The Physics of Modern Brachytherapy for Oncology

Series in
Medical Physics and Biomedical Engineering

Series Editor: **J G Webster**, University of Wisconsin-Madison, USA
Consultant Editor: **R F Mould**

Series in
Medical Physics and Biomedical Engineering

The Physics of Modern Brachytherapy for Oncology

D Baltas
Klinikum Offenbach, Germany
and
University of Athens, Greece

L Sakelliou
University of Athens, Greece

N Zamboglou
Klinikum Offenbach, Germany

CRC Press
Taylor & Francis Group
Boca Raton London New York

CRC Press is an imprint of the
Taylor & Francis Group, an **informa** business
A TAYLOR & FRANCIS BOOK

CRC Press
Taylor & Francis Group
6000 Broken Sound Parkway NW, Suite 300
Boca Raton, FL 33487-2742

First issued in paperback 2019

ISBN-13: 978-0-7503-0708-6 (hbk)
ISBN-13: 978-0-367-86421-7 (pbk)

Library of Congress Cataloging-in-Publication Data

Baltas, Dimos.
   The physics of modern brachytherapy for oncology / Dimos Baltas, Loukas Sakelliou, and Nikolaos Zamboglou.
      p. cm. -- (Series in medical physics and biomedical engineering)
   Includes bibliographical references and index.
   ISBN-13: 978-0-7503-0708-6 (alk. paper)
   ISBN-10: 0-7503-0708-0 (alk. paper)
   1. Radioisotope brachytherapy. I. Zamboglou, N. (Nickolaos) II. Sakelliou, Loukas. III. Title. IV. Series.

RC271.R27B35 2007
615.8'424--dc22   2006005610

Visit the Taylor & Francis Web site at
http://www.taylorandfrancis.com

and the CRC Press Web site at
http://www.crcpress.com

Imperturbable wisdom is worth everything. To a wise man, the whole earth is open; for the native land of a good soul is the whole earth. Medicine heals diseases of the body; wisdom frees the soul from passions. Neither skill nor wisdom is attainable unless one learns. Beautiful objects are wrought by study through effort, but ugly things are reaped automatically without toil.

On Learning, Democritus (460–370 B.C.)

Preface

This *Physics of Modern Brachytherapy for Oncology* is the most comprehensive brachytherapy textbook for physicists that has so far been written and intentionally includes chapters on basic physics that are necessary for an understanding of modern brachytherapy. The book therefore stands alone as a *total* reference book, with readers not having to consult other texts to obtain information on the basics, which are presented in Chapter 2 through Chapter 4 and are set out in a logical fashion starting with quantities and units, followed by basic atomic and nuclear physics.

It was only after the development of the atomic bomb in the Manhattan project in the late 1940s that the peaceful uses of ionizing radiation (other than radium and radon) were harnessed in medicine. This led to the introduction of ^{60}cobalt and ^{137}cesium in the late 1950s and these radionuclides replaced radium for the manufacture of tubes and needles and ^{198}gold seeds replaced the radon seeds. However, there was still a problem with the tubes and needles because a sufficiently high specific activity could not be achieved and thin malleable wire sources were impossible to make using ^{60}Co and ^{137}Cs. This was eventually overcome in the early 1960s by initially using ^{182}tantalum wire and later ^{192}iridium wire, with the latter being the standard brachytherapy source today not only for use with the Paris system but also in the form of miniature sources for use with high dose rate (HDR) *afterloading* machines. With radium and ^{137}Cs *only*, low dose rate (LDR) brachytherapy could be performed, again because of specific activity limitations. ^{60}Co was first used for HDR afterloaders but was too expensive because of the need for regular source replacements and for extended radiation protection infrastructure. ^{192}Ir was a cheaper option. Work continues on possibilities for the use of new radionuclides such as ^{241}americium, ^{169}ytterbium and ^{145}samarium, but currently the three mainstay radionuclides are ^{192}Ir, ^{125}iodine and ^{103}palladium. This is reflected in the contents of Chapter 5 and Chapter 6.

Dosimetry is an essential part of brachytherapy physics and has been from the earliest days when the first measurements were made by Marie Curie using the piezo electrometer designed by Pierre Curie and his brother Jacques. However, international agreement on radium dosimetry units took many years and, for example, the roentgen unit of exposure was not recommended by the ICRU as a unit of measurement for both x-rays and radium gamma rays until 1937. Previously there was a differentiation between x-ray roentgens and gamma ray roentgens. Also, prior to 1937, more than 50 proposals for radiation units had been made, depending not only on the ionization effect, but also on, for example, silver bromide photographic film blackening, fluorescence and skin erythema. The latter formed the basis of the radium dosage system developed by the physicist Edith Quimby at Memorial Hospital, New York.

The most popular unit used was the milligram-hour, which was merely the product of the milligrams of radium and the duration of the treatment. For radon, the analogous unit was millicurie-destroyed since radon had a half-life of only some 3.5 days. However, the best experimental results were made using ionization chambers, particularly those developed in the late 1920s by the physicist Rolf Sievert of the Radiumhemmet in Stockholm, who was also responsible for the Sievert integral. Source calibration and dosimetry protocols are considered in Chapter 7 and Chapter 8. Monte Carlo aided and experimental dosimetry, including gel dosimetry, which has only been available in the last decade, are the topics in Chapter 9 and Chapter 10.

Dimos Baltas
Nikolaos Zamboglou
Loukas Sakelliou

Acknowledgments

This book could not be completed without the contributions, simulations, discussions and efforts of several colleagues, collaborators and friends. There are a lot of people who played specific roles in the book itself. We cannot mention all but would particularly like to single out, for special thanks and gratitude, the following:

R.F. Mould, for creating with us the idea for this book and providing the excellent review on the early history of brachytherapy physics presented in Chapter 1.

A.B. Lahanas, for contributing the comprehensive review of elementary particles and the Standard Model in Chapter 3.

J. Heijn, for enabling a detailed insight into the production methods and processes of the most important radionuclides by providing the material in Chapter 6.

P. Papagiannis, for his essential contribution to the TLD and gel dosimetry sections in Chapter 10 as well as to the Monte Carlo sections in Chapter 9 of the book. We are very grateful for his comments, clarifications, discussions, collaboration and friendship.

E. Pantelis, for the production of the original data and Monte Carlo simulation results that facilitated the discussion in Chapter 4 and Chapter 9.

G. Anagnostopoulos, for his gracious help in providing Monte Carlo simulation results for the different phantom materials and his contribution to the TLD section in the experimental dosimetry, Chapter 10.

Authors

Dr. Dimos Baltas is the director of the Department of Medical Physics and Engineering of the State Hospital in Offenbach, Germany. Since 1996, he has been associate adjunct research professor for Medical Physics and Engineering of the Institute of Communication and Computer Systems (ICCS), at the Department of Electrical and Computer Engineering, National Technical University of Athens and since 2005 he has been an honorary scientific associate of the Nuclear Physics and Elementary Particles Section, at the Physics Department, University of Athens, Greece. He received his PhD in medical physics in 1989 at the University of Heidelberg, Germany and did research in applied radiation biology, modeling and radiobiology-based planning. He has pursued research in the field of dosimetry, quality assurance and technology in brachytherapy since the late 1980s. His contributions to brachytherapy physics and technology include experimental dosimetry of high activity sources, development and establishment of dosimetry protocols, advanced quality assurance procedures and systems, experimental and Monte Carlo–based dosimetry of new high and low activity sources, development of new algorithms for dose calculation and for imaging-based treatment planning in brachytherapy as well as development of new radionuclides for brachytherapy. He has written more than 80 refereed publications and ten chapter contributions in published books. Ten diploma theses and 12 PhD dissertations have been completed under his supervision. He is member of the German Association of Medical Physics (DGMP), of the German Association for Radiation Oncology (DEGRO), of the European Society for Therapeutic Radiology and Oncology (ESTRO), of the American Association of Physicists in Medicine (AAPM) and of the Greek Association of Physicists in Medicine (EFIE). Since 2000, he has been the Chairman of the Task Group on Afterloading Dosimetry of the German Association of Medical Physics (DGMP). Since 2004, he has been a member of the Medical Physics Experts Group for Reference Dosimetry Data of Brachytherapy Sources (BRAPHYQS group of ESTRO) and since 2005, a member of the Scientific Advisory Board of the *Journal Strahlentherapie und Onkologie, Journal of Radiation Oncology Biology Physics.*

Dr. Loukas Sakelliou is associate professor in the Nuclear Physics and Elementary Particles Section, at the Physics Department, University of Athens, Greece. He received his PhD in physics in 1982, then did research in experimental particle physics and participated in a number of experiments at CERN (including CPLEAR, Energy Amplifier and TARC). He pursued research in the field of medical physics in the early 1990s. His contribution to the field of brachytherapy includes, mainly although not exclusively, the Monte Carlo modeling of medical radiation sources for the generation of dosimetry data for use in clinical practice. He has written over 100 refereed publications and 10 PhD dissertations have been completed under his supervision. Dr. Sakelliou is presently the head of the Dosimetry Laboratory of the Institute for Accelerating Systems and Applications (IASA, Athens, Greece) and a member of the founding board of an MSc course in medical physics. He teaches atomic and nuclear physics, modern physics and medical physics and his current research interests include Monte Carlo simulation and analytical dosimetry methods with a focus on the amendment of radiation therapy treatment planning methods, experimental brachytherapy dosimetry using polymer gels and treatment verification in contemporary radiation therapy techniques.

Dr. Nikolaos Zamboglou is the chairman of the Radiation Oncology Clinic of the State Hospital in Offenbach, Germany. Since 1990, he has been a member of the Faculty of Medicine in Radiation Oncology of the University of Düsseldorf Germany, and since 1993 adjunct research professor at the Institute of Communication and Computer Systems (ICCS) of the National Technical University of Athens. He received his PhD in physics in 1977 at the Faculty of Science of the University of Düsseldorf and in medicine, MD, in 1987 at the Faculty of Medicine, University of Essen, Germany. He did research in biological dosimetry, radiation protection, radiation biology focused on the radio-sensitivity of cells and clinical research in simultaneous radio-chemotherapy and brachytherapy. His current research area is the clinical implementation of advanced technologies and imaging-based techniques in radiation oncology especially focused on high dose rate brachytherapy, in which he has been active since the late 1980s. He has written more than 110 refereed publications and more than 30 chapter contributions in published books. Twenty-five PhD and MD dissertations have been completed under his supervision. He is a member of the German Association of Medical Physics (DGMP), of the German Association for Radiation Oncology (DEGRO), of the European Society for Therapeutic Radiology and Oncology (ESTRO), and of the Hellenic Society for Radiation Oncology (HSRO). Since 2000, he has been the Chairman of the Task Group on 3D Treatment Planning in Brachytherapy of the German Association for Radiation Oncology (DEGRO). Since 2001 he has been a member of the Steering Committee of the German Association for Radiation Oncology (DEGRO), where for the period of 2003 to 2005 he was President. Since 1996, he has been a member of the Scientific Advisory Board of the *Journal Strahlentherapie und Onkologie, Journal of Radiation Oncology Biology Physics*.

Contents

1

The Early History of Brachytherapy Physics

1.1 Introduction

The textbook written by Paul Davies in 1989[1] entitled, *The New Physics*, commenced with the following opinion.

> Many elderly scientists look back nostalgically at the first 30 years of the 20th century, and refer to it as the golden age of physics. Historians, however, may come to regard those years as the dawning of the New Physics. The events which the quantum and relativity theories set in train are only now impinging on science, and many physicists believe that the golden age was only the beginning of the revolution.

This is particularly true for the start of the 21st century in the field of modern brachytherapy which is a treatment modality within the framework of oncology, and which encompasses the entire spectrum of cancer diagnosis and treatment. Complex software programs are now an integral part of the brachytherapy process and links to CT, MR, and ultrasound imaging are now the norm rather than the unusual. Indeed, modern brachytherapy has changed out of all recognition to what it was even 50 years ago. Then, brachytherapy had been more or less standardized for a couple of decades with geometric-based dosimetry systems, no radium and radon replacement radionuclides such as ^{60}Co and ^{137}Cs, no digital computers, with all body tissues assumed to be of unit density, no routine consideration of radiobiology, no remote afterloading and very little manual afterloading and with many institutions relying on the milligram–hour unit for radium and the millicurie destroyed per square centimeter for radon, even though the roentgen unit had been introduced for both x-rays and gamma rays in 1937.[2,3]

The golden age of brachytherapy is therefore still ahead of us and certainly not behind us in the 20th century, when in the 1940s and 1950s it might even be said to have stagnated for a period of time. If there was anything golden about such a time it was a misnomer in that dosimetry and planning had been unchanged for so long that physicists did not have to think very hard in terms of calculations. For example, it was very easy to plan a radium implant delivering 1000 R per day at 0.5 or 1.0 cm distance; and for intracavitary gynecological insertions there was often only a choice of vaginal ovoids which were either large, medium, or small and of an intrauterine tube which was standardized as short, medium, or long. Simulators had not been developed and a three-dimensional distribution using an analog-computing device could take a physicist 24 h to perform, which in practice meant that very few hospitals bothered to consider this. The planning/verification images were mainly only lateral and AP orthogonal radiographs, obtained to determine whether or not the radium sources were correctly positioned.

However, history should not be ignored because one can always learn from the past and it is interesting to record that by the year 1910 the principal of afterloading had been enunciated both in Europe[4] and the United States.[5] Also, many radiobiological experiments had been performed: one of the earliest self-exposure studies being by Pierre Curie working with Henri Becquerel in 1901.[6] This first chapter in *The Physics of Modern Brachytherapy for Oncology* therefore looks back to some of the physics discoveries, concepts, and ideas that have formed the foundations of today's modern brachytherapy practice.

1.2 Discoveries

The three seminal discoveries that occurred at the end of the 19th century were, of course, those of x-rays by Röntgen,[7,8] radioactivity by Becquerel,[9] and radium by the Curies.[10] Also, between the discoveries of radioactivity and radium, there occurred the discoveries of thorium by Schmidt[11] and of polonium by the Curies.[12] These set the scene for the work of Rutherford and for leading to the nuclear transformation theory of Rutherford and Soddy in 1902 and later the Rutherford–Bohr atom.

1.3 Ionization and X-Rays

Ernest Rutherford's research career began in 1893 and his first paper was on the magnetization of iron by high-frequency discharges.[13] His work on magnetism continued in Cambridge[14] until 1896 but then at Thomson's suggestion he abandoned this work to concentrate on research using the newly discovered x-rays. In particular, this was to study the power of x-rays to render air conductive by the process of ionization.

The first paper on ionization was published in 1896[15] and this was followed by one[16] in which Rutherford introduced the concept of the coefficient of absorption and demonstrated that the leak of a gas (i.e., the ionization current) is proportional to the intensity of radiation at any point in time. His third paper[17] on ionization contained at the end of the paper his first reference to the use of "the radiation given out by uranium and its salts." However, as emphasized by Cohen,[13] Rutherford at this time considered radiation to be a tool for investigating, and not the other way around.

1.4   α, β, γ, and Half-Life

The work for the papers on ionization[15-19] was undertaken while Rutherford was still in Cambridge, but the last paper in this series[19] was not published until January 1899 when he was then in Montreal. His opinion on the photographic method of investigation vs. the ionization method mirrors that of Marie and Pierre Curie.

> The properties of uranium radiation may be investigated by two methods, one depending on the action on a photographic plate and the other on the discharge of electrification. The photographic method is very slow and tedious, and admits of only the roughest measurements. Two or three days exposure to the radiation is generally required to produce any marked effect on the photographic plate. In addition, when we are dealing with very slight photographic action, the fogging of the plate, during the long exposures required, by the vapors of the substances is liable to obscure the results. On the other hand the method of testing the electrical discharge caused by the radiation is much more rapid than the photographic method, and also admits of fairly accurate quantitative determinations.[13]

It was in this 1899 paper[19] that Rutherford proposed the terms α and β for two of the uranium radiations. The discovery of the third type, the γ radiation, is usually attributed to Paul Villard in 1900.[20]

> There are present at least two distinct types of radiation, one that is very readily absorbed, which will be termed for convenience the α radiation, and the other of a more penetrative character, which will be termed the β radiation.

Rutherford was also responsible for introducing the concept of half-life.[21] This was in a paper reporting studies on thorium compounds, which also illustrated for the first time, the growth and decay of a radioactive substance, in this instance thorium emanation (now termed ^{220}Rn) which has a half-life of 55.6 sec. Rutherford, by measuring current with time, showed that the curve of the rise of the current intersected the curve of the decay of the current at the time equal to the half-life. Furthermore, he reported that if all the emanation from a layer of thorium oxide is removed by a current of air, it builds up again with the same half-period.

1.5 Nuclear Transformation

By 1901, Rutherford had obtained a reasonably pure radium sample from Germany that enabled him to work with radium emanation (although the amount was still too small for chemical analysis) rather than thorium emanation, as the former had the experimental advantage over the latter of a half-life of approximately 4 d compared to approximately 1 min. Rutherford attempted to estimate the molecular (or atomic) weight by measuring its rate of diffusion in air, using a long cylindrical ionization chamber divided into two by a moveable metal shutter.

The experiment[22] gave a molecular weight in the range 40 to 100 (which was actually far too low because of measurement error). This convinced Rutherford that the emanation was not radium in vapor form because Marie Curie had already shown[23] that radium had an atomic weight of at least 140. This evidence supported the transformation theory which he published in 1902[24] with Frederick Soddy (1877–1956, Nobel Prize for Chemistry 1921) during a collaboration which lasted only 18 months, from October 1901 to March 1903. The theory[24,25] was a radical departure from the conviction that the atoms of each element are permanent and indestructible. Rutherford then showed in 1903[26] that alpha rays were actually, heavy, positively charged particles, and in 1906 argued[27] that their precise nature was "a helium atom carrying twice the ionic charge of hydrogen."

The exponential law of transformation can be described in the following terms. If N_0 is the number of atoms breaking up per second at time $t = 0$, the corresponding number after time t is given by

$$N_t = N_0 e^{-\lambda t}$$

where λ is a constant. The number n_t of atoms which are unchanged after an interval t is equal to the number which change between t and infinity. That is

$$n_t = \int_t^\infty N_0 e^{-\lambda t} dt = \frac{N_0 e^{-\lambda t}}{\lambda} = \frac{N_t}{\lambda}$$

and the number of atoms present at time $t = 0$ is $n_0 = N_0/\lambda$ and thus $n_t = n_0 e^{-\lambda t}$ or the number of atoms of a radioactive substance decreases with time according to an exponential law. Since $N_t = \lambda n_t$ we can say that λ is the fraction of the total number of atoms present which break up per unit time (e.g., second) and thus λ has a distinct physical meaning. It is called the transformation constant and its reciprocal $1/\lambda$ is called the average life.

Also, if the half-life is $T_{1/2}$ then

$$T_{1/2} = 1/\lambda \ln 2 = 0.693/\lambda$$

Because of the statistical character of the transformation law, λ is the average fraction of atoms which break up in unit time. The number which break up in unit time is subject to fluctuations around this average value and the magnitude of these fluctuations can be calculated from the laws of chance.

1.6 Rutherford–Bohr Atom

J. J. Thomson's view of an atom was that it was a positively charged evenly distributed substance in which corpuscles (later called electrons) were thought to be embedded. George Gamow's[28] analogy was "as raisins in a round loaf of raisin bread." Rutherford's concept was that an atom had a large positive central charge (the term nucleus was not used initially) surrounded by a sphere of negatively charged electricity.[29,30] His 1920 Bakerian Lecture to the Royal Society[31] was on the subject of the nuclear constitution of atoms and the introduction to this important lecture is given below.

> The conception of the nuclear constitution of atoms arose initially from attempts to account for the scattering of α-particles through large angles in traversing thin sheets of matter.[32] Taking into account the large mass and velocity of the α-particles, these large deflections were very remarkable, and indicated that very intense electric or magnetic fields exist within the atom. To account for these results, it was found necessary to assume[29,30] that the atom consists of a charged massive nucleus of dimensions very small compared with the ordinarily accepted magnitude of the diameter of the atom. This positively charged nucleus contains most of the mass of the atom, and is surrounded at a distance by a distribution of negative electrons equal in number to the resultant positive charge on the nucleus. Under those conditions a very intense electric field exists close to the nucleus, and the large deflection of the α-particle in an encounter with a single atom happens when the particle passes close to the nucleus. Assuming that the electric forces between the α-particle and the nucleus varied according to the inverse square law in the region close to the nucleus, the writer worked out the relations connecting the number of α-particles scattered through any angle with the charge on the nucleus and the energy of the α-particle. Under the central field of force, the α-particle describes a hyperbolic orbit round the nucleus, and the magnitude of the deflection depends on

the closeness of approach to the nucleus. From the data of scattering of α-particles then available, it was deduced that the resultant charge on the nucleus was about (1/2)*Ae*, where *A* is the atomic weight and *e* the fundamental unit of charge. Geiger and Marsden[32,33] made an elaborate series of experiments to test the correctness of the theory, and confirmed the main conclusion. They found the nucleus was about (1/2)*Ae*, but, from the nature of the experiments, it was difficult to fix the actual value within about 20 percent. C. G. Darwin[34] worked out completely the deflection of the α-particle and of the nucleus, taking into account the mass of the latter, and showed that the scattering experiments of Geiger and Marsden could not be reconciled with any law of central force, except the inverse square. The nuclear constitution of the atom was thus very strongly supported by the scattering of α-rays.

Since the atom is electrically neutral, the number of external electrons surrounding the nucleus must be equal to the number of units of resultant charge on the nucleus. It should be noted that, from consideration of the scattering of x-rays by light elements, Barkla[35] had shown in 1911 that the number of electrons was equal to about half the atomic weight. This was deduced from the theory of scattering of J. J. Thomson, in which it was assumed that each of the external electrons in an atom acted as an independent scattering unit.

Two entirely different methods had thus given similar results with regard to the number of external electrons in the atom, but the scattering of α-rays had shown in addition that the positive charge must be concentrated on a massive nucleus of small dimensions. It was suggested by van den Broek[36] that the scattering of α-particles by the atoms was not inconsistent with the possibility that the charge on the nucleus was equal to the atomic number of the atom, i.e., to the number of the atom when arranged in order of increasing atomic weight. The importance of the atomic number in fixing the properties of an atom was shown by the remarkable work of Moseley[37,38] on the x-ray spectra of the elements. He showed that the frequency of vibration of corresponding lines in the x-ray spectra of the elements depended on the square of a number which varied by unity in successive elements. This relation received an interpretation by supposing that the nuclear charge varied by unity in passing from atom to atom, and was given numerically by the atomic weight. I can only emphasize in passing the great importance of Moseley's work, not only from fixing the number of possible elements, and the position of undetermined elements, but in showing the properties of an atom were defined by a number which varied by unity in successive atoms. This gives a new method of regarding the periodic classification of the elements, for the atomic number, or its equivalent the nuclear charge, is of more fundamental importance than its atomic weight. In Moseley's work, the frequency of vibration of the atom was not exactly proportional to N, where N is the atomic number, but to $(N - a)^2$ where a was a constant which had different values, depending on whether the K or L series of characteristic radiation were measured. It was supposed that this constant depended on the number and position of the electrons close to the nucleus.

Niels Bohr (1885–1962) realized that the electronic constitution of the atom was quantum in nature and using the optical spectrum as a basis, in 1913 proposed his modifications to the Rutherford concept of the atom, but retaining the positively charged atomic nucleus.[39–42] Then in 1916, Arnold Sommerfeld (1868–1961) suggested[43] added elliptical paths for the electrons in addition to the circular ones proposed by Bohr. The concept of isotopes was enunciated in the same year by Soddy.[44,45] Further major advances then had to wait for the discovery of the neutron in 1932[46] by James Chadwick (1891–1974, Nobel Prize for Physics in 1935) and the knowledge that the constituents of a nucleus consisted of a combination of protons and neutrons. However, it is interesting to note that 12 years before the discovery of the neutron, Rutherford in his 1920 Bakerian lecture[25] stated

> Under some conditions … it may be possible for an electron to combine much more closely with the hydrogen nucleus [i.e., a proton], forming a kind of neutral doublet. Such an atom would have novel properties. Its external field would be practically zero … and consequently it should be able to move freely through matter. Its presence would probably be difficult to detect.

1.7 The Start of Brachytherapy

It can be argued that the true start of brachytherapy was Röntgen's discovery of x-rays in 1895, since that was directly linked to Becquerel's discovery of radioactivity in 1896, which in turn led to the discovery of radium by Marie and Pierre Curie in 1898. This was followed in 1901 by Pierre Curie's self-exposure experiment.[6] This early radiobiology experiment was planned together with Becquerel after a German chemist, Friedrich Giesel, reported a burn following a 2-h self-exposure on his arm. Pierre Curie kept a daily record for 52 d of the status of the radiation reaction following a 10-h exposure on his forearm.[6,47]

The first successful radium brachytherapy treatment for cancer followed in 1903 in St. Petersburg where two patients were treated for facial basal cell carcinoma.[2,48] Such surface mold and plaque treatments were followed by intracavitary techniques for cancers of the cervix, uteri and endometrium, because these sites were readily accessible compared to deeper seated sites. Then, only a few years later, interstitial radium brachytherapy techniques were implemented and by the end of the first decade of the 20th century most body sites which are treated today had been, with varying degrees of success, treated by radium brachytherapy. The only exceptions were sites requiring a catheter to be passed via a narrow and markedly curved lumen, such as tumors of the lung and bile duct.

1.8 Dose Rates

The dose rates using radium were low and treatment times were measured in days, and it was not until some 80 years later when technology had vastly improved that the ICRU proposed[49] definitions for low (LDR), medium (MDR), and high (HDR) dose rates. These are, respectively, 0.4 to 2.0 Gy/h (LDR), 2 to 12 Gy/h (MDR) and >0.2 Gy/min (HDR). LDR techniques are still used, particularly by the French School[50–52] but HDR has now largely superseded the use of LDR for the majority of cancer treatments using brachytherapy. This is reflected in the current availability of remote afterloading machines, with those such as the Selectron-LDR, which formerly were the most widely used, now no longer being manufactured.

1.9 Dosimetry Systems

Dosimetry techniques and systems have also changed beyond all recognition over the last century. For example, it was originally thought[53] that a uniform distribution of sources on a surface applicator or for a single planar interstitial implant would produce a uniform dose distribution at the treating distance of usually 0.5 or 1.0 cm. In clinical practice this was the opinion until the late 1920s, but by the 1930s it had been unequivocally shown that this was not true and that a nonuniform distribution of sources produced a uniform dose distribution. This formed the basis of the Manchester System of Paterson and Parker,[54] which was developed in the 1930s. The rules of the system were formulated from theoretical studies of the distribution of exposure dose expressed in roentgens around sources of simple geometrical shape (line, disk, annulus, sphere, cylinder); the line source having been studied much earlier and the Sievert integral[55] by then being well known. The Manchester System was widely used until the 1970s and the advent of computer treatment planning, and together with the

Quimby System, which had its origins in the early 1920s and was the system of choice in the U.S., were the standard methods of brachytherapy dosimetry for many years.

In both systems the planning of an interstitial implant consisted of determining the area or volume of a target region and then referring to a table or graph for the required total source strength (milligram–hours) per unit peripheral dose (1000 cGy) or, alternatively the source activity for a given peripheral dose rate. The objective of the Manchester was to deliver a uniform dose to within ±10% of the prescribed dose throughout the implanted region or in the treatment plane.[54] Quimby called for a uniform distribution of source strength and accepted the hot spots in the central region of the implant. The Quimby prescribed dose was the same as the peripheral dose.[56–59]

The Paris System was developed much later[50–52] than the Manchester and Quimby, and was designed for use with ^{192}Ir wires and not radium sources. The dosimetry is based on the basal dose rate, which is a measure of the dose rate at the center of the treated volume. It is calculated from the position of the sources in the central plane and is the minimum dose rate between a pair or group of sources. It must be recognized that the relationship between the basal dose and the spacing of the ^{192}Ir wires is critical.

There are now international recommendations[60] for dose and volume specification in interstitial brachytherapy which replace the earlier concepts and terminology such that for volumes and planes we now have: gross tumor volume (GTV), clinical target volume (CTV), planning target volume (PTV), treated volume, and central plane. For a description of the dose distribution we now have prescribed dose, minimum target dose, mean central dose, high dose volumes, and low dose volumes.

Previously, apart from the Manchester ±10% dose homogeneity objective there were no real attempts to specify implant quality until the proposal in 1986 for dose–volume histograms.[61] Since then, many indices have been proposed, for example, that from the Offenbach Radiotherapy Clinic for the COIN index.[62]

Other early work which has so far not been mentioned, but which had an impact on the development of dosage systems for surface and interstitial applications, including theoretical calculations of dose around sources of standard geometrical shape, are described in the next paragraph.

In 1931, Murdoch from the Brussels Cancer Center described a series of radium surface brachytherapy treatments,[63] and this work directly preceded the papers of Paterson and Parker[64–66] which formed the basis of the Manchester system.[54] Theoretical calculations of dose distributions around radium sources began in 1916[67] some 5 years before the Sievert integral was published for linear radium sources[55] and continued throughout the 1930s.[64–66,68,69] Experimental measurements were also undertaken during the 1920s,[70–72] with in some cases, rather unusual choices to mimic the surrounding tissues for a real patient. These included butter, rabbit muscle, and rat tissues. The underlying principle of the use of the butter phantom[72] was that by sectioning the butter the bleaching of its color by the radium gamma-rays would effectively produce isodose curves.

Dosimetry systems were also developed for intracavitary gynecological applications for either the cervix uteri[73] or the corpus uteri, the endometrium.[74] For the cervix the Christie Hospital, Manchester system was most widely used with a choice of vaginal ovoid size and radium loading and of the intrauterine tube. The corpus system was developed at the Radiumhemmet, Stockholm and became known as the Heyman packing method because several small radium capsules were tightly packed into the uterine cavity. The Paris system of Regaud from the Institut du Radium[75] used corks for the vaginal applicators and predated the Manchester system, which was adapted from this earlier method. Later, the Fletcher system,[76] which was developed at the M.D. Anderson Hospital in Houston was directly influenced by the Manchester system. There were, however, several other systems developed using different designs of vaginal applicator to the ovoid-type of Paris, Manchester, and Fletcher. These included a ring applicator[77] and a pin and plate method developed in German centers[78] which was the first example of the intrauterine source and vaginal source being mechanically attached to each other. This overcame the problem of the vaginal source moving from the correct location, which often occurred because it was fixed into position by only using gauze packing.

1.10   Marie Curie

It is appropriate that the discoverer of radium should be given the last word in this historical review on the early history of radium brachytherapy. It is often thought that Marie Curie limited her scientific work to the chemistry of radioactive materials (except for her sojourn in World War I in the organization and operation of mobile x-ray lorries, known as "Little Curies") and that it was left to her Institut du Radium colleague Claudius Regaud to investigate brachytherapy (known in France as curietherapie). This is a misunderstanding as seen from two of her quotations,[79] which clearly show that she was fully aware of the need for a good scientific basis for brachytherapy clinical practice.

> It is easy to understand how important for me is the conviction that our discovery is a blessing for humankind not only by its scientific importance but also because it permits to reduce human suffering and to treat a terrible disease. This is indeed, a great reward for the years of our enormous effort.

> Treatment in such a new specialty requires a sound basis to be provided by physical and chemical studies on new substances .... wherever this basis is not ensured, theory acquires the form of

empiricism and routine by the uncritical application of the popular method which, sometimes, includes major mistakes.

References

1. P. Davies, ed. *The New Physics*, Cambridge University Press, Cambridge, 1989.
2. Mould, R.F. *A Century of X-Rays and Radioactivity in Medicine*, Institute of Physics Publishing, Bristol, 1993.
3. Mould, R.F. Radium mosaic: A scientific history of radium, *Nowotwory J. Oncol.*, Special Suppl., Nowotwory, Warsaw, 2005.
4. Strebel, H. Vorschlaege zur Radiumtherapie, *Deut. Med. Zeit.*, 24, 11, 1903.
5. Abbe, R. News item. A very ingenious method of introducing radium into the substance of a tumour, *Arch. Roentgenol. Ray*, 15, 74, 1910.
6. Becquerel, H. and Curie, P. Action physiologiques des rayons du radium, *Comptes Rendus de l'Académie des Sciences*, Paris, 132, 1289–1291, 1901.
7. Röntgen, W.C. Ueber eine neue Art von Strahlen [Vorläufige Mittheilung], *Sitzungsberichte der Physikalische-medizinischen Gessellschaft zu Würzburg*, 9, 132–141, 1895, Reprinted in English: On a new kind of rays: preliminary communication. See Mould, R.F., Röntgen and the discovery of x-rays, *Br. J. Radiol.*, 68, 1145–1176, 1995.
8. Röntgen, W.C. Ueber eine neue Art von Strahlen [2. Mittheilung], *Sitzungsberichte der Physikalische-medizinischen Gessellschaft zu Würzburg*, 2, 11–17, 1896.
9. Becquerel, H. Sur les radiations invisibles émises par les corps phosphorescents, *Comptes Rendus de l'Académie des Sciences*, Paris, 122, 501–503, 1896.
10. Curie, P., Curie, M., and Bémont, G. Sur une nouvelle substance fortement radio-active contenue dans la pechblende, *Comptes Rendus de l'Académie des Sciences*, Paris, 127, 1215–1218, 1898.
11. Schmidt, C.G. Ueber die von den Thorverbindungen und einigen anderen Substanzen ausgehenden Strahlung. {On the rays emitted by thorium compounds and some other substances} *Verh. Phys. Ges.* Berlin, 17 and *Ann. Phys. Chem.*, 65, 141–151, 1898.
12. Curie, P. and Curie, M. Sur une substance nouvelle radioactive contenue dans la pechblende. {On a new radioactive substance contained in pitchblende}, *Comptes Rendus de l'Académie des Sciences*, 127, 175–178, 1898.
13. Cohen, M. Rutherford's curriculum vitae 1894–1907, *Med. Phys.*, 22, 841–859, 1995.
14. Rutherford, E. A magnetic detector of electrical waves and some of its applications, *Philos. Trans. R. Soc. Ser. A*, 189, 1–24, 1897.
15. Thomson, J.J. and Rutherford, E. On the passage of electricity through gases exposed to Röntgen rays, *Philos. Mag. Ser. 5*, 42, 392–407, 1896.
16. Rutherford, E. On the electrification of gases exposed to Röntgen radiation by gases and vapours, *Philos. Mag. Ser. 5*, 43, 241–255, 1897.
17. Rutherford, E. The velocity and rate of recombination of the ions of gases exposed to Röntgen radiation, *Philos. Mag. Ser. 5*, 44, 422–440, 1897.
18. Rutherford, E. The discharge of electrification by ultraviolet light, *Proc. Cambridge Philos. Soc.*, 9, 401–416, 1898.

19.   Rutherford, E. Uranium radiation and the electrical conduction produced by it, *Philos. Mag. Ser. 5*, 47, 109–163, 1899.

20.   Villard, P. Sur la réflexion et la réfraction de rayons cathodiques et des rayons déviables du radium, *Comptes Rendus de l'Académie des Sciences*, 130, 1010–1011, 1900.

21.   Rutherford, E. A radioactive substance emitted from thorium compounds, *Philos. Mag. Ser. 5*, 49, 1–14, 1900.

22.   Rutherford, E. and Brooks, H.T. The new gas from radium, *Trans. Roy. Soc. Can. Sec. iii Ser. ii*, 7, 21–25, 1901.

23.   Curie, M. Sur le poids atomique du métal dans le chlorure de baryum radifère, *Comptes Rendus de l'Académie des Sciences*, 129, 760–762, 1899.

24.   Rutherford, E. and Soddy, F. The cause and nature of radioactivity, *Philos. Mag. Ser. 6*, 4, 370–396, See also pages 569–585, 1902.

25.   Rutherford, E. and Soddy, F. Radioactive change, *Philos. Mag. Ser. 6*, 5, 576–591, 1903.

26.   Rutherford, E. The magnetic and electric deviation of the easily absorbed rays from radium, *Philos. Mag. Ser. 6*, 5, 177–187, 1903.

27.   Rutherford, E. The mass and velocity of the α particles expelled from radium and actinium, *Philos. Mag. Ser. 6*, 12, 348–371, 1906.

28.   Gamow, G. *Thirty Years that Shook Physics: The Story of the Quantum Theory*, Doubleday Anchor Science Series, New York, 1966.

29.   Rutherford, E. The scattering of alpha and beta particles by matter and the structure of the atom, *Philos. Mag. Ser. 6*, 21, 669–688, 1911.

30.   Rutherford, E. The structure of the atom, *Philos. Mag. Ser. 6*, 27, 488–498, 1914.

31.   Rutherford, E. Nuclear constitution of atoms. Royal Society Bakerian Lecture, *Proc. Roy. Soc. A*, 97, 374–400, 1920.

32.   Geiger, H. and Marsden, E. On a diffuse reflection of the α-particles, *Proc. Roy. Soc. A*, 82, 495–500, 1909.

33.   Geiger, H. and Marsden, E. The laws of deflexion of α particles through large angles, *Philos. Mag. Ser. 6*, 25, 604–623, 1913.

34.   Darwin, C.G. Collision of alpha particles with light atoms, *Philos. Mag. Ser. 6*, 27, 499–506, 1914.

35.   Barkla, C.G. Note on the energy of scattered x-radiation, *Philos. Mag. Ser. 6*, 21, 648–652, 1911.

36.   van den Broek, A. Radioactive elements and the periodic system, *Phys. Z.*, 14, 32–41, 1913.

37.   Moseley, H.G.J. The high frequency spectra of the elements I, *Philos. Mag. Ser. 6*, 26, 1024–1034, 1913.

38.   Moseley, H.G.J. The high frequency spectra of the elements II, *Philos. Mag. Ser. 6*, 27, 703–713, 1914.

39.   Bohr, N. On the constitution of atoms and molecules, *Philos. Mag. Ser. 6*, 26, 1–25, 1913.

40.   Bohr, N. On the constitution of atoms and molecules II. Systems containing only a single nucleus, *Philos. Mag. Ser. 6*, 26, 476–502, 1913.

41.   Bohr, N. On the constitution of atoms and molecules III. Systems containing several nucleii, *Philos. Mag. Ser. 6*, 26, 857–875, 1913.

42.   Bohr, N. On the series spectrum of hydrogen and the structure of the atom, *Philos. Mag. Ser. 6*, 29, 332–335, 1915.

43.   Sommerfeld, A. *Atom und Spektrallinien*, Friedrich Vieweg und Sohn, Braunschweig, 1919.

44. Soddy, F. Intra-atomic charge, *Nature*, 92, 399–400, 1913.
45. Soddy, F. The radio-elements and the periodic law, *Nature*, 91, 57–58, 1913.
46. Chadwick, J. The existence of a neutron, *Proc. Roy. Soc. A*, 136, 692–708, 1932.
47. Mould, R.F. The discovery in 1898 by Maria Sklodowska-Curie (1867–1934) and Pierre Curie (1859–1906) with commentary on their life and times, *Br. J. Radiol.*, 71, 1229–1254, 1998.
48. Goldberg, S.W. and London, E.S. Zur Frage der Beziehungen zwischen Becquerelstrahlen und Hautaffectionen, *Dermatologische Zeit*, 10, 457, 1903.
49. International Commission on Radiation Units and Measurements, Dose and volume specification for reporting intracavitary therapy in gynecology, ICRU Report 38. ICRU, Bethesda, 1985.
50. Pierquin, B., Chassagne, D., and Perez, R. *Precis de curietherapie*, Masson, Paris, 1964.
51. Dutreix, A., Marinello, G., and Wambersie, A. *Dosimetrie en curietherapie*, Masson, Paris, 1982.
52. Pierquin, B., Wilson, J.F., and Chassagne, D. *Modern Brachytherapy*, Masson, New York, 1987.
53. Wickham, L. and Degrais, P. *Radiumtherapy*, English ed., Cassell, London, 1910.
54. W.J. Meredith, ed., *The Manchester System*, Livingstone, Edinburgh, 1947.
55. Sievert, R. Die Intensitätsverteilung der primaeren Gammastrahlung in der Naehe medizinscher Radiumpräparate, *Acta. Radiol.*, 1, 89–128, 1921.
56. Quimby, E.H. The effect of the size of radium applicators on skin doses, *Am. J. Roentgenol.*, 9, 671–683, 1922.
57. Quimby, E.H. The grouping of radium tubes in packs and plaques to produce the desired distribution of radiation, *Am. J. Roentgenol.*, 27, 18–39, 1932.
58. Quimby, E.H. Dosage tables for linear radium sources, *Radiology*, 43, 572–577, 1944.
59. Hilaris, B.S., Nori, D., and Anderson, L.L. *Atlas of Brachytherapy*, Macmillan, New York, 1988.
60. International Commission on Radiation Units and Measurements, Dose and volume specification for reporting interstitial therapy, ICRU Report 58. ICRU, Bethesda, 1997.
61. Anderson, L.L. A natural volume–dose histogram for brachytherapy, *Med. Phys.*, 13, 898–903, 1986.
62. Baltas, D., Kolotas, C., Geramani, K., Mould, R.F., Ioannides, G., Kekchidi, M., and Zamboglou, N. A conformal index (COIN) to evaluate implant quality and dose specification in brachytherapy, *Int. J. Radiat. Oncol. Biol. Phys.*, 40, 515–524, 1998.
63. Murdoch, J. Dosage in radium therapy, *Br. J. Radiol.*, 4, 256–284, 1931.
64. Paterson, R. and Parker, H.M. A dosage system for gamma-ray therapy, *Br. J. Radiol.*, 7, 592–632, 1934.
65. Paterson, R., Parker, H.M., and Spiers, F.W. A system of dosage for cylindrical distributions of radium, *Br. J. Radiol.*, 9, 487–508, 1936.
66. Paterson, R. and Parker, H.M. A dosage system for interstitial radium therapy, *Br. J. Radiol.*, 11, 252–266, see also pages 313–340, 1938.
67. Meyer, S. and Schweidler, E. *Raioactivität*, Teubner, Berlin, pp. 70–76, 1966.
68. Mayneord, W.V. The distribution of radiation around simple radioactive sources, *Br. J. Radiol.*, 5, 677–716, 1932.
69. Souttar, H.S. *Radium and its Surgical Applications*, Heinemann, London, 1929.

70.   Bagg, H. The action of buried tubes of radium emanation upon normal and neoplastic tissues, *Am. J. Roentgenol.*, 7, 535–543, 1920.

71.   Cutler, M. Comparison of the effects of filtered and unfiltered tubes buried in rabbit muscle, *Am. J. Roentgenol.*, 15, 1–35, 1926.

72.   Failla, G., et al., Dosage study of the therapeutic use of unfiltered radium, *Am. J. Roentgenol.*, 15, 1–35, 1926.

73.   Tod, M. and Meredith, W.J. A dosage system for use in the treatment of cancer of the uterine cervix, *Br. J. Radiol.*, 11, 809, 1938.

74.   Heyman, J., Reuterwall, O., and Benner, S. The Radiumhemmet experience with radiotherapy in cancer of the corpus of the uterus: Classification, method and treatment results, *Acta Radiol.*, 22, 31, 1941.

75.   Regaud, C. Services de curietherapie, *Radiophysiologie et Radiotherapie Recuil de Traveaux Biologiques, Techniques et Therapeutiques*, Vol. 2, C. Regaud, A. Lacassagne and R. Ferroux, eds., Institut du Radium, Paris, p. 218, 1922.

76.   Fletcher, G.H., Stovall, M., and Sampiere, V. *Carcinoma of the Uterine Cervix, Endometrium and Ovary*, Year Book Medical Publishers, Chicago, 1962.

77.   von Seuffert, E. Die Radiumbehanglung maligner Neubildungen in der Gynakologie, *Lehrbuch der Strahlentherapie: Gynakologie*, H. Meyer, ed., Urban and Schwarzenberg, Berlin, p. 940, 1929.

78.   Rotte, K. and Sauer, O. From radium to remote afterloading: German gynaecological experience 1903–1992 with special reference to Wurzburg, *International Brachytherapy*, R.F. Mould, ed., Nucletron, Veenendaal, pp. 581–587, 1992.

79.   Towpik, E. and Mould, R.F. eds. Maria Sklodowska-Curie memorial issue of the Polish, *Nowotwory J. Oncol.*, Nowotwory, Warsaw, 1998.

2

Radiation Quantities and Units

2.1 Introduction

This chapter considers quantities and units that are required for the practice of medical physics in radiation therapy, including those that are solely relevant to brachytherapy. Measurements are an essential part of brachytherapy physics both for routine and research aspects. Precision and accuracy of the spectrum of brachytherapy measurements depend on the particular application, which may for example involve equipment calibration or measurements on patients.

Unequivocal definitions of the quantities and units involved are essential for interchanges of results among professional scientists and must be made according to international standards[1,2] of the Système Internationale (SI), the International Bureau of Weights and Measures (BIPM),[3] International Organization for Standardization (ISO),[4] and International Commission on Radiation Units and Measurements (ICRU).[5]

According to ICRU Report 60[5] the term *physical quantity*[*] is used to describe quantitatively a physical phenomenon or a physical object. A unit is a selected reference sample of a quantity with which other quantities of the same kind are compared. Every quantity can be expressed as the product of a numerical value and a unit.

The set of quantities that can be used to derive other quantities are termed *base quantities*. All other quantities derived by multiplying or dividing base quantities are termed *derived quantities*.

A system of units is constructed by first defining the units for the base quantities: these are the base units. Second, the units for the derived quantities are defined: these are the derived units. Table 2.1 summarizes base units according to the SI system. In addition, there are two SI supplementary units that are relevant in brachytherapy physics. These are the quantity plane angle, which is given the name *radian* and the symbol rad, and the quantity solid angle, which is given the name *steradian* and the symbol sr.

---

[*] For reasons of simplicity, the term *quantity* will be used instead of *physical quantity*.

TABLE 2.1

SI Base and Supplementary Units

Category	Quantity	Unit Name	Unit Symbol
Base units	Length	Meter	m
	Mass	Kilogram	kg
	Time	Second	s
	Amount of substance	Mole	mol
	Electric current	Ampere	A
	Thermodynamic temperature	Kelvin	K
Supplementary units	Plane angle	Radian	rad
	Solid angle	Steradian	sr

Source: From Bureau International des Poids et Mesures (BIPM), *Le Système International d'Unités (SI)*, BIPM, Sèvre, 1998; International Organization for Standardization (ISO), *Handbook: Quantities and Units*, ISO, Geneva, 1993; International Commission on Radiation Units and Measurements, *ICRU Report 60*, ICRU, Bethesda, 1998. With permission.

Some derived SI units are given special names, such as coulomb for ampere second. Other derived units are given special names only when they are used with certain derived quantities. Special names currently in use in this restricted category are becquerel (equal to reciprocal second for activity of a radionuclide) and gray (equal to joule per kilogram for absorbed dose, kerma, cema, and specific energy) (Table 2.2).

There are also a few units outside of SI and some of their values in terms of SI units are obtained experimentally. Two examples in current use are electronvolt (eV) and (unified) atomic mass unit (u). Others, such as day, hour, and minute are not coherent with SI but, because of long usage, are allowed to be used with SI[5] (Table 2.3).

Table 2.4 gives SI prefixes for factors in the range 10^{24} to 10^{-24}.

The conversion factors for units that are no longer recommended but are still encountered are given in Table A.2.1 of Appendix 2.

The effects of radiation on matter depend on the radiation field and on the interactions between the radiation and the matter. The radiation field is defined by the radiometric quantities depicted in Section 2.3, where the interactions are described by the interaction quantities (coefficients) illustrated in Section 2.4. In order to be able to provide a physical measure that correlates with actual or potential effects of radiation on matter, a third group, the dosimetric quantities, are defined in Section 2.5. Finally, in Section 2.6 quantities that are needed to describe radioactivity are defined.

2.2   Ionization and Excitation

The process by which a neutral atom (in a natural state) gets a positive or negative charge is known as ionization. Processes that lead to the removal

TABLE 2.2

SI-Derived Units

Category	Quantity	Unit Name	Unit Symbol	Expression in Terms of Other Units
Special names in general use	Frequency	Hertz	Hz	s^{-1}
	Force	Newton	N	$kg\, m\, s^{-2}$
	Pressure	Pascal	Pa	$kg\, m^{-1}\, s^{-2}$
	Energy	Joule	J	N m
	Power	Watt	W	$J\, s^{-1}$
	Electric charge	Coulomb	C	A s
	Electric potential	Volt	V	$W\, A^{-1}$
	Celsius temperature	Degree Celsius	°C	K^a
Special names in restricted use	Activity	Becquerel	Bq	s^{-1}
	Absorbed dose, kerma, cema, specific energy	Gray	Gy	$J\, kg^{-1}$

[a] *Degree Celsius* is a special name for the Kelvin unit for use in stating values of temperature[5] (see also Appendix 2, Table A.2.1).

Source: From Bureau International des Poids et Mesures (BIPM), *Le Système International d'Unités (SI)*, BIPM, Sèvre, 1998; International Organization for Standardization (ISO), *Handbook: Quantities and Units*, ISO, Geneva, 1993; International Commission on Radiation Units and Measurements, *ICRU Report 60*, ICRU, Bethesda, 1998. With permission.

of an orbital electron from the atom result in a positively charged atom and a liberated electron: an ion pair. There are also cases where an electron is captured by a neutral atom which results in a negatively charged atom: a single negative ion. On the other hand, the process by which an orbital

TABLE 2.3

Units Used in SI

Category	Quantity	Unit Name	Unit Symbol	Expression in Terms of Other Units
Experimentally obtained	Energy	Electron volt	eV	$1.60217733 \times 10^{-19}$ C V
	Mass	(Unified) atomic mass unit	u	$1.6605402 \times 10^{-27}$ kg
Widely used	Time	Minute	min	60 s
		Hour	h	3600 s
		Day	d	8.640×10^4 s

Source: From Bureau International des Poids et Mesures (BIPM), *Le Système International d'Unités (SI)*, BIPM, Sèvre, 1998; International Organization for Standardization (ISO), *Handbook: Quantities and Units*, ISO, Geneva, 1993; International Commission on Radiation Units and Measurements, *ICRU Report 60*, ICRU, Bethesda, 1998. With permission.

TABLE 2.4

SI Prefixes

Factor	Prefix	Symbol	Factor	Prefix	Symbol
10^{24}	yotta	Y	10^{-1}	deci	d
10^{21}	zetta	Z	10^{-2}	centi	c
10^{18}	exa	E	10^{-3}	milli	m
10^{15}	peta	P	10^{-6}	micro	μ
10^{12}	tera	T	10^{-9}	nano	n
10^{9}	giga	G	10^{-12}	pico	p
10^{6}	mega	M	10^{-15}	femto	f
10^{3}	kilo	k	10^{-18}	atto	a
10^{2}	hecto	h	10^{-21}	zepto	z
10^{1}	deca	da	10^{-24}	yocto	y

Source: From Bureau International des Poids et Mesures (BIPM), *Le Système International d'Unités (SI)*, BIPM, Sèvre, 1998; International Organization for Standardization (ISO), *Handbook: Quantities and Units*, ISO, Geneva, 1993; International Commission on Radiation Units and Measurements, *ICRU Report 60*, ICRU, Bethesda, 1998. With permission.

electron is simply transferred to a higher energy level than that of its natural state in the atom is called excitation. Both ionization and excitation can occur when particles undergo collisions with the atoms or molecules.

2.2.1   Ionizing Radiation

The word *radiation* was used until the end of the 19th century to describe electromagnetic waves. Since the discovery of x-rays and radioactivity and the establishment of the theory of the duality of matter formulated by DeBroglie (1892–1987), radiation refers to the entire electromagnetic spectrum as well as to all atomic and subatomic particles. The term *ionizing radiation* refers to charged and uncharged particles that can produce ionization when penetrating matter.

Charged particles such as electrons and protons are usually characterized as directly ionizing radiation when they have sufficient kinetic energy to produce ionization by collision when they penetrate matter. Collision here does not necessarily mean an actual physical contact of the charge particle and the orbital electron; it is generally the Coulomb-force interaction between the electromagnetic fields associated with the charged particle and the orbital electron.

Charged particles are, of course, slowed down during this process and when their kinetic energy is reduced sufficiently, ionization becomes unlikely or impossible. From that point, charged particles dissipate their remaining energy, mainly in excitation or elastic scattering processes, and the initially ionizing particles become non-ionizing.

For the uncharged particles such as photons or neutrons, the term *indirectly ionizing radiation* is used. This is because uncharged particles produce directly ionizing particles in the matter when penetrating and interacting with matter. Based on this fact, the deposition of energy in matter by indirectly ionizing radiation is a two-step process.

Since the energy needed to cause an electron to escape from an atom can be as low as 4 eV, electromagnetic radiation of up to about 310 nm can be theoretically considered as ionizing radiation. This includes the majority of the ultraviolet light (UV) with wavelength $\lambda = 10-390$ nm. Due to the negligible penetration capability in matter and the restricted ionization of UV light, this is usually excluded from radiation physics. Consequently, ionizing radiation includes only electromagnetic radiation with wavelength up to 10 nm. Electromagnetic radiation with wavelength above 10 nm is non-ionizing radiation and includes radiowaves, microwaves, visible light ($\lambda = 390-770$ nm), and UV light.

2.2.2 Types and Sources of Ionizing Radiation

The types of ionizing radiation most relevant to radiation oncology can be divided into the following two categories.

2.2.2.1 *Electromagnetic Radiation*

Robert Maxwell (1831–1879) first used the term *electromagnetic* as a descriptive for oscillating electric and magnetic fields. The two kinds of electromagnetic radiation are γ-rays (gamma rays) and x-rays.

2.2.2.2 *γ-Rays*

γ-Rays are electromagnetic radiation emitted either from a nucleus or in annihilation reactions between matter and antimatter.

2.2.2.3 *X-Rays*

These are electromagnetic radiations emitted either by charged particles, usually electrons, when changing atomic energy levels (characteristic or fluorescence x-rays) or in slowing down processes in Coulomb-force interactions (bremsstrahlung).

The quantum energy of a γ or x electromagnetic photon is given by

$$E_\gamma = h\nu = \frac{hc}{\lambda} \tag{2.1}$$

where h is Planck's constant (see Appendix 2), ν is the frequency, λ is the wavelength of the electromagnetic radiation, and c is the speed of light in vacuum (Appendix 2). According to the above, a γ-ray photon and an x-ray photon of the same energy differ only in their origins. Furthermore, and in contrast to γ-rays, x-rays can also be produced artificially either using x-ray tubes with accelerating potentials up to 300 kV or linear accelerators with accelerating potentials usually in the range 4 to 25 MV.

2.2.2.4   Particulate Radiation

According to the particles comprising the radiation, the following three common types of particulate radiation exist.

2.2.2.5   Electrons

These are emitted either from the nucleus, result from a collision process by a charged particle, or are produced artificially by electron accelerators. The first type are called β-rays and can be positive, β^+, or negative, β^-, in charge (electrons and positrons). The second type are called δ-rays. Electron accelerators create pulsed electron beams of high energy, usually in the range 5 to 25 MeV.

2.2.2.6   Neutrons

Since neutrons have no charge, neutron radiation is generated only in nuclear reactions.

2.2.2.7   Heavy Charged Particles

We understand heavy charged particles generally to mean protons, deuterons (a nucleus of deuterium consists of one neutron and one proton), tritons (a nucleus of tritium consists of two neutrons and one proton), alpha particles (a helium nucleus consists of two neutrons and two protons), pions, and nuclei of other heavier atoms, such as carbon and nitrogen. With the exception of α particles, which can also be emitted by radioactive nuclei, all other types of heavy charged particle radiation are generated by accelerators.

2.3   Radiometry

Radiation fields of various types of ionizing particles, such as photons or electrons, are characterized by radiometric quantities that specify and describe the radiation field at a specific point of interest in free space or in material.

2.3.1   Radiant Energy

The number of particles emitted, transferred, or received at a point in a radiation field is called the particle number N. The energy of the particles emitted, transferred, or received is called the *radiant energy R* (J); see Table 2.5 for a summary of radiometric quantities and units. In this definition of *radiant energy* the rest energy of the particles considered is excluded.

Considering particles of energy E, then $R = NE$, where for E the rest energy of the particles is excluded. The distributions of the particle number N_E (J^{-1}) and of the radiant energy R_E with respect to energy (energy distribution of particle number and energy distribution of radiant

TABLE 2.5

Summary of Radiometric Quantities

Name	Symbol	Unit	Definition
Particle number	N	1	
Radiant energy	R	J	
Volumic particle number	N	m^{-3}	dN/dV
Energy distribution of particle number	N_E	J^{-1}	dN/dE
Energy distribution of radiant energy	R_E	1	dR/dE
Flux	$\dot{N}$	s^{-1}	dN/dt
Energy flux	$\dot{R}$	W	dR/dt
Energy distribution of flux	$\dot{N}_E$	$J^{-1}s^{-1}$	$d\dot{N}/dE$
Energy distribution of energy flux	$\dot{R}_E$	s^{-1}	$d\dot{R}/dE$
Fluence	Φ	m^{-2}	$dN/d\alpha$
Energy fluence	ψ	$J\,m^{-2}$	$dR/d\alpha$
Energy distribution of fluence	Φ_E	$J^{-1}m^{-2}$	$d\Phi/dE$
Energy distribution of energy fluence	ψ_E	m^{-2}	$d\Psi/dE$
Fluence rate	$\dot{\Phi}$	$m^{-2}s^{-1}$	$d\Phi/dt$
Energy fluence rate	$\dot{\Psi}$	$W\,m^{-2}$	$d\Psi/dt$
Particle radiance	$\dot{\Phi}_\Omega$	$m^{-2}s^{-1}sr^{-1}$	$d\dot{\Phi}/d\Omega$
Energy radiance	$\dot{\Psi}_\Omega$	$W\,m^{-2}sr^{-1}$	$d\dot{\Psi}/d\Omega$

energy) are

$$N_E = \frac{dN}{dE}, \qquad R_E = \frac{dR}{dE} \qquad \text{and} \qquad R_E = N_E E \qquad (2.2)$$

where dN is the number of particles with energy in the interval E to $E + dE$ and dR is the radiant energy of these particles.

The *volumic particle number* n (m^{-3}) is defined as the quotient of the number of particles dN in the volume element dV and the volume element dV; thus $n = dN/dV$.

2.3.2 Flux and Energy Flux

If dN is the incremental change of the particle number in the time interval dt, the *flux* $\dot{N}$ (s^{-1}) is defined as

$$\dot{N} = \frac{dN}{dt} \qquad (2.3)$$

Taking dR to be the incremental change of the radiant energy related to the particle number change dN within the time interval dt, the *energy flux* $\dot{R}$ (W) is defined as

$$\dot{R} = \frac{\mathrm{d}R}{\mathrm{d}t} \tag{2.4}$$

2.3.3   Fluence and Energy Fluence

If $\mathrm{d}N$ is the number of particles incident on a sphere of cross-sectional area $\mathrm{d}\alpha$, the fluence Φ (m^{-2}) is defined as

$$\Phi = \frac{\mathrm{d}N}{\mathrm{d}\alpha} \tag{2.5}$$

where $\mathrm{d}\alpha$ is *actually* considered as an area perpendicular to the direction of each particle. This is most simply described in terms of a sphere of cross-sectional area $\mathrm{d}\alpha$. If $\mathrm{d}R$ is the radiant energy corresponding to the number of particles $\mathrm{d}N$, the *energy fluence* Ψ $(\mathrm{J}\,\mathrm{m}^{-2})$ is given by

$$\Psi = \frac{\mathrm{d}R}{\mathrm{d}\alpha} \tag{2.6}$$

The energy distribution of fluence, Φ_{E} $(\mathrm{J}^{-1}\,\mathrm{m}^{-2})$, and the energy distribution of energy fluence, Ψ_{E} (m^{-2}), are given by

$$\Phi_{\mathrm{E}} = \frac{\mathrm{d}\Phi}{\mathrm{d}E} \qquad \Psi_{\mathrm{E}} = \frac{\mathrm{d}\Psi}{\mathrm{d}E} \qquad \text{and} \qquad \Psi_{\mathrm{E}} = \Phi_{\mathrm{E}}E \tag{2.7}$$

where $\mathrm{d}\Phi$ is the fluence of particles with energy in the interval E to $E + \mathrm{d}E$ and $\mathrm{d}\Psi$ is the corresponding energy fluence of these particles.

2.3.4   Fluence Rate and Energy Fluence Rate

Taking $\mathrm{d}\Phi$ to be the incremental change of the fluence observed in the time interval $\mathrm{d}t$, the *fluence rate* $\dot{\Phi}$ $(\mathrm{m}^{-2}\,\mathrm{s}^{-1})$ is

$$\dot{\Phi} = \frac{\mathrm{d}\Phi}{\mathrm{d}t} \tag{2.8}$$

If $\mathrm{d}\Psi$ is the incremental change of the energy fluence in that time interval $\mathrm{d}t$, the energy fluence rate $\dot{\Psi}$ $(\mathrm{W}\,\mathrm{m}^{-2})$ is

$$\dot{\Psi} = \frac{\mathrm{d}\Psi}{\mathrm{d}t} \tag{2.9}$$

2.3.5   Particle Radiance and Energy Radiance

If $\mathrm{d}\dot{\Phi}$ is the fluence rate of particles propagating within a solid angle $\mathrm{d}\Omega$ around a specific direction, the *particle radiance* $\dot{\Phi}_{\Omega}$ $(\mathrm{m}^{-2}\,\mathrm{s}^{-1}\,\mathrm{sr}^{-1})$ is defined

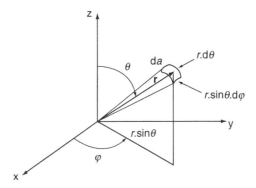

FIGURE 2.1
Schematic representation of the solid angle $d\Omega$ around a specific direction in space as defined by the vector **r**. Using spherical coordinates, **r** is expressed as (r,θ,φ). The small surface element da perpendicular to **r** and lying on a sphere of radius r is given by $da = r^2\sin\theta\, d\theta\, d\varphi$. Consequently, the solid angle $d\Omega$ is given by $d\Omega = \sin\theta\, d\theta\, d\varphi$.

as (see also Figure 2.1)

$$\dot{\Phi}_\Omega = \frac{d\dot{\Phi}}{d\Omega} \tag{2.10}$$

The *energy radiance* $\dot{\Psi}_\Omega (\mathrm{W\ m^{-2}\ sr^{-1}})$ is

$$\dot{\Psi}_\Omega = \frac{d\dot{\Psi}}{d\Omega} \tag{2.11}$$

where $d\dot{\Psi}$ is the corresponding energy fluence rate.

2.4   Interaction Coefficients

Various interaction processes may occur as particles pass through material. For a specific interaction the following are possible:

1. The energy of the incident particle is changed.
2. The propagation direction of the particle is changed.
3. Both the energy and the direction of the incident particle are changed.
4. The particle is absorbed.

In addition, this interaction process can be followed by the emission of one or more secondary particles. The probability of occurrence of a specific interaction process for a specific type of radiation (particle) in a specific target material is characterized by *interaction coefficients*. The fundamental interaction coefficient is the *cross-section*.

2.4.1 Total and Differential Cross-Section

The cross-section σ (m^2) of a target entity for a particular type of interaction with a particular type of particle is defined as

$$\sigma = \frac{\langle N_{\text{int}} \rangle}{\Phi} \tag{2.12}$$

where $\langle N_{\text{int}} \rangle$ is the expected number of this type of interaction with a single target entity and Φ is the particle fluence. For a unit fluence Φ, $\langle N_{\text{int}} \rangle$ can be interpreted as the probability P of occurrence of this type of interaction with a single target entity.

Although the SI unit for cross-section is m^2, there is a special unit, the barn (b), that is still in widespread use: $1 \, b = 10^{-28} \, m^2$.

In general, a full description of an interaction process requires knowledge of the distributions of the cross-sections in terms of the energy and direction of all particles emerging as a result of the interaction. Such distributions are obtained by differentiations of the cross-section σ with respect to energy E and solid angle Ω and are usually called *differential cross-sections*.

Taking the case where the incident particles undergo different types of interaction in a target, the resulting cross-section is usually called the *total cross-section* σ and is defined as the sum of all cross-sections σ_i of the partial specific interactions:

$$\sigma = \sum_i \sigma_i = \frac{1}{\Phi} \sum_i \langle N_{\text{int}} \rangle_i \tag{2.13}$$

where i is the index over all types of interactions that occurred in the target when it was subjected to the particle fluence Φ. $\langle N_{\text{int}} \rangle_i$ is the expected number of the ith interaction with the single target entity under consideration.

2.4.2 Linear and Mass Attenuation Coefficient

If we consider N *uncharged* particles, such as photons traversing a distance dl in a material of density ρ, then if dN particles experience interactions with the material, we can define the *linear attenuation coefficient* μ (m^{-1}) (Figure 2.2) as

$$\mu = \frac{1}{dl} \frac{dN}{N} \tag{2.14}$$

According to this definition of μ, the probability that a particle at a normal incidence in a layer of material of thickness dl undergoes an interaction is μdl.

The linear attenuation coefficient depends on the density ρ of the material and, in order to overcome this dependence, the *mass attenuation coefficient* μ/ρ

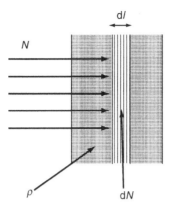

FIGURE 2.2
N particles traverse a distance $\mathrm{d}l$ in a material of density ρ, where $\mathrm{d}N$ particles out of N initial particles experience interactions with that material.

$(\mathrm{m}^2\,\mathrm{kg}^{-1})$ is defined as

$$\frac{\mu}{\rho} = \frac{1}{\rho}\frac{1}{\mathrm{d}l}\frac{\mathrm{d}N}{N} \tag{2.15}$$

If σ is the total cross-section for the specific uncharged particles and the specific material, by combining Equation 2.5, Equation 2.12, and Equation 2.15 it can be easily shown that

$$\frac{\mu}{\rho} = \frac{N_A}{M}\sigma = \frac{N_A}{M}\sum_i \sigma_i \tag{2.16}$$

where N_A is the Avogadro constant (see Table A.2.2 of Appendix 2) and M is the molar mass of the target material, N_A/M is the number of target entities per unit mass of the target material, and i is the index over all types of interactions and the summation is extended over all cross-sections σ_i of the specific interactions (see also Equation 2.12). Taking the *volumic number of target entities* n_t (m^{-3}) to be the number of target entities in a volume element of the target divided by its volume, the above expression can also be written as

$$\frac{\mu}{\rho} = \frac{n_t}{\rho}\sigma \tag{2.17}$$

If the material under consideration is a compound, usually the approximation is made that consists of independent atoms and the mass attenuation coefficient of that compound material is given by

$$\frac{\mu}{\rho} = \frac{1}{\rho}\sum_j \left[(n_t)_j\sigma_j\right] = \frac{1}{\rho}\sum_j \left[(n_t)_j\sum_i \sigma_{i,j}\right] \tag{2.18}$$

where $(n_t)_j$ is the volumic number of target entities of the jth type, σ_j is the total cross-section for the jth target entity, and $\sigma_{i,j}$ is the cross-section for the ith specific interaction for the jth single target entity. Since the above expression ignores the changes in the molecular, chemical, or crystalline environment of an atom, for some cases (e.g., interaction of low energy photons with molecules) it can lead to errors.

2.4.3 Atomic and Electronic Attenuation Coefficient

The attenuation capacity per atom of the target material is given by the *atomic attenuation coefficient* μ_{atom} expressed in cm^2 (usually). Taking A as the atomic mass of the target material, the mass of one atom is A/N_A with N_A the Avogadro constant. In this way μ_{atom} is given by

$$\mu_{atom} = \frac{\mu}{\rho} \frac{A}{N_A} \tag{2.19}$$

Similarly, the attenuation capacity per electron in the target material is given by the *electronic attenuation coefficient* $\mu_{electron}$ (also usually expressed in cm^2). For neutral atoms and with Z the atomic number of the target material, $\mu_{electron}$ is

$$\mu_{electron} = \frac{\mu_{atom}}{Z} = \frac{1}{Z} \frac{\mu}{\rho} \frac{A}{N_A} \tag{2.20}$$

where $(N_A/A)Z$ is the number of electrons per unit mass of the target material.

2.4.4 Mass Energy Transfer Coefficient

If we consider N *uncharged* particles each of a given energy E (excluding rest energy) traversing a distance dl in a material of density ρ, the incident radiant energy R for this target is $R = NE$. If dR_{tr} is the radiant energy that is transferred through interactions along the distance dl to the kinetic energy of charged particles, the mass energy transfer coefficient μ_{tr}/ρ (m^2 kg^{-1}) is defined as

$$\frac{\mu_{tr}}{\rho} = \frac{1}{\rho} \frac{1}{dl} \frac{dR_{tr}}{R} \tag{2.21}$$

For photon radiation, the binding energy is usually included in μ_{tr}/ρ.

In general, the incident particles (of specific type and given energy) can produce several types of interactions in a target entity. In this case the mass energy transfer coefficient μ_{tr}/ρ is given by

$$\frac{\mu_{tr}}{\rho} = \frac{N_A}{M} \sum_i [f_i \sigma_i] \tag{2.22}$$

where N_A is the Avogadro constant (see Table A.2.2 of Appendix 2), M is the molar mass of the target material, i is the index over all types of interactions, and the summation is extended over all cross-sections σ_i of the specific interactions (see also Equation 2.13). f_i is the average fraction of the incident particle energy that is transferred to the kinetic energy of charged particles in the ith type of interaction: $f_i = \bar{E}_{tr,i}/E$ with $\bar{E}_{tr,i}$ the average energy transferred to the kinetic energy of the charged particles for that ith interaction.

Combining Equation 2.16 and Equation 2.22, the following relation between mass energy transfer and mass attenuation coefficients is derived:

$$\frac{\mu_{tr}}{\rho} = \frac{\mu}{\rho} \frac{\sum_i [f_i \sigma_i]}{\sum_i f_i} \tag{2.23}$$

or, by defining $f = \sum_i [f_i \sigma_i]/\sum_i f_i$, we have

$$\frac{\mu_{tr}}{\rho} = \frac{\mu}{\rho} f \tag{2.24}$$

For a compound material, by making the same approximation as for the mass attenuation coefficient (Equation 2.18), namely that the material consists of independent atoms, and in this way the mass attenuation coefficient of that compound material, the mass energy transfer coefficient μ_{tr}/ρ is given by

$$\frac{\mu_{tr}}{\rho} = \frac{1}{\rho} \sum_j \left[(n_t)_j \cdot \sum_i [f_{i,j} \sigma_{i,j}] \right] \tag{2.25}$$

where $(n_t)_j$ is the volumic number of target entities of the jth type and $\sigma_{i,j}$ and $f_{i,j}$ are the cross-section and the average fraction of the incident particle energy that is transferred to the kinetic energy of charged particles for the ith specific interaction for the jth single target entity, respectively. As in Equation 2.18, the above expression ignores the changes in the molecular, chemical, or crystalline environment of an atom in some cases (e.g., interaction of low energy photons with molecules) and it can lead to errors.

2.4.5 Mass Energy Absorption Coefficient

If we consider again uncharged particles traversing through a target material of density ρ, and if g is the fraction of the energy of the charged particles rebated in this material as a result of the interactions that have occurred, the *mass energy absorption coefficient* μ_{en}/ρ ($m^2 \ kg^{-1}$) is defined as

$$\frac{\mu_{en}}{\rho} = \frac{\mu_{tr}}{\rho}(1 - g) \tag{2.26}$$

For compound materials, unless g has a sufficiently small value, the mass energy absorption coefficient cannot be calculated simply by a summation of the mass absorption coefficients of the atomic constituents as in the case of the mass energy transfer coefficient (Equation 2.25).

2.4.6 Linear and Mass Stopping Power

If we consider *charged* particles such as electrons traversing a distance dl in a material of density ρ, then if dE is the energy lost by such a particle along this distance we define the *linear stopping power* S (J m^{-1}) as

$$S = \frac{\mathrm{d}E}{\mathrm{d}l} \tag{2.27}$$

The *mass stopping power* S/ρ is, according to the above equation, defined as

$$\frac{S}{\rho} = \frac{1}{\rho}\frac{\mathrm{d}E}{\mathrm{d}l} \tag{2.28}$$

The SI unit for S/ρ is J m^2 kg^{-1}. Since energy can also be expressed in eV, the mass stopping power can be expressed in terms of eV m^2 kg^{-1} or multiples or submultiples of eV m^2 kg^{-1}.

If we consider separately the three components of the energy loss of charged particles traversing the target material; namely:

1. The energy lost due to collisions with electrons
2. Energy losses due to emission of bremsstrahlung in the electric field of atomic nuclei or atomic electrons
3. Energy losses due to elastic Coulomb collisions in which recoil energy is imparted to atoms

we can define the following three independent mass stopping power components:

Mass electronic (or collision) stopping power

$$\frac{S_{\mathrm{el}}}{\rho} = \frac{1}{\rho}\left(\frac{\mathrm{d}E}{\mathrm{d}l}\right)_{\mathrm{el}}$$

Mass radiative stopping power

$$\frac{S_{\mathrm{rad}}}{\rho} = \frac{1}{\rho}\left(\frac{\mathrm{d}E}{\mathrm{d}l}\right)_{\mathrm{rad}}$$

Mass nuclear stopping power

$$\frac{S_{\mathrm{nuc}}}{\rho} = \frac{1}{\rho}\left(\frac{\mathrm{d}E}{\mathrm{d}l}\right)_{\mathrm{nuc}}$$

This means that the mass stopping power is

$$\frac{S}{\rho} = \frac{S_{\mathrm{el}}}{\rho} + \frac{S_{\mathrm{rad}}}{\rho} + \frac{S_{\mathrm{nuc}}}{\rho} \tag{2.29}$$

The separate mass stopping power components can be expressed in terms of cross-sections. Taking the mass electronic or mass collision stopping power S_{el}/ρ, it can be described by the following expression:

$$\frac{S_{el}}{\rho} = \frac{N_A}{M} Z \int w \frac{d\sigma}{dw} dw \qquad (2.30)$$

where N_A is the Avogadro number, M is the molar mass of the target material, Z is its atomic number, $d\sigma/dw$ is the differential cross-section per atomic electron for collisions, and w is the energy loss.

2.4.7 Linear Energy Transfer (LET)

The *linear energy transfer* (LET) (also called *restricted linear electronic stopping power*) L_Δ (J m^{-1}) of a material and for charged particles is defined as

$$L_\Delta = \frac{dE_\Delta}{dl} \qquad (2.31)$$

where dE_Δ is the energy lost by a charged particle due to electronic collisions in traversing a distance dl in the material reduced by the sum of the kinetic energies of all electrons released by the charged particle along dl with kinetic energies higher than the value Δ. Since energy can also be expressed in eV, L_Δ can be expressed in terms of eV m^{-1} or multiples or submultiples of eV m^{-1} (e.g., keV μm^{-1}).

Taking Equation 2.29 for the definition of the electronic stopping power, the linear energy transfer L_Δ can also be given by

$$L_\Delta = S_{el} - \frac{dE_{ke,\Delta}}{dl} \qquad (2.32)$$

where S_{el} is the linear electronic stopping power and $dE_{ke,\Delta}$ is the sum of the kinetic energies greater than Δ of all electrons released by the charged particle along its travel distance dl in the material in question. The lower the value of Δ, the stricter the applicability of L_Δ to the track of the primary particle. For simplicity, Δ is expressed in eV. Thus, L_{100} is the linear energy transfer for an energy cut-off $\Delta = 100$ eV.

According to the above definition, L_0 considers the energy lost by the charged particle that does not reappear as kinetic energy of electrons released along the particle path dl. On the other hand, $L_\infty = S_{el}$ and is also called *unrestricted linear energy transfer*.

2.4.8 Mean Energy Expended in a Gas per Ion Pair Formed

If we consider charged particles traversing a volume filled by a gas, and if N is the mean number of ion pairs formed when the initial kinetic energy E of a charged particle is completely dissipated in the gas, the *mean energy expended in a gas per ion pair formed* W (J) is defined as

$$W = \frac{E}{N} \tag{2.33}$$

Electronvolts (eV) can also be used as a unit for W. According to the above definition, N also includes ion pairs that are produced by bremsstrahlung or other secondary radiation produced by the charged particle.

Table 2.6 gives a summary of all interaction coefficients discussed in this chapter.

TABLE 2.6

Summary of Interaction Coefficients

Name	Symbol	Unit	Definition
Cross-section	σ	m^2	P/Φ
Linear attenuation coefficient	μ	m^{-1}	$(1/dl)(dN/N)$
Mass attenuation coefficient	μ/ρ	$m^2\,kg^{-1}$	$(1/\rho)(1/dl)(dN/N)$
Atomic attenuation coefficient	μ_{atom}	m^2	$(\mu/\rho)(A/N_A)$
Electronic attenuation coefficient	$\mu_{electron}$	m^2	$(1/Z)(\mu/\rho)(A/N_A)$
Mass energy transfer coefficient	μ_{tr}/ρ	$m^2\,kg^{-1}$	$(1/\rho)(1/dl)(dR_{tr}/R)$
Mass energy absorption coefficient	μ_{en}/ρ	$m^2\,kg^{-1}$	$(\mu_{tr}/\rho)(1-g)$
Linear stopping power	S	$J\,m^{-1}$	dE/dl
Mass stopping power	S/ρ	$J\,m^2\,kg^{-1}$	$(1/\rho)(dE/dl)$
Linear energy transfer	L_Δ	$J\,m^{-1}$	dE_Δ/dl
Mean energy expended in a gas per ion pair formed	W	J	E/N

2.5 Dosimetry

Dosimetric quantities are quantities defined in order to provide a physical measure that correlates with the effects of radiation on matter: actual or potential effects. They describe the processes of interaction of radiation with matter: conversion/transfer of particle energy and energy deposition in matter. There are dosimetric quantities describing the conversion part of the radiation interaction processes and the energy deposition processes. It is obvious that the dosimetric quantities result directly from the radiometric quantities and the interaction coefficients.

2.5.1 Energy Conversion

Energy conversion means transfer of energy from ionizing particles to secondary ionizing particles.

2.5.1.1 *Kerma*

The quantity *kerma* (from the *kinetic energy released* per unit *mass*) refers to the kinetic energy of charged particles, e.g., electrons and positrons, that have been liberated by uncharged particles such as photons. Kerma does not include the energy that has been expended against the binding energies of these charged particles, even if this is usually a relatively small component.

If dE_{tr} is the sum of the initial kinetic energies of all charged particles liberated by uncharged particles within a volume element dV of a material containing a mass $dm = \rho dV$ of that material, the kerma K ($J\,kg^{-1}$) is given by

$$K = \frac{dE_{tr}}{dm} = \frac{1}{\rho}\frac{dE_{tr}}{dV} \tag{2.34}$$

The name of the SI unit for kerma is gray (Gy): $1\,Gy = 1\,J\,kg^{-1}$. It should be noted that dE_{tr} also includes the kinetic energy of Auger electrons.

If we consider uncharged particles of a given energy E (monoenergetic particles), and if the fluence of these uncharged particles at the position of the volume element dV is Φ, the kerma K is related to Φ according to

$$K = \Phi E\left(\frac{\mu_{tr}}{\rho}\right) \tag{2.35}$$

where μ_{tr}/ρ is the mass energy transfer coefficient of the material for these particles of energy E.

The kerma per unit fluence K/Φ is called the kerma coefficient for uncharged particles of energy E in a specific material.

For the monoenergetic uncharged particles of energy E, the energy fluence ψ is given by $\psi = \Phi E$, and thus

$$K = \Psi\left(\frac{\mu_{tr}}{\rho}\right) \tag{2.36}$$

Now for uncharged particles of a given energy distribution, the kerma K is given according to

$$K = \int_E \Phi_E E\left(\frac{\mu_{tr}}{\rho}\right)_E dE$$

or

$$\tag{2.37}$$

$$K = \int_E \Psi_E \left(\frac{\mu_{tr}}{\rho}\right)_E dE$$

where Φ_E and ψ_E are the energy distribution of fluence and energy fluence of the uncharged particles, respectively, and $(\mu_{tr}/\rho)_E$ is the mass energy transfer coefficient of the material under consideration for uncharged particles of energy E.

According to the above expressions, kerma can be defined for a specific material (that is enclosed in the volume element dV in Equation 2.34) at a point inside a different material. Thus, one can speak about the air kerma in air or air kerma in water, which are the kerma in air-filled volume element dV at a point in air medium and at a point in water medium, respectively.

2.5.1.2   Kerma Rate

The kerma rate $\dot{K}$ (J kg^{-1} s^{-1}) is the quotient of the kerma increment dK occurring in the time interval dt by the time interval dt:

$$\dot{K} = \frac{dK}{dt} \tag{2.38}$$

If the special name gray is used, the unit of kerma rate is gray per second: $1 \, \text{Gy s}^{-1} = 1 \, \text{J kg}^{-1} \text{s}^{-1}$.

   As with the case of kerma, kerma rate can be defined for a specific material at a point inside a different material (medium).

2.5.1.3   Exposure

Let us consider again a volume element dV filled by air of mass dm at a point in a photon radiation field. If dQ is the absolute value of the total charge of the ions of one sign produced in air when all the electrons and positrons liberated or created by photons in the air-filled volume element dV are completely stopped in this volume, then the *exposure* X (C kg^{-1}) is defined as

$$X = \frac{dQ}{dm} = \frac{1}{\rho_a} \frac{dQ}{dV} \tag{2.39}$$

where ρ_α is the density of air in volume dV. Note that, according to this definition, ionization produced by Auger electrons is included in dQ whereas ionization produced by radiative processes such as bremsstrahlung and fluorescence photons is not included in dQ.

   The exposure X can be expressed in terms of the energy distribution of fluence Φ_E or of the energy distribution of energy fluence ψ_E of the photons at that point of consideration (at dV) in analogy to Equation 2.37 for the kerma K:

$$X = \frac{e}{W} \int_E \Phi_E E \left(\frac{\mu_{tr}}{\rho} \right)_E (1 - g_a) dE$$

or   (2.40)

$$X = \frac{e}{W} \int_E \psi_E \cdot \left(\frac{\mu_{tr}}{\rho} \right)_E (1 - g_a) dE$$

where e is the elementary charge and W is the mean energy expended in air per ion pair formed (see also Section 2.4.8).

   g_α is the fraction of the energy of the electrons liberated by photons in air that are lost to radiative processes (bremsstrahlung and fluorescence).

Finally, $(\mu_{tr}/\rho)_E$ is the mass energy transfer coefficient of air for photons of energy E. The integral in the two equations above is carried out over the energy spectrum of the photon radiation field at the position of the air-filled volume element dV.

For photon energies of the order of 1 MeV or below, g has a small value (for the air-filled volume element dV considered here, g_α is less than 0.3% for 1.0 MeV photons and less than 0.03% for 0.1 MeV photons[6]). In this case and by taking into account Equation 2.37, the above expressions for the exposure X can be approximated by

$$X = \frac{e}{W} K(1 - \bar{g}_a) \tag{2.41}$$

where K is the air kerma of the primary photons in the volume element dV and $\bar{g}_a$ is the mean value of g_α calculated over the distribution of the air kerma with respect to the liberated electron energy.

In many publications the product of the last two terms on the right-hand side of the above equation is called the *collision kerma* K_{col}:

$$K_{col} = K(1 - \bar{g}_a) \tag{2.42}$$

According to this, the collision kerma K_{col} is the part of the air kerma K which is defined by that part of the initial kinetic energy of the electrons liberated by photons in air which is expended totally by inelastic collisions with atomic electrons (see also *mass electronic stopping power* in Section 2.4.6). For photon energies of 1 MeV or below K_{col} is the major part of the air kerma K: only a small part of the initial kinetic energy of the electrons in air is expended by radiative processes ($g_\alpha < 0.3\%$ bremsstrahlung).

Like kerma, exposure can be defined for different materials (medium) surrounding the air-filled volume element dV; one can speak about exposure X in air or in water.

2.5.1.4 Exposure Rate

The exposure rate $\dot{X}$ (C kg^{-1} s^{-1}) is defined as the quotient of the increment of exposure dX observed in the time interval dt by this time interval:

$$\dot{X} = \frac{dX}{dt} \tag{2.43}$$

2.5.1.5 Cema

The quantity *cema* (from *Converted Energy* per unit *Mass*) refers to the energy lost by charged particles in electronic collisions in a specific material. This energy loss includes the charged particles energy expended against binding energy and any kinetic energy of the liberated electrons. These latest are called *secondary electrons*. The secondary electrons are not considered for the determination of cema.

If dE_c is that energy lost by the charged particles in electronic collisions within a volume element dV of a material containing a mass $dm = \rho dV$

of that material, the cema C (J kg^{-1}) is given by

$$C = \frac{dE_c}{dm} = \frac{1}{\rho}\frac{dE_c}{dV} \tag{2.44}$$

The special name of the SI unit for cema is gray (Gy): 1 Gy $= 1$ J kg^{-1}.

In the above definition the energy subsequently lost by all secondary electrons within the volume dV is excluded from dE_c.

If we consider now the mass electronic (collision) stopping power (see also Section 2.4.6) of the specified material that fills the volume element dV for the considered charged particles of energy E, $(S_{el}/\rho)_E$, the above equation for cema becomes

$$C = \int_E \Phi_E \left(\frac{S_{el}}{\rho}\right)_E dE \tag{2.45}$$

where Φ_E is the energy distribution of the charged particle fluence. According to the definition of cema, Φ_E does not consider any contribution from secondary electrons.

In this expression for the cema C the electronic stopping power S_{el} can be replaced by the corresponding unrestricted linear energy transfer L_∞ (see also Section 2.4.7).

According to the above expressions, cema can be defined for a specific material (that is enclosed in the volume element dV) at a point inside a different material (medium). Thus, one can speak about the air cema in air or the water cema in air, which are the cema in air-filled volume element dV at a point in air medium and the cema in water-filled volume element dV at a point in air medium, respectively.

2.5.1.6 Cema Rate

The cema rate $\dot{C}$ (J kg^{-1} s^{-1}) is defined as the quotient of the increment of cema dC observed in the time interval dt by this time interval:

$$\dot{C} = \frac{dC}{dt} \tag{2.46}$$

If the special name gray is used, the unit of cema rate is gray per second: 1 Gy s$^{-1} = 1$ J kg^{-1} s^{-1}.

Furthermore, as in the case of cema, cema rate can be defined for a specific material at a point inside a different material (medium).

2.5.2 Deposition of Energy

The energy deposition in matter is a stochastic process and the quantities discussed in the following for describing this are stochastic: their values are not unique but they follow a probability distribution.

2.5.2.1 Energy Deposit

Let us consider an ionizing particle that undergoes a single interaction i in matter. The energy deposited by this particle in this single interaction is called *energy deposit* ε_i (J) and is defined according to

$$\varepsilon_i = \varepsilon_{in} - \varepsilon_{out} + Q \tag{2.47}$$

ε_{in} is the energy of the ionizing particle before the interaction with its rest energy is excluded and ε_{out} is the sum of the energies of all ionizing particles leaving this interaction where their rest energies are again excluded. Q is the change occurring in the rest energies (masses) of the nucleus and of all particles that are involved in the interaction. If there is a decrease of rest energy, Q has a positive value; otherwise Q has a negative value.

The point where the interaction takes place and where the ionizing particle loses kinetic energy is termed the *transfer point*; the energy deposit ε_i is the energy deposited by the ionizing particle at the transfer point.

2.5.2.2 Energy Imparted

If we consider now all energy deposits ε_i that take place within a given volume in matter, the *energy imparted* ε (J) is defined as the total energy deposited in that volume expressed as the sum of all these energy deposits:

$$\varepsilon = \sum_i \varepsilon_i \tag{2.48}$$

Given that the energy deposit is the result of a single interaction of an ionizing particle at some transfer point within the volume under consideration, we can express the *mean energy imparted* $\bar{\varepsilon}$ to that volume in matter in terms of radiant energy (see Section 2.3.1) as

$$\varepsilon = R_{in} - R_{out} + \sum Q \tag{2.49}$$

where R_{in} is the radiant energy of all ionizing particles entering the volume and R_{out} is the radiant energy of all ionizing particles leaving that volume. The summation term in the above equation extends over all changes Q of the rest energy of nuclei and particles occurring in the volume. Here, the same notation is used as for the case of energy deposit in Equation 2.47: if there is a decrease of rest energy, Q has a positive value; otherwise Q has a negative value.

It is common to use the term *event* when dealing with energy deposition and interactions in matter. Event characterizes the imparting of energy to matter by statistically correlated particles. Thus, the primary photon and the electrons are liberated by it or by the electron–positron particle pair.

2.5.2.3 Absorbed Dose

If $d\bar{\varepsilon}$ is the mean energy imparted to matter in a volume dV of mass $dm = \rho dV$ then the *absorbed dose D* ($J\,kg^{-1}$) is defined as

$$D = \frac{d\bar{\varepsilon}}{dm} = \frac{1}{\rho}\frac{d\bar{\varepsilon}}{dV} \qquad (2.50)$$

The name of the SI unit for absorbed dose is gray (Gy): $1\,Gy = 1\,J\,kg^{-1}$.

2.5.2.4 Absorbed Dose Rate

The *absorbed dose rate* $\dot{D}$ ($J\,kg^{-1}\,s^{-1}$) is the quotient of the increment of absorbed dose dD observed in the time interval dt by this time interval:

$$\dot{D} = \frac{dD}{dt} \qquad (2.51)$$

As in the case of kerma rate, if the special name gray is used, the unit of absorbed dose rate is gray per second: $1\,Gy\,s^{-1} = 1\,J\,kg^{-1}\,s^{-1}$.

Table 2.7 summarizes all dosimetric quantities and their units as described in this chapter.

TABLE 2.7

Summary of Dosimetric Quantities

Category	Name	Symbol	Unit	Definition
Energy conversion	Kerma	K	$J\,kg^{-1} = Gy$	dE_{tr}/dm
	Kerma coefficient		$J\,m^2\,kg^{-1} = Gy\,m^2$	K/Φ
	Kerma rate	$\dot{K}$	$J\,kg^{-1}\,s^{-1} = Gy\,s^{-1}$	dK/dt
	Exposure	X	$C\,kg^{-1}$	dQ/dm
	Exposure rate	$\dot{X}$	$C\,kg^{-1}\,s^{-1}$	dX/dt
	Cema	C	$J\,kg^{-1} = Gy$	dE_c/dm
	Cema rate	$\dot{C}$	$J\,kg^{-1}\,s^{-1} = Gy\,s^{-1}$	dC/dt
Energy deposition	Energy deposit	ε_i	J	$\varepsilon_{in} - \varepsilon_{out} + Q$
	Energy imparted	ε	J	$\sum \varepsilon_i$
	Absorbed dose	D	$J\,kg^{-1} = Gy$	$d\varepsilon/dm$
	Absorbed dose rate	$\dot{D}$	$J\,kg^{-1}\,s^{-1} = Gy\,s^{-1}$	dD/dt

2.6 Radioactivity

A *nuclide* is a species of atoms that have a specified number of protons (atomic number Z) and a specified number of neutrons (N) in their nucleus. The property of several nuclides (unstable nuclides) spontaneously, without any external influence such as temperature or pressure, and under emission of irradiation to be transformed (disintegrated) in other stable nuclides is described as *radioactivity*. In this process, the whole atom (nuclei and orbital electrons) is involved. The unstable nuclides are called *radionuclides*. Radioactivity is a *stochastic* process; its values follow a specific probability distribution. In a number of nuclei of a specific radionuclide it is not possible to predict which of them will undergo a spontaneous nuclear transformation (i.e., will *decay*) and when. Any nucleus may decay in any given interval with a certain probability.

2.6.1 Decay Constant

If dP is the probability that a given nucleus of a radionuclide in a particular energy state (i.e., the ground energy state of the radionuclide) undergoes a spontaneous nuclear transformation from that energy state in the time interval dt, the *decay constant* λ (s^{-1}) is defined as

$$\lambda = \frac{dP}{dt} \qquad (2.52)$$

According to this the *half-life* $T_{1/2}$ (s) of a radionuclide is defined as

$$T_{1/2} = \frac{\ln(2)}{\lambda} \qquad (2.53)$$

$T_{1/2}$ is the time increment required for the number of radionuclides in the particular energy state to be reduced to one half of this number.

Another quantity of great practical importance is the *mean life* τ of a radionuclide. It is defined as the inverse of the *decay constant* and represents the mean lifetime of an individual nucleus:

$$\tau = \frac{1}{\lambda} = \frac{T_{1/2}}{\ln(2)} \qquad (2.54)$$

2.6.2 Activity

In order to describe quantitatively the time rhythm of the occurrence of spontaneous nuclear transformations (decays) of an amount of a specific radionuclide, the *activity* A (s^{-1}) is defined as

$$A = \frac{dN}{dt} \qquad (2.55)$$

where dN is the number of decays observed during the time interval dt. The special name for the unit of activity is becquerel (Bq). Until recently, the unit used to describe activity has been the *curie* (Ci): 1 Ci $= 3.7 \times 10^{10}$ Bq, but the use of this unit is no longer recommended even if it is still present in several publications.

dP in Equation 2.52 is given by $dP = dN/N$. Combining Equation 2.52 and Equation 2.55 and assuming that a number $N(t)$ of nuclei of the specific radionuclide at this particular energy state are available in an amount of that radionuclide at time t, then the activity $A(t)$ of that radionuclide amount at that time t is

$$A(t) = \lambda N(t) \qquad (2.56)$$

Taking dN to be the change (reduction, $dN < 0$) of the number of atoms of that radionuclide at a specific time t, $dP = -dN/N$. Applying this in Equation 2.52 and using Equation 2.55, we obtain

$$\frac{dN}{N(t)} = -\lambda dt$$

and thus

$$N(t) = N_0 e^{-\lambda t} \tag{2.57}$$

where N_0 is the initial number of the atoms (nuclei) of that radionuclide: $N_0 = N(t = 0)$.

Combining Equation 2.56 and Equation 2.57 we obtain for the activity $A(t)$:

$$A(t) = A_0 e^{-\lambda t} \tag{2.58}$$

with $A_0 = \lambda N_0$. This is the initial activity of the radionuclide.

2.6.2.1 Specific Activity

The *specific activity* A_{specific} (Bq kg^{-1}) of a radioactive source of mass m containing a single radionuclide of activity A is given by

$$A_{\text{specific}} = \frac{A}{m} \tag{2.59}$$

Using the decay constant λ, the specific activity A_{specific} for a pure radionuclide can be calculated by

$$A_{\text{specific}} = \lambda \frac{N_A}{M} \tag{2.60}$$

where N_A is the Avogadro constant (see Table A.2.2 of Appendix 2) and M is the molar mass of the radionuclide.

2.6.3 Air Kerma-Rate Constant

Let us consider a source of irradiation being a point in free space (vacuum), an *ideal point source*, consisting of a specific radionuclide and having an activity A. If $\dot{K}_\delta$ is the air kerma rate due to photons of energy greater than a certain cut-off value δ at a distance r from that point source, the *air kerma-rate constant* Γ_δ (J kg^{-1} m^2) of that radionuclide is defined as

$$\Gamma_\delta = \frac{r^2 \dot{K}_\delta}{A} \tag{2.61}$$

Using the special names gray (Gy) for the kerma and becquerel (Bq) for the activity, the unit of the air kerma-rate constant is Gy s^{-1} Bq^{-1} m^2.

Photons are here considered to be γ-rays, characteristic x-rays, and internal bremsstrahlung. The air kerma-rate constant is defined for an ideal point source and is characteristic for each radionuclide. For a real source of finite size, attenuation and scattering in the source itself occur and annihilation radiation and external bremsstrahlung may be produced. To account for these processes, specific corrections may be required. In addition, if any medium such as air or water intervenes between the source and the point of measurement, this will result in additional absorption and scattering of the radiation and thus additional corrections have to be applied.

The value of the energy cut-off δ depends on the specific application. For simplicity and standardization it is recommended that δ is expressed in keV. The notation Γ_{10} for the air kerma-rate constant means that a value of 10 keV for the energy cut-off δ is used.

Let us now consider an ideal point source S of a specific radionuclide positioned at a given position in free space (vacuum; see also Figure 2.3). Let A be the activity of that point source and $\dot{K}_\delta$ the air kerma rate due to photons of energy greater than the cut-off value δ in the air-filled volume element dV at point P in a radial distance r from the point source. Let us also assume that the source is emitting a single photon of energy E per disintegration (monoenergetic point source, $E > \delta$).

Based on Figure 2.3, and according to Equation 2.5, Equation 2.7, and Equation 2.55, the energy fluence rate $\dot{\Psi}$ of photons at P is given by

$$\dot{\Psi} = \left(\frac{da}{4\pi r^2}\right)\left(\frac{1}{da}\right)AE = \left(\frac{1}{4\pi r^2}\right)AE$$

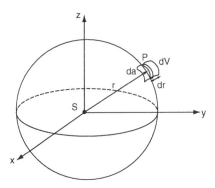

FIGURE 2.3
Schematic representation of an ideal point source S of activity A which is positioned at the center of a sphere of radius r, emitting a single photon of energy E per disintegration. The point P is at the surface of that sphere and a small spherical surface element da is centered at P. The spherical coordinates of P are expressed as (r,θ,φ) and according to Figure 2.1 we have $da = r^2\sin\theta\,d\theta\,d\varphi$. The volume element dV at P can be expressed as $dV = da\,dr = r^2\sin\theta\,d\theta\,d\varphi\,dr$. The number of photons emitted by the source that pass through the surface element da per unit time interval is $[da/(4\pi r^2)A]$.

Consequently, and according to Equation 2.36 and Equation 2.38, $\dot{K}_\delta$ at point P is

$$\dot{K}_\delta = \dot{\Psi}\left(\frac{\mu_{tr}}{\rho}\right)_{\alpha,E} = \left(\frac{1}{4\pi r^2}\right)AE\left(\frac{\mu_{tr}}{\rho}\right)_{\alpha,E} \tag{2.62}$$

where $(\mu_{tr}/\rho)_{\alpha,E}$ is the mass energy transfer coefficient of air for the photon energy E.

Combining Equation 2.61 and Equation 2.62, the air kerma-rate *constant* Γ_δ for the monoenergetic point source is given by

$$\Gamma_\delta = \left(\frac{1}{4\pi}\right)E\left(\frac{\mu_{tr}}{\rho}\right)_{\alpha,E} \tag{2.63}$$

For the case of a point source with a known photon spectrum, the above expression takes the following form:

$$\Gamma_\delta = \left(\frac{1}{4\pi}\right)\sum_i\left\{n_iE_i\left(\frac{\mu_{tr}}{\rho}\right)_{\alpha,E_i}\right\} \tag{2.64}$$

where n_i is the number of photons emitted with energy E_i per disintegration, $(\mu_{tr}/\rho)_{\alpha,E_i}$ is the mass energy transfer coefficient of air for the photon energy E_i, and i is the index over all emitted photon energies E_i with $E_i > \delta$.

A summary of all quantities related to radioactivity and of the corresponding units is given in Table 2.8.

TABLE 2.8

Summary of Quantities Related to Radioactivity

Name	Symbol	Unit	Definition
Decay constant	λ	s^{-1}	dP/dt
Half-life	$T_{1/2}$	s	$\ln(2)/\lambda$
Mean life	τ	s	$1/\lambda$
Activity	A	$Bq = s^{-1}$	dN/dt
Specific activity	$A_{specific}$	$Bq\,kg^{-1}$	A/m
Air kerma-rate constant	Γ_δ	$J\,kg^{-1}\,m^2$ or $Gy\,s^{-1}\,Bq^{-1}\,m^2$	$r^2\dot{K}_\delta/A$

References

1.   Freim, J. and Feldman, A. *Medical Physics Handbook of Units and Measures*, Medical Physics Publishing, Madison, 1992.
2.   Wright, A.E. *Medical Physics Handbook of Radiation Therapy*, Medical Physics Publishing, Madison, 1992.
3.   Bureau International des Poids et Mesures (BIPM), *Le Système International d'Unités (SI)*, 7th ed., BIPM, Sèvre, 1998.

4. International Organization for Standardization (ISO), *Handbook: Quantities and Units*, 3rd ed., ISO, Geneva, 1993, ISO 31-0:1992(E).
5. International Commission on Radiation Units and Measurements, *Fundamental Quantities and Units for Ionizing Radiation ICRU Report 60*, ICRU, Bethesda, 1998.
6. Deutsches Institut für Normung e.V. [DIN], *Terms in the field of radiological technique. Part 3. Dose quantities and units*, DIN, Berlin, 2001 (Begriffe in der radiologischen Technik. Teil 3. Dosisgrössen und Dosiseinheiten) DIN 6814-3.

3

Atoms, Nuclei, Elementary Particles, and Radiations

3.1 Atoms

In ancient Greek philosophy the term *atom* (the indivisible) was used to describe the small indivisible pieces of which matter consists. Father of the so-called *Atomism* theory was the philosopher Leucippus (450–370 B.C.) and his student Democritus (460–370 B.C.). According to them, matter is built of identical, invisible, and indivisible particles, the atoms. Atoms are continuously moving in the infinite empty space. This infinite empty space exists without itself being made of atoms. Atoms show variations in their form and size and they tend to be bound with other atoms. This behavior of the atoms results in the building of the material world. According to Democritus, the origin of the universe was the result of the incessant movement of atoms in space. In this sense atoms were the elementary particles of nature.

It was in 1803, more than 2000 years later, when John Dalton (1766–1844) first provided evidence of the existence of atoms by applying chemical methods. In his theory, Dalton supported the concept that matter is built of indivisible atoms of different weights. All atoms of a specific chemical element are identical in respect of their mass (weight) and their chemical behavior. Atoms are able to build compounds keeping their proportion to each other in the form of simple integers. If these compounds decompose, the atoms involved emerge unchanged from this reaction. In 1896, Dmitrij Iwanowitsch Mendelejew (1834–1907) proposed the periodic law, according to which the properties of the elements are a periodic function of their atomic masses. The definitive arrangement of the elements in the periodic table according to their atomic number and not according to their atomic masses was accomplished in 1914 by Henry Moseley (1887–1915).

In 1904, Sir Joseph John Thomson (1856–1940) proposed the first model of the atom, according to which, the atom is a positively charged sphere of radius about 10^{-10} m with electrons interspersed over its volume.

The Thomson model incorporates many of the known properties of the atom, i.e., size, mass, number of electrons, and electrical neutrality. In 1911, Ernest Rutherford (1871–1937) and his group disproved Thomson's atom model in a series of scatter experiments using alpha particles and thin metal foils. These experiments revealed that besides scattering at small angles corresponding to the Coulomb interaction of α-particles with a Thomson-atom, some α-particles were scattered at very large angles. To explain these results, Rutherford suggested that the positive atomic charge and almost all the atomic mass is concentrated in a very small nucleus of about 10^{-14} m, instead of being distributed over the whole volume of the atom, with electrons distributed around it at comparatively large distances of about 10^{-10} m.

Niels Hendrick David Bohr (1885–1962) formulated his model for the hydrogen atom in 1913, during his stay in Rutherford's laboratory. It was based on the results of Rutherford and the works of Max Karl Ernst Ludwig Planck (1858–1947) and Albert Einstein (1879–1955). According to Bohr's model, electrons move in orbits around the positive charged nucleus similar to the planetary movement around the sun. The role of the gravitational force in the solar system is undertaken by the Coulomb attractive force. In 1926, Werner Karl Heisenberg (1901–1976) and Erwin Schrödinger (1887–1961) founded a new approach, termed quantum mechanics, for describing microscopic phenomena where the deterministic approach of classical mechanics was replaced by probabilistic theory.

3.1.1 The Bohr Hydrogen Atom Model

Bohr's model of the atom is reminiscent of the solar system. The stability of the solar system is ensured by equilibrium of the gravitational and centrifugal forces. In the hydrogen atom a single electron (charge e^-) circulates about a proton (charge e^+). The radius of the circular orbit is r and the electron of rest mass m_e moves with constant tangential speed v. The proton is assumed to be at rest. The attractive Coulomb force provides the centripetal acceleration v^2/r, so:

$$F = \frac{e^2}{4\pi\varepsilon_0}\frac{1}{r^2} = \frac{m_e v^2}{r} \tag{3.1}$$

where ε_0 is the permittivity of free space (see Table A.2.2).

Since an electron undergoes an accelerated movement in its orbit, according to the classical electrodynamics theory it should continuously radiate energy in the form of electromagnetic waves. This, in turn, should result in the collapse of the atom since by continuously losing energy the electron should spiral on to the nucleus. Obviously, this atomic model contradicts two of the most significant experimental results, namely the stability of the atom and the discrete nature of atomic spectra. To overcome these difficulties Bohr postulated the existence of stationary electron orbits in the atom wherein the electron does not radiate energy. These orbits are characterized by definite values of angular momentum L, which is an integer multiplicate of Planck's constant h:

$$L = m_e vr = n \frac{h}{2\pi} = n\hbar \tag{3.2}$$

where n, called the *quantum number*, is an integer of value $n = 1,2,...,\infty$.

Manipulating Equation 3.1 and Equation 3.2 it is found that electron orbits lie at certain distances,

$$r_n = n^2 \frac{4\pi\varepsilon_0\hbar^2}{m_e e^2} = n^2 \times 0.529 \times 10^{-10} \text{ m} \tag{3.3}$$

and are characterized by definite energies:

$$E_n = -\frac{1}{n^2} \frac{m_e e^4}{32\pi^2\varepsilon_0^2\hbar^2} = -\frac{1}{n^2} 13.606 \text{ eV} \tag{3.4}$$

These important results are very different from those we expect from classical physics. For example, a satellite may be placed into Earth's orbit at any desired altitude when supplied with the proper speed. This is not true for an electron's orbit r_n, for which, according to Equation 3.3 there are only certain, discrete allowed orbits. The same applies for the energy levels E_n, which appear in Equation 3.4; only certain values are allowed. They are quantized. The new theory proposed by Bohr is termed the quantum theory of atomic processes.

Bohr postulated that, even though the electron does not radiate when it is in a certain energy state, it may make transitions from one state to a lower energy state. The energy difference between the two states is emitted as a quantum of radiation (photon) whose energy E_{photon} is equal to this energy difference:

$$E_{photon} = E_{initial} - E_{final} \tag{3.5}$$

For example, in a transition from initial state $n = 3$ to final state $n = 2$, the energy E_{photon} emitted is given by Equation 3.4 and Equation 3.5:

$$E_{photon} = E_2 - E_3 = \left(\frac{-1}{3^2} - \frac{-1}{2^2} \right) 13.606 \text{ eV} = 2.520 \text{ eV}$$

This photon has a wavelength $\lambda = hc/E_{photon} = 656.1$ nm, which is exactly the measured wavelength of the visible, red line of the Balmer series. Likewise the wavelengths of all spectral lines were accurately predicted.

The absolute value of energy E_n in a state n is known as the binding energy, b, of the state (i.e., $b = |E_n|$ since E_n is negative, indicating a bound electron–proton system). The state $n = 1$, corresponds to the lowest energy $E_1 = -13.606$ eV. This is the ground state of the hydrogen atom. The atom can remain unchanged over infinitely long periods of time in this ground state. It is a stable state, from which the atom can *only* depart by energy absorption. An energy equal to binding energy b is just sufficient to break up the hydrogen atom into a free electron and a proton, a process termed ionization. If $E > b$ is supplied to the hydrogen atom (e.g., photoelectric absorption, Compton scattering, and other processes discussed in Chapter 4) the electron will leave the atom with a kinetic energy $K = E - b$.

Higher energy states $n \geq 2$ are *excited* states where the atom arrives upon absorption of energy equal to $E_n - E_1$. Excited states are usually short living (half-life $\sim 10^{-8}$ sec) and therefore unstable. The atom spontaneously decays from an excited state to lower energy states, eventually returning to the ground state, emitting upon the process quanta of radiation (photons) termed characteristic fluorescence radiation (see Section 3.1.4).

It is through these processes of excitation and ionization that radiation deposits energy to matter.

The Bohr model provides a coherent picture of the hydrogen atom and its size, explains the discrete nature of the atomic spectra, and accurately predicts the wavelengths of the emitted radiations (lacking, however, the potential to make any prediction for their intensities). Bohr's theory was later modified and perfected; ellipsoidal orbits and motion of proton about the common center of mass were considered and new quantum numbers were introduced to explain phenomena such as the fine structure (many spectral lines are not single but actually composed of two closely spaced lines). However, the Bohr model remains a phenomenological and incomplete model applicable to atoms containing one electron and yet with serious deficiencies (it violates the uncertainty principle and angular momentum conservation, as discussed in Section 3.1.2). More generally, it did not provide any insight as to why the concepts of classical mechanics must be renounced in order to describe the atomic processes. These difficulties were overcome in 1826 when Heisenberg and Schrödinger proposed a quite new approach, termed *quantum mechanics*, for describing microscopic phenomena.

3.1.2   The Quantum Mechanical Atomic Model

The quantum theory was introduced by Planck in 1900 in order to describe the quantization of the energy emitted by a black body: the quantum hypothesis. In 1905, Einstein, based on Planck's quantum hypothesis, succeeded in explaining the photoelectric effect (see Section 4.2.1) by assuming that the energy of light is bounded in light particles, photons, whose energy is quantized as described by Planck. The particle nature of the electromagnetic radiation was further needed to explain the observation made in 1922 by Arthur Holly Compton (1892–1962), known as the Compton effect or Compton scattering (see Section 4.2.3). In summary, it was founded that some effects, such as interference could be explained on the basis of the wave nature of light whereas the explanation of others, such as the photoelectric effect, required the assumption of the particle nature of light. This was termed the wave-particle duality.

In 1924, Louis Victor Duc de Broglie (1892–1987) postulated in his Ph.D. thesis that all forms of matter exhibit wave as well as particle properties, as photons do. According to this conception, an electron has a dual particle-wave nature. Accompanying the electron is a kind of wave that guides the electron through space. The corresponding wavelength, called the De Broglie wavelength λ of the particle, is given by

$$\lambda = \frac{h}{p} \tag{3.6}$$

where h is Planck's constant and p is the momentum of the electron. Clinton Joseph Davisson (1881–1958) and Lester Halbert Germer (1896–1971), in 1927, succeeded in providing experimental proof of de Broglie's assumption.

Werner Karl Heisenberg (1901–1976), who had worked with Bohr in his laboratory in Copenhagen during the period 1924–1927, began his work on the development of quantum mechanics in 1925, focusing on the mathematical description of the frequencies and intensities (amplitudes) of the radiation emitted or absorbed by atoms. In 1927, Heisenberg formulated the uncertainty principle, which states that no experiment can ever be performed yielding uncertainties below the limits expressed by the following uncertainty relations:

$$\Delta x \Delta p_x \sim \hbar \tag{3.7}$$

$$\Delta E \Delta \tau \sim \hbar \tag{3.8}$$

or any other relation combining two physical quantities whose product can be expressed in $\hbar$ dimensions. According to the uncertainty principle, the simultaneous definition of neither the exact position and the exact momentum of an electron in a given state, nor the exact energy of a state and its exact lifetime, are possible. For example, Bohr's theory violates the uncertainty principle of Equation 3.7 since it allows for the simultaneous definition of radial distance r (thus $\Delta r = 0$) and the momentum p_r at the radial direction (it is actually zero and thus $\Delta p_r = 0$). Another example is given by Equation 3.8; it shows that the energy of stable ground state, that can remain unchanged for ever (thus $\Delta \tau = \infty$), can be determined with high accuracy ($\Delta E = 0$). An excited state, however, usually presents a lifetime of the order of 10^{-8} sec (thus $\Delta \tau \approx 10^{-8}$ sec) and therefore $\Delta E \approx \hbar / \Delta \tau \approx 6.6 \times 10^{-8}$ eV which corresponds to the natural width of spectral lines. Note that ΔE is the uncertainty in the energy of the excited state and also the uncertainty in the energy of the photon emitted during de-excitation. In practice, the observed width of spectral lines is wider than the natural width due to Doppler broadening caused by thermal motion.

Particle-waves are described by a complex-value wave function, usually denoted by the Greek letter ψ whose absolute square $|\psi|^2$ yields the probability of finding the particle at a given point at some instant. In 1926, Erwin Schrödinger (1887–1961) proposed a wave equation that describes the manner in which particle-waves change in space and time.

The time-independent Schrödinger equation for the hydrogen atom reads:

$$-\frac{\hbar^2}{2m}\Delta\psi(\vec{r}) + V(\vec{r})\psi(\vec{r}) = E\psi(\vec{r}) \tag{3.9}$$

where

$$V(\vec{r}) = \frac{e^2}{4\pi\varepsilon_0}\frac{1}{r} \tag{3.10}$$

is the potential and r is the radial distance of the electron from the proton. The solution of the Schrödinger equation for the hydrogen atom is a wave function $\psi_{nlm_l}(\vec{r})$ where the three quantum numbers n, l, m_l which emerge from the theory are called:

- Principal quantum number n that takes the values $n = 1,2,3,\ldots,\infty$.
- Angular momentum quantum number l that, for given n, takes the values $l = 0,1,2,\ldots,n-1$.
- Magnetic quantum number m_l that, for given l, takes the values $m_l = -l, -l+1, -l+2,\ldots, +l$.

The quantized energy levels in the hydrogen atom are found to depend *only* on the principal quantum number n and are given by Equation 3.4, i.e., they are exactly the same as predicted by Bohr. Note, however, that this only applies for the hydrogen atom due to the special form of the potential of Equation 3.10 and its r^{-1} dependence (see also Section 3.1.3).

The angular momentum, however, is totally different from that postulated by Bohr (Equation 3.2). According to quantum mechanics the orbital angular momentum is a vector, $\vec{L}$, for which we can *only* determine its magnitude $|L|$ and possible projections L_z along a given z-direction (e.g., a magnetic field applied along the z-axis). The angular momentum quantum number l specifies the magnitude L according to

$$|L| = \sqrt{l(l+1)}\,\hbar \tag{3.11}$$

while the magnetic quantum number m_l describes the orientation of the angular momentum in space; for a given l there are $(2l+1)$ discrete values of the projection L_z of the angular momentum vector into a specific direction-z in space:

$$L_z = m_l\hbar \tag{3.12}$$

In a three-dimensional representation, the angular momentum vector $\vec{L}$ must lie on the surface of a cone which forms an angle θ with the z-axis. The values of θ are also quantized and given by

$$\cos\theta = \frac{m_l}{\sqrt{l(l+1)}} \tag{3.13}$$

Summarizing, the electron angular momentum is a vector $\vec{L}$ quantized in space: for a given n there are n discrete values of the magnitude $|L|$ given by Equation 3.11. For each of these n discrete values of l, there are $(2l+1)$

(i.e., a total of $n(2l + 1)$) discrete orientations in space given by Equation 3.12 and/or Equation 3.13.

The next plausible step was the modification of Schrödinger's equation for consistency with the theory of special relativity. This was achieved in 1928 by Dirac (1902–1984), who showed that the requirements imposed by relativity on the quantum theory have the following consequences: the electron has intrinsic angular momentum and associated magnetic dipole moment and there are fine structure corrections to the Bohr formula for the energy levels which do not depend only on the principal quantum number n. The intrinsic angular momentum or spin, $\vec{S}$, is a quantum property of the electron. As for the angular momentum, $\vec{L}$, the spin vector $\vec{S}$ has a quantized magnitude $|S|$ which in units of $\hbar$ is given by

$$|S| = \sqrt{s(s + 1)}\hbar \qquad (3.14)$$

possible projections S_z along a given z-direction given by

$$S_z = m_s\hbar \qquad (3.15)$$

and corresponding possible values of angle θ with the z-axis, given by

$$\cos\theta = \frac{m_s}{\sqrt{l(l + 1)}} \qquad (3.16)$$

For electrons (as well as protons and neutrons) the spin quantum number s has a single value of $s = 1/2$, that specifies the magnitude $|S|$. As for the orbital angular momentum, there is a spin magnetic quantum number m_s that takes the two quantized values $m_s = \pm 1/2$ that specify the orientation of the spin vector. Thus, Equation 3.14 through Equation 3.16 for $s = 1/2$ yield:

$$|S| = \sqrt{\frac{1}{2}\left(\frac{1}{2} + 1\right)}\hbar = \frac{\sqrt{3}}{2}\hbar, \qquad S_z = \pm\frac{1}{2}\hbar \ \text{ and } \ \cos\theta = \pm\frac{\sqrt{3}}{3}$$

A vector model describing space quantization of an electron's spin is shown in Figure 3.1. According to quantum mechanics there are two equally probable orientations of spin; either parallel or antiparallel to, for instance, an external magnetic field along the z-axis. These states are usually called spin up and spin down. (Associated with the spin of electrons, protons as well as neutrons, is a magnetic dipole moment and an interaction energy in the presence of an external magnetic field and hence these states form two separate energy levels. Despite the equal quantum mechanical probability in spin orientations, the population of these two levels is not equal in ambient temperature and this constitutes the physical basis for spectrometric or imaging techniques such as electron spin resonance (ESR) nuclear magnetic resonance (NMR) and magnetic resonance imaging (MRI).)

A complete description of an electron state is achieved through the four quantum numbers (n, l, m_l, m_s). For the ground state of hydrogen $n = 1$, and therefore only $l = 0$ and $m_l = 0$ are permitted. With the addition of

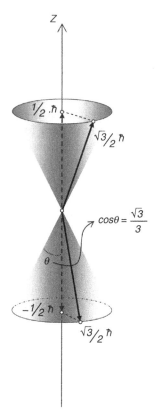

FIGURE 3.1
A vector model describing space quantization of electron spin.

spin ($m_s = \pm 1/2$) the ground state is therefore either $(1,0,0, + 1/2)$ or $(1,0,0, - 1/2)$. These two states have the same energy, given by Equation 3.4 for $n = 1$. They are thus twofold degenerate (degeneracy, in general, disappears in the presence of an external magnetic field; e.g., Zeeman effect, MRI). For the first excited state, $n = 2$ and thus the allowed values of l are $l = 0$ or $l = 1$ (equally probable). For $l = 0$, only $m_l = 0$ is permitted. For $l = 1$, m_l can take the values of -1, 0, and $+1$. There are, therefore, $2n^2 = 2(2^2)$ eightfold degenerate states, namely $(2,0,0, + 1/2)$, $(2,0,0, - 1/2)$, $(2,1,1, + 1/2)$, $(2,0,0, - 1/2)$, $(2,1,0, + 1/2)$, $(2,1,0, - 1/2)$, $(2,1, - 1, + 1/2)$, $(2,1, - 1, - 1/2)$. For $n = 3$ there are $2n^2 = 2(3^2) = 18$ states.

For historical reasons, all states with the same principal quantum number, n, are said to form a shell. These shells are designated with the following capital letters:

Value of n	1	2	3	4	5	6...
Shell symbol	K	L	M	N	O	P...

Likewise, there is a spectroscopic notation in which small-case letters are used to identify different l values:

Value of l	0	1	2	3	4	5	6
Notation	s	p	d	f	g	h	i

(the first four letters stand for sharp, principal, diffuse, and fundamental, which were terms originally used to describe atomic spectra before atomic theory was developed).

In spectroscopic notation the ground state of hydrogen is 1s where the value of $n = 1$ is specified before the s. The first excited state is either 2s or 2p (the 2s state, as all s states, is single while the 2p state, as all p, d, f, and subsequent states, is double due to spin–orbital momentum interaction in the presence of the internal magnetic field, causing the effect of fine structure). For $n = 3$ there are 3s, 3p, or 3d states. Not all transitions between different states are allowed. The transitions most likely to occur are those resulting to a change of l by one unit (i.e., those that conserve angular momentum since the emitted photon has spin 1) and thus the selection rule for allowed transitions is

$$\Delta l = \pm 1 \tag{3.17}$$

Thus, a transition from $n = 3$ to the $n = 1$ state, that according to the Bohr model does not conserve angular momentum and therefore is not allowed, quantum mechanically could only be from 3p state to the 1s state. Transitions such as 3s → 1s or 3d → 1s are not allowed due to the selection rule of Equation 3.17.

3.1.3 Multielectron Atoms and Pauli's Exclusion Principle

An electronic state is completely specified by the four quantum numbers (n, l, m_l, m_s). In addition, we have seen that in the hydrogen atom the energy E_n depends only on the principle quantum number n in Equation 3.4, and not on the angular momentum quantum number l as a consequence of the special form of the potential (r^{-1} dependence). For atoms with more than one electron ($Z > 1$) the energy of a quantum state depends on n as well as on the angular momentum quantum number l and, in general, the lower the value of l the lower the energy of the electron. For example a 4s state is of lower energy than a 3d state, as naturally provided quantum mechanically. States having the same values of n and l (for instance 4s or 3d) are known as subshells. The number of quantum states in a subshell is $2(2l + 1)$. The $(2l + 1)$ factor comes from the number of different m_l values for each l. The extra factor of 2, comes from the two possible values of $m_s = \pm 1/2$. Thus, subshell 4s has two quantum states (as all s states) and 3d has ten quantum states (as all d states). The way in which these different quantum states are occupied by the electrons in a multielectron atom is based on the exclusion principle formulated in 1925 by Wolfgang Pauli (1900–1958): no two electrons in a single atom can ever be in the same quantum state; that is, no two electrons in the same atom can have the same set of the four quantum numbers n, l, m_l, and m_s. In fact, particles that obey the exclusion principle are those of half-integer spin quantum numbers called fermions (such as electrons, protons, and neutrons for which $s = 1/2$) and not those of integer spin quantum numbers called bosons (such as photons of $s = 1$).

One of the most important consequences of the exclusion principle is that it explains the periodic table of the elements which constitutes the basis for studying chemical behavior. The ordering of the energy levels in many electron atoms is well known. The two $1s$ states of the K-shell are the ones to be filled first, since these states correspond to the lowest energy. For the L-shell, the two $2s$ states are of lower energy than the six $2p$ states. Therefore, the electron configuration of the fluorine ($Z = 9$) is: $1s^2 2s^2 2p^5$, i.e., there are two electrons in the $1s$ states, two more in the $2s$ state and five electrons occupy the $2p$ states. For neon ($Z = 10$) the configuration is $1s^2 2s^2 2p^6$ and for sodium ($Z = 11$) $1s^2 2s^2 2p^6 3s^1$. Neon which has all its subshells filled is an inert gas (practically nonreactive) while its neighbors, fluorine and sodium, are among the most reactive elements.

3.1.4 Characteristic X-Rays. Fluorescence Radiation

Owing to historical reasons, x-rays immediately bring to mind the electromagnetic radiation produced in an x-ray tube or linear accelerator where electrons are rapidly decelerated in the anode (see Section 4.5). These x-rays present a continuous energy spectrum (bremsstrahlung) spreading from zero up to a maximum energy, which depends on the applied acceleration potential (e.g., for 100 kV potential the maximum energy of the spectrum is 100 keV). Superimposed on this continuous spectrum are discrete x-ray line spectra, called characteristic x-rays, since they are emitted by the atoms of the anode (in general characteristic x-rays are those emitted by atoms while γ-rays are those emitted by nuclei).

Since all the inner shells of an atom are filled, x-ray transitions do not normally occur between these levels. However, once an inner electron is removed by an atom (as in the case of photo-absorption which is of higher probability for K-shell electrons, as discussed in Section 4.2.1) the vacancy created is filled by outer electrons falling into it, and this process may be accompanied by emission of fluorescent radiation or Auger electron emission. The fluorescent x-rays emitted in the process of filling a K-shell vacancy are known as K x-rays and may be of significant impact on the dosimetry of low-energy photon emitters such as ^{125}I and [103]Pd (see Section 4.2.1, Chapter 5, and Chapter 9).

Consider the K_α x-rays, originating by the electron transition from the L to the K-shell. An electron in the L-shell is screened by the two $1s$ electrons and so it faces an effective nuclear charge of $Z_{effective} \approx Z - 2$. When one of these $1s$ electrons is removed and a K-shell vacancy is created, only the remaining $1s$ electron screens the L-shell, and so $Z_{effective} \approx Z - 1$. Bohr's theory for the hydrogen atom includes the nuclear electric charge only in Equation 3.1. Therefore the allowed energies (Equation 3.4) for $Z_{\text{effective}} \approx Z - 1$ are given by

$$E_n = -\frac{1}{n^2} \frac{m_e e^4 (Z-1)^2}{32\pi^2 \varepsilon_0^2 \hbar^2} = -\frac{1}{n^2}(Z-1)^2 13.606 \text{ eV} \qquad (3.18)$$

For $n = 1$, this equation provides a crude yet useful approximation for the K-shell binding energy (see also Equation 4.15).

The energy $E(K_\alpha)$ of the K_α x-ray, i.e., transition from $n = 2$ to $n = 1$, is $E_2 - E_1$:

$$E(K_\alpha) = \frac{3}{4}(Z-1)^2 13.606 \text{ eV} \qquad (3.19)$$

A plot of the square root of the K_α x-ray energy, i.e., $\{E(K_\alpha)\}^{1/2}$ as a function of electric charge Z (and not $Z_{\text{effective}}$) is a straight line (the approximation $Z_{\text{effective}} \approx Z - 1$ is not crucial; it could easily be $Z_{\text{effective}} \approx Z - k$, where k is an unknown number to be evaluated from the intercept of the straight line).

This is the Moseley (1887–1915) law, formulated in 1913. It is a simple, yet powerful way, to determine the atomic number Z of the atom, hence its specific place in the periodical table of the elements, which were previously ordered according to increasing mass causing some abnormalities. While an element is exclusively specified by its atomic number (number of protons), elemental atomic weight depends on the mass number (sum of protons and neutrons in its nucleus).

3.2 Atomic Nucleus

Rutherford suggested, in 1911, that the positive atomic charge and almost all the atomic mass are concentrated in a very small central body, the nucleus. Continuing his experiments, in 1919, Rutherford discovered that there were hydrogen nuclei ejected from materials upon being bombarded by fast alpha-particles. He identified the ejected hydrogen nuclei, the protons, as nuclear constituents; the atomic nuclei contain protons.

Until 1932, physicists assumed that atomic nuclei are constructed of protons, alpha-particles, and electrons. In 1932, Sir James Chadwick (1891–1974) identified the neutron by interpreting correctly the results of the experiments carried out mainly by Jean Frédéric (1897–1958) and Irène Joliot-Curie (1900–1956): neutrons, uncharged particles were ejected out of beryllium nuclei after their bombardment with alpha-particles. Chadwick considered neutrons to be an electron–proton compound and added it to the nuclear mix.

In July 1932, Heisenberg published his neutron–proton nuclear model by assuming that neutrons and protons are constituents of the nucleus. His model of the nucleus also contained electrons (nuclear electrons) either bound or unbound. The assumption of the existence of nuclear electrons was definitively rejected in the late 1930s after the introduction of the neutrino by Pauli in 1931 and the establishment of Enrico Fermi's (1901–1954) theory of beta decay, published in 1933.

Contrary to the original idea of indivisible atom, it has been proved that the atomic nucleus has an internal structure. The nucleus contains two kinds of particles, the nucleons: the positive charged protons with a charge of $+e$ and the uncharged neutrons.

3.2.1   Chart of the Nuclides

All nuclei are composed of protons and neutrons. The atomic number Z denotes the number of protons and the neutron number N denotes the number of neutrons in a nucleus. The mass number $A = Z + N$ indicates the total number of nucleons (protons and neutrons). A nucleus X specified by atomic number Z, neutron number N and mass number A is called a nuclide and it is symbolized by

$$_Z^A X_N$$

Usually, however, the atomic number Z (implied by the chemical symbol) and the neutron number N $(N = A - Z)$ are omitted and a nuclide is identified by its chemical symbol and its mass number A, e.g., ^{12}C instead of $_6^{12}C_6$, ^{192}Ir instead of $_{77}^{192}Ir_{115}$, etc.

A convenient way of depicting nuclear information is offered by the chart of the nuclides, first used by Segre, where each nuclide is represented by a unit square in a plot of atomic number Z vs. the neutron number N. A small part of such a chart is shown in Figure 3.2 (a complete chart can be found at references 1 and 2). In this chart isotopes (nuclides of the same atomic number Z) are arranged in horizontal lines; e.g., ^1H, ^2H, and ^3H are isotopes of hydrogen. Isotopes have identical chemical properties, they are chemically undistinguishable but they have significantly different nuclear properties. Isotones (nuclides of the same N) are arranged in vertical lines (e.g., ^3H, ^4He, ^5Li…) and isobars (nuclides of the same A) fall along descending diagonals from left to right (e.g., ^6Be, ^6Li, ^6He).

Figure 3.3 shows the general layout of a complete chart of nuclides. There are about 3000 nuclides known to exist (most of them artificial), divided into two broad classes: stable and radioactive. Stable nuclides remain unchanged over an infinitely long time, while radioactive nuclides are unstable, undergoing spontaneous transformations known as radioactive decays (see Section 3.3.1). Filled squares in Figure 3.2 and/or Figure 3.3 denote natural nuclides, i.e., primordial nuclides (formed before the creation of Earth) occurring in nature. These are either stable (there are only 274 stable nuclides in nature) or long-living radioactive nuclides surviving since the creation of the Earth about five billion years ago (there are 14 including ^{40}K, ^{87}Ru, ^{232}Th, ^{234}U, ^{235}U, ^{238}U). Natural nuclides form the stability curve or stability valley, discerned in Figure 3.3 (see also Section 3.4.2 and Figure 3.9 in that section). For light nuclei, $Z < 20$, the greatest stability is achieved

Z \ N	0	1	2	3	4	5	6	7	8	9	10	11	12	13	14	15	16
8						^{13}O 13.0248104 8.58 ms EC 17.765	^{14}O 14.0085953 70.606 s EC 5.143	^{15}O 15.0030654 122.24 s EC 2.754	^{16}O 15.9949146 Stable 99.762%	^{17}O 16.9991315 Stable 0.038%	^{18}O 17.9991604 Stable 0.200%	^{19}O 19.0035787 26.91 s β^- 4.821	^{20}O 20.0040761 13.51 s β^- 3.814	^{21}O 21.0086546 3.42 s β^- 8.109	^{22}O 22.0099672 2.25 s β^- 6.490	^{23}O 23.0156913 82 ms β^- 11.290	^{24}O 24.0203699 61 ms β^- 11.400
7					^{11}N 11.0267962 3 E-22 s p 2.290	^{12}N 12.0186132 11.000 ms EC 17.338	^{13}N 13.0057386 9.965 min EC 2.220	^{14}N 14.0030740 Stable 99.634%	^{15}N 15.0001089 Stable 3.66%	^{16}N 16.0061014 7.13 s β^- 10.419	^{17}N 17.0084497 4.173 s β^- 8.680	^{18}N 18.0140818 624 ms β^- 13.899	^{19}N 19.0170269 290 ms β^- 12.527	^{20}N 20.0233673 142 ms β^- 17.970	^{21}N 21.0270876 87 ms β^- 17.170	^{22}N 22.0344403 18 ms β^- 22.800	
6			^{8}C 8.037675 2 E-21 s 2p 2.142	^{9}C 9.0310401 126.5 ms EC 16.498	^{10}C 10.0168531 19.255 s EC 3.648	^{11}C 11.0114338 20.39 mon EC 1.982	^{12}C 12.0000000 Stable 98.93%	^{13}C 12.0033548 Stable 1.07%	^{14}C 14.003242 5730 years β^- 0.156	^{15}C 15.0105993 2.449 s β^- 9.772	^{16}C 16.0147012 0.747 s β^- 8.012	^{17}C 17.0225837 193 ms β^- 13.166	^{18}C 18.0267571 95 ms β^- 11.810	^{19}C 19.0352481 49 ms β^- 16.970	^{20}C 20.0403224 14 ms β^- 15.790	^{21}C 21.04934 Unknown	
5			^{7}B 7.0299174 3 E-22 s 2p 2.200	^{8}B 8.0246067 770 ms EC 17.979	^{9}B 9.0133288 8 E-19 s 2α 0.277	^{10}B 10.0129370 Stable 19.9%	^{11}B 11.0093055 Stable 80.1%	^{12}B 12.0143521 20.20 ms β^- 13.369	^{13}B 13.0177803 17.36 ms β^- 13.437	^{14}B 14.0254041 13.8 ms β^- 20.664	^{15}B 15.0310973 10.5 ms β^- 19.094	^{16}B 16.0398088 <190 ps n 0.040	^{17}B 17.0469314 5.08 ms β^- 22.680				
4			^{6}Be 6.0197258 5 E-21 s 2p 1.372	^{7}Be 7.0169292 53. 29d EC 0.862	^{8}Be 8.0053051 3 E-16 s 2α 0.092	^{9}Be 9.0121821 Stable 100%	^{10}Be 10.0135337 1510000 y β^- 0.566	^{11}Be 11.0216577 13.81 s β^- 11.506	^{12}Be 12.0269206 23.6 ms β^- 11.708	^{13}Be 13.0361338 3 E-21 s n 0.500	^{14}Be 14.0428155 3 E-21 s β^- 16.220						
3		^{4}Li 4.0271823 8 E-23 s p 3.100	^{5}Li 5.0125378 3 E-22 s β^- 1.970	^{6}Li 6.0151223 Stable 7.5%	^{7}Li 7.0160040 Stable 92.5%	^{8}Li 8.0224867 138 ms β^- 16.004	^{9}Li 9.0267891 178.3 ms β^- 13.606	^{10}Li 10.0354809 4 E-22 s n 0.420	^{11}Li 11.0437962 8.5 ms β^- 20.610								
2		^{3}He 3.0160293 Stable 0.000137%	^{4}He 4.0026032 Stable 99.99986%	^{5}He 5.0122236 8 E-22 s α0.890	^{6}He 6.0188881 806.7 ms β^- 3.508	^{7}He 7.0280305 3 E-21 s n 0.440	^{8}He 8.0339218 119.0 ms β^- 10.653										
1	^{1}H 1.0078250 Stable 99.985%	^{2}H 2.0141018 Stable 0.015%	^{3}H 3.0160493 12.33 y β^- 0.019														
0		^{0}n 1.0086649 10.4 min β^- 0.782															

FIGURE 3.2
Part of the chart of nuclides. In this chart each nuclide is represented by a unit square in a plot of atomic number, *Z*, vs. neutron number, *N*. Shaded squares correspond to stable nuclides for which atomic mass in units of u and natural isotopic abundance are provided. For the remaining nuclides, which are unstable, atomic mass, half-life, decay mode, and decay energy in units of MeV are provided.

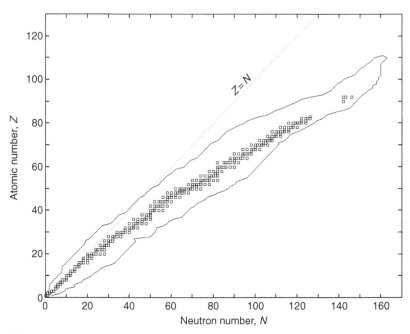

FIGURE 3.3
An overview of the complete chart of nuclides. Shaded squares correspond to natural nuclides (forming the stability valley) while the remaining area bounded by the full line is occupied by radioactive nuclides.

when the number of protons equals the number of neutrons, $Z = N$. For heavier nuclei, the instability caused by the Coulomb repulsion between the protons is counterbalanced by a high-neutron excess, $Z < N$ (see Section 3.2.3). Neighboring the natural nuclides are radioactive nuclides which occupy the rest of the area bounded by the full line in Figure 3.3. These nuclides decay in such a way that they finally reach a stable nuclide lying on the stability curve. The nuclides below the stability curve are β^--unstable while those above the stability curve are β^+-unstable (see Section 3.4.2).

From the three hydrogen isotopes ^{1}H, ^{2}H, and ^{3}H, only ^{1}H and ^{2}H are natural (actually stable nuclides, see Figure 3.2). ^{3}H is a relatively short-living radioactive nuclide which has not survived since the creation of the Earth approximately 5 billion years ago. For carbon ($Z = 6$) there are two natural isotopes, ^{12}C and ^{13}C, with corresponding natural isotopic abundance of 98.89 and 1.11%. The short-living nuclide ^{14}C is not considered a natural isotope although it is being constantly produced by cosmic ray neutron bombardment of ^{14}N and this activity is continuously replenished (carbon dating is based on this isotope). Since 1961, the atomic weight has been based on a ^{12}C-scale; the atomic weight of this carbon isotope is exactly 12. However, the atomic weight of carbon presented in the periodic table is not exactly 12, but $0.9893 \times 12 + 0.0107 \times 13.00033548 = 12.0107$ due to the natural isotopic abundance of ^{13}C. ^{238}U is not stable but is long-living (half-life $T_{1/2} = 4.51 \times 10^9$ years) and is considered a natural nuclide with natural isotope abundance of 99.27%. Through successive transformations, ^{238}U eventually arrives at the stable end product ^{206}Pb. None of the members of the corresponding radioactive series (14 overall) are considered as natural nuclides although they are found in nature and sometimes are very important, as was ^{226}Ra in early brachytherapy practice. This is because they do not have characteristic terrestrial compositions.

The chart of the nuclides provides some of the most important information on each nuclide including its mass and binding energy, spin and magnetic moment, natural isotopic abundance, mode of decay, and decay constant if it is unstable, etc.

3.2.2 Atomic and Nuclear Masses and Binding Energies

Published tabulations present atomic rather than nuclear masses. Atomic masses are known with high accuracy. This accuracy is achieved by measuring atomic masses relative to each other, and in particular, relative to the neutral unexcited ^{12}C atom which has arbitrarily been assigned 12.00000 atomic mass units. The unit employed is the unified atomic mass unit, u, defined as the $1/12$ of the mass of the neutral unexcited ^{12}C atom. In SI units, the unified atomic mass unit u is therefore:

$$1\,u = \frac{1}{12} \times \frac{12 \times 10^{-3}\ \text{kg}}{N_A} = 1.66053873 \times 10^{-27}\ \text{kg} \tag{3.20}$$

where N_A is Avogadro's number.

The energy equivalence of the unified mass unit in MeV is, according to Einstein's mass–energy relation, $E = 1\,u \times c^2 = 931.494013$ MeV, thus:

$$1\,u = 931.494013\ \text{MeV}/c^2 \tag{3.21}$$

The rest mass m in units of u and the rest energy mc^2 in units of MeV, for electron, proton, and neutron are (see also Appendix 2):

Electron	$m_e = 5.48597 \times 10^{-4}\,u$	$m_e c^2 = 0.511$ MeV
Proton	$m_p = 1.008665\,u$	$m_p c^2 = 938.28$ MeV
Neutron	$m_n = 1.007277\,u$	$m_n c^2 = 939.57$ MeV

The rest energy of the hydrogen atom in its ground state is

$$m_{\text{Hydrogen}} c^2 = m_e c^2 + m_p c^2 - 13.606\ \text{eV}$$

that is, the binding energy $b = 13.606$ eV is subtracted from the sum of the rest energies of the electron and proton composing the hydrogen atom. The electron binding energy is only $a \sim 10^{-8}$ fraction of the atomic rest energy which is 938.783 MeV. The rest mass of the hydrogen atom is 1.007825 u.

To a first approximation the atomic mass $m_{\text{atom}}(Z,A)$ of an atom with Z protons, $A - Z$ neutrons and Z electrons in units of u or GeV/c^2 is

$$m_{\text{atom}}(Z, A) \approx A\, u \approx A\, \text{GeV}/c^2 \tag{3.22}$$

Some times, instead of atomic mass the mass excess, Δ, defined as the difference of the mass number, A, by the atomic mass, i.e. $\Delta = m_{\text{atom}}(Z,A) - A$, is used (see Appendix 1).

The nucleus contains almost all the mass of the atom. The nuclear mass $m_{\text{nucleus}}(Z,A)$ of a certain nuclide of atomic number Z and mass number A, can be found from the atomic mass $m_{\text{atom}}(Z,A)$ of the corresponding atom, by subtracting the mass of the Z electrons with due consideration of the mass reduction associated with the total binding energies b_e of all electrons in the atom. The equation for the rest energies is therefore:

$$m_{\text{nucleus}}(Z, A)c^2 = m_{\text{atom}}(Z, A)c^2 - Zm_ec^2 - b_e \tag{3.23}$$

The total binding energy b_e of all Z electrons expressed in eV, can be adequately approximated by the Thomas–Fermi empirical expression:

$$b_e = 15.73 \times Z^{7/3}\,\text{eV} \tag{3.24}$$

This is negligible when compared with the atomic masses and it is usually neglected in the calculations of the nuclear masses. Moreover, it usually makes no difference which of the atomic or nuclear mass is used, because the number of electrons, and hence electron masses, cancel out in most nuclear reactions (see for example Section 3.4.1).

The binding energy B of a nucleus $^A_Z X_N$ is defined as

$$B = \left\{ Zm_p + Nm_n - m_{\text{nucleus}}(Z, A) \right\} c^2 \tag{3.25}$$

That is, as the difference in rest energies between the nucleus and its constituent Z protons and N neutrons. The binding energy is in that sense the amount of energy that has to be supplied in order to break up the nucleus into free neutrons and protons. Equivalently, the nuclear binding energy can be calculated from atomic mass ignoring the total binding energy b_e of the atomic electrons:

$$B = \left\{ Z(m_p + m_e) + (A - Z)m_n - m_{\text{atom}}(Z, A) \right\} c^2 \tag{3.26}$$

Figure 3.4 presents the variation of binding energy per nucleon, B/A, with mass number, A, for the natural nuclides presented in Figure 3.3. The binding energy per nucleon starts at small values of 1.11 for ^2H ($B = 2.22$ MeV), rises to a maximum of 8.79 for ^{56}Fe ($B = 492.25$ MeV) and then falls to a value of 7.57 for ^{238}U ($B = 1801.69$ MeV). A number of interesting conclusions can be drawn by noting the features of the B/A vs. A plot:

1. B/A is approximately constant and equal to about 8 MeV/nucleon. This implies that the nuclear force acting between nucleons is due to a short-range interaction. Indeed, if each nucleon interacted with

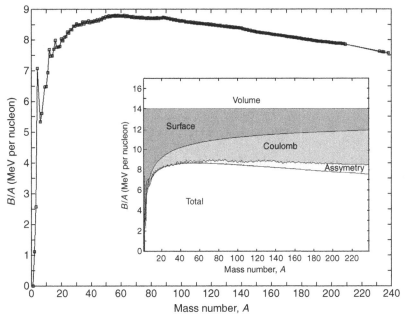

FIGURE 3.4
The binding energy per nucleon, B/A, vs. mass number, A, for natural nuclides. The inset presents corresponding data calculated using the semiempirical liquid drop model (Equation 3.27) as well as the relative magnitude of the terms in the model.

all the remaining $(A - 1)$ nucleons, the total binding energy B would be proportional to $A(A - 1) \approx A^2$ and not to A.

2. B/A falls of at large A. This is due to Coulomb repulsion between protons. Coulomb is a long-range interaction and therefore each proton interacts with the remaining $(Z - 1)$ protons in the nucleus, resulting in a Coulomb repulsion energy which increases similarly to $Z(Z - 1) \approx Z^2$. Since Z^2 increases faster than A, heavy nuclei, as shown by Figure 3.3, have more neutrons than protons. This compensation of the Coulomb repulsion between all protons by the strong nuclear attraction between neighboring nucleons could not last for ever; elements heavier than uranium do not occur in nature.

3. The B/A vs. A plot peaks at $A \approx 60$. This implies that binding energies can be increased by either splitting a heavy nucleus into two lighter nuclei or fusing two light nuclei together, highlighting the importance of fission and fusion reactions in the production of nuclear energy.

3.2.3 The Semiempirical Mass Formula

The general features of the B/A vs. A plot (Figure 3.4) indicate that nuclear behavior resembles that of a drop of liquid. Using a semiempirical approach,

Weizsacher demonstrated, in 1935, that it is possible to achieve a quantitative interpretation of binding energy by regarding B as akin to a latent energy of condensation. According to this idea, the binding energy $B(Z,A)$ of a nucleus $^A_Z X$ is given by a number of terms that are functions of A and Z:

$$B(Z, A) = \alpha_{\text{Volume}}A - \alpha_{\text{Surface}}A^{2/3} - \alpha_{\text{Coulomb}}\frac{Z(Z-1)}{A^{1/3}}$$

$$- \alpha_{\text{Asymmetry}}\frac{(A-2Z)^2}{A} - \alpha_{\text{Pairing}}\delta A^{-3/4} \qquad (3.27)$$

The first two terms in this formula indicate that there is an analog of a nucleus with a drop of liquid. A liquid drop has volume and surface energy due to the fact that the attractive forces between molecules are short-ranged forces (intermolecular attractions—van der Waals forces). Each molecule interacts only with its neighbors, and the number of neighbors surrounding a molecule is independent of the overall size of the liquid drop. Therefore, the energy required to overcome these short-range forces and completely evaporate the drop is, on average, the same for each molecule (volume energy). Molecules, however that lie on the surface of a drop are not surrounded by neighbors on all sides, and consequently, they are not bound as tightly as the molecules in the interior of the drop. Hence, the contribution of these surface molecules to the total energy of the drop is, on average, diminished by a factor approximately equal to the surface of the drop (surface energy). The attractive force between nucleons in a nucleus (regardless of the nucleon being a proton or a neutron), i.e., the strong nuclear force, is a short-range one. Nuclei are roughly spherical, with the nuclear radius R given approximately by

$$R = R_0 A^{1/3} \text{ where } R_0 \approx 1.2 \times 10^{-15} \text{ m} = 1.2 \text{ fm} \qquad (3.28)$$

Nuclear volume V_{nucleus} is

$$V_{\text{nucleus}} = \frac{4}{3}\pi R_0^3 A \rightarrow V_{\text{nucleus}} \propto A \qquad (3.29)$$

and thus nuclear volume is directly proportional to the mass number A. Nuclear mass is also proportional to A (see Equation 3.22). In this way the nucleus can be seen to be analogous to a drop of incompressible liquid, which has a constant and very high-density ($\sim 10^{12}$ times greater than that of ordinary matter) independent of its size. The term ($\alpha_{\text{Volume}}A$) in Equation 3.27 represents a constant, bulk-binding energy per nucleon B/A (see inset of Figure 3.4), similar to the cohesive energy of a simple liquid drop. The second term ($\alpha_{\text{Surface}}A^{2/3}$), representing the diminished contribution of the surface nucleons to the total binding energy of the nucleus, is proportional to the surface area of the nucleus ($\propto 4\pi A^{2/3}$) and this surface term is subtracted by the bulk-binding energy (see inset of Figure 3.4). As shown in Figure 3.4, for light nuclei the fraction of nucleons on the surface is quite large and hence there is a sharp fall in B/A. For heavy nuclei this term is less important.

The third term in Equation 3.27 represents the Coulomb repulsion between protons and has a simple explanation; it is the electrostatic energy of the nuclear charge distribution. Assuming that the nucleus is a uniformly charged sphere of radius $R_0 A^{1/3}$ as in Equation 3.28, and total charge Ze, its energy would be

$$E_C = \frac{3}{5} \frac{(Ze)^2}{(4\pi\varepsilon_0)R_0 A^{1/3}} \tag{3.30}$$

Since each proton interacts with all other protons in the nucleus but itself, the Z^2 term in Equation 3.30 has been replaced by the term $Z(Z-1)$ in Equation 3.27. The Coulomb term, similar to the surface term, has a negative sign and thus is subtracted from the total binding energy (see inset of Figure 3.4). As shown in this figure, the Coulomb energy becomes very significant for heavy nuclei, since $Z(Z-1)$ increases more rapidly than A (Coulomb is a long-range interaction, in contrast to the short-range nuclear interaction). Note that it is the balance between the increase of Coulomb energy and the decrease of surface energy for heavier nuclei that produces the maximum B/A at around $A = 60$, shown in the plot of B/A vs. A.

The asymmetry term in Equation 3.27 originates from Pauli's exclusion principle to which neutrons and protons, as fermions, obey. For a given A, it is energetically advantageous to maximize the number of neutron–proton pairs. The term $(A - 2Z)^2/A$ is a simple empirical expression which, if considered alone, indicates maximum stability is achieved when the neutron excess $N - Z = A - 2Z$ is minimum for a given A. It is sometimes called the symmetry term since it tends to make nuclei symmetric with respect to the number of neutrons and protons. The Segre plot (Figure 3.3) shows that for light nuclei the maximum stability is achieved when $Z \approx N$, while for heavier nuclei stability ensues only if there is a neutron excess, i.e., $Z < N$, due to the relatively higher importance of the Coulomb energy term. The $(A - 2Z)$ excess neutrons occupy higher energy quantum states and consequently they are less tightly bound than the first $2Z$ nucleons which occupy the lower energy states.

The last term in Equation 3.27 is purely phenomenological in form and it's A dependence. Nuclei display a systematic trend for pairing: those having even number of protons and neutrons (even–even) tend to be very stable (there are 165 even–even stable nuclides); those with even-Z and odd-N (55 nuclides) or odd-Z and even-N (50 nuclides) are somewhat less stable; and those with an odd number of Z and N (odd–odd) are mainly unstable (there are only four stable odd–odd nuclides known: ^{2}H, ^{6}Li, ^{10}B, and ^{14}N). Also, whenever Z or N becomes equal to the so-called magic-numbers: 2, 4, 8, 20, 50, 82, and 126, the corresponding nuclides have large binding energies. To account for the pairing energy, the parameter δ in Equation 3.27 takes the values of

$$\delta = -1 \text{ for even–even nuclei}$$

$$\delta = 0 \text{ for even–odd nuclei}$$

$$\delta = +1 \text{ for odd–odd nuclei}$$

indicating that the largest stability (largest binding energy) is achieved for even–even nuclei.

The five parameters in Equation 3.27 are evaluated by fitting the formula to known binding energies (hence the name semiempirical). Typical fitted values for these parameters are

$$\alpha_{\text{Volume}} = 14 \text{ MeV}$$

$$\alpha_{\text{Surface}} = 13 \text{ MeV}$$

$$\alpha_{\text{Coulomb}} = 0.6 \text{ MeV}$$

$$\alpha_{\text{Asymmetry}} = 19 \text{ MeV}$$

$$\alpha_{\text{Pairing}} = 33.5 \text{ MeV}$$

For example, employing the above set of the five parameters yields a binding energy calculation of $B = 486.65$ MeV for ^{56}Fe (to be compared with the actual value of 492.25 MeV) and $B = 1799.2$ MeV for ^{238}U (to be compared with the actual value of 1801.69 MeV).

Incorporating Equation 3.27 into Equation 3.25, the semiempirical mass formula reads:

$$m_{\text{nucleus}}(Z, A)c^2 = \left\{ Zm_{\text{p}} + (A - 2Z)m_{\text{n}} \right\} c^2$$

$$+ \left\{ - \alpha_{\text{Volume}} A + \alpha_{\text{Surface}} A^{2/3} + \alpha_{\text{Coulomb}} \frac{Z(Z - 1)}{A^{1/3}} \right.$$

$$\left. + \alpha_{\text{Asymmetry}} \frac{(A - 2Z)^2}{A} + \alpha_{\text{Pairing}} \delta A^{-3/4} \right\}$$

$$(3.31)$$

3.3 Nuclear Transformation Processes

Nuclear transformation processes are those inducing transitions from one nuclear state to another. They can fall into two categories: those which occur spontaneously, referred to as decays and those which are initiated by bombardment with a particle from outside, called reactions. When a process occurs spontaneously, conservation of energy requires that the final state be of lower energy than the initial state and the difference in these energies, called Q value, is liberated as kinetic energy of energetic particles being emitted (see Section 3.3.3). These particles are rather easy to observe experimentally, and, in fact, it was their discovery in the 1890s that initiated research in nuclear physics.

3.3.1   Radioactive Decay

As mentioned in Section 3.2.1 there are only 274 stable nuclides (forming the stability valley in Figure 3.3) and approximately 2800 unstable nuclides. All unstable nuclei spontaneously transform into other nuclear species by means of different decay processes that change the Z and N numbers of nucleus, until stability is reached. Such spontaneous nuclear processes are called radioactive decays. Among well-known radioactive decays are alpha decay, beta decay (including electron capture (EC)), and spontaneous fission of heavy nuclei. Excited states of nuclei are also unstable (usually when a nucleus decays by alpha- or beta-emission it is left in an excited state) and eventually decay to the ground state by emission of gamma radiation (no change of Z or N occurs). All the above processes follow the laws of radioactive decay. Radioactivity is thus a property of the nucleus or, to be more precise, of a state of the nucleus, according to which the nucleus decays to a more stable state.

The probability for a radioactive decay per unit time for a specific nuclide is constant and called the decay constant, λ (see Section 2.6.1). Since radioactive decay is a stochastic, spontaneous process it is not possible to identify which particular atoms out of an amount of a specific radionuclide will undergo such decay at a specific time. It is only possible to predict the mean number of disintegrated nuclei at a specific time, i.e., the activity $A(t)$ defined as

$$A(t) = -\frac{dN(t)}{dt} \tag{3.32}$$

where $dN(t)$ is the number of decays observed during the time interval dt (the minus sign is included since $dN(t)/dt$ is negative due to the decrease of $N(t)$ with time while activity, $A(t)$, is a positive number). Experimentally, it is found that the activity, $A(t)$, at any instant of time t is directly proportional to the number, $N(t)$, of the radioactive parent nuclei present at that time:

$$A(t) = \lambda N(t) \tag{3.33}$$

where λ is the decay constant. The SI unit of activity is the Becquerel (Bq, named after the discoverer of radioactivity): 1 Bq = 1 disintegration per second = 1 sec^{-1}. Activity was traditionally measured in units of Curies (Ci) with one Ci originally defined as the activity of 1 g pure ^{226}Ra (1 Ci = 3.7×10^{10} Bq).

Combining Equation 3.32 and Equation 3.33, gives

$$-\frac{dN(t)}{dt} = \lambda N(t) \tag{3.34}$$

Integrating this differential equation results in

$$N(t) = N_0 \exp(-\lambda t) \tag{3.35}$$

where N_0 is the (initial) number of radioactive nuclei at $t = 0$, i.e., $N_0 = N(0)$. Multiplying both sides of Equation 3.35 by the decay constant λ and recalling

Equation 3.33, results in the following equation for the activity:

$$A(t) = A_0 \exp(-\lambda t) \tag{3.36}$$

where A_0 is the (initial) activity at $t = 0$, i.e., $A_0 = A(0)$.

Both Equation 3.35 and Equation 3.36 present the exponential law of radioactive decay, which states that the number of nuclei that have not decayed in the sample as well as the activity of the sample, both decrease exponentially with time.

The time needed for half of the radionuclides to decay (or equivalently the activity of a sample to be reduced to half its initial value) is called half-life, $T_{1/2}$, and it can be calculated using Equation 3.35 for $N(T_{1/2}) = N_0/2$ (or equivalent to Equation 3.36 for $A(T_{1/2}) = A_0/2$):

$$\frac{N_0}{2} = N_0 \exp(-\lambda T_{1/2}) \rightarrow T_{1/2} = \frac{\ln 2}{\lambda} \tag{3.37}$$

The mean lifetime τ, i.e., the average lifetime of a given radioactive nucleus is the average value of t calculated as

$$\tau = \langle t \rangle = \frac{\int_0^\infty t \, dN(t)}{\int_0^\infty dN(t)} = \frac{N_0 \int_0^\infty \lambda t \exp(-\lambda t) dt}{N_0} = \frac{1}{\lambda} \tag{3.38}$$

The mean lifetime τ, is the reciprocal of the decay constant λ, and this result is natural since the decay constant has the physical meaning of the disintegration probability, i.e., the fraction of decays taking place per unit time. Apparently, within time τ the initial number of nuclei decreases by a factor of e.

Figure 3.5 presents the exponential decrease of unit activity with time for various radionuclides used in brachytherapy which are characterized by half-lives spanning from a couple of days to 30 years.

3.3.2 Radioactive Growth and Decay

Activity calculations in successive radioactive decays are more complicated. Suppose a chain decay of the form:

$$N_1 \xrightarrow{\lambda_1} N_2 \xrightarrow{\lambda_2} N_3 \tag{3.39}$$

In this chain, the parent radionuclide N_1 with decay constant λ_1 decays to daughter nuclide N_2 which is also radioactive and has a decay constant λ_2 and therefore decays to N_3 (assumed for simplicity to be stable). It is clear that there is a growth of N_2 with time due to decay of N_1, as well as a decay of N_2 since it is itself radioactive. This is also the case of growth and decay of the excited states usually created when a parent nucleus decays by alpha- or beta-emission.

A system of two differential equations can be written to describe the two successive decays in Equation 3.39. For simplicity let the numbers N_1, N_2,

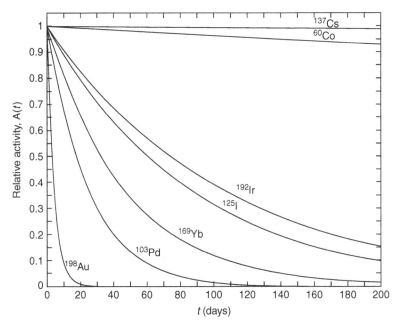

FIGURE 3.5
The exponential time decrease for a unit of initial activity of radionuclides used in brachytherapy: ^{137}Cs $(T_{1/2} = 30.20\text{y})$, ^{60}Co $(T_{1/2} = 5.27\text{y})$, ^{192}Ir $(T_{1/2} = 73.81\text{d})$, ^{125}I $(T_{1/2} = 59.49\text{d})$, ^{169}Yb $(T_{1/2} = 32.02\text{d})$, ^{103}Pd $(T_{1/2} = 16.99\text{d})$, and ^{198}Au $(T_{1/2} = 2.70\text{d})$.

and N_3 represent also the number of nuclei from each nuclide. If originally only the nuclide N_1 is present, i.e., $N_1(0) \neq 0$ is the number of parent nuclei at time $t = 0$, then $N_2(0) = N_3(0) = 0$ and the differential equation for N_1 is (see Equation 3.34):

$$-\frac{dN_1(t)}{dt} = \lambda_1 N_1(t) \tag{3.40}$$

which integrated results to

$$N_1(t) = N_1(0)\exp(-\lambda_1 t) \tag{3.41}$$

for the number of nuclei $N_1(t)$ present at time t. The parent activity $A_1(t)$ at time t is given by

$$A_1(t) = A_1(0)\exp(-\lambda_1 t) \tag{3.42}$$

The differential equation describing the growth and decay of the daughter nuclei N_2 is

$$\frac{dN_2(t)}{dt} = \lambda_1 N_1(t) - \lambda_2 N_2(t) \tag{3.43}$$

where the first term $\lambda_1 N_1$ on the right side of the equation corresponds to the growth of N_2 due to the parent decay and the second term $\lambda_2 N_2$ corresponds to the daughter decay.

Substituting Equation 3.41 into Equation 3.43, results in

$$\frac{dN_2(t)}{dt} + \lambda_2 N_2 = \lambda_1 N_1(0) \exp(-\lambda_1 t) \tag{3.44}$$

which integrated gives

$$N_2(t) = \frac{\lambda_1}{\lambda_2 - \lambda_1} N_1(0) \{(\exp(-\lambda_1 t) - \exp(-\lambda_2 t)\} \tag{3.45}$$

for the number of daughter nuclei $N_2(t)$ present at time t. The daughter activity $A_2(t)$ at time t is given by

$$A_2(t) = A_1(0) \frac{\lambda_2}{\lambda_2 - \lambda_1} \{(\exp(-\lambda_1 t) - \exp(-\lambda_2 t)\} \tag{3.46}$$

Daughter activity $A_2(t)$ reaches its maximum value when its time derivative vanishes, i.e., at time t for which $dA_2/dt = 0$ (or $dN_2/dt = 0$). This time is calculated as

$$t(A_2 = \text{max}) = \frac{\ln(\lambda_1/\lambda_2)}{\lambda_1 - \lambda_2} \tag{3.47}$$

While the parent activity $A_1(t)$ decreases exponentially with time t as in Equation 3.42, the daughter activity $A_2(t)$ in Equation 3.46 starts from zero at $t = 0$, increasing to its maximum value at time $t(A_2 = \text{max})$ as in Equation 3.47. At that time, and only at that time, the parent activity and the accumulated daughter activity are equal. This can be easily deduced from Equation 3.43, since $dN_2/dt = 0$.

There are many cases in nature for which $\lambda_1 \ll \lambda_2$, i.e., the daughter is shorter living than the parent. In such cases some approximations can be used. These are

$$\lambda_2 - \lambda_1 \equiv \lambda_2 \text{ and } \exp(-\lambda_2 t) \equiv 0 \text{ when } t > t(A_2 = \text{max}) \tag{3.48}$$

Then it follows:

$$\frac{A_2(t)}{A_1(t)} = \frac{\lambda_2}{\lambda_2 - \lambda_1} \left\{ \frac{\exp(-\lambda_1 t) - \exp(-\lambda_2 t)}{\exp(-\lambda_1 t)} \right\} \approx 1$$

That is

$$A_2(t) \equiv A_1(t) = A_1(0) \exp(-\lambda_1 t) \tag{3.49}$$

This is a very interesting result since it allows for simple calculations of daughter activity when the daughter nuclear state is shorter living than the parent one. When the activities of parent and daughter are equal, the situation is called *ideal equilibrium*. Figure 3.6 illustrates an example of the decay of unitary activity of ^{226}Ra and the corresponding growth and decay

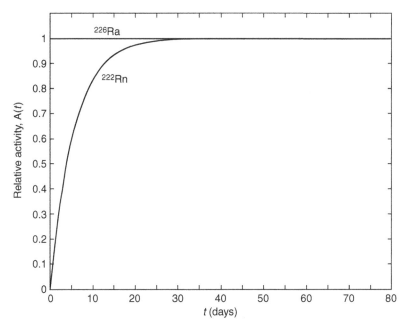

FIGURE 3.6
Relative activity plotted versus time for ^{226}Ra and ^{222}Rn.

of ^{222}Rn versus time, in the alpha decay of ^{226}Ra ($T_{1/2} = 1600$ years) to ^{222}Rn ($T_{1/2} = 91.8$ hours) (see also Equation 3.68 and Figure 3.8 in Section 3.4.1).

Radioactive decay usually leaves daughter nuclides in an excited state. Excited states are usually short living. These states de-excite emitting gamma radiation. It is such gamma radiations that are mainly utilized in brachytherapy. Chapter 5 presents brachytherapy-related radionuclides and their decay modes. For example, ^{137}Cs is beta minus radioactive, decaying to the stable nuclide ^{137}Ba (see also Equation 3.76 and Figure 3.10 in Section 3.4.2.1). However, this decay does not lead necessarily to the ground state of ^{137}Ba, but also to an excited state from which the 662 keV gamma radiation useful to brachytherapy arises. In computing how gamma radiation output of a ^{137}Cs brachytherapy source changes with time, the decay constant of the parent nuclide ^{137}Cs is used. In this example the half-time of ^{137}Cs is approximately 30 years. That means that in 30 years not only will the activity of a ^{137}Cs source drop to half of its initial value, but also its gamma radiation output.

3.3.3 Nuclear Reactions

In a nuclear reaction, two nuclei or a nucleon and a nucleus come together at such close approach (of the order of 10^{-15} m) that they interact through the strong force. A nuclear reaction is accompanied by a redistribution of energy and momentum between both particles and this may lead to the formation of new particles. There are many different types of nuclear reactions. Depending on the particles responsible for these reactions they are usually classified as neutron-induced reactions, reactions induced by charged particles or even by gamma radiation. The latter are associated with electromagnetic interaction, but are usually referred to as nuclear since the interaction takes place in the vicinity of the nucleus and results in its transformation. Radioactive decays are also considered as nuclear reactions.

The most commonly encountered type of nuclear reaction involves a light particle a and a nucleus A, resulting in the formation of a light particle b and a nucleus B. This is called an (a,b) reaction and can be written in the following general form:

$$a + A \rightarrow b + B \Leftrightarrow A(a, b)B \tag{3.50}$$

Suppose an incident particle a of rest mass m_a and kinetic energy K_a is in collision with a target nucleus A of rest mass m_A and kinetic energy $K_A = 0$. After the collision, the particle b has rest mass m_b and kinetic energy K_b while the residual nucleus has rest mass m_B and kinetic energy K_B. Energy conservation for the reaction 3.50 implies that

$$m_a c^2 + K_a + m_A c^2 = m_b c^2 + K_b + m_B c^2 + K_B \tag{3.51}$$

An important aspect of a nuclear reaction is its energy balance called the Q value or reaction energy Q. The Q value for the reaction in Equation 3.50, taking into account Equation 3.51, is defined as

$$Q = (m_a + m_A)c^2 - (m_b - m_B)c^2 = K_b + K_B - K_a \tag{3.52}$$

This equation shows that the rest energy balance equals the kinetic energy balance.

If $Q > 0$, the reaction is called exoergic and it is accompanied by a liberation of kinetic energy at the expense of the rest energy. Radioactive decays are all exoergic reactions.

If $Q < 0$, the reaction is called endoergic and involves an increase in the rest energy at the expense of the kinetic energy. Endoergic reactions can take place only when the incident particle has sufficient kinetic energy.

Finally, $Q = 0$ corresponds to elastic scattering, denoted as $A(a,a)A$, in which there is no production of new particles and the kinetic energy is conserved.

Radionuclides used for the construction of brachytherapy sources (see Chapter 6) are either natural (e.g., radium ^{226}Ra, see Section 3.1) or artificial. They can be fission products (e.g., ^{137}Cs, 6.15 atoms formed per 100 undergoing fission) or they can be obtained as a result of a neutron capture reaction of a stable nuclide, i.e., (n, γ) reaction.

When a beam of particles is incident normally upon a thin sheet of material containing target nuclei, the probability of reaction is proportional to the number of target nuclei per unit area of the sheet. The proportionality constant has the units of area and is called the cross-section σ. A cross-section may be visualized as the effective area a target nucleus presents to the incident particles, for undergoing the reaction; the probability of reaction is just equal to the probability that the incident particle strikes within the effective target area. Since the number of target nuclei per unit area is $n_i x$, where n_i is the number of target per unit volume in the material and x is the thickness of the sheet, the probability of the reaction occurring is equal to the product $n_i x \sigma$.

Assuming that the beam of particles has a fluence rate $\dot{\Phi}$ in particles $cm^{-2} sec^{-1}$, the rate at which the reaction proceeds is

$$\text{reaction rate } \dot{R} = n_i x \dot{\Phi} \sigma \tag{3.53}$$

Suppose that a certain radionuclide with decay constant λ is produced in a nuclear reactor with a constant rate, $\dot{R}$. The differential equation describing the production and decay of the radionuclide is

$$\frac{dN(t)}{dt} = \dot{R} - \lambda N(t) \tag{3.54}$$

which solved for the number N gives

$$N(t) = \frac{\dot{R}}{\lambda}\{1 - \exp(-\lambda t)\} \tag{3.55}$$

The accumulated activity $A(t)$ at a time t is

$$A(t) = \dot{R}\{1 - \exp(-\lambda t)\} \tag{3.56}$$

The last equation shows that the accumulated activity $A(t)$ depends on both the reaction rate $\dot{R}$ and the decay constant λ through the term $[1 - \exp(-\lambda t)]$ (see also Section 6.1).

A more complicated situation is when the radionuclide of interest is a daughter product of a shorter-living nuclide as is the case of ^{125}I which is the daughter product of ^{125}Xe. This decay chain and associated $T_{1/2}$ values are

$$^{125}_{54}Xe \xrightarrow{16.9 \text{ h}} {}^{125}_{53}I \xrightarrow{59.49 \text{ d}} {}^{125}_{52}Te \tag{3.57}$$

Assuming that ^{125}Xe is produced in a nuclear reactor at a constant rate $\dot{R}$, the accumulated activity $A_{Xe}(t)$ is given from 3.56:

$$A_{Xe}(t) = \dot{R}\{1 - \exp(-\lambda_{Xe}t)\} \tag{3.58}$$

The differential equation for ^{125}I is

$$\frac{dN_I(t)}{dt} = \lambda_{Xe}N_{Xe}(t) - \lambda_I N_I(t) = \dot{R}\{1 - \exp(-\lambda_{Xe}t)\} - \lambda_I N_I(t) \tag{3.59}$$

If $N_I(0) = 0$ then

$$N_I(t) = \dot{R}\frac{1 - \exp(-\lambda_I t)}{\lambda_I} - \frac{\exp(-\lambda_{Xe}t) - \exp(-\lambda_I t)}{\lambda_I - \lambda_{Xe}} \tag{3.60}$$

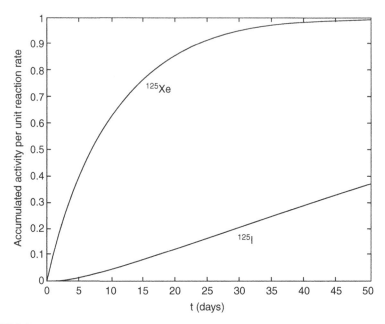

FIGURE 3.7
A plot of accumulated activity per unit reaction rate vs. time for ^{125}Xe ($T_{1/2} = 16.9$ h) and ^{125}I ($T_{1/2} = 59.49$d).

and the accumulated activity $A_I(t)$ for a time t, becomes

$$A_I(t) = \dot{R}\left\{1 - \exp(-\lambda_I t) - \lambda_I \frac{\exp(-\lambda_{Xe}t) - \exp(-\lambda_I t)}{\lambda_I - \lambda_{Xe}}\right\} \qquad (3.61)$$

Figure 3.7 presents the increase of accumulated activity per unit reaction rate with time for ^{125}Xe and ^{125}I (see also Chapter 6).

3.4 Modes of Decay

The laws of radioactivity discussed so far, describe the rates at which unstable nuclear states decay, and have nothing to do with the modes of decay themselves as well as the kind, energy, or intensity of emitted radiation. All these properties, including the decay constant, are characteristic for every nuclide and they univocally designate it.

The mode of decay is named according to the kind of radiation emitted, for which, historically, the Greek alphabet is used, i.e., α decay, β decay, and γ decay. The strong nuclear force is responsible for the alpha decay, the weak nuclear force for beta decay, and the electromagnetic for gamma decay. All these decay modes are spontaneous and the energy liberated, disintegration energy or Q-value (see Section 3.3.3), is distributed as kinetic energy among the decay products, i.e., the residual nucleus and the emitted particles. Conservation of energy and momentum specifies how this energy is distributed, conservation of charge, nucleon, and lepton numbers specify the particles emitted, and other conservation laws (parity, isospin) impose certain restrictions on allowed transitions. These, however, are beyond the scope of this book, which focuses on the essential.

3.4.1 Alpha Decay

The peak of the mean binding energy per nucleon (B/A vs. A curve in Figure 3.4) at $A \sim 60$, indicates that it is advantageous for a heavy nucleus to split into two smaller nuclei, which together have a greater net binding energy. There are two processes of spontaneous heavy nuclei splitting, namely α decay and fission. In α decay, which is the principal mode of nuclear splitting, a heavy nucleus splits into a light $^4_2$He nucleus (the α-particle) and another heavy nucleus. $^4_2$He has the largest binding energy among all light nuclides: $B = 28.29$ MeV. In spontaneous fission, the nucleus splits into two more or less equal nuclei.

In α decay a parent nuclide, $^A_Z X$, decays spontaneously to the daughter nuclide, $^{A-4}_{Z-2}Y$, and an α-particle is emitted, while energy Q_α is liberated:

$$^A_Z X \rightarrow {}^{A-4}_{Z-2}Y + \alpha + Q_\alpha \tag{3.62}$$

Conservation of energy demands:

$$m_X c^2 + K_X = m_Y c^2 + K_Y + m_\alpha c^2 + K_\alpha \tag{3.63}$$

where rest masses are denoted by m and kinetic energies by K. Rewriting the last equation and given zero kinetic energy of the parent nucleus ($K_X = 0$), yields:

$$Q_\alpha \equiv m_X c^2 - m_Y c^2 - m_\alpha c^2 = K_Y + K_\alpha \tag{3.64}$$

The last equation shows that the rest energy balance equals the kinetic energy balance. The disintegration energy Q_α, as the rest energy balance (left side of Equation 3.64), can be accurately calculated from the known masses. Atomic masses can be readily used in these calculations instead of calculating the nuclear masses since the number of electrons, and hence electron masses, cancels out as seen in Equation 3.62, and moreover, there is good approximation to a balance of the electrons' binding energies since the most tightly bound, K electrons practically remain unchanged.

The disintegration energy Q_α, is distributed as kinetic energy of the α-particle and recoil energy of the daughter nucleus. Conservation of momentum in reaction 3.62 demands that the momentum p_α of the α-particle is equal in magnitude to the momentum p_Y of the daughter nucleus, since the parent nucleus spontaneously decays at rest:

$$p_\alpha = p_Y \tag{3.65}$$

For the energies considered herein, both the α-particle and daughter nucleus can be treated as nonrelativistic ($p^2 = 2\,mK$). Thus, Equation 3.65 gives

$$\frac{K_\alpha}{K_Y} = \frac{m_Y}{m_\alpha} \approx \frac{A-4}{4} \tag{3.66}$$

where the mass ratio is approximated by mass number ratio as in Equation 3.22. From Equation 3.64 and Equation 3.66 the kinetic energy K_α of the emitted α-particle and the recoil energy K_Y of the daughter nucleus are obtained:

$$K_\alpha \approx Q_\alpha\left(1 - \frac{4}{A}\right) \text{ and } K_Y \approx Q_\alpha\frac{4}{A} \tag{3.67}$$

That is, the emitted α-particles have discrete (kinetic) energy which constitutes a considerable part of the disintegration energy Q_α.

In the above calculations the daughter nucleus is assumed to be created in its ground state. This is not however mandatory; the daughter nucleus can be left in an excited state. If this excited state has (excitation) energy E^* above the ground state, the rest energy of the daughter nucleus is less than this value, and consequently, the disintegration energy Q_α in Equation 3.67 has to be replaced by $Q_\alpha - E^*$.

A typical example of an alpha decay is that of ^{226}Ra (half-life $T_{1/2} = 1600$ years):

$$^{226}_{88}\text{Ra} \rightarrow {}^{222}_{86}\text{Rn} + \alpha + 4.8706\,\text{MeV} \tag{3.68}$$

The atomic masses are: $m_{\text{Ra}} = 226.025403\,u$, $m_{\text{Rn}} = 222.017571\,u$, and $m_\alpha = 4.0026032\,u$ yielding (according to Equation 3.64 and Equation 3.21) disintegration energy $Q_\alpha = 4.8706\,\text{MeV}$. This disintegration energy is distributed as in Equation 3.67, resulting in $K_\alpha = 4.7843\,\text{MeV}$ and $K_{\text{Rn}} = 0.0863\,\text{MeV}$.

The above disintegration scheme occurs with 94.5% probability. In the remaining 5.5% of cases, the radon daughter nucleus is in an excited state ($^{222}_{86}\text{Rn}^*$) and decays to its ground state by emitting γ-rays of 0.1862 MeV. In this case, the kinetic energy of the emitted α-particle is $(4.7843 - 0.1862)$ MeV $= 4.60$ MeV. Figure 3.8 summarizes the simplified decay scheme of ^{226}Ra.

3.4.2 Beta Decay

The semiempirical mass formula presented in Equation 3.31 reveals that when nuclear masses $m(Z,A)$ for a given mass number A are plotted vs. the atomic number Z, they form a parabola when A is odd or two parabolas displaced in mass by $2 \times \alpha_{\text{Pairing}}\delta A^{-3/4}$ when A is even. The most stable nuclides lie at the base of the parabola. All nuclides belong to a corresponding A-value parabola. All these parabolas form a three-dimensional presentation of the two-dimensional Segre plot presented in

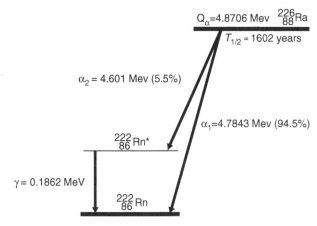

FIGURE 3.8
The decay scheme of ^{226}Ra.

Figure 3.3 with nuclear mass as the third dimension and warranting the name stability valley.

For example, the odd-A parabolas for the isobaric families of $A = 125$ and $A = 137$ are presented in Figure 3.9. For $A = 125$, only ^{125}Te, which lies close to the parabola's minimum, is stable. All other members of the isobaric

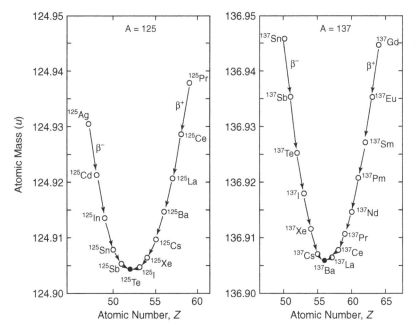

FIGURE 3.9
A plot of the atomic mass of the isobaric families with $A = 125$ and $A = 137$ vs. atomic number Z (i.e., the parabola on which these two odd-A isobaric families lie).

family are radioactive, and by successive decays they finally reach the stable nuclide ^{125}Te. For $A = 137$, the stable nuclide is ^{137}Ba. The mode of decay between isobars is named β decay from the Greek letter β (beta); weak nuclear force is responsible for β decay. There are three types of β decay: $β^-$ (beta minus), $β^+$ (beta plus), and electron capture. Members of the isobaric family that lie on the left arm of the parabola ($Z < Z_{minimum}$), as those presented in Figure 3.9, are $β^-$ radioactive while members on the right-arm ($Z > Z_{minimum}$) are $β^+$ radioactive and/or decay by EC.

For even A, the two existing parabolas correspond to one with even Z and even N (the lower of the two) and one with odd Z and odd N. In this case, the number of stable nuclides of an isobaric family could be one, two or even three in rare cases. An example of an isobaric family with two stable nuclides is that with $A = 192$ (the two stable nuclides are ^{192}Pt and ^{192}Os). ^{192}Ir, which is the most commonly used nuclide in current brachytherapy practice lies on the upper parabola (since both Z and N are odd) and decays with either $β^-$ to ^{192}Pt (95%) or to ^{192}Os by EC (5%) (see Chapter 5).

3.4.2.1 β⁻ Decay

In $β^-$ decay a parent nuclide, ^A_ZX, decays spontaneously to the daughter nuclide, $^{A}_{Z+1}Y$, a $β^-$-particle and an antineutrino, $\overline{ν}_e$ are emitted and energy $Q_{β^-}$ is liberated:

$$^A_ZX \rightarrow ^A_{Z+1}Y + β^- + \overline{ν}_e + Q_{β^-} \tag{3.69}$$

The $β^-$-particle is an electron e^- whose presence in the reaction satisfies charge conservation (since nucleons are conserved and there is a proton excess in the right side of the reaction). The antineutrino $\overline{ν}_e$ has zero mass (or very small), has no charge, and its presence in the reaction is to ensure lepton conservation.

Conservation of energy in reaction 3.69 demands:

$$m_Xc^2 + K_X = m_Yc^2 + K_Y + m_ec^2 + K_{β^-} + K_{\overline{ν}} \tag{3.70}$$

where rest masses are denoted by m (excluding the massless $\overline{ν}_e$) and kinetic energies by K ($\overline{ν}_e$ does have kinetic energy). Rewriting the last equation, and since the kinetic energy of the parent nucleus is $K_X = 0$, gives

$$Q_{β^-} \equiv m_Xc^2 - m_Yc^2 - m_ec^2 = K_Y + K_{β^-} + K_{\overline{ν}} \tag{3.71}$$

The masses m_X and m_Y in Equation 3.71 are nuclear. In terms of atomic masses M_X and M_Y, if Equation 3.23 is recalled the right side of Equation 3.71 gives

$$Q_{β^-} \equiv M_Xc^2 - M_Yc^2 \tag{3.72}$$

This is the energy condition for $β^-$ decay; the atomic mass of parent nuclide must be greater than the atomic mass of daughter nuclide, since $Q_{β^-} > 0$ for a spontaneous decay.

Equation 3.71 shows that the disintegration energy, Q_{β^-}, is distributed as kinetic energy among three particles: the daughter nucleus, the β^- particle, and the antineutrino $\overline{\nu}_e$. Since there is only one additional restriction due to momentum conservation:

$$\vec{p}_X = 0 = \vec{p}_Y + \vec{p}_{\beta^-} + \vec{p}_{\overline{\nu}} \tag{3.73}$$

there is no unique way of distributing the disintegration energy. Only a minute proportion of this energy can be carried away by the daughter nucleus due to its huge mass relative to the electron-antineutrino pair. Therefore, almost all the energy goes to the lepton pair, the electron and the antineutrino, which present continuous and not discrete energy spectrum. Thus, the β^--particle can have any kinetic energy up to the maximum possible value of Q_{β^-}, that is

$$K_{\beta^-,max} = Q_{\beta^-} \equiv M_X c^2 - M_Y c^2 \tag{3.74}$$

It is customary to tabulate this maximum kinetic energy $K_{\beta^-,max}$ of the β^--particle (see for example Figure 3.2).

The simplest example of β^- decay is the decay of free neutron (see Figure 3.2):

$$n \rightarrow p + \beta^- + \overline{\nu}_e + 0.782 \, \text{MeV} \tag{3.75}$$

The free neutron is not stable. It has a lifetime of $T_{1/2} \approx 10.4$ min!

A typical example of β^- decay is that of $^{137}_{55}\text{Cs}$ (half-life $T_{1/2} = 30.07$ years) to $^{137}_{56}\text{Ba}$ (see corresponding isobaric parabola in Figure 3.9):

$$^{137}_{55}\text{Cs} \rightarrow \, ^{137}_{56}\text{Ba} + \beta^- + \overline{\nu}_e + 1.1756 \, \text{MeV} \tag{3.76}$$

Substituting the atomic masses of ^{137}Cs (136.9070835 u) and ^{137}Ba (136.9058214 u) into Equation 3.72 and accounting for the conversion factor of Equation 3.21, yields the disintegration energy of $Q_{\beta^-} = 1.1756$ MeV and hence $K_{\beta^-,max} = 1.1756$ MeV.

The above disintegration scheme occurs with 5.6% probability. In the other 94.4% of cases the daughter nucleus is in an excited state ($^{137}_{56}\text{Ba}^*$) and further decays to its ground state by emitting γ-rays of 0.6617 MeV. In this case, the maximum kinetic energy of the emitted β^--particle is $(1.1756 - 0.6617)$ MeV $= 0.514$ MeV. Figure 3.10 summarizes the decay scheme of ^{137}Cs.

3.4.2.2 β^+ Decay

In β^+ decay a parent nuclide, $^A_Z X$, decays spontaneously to the daughter nuclide $_{Z-1}^{A}Y$, a β^+-particle and a neutrino, ν_e are emitted, while energy Q_{β^+} is liberated:

$$^A_Z X \rightarrow \, ^A_{Z-1}Y + \beta^+ + \nu_e + Q_{\beta^+} \tag{3.77}$$

Conservation of energy in reaction 3.77 demands that

$$m_X c^2 + K_X = m_Y c^2 + K_Y + m_e c^2 + K_{\beta^+} + K_\nu \tag{3.78}$$

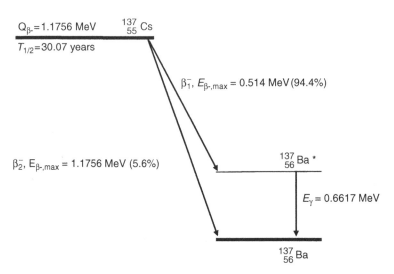

FIGURE 3.10
The β^- decay scheme of ^{137}Cs.

where rest masses are denoted by m (excluding the massless ν_e) and kinetic energies by K (ν_e does have kinetic energy). Rewriting the last equation, given that the kinetic energy of the parent nucleus is $K_X = 0$, gives

$$Q_{\beta^+} \equiv m_X c^2 - m_Y c^2 - m_e c^2 = K_Y + K_{\beta^+} + K_\nu \qquad (3.79)$$

In terms of atomic masses M_X and M_Y, if Equation 3.23 is recalled the right side of Equation 3.79 gives

$$Q_{\beta^+} \equiv M_X c^2 - M_Y c^2 - 2m_e c^2 \qquad (3.80)$$

This is the energy condition for β^+ decay; the atomic mass of the parent nuclide must be greater by at least two times the electrons' rest energy than the atomic mass of the daughter nuclide, since $Q_{\beta^+} > 0$ for a spontaneous decay. That is

$$M_X c^2 - M_Y c^2 > 2m_e c^2 \approx 1.02 \text{ MeV} \qquad (3.81)$$

As for β^- decay, the disintegration energy Q_{β^+} is practically distributed between the β^+-particle and the neutrino ν_e, both of which present a continuous energy spectrum. Thus, the β^+-particle can have any kinetic energy up to a maximum value of Q_{β^+}. That is

$$K_{\beta^+,\text{max}} = Q_{\beta^+} \equiv M_X c^2 - M_Y c^2 - 1.02 \text{ MeV} \qquad (3.82)$$

It is customary to tabulate this maximum kinetic energy $K_{\beta^+,\text{max}}$ of the β^+-particle.

A typical example of β^+ decay is that of ^{22}Na (half-life of $T_{1/2} = 2.609$ years) to ^{22}Ne (0.06% to the ground state of ^{22}Ne and 89.84% to an excited

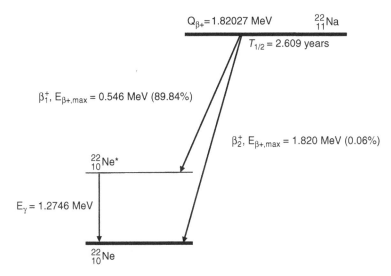

FIGURE 3.11
The β^+ decay scheme of ^{22}Na (see also Figure 3.12).

state of ^{22}Ne* with excitation energy $E^* = 1.2746$ MeV):

$$^{22}_{11}\text{Na} \rightarrow \,^{22}_{10}\text{Ne} + \beta^+ + \nu_e + 1.82027 \text{ MeV} \tag{3.83}$$

Substituting the atomic masses of ^{22}Na (21.9944368 u) and ^{22}Ne (21.9913855) into Equation 3.82 and accounting for the conversion factor of Equation 3.21 yields the disintegration energy of $Q_{\beta^+} = 1.82027$ MeV and hence $K_{\beta^+,\text{max}} = 1.82027$ MeV for decays to the ground state of ^{22}Ne and $Q_{\beta^+} = 0.546$ MeV and hence $K_{\beta^+,\text{max}} = 0.546$ MeV for decays leading to the first excited state of ^{22}Ne. Figure 3.11 summarizes the β^+ decay scheme of ^{22}Na (see also Figure 3.12).

3.4.2.3 Electron Capture

Electron capture (EC) is competitive to β^+ decay. In EC, a parent nucleus, $^A_Z X$, captures an electron from its own atomic electron shells and decays spontaneously to the daughter nucleus $_{Z-1}^A Y$, a neutrino, ν_e is emitted, while energy Q_{EC} is liberated:

$$^A_Z X + e^- \rightarrow \,_{Z-1}^A Y + \nu_e + Q_{\text{EC}} \tag{3.84}$$

Conservation of energy in reaction 3.84 demands that

$$m_X c^2 + K_X + m_e c^2 + K_e = m_Y c^2 + K_Y + K_\nu \tag{3.85}$$

where rest masses are denoted by m (excluding the massless ν_e) and kinetic energies by K (ν_e does have kinetic energy). Rewriting the last equation and ignoring kinetic energies of both the parent nucleus and the atomic electron,

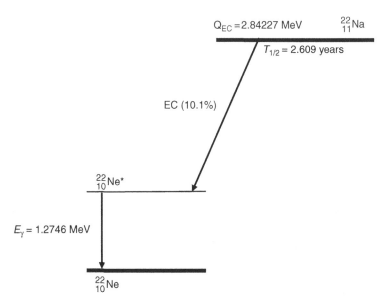

FIGURE 3.12
The EC decay scheme of ^{22}Na.

gives

$$Q_{EC} \equiv m_X c^2 + m_e c^2 - m_Y c^2 = K_Y + K_\nu \qquad (3.86)$$

The masses m_X and m_Y appearing in 3.86 are nuclear. In terms of atomic masses M_X and M_Y, if Equation 3.23 is recalled the right side of Equation 3.86 gives

$$Q_{EC} \equiv M_X c^2 - M_Y c^2 \qquad (3.87)$$

This is the energy condition for electron capture; the atomic mass of the parent nuclide must be greater than the atomic mass of the daughter nuclide, since $Q_{EC} > 0$ for a spontaneous decay. That is

$$M_X c^2 - M_Y c^2 > 0 \quad \text{or} \quad M_X > M_Y \qquad (3.88)$$

The latter equation, in view of the energy condition for β^+ decay of Equation 3.81, implies that besides being competitive to β^+ decay, EC provides an alternative path of decay between two neighboring isobar nuclei lying on the right arm of the isobar parabola and presenting a positive, yet less than $2m_e c^2 \approx 1.02$ MeV atomic rest energy difference.

A characteristic example (see Figure 3.9) is that of ^{125}I (atomic mass = 124.9046242 u) which decays to ^{125}Te (atomic mass = 124.9044247 u). In this example, the atomic rest energy difference is 0.186 MeV which fulfils condition 3.88 for EC but not that for β^+ decay in Equation 3.81. Therefore, as discussed in Chapter 5, ^{125}I decays to ^{125}Te exclusively by EC.

Equation 3.86 shows that the disintegration energy Q_{EC}, is distributed as kinetic energy between two particles: the daughter nucleus $^A_{Z-1}Y$ and the neutrino ν_e. Owing to momentum conservation:

$$\vec{p}_Y + \vec{p}_\nu = 0 \tag{3.89}$$

that is, the recoil momentum p_Y of the daughter nucleus is equal in magnitude to the neutrinos' momentum p_ν. Thus, the neutrino carries away almost the entire disintegration energy Q_{EC} in the form of kinetic energy. It therefore has a discrete and not a continuous energy spectrum:

$$K_\nu = Q_{EC} \equiv M_X c^2 - M_Y c^2 \tag{3.90}$$

Given the high accuracy in Q_{EC} calculations from atomic masses, measurement of the daughter nucleus recoil energy was used by Davis (1952) to provide an estimate of the upper limit for the neutrino mass.

The captured electron is an inner-shell electron, usually from the K-shell (the electron capture process is alternatively named as K capture). The vacancy created is filled by outer electrons falling into it and the process is accompanied by emission of fluorescent radiation or Auger electron emission. This has significant impact on the dosimetry of low-energy emitters such as ^{125}I and ^{103}Pd (see Chapter 4 and Chapter 5).

The β^+ decay of ^{22}Na discussed in the previous section and presented in Figure 3.11, reveals that there is an alternative 10.1% possibility for EC (β^+ decay appears with a frequency of 89.9%). The rest energy difference is

$$Q_{EC} \equiv M_{^{22}Na} c^2 - M_{^{22}Ne} c^2 = 2.842268 \text{ MeV} \tag{3.91}$$

Figure 3.12 summarizes the EC decay for ^{22}Na.

3.4.3 Gamma Decay and Internal Conversion

Alpha and beta decays discussed in previous sections (Section 3.4.1 and Section 3.4.2) reveal that the daughter nucleus in such decays is usually created in an excited state. These states are usually short living and de-excite spontaneously emitting electromagnetic radiation named γ-rays from the Greek letter γ (gamma); the decay is called γ decay. There are single transitions when a nucleus emits a single γ-ray and at once falls to the ground state (see Figure 3.8 and Figure 3.10 through Figure 3.12) or cascade transitions when excitation is removed by a successive emission of several γ-rays (see for example the decay scheme of ^{60}Co and ^{192}Ir in Chapter 5). The energy, $E\gamma$, of the γ-rays is determined by the difference ΔE in the nuclear rest energy of the two levels involved:

$$\Delta E = E_{\text{initial}} - E_{\text{final}} = E_\gamma + K_{\text{recoil}} \tag{3.92}$$

where K_{recoil} is the kinetic energy of the recoil nucleus. Since the momentum of a γ-ray (E_γ / c) should be equal in magnitude to the recoil momentum due to momentum conservation, the recoil energy K_{recoil} is

$$K_{\text{recoil}} = \frac{E_\gamma^2}{2mc^2} \approx \frac{(\Delta E)^2}{2mc^2} \qquad (3.93)$$

Since ΔE is in the energy range of 10 keV to 5 Mev while the remaining energy of the atom accounts for many GeV, the recoil energy is in the eV range. Thus, the γ-ray carries away an overwhelming part of the nuclear excitation energy.

In addition to the γ-ray emission, there is another mechanism by which the excited daughter nucleus may de-excite; by transferring its energy excess directly to an inner orbital electron. The electron is then ejected by the atom with a kinetic energy equal to the difference of the nucleus excess energy ΔE and the binding energy of the involved electron. This process is called internal conversion (IC) and the ejected electron is called an internal conversion electron. Although ^{137}Cs decays to the excited state of ^{137}Ba with a probability of 94.4% (see Figure 3.10) the 0.6627 MeV γ-ray is emitted with a probability of 85.9% and not 94.4%. The remaining probability corresponds to de-excitation of the daughter nuclide by IC.

Both internal conversion electrons and γ-rays present discrete line spectra which are characteristic for each nuclide. As in the case of the electron capture process, IC is accompanied by atomic characteristic fluorescent radiation and Auger electrons. Obviously, neither gamma decay nor internal conversion change the atomic number Z, the mass number A, and the neutron number N of the nucleus.

3.5 Elementary Particles and the Standard Model

In the history of physics, during the last 85 years, we have witnessed tremendous progress in the area of atomic and subatomic physics which has had a major impact on the frontier of technological advancement. In 1920, the laws of nature governing the behavior of the smallest matter entity, at that time, the atom, started to be understood with the establishment of quantum mechanics, followed by a breakthrough almost 50 years later, in 1969, with paramount discoveries that established the Standard Model, proposed by Weinberg, Salam, and Glashow, as a true theory of nature.

With the advent of quantum mechanics we were able to interpret, in a probabilistic way, the motion of small-sized objects, such as that of an atom for instance, whose size is of the order of 10^{-10} m or almost ten billion times smaller than the height of a human being. At this microscopic scale the laws of nature are not the same as those governing the motion of macroscopic objects. We are not in a position to know the trajectory of such a small-sized object, of atomic or subatomic scale, merely because we cannot measure its position and its velocity simultaneously. This is actually the content of the uncertainty principle (see Section 3.1.2), an inherent property of nature. We are therefore unable to know the position and velocity of a microscopic particle but we are in position to speak of the probability to find it here or there. This probability is expressed through a function, known as wave function, which satisfies a certain equation, known as the Schroedinger equation, which is the fundamental law in quantum mechanics. The motion and interaction of atoms, the formation of molecules and chemical bonds are all explained with the rules of quantum mechanics. Knowing the laws of nature at this microscopic level we have been able to understand the mechanisms responsible for the behavior of matter at small scales, which has been of paramount importance for technological developments.

The construction of new materials, for instance in contemporary times the silicon chip which is the heart of any computer machine, as well as other inventions used in modern medicine, such as MRI, positron emission tomography, and other diagnostic tools, are based on the knowledge of our microcosmos.

The present knowledge of the microcosm is based on research, conducted from 1930 to 1970, of a physics theory that completes quantum mechanics, in the sense that it can explain phenomena that quantum mechanics was unable to interpret, and that occurs at a smaller scale, much smaller than that of an atom or a nucleus. The Standard Model, which is the new landmark, was established in 1969 and it succeeds in explaining with unprecedented accuracy, all phenomena occurring at distances 10^{-18} m, or equivalently one hundred million times smaller than the dimensions of an atom! But how can we see at such small distances? The physics laws and the validity of the proposed theories can be tested with large accelerator machines, running at very high energies, which are the eyes of the particle physicists, capable of probing such tiny distances.

In order to conceive the magnitude of such machines, the Large Hadron Collider (LHC) accelerator being built at CERN (Switzerland), which will start working in 2007, is a circular accelerator with a circumference of about 27 km. In it, protons will be accelerated to total energies reaching 14 TeV (14 thousand billions eV!). With this accelerator we will be able to go deeper and probe at even smaller distances.

Perhaps one of the spectacular discoveries of past accelerators was the fact that protons and neutrons, the ingredients of the nuclei of the atoms, are not fundamental but are composed of other particles which are named quarks. The atom consists of electrons and a nucleus, the nucleus consists of protons and neutrons, which are bound together by a sort of force known as the strong force which must be stronger than the electromagnetic force to overcome the repulsion of protons within the nuclei due to electric forces.

The major discovery that protons, neutrons, as well as other particles discovered since 1930 (which collectively we call hadrons), are not fundamental, had been theoretically founded and was experimentally verified and it is at the heart of the Standard Model. According to it, the fundamental building blocks of matter are the leptons and the quarks. There are six leptons and one of them is the well-known electron, one of the ingredients of the atom. Quarks are the ingredients of protons, neutrons, and in general, of all hadrons. These fundamental particles carry spin equal to 1/2 and they are grouped in three families which have exactly the same characteristics, such as electric charges, spins, and other quantum numbers, but different masses. These families are also called generations. The first family, or generation, accommodates two leptons, the electron, e^-, and its neutrino, ν_e, and six quarks denoted by u_a, d_a existing in three different species, or colors, labeled by $a = 1, 2, 3$. These quarks are called up and down, respectively. Their names reflect the way they are grouped in mathematical entities used to describe the theory and will not concern us here. The electric charge of the neutrino is zero and for this reason it does not interact electromagnetically, that is it does not interact with other charged particles. Its mass is very small, almost vanishing, being therefore very fast almost as fast as light. The only sort of interaction neutrinos can have is a feeble interaction, known as weak, which is responsible for the radioactivity of some particles or nuclei.

The neutron, for instance is unstable due to this kind of interaction and it decays into a proton, an electron, and a neutrino (see Section 3.4.2.1). This process is known as β^- decay. Owing to the fact that neutrinos interact only weakly they can penetrate the Earth and travel galactic distances without being captured by matter, carrying information from remote parts of our Universe. Recently, great efforts have been made towards building neutrino telescopes, which can see and study the properties of such cosmic neutrinos arriving at Earth, that could possibly reveal information concerning the creation of our Universe.

Unlike neutrinos, the electric charge of the electron is not zero; it is actually just opposite to that of the proton which conventionally is taken equal to $+1$. Then, the charges of the u_a, d_a quarks are fractional and equal to 2/3 and $-1/3$ times the proton's charge, respectively. The proton consists of three quarks, namely two up quarks u and one down quark d, whose total charge is $2 \times 2/3 - 1/3 = 1$, the charge of the proton. Similarly, the neutron consists of two d quarks and one u and their total charge is zero, which is actually the charge of the neutron. The color $a = 1, 2, 3$, labeling the quarks, is a quantum number analogous to the electric charge. In order for protons, neutrons, as well as other hadrons to be composed of quarks, the latter have to be bound or glued together by some sort of force. This force feels not the electric charge but the color instead and the part of the theory describing this kind of interaction is called quantum chromodynamics or QCD for short.

For every one of the three generations of the fundamental fermions there corresponds a generation that includes their antiparticles. For instance, for the first generation, which accommodates the electron e^-, there is a generation which includes its antiparticle e^+, the so-called positron having opposite electric charge but exactly the same mass and spin, and also the antiparticles of the neutrino and up and down quarks. The leptons and quarks of the three generations as well as their antiparticles are shown

TABLE 3.1

Leptons, Quarks and Their Antiparticles

Leptons			Quarks		
Particle Name	Symbol	Antiparticle	Particle Name	Symbol	Antiquark
Electron	e^-	e^+	Up	u	$\bar{u}$
Neutrino (e)	ν_e	$\bar{\nu}_e$	Down	d	$\bar{d}$
Muon	μ^-	μ^+	Strange	s	$\bar{s}$
Neutrino (μ)	ν_μ	$\bar{\nu}_\mu$	Charmed	c	$\bar{c}$
Tau	τ^-	τ^+	Bottom	b	$\bar{b}$
Neutrino (τ)	ν_τ	$\bar{\nu}_\tau$	Top	t	$\bar{t}$

in Table 3.1 and Figure 3.13. Detailed information can be found in the review of particle properties.[3] Anything we know in nature consists of these fundamental building blocks accommodated in the three families of quarks and leptons. There is no experimental evidence, up to the time of the writing, for the existence of any other form of matter although at the theoretical level there are proposals advocating the existence of new forms of matter undiscovered in the laboratory as yet. New experiments scheduled to run in the near future, starting with the LHC in 2007, will test some of these theoretical proposals and may discover new massive ingredients providing vital information to particle physics. It is worth noting that some of these ingredients are eagerly awaited in order to explain the missing mass of the Universe, the so-called dark matter, which is one of the biggest mysteries of modern cosmology.

Except for the fundamental fermions accomodated in the three generations the Standard Model also includes the carriers of the forces. With the exception of gravity, whose energy here plays little role, there are three

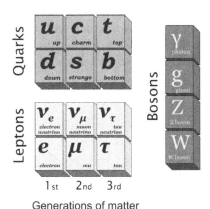

Generations of matter

FIGURE 3.13

Quarks and Leptons (left side) and the Force Carriers (right side).

forces: the strong, the electromagnetic, and the weak. The strong force is responsible for the binding of the protons and neutrons to form nuclei. The electromagnetic force is the one felt by charged matter, which at the classical level and for two charged particles of charges q_1, q_2, is given by the well-known Coulomb's law $F = q_1 q_2 / r^2$. The weak force, as well as the strong, is not manifested as a classical force, and is responsible for a variety of processes occurring in nuclear physics, among these being the β decay discussed earlier (see Section 3.4.2). Concerning the strength of these forces, if that of the strong force is taken equal to 1, the electromagnetic is roughly 0.01, and that of the weak even smaller, 0.00001. The gravitational force is left out of our discussion since it is imperceptibly small, 10^{40} times smaller than the strong force. Particles interact with each other because they exchange the carriers, or mediators, of the corresponding force. For instance, two electrically charged particles interact with each other since they exchange a photon which is the carrier of the electromagnetic force. In order to conceive of this exchange mechanism, imagine two people ice-skating and one throwing a ball to the other when they approach at a close distance. The trajectory of the skater who throws the ball will change since the ball leaving his hands carries momentum. The trajectory of the other skater who catches the ball will change as well for the same reason. This is how the interaction takes place. The trajectories of the skaters change as a result of the exchange of the ball which is the carrier, or mediator, of the interaction in this example. The carrier of the electromagnetic force is called a photon, it is usually denoted by the symbol γ, and it has zero mass! The lack of mass is intimately connected with the fact that the electromagnetic force is long range following the classical law $F \sim 1/r^2$. It should be noted that the light we see is a bunch of an enormous number of photons! In Figure 3.14, two electrons are depicted interacting electromagnetically by exchanging a photon (wavy line). The weak force has three mediators named W^+, W^-, Z, which unlike the photon are quite massive, each weighing roughly 100 times the mass of the proton. Owing to this, the weak force is of short range and cannot manifest itself as a classical force. The carriers of the strong (QCD) forces are eight massless particles called gluons which will be collectively denoted by the symbol g. Although they have zero mass, as the photon does, they do not induce long-range forces, due to a special property inherent in the dynamics

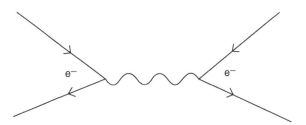

FIGURE 3.14
Two electrons interact by exchanging a photon.

of the QCD, which prohibits the propagation of the force at large distances. All of these carriers have spin equal to 1, and accordingly in the terminology of particle physics they are called gauge bosons. In Figure 3.13 all quarks and leptons are displayed along with the mediators of the three forces.

The Standard Model explains perfectly well all particle interactions to an unprecedented accuracy, as laboratory experiments have shown, for energies less than about 200 GeV, that is two hundred billion eV, equivalent to distances 10^{-18} m. However, there is still a missing link, escaping detection as yet, which is a fundamental particle needed for the correct mathematical description of the theory. From the physics point of view, this particle, which bears the name Higgs boson, is spinless and electrically neutral, and its presence is theoretically needed in order for the leptons and quarks, and also the carriers of the weak interactions W, Z, to acquire masses. Thus, its role is extremely important and searching for it will be one of the primary tasks in the new accelerators that will run in the near future. The new accelerators will probe at distances smaller than 10^{-18} m and will be able to provide us with new information concerning the physics laws that govern the interactions of particles at smaller scales. Theoretical proposals predict that when we probe more deeply, new species of particles will be discovered which at present escape detection because of lack of energy. With new accelerators, of sufficiently high energy, these new degrees of freedom will be produced in the laboratory, provided they exist, and their detection will open a new era in particle physics.

References

1. National Institute of Standards and Technology Physics Laboratory Physical Reference Data Electronic Version, available online at http://physics.nist.gov/PhysRefData (Gaithersburg: National Institute of Standards and Technology), August 2004.
2. Nuclear Data Evaluation La, Table of Nuclides (http://atom.kaeri.re.kr). Korea Atomic Energy Research Institute, 2000.
3. The Review of Particle Physics by Particle Data Group (PDG). Available online at http://pdg.lbl.gov/

4

Interaction Properties of Photons and Electrons

4.1 Introduction

This chapter provides an understanding on how photons in the energy range relevant to brachytherapy, interact with matter. Photons are indirectly ionizing particles, which eventually interact with the material through which they travel and appropriate interaction coefficients describe the probability of such interactions. Consequently, they also describe radiation transmission, energy transfer, and energy deposition in matter.

Let us consider two of these coefficients, the linear attenuation coefficient, μ, and the linear energy absorption coefficient, μ_{en}, (see also Section 2.4) that are essential when describing radiation transport and predicting absorbed dose in media subjected to photon irradiation. Figure 4.1 shows the energy dependence of these coefficients for two materials of radiobiological interest, soft tissue and cortical bone. In SI units these two coefficients are normally expressed in m^{-1} but in brachytherapy applications where distances of dosimetric interest are on the centimeter scale, they are more conveniently expressed in cm^{-1}. Inspection of Figure 4.1 reveals that there is strong dependence of these two coefficients on both the photon energy and the target material. Particularly, in the low energy region, changes of four orders of magnitude can be observed.

The linear attenuation coefficient, μ, of a material determines the probability that a photon will interact per unit distance traveled in the medium. Hence, the probability $P(r)$ that a photon will travel a distance r within that material, called the probability function, is given by

$$P(r) = \exp(-\mu r) \tag{4.1}$$

The average distance traversed by the photon before interacting with the material is called the mean free path (mfp), is designated usually by the

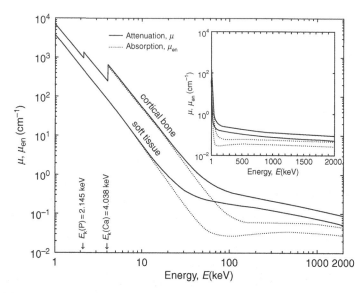

FIGURE 4.1
Linear attenuation coefficients, μ, and linear energy absorption coefficients, μ_{en}, for soft tissue and cortical bone, plotted as a function of incident photon energy, E, using logarithmic scales in both axes. Absorption K-edges of calcium and phosphorus, which are constituents of cortical bone, are indicated on the energy scale. These data are also given in the inset using a linear energy scale.

Greek letter λ and is obtained by

$$\lambda = \int_0^\infty rP(r)\mathrm{d}r = \int_0^\infty r\exp(-\mu r)\mathrm{d}r = \frac{1}{\mu} \tag{4.2}$$

The energy range relevant to brachytherapy, extents from about 20 keV (for the low energy emitters ^{125}I and ^{103}Pd) to about 1 MeV (for the higher energy emitters ^{192}Ir, ^{137}Cs, and ^{60}Co, see Chapter 5). This energy range corresponds to a mean free path range of about 1 to 13 cm in soft tissue. This can be deduced from Equation 4.2 and the data in Figure 4.1. Photons outside this energy range are not appropriate for brachytherapy applications, since they either have too small or too large mean free paths.

In the case of a brachytherapy source placed within a target material, a large number of photons are emitted in all directions, depending on the source strength. Assume for simplicity a monoenergetic point source. Such a source will emit photons of the same energy E, with the fraction of the emitted photons that arrive at a distance r from the source determined by the exponential term $e^{-\mu r}$. Given the fact that the source occupies a single point and that photon emission is isotropic, photons arriving at distance r will be spatially distributed over the surface of a sphere of area $4\pi r^2$. Thus, photon fluence at a distance r from the source is proportional to $\exp(-\mu r)$ and inversely proportional to $(1/4\pi r^2)$ and consequently the energy fluence is obtained by

$$\text{primary energy fluence}: \quad \Psi(E, r) \propto \frac{\exp(-\mu r)}{4\pi r^2} E \qquad (4.3)$$

The factor $(1/4\pi r^2)$ is called the point source geometry factor and it accounts for the reduction of photon fluence with distance, purely due to geometry and in the absence of any photon interactions. At small radial distances the r^{-2} factor predominates over the exponential factor $\exp(-\mu r)$ and consequently the geometry factor determines the energy fluence close to the source.

The linear energy absorption coefficient μ_{en}, determines the average fraction of photon energy available for deposition per photon interaction. This is equal to μ_{en}/μ. Since μ defines the photon interaction probability, the energy available for deposition per unit volume of material, at a distance r from a source emitting photons of energy E, is proportional to the product of

1. The primary energy fluence at distance r as given by Equation 4.3
2. The interaction probability μ
3. The average fraction of energy available for deposition per interaction given by (μ_{en}/μ)

The energy deposition per unit mass is obtained by dividing the energy available for deposition per unit volume of material by the material density. Therefore the primary absorbed dose at r is described by

$$\text{primary dose} \propto \frac{\exp(-\mu r)}{4\pi r^2} E \frac{\mu_{en}}{\rho} \qquad (4.4)$$

The linear attenuation coefficient μ, is included only in the exponential term which determines the energy fluence. Equation 4.4 shows that the energy fluence when multiplied by the mass energy absorption coefficient μ_{en}/ρ, gives the absorbed dose.

There are three principal effects which lead to the attenuation of photons for the brachytherapy range of energies. These are, the photoelectric effect or photoabsorption, elastic scattering from atoms, which is also known as Rayleigh or coherent scattering, and inelastic scattering also known as Compton or incoherent scattering. Pair production which has threshold energy of 1.022 MeV can be ignored for brachytherapy, even for the high energy emitter ^{60}Co.

Complete absorption of the photon occurs only in the photoabsorption process, where the photon's energy is transferred to a bound electron, which is then released from the atom. This is a primary ionization and the ejected electron, the photoelectron, deposits the energy gained as it travels through the medium. This causes thousands of secondary ionizations and excitations. In Compton scattering, the photon is not absorbed, but it is scattered with reduced energy in a different direction, transferring only a part of its energy to the released electron, the Compton recoil electron. There is no ionization, electron release, and energy deposition in Rayleigh scattering where the photon is simply scattered in a different direction. It is evident therefore that the above equations, ignore any scattered radiation. Moreover, they assume that any energy transfer from photons to electrons during an interaction is deposited on spot.

Actual brachytherapy sources are not points, but have finite dimensions. They usually consist of a radioactive core encapsulated in a metallic housing. Photon interactions within radioactive core and encapsulation material produce a photon energy spectrum different from that emitted by the radioisotope without any filtration. Moreover, sources are not usually monoenergetic. For example, ^{192}Ir emits more than 50 different photon energies, ranging from a few keV to approximately 1 MeV (see Section 5.6).

In the following section photon interactions with matter are presented in some detail, emphasizing factors that are essential from the dosimetry point of view in brachytherapy. In this context, not only tissue materials, but also phantom and detector as well as encapsulation and shielding materials are considered. Corresponding interaction coefficients are defined and their energy and material dependencies discussed.

4.2 Photon Interaction Processes

Essentially, there are numerous possible processes in which a photon may interact with matter. However, for the energies relevant to brachytherapy, only the photoelectric effect, coherent (Rayleigh), and incoherent (Compton) scattering contribute to the photon attenuation and scatter. For pure elements, the probability that each one of these interactions occurs is described by the corresponding atomic cross-section, designated respectively as $_a\sigma^{ph}$, $_a\sigma^{coh}$, and $_a\sigma^{incoh}$, where the subscript a stands for atom, i.e., they are stated per atom. Since these are independent probabilities, the total atomic cross-section $_a\sigma$, i.e., the total interaction probability per atom, is written:

$$_a\sigma = {_a\sigma^{ph}} + {_a\sigma^{coh}} + {_a\sigma^{incoh}} \tag{4.5}$$

The atomic cross-section is a quantum mechanical property attributed jointly to the photon and the atom (in general, for the incoming particle and the target, which could be another particle, nucleus, atom, or molecule). It is a function sensitive to both the photon energy E and the atomic number Z and represents the effective size of the atom, which may be very different from its geometrical size. Accordingly, electronic cross-sections $_e\sigma$ and molecular cross-sections $_m\sigma$ are defined, representing the probability interactions per electron or per molecule, respectively. Cross-sections have the dimensions of an area and since they are usually a small number, they are expressed in units of barn/atom, where 1 b $= 10^{-28}$ m$^2 = 10^{-24}$ cm^2.

The linear attenuation coefficient μ, which is a macroscopic cross-section, is related to the atomic cross-section $_a\sigma$ by the expression:

$$\mu = \frac{\rho}{uA}\,_a\sigma \tag{4.6}$$

where ρ and A are the density and the atomic weight of the target material respectively, and u is the unified mass unit. The quantity ρ/uA, represents the number of atoms per unit volume. Since μ depends on the material density ρ, which can vary considerably, for compilation purposes this dependency is removed by tabulating the mass attenuation coefficient μ/ρ:

$$\frac{\mu}{\rho} = \frac{1}{uA}\,_a\sigma \tag{4.7}$$

The unified mass unit u is defined as the $1/12$ of the mass of the atom of the nuclide ^{12}C and equals to $1u = (1\ g)/(N_A\ mol)$ where N_A is Avogadro's number, expressed as $0.6022\ 10^{24}\ mol^{-1}$ (see Section 3.2.2, Equation 3.20, and Equation 3.21). The reason for writing Avogadro's number as 10^{24}, is that it cancels out the 10^{-24} factor, obtained from the conversion of barns to cm^{-2}. Therefore when atomic cross-sections are given in units of barn/atom, the mass attenuation coefficient μ/ρ is given in units of cm^2/g by

$$\frac{\mu}{\rho}\left(\frac{cm^2}{g}\right) = \frac{0.6022}{A}\,_a\sigma\left(\frac{barn}{atom}\right) \tag{4.8}$$

Equation 4.5 can be rewritten in terms of the linear attenuation coefficient μ (using Equation 4.6), as

$$\mu = \mu^{ph} + \mu^{coh} + \mu^{incoh} \tag{4.9}$$

where μ^{ph}, μ^{coh}, and μ^{incoh} are the linear attenuation coefficients due to photoabsorption, coherent, and incoherent scattering. Accordingly:

$$\frac{\mu}{\rho} = \frac{\mu^{ph}}{\rho} + \frac{\mu^{coh}}{\rho} + \frac{\mu^{incoh}}{\rho} \tag{4.10}$$

Current compilations of the mass attenuation coefficients μ/ρ for pure elements are derived from theoretical or semiempirical values of the atomic cross-sections for the individual processes, described in Equation 4.5. As reviewed in detail by Hubbell,[1] these are the product of extensive theoretical and experimental research. Tables and graphs of the photon mass attenuation coefficient μ/ρ for all of the elements $Z = 1$ to 92, can be found in the XCOM database[2] available online. For the photon energy range of interest in brachytherapy applications, the envelope of uncertainty on these μ/ρ values is of the order of 1 to 2%.

For homogeneous mixtures and compounds, the Bragg rule for mixing elements postulates that radiation interacts with atoms individually, and that atoms do not influence each other's interaction probability. Accordingly, the molecular cross-section $_m\sigma$ is equal to the sum of the atomic cross-sections of individual atoms which constitute the molecule. For example, the molecular cross-section $_m\sigma_{H_2O}$ for the water molecule is given by

$$_m\sigma_{H_2O} = 2\,_a\sigma_H +\,_a\sigma_O \qquad (4.11)$$

where $_a\sigma_H$ and $_a\sigma_O$ are the atomic cross-sections for the hydrogen and the oxygen atoms. The factor 2 in this equation results from the fact that there are two hydrogen atoms in a water molecule.

Equation 4.5 through Equation 4.10 do not only apply for pure elements but can also be used for chemical compounds by replacing the atomic cross-section $_a\sigma$ by the molecular cross-section $_m\sigma$ and the atomic weight A by the molecular weight M. Substituting cross-sections in Equation 4.11 using attenuation coefficients from Equation 4.6, gives the mass attenuation coefficient for water as follows:

$$\left(\frac{\mu}{\rho}\right)_{H_2O} = \frac{2A_H}{M_{H_2O}}\left(\frac{\mu}{\rho}\right)_H + \frac{A_O}{M_{H_2O}}\left(\frac{\mu}{\rho}\right)_O$$

where A_H and A_O are the atomic weights of hydrogen and oxygen and M_{H_2O} is the molecular weight of water. Hence:

$$\left(\frac{\mu}{\rho}\right)_{H_2O} = 0.1119\left(\frac{\mu}{\rho}\right)_H +0.8881\left(\frac{\mu}{\rho}\right)_O \qquad (4.12)$$

where the fractions by weight are 0.1119 for hydrogen and 0.8881 for oxygen.

In general, values of the mass attenuation coefficient, μ/ρ, for homogeneous mixtures and compounds are obtained according to simple additivity:

$$\left(\frac{\mu}{\rho}\right)_{mix} = \sum_i w_i\left(\frac{\mu}{\rho}\right)_i \qquad (4.13)$$

where $(\mu/\rho)_{mix}$ is the mass attenuation coefficient of the mixture or compound, $(\mu/\rho)_i$ is the mass attenuation coefficient of the ith component element and w_i is its fraction by weight in the compound or mixture, with the summation over all constituent elements. The Bragg mixture rule is a safe assumption for brachytherapy energies and is also used in calculating the partial mass attenuation coefficients μ^{ph}/ρ, μ^{coh}/ρ, and μ^{incoh}/ρ of a mixture or compound. The XCOM database,[2] includes 48 compounds and mixtures of radiological interest. Additionally, it enables compilation of any mixture and compound provided that the corresponding elemental composition is given.

Table 4.1a, gives the percentage elemental composition of selected tissue, dosimeter, and phantom media. The atomic number Z for each element is denoted by a subscript on the left of the element symbol. For tissue materials, the elemental composition by weight is taken from ICRU 44.[3] Solid water is a phantom material used in TLD dosimetry (see Section 10.4), and its

TABLE 4.1a

Percentage Elemental Compositions by Weight of Selected Tissue, Dosimeter, and Phantom Media of Interest in Brachytherapy Applications

Medium	Density (g/cm³)	Elements														
		1H	3Li	6C	7N	8O	9F	11Na	12Mg	14Si	15P	16S	17Cl	18Ar	19K	20Ca
Water	1.000	11.19	—	—	—	88.81	—	—	—	—	—	—	—	—	—	—
Soft tissue	1.060	10.2	—	14.3	3.4	70.8	—	0.2	—	—	—	0.3	0.2	—	0.3	—
Cortical bone	1.920	3.4	—	15.5	4.2	43.5	—	0.1	0.2	—	10.3	0.3	—	—	—	22.5
Breast tissue	1.020	10.6	—	33.2	3.0	52.7	—	0.1	—	—	0.1	0.2	0.1	—	—	—
Air	1.205×10^{-3}	—	—	0.01	75.53	23.18	—	—	—	—	—	—	—	1.28	—	—
LiF (TLD-100)	2.635	—	26.76	—	—	—	73.24	—	—	—	—	—	—	—	—	—
SiO2	2.650	—	—	—	—	53.26	—	—	—	46.74	—	—	—	—	—	—
Radiochromic dye film, nylon base	1.080	10.20	—	65.44	9.89	14.47	—	—	—	—	—	—	—	—	—	—
GaF chromic sensor	1.300	8.97	—	60.58	11.22	19.23	—	—	—	—	—	—	—	—	—	—
VIPAR gel	1.018	10.74	—	7.18	2.06	80.01	—	—	—	—	—	—	—	—	—	—
Solid water	1.015	8.09	—	67.17	2.42	19.87	—	—	—	—	—	—	0.13	—	—	2.32
Polymethyl methacrylate (PMMA)	1.190	8.05	—	59.99	—	31.96	—	—	—	—	—	—	—	—	—	—
Polystyrene	1.060	7.74	—	92.26	—	—	—	—	—	—	—	—	—	—	—	—

elemental composition is taken from Anagnostopoulos et al.[4] VIPAR is a polymer gel formulation (see Section 10.5) that has evolved as a promising new three-dimensional dosimetry tool in brachytherapy. Its elemental composition is taken from Kipouros et al.[5] All other elemental compositions are as adopted in the XCOM database[2] available on line.

Table 4.1b, gives the corresponding percentage elemental composition of selected shielding materials including stainless steel, which is a typical encapsulation material for brachytherapy sources (the stated composition refers to AISI 304)[6] and tungsten alloy, which is a shielding material used in brachytherapy for high energy emitters such as ^{192}Ir.[7-9]

4.2.1 Photoelectric Effect (Photoabsorption)

The photoelectric effect is the dominant interaction at low photon energies. It is an interaction between the photon and the atom and it cannot occur between a photon and a free electron, since the latter process does not simultaneously conserve energy and momentum. Photoabsorption is therefore possible only for bound electrons. The energy of the absorbed photon is transferred to the ejected electron and the atom. The energy transferred to the atom, as recoil energy, is negligible due to its large mass. In fact, even if the entire photon momentum E/c was to be transferred to the atom, its recoil energy $E^2/2mc^2$, where mc^2 is the rest energy of the entire atom, could not exceed a few eV. The energy E of the photon minus the energy E_{shell} binding the electron to the atom, is taken by the ejected electron, whose kinetic energy T_e is

$$T_e = E - E_{shell} \qquad (4.14)$$

Equation 4.14 shows that photoabsorption can occur only if the photon has energy higher than the binding energy of the electron to be removed. The most tightly bound electrons are the K-shell electrons, for which the binding energies E_K for medium Z materials are approximately given by Moseley's law (see Section 3.1.4):

$$\text{K-edge}: \qquad E_K = 13.6\,\text{eV}(Z - 3)^2 \qquad (4.15)$$

where Z is the atomic number of the atom. This gives K-edge values of 3.9 and 84.9 keV for $_{20}$Ca and $_{82}$Pb, respectively, compared with actual values of 4.04 and 88.00 keV.

The probability of photoabsorption is highest for K-shell electrons; about 80% of the interactions involve the K-shell electrons, and consequently, these electrons are the most important in the brachytherapy energy range. When the photon energy falls below the binding energy of the K shell, a new channel, e.g., L-shell electrons, becomes energetically allowed. The probability of photoabsorption falls abruptly at the binding energy of the K shell, and the curve of the attenuation coefficient presents a sawtooth shape, as the one observed in Figure 4.1 for bone. In this figure, the two

TABLE 4.1b

Percentage Elemental Compositions by Weight of Selected Shielding Materials of Interest in Brachytherapy Applications

Medium	Density (g/cm^3)	$_1$H	$_6$C	$_8$O	$_9$F	$_{11}$Na	$_{12}$Mg	$_{13}$Al	$_{14}$Si	$_{19}$K	$_{20}$Ca	$_{22}$Ti	$_{24}$Cr	$_{25}$Mn	$_{26}$Fe	$_{28}$Ni	$_{33}$As	$_{74}$W	$_{82}$Pb
Stainless steel	8.02	—	—	—	—	—	—	—	1	—	—	—	19	2	68	10	—	—	—
Tungsten alloy	17.00	—	—	—	—	—	—	—	—	—	—	—	—	—	3	6.5	—	90.5	—
Lead	11.35	—	—	—	—	—	—	—	—	—	—	—	—	—	—	—	—	—	100
Glass lead	6.22	—	—	15.65	—	—	—	—	8.09	—	—	0.81	—	—	—	—	—	—	75.19
Ordinary concrete	2.30	2.21	0.25	57.49	0.64	1.52	0.13	2.00	30.46	1.00	4.30	—	—	—	0.64	—	0.26	—	—

sharp discontinuities observed at 2.145 and 4.038 keV are the K-edges of $_{15}$P and $_{20}$Ca.

Following photoabsorption the atom is left excited. The vacancy created by the ejection of an electron from the inner shells is filled by outer electrons falling into it (de-excitation) and this process may be accompanied by emission of fluorescent radiation or Auger electron emission. The binding energy E_{shell} is thus emitted promptly by the residual atom.

Figure 4.2 shows the mass attenuation coefficient due to photoabsorption, μ^{ph}/ρ, for selected pure elements as a function of the photon energy E. K-edge energies are shown on the energy scale for each element. A strong energy and/or material dependence is observed. Specifically, when the photon energy is much larger than the K-shell energy E_K, which according to Equation 4.15 depends largely on Z, the probability of photoabsorption is quite small. As the photon energy E decreases and approaches E_K this probability rapidly increases. For photon energies $E < E_K$ photoabsorption is impossible for K-shell electrons and the probability is determined by the interactions with L, M,..., shells. As a result, a series of sawtooth discontinuities are observed in the curve of the absorption coefficient, corresponding to the binding energies of these different shells. Data suggest that photoabsorption is significant at low photon energies, where the term low refers to photon energies not much larger than the K-shell energies of elements.

No single formula accurately describes the photoelectric effect over a wide range of photon energies E and/or atomic numbers Z. What is more,

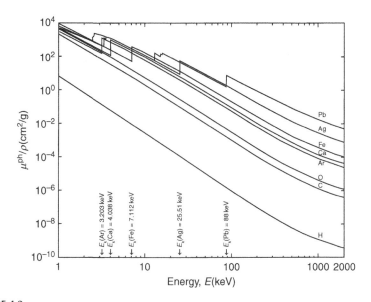

FIGURE 4.2
Mass attenuation coefficients due to photoabsorption, μ^{ph}/ρ, for selected pure elements, plotted as a function of incident photon energy, E. Absorption K-edges are indicated on the energy scale.

photoabsorption does not have the same energy dependence for all Z, or the same Z dependence for all photon energies. A crude but useful guide which gives the order of magnitude for the photoelectric atomic cross-section, at energies above K-edges, is

$$\sigma^{\text{ph}} \propto \frac{Z^4}{E^3}$$

Hence, according to Equation 4.7, the energy and material dependence of the mass attenuation coefficient due to photoabsorption is approximately described by

$$\frac{\mu^{\text{ph}}}{\rho} \propto \frac{Z}{A} \frac{Z^3}{E^3} \approx \frac{Z^3}{E^3} \qquad (4.16)$$

since the Z/A ratio is approximately equal to one half for most elements.

Mass attenuation coefficients due to photoabsorption, μ^{ph}/ρ, for homogeneous mixtures and chemical compounds are calculated from the individual atoms which form the molecule, according to the Bragg additivity rule of Equation 4.13. It is expected therefore, that any high-Z element in the molecule or mixture effectively determines the photoabsorption probability of the compound, due to the strong Z dependence. The characteristic absorption edges of the compound or mixture relate directly to those corresponding to the participating atoms, since molecular bonds do not disturb inner atomic shell-energies in a molecule. This is illustrated using water as an example, for which Equation 4.13 is written:

$$\left(\frac{\mu^{\text{ph}}}{\rho}\right)_{\text{H}_2\text{O}} = 0.1119\left(\frac{\mu^{\text{ph}}}{\rho}\right)_{\text{H}} + 0.8881\left(\frac{\mu^{\text{ph}}}{\rho}\right)_{\text{O}} \approx 0.8881\left(\frac{\mu^{\text{ph}}}{\rho}\right)_{\text{O}} \qquad (4.17)$$

The contribution of hydrogen is negligible as seen in Figure 4.2. The last approximation is within 1% of the actual value. The last two equations reveal that for the photoabsorption, water presents an effective atomic number close to that of oxygen ($Z = 8$).

Figure 4.3 shows the mass attenuation coefficient due to photoabsorption, μ^{ph}/ρ, for selected compounds and mixtures as a function of the photon energy E. Absorption K edges are indicated on the energy scale as are the constituent elements from which they originate. It is observed that attenuation curves for soft tissue, water, air, solid water, and VIPAR gel almost coincide, with an effective atomic number close to that of oxygen. Moreover, water proves to be tissue equivalent in all the energy range given in the figure, justifying the wide use of water as a phantom material. VIPAR and solid water can be used equivalently as phantom materials. Noticeable deviations are observed between solid water and soft tissue or water for energies lower than 5 keV, as shown in the inset. These deviations are due to Ca, one of the elements comprising solid water, and necessitate corrections (that are sensitive to the percentage contribution of Ca) to be applied in the experimental dosimetry of low energy emitters, such as ^{125}I and [103]Pd.[4,10]

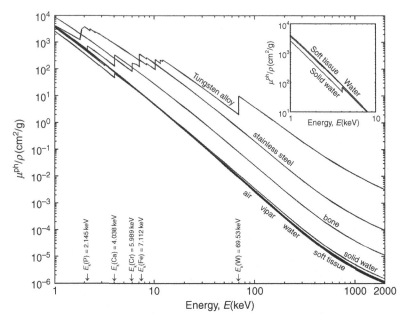

FIGURE 4.3

Mass attenuation coefficients due to photoabsorption, μ^{ph}/ρ, for selected compounds and mixtures, plotted as a function of incident photon energy, E. Absorption K-edges are indicated on the energy scale. The low energy region of 1 to 10 keV is given in the inset for water, soft tissue, and solid water.

Corresponding corrections are significantly lower for VIPAR as well as other polymer gel dosimetry formulations which present comparable attenuation properties to water.[11]

Data shown in Figure 4.2 and Figure 4.3 indicate that the atomic photoelectric effect is the dominant interaction at low photon energies, especially for high-Z materials, such as are usually incorporated in a brachytherapy source. Photoabsorption in these high-Z materials generates characteristic x-ray fluorescence radiation, which is important for the low-energy photon transport computations. Figure 4.4 presents the fluorescence yield for the K shell, which is defined as the fraction of K-shell vacancies that are filled by photon emission rather than Auger electron ejection, as a function of atomic number Z, and shows how the fluorescence yield for the K shell rises from essentially zero for the low-Z elements to almost unity for high-Z elements. The probability for fluorescence emission is given by the product of the fluorescence yield pertinent to the atomic shell and the probability of participation of the particular atomic shell in the photoeffect. Both these probabilities are small for shells other than the K shell.

^{125}I and ^{103}Pd are low energy photon emitters which have an increased probability of photoelectric absorption in the source core and in the

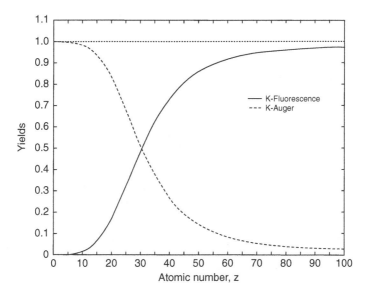

FIGURE 4.4
K-fluorescence and K-Auger yields plotted as a function of atomic number, Z.

encapsulation material. The fluorescence radiation following photo-absorption may be of significant impact on ^{125}I and ^{103}Pd dosimetry. Characteristic examples are: (1) the fluorescence x-ray radiation of approximately 4.5 keV emerging from the commonly used Ti encapsulation which led to the implementation of a new primary air kerma strength standard;[12] and (2) the fluorescence x-ray radiation of approximately 22 keV emerging from the commonly used Ag marker in low energy interstitial brachytherapy seeds which results to different dosimetric characteristics of the sources utilizing this marker. These are discussed in detail in Chapter 9.

The XCOM database[2] incorporates the photoelectric cross-section, σ^{ph}, values of Scofield.[13] These photoelectric cross-sections, that are of major importance for the low energy brachytherapy range, correspond to nonrelativistic calculations based on the Hartree-Slater atomic model that were found to be in better agreement with experimental results than those renormalized to the relativistic Hartree-Fock model values.[1] Table 4.2 through Table 4.19 give mass attenuation coefficients due to photoabsorption, μ^{ph}/ρ, for all media given in Table 4.1 and the principal photon energies emitted by radionuclides currently used or proposed for use in brachytherapy sources of ^{103}Pd, ^{125}I, ^{170}Tm, ^{169}Yb, ^{137}Cs, and ^{60}Co. The attenuation data were derived based on the mixture rule of Equation 4.13 and the elemental composition of Table 4.1, by log–log interpolation at the principal photon energies on the values of the atomic photoelectric cross-sections of Scofield.[13]

TABLE 4.2

Total and Partial (Photoabsorption, Coherent, and Incoherent) Mass Attenuation Coefficients, as Well as Mass Energy Absorption Coefficients, for the Principal Energies of ^{103}Pd, ^{125}I, ^{170}Tm, ^{169}Yb, ^{192}Ir, ^{137}Cs, and ^{60}Co, for Water

	E (keV) Intensity (%)	μ^{ph}/ρ (cm²/g)	μ^{coh}/ρ (cm²/g)	μ^{incoh}/ρ (cm²/g)	μ/ρ (cm²/g)	μ_{en}/ρ (cm²/g)
^{103}Pd	20.07 (22.06)	$0.579 \times 10^{+00}$	0.881×10^{-01}	$0.177 \times 10^{+00}$	$0.805 \times 10^{+00}$	$0.544 \times 10^{+00}$
	20.22 (41.93)	$0.525 \times 10^{+00}$	0.871×10^{-01}	$0.177 \times 10^{+00}$	$0.793 \times 10^{+00}$	$0.532 \times 10^{+00}$
	22.70 (13.05)	$0.361 \times 10^{+00}$	0.726×10^{-01}	$0.179 \times 10^{+00}$	$0.637 \times 10^{+00}$	$0.371 \times 10^{+00}$
^{125}I	27.20 (39.90)	$0.201 \times 10^{+00}$	0.547×10^{-01}	$0.182 \times 10^{+00}$	$0.453 \times 10^{+00}$	$0.212 \times 10^{+00}$
	27.47 (74.50)	$0.194 \times 10^{+00}$	0.539×10^{-01}	$0.182 \times 10^{+00}$	$0.444 \times 10^{+00}$	$0.205 \times 10^{+00}$
	31.00 (25.90)	$0.131 \times 10^{+00}$	0.443×10^{-01}	$0.183 \times 10^{+00}$	$0.362 \times 10^{+00}$	$0.142 \times 10^{+00}$
	35.50 (6.68)	0.840×10^{-01}	0.352×10^{-01}	$0.183 \times 10^{+00}$	$0.308 \times 10^{+00}$	0.972×10^{-01}
^{170}Tm	51.35 (0.97)	0.249×10^{-01}	0.185×10^{-01}	$0.180 \times 10^{+00}$	$0.223 \times 10^{+00}$	0.405×10^{-01}
	52.39 (1.69)	0.234×10^{-01}	0.178×10^{-01}	$0.180 \times 10^{+00}$	$0.221 \times 10^{+00}$	0.393×10^{-01}
	59.38 (0.36)	0.154×10^{-01}	0.142×10^{-01}	$0.180 \times 10^{+00}$	$0.207 \times 10^{+00}$	0.328×10^{-01}
	84.25 (2.48)	0.486×10^{-02}	0.741×10^{-02}	$0.177 \times 10^{+00}$	$0.180 \times 10^{+00}$	0.259×10^{-01}
^{169}Yb	49.77 (53)	0.276×10^{-01}	0.196×10^{-01}	$0.180 \times 10^{+00}$	$0.228 \times 10^{+00}$	0.426×10^{-01}
	50.74 (93.8)	0.259×10^{-01}	0.189×10^{-01}	$0.180 \times 10^{+00}$	$0.225 \times 10^{+00}$	0.413×10^{-01}
	57.50 (38.5)	0.172×10^{-01}	0.150×10^{-01}	$0.178 \times 10^{+00}$	$0.211 \times 10^{+00}$	0.341×10^{-01}
	63.12 (44.2)	0.126×10^{-01}	0.127×10^{-01}	$0.176 \times 10^{+00}$	$0.202 \times 10^{+00}$	0.308×10^{-01}
	109.78 (17.5)	0.203×10^{-02}	0.447×10^{-02}	$0.159 \times 10^{+00}$	$0.166 \times 10^{+00}$	0.260×10^{-01}
	177.98 (35.8)	0.421×10^{-03}	0.175×10^{-02}	$0.140 \times 10^{+00}$	$0.143 \times 10^{+00}$	0.288×10^{-01}
	307.74 (10.05)	0.757×10^{-04}	0.591×10^{-03}	$0.117 \times 10^{+00}$	$0.118 \times 10^{+00}$	0.320×10^{-01}
^{198}Au	411.80 (95.58)	0.324×10^{-04}	0.331×10^{-03}	$0.104 \times 10^{+00}$	$0.105 \times 10^{+00}$	0.326×10^{-01}
^{192}Ir	295.96 (28.72)	0.851×10^{-04}	0.639×10^{-03}	$0.119 \times 10^{+00}$	$0.120 \times 10^{+00}$	0.318×10^{-01}
	308.46 (29.68)	0.752×10^{-04}	0.589×10^{-03}	$0.117 \times 10^{+00}$	$0.118 \times 10^{+00}$	0.320×10^{-01}
	316.51 (82.71)	0.697×10^{-04}	0.559×10^{-03}	$0.116 \times 10^{+00}$	$0.116 \times 10^{+00}$	0.321×10^{-01}
	468.07 (47.81)	0.223×10^{-04}	0.257×10^{-03}	0.993×10^{-01}	0.995×10^{-01}	0.329×10^{-01}
	588.58 (4.52)	0.123×10^{-04}	0.162×10^{-03}	0.901×10^{-01}	0.903×10^{-01}	0.328×10^{-01}
	604.41 (8.20)	0.115×10^{-04}	0.154×10^{-03}	0.891×10^{-01}	0.893×10^{-01}	0.328×10^{-01}
	612.46 (5.34)	0.111×10^{-04}	0.150×10^{-03}	0.886×10^{-01}	0.888×10^{-01}	0.327×10^{-01}
^{137}Cs	662 (85.1)	0.924×10^{-05}	0.128×10^{-03}	0.856×10^{-01}	0.857×10^{-01}	0.326×10^{-01}
^{60}Co	1173 (100)	0.263×10^{-05}	0.409×10^{-04}	0.652×10^{-01}	0.653×10^{-01}	0.301×10^{-01}
	1332 (100)	0.208×10^{-05}	0.317×10^{-04}	0.611×10^{-01}	0.612×10^{-01}	0.292×10^{-01}

TABLE 4.3

Total and Partial (Photoabsorption, Coherent, and Incoherent) Mass Attenuation Coefficients, as Well as Mass Energy Absorption Coefficients, for the Principal Energies of ^{103}Pd, ^{125}I, ^{170}Tm, ^{169}Yb, ^{198}Au, ^{192}Ir, ^{137}Cs, and ^{60}Co, for Soft Tissue

	E (keV) Intensity (%)	μ^{ph}/ρ (cm²/g)	μ^{coh}/ρ (cm²/g)	μ^{incoh}/ρ (cm²/g)	μ/ρ (cm²/g)	μ_{en}/ρ (cm²/g)
^{103}Pd	20.07 (22.06)	$0.555 \times 10^{+00}$	0.857×10^{-01}	$0.176 \times 10^{+00}$	$0.818 \times 10^{+00}$	$0.560 \times 10^{+00}$
	20.22 (41.93)	$0.542 \times 10^{+00}$	0.847×10^{-01}	$0.176 \times 10^{+00}$	$0.806 \times 10^{+00}$	$0.547 \times 10^{+00}$
	22.70 (13.05)	$0.373 \times 10^{+00}$	0.707×10^{-01}	$0.178 \times 10^{+00}$	$0.646 \times 10^{+00}$	$0.383 \times 10^{+00}$
^{125}I	27.20 (39.90)	$0.208 \times 10^{+00}$	0.532×10^{-01}	$0.180 \times 10^{+00}$	$0.457 \times 10^{+00}$	$0.219 \times 10^{+00}$
	27.47 (74.50)	$0.202 \times 10^{+00}$	0.524×10^{-01}	$0.180 \times 10^{+00}$	$0.449 \times 10^{+00}$	$0.213 \times 10^{+00}$
	31.00 (25.90)	$0.137 \times 10^{+00}$	0.431×10^{-01}	$0.181 \times 10^{+00}$	$0.364 \times 10^{+00}$	$0.148 \times 10^{+00}$
	35.50 (6.68)	0.880×10^{-01}	0.342×10^{-01}	$0.181 \times 10^{+00}$	$0.310 \times 10^{+00}$	$0.101 \times 10^{+00}$
^{170}Tm	51.35 (0.97)	0.257×10^{-01}	0.178×10^{-01}	$0.178 \times 10^{+00}$	$0.222 \times 10^{+00}$	0.411×10^{-01}
	52.39 (1.69)	0.240×10^{-01}	0.172×10^{-01}	$0.178 \times 10^{+00}$	$0.219 \times 10^{+00}$	0.398×10^{-01}
	59.38 (0.36)	0.159×10^{-01}	0.137×10^{-01}	$0.176 \times 10^{+00}$	$0.205 \times 10^{+00}$	0.331×10^{-01}
	84.25 (2.48)	0.505×10^{-02}	0.715×10^{-02}	$0.167 \times 10^{+00}$	$0.179 \times 10^{+00}$	0.259×10^{-01}
^{169}Yb	49.77 (53)	0.292×10^{-01}	0.190×10^{-01}	$0.179 \times 10^{+00}$	$0.227 \times 10^{+00}$	0.441×10^{-01}
	50.74 (93.8)	0.274×10^{-01}	0.183×10^{-01}	$0.179 \times 10^{+00}$	$0.224 \times 10^{+00}$	0.426×10^{-01}
	57.50 (38.5)	0.182×10^{-01}	0.146×10^{-01}	$0.176 \times 10^{+00}$	$0.210 \times 10^{+00}$	0.349×10^{-01}
	63.12 (44.2)	0.134×10^{-01}	0.123×10^{-01}	$0.174 \times 10^{+00}$	$0.201 \times 10^{+00}$	0.314×10^{-01}
	109.78 (17.5)	0.218×10^{-02}	0.434×10^{-02}	$0.157 \times 10^{+00}$	$0.164 \times 10^{+00}$	0.259×10^{-01}
	177.98 (35.8)	0.456×10^{-03}	0.170×10^{-02}	$0.139 \times 10^{+00}$	$0.141 \times 10^{+00}$	0.286×10^{-01}
	307.74 (10.05)	0.827×10^{-04}	0.575×10^{-03}	$0.116 \times 10^{+00}$	$0.117 \times 10^{+00}$	0.317×10^{-01}
^{198}Au	411.80 (95.58)	0.358×10^{-04}	0.322×10^{-03}	$0.103 \times 10^{+00}$	$0.104 \times 10^{+00}$	0.323×10^{-01}
^{192}Ir	295.96 (28.72)	0.929×10^{-04}	0.621×10^{-03}	$0.118 \times 10^{+00}$	$0.119 \times 10^{+00}$	0.315×10^{-01}
	308.46 (29.68)	0.821×10^{-04}	0.572×10^{-03}	$0.116 \times 10^{+00}$	$0.117 \times 10^{+00}$	0.317×10^{-01}
	316.51 (82.71)	0.761×10^{-04}	0.544×10^{-03}	$0.115 \times 10^{+00}$	$0.115 \times 10^{+00}$	0.318×10^{-01}
	468.07 (47.81)	0.248×10^{-04}	0.250×10^{-03}	0.984×10^{-01}	0.986×10^{-01}	0.326×10^{-01}
	588.58 (4.52)	0.136×10^{-04}	0.158×10^{-03}	0.893×10^{-01}	0.894×10^{-01}	0.325×10^{-01}
	604.41 (8.20)	0.127×10^{-04}	0.150×10^{-03}	0.883×10^{-01}	0.884×10^{-01}	0.325×10^{-01}
	612.46 (5.34)	0.123×10^{-04}	0.146×10^{-03}	0.878×10^{-01}	0.879×10^{-01}	0.324×10^{-01}
^{137}Cs	662 (85.1)	0.102×10^{-04}	0.125×10^{-03}	0.847×10^{-01}	0.849×10^{-01}	0.323×10^{-01}
^{60}Co	1173 (100)	0.289×10^{-05}	0.398×10^{-04}	0.646×10^{-01}	0.647×10^{-01}	0.298×10^{-01}
	1332 (100)	0.229×10^{-05}	0.309×10^{-04}	0.606×10^{-01}	0.607×10^{-01}	0.289×10^{-01}

TABLE 4.4

Total and Partial (Photoabsorption, Coherent, and Incoherent) Mass Attenuation Coefficients, as Well as Mass Energy Absorption Coefficients, for the Principal Energies of ^{103}Pd, ^{125}I, ^{169}Yb, ^{170}Tm, ^{198}Au, ^{192}Ir, ^{137}Cs, and ^{60}Co, for Cortical Bone

	E (keV) Intensity (%)	μ^{ph}/ρ (cm²/g)	μ^{coh}/ρ (cm²/g)	μ^{incoh}/ρ (cm²/g)	μ/ρ (cm²/g)	μ_{en}/ρ (cm²/g)
^{103}Pd	20.07 (22.06)	$0.364 \times 10^{+01}$	$0.169 \times 10^{+00}$	$0.155 \times 10^{+00}$	$0.396 \times 10^{+01}$	$0.356 \times 10^{+01}$
	20.22 (41.93)	$0.356 \times 10^{+01}$	$0.167 \times 10^{+00}$	$0.155 \times 10^{+00}$	$0.388 \times 10^{+01}$	$0.348 \times 10^{+01}$
	22.70 (13.05)	$0.251 \times 10^{+01}$	$0.140 \times 10^{+00}$	$0.157 \times 10^{+00}$	$0.284 \times 10^{+01}$	$0.246 \times 10^{+01}$
^{125}I	27.20 (39.90)	$0.145 \times 10^{+01}$	$0.106 \times 10^{+00}$	$0.161 \times 10^{+00}$	$0.174 \times 10^{+01}$	$0.143 \times 10^{+01}$
	27.47 (74.50)	$0.141 \times 10^{+01}$	$0.104 \times 10^{+00}$	$0.161 \times 10^{+00}$	$0.169 \times 10^{+01}$	$0.139 \times 10^{+01}$
	31.00 (25.90)	$0.976 \times 10^{+00}$	0.861×10^{-01}	$0.163 \times 10^{+00}$	$0.123 \times 10^{+01}$	$0.970 \times 10^{+00}$
	35.50 (6.68)	$0.642 \times 10^{+00}$	0.690×10^{-01}	$0.164 \times 10^{+00}$	$0.887 \times 10^{+00}$	$0.645 \times 10^{+00}$
^{170}Tm	51.35 (0.97)	$0.204 \times 10^{+00}$	0.372×10^{-01}	$0.163 \times 10^{+00}$	$0.404 \times 10^{+00}$	$0.216 \times 10^{+00}$
	52.39 (1.69)	$0.192 \times 10^{+00}$	0.359×10^{-01}	$0.163 \times 10^{+00}$	$0.391 \times 10^{+00}$	$0.205 \times 10^{+00}$
	59.38 (0.36)	$0.130 \times 10^{+00}$	0.289×10^{-01}	$0.161 \times 10^{+00}$	$0.320 \times 10^{+00}$	$0.144 \times 10^{+00}$
	84.25 (2.48)	0.431×10^{-01}	0.154×10^{-01}	$0.154 \times 10^{+00}$	$0.212 \times 10^{+00}$	0.627×10^{-01}
^{169}Yb	49.77 (53)	$0.225 \times 10^{+00}$	0.393×10^{-01}	$0.163 \times 10^{+00}$	$0.428 \times 10^{+00}$	$0.237 \times 10^{+00}$
	50.74 (93.8)	$0.212 \times 10^{+00}$	0.380×10^{-01}	$0.163 \times 10^{+00}$	$0.414 \times 10^{+00}$	$0.225 \times 10^{+00}$
	57.50 (38.5)	$0.144 \times 10^{+00}$	0.306×10^{-01}	$0.161 \times 10^{+00}$	$0.338 \times 10^{+00}$	$0.158 \times 10^{+00}$
	63.12 (44.2)	$0.107 \times 10^{+00}$	0.259×10^{-01}	$0.160 \times 10^{+00}$	$0.296 \times 10^{+00}$	$0.124 \times 10^{+00}$
	109.78 (17.5)	0.187×10^{-01}	0.939×10^{-02}	$0.146 \times 10^{+00}$	$0.176 \times 10^{+00}$	0.422×10^{-01}
	177.98 (35.8)	0.409×10^{-02}	0.373×10^{-02}	$0.129 \times 10^{+00}$	$0.138 \times 10^{+00}$	0.307×10^{-01}
	307.74 (10.05)	0.770×10^{-03}	0.128×10^{-02}	$0.108 \times 10^{+00}$	$0.110 \times 10^{+00}$	0.303×10^{-01}
^{198}Au	411.80 (95.58)	0.339×10^{-03}	0.721×10^{-03}	0.966×10^{-01}	0.977×10^{-01}	0.306×10^{-01}
^{192}Ir	295.96 (28.72)	0.864×10^{-03}	0.139×10^{-02}	$0.110 \times 10^{+00}$	$0.112 \times 10^{+00}$	0.303×10^{-01}
	308.46 (29.68)	0.765×10^{-03}	0.128×10^{-02}	$0.108 \times 10^{+00}$	$0.110 \times 10^{+00}$	0.303×10^{-01}
	316.51 (82.71)	0.710×10^{-03}	0.121×10^{-02}	$0.107 \times 10^{+00}$	$0.109 \times 10^{+00}$	0.304×10^{-01}
	468.07 (47.81)	0.236×10^{-03}	0.560×10^{-03}	0.919×10^{-01}	0.927×10^{-01}	0.307×10^{-01}
	588.58 (4.52)	0.130×10^{-03}	0.355×10^{-03}	0.836×10^{-01}	0.840×10^{-01}	0.305×10^{-01}
	604.41 (8.20)	0.122×10^{-03}	0.337×10^{-03}	0.826×10^{-01}	0.830×10^{-01}	0.305×10^{-01}
	612.46 (5.34)	0.118×10^{-03}	0.328×10^{-03}	0.821×10^{-01}	0.825×10^{-01}	0.304×10^{-01}
^{137}Cs	662 (85.1)	0.982×10^{-04}	0.281×10^{-03}	0.793×10^{-01}	0.797×10^{-01}	0.302×10^{-01}
^{60}Co	1173 (100)	0.283×10^{-04}	0.897×10^{-04}	0.605×10^{-01}	0.606×10^{-01}	0.279×10^{-01}
	1332 (100)	0.223×10^{-04}	0.696×10^{-04}	0.567×10^{-01}	0.568×10^{-01}	0.270×10^{-01}

TABLE 4.5

Total and Partial (Photoabsorption, Coherent, and Incoherent) Mass Attenuation Coefficients, as Well as Mass Energy Absorption Coefficients, for the Principal Energies of ^{103}Pd, ^{125}I, ^{170}Tm, ^{169}Yb, ^{198}Au, ^{192}Ir, ^{137}Cs, and ^{60}Co, for Breast Tissue

	E (keV) Intensity (%)	μ^{ph}/ρ (cm²/g)	μ^{coh}/ρ (cm²/g)	μ^{incoh}/ρ (cm²/g)	μ/ρ (cm²/g)	μ_{en}/ρ (cm²/g)
^{103}Pd	20.07 (22.06)	$0.428 \times 10^{+00}$	0.776×10^{-01}	$0.178 \times 10^{+00}$	$0.685 \times 10^{+00}$	$0.434 \times 10^{+00}$
	20.22 (41.93)	$0.418 \times 10^{+00}$	0.767×10^{-01}	$0.178 \times 10^{+00}$	$0.676 \times 10^{+00}$	$0.424 \times 10^{+00}$
	22.70 (13.05)	$0.287 \times 10^{+00}$	0.639×10^{-01}	$0.180 \times 10^{+00}$	$0.553 \times 10^{+00}$	$0.297 \times 10^{+00}$
^{125}I	27.20 (39.90)	$0.159 \times 10^{+00}$	0.480×10^{-01}	$0.182 \times 10^{+00}$	$0.403 \times 10^{+00}$	$0.170 \times 10^{+00}$
	27.47 (74.50)	$0.154 \times 10^{+00}$	0.472×10^{-01}	$0.182 \times 10^{+00}$	$0.396 \times 10^{+00}$	$0.165 \times 10^{+00}$
	31.00 (25.90)	$0.104 \times 10^{+00}$	0.389×10^{-01}	$0.183 \times 10^{+00}$	$0.329 \times 10^{+00}$	$0.115 \times 10^{+00}$
	35.50 (6.68)	0.669×10^{-01}	0.308×10^{-01}	$0.183 \times 10^{+00}$	$0.286 \times 10^{+00}$	0.799×10^{-01}
^{170}Tm	51.35 (0.97)	0.199×10^{-01}	0.161×10^{-01}	$0.180 \times 10^{+00}$	$0.216 \times 10^{+00}$	0.354×10^{-01}
	52.39 (1.69)	0.186×10^{-01}	0.155×10^{-01}	$0.179 \times 10^{+00}$	$0.213 \times 10^{+00}$	0.345×10^{-01}
	59.38 (0.36)	0.123×10^{-01}	0.124×10^{-01}	$0.177 \times 10^{+00}$	$0.201 \times 10^{+00}$	0.295×10^{-01}
	84.25 (2.48)	0.388×10^{-02}	0.644×10^{-02}	$0.168 \times 10^{+00}$	$0.178 \times 10^{+00}$	0.247×10^{-01}
^{169}Yb	49.77 (53)	0.220×10^{-01}	0.170×10^{-01}	$0.180 \times 10^{+00}$	$0.220 \times 10^{+00}$	0.370×10^{-01}
	50.74 (93.8)	0.207×10^{-01}	0.165×10^{-01}	$0.180 \times 10^{+00}$	$0.217 \times 10^{+00}$	0.360×10^{-01}
	57.50 (38.5)	0.137×10^{-01}	0.131×10^{-01}	$0.178 \times 10^{+00}$	$0.205 \times 10^{+00}$	0.305×10^{-01}
	63.12 (44.2)	0.101×10^{-01}	0.110×10^{-01}	$0.176 \times 10^{+00}$	$0.197 \times 10^{+00}$	0.280×10^{-01}
	109.78 (17.5)	0.163×10^{-02}	0.388×10^{-02}	$0.158 \times 10^{+00}$	$0.164 \times 10^{+00}$	0.254×10^{-01}
	177.98 (35.8)	0.337×10^{-03}	0.151×10^{-02}	$0.140 \times 10^{+00}$	$0.141 \times 10^{+00}$	0.286×10^{-01}
	307.74 (10.05)	0.607×10^{-04}	0.511×10^{-03}	$0.116 \times 10^{+00}$	$0.117 \times 10^{+00}$	0.318×10^{-01}
^{198}Au	411.80 (95.58)	0.262×10^{-04}	0.286×10^{-03}	$0.104 \times 10^{+00}$	$0.104 \times 10^{+00}$	0.324×10^{-01}
^{192}Ir	295.96 (28.72)	0.682×10^{-04}	0.553×10^{-03}	$0.118 \times 10^{+00}$	$0.119 \times 10^{+00}$	0.316×10^{-01}
	308.46 (29.68)	0.603×10^{-04}	0.509×10^{-03}	$0.116 \times 10^{+00}$	$0.117 \times 10^{+00}$	0.318×10^{-01}
	316.51 (82.71)	0.558×10^{-04}	0.484×10^{-03}	$0.115 \times 10^{+00}$	$0.116 \times 10^{+00}$	0.319×10^{-01}
	468.07 (47.81)	0.181×10^{-04}	0.222×10^{-03}	0.987×10^{-01}	0.991×10^{-01}	0.327×10^{-01}
	588.58 (4.52)	0.990×10^{-05}	0.140×10^{-03}	0.896×10^{-01}	0.897×10^{-01}	0.327×10^{-01}
	604.41 (8.20)	0.926×10^{-05}	0.133×10^{-03}	0.886×10^{-01}	0.887×10^{-01}	0.327×10^{-01}
	612.46 (5.34)	0.897×10^{-05}	0.130×10^{-03}	0.881×10^{-01}	0.882×10^{-01}	0.326×10^{-01}
^{137}Cs	662 (85.1)	0.745×10^{-05}	0.111×10^{-03}	0.851×10^{-01}	0.852×10^{-01}	0.324×10^{-01}
^{60}Co	1173 (100)	0.211×10^{-05}	0.354×10^{-04}	0.648×10^{-01}	0.649×10^{-01}	0.299×10^{-01}
	1332 (100)	0.167×10^{-05}	0.275×10^{-04}	0.608×10^{-01}	0.609×10^{-01}	0.290×10^{-01}

TABLE 4.6

Total and Partial (Photoabsorption, Coherent, and Incoherent) Mass Attenuation Coefficients, as Well as Mass Energy Absorption Coefficients, for the Principal Energies of ^{103}Pd, ^{125}I, ^{170}Tm, ^{169}Yb, ^{198}Au, ^{192}Ir, ^{137}Cs, and ^{60}Co, for Air

	E (keV) Intensity (%)	μ^{ph}/ρ (cm²/g)	μ^{coh}/ρ (cm²/g)	μ^{incoh}/ρ (cm²/g)	$\mu_{/}\rho$ (cm²/g)	μ_{en}/ρ (cm²/g)
^{103}Pd	20.07 (22.06)	$0.529 \times 10^{+00}$	0.870×10^{-01}	$0.156 \times 10^{+00}$	$0.773 \times 10^{+00}$	$0.533 \times 10^{+00}$
	20.22 (41.93)	$0.516 \times 10^{+00}$	0.860×10^{-01}	$0.156 \times 10^{+00}$	$0.762 \times 10^{+00}$	$0.521 \times 10^{+00}$
	22.70 (13.05)	$0.356 \times 10^{+00}$	0.717×10^{-01}	$0.158 \times 10^{+00}$	$0.608 \times 10^{+00}$	$0.364 \times 10^{+00}$
^{125}I	27.20 (39.90)	$0.199 \times 10^{+00}$	0.539×10^{-01}	$0.161 \times 10^{+00}$	$0.428 \times 10^{+00}$	$0.208 \times 10^{+00}$
	27.47 (74.50)	$0.193 \times 10^{+00}$	0.531×10^{-01}	$0.161 \times 10^{+00}$	$0.420 \times 10^{+00}$	$0.202 \times 10^{+00}$
	31.00 (25.90)	$0.130 \times 10^{+00}$	0.437×10^{-01}	$0.162 \times 10^{+00}$	$0.340 \times 10^{+00}$	$0.140 \times 10^{+00}$
	35.50 (6.68)	0.840×10^{-01}	0.347×10^{-01}	$0.163 \times 10^{+00}$	$0.287 \times 10^{+00}$	0.957×10^{-01}
^{170}Tm	51.35 (0.97)	0.253×10^{-01}	0.182×10^{-01}	$0.161 \times 10^{+00}$	$0.204 \times 10^{+00}$	0.392×10^{-01}
	52.39 (1.69)	0.237×10^{-01}	0.175×10^{-01}	$0.161 \times 10^{+00}$	$0.202 \times 10^{+00}$	0.380×10^{-01}
	59.38 (0.36)	0.157×10^{-01}	0.139×10^{-01}	$0.159 \times 10^{+00}$	$0.180 \times 10^{+00}$	0.313×10^{-01}
	84.25 (2.48)	0.499×10^{-02}	0.728×10^{-02}	$0.151 \times 10^{+00}$	$0.163 \times 10^{+00}$	0.239×10^{-01}
^{169}Yb	49.77 (53)	0.279×10^{-01}	0.193×10^{-01}	$0.161 \times 10^{+00}$	$0.209 \times 10^{+00}$	0.414×10^{-01}
	50.74 (93.8)	0.262×10^{-01}	0.186×10^{-01}	$0.161 \times 10^{+00}$	$0.206 \times 10^{+00}$	0.400×10^{-01}
	57.50 (38.5)	0.175×10^{-01}	0.148×10^{-01}	$0.159 \times 10^{+00}$	$0.192 \times 10^{+00}$	0.326×10^{-01}
	63.12 (44.2)	0.129×10^{-01}	0.125×10^{-01}	$0.158 \times 10^{+00}$	$0.183 \times 10^{+00}$	0.292×10^{-01}
	109.78 (17.5)	0.210×10^{-02}	0.438×10^{-02}	$0.143 \times 10^{+00}$	$0.150 \times 10^{+00}$	0.237×10^{-01}
	177.98 (35.8)	0.440×10^{-03}	0.171×10^{-02}	$0.126 \times 10^{+00}$	$0.128 \times 10^{+00}$	0.260×10^{-01}
	307.74 (10.05)	0.798×10^{-4}	0.580×10^{-03}	$0.105 \times 10^{+00}$	$0.106 \times 10^{+00}$	0.288×10^{-01}
^{198}Au	411.80 (95.58)	0.346×10^{-04}	0.324×10^{-03}	0.937×10^{-01}	0.942×10^{-01}	0.293×10^{-01}
^{192}Ir	295.96 (28.72)	0.897×10^{-04}	0.627×10^{-03}	$0.107 \times 10^{+00}$	$0.108 \times 10^{+00}$	0.286×10^{-01}
	308.46 (29.68)	0.793×10^{-04}	0.577×10^{-03}	$0.105 \times 10^{+00}$	$0.106 \times 10^{+00}$	0.288×10^{-01}
	316.51 (82.71)	0.735×10^{-04}	0.548×10^{-03}	$0.104 \times 10^{+00}$	$0.105 \times 10^{+00}$	0.288×10^{-01}
	468.07 (47.81)	0.240×10^{-04}	0.251×10^{-03}	0.892×10^{-01}	0.895×10^{-01}	0.296×10^{-01}
	588.58 (4.52)	0.131×10^{-04}	0.159×10^{-03}	0.811×10^{-01}	0.813×10^{-01}	0.295×10^{-01}
	604.41 (8.20)	0.123×10^{-04}	0.151×10^{-03}	0.801×10^{-01}	0.803×10^{-01}	0.295×10^{-01}
	612.46 (5.34)	0.119×10^{-04}	0.147×10^{-03}	0.797×10^{-01}	0.798×10^{-01}	0.294×10^{-01}
^{137}Cs	662 (85.1)	0.989×10^{-05}	0.126×10^{-03}	0.769×10^{-01}	0.771×10^{-01}	0.293×10^{-01}
^{60}Co	1173 (100)	0.280×10^{-05}	0.400×10^{-04}	0.586×10^{-01}	0.587×10^{-01}	0.270×10^{-01}
	1332 (100)	0.222×10^{-05}	0.311×10^{-04}	0.549×10^{-01}	0.550×10^{-01}	0.263×10^{-01}

TABLE 4.7

Total and Partial (Photoabsorption, Coherent, and Incoherent) Mass Attenuation Coefficients, as Well as Mass Energy Absorption Coefficients, for the Principal Energies of ^{103}Pd, ^{125}I, ^{170}Tm, ^{169}Yb, ^{198}Au, ^{192}Ir, ^{137}Cs, and ^{60}Co, for LiF

	E (keV) Intensity (%)	μ^{ph}/ρ (cm²/g)	μ^{coh}/ρ (cm²/g)	μ^{incoh}/ρ (cm²/g)	μ/ρ (cm²/g)	μ_{en}/ρ (cm²/g)
^{103}Pd	20.07 (22.06)	$0.636 \times 10^{+00}$	0.900×10^{-01}	$0.145 \times 10^{+00}$	$0.871 \times 10^{+00}$	$0.642 \times 10^{+00}$
	20.22 (41.93)	$0.622 \times 10^{+00}$	0.890×10^{-01}	$0.145 \times 10^{+00}$	$0.856 \times 10^{+00}$	$0.628 \times 10^{+00}$
	22.70 (13.05)	$0.429 \times 10^{+00}$	0.746×10^{-01}	$0.147 \times 10^{+00}$	$0.651 \times 10^{+00}$	$0.435 \times 10^{+00}$
^{125}I	27.20 (39.90)	$0.239 \times 10^{+00}$	0.561×10^{-01}	$0.150 \times 10^{+00}$	$0.445 \times 10^{+00}$	$0.247 \times 10^{+00}$
	27.47 (74.50)	$0.232 \times 10^{+00}$	0.552×10^{-01}	$0.150 \times 10^{+00}$	$0.437 \times 10^{+00}$	$0.239 \times 10^{+00}$
	31.00 (25.90)	$0.157 \times 10^{+00}$	0.453×10^{-01}	$0.151 \times 10^{+00}$	$0.353 \times 10^{+00}$	$0.165 \times 10^{+00}$
	35.50 (6.68)	$0.101 \times 10^{+00}$	0.360×10^{-01}	$0.151 \times 10^{+00}$	$0.288 \times 10^{+00}$	$0.111 \times 10^{+00}$
^{170}Tm	51.35 (0.97)	0.302×10^{-01}	0.188×10^{-01}	$0.149 \times 10^{+00}$	$0.198 \times 10^{+00}$	0.432×10^{-01}
	52.39 (1.69)	0.283×10^{-01}	0.181×10^{-01}	$0.149 \times 10^{+00}$	$0.195 \times 10^{+00}$	0.416×10^{-01}
	59.38 (0.36)	0.188×10^{-01}	0.144×10^{-01}	$0.147 \times 10^{+00}$	$0.180 \times 10^{+00}$	0.332×10^{-01}
	84.25 (2.48)	0.593×10^{-02}	0.754×10^{-02}	$0.140 \times 10^{+00}$	$0.153 \times 10^{+00}$	0.253×10^{-01}
^{169}Yb	49.77 (53)	0.335×10^{-01}	0.199×10^{-01}	$0.149 \times 10^{+00}$	$0.203 \times 10^{+00}$	0.459×10^{-01}
	50.74 (93.8)	0.314×10^{-01}	0.192×10^{-01}	$0.149 \times 10^{+00}$	$0.200 \times 10^{+00}$	0.440×10^{-01}
	57.50 (38.5)	0.208×10^{-01}	0.153×10^{-01}	$0.148 \times 10^{+00}$	$0.184 \times 10^{+00}$	0.346×10^{-01}
	63.12 (44.2)	0.153×10^{-01}	0.129×10^{-01}	$0.146 \times 10^{+00}$	$0.174 \times 10^{+00}$	0.299×10^{-01}
	109.78 (17.5)	0.249×10^{-02}	0.456×10^{-02}	$0.133 \times 10^{+00}$	$0.140 \times 10^{+00}$	0.222×10^{-01}
	177.98 (35.8)	0.524×10^{-03}	0.180×10^{-02}	$0.117 \times 10^{+00}$	$0.119 \times 10^{+00}$	0.242×10^{-01}
	307.74 (10.05)	0.935×10^{-04}	0.604×10^{-03}	0.973×10^{-01}	0.980×10^{-01}	0.267×10^{-01}
^{198}Au	411.80 (95.58)	0.399×10^{-04}	0.338×10^{-03}	0.871×10^{-01}	0.875×10^{-01}	0.274×10^{-01}
^{192}Ir	295.96 (28.72)	0.105×10^{-03}	0.653×10^{-03}	0.987×10^{-01}	0.995×10^{-01}	0.266×10^{-01}
	308.46 (29.68)	0.928×10^{-04}	0.601×10^{-03}	0.973×10^{-01}	0.980×10^{-01}	0.267×10^{-01}
	316.51 (82.71)	0.859×10^{-04}	0.571×10^{-03}	0.964×10^{-01}	0.970×10^{-01}	0.268×10^{-01}
	468.07 (47.81)	0.279×10^{-04}	0.262×10^{-03}	0.827×10^{-01}	0.830×10^{-01}	0.275×10^{-01}
	588.58 (4.52)	0.153×10^{-04}	0.166×10^{-03}	0.751×10^{-01}	0.753×10^{-01}	0.274×10^{-01}
	604.41 (8.20)	0.143×10^{-04}	0.157×10^{-03}	0.743×10^{-01}	0.744×10^{-01}	0.274×10^{-01}
	612.46 (5.34)	0.138×10^{-04}	0.153×10^{-03}	0.738×10^{-01}	0.740×10^{-01}	0.273×10^{-01}
^{137}Cs	662 (85.1)	0.113×10^{-04}	0.131×10^{-03}	0.714×10^{-01}	0.715×10^{-01}	0.272×10^{-01}
^{60}Co	1173 (100)	0.320×10^{-05}	0.418×10^{-04}	0.544×10^{-01}	0.544×10^{-01}	0.251×10^{-01}
	1332 (100)	0.260×10^{-05}	0.324×10^{-04}	0.509×10^{-01}	0.510×10^{-01}	0.243×10^{-01}

TABLE 4.8

Total and Partial (Photoabsorption, Coherent, and Incoherent) Mass Attenuation Coefficients, as Well as Mass Energy Absorption Coefficients, for the Principal Energies of ^{103}Pd, ^{125}I, ^{170}Tm, ^{169}Yb, ^{198}Au, ^{192}Ir, ^{137}Cs, and ^{60}Co, for SiO$_2$

	E (keV) Intensity (%)	μ^{ph}/ρ (cm²/g)	μ^{coh}/ρ (cm²/g)	μ^{incoh}/ρ (cm²/g)	μ/ρ (cm²/g)	μ_{en}/ρ (cm²/g)
^{103}Pd	20.07 (22.06)	$0.221 \times 10^{+01}$	$0.161 \times 10^{+00}$	$0.148 \times 10^{+00}$	$0.252 \times 10^{+01}$	$0.221 \times 10^{+01}$
	20.22 (41.93)	$0.216 \times 10^{+01}$	$0.160 \times 10^{+00}$	$0.148 \times 10^{+00}$	$0.247 \times 10^{+01}$	$0.216 \times 10^{+01}$
	22.70 (13.05)	$0.151 \times 10^{+01}$	$0.134 \times 10^{+00}$	$0.151 \times 10^{+00}$	$0.180 \times 10^{+01}$	$0.151 \times 10^{+01}$
^{125}I	27.20 (39.90)	$0.857 \times 10^{+00}$	$0.101 \times 10^{+00}$	$0.155 \times 10^{+00}$	$0.111 \times 10^{+01}$	$0.862 \times 10^{+00}$
	27.47 (74.50)	$0.831 \times 10^{+00}$	0.996×10^{-01}	$0.155 \times 10^{+00}$	$0.109 \times 10^{+01}$	$0.836 \times 10^{+00}$
	31.00 (25.90)	$0.568 \times 10^{+00}$	0.822×10^{-01}	$0.157 \times 10^{+00}$	$0.807 \times 10^{+00}$	$0.575 \times 10^{+00}$
	35.50 (6.68)	$0.370 \times 10^{+00}$	0.659×10^{-01}	$0.158 \times 10^{+00}$	$0.594 \times 10^{+00}$	$0.379 \times 10^{+00}$
^{170}Tm	51.35 (0.97)	$0.114 \times 10^{+00}$	0.351×10^{-01}	$0.157 \times 10^{+00}$	$0.306 \times 10^{+00}$	$0.128 \times 10^{+00}$
	52.39 (1.69)	$0.107 \times 10^{+00}$	0.339×10^{-01}	$0.157 \times 10^{+00}$	$0.298 \times 10^{+00}$	$0.121 \times 10^{+00}$
	59.38 (0.36)	0.715×10^{-01}	0.271×10^{-01}	$0.156 \times 10^{+00}$	$0.254 \times 10^{+00}$	0.869×10^{-01}
	84.25 (2.48)	0.232×10^{-01}	0.143×10^{-01}	$0.149 \times 10^{+00}$	$0.187 \times 10^{+00}$	0.424×10^{-01}
^{169}Yb	49.77 (53)	$0.126 \times 10^{+00}$	0.371×10^{-01}	$0.158 \times 10^{+00}$	$0.321 \times 10^{+00}$	$0.139 \times 10^{+00}$
	50.74 (93.8)	$0.119 \times 10^{+00}$	0.358×10^{-01}	$0.158 \times 10^{+00}$	$0.312 \times 10^{+00}$	$0.132 \times 10^{+00}$
	57.50 (38.5)	0.793×10^{-01}	0.287×10^{-01}	$0.156 \times 10^{+00}$	$0.264 \times 10^{+00}$	0.938×10^{-01}
	63.12 (44.2)	0.588×10^{-01}	0.243×10^{-01}	$0.155 \times 10^{+00}$	$0.238 \times 10^{+00}$	0.743×10^{-01}
	109.78 (17.5)	0.987×10^{-02}	0.870×10^{-02}	$0.142 \times 10^{+00}$	$0.161 \times 10^{+00}$	0.311×10^{-01}
	177.98 (35.8)	0.213×10^{-02}	0.347×10^{-02}	$0.126 \times 10^{+00}$	$0.131 \times 10^{+00}$	0.276×10^{-01}
	307.74 (10.05)	0.387×10^{-03}	0.117×10^{-02}	$0.105 \times 10^{+00}$	$0.107 \times 10^{+00}$	0.291×10^{-01}
^{198}Au	411.80 (95.58)	0.166×10^{-03}	0.658×10^{-03}	0.939×10^{-01}	0.948×10^{-01}	0.296×10^{-01}
^{192}Ir	295.96 (28.72)	0.435×10^{-03}	0.127×10^{-02}	$0.106 \times 10^{+00}$	$0.108 \times 10^{+00}$	0.290×10^{-01}
	308.46 (29.68)	0.384×10^{-03}	0.111×10^{-02}	$0.105 \times 10^{+00}$	$0.106 \times 10^{+00}$	0.291×10^{-01}
	316.51 (82.71)	0.356×10^{-03}	0.117×10^{-02}	$0.104 \times 10^{+00}$	$0.105 \times 10^{+00}$	0.292×10^{-01}
	468.07 (47.81)	0.117×10^{-03}	0.510×10^{-03}	0.892×10^{-01}	0.899×10^{-01}	0.297×10^{-01}
	588.58 (4.52)	0.642×10^{-04}	0.323×10^{-03}	0.810×10^{-01}	0.814×10^{-01}	0.296×10^{-01}
	604.41 (8.20)	0.600×10^{-04}	0.306×10^{-03}	0.801×10^{-01}	0.805×10^{-01}	0.295×10^{-01}
	612.46 (5.34)	0.580×10^{-04}	0.298×10^{-03}	0.797×10^{-01}	0.800×10^{-01}	0.295×10^{-01}
^{137}Cs	662 (85.1)	0.478×10^{-04}	0.256×10^{-03}	0.770×10^{-01}	0.773×10^{-01}	0.294×10^{-01}
^{60}Co	1173 (100)	0.136×10^{-04}	0.815×10^{-04}	0.587×10^{-01}	0.588×10^{-01}	0.270×10^{-01}
	1332 (100)	0.110×10^{-04}	0.632×10^{-04}	0.550×10^{-01}	0.551×10^{-01}	0.262×10^{-01}

TABLE 4.9

Total and Partial (Photoabsorption, Coherent, and Incoherent) Mass Attenuation Coefficients, as Well as Mass Energy Absorption Coefficients, for the Principal Energies of ^{103}Pd, ^{125}I, ^{170}Tm, ^{169}Yb, ^{198}Au, ^{192}Ir, ^{137}Cs, and ^{60}Co, for Radiochromic Dye Film, Nylon Base

	E (keV) Intensity (%)	μ^{ph}/ρ (cm²/g)	μ^{coh}/ρ (cm²/g)	μ^{incoh}/ρ (cm²/g)	μ/ρ (cm²/g)	μ_{en}/ρ (cm²/g)
^{103}Pd	20.07 (22.06)	$0.266 \times 10^{+00}$	0.650×10^{-01}	$0.179 \times 10^{+00}$	$0.510 \times 10^{+00}$	$0.272 \times 10^{+00}$
	20.22 (41.93)	$0.260 \times 10^{+00}$	0.643×10^{-01}	$0.179 \times 10^{+00}$	$0.503 \times 10^{+00}$	$0.266 \times 10^{+00}$
	22.70 (13.05)	$0.178 \times 10^{+00}$	0.538×10^{-01}	$0.181 \times 10^{+00}$	$0.413 \times 10^{+00}$	$0.185 \times 10^{+00}$
^{125}I	27.20 (39.90)	0.981×10^{-01}	0.402×10^{-01}	$0.183 \times 10^{+00}$	$0.322 \times 10^{+00}$	$0.107 \times 10^{+00}$
	27.47 (74.50)	0.950×10^{-01}	0.395×10^{-01}	$0.184 \times 10^{+00}$	$0.318 \times 10^{+00}$	$0.104 \times 10^{+00}$
	31.00 (25.90)	0.638×10^{-01}	0.323×10^{-01}	$0.184 \times 10^{+00}$	$0.280 \times 10^{+00}$	0.741×10^{-01}
	35.50 (6.68)	0.408×10^{-01}	0.256×10^{-01}	$0.184 \times 10^{+00}$	$0.250 \times 10^{+00}$	0.523×10^{-01}
^{170}Tm	51.35 (0.97)	0.120×10^{-01}	0.133×10^{-01}	$0.180 \times 10^{+00}$	$0.205 \times 10^{+00}$	0.274×10^{-01}
	52.39 (1.69)	0.112×10^{-01}	0.128×10^{-01}	$0.179 \times 10^{+00}$	$0.203 \times 10^{+00}$	0.269×10^{-01}
	59.38 (0.36)	0.739×10^{-02}	0.102×10^{-02}	$0.176 \times 10^{+00}$	$0.194 \times 10^{+00}$	0.245×10^{-01}
	84.25 (2.48)	0.231×10^{-02}	0.528×10^{-02}	$0.167 \times 10^{+00}$	$0.175 \times 10^{+00}$	0.230×10^{-01}
^{169}Yb	49.77 (53)	0.133×10^{-01}	0.141×10^{-01}	$0.180 \times 10^{+00}$	$0.208 \times 10^{+00}$	0.282×10^{-01}
	50.74 (93.8)	0.125×10^{-01}	0.136×10^{-01}	$0.180 \times 10^{+00}$	$0.206 \times 10^{+00}$	0.276×10^{-01}
	57.50 (38.5)	0.824×10^{-02}	0.108×10^{-01}	$0.177 \times 10^{+00}$	$0.196 \times 10^{+00}$	0.246×10^{-01}
	63.12 (44.2)	0.604×10^{-02}	0.909×10^{-02}	$0.175 \times 10^{+00}$	$0.190 \times 10^{+00}$	0.235×10^{-01}
	109.78 (17.5)	0.962×10^{-03}	0.317×10^{-02}	$0.158 \times 10^{+00}$	$0.162 \times 10^{+00}$	0.245×10^{-01}
	177.98 (35.8)	0.201×10^{-03}	0.124×10^{-02}	$0.139 \times 10^{+00}$	$0.141 \times 10^{+00}$	0.284×10^{-01}
	307.74 (10.05)	0.354×10^{-04}	0.416×10^{-03}	$0.116 \times 10^{+00}$	$0.116 \times 10^{+00}$	0.317×10^{-01}
^{198}Au	411.80 (95.58)	0.150×10^{-04}	0.233×10^{-03}	$0.104 \times 10^{+00}$	$0.104 \times 10^{+00}$	0.325×10^{-01}
^{192}Ir	295.96 (28.72)	0.398×10^{-04}	0.449×10^{-03}	$0.118 \times 10^{+00}$	$0.118 \times 10^{+00}$	0.315×10^{-01}
	308.46 (29.68)	0.351×10^{-04}	0.414×10^{-03}	$0.116 \times 10^{+00}$	$0.116 \times 10^{+00}$	0.317×10^{-01}
	316.51 (82.71)	0.325×10^{-04}	0.393×10^{-03}	$0.115 \times 10^{+00}$	$0.115 \times 10^{+00}$	0.318×10^{-01}
	468.07 (47.81)	0.105×10^{-04}	0.180×10^{-03}	0.984×10^{-01}	0.986×10^{-01}	0.327×10^{-01}
	588.58 (4.52)	0.575×10^{-05}	0.114×10^{-03}	0.893×10^{-01}	0.894×10^{-01}	0.326×10^{-01}
	604.41 (8.20)	0.536×10^{-05}	0.108×10^{-03}	0.883×10^{-01}	0.884×10^{-01}	0.325×10^{-01}
	612.46 (5.34)	0.518×10^{-05}	0.105×10^{-03}	0.878×10^{-01}	0.879×10^{-01}	0.325×10^{-01}
^{137}Cs	662 (85.1)	0.424×10^{-05}	0.903×10^{-04}	0.849×10^{-01}	0.849×10^{-01}	0.324×10^{-01}
^{60}Co	1173 (100)	0.118×10^{-05}	0.287×10^{-04}	0.647×10^{-01}	0.647×10^{-01}	0.298×10^{-01}
	1332 (100)	0.972×10^{-06}	0.223×10^{-04}	0.606×10^{-01}	0.606×10^{-01}	0.290×10^{-01}

TABLE 4.10

Total and Partial (Photoabsorption, Coherent, and Incoherent) Mass Attenuation Coefficients, as Well as Mass Energy Absorption Coefficients, for the Principal Energies of ^{103}Pd, ^{125}I, ^{170}Tm, ^{169}Yb, ^{198}Au, ^{192}Ir, ^{137}Cs, and ^{60}Co, for Gafchromic Sensor

	E (keV) Intensity (%)	μ^{ph}/ρ (cm²/g)	μ^{coh}/ρ (cm²/g)	μ^{incoh}/ρ (cm²/g)	μ/ρ (cm²/g)	μ_{en}/ρ (cm²/g)
^{103}Pd	20.07 (22.06)	$0.289 \times 10^{+00}$	0.675×10^{-01}	$0.176 \times 10^{+00}$	$0.533 \times 10^{+00}$	$0.296 \times 10^{+00}$
	20.22 (41.93)	$0.282 \times 10^{+00}$	0.668×10^{-01}	$0.177 \times 10^{+00}$	$0.526 \times 10^{+00}$	$0.289 \times 10^{+00}$
	22.70 (13.05)	$0.193 \times 10^{+00}$	0.559×10^{-01}	$0.179 \times 10^{+00}$	$0.428 \times 10^{+00}$	$0.201 \times 10^{+00}$
^{125}I	27.20 (39.90)	$0.107 \times 10^{+00}$	0.418×10^{-01}	$0.181 \times 10^{+00}$	$0.329 \times 10^{+00}$	$0.116 \times 10^{+00}$
	27.47 (74.50)	$0.104 \times 10^{+00}$	0.411×10^{-01}	$0.181 \times 10^{+00}$	$0.325 \times 10^{+00}$	$0.113 \times 10^{+00}$
	31.00 (25.90)	0.696×10^{-01}	0.336×10^{-01}	$0.181 \times 10^{+00}$	$0.285 \times 10^{+00}$	0.797×10^{-01}
	35.50 (6.68)	0.445×10^{-01}	0.266×10^{-01}	$0.181 \times 10^{+00}$	$0.252 \times 10^{+00}$	0.559×10^{-01}
^{170}Tm	51.35 (0.97)	0.131×10^{-01}	0.138×10^{-01}	$0.177 \times 10^{+00}$	$0.204 \times 10^{+00}$	0.283×10^{-01}
	52.39 (1.69)	0.123×10^{-01}	0.133×10^{-01}	$0.177 \times 10^{+00}$	$0.203 \times 10^{+00}$	0.278×10^{-01}
	59.38 (0.36)	0.808×10^{-02}	0.106×10^{-01}	$0.175 \times 10^{+00}$	$0.193 \times 10^{+00}$	0.250×10^{-01}
	84.25 (2.48)	0.253×10^{-02}	0.550×10^{-02}	$0.165 \times 10^{+00}$	$0.173 \times 10^{+00}$	0.230×10^{-01}
^{169}Yb	49.77 (53)	0.145×10^{-01}	0.146×10^{-01}	$0.178 \times 10^{+00}$	$0.207 \times 10^{+00}$	0.292×10^{-01}
	50.74 (93.8)	0.136×10^{-01}	0.141×10^{-01}	$0.178 \times 10^{+00}$	$0.205 \times 10^{+00}$	0.285×10^{-01}
	57.50 (38.5)	0.899×10^{-02}	0.113×10^{-01}	$0.175 \times 10^{+00}$	$0.196 \times 10^{+00}$	0.252×10^{-01}
	63.12 (44.2)	0.660×10^{-02}	0.947×10^{-02}	$0.173 \times 10^{+00}$	$0.189 \times 10^{+00}$	0.238×10^{-01}
	109.78 (17.5)	0.105×10^{-02}	0.331×10^{-02}	$0.156 \times 10^{+00}$	$0.161 \times 10^{+00}$	0.243×10^{-01}
	177.98 (35.8)	0.220×10^{-03}	0.130×10^{-02}	$0.138 \times 10^{+00}$	$0.139 \times 10^{+00}$	0.281×10^{-01}
	307.74 (10.05)	0.387×10^{-04}	0.434×10^{-03}	$0.115 \times 10^{+00}$	$0.115 \times 10^{+00}$	0.314×10^{-01}
^{198}Au	411.80 (95.58)	0.165×10^{-04}	0.243×10^{-03}	$0.103 \times 10^{+00}$	$0.103 \times 10^{+00}$	0.322×10^{-01}
^{192}Ir	295.96 (28.72)	0.436×10^{-04}	0.469×10^{-03}	$0.116 \times 10^{+00}$	$0.117 \times 10^{+00}$	0.312×10^{-01}
	308.46 (29.68)	0.385×10^{-04}	0.432×10^{-03}	$0.114 \times 10^{+00}$	$0.115 \times 10^{+00}$	0.314×10^{-01}
	316.51 (82.71)	0.356×10^{-04}	0.411×10^{-03}	$0.113 \times 10^{+00}$	$0.114 \times 10^{+00}$	0.315×10^{-01}
	468.07 (47.81)	0.115×10^{-04}	0.188×10^{-03}	0.973×10^{-01}	0.975×10^{-01}	0.323×10^{-01}
	588.58 (4.52)	0.630×10^{-05}	0.119×10^{-03}	0.883×10^{-01}	0.885×10^{-01}	0.322×10^{-01}
	604.41 (8.20)	0.588×10^{-05}	0.113×10^{-03}	0.873×10^{-01}	0.874×10^{-01}	0.322×10^{-01}
	612.46 (5.34)	0.568×10^{-05}	0.110×10^{-03}	0.868×10^{-01}	0.869×10^{-01}	0.322×10^{-01}
^{137}Cs	662 (85.1)	0.465×10^{-05}	0.943×10^{-04}	0.839×10^{-01}	0.840×10^{-01}	0.320×10^{-01}
^{60}Co	1173 (100)	0.129×10^{-05}	0.300×10^{-04}	0.640×10^{-01}	0.640×10^{-01}	0.295×10^{-01}
	1332 (100)	0.107×10^{-05}	0.233×10^{-04}	0.599×10^{-01}	0.599×10^{-01}	0.286×10^{-01}

TABLE 4.11

Total and Partial (Photoabsorption, Coherent, and Incoherent) Mass Attenuation Coefficients, as Well as Mass Energy Absorption Coefficients, for the Principal Energies of ^{103}Pd, ^{125}I, ^{170}Tm, ^{169}Yb, ^{198}Au, ^{192}Ir, ^{137}Cs, and ^{60}Co, for VIPAR Gel

	E (keV) Intensity (%)	μ^{ph}/ρ (cm²/g)	μ^{coh}/ρ (cm²/g)	μ^{incoh}/ρ (cm²/g)	μ/ρ (cm²/g)	μ_{en}/ρ (cm²/g)
^{103}Pd	20.07 (22.06)	$0.508 \times 10^{+00}$	0.856×10^{-01}	$0.177 \times 10^{+00}$	$0.772 \times 10^{+00}$	$0.514 \times 10^{+00}$
	20.22 (41.93)	$0.496 \times 10^{+00}$	0.846×10^{-01}	$0.177 \times 10^{+00}$	$0.761 \times 10^{+00}$	$0.502 \times 10^{+00}$
	22.70 (13.05)	$0.341 \times 10^{+00}$	0.706×10^{-01}	$0.179 \times 10^{+00}$	$0.614 \times 10^{+00}$	$0.351 \times 10^{+00}$
^{125}I	27.20 (39.90)	$0.190 \times 10^{+00}$	0.532×10^{-01}	$0.181 \times 10^{+00}$	$0.438 \times 10^{+00}$	$0.200 \times 10^{+00}$
	27.47 (74.50)	$0.184 \times 10^{+00}$	0.524×10^{-01}	$0.181 \times 10^{+00}$	$0.430 \times 10^{+00}$	$0.194 \times 10^{+00}$
	31.00 (25.90)	$0.124 \times 10^{+00}$	0.431×10^{-01}	$0.182 \times 10^{+00}$	$0.352 \times 10^{+00}$	$0.135 \times 10^{+00}$
	35.50 (6.68)	0.794×10^{-01}	0.342×10^{-01}	$0.182 \times 10^{+00}$	$0.302 \times 10^{+00}$	0.922×10^{-01}
^{170}Tm	51.35 (0.97)	0.235×10^{-01}	0.179×10^{-01}	$0.179 \times 10^{+00}$	$0.221 \times 10^{+00}$	0.390×10^{-01}
	52.39 (1.69)	0.220×10^{-01}	0.173×10^{-01}	$0.179 \times 10^{+00}$	$0.218 \times 10^{+00}$	0.379×10^{-01}
	59.38 (0.36)	0.145×10^{-01}	0.138×10^{-01}	$0.177 \times 10^{+00}$	$0.205 \times 10^{+00}$	0.318×10^{-01}
	84.25 (2.48)	0.458×10^{-02}	0.718×10^{-02}	$0.168 \times 10^{+00}$	$0.179 \times 10^{+00}$	0.255×10^{-01}
^{169}Yb	49.77 (53)	0.261×10^{-01}	0.190×10^{-01}	$0.180 \times 10^{+00}$	$0.225 \times 10^{+00}$	0.410×10^{-01}
	50.74 (93.8)	0.245×10^{-01}	0.183×10^{-01}	$0.180 \times 10^{+00}$	$0.222 \times 10^{+00}$	0.397×10^{-01}
	57.50 (38.5)	0.162×10^{-01}	0.146×10^{-01}	$0.177 \times 10^{+00}$	$0.209 \times 10^{+00}$	0.330×10^{-01}
	63.12 (44.2)	0.119×10^{-01}	0.123×10^{-01}	$0.175 \times 10^{+00}$	$0.200 \times 10^{+00}$	0.299×10^{-01}
	109.78 (17.5)	0.191×10^{-02}	0.433×10^{-02}	$0.158 \times 10^{+00}$	$0.165 \times 10^{+00}$	0.257×10^{-01}
	177.98 (35.8)	0.396×10^{-03}	0.169×10^{-02}	$0.140 \times 10^{+00}$	$0.142 \times 10^{+00}$	0.287×10^{-01}
	307.74 (10.05)	0.712×10^{-04}	0.572×10^{-03}	$0.116 \times 10^{+00}$	$0.117 \times 10^{+00}$	0.319×10^{-01}
^{198}Au	411.80 (95.58)	0.308×10^{-04}	0.321×10^{-03}	$0.104 \times 10^{+00}$	$0.104 \times 10^{+00}$	0.325×10^{-01}
^{192}Ir	295.96 (28.72)	0.801×10^{-04}	0.618×10^{-03}	$0.118 \times 10^{+00}$	$0.119 \times 10^{+00}$	0.317×10^{-01}
	308.46 (29.68)	0.708×10^{-04}	0.570×10^{-03}	$0.116 \times 10^{+00}$	$0.117 \times 10^{+00}$	0.319×10^{-01}
	316.51 (82.71)	0.656×10^{-04}	0.541×10^{-03}	$0.115 \times 10^{+00}$	$0.116 \times 10^{+00}$	0.319×10^{-01}
	468.07 (47.81)	0.213×10^{-04}	0.249×10^{-03}	0.987×10^{-01}	0.992×10^{-01}	0.328×10^{-01}
	588.58 (4.52)	0.116×10^{-04}	0.157×10^{-03}	0.897×10^{-01}	0.899×10^{-01}	0.327×10^{-01}
	604.41 (8.20)	0.108×10^{-04}	0.149×10^{-03}	0.887×10^{-01}	0.889×10^{-01}	0.327×10^{-01}
	612.46 (5.34)	0.105×10^{-04}	0.145×10^{-03}	0.882×10^{-01}	0.884×10^{-01}	0.326×10^{-01}
^{137}Cs	662 (85.1)	0.872×10^{-05}	0.124×10^{-03}	0.852×10^{-01}	0.853×10^{-01}	0.324×10^{-01}
^{60}Co	1173 (100)	0.247×10^{-05}	0.396×10^{-04}	0.649×10^{-01}	0.650×10^{-01}	0.299×10^{-01}
	1332 (100)	0.196×10^{-05}	0.307×10^{-04}	0.609×10^{-01}	0.610×10^{-01}	0.291×10^{-01}

TABLE 4.12

Total and Partial (Photoabsorption, Coherent, and Incoherent) Mass Attenuation Coefficients, as Well as Mass Energy Absorption Coefficients, for the Principal Energies of ^{103}Pd, ^{125}I, ^{170}Tm, ^{169}Yb, ^{198}Au, ^{192}Ir, ^{137}Cs, and ^{60}Co, for Solid Water

Isotope	E (keV) Intensity (%)	μ^{ph}/ρ (cm²/g)	μ^{coh}/ρ (cm²/g)	μ^{incoh}/ρ (cm²/g)	μ/ρ (cm²/g)	μ_{en}/ρ (cm²/g)
^{103}Pd	20.07 (22.06)	$0.572 \times 10^{+00}$	0.745×10^{-01}	$0.174 \times 10^{+00}$	$0.822 \times 10^{+00}$	$0.570 \times 10^{+00}$
	20.22 (41.93)	$0.559 \times 10^{+00}$	0.736×10^{-01}	$0.174 \times 10^{+00}$	$0.810 \times 10^{+00}$	$0.557 \times 10^{+00}$
	22.70 (13.05)	$0.389 \times 10^{+00}$	0.612×10^{-01}	$0.176 \times 10^{+00}$	$0.649 \times 10^{+00}$	$0.394 \times 10^{+00}$
^{125}I	27.20 (39.90)	$0.221 \times 10^{+00}$	0.459×10^{-01}	$0.178 \times 10^{+00}$	$0.459 \times 10^{+00}$	$0.229 \times 10^{+00}$
	27.47 (74.50)	$0.215 \times 10^{+00}$	0.452×10^{-01}	$0.178 \times 10^{+00}$	$0.451 \times 10^{+00}$	$0.222 \times 10^{+00}$
	31.00 (25.90)	$0.147 \times 10^{+00}$	0.372×10^{-01}	$0.179 \times 10^{+00}$	$0.366 \times 10^{+00}$	$0.156 \times 10^{+00}$
	35.50 (6.68)	0.956×10^{-01}	0.294×10^{-01}	$0.179 \times 10^{+00}$	$0.310 \times 10^{+00}$	$0.107 \times 10^{+00}$
^{170}Tm	51.35 (0.97)	0.296×10^{-01}	0.154×10^{-01}	$0.176 \times 10^{+00}$	$0.221 \times 10^{+00}$	0.445×10^{-01}
	52.39 (1.69)	0.278×10^{-01}	0.149×10^{-01}	$0.175 \times 10^{+00}$	$0.218 \times 10^{+00}$	0.431×10^{-01}
	59.38 (0.36)	0.184×10^{-01}	0.105×10^{-01}	$0.173 \times 10^{+00}$	$0.192 \times 10^{+00}$	0.353×10^{-01}
	84.25 (2.48)	0.605×10^{-02}	0.620×10^{-02}	$0.164 \times 10^{+00}$	$0.176 \times 10^{+00}$	0.265×10^{-01}
^{169}Yb	49.77 (53)	0.327×10^{-01}	0.163×10^{-01}	$0.176 \times 10^{+00}$	$0.225 \times 10^{+00}$	0.471×10^{-01}
	50.74 (93.8)	0.307×10^{-01}	0.158×10^{-01}	$0.176 \times 10^{+00}$	$0.222 \times 10^{+00}$	0.454×10^{-01}
	57.50 (38.5)	0.206×10^{-01}	0.126×10^{-01}	$0.174 \times 10^{+00}$	$0.207 \times 10^{+00}$	0.369×10^{-01}
	63.12 (44.2)	0.153×10^{-01}	0.106×10^{-01}	$0.172 \times 10^{+00}$	$0.198 \times 10^{+00}$	0.329×10^{-01}
	109.78 (17.5)	0.259×10^{-02}	0.373×10^{-02}	$0.154 \times 10^{+00}$	$0.161 \times 10^{+00}$	0.259×10^{-01}
	177.98 (35.8)	0.556×10^{-03}	0.146×10^{-02}	$0.136 \times 10^{+00}$	$0.138 \times 10^{+00}$	0.282×10^{-01}
	307.74 (10.05)	0.103×10^{-03}	0.494×10^{-03}	$0.114 \times 10^{+00}$	$0.114 \times 10^{+00}$	0.311×10^{-01}
^{198}Au	411.80 (95.58)	0.451×10^{-04}	0.276×10^{-03}	$0.102 \times 10^{+00}$	$0.102 \times 10^{+00}$	0.317×10^{-01}
^{192}Ir	295.96 (28.72)	0.116×10^{-03}	0.534×10^{-03}	$0.116 \times 10^{+00}$	$0.116 \times 10^{+00}$	0.310×10^{-01}
	308.46 (29.68)	0.102×10^{-03}	0.492×10^{-03}	$0.114 \times 10^{+00}$	$0.114 \times 10^{+00}$	0.311×10^{-01}
	316.51 (82.71)	0.950×10^{-04}	0.467×10^{-03}	$0.113 \times 10^{+00}$	$0.113 \times 10^{+00}$	0.312×10^{-01}
	468.07 (47.81)	0.314×10^{-04}	0.214×10^{-03}	0.965×10^{-01}	0.966×10^{-01}	0.320×10^{-01}
	588.58 (4.52)	0.172×10^{-04}	0.136×10^{-03}	0.876×10^{-01}	0.877×10^{-01}	0.319×10^{-01}
	604.41 (8.20)	0.161×10^{-04}	0.129×10^{-03}	0.866×10^{-01}	0.867×10^{-01}	0.319×10^{-01}
	612.46 (5.34)	0.156×10^{-04}	0.126×10^{-03}	0.861×10^{-01}	0.862×10^{-01}	0.319×10^{-01}
^{137}Cs	662 (85.1)	0.130×10^{-04}	0.108×10^{-03}	0.831×10^{-01}	0.832×10^{-01}	0.317×10^{-01}
^{60}Co	1173 (100)	0.371×10^{-05}	0.342×10^{-04}	0.634×10^{-01}	0.634×10^{-01}	0.292×10^{-01}
	1332 (100)	0.294×10^{-05}	0.266×10^{-04}	0.594×10^{-01}	0.594×10^{-01}	0.284×10^{-01}

TABLE 4.13

Total and Partial (Photoabsorption, Coherent, and Incoherent) Mass Attenuation Coefficients, as Well as Mass Energy Absorption Coefficients, for the Principal Energies of ^{103}Pd, ^{125}I, ^{169}Yb, ^{170}Tm, ^{198}Au, ^{192}Ir, ^{137}Cs, and ^{60}Co, for PMMA

	E (keV) Intensity (%)	μ^{ph}/ρ (cm²/g)	μ^{coh}/ρ (cm²/g)	μ^{incoh}/ρ (cm²/g)	μ/ρ (cm²/g)	μ_{en}/ρ (cm²/g)
^{103}Pd	20.07 (22.06)	$0.322 \times 10^{+00}$	0.706×10^{-01}	$0.174 \times 10^{+00}$	$0.567 \times 10^{+00}$	$0.329 \times 10^{+00}$
	20.22 (41.93)	$0.315 \times 10^{+00}$	0.699×10^{-01}	$0.174 \times 10^{+00}$	$0.559 \times 10^{+00}$	$0.322 \times 10^{+00}$
	22.70 (13.05)	$0.216 \times 10^{+00}$	0.585×10^{-01}	$0.177 \times 10^{+00}$	$0.451 \times 10^{+00}$	$0.223 \times 10^{+00}$
^{125}I	27.20 (39.90)	$0.120 \times 10^{+00}$	0.438×10^{-01}	$0.179 \times 10^{+00}$	$0.342 \times 10^{+00}$	$0.128 \times 10^{+00}$
	27.47 (74.50)	$0.116 \times 10^{+00}$	0.431×10^{-01}	$0.179 \times 10^{+00}$	$0.338 \times 10^{+00}$	$0.125 \times 10^{+00}$
	31.00 (25.90)	0.779×10^{-01}	0.352×10^{-01}	$0.179 \times 10^{+00}$	$0.293 \times 10^{+00}$	0.879×10^{-01}
	35.50 (6.68)	0.499×10^{-01}	0.279×10^{-01}	$0.179 \times 10^{+00}$	$0.257 \times 10^{+00}$	0.611×10^{-01}
^{170}Tm	51.35 (0.97)	0.147×10^{-01}	0.145×10^{-01}	$0.176 \times 10^{+00}$	$0.205 \times 10^{+00}$	0.298×10^{-01}
	52.39 (1.69)	0.138×10^{-01}	0.140×10^{-01}	$0.175 \times 10^{+00}$	$0.203 \times 10^{+00}$	0.292×10^{-01}
	59.38 (0.36)	0.907×10^{-02}	0.111×10^{-01}	$0.173 \times 10^{+00}$	$0.193 \times 10^{+00}$	0.259×10^{-01}
	84.25 (2.48)	0.284×10^{-02}	0.578×10^{-02}	$0.164 \times 10^{+00}$	$0.172 \times 10^{+00}$	0.232×10^{-01}
^{169}Yb	49.77 (53)	0.163×10^{-01}	0.154×10^{-01}	$0.176 \times 10^{+00}$	$0.208 \times 10^{+00}$	0.309×10^{-01}
	50.74 (93.8)	0.153×10^{-01}	0.148×10^{-01}	$0.176 \times 10^{+00}$	$0.206 \times 10^{+00}$	0.301×10^{-01}
	57.50 (38.5)	0.101×10^{-01}	0.118×10^{-01}	$0.174 \times 10^{+00}$	$0.196 \times 10^{+00}$	0.262×10^{-01}
	63.12 (44.2)	0.742×10^{-02}	0.995×10^{-02}	$0.172 \times 10^{+00}$	$0.189 \times 10^{+00}$	0.245×10^{-01}
	109.78 (17.5)	0.119×10^{-02}	0.348×10^{-02}	$0.155 \times 10^{+00}$	$0.160 \times 10^{+00}$	0.242×10^{-01}
	177.98 (35.8)	0.248×10^{-03}	0.137×10^{-02}	$0.137 \times 10^{+00}$	$0.138 \times 10^{+00}$	0.279×10^{-01}
	307.74 (10.05)	0.437×10^{-04}	0.457×10^{-03}	$0.114 \times 10^{+00}$	$0.114 \times 10^{+00}$	0.311×10^{-01}
^{198}Au	411.80 (95.58)	0.399×10^{-04}	0.338×10^{-03}	0.871×10^{-01}	0.875×10^{-01}	0.319×10^{-01}
^{192}Ir	295.96 (28.72)	0.492×10^{-04}	0.494×10^{-03}	$0.115 \times 10^{+00}$	$0.116 \times 10^{+00}$	0.309×10^{-01}
	308.46 (29.68)	0.434×10^{-04}	0.455×10^{-03}	$0.114 \times 10^{+00}$	$0.114 \times 10^{+00}$	0.311×10^{-01}
	316.51 (82.71)	0.402×10^{-04}	0.432×10^{-03}	$0.112 \times 10^{+00}$	$0.113 \times 10^{+00}$	0.312×10^{-01}
	468.07 (47.81)	0.130×10^{-04}	0.198×10^{-03}	0.965×10^{-01}	0.967×10^{-01}	0.320×10^{-01}
	588.58 (4.52)	0.712×10^{-05}	0.125×10^{-03}	0.866×10^{-01}	0.877×10^{-01}	0.319×10^{-01}
	604.41 (8.20)	0.665×10^{-05}	0.119×10^{-03}	0.866×10^{-01}	0.867×10^{-01}	0.319×10^{-01}
	612.46 (5.34)	0.642×10^{-05}	0.116×10^{-03}	0.861×10^{-01}	0.862×10^{-01}	0.319×10^{-01}
^{137}Cs	662 (85.1)	0.525×10^{-05}	0.992×10^{-04}	0.832×10^{-01}	0.833×10^{-01}	0.317×10^{-01}
^{60}Co	1173 (100)	0.147×10^{-05}	0.316×10^{-04}	0.634×10^{-01}	0.635×10^{-01}	0.292×10^{-01}
	1332 (100)	0.121×10^{-05}	0.245×10^{-04}	0.594×10^{-01}	0.595×10^{-01}	0.284×10^{-01}

TABLE 4.14

Total and Partial (Photoabsorption, Coherent, and Incoherent) Mass Attenuation Coefficients, as Well as Mass Energy Absorption Coefficients, for the Principal Energies of ^{103}Pd, ^{125}I, ^{169}Yb, ^{170}Tm, ^{198}Au, ^{192}Ir, ^{137}Cs, and ^{60}Co, for Polystyrene

	E (keV) Intensity (%)	μ^{ph}/ρ (cm²/g)	μ^{coh}/ρ (cm²/g)	μ^{incoh}/ρ (cm²/g)	μ/ρ (cm²/g)	μ_{en}/ρ (cm²/g)
^{103}Pd	20.07 (22.06)	$0.198 \times 10^{+00}$	0.599×10^{-01}	$0.175 \times 10^{+00}$	$0.434 \times 10^{+00}$	$0.205 \times 10^{+00}$
	20.22 (41.93)	$0.194 \times 10^{+00}$	0.593×10^{-01}	$0.176 \times 10^{+00}$	$0.429 \times 10^{+00}$	$0.201 \times 10^{+00}$
	22.70 (13.05)	$0.132 \times 10^{+00}$	0.495×10^{-01}	$0.178 \times 10^{+00}$	$0.359 \times 10^{+00}$	$0.140 \times 10^{+00}$
^{125}I	27.20 (39.90)	0.728×10^{-01}	0.369×10^{-01}	$0.180 \times 10^{+00}$	$0.289 \times 10^{+00}$	0.817×10^{-01}
	27.47 (74.50)	0.705×10^{-01}	0.363×10^{-01}	$0.180 \times 10^{+00}$	$0.286 \times 10^{+00}$	0.795×10^{-01}
	31.00 (25.90)	0.472×10^{-01}	0.296×10^{-01}	$0.180 \times 10^{+00}$	$0.257 \times 10^{+00}$	0.574×10^{-01}
	35.50 (6.68)	0.301×10^{-01}	0.234×10^{-01}	$0.180 \times 10^{+00}$	$0.234 \times 10^{+00}$	0.414×10^{-01}
^{170}Tm	51.35 (0.97)	0.880×10^{-02}	0.121×10^{-01}	$0.176 \times 10^{+00}$	$0.197 \times 10^{+00}$	0.239×10^{-01}
	52.39 (1.69)	0.823×10^{-02}	0.117×10^{-01}	$0.175 \times 10^{+00}$	$0.195 \times 10^{+00}$	0.236×10^{-01}
	59.38 (0.36)	0.541×10^{-02}	0.928×10^{-02}	$0.173 \times 10^{+00}$	$0.188 \times 10^{+00}$	0.220×10^{-01}
	84.25 (2.48)	0.168×10^{-02}	0.480×10^{-02}	$0.164 \times 10^{+00}$	$0.170 \times 10^{+00}$	0.219×10^{-01}
^{169}Yb	49.77 (53)	0.977×10^{-02}	0.128×10^{-01}	$0.176 \times 10^{+00}$	$0.199 \times 10^{+00}$	0.243×10^{-01}
	50.74 (93.8)	0.916×10^{-02}	0.124×10^{-01}	$0.176 \times 10^{+00}$	$0.198 \times 10^{+00}$	0.239×10^{-01}
	57.50 (38.5)	0.603×10^{-02}	0.985×10^{-02}	$0.174 \times 10^{+00}$	$0.190 \times 10^{+00}$	0.221×10^{-01}
	63.12 (44.2)	0.442×10^{-02}	0.829×10^{-02}	$0.172 \times 10^{+00}$	$0.184 \times 10^{+00}$	0.214×10^{-01}
	109.78 (17.5)	0.698×10^{-03}	0.288×10^{-02}	$0.155 \times 10^{+00}$	$0.158 \times 10^{+00}$	0.237×10^{-01}
	177.98 (35.8)	0.145×10^{-03}	0.113×10^{-02}	$0.136 \times 10^{+00}$	$0.138 \times 10^{+00}$	0.277×10^{-01}
^{198}Au	307.74 (10.05)	0.255×10^{-04}	0.376×10^{-03}	$0.113 \times 10^{+00}$	$0.114 \times 10^{+00}$	0.310×10^{-01}
	411.80 (95.58)	0.108×10^{-04}	0.210×10^{-03}	$0.101 \times 10^{+00}$	$0.102 \times 10^{+00}$	0.318×10^{-01}
^{192}Ir	295.96 (28.72)	0.287×10^{-04}	0.407×10^{-03}	$0.115 \times 10^{+00}$	$0.115 \times 10^{+00}$	0.308×10^{-01}
	308.46 (29.68)	0.253×10^{-04}	0.374×10^{-03}	$0.113 \times 10^{+00}$	$0.114 \times 10^{+00}$	0.310×10^{-01}
	316.51 (82.71)	0.234×10^{-04}	0.356×10^{-03}	$0.112 \times 10^{+00}$	$0.113 \times 10^{+00}$	0.311×10^{-01}
	468.07 (47.81)	0.754×10^{-05}	0.163×10^{-03}	0.962×10^{-01}	0.964×10^{-01}	0.319×10^{-01}
	588.58 (4.52)	0.413×10^{-05}	0.103×10^{-03}	0.874×10^{-01}	0.875×10^{-01}	0.319×10^{-01}
	604.41 (8.20)	0.385×10^{-05}	0.978×10^{-04}	0.863×10^{-01}	0.865×10^{-01}	0.318×10^{-01}
	612.46 (5.34)	0.372×10^{-05}	0.953×10^{-04}	0.859×10^{-01}	0.860×10^{-01}	0.318×10^{-01}
^{137}Cs	662 (85.1)	0.304×10^{-05}	0.816×10^{-04}	0.830×10^{-01}	0.831×10^{-01}	0.316×10^{-01}
^{60}Co	1173 (100)	0.838×10^{-06}	0.260×10^{-04}	0.632×10^{-01}	0.633×10^{-01}	0.291×10^{-01}
	1332 (100)	0.695×10^{-06}	0.201×10^{-04}	0.592×10^{-01}	0.593×10^{-01}	0.283×10^{-01}

TABLE 4.15

Total and Partial (Photoabsorption, Coherent, and Incoherent) Mass Attenuation Coefficients, as Well as Mass Energy Absorption Coefficients, for the Principal Energies of ^{103}Pd, ^{125}I, ^{169}Yb, ^{170}Tm, ^{198}Au, ^{192}Ir, ^{137}Cs, and ^{60}Co, for Stainless Steel

	E (keV) Intensity (%)	μ^{ph}/ρ (cm²/g)	μ^{coh}/ρ (cm²/g)	μ^{incoh}/ρ (cm²/g)	μ/ρ (cm²/g)	μ_{en}/ρ (cm²/g)
^{103}Pd	20.07 (22.06)	$0.242 \times 10^{+02}$	$0.507 \times 10^{+00}$	$0.117 \times 10^{+00}$	$0.249 \times 10^{+02}$	$0.218 \times 10^{+02}$
	20.22 (41.93)	$0.236 \times 10^{+02}$	$0.502 \times 10^{+00}$	$0.117 \times 10^{+00}$	$0.243 \times 10^{+02}$	$0.214 \times 10^{+02}$
	22.70 (13.05)	$0.169 \times 10^{+02}$	$0.423 \times 10^{+00}$	$0.121 \times 10^{+00}$	$0.175 \times 10^{+02}$	$0.155 \times 10^{+02}$
^{125}I	27.20 (39.90)	$0.100 \times 10^{+02}$	$0.325 \times 10^{+00}$	$0.126 \times 10^{+00}$	$0.105 \times 10^{+02}$	$0.930 \times 10^{+01}$
	27.47 (74.50)	$0.975 \times 10^{+01}$	$0.320 \times 10^{+00}$	$0.126 \times 10^{+00}$	$0.102 \times 10^{+02}$	$0.905 \times 10^{+01}$
	31.00 (25.90)	$0.686 \times 10^{+01}$	$0.267 \times 10^{+00}$	$0.130 \times 10^{+00}$	$0.727 \times 10^{+01}$	$0.643 \times 10^{+01}$
	35.50 (6.68)	$0.460 \times 10^{+01}$	$0.214 \times 10^{+00}$	$0.132 \times 10^{+00}$	$0.496 \times 10^{+01}$	$0.434 \times 10^{+01}$
^{170}Tm	51.35 (0.97)	$0.153 \times 10^{+01}$	$0.117 \times 10^{+00}$	$0.136 \times 10^{+00}$	$0.178 \times 10^{+01}$	$0.148 \times 10^{+01}$
	52.39 (1.69)	$0.144 \times 10^{+01}$	$0.113 \times 10^{+00}$	$0.136 \times 10^{+00}$	$0.169 \times 10^{+01}$	$0.139 \times 10^{+01}$
	59.38 (0.36)	$0.928 \times 10^{+00}$	0.887×10^{-01}	$0.137 \times 10^{+00}$	$0.115 \times 10^{+01}$	$0.961 \times 10^{+00}$
	84.25 (2.48)	$0.337 \times 10^{+00}$	0.504×10^{-01}	$0.133 \times 10^{+00}$	$0.520 \times 10^{+00}$	$0.345 \times 10^{+00}$
^{169}Yb	49.77 (53)	$0.167 \times 10^{+01}$	$0.124 \times 10^{+00}$	$0.136 \times 10^{+00}$	$0.193 \times 10^{+01}$	$0.162 \times 10^{+01}$
	50.74 (93.8)	$0.158 \times 10^{+01}$	$0.120 \times 10^{+00}$	$0.136 \times 10^{+00}$	$0.184 \times 10^{+01}$	$0.153 \times 10^{+01}$
	57.50 (38.5)	$0.108 \times 10^{+01}$	0.971×10^{-01}	$0.136 \times 10^{+00}$	$0.132 \times 10^{+01}$	$0.106 \times 10^{+01}$
	63.12 (44.2)	$0.815 \times 10^{+00}$	0.829×10^{-01}	$0.135 \times 10^{+00}$	$0.104 \times 10^{+01}$	$0.802 \times 10^{+00}$
	109.78 (17.5)	$0.149 \times 10^{+00}$	0.312×10^{-01}	$0.128 \times 10^{+00}$	$0.317 \times 10^{+00}$	$0.169 \times 10^{+00}$
	177.98 (35.8)	0.338×10^{-01}	0.127×10^{-01}	$0.115 \times 10^{+00}$	$0.164 \times 10^{+00}$	0.582×10^{-01}
	307.74 (10.05)	0.658×10^{-02}	0.442×10^{-02}	0.971×10^{-01}	$0.108 \times 10^{+00}$	0.332×10^{-01}
^{198}Au	411.80 (95.58)	0.292×10^{-02}	0.250×10^{-02}	0.869×10^{-01}	0.929×10^{-01}	0.306×10^{-01}
^{192}Ir	295.96 (28.72)	0.736×10^{-02}	0.477×10^{-02}	0.984×10^{-01}	$0.111 \times 10^{+00}$	0.338×10^{-01}
	308.46 (29.68)	0.653×10^{-02}	0.440×10^{-02}	0.970×10^{-01}	$0.108 \times 10^{+00}$	0.331×10^{-01}
	316.51 (82.71)	0.607×10^{-02}	0.419×10^{-02}	0.961×10^{-01}	$0.107 \times 10^{+00}$	0.328×10^{-01}
	468.07 (47.81)	0.204×10^{-02}	0.195×10^{-02}	0.828×10^{-01}	0.870×10^{-01}	0.295×10^{-01}
	588.58 (4.52)	0.113×10^{-02}	0.124×10^{-02}	0.754×10^{-01}	0.778×10^{-01}	0.284×10^{-01}
	604.41 (8.20)	0.106×10^{-02}	0.117×10^{-02}	0.746×10^{-01}	0.768×10^{-01}	0.283×10^{-01}
	612.46 (5.34)	0.103×10^{-02}	0.114×10^{-02}	0.741×10^{-01}	0.763×10^{-01}	0.283×10^{-01}
^{137}Cs	662 (85.1)	0.858×10^{-03}	0.979×10^{-03}	0.716×10^{-01}	0.735×10^{-01}	0.279×10^{-01}
^{60}Co	1173 (100)	0.248×10^{-03}	0.313×10^{-03}	0.547×10^{-01}	0.553×10^{-01}	0.251×10^{-01}
	1332 (100)	0.195×10^{-03}	0.243×10^{-03}	0.513×10^{-01}	0.519×10^{-01}	0.244×10^{-01}

TABLE 4.16

Total and Partial (Photoabsorption, Coherent, and Incoherent) Mass Attenuation Coefficients, as Well as Mass Energy Absorption Coefficients, for the Principal Energies of ^{103}Pd, ^{125}I, ^{170}Tm, ^{169}Yb, ^{198}Au, ^{192}Ir, ^{137}Cs, and ^{60}Co, for Tungsten Alloy

	E (keV) Intensity (%)	μ^{ph}/ρ (cm²/g)	μ^{coh}/ρ (cm²/g)	μ^{incoh}/ρ (cm²/g)	μ/ρ (cm²/g)	μ_{en}/ρ (cm²/g)
^{103}Pd	20.07 (22.06)	$0.598 \times 10^{+02}$	$0.189 \times 10^{+01}$	0.767×10^{-01}	$0.617 \times 10^{+02}$	$0.535 \times 10^{+02}$
	20.22 (41.93)	$0.586 \times 10^{+02}$	$0.187 \times 10^{+01}$	0.769×10^{-01}	$0.605 \times 10^{+02}$	$0.525 \times 10^{+02}$
	22.70 (13.05)	$0.430 \times 10^{+02}$	$0.161 \times 10^{+01}$	0.807×10^{-01}	$0.447 \times 10^{+02}$	$0.389 \times 10^{+02}$
^{125}I	27.20 (39.90)	$0.264 \times 10^{+02}$	$0.127 \times 10^{+01}$	0.869×10^{-01}	$0.278 \times 10^{+02}$	$0.243 \times 10^{+02}$
	27.47 (74.50)	$0.257 \times 10^{+02}$	$0.126 \times 10^{+01}$	0.873×10^{-01}	$0.271 \times 10^{+02}$	$0.237 \times 10^{+02}$
	31.00 (25.90)	$0.186 \times 10^{+02}$	$0.107 \times 10^{+01}$	0.913×10^{-01}	$0.197 \times 10^{+02}$	$0.172 \times 10^{+02}$
	35.50 (6.68)	$0.128 \times 10^{+02}$	$0.877 \times 10^{+00}$	0.949×10^{-01}	$0.138 \times 10^{+02}$	$0.120 \times 10^{+02}$
^{170}Tm	51.35 (0.97)	$0.462 \times 10^{+01}$	$0.500 \times 10^{+00}$	$0.103 \times 10^{+00}$	$0.523 \times 10^{+01}$	$0.442 \times 10^{+01}$
	52.39 (1.69)	$0.437 \times 10^{+01}$	$0.484 \times 10^{+00}$	$0.103 \times 10^{+00}$	$0.496 \times 10^{+01}$	$0.419 \times 10^{+01}$
	59.38 (0.36)	$0.308 \times 10^{+01}$	$0.397 \times 10^{+00}$	$0.104 \times 10^{+00}$	$0.359 \times 10^{+01}$	$0.294 \times 10^{+01}$
	84.25 (2.48)	$0.592 \times 10^{+01}$	$0.225 \times 10^{+00}$	$0.106 \times 10^{+00}$	$0.625 \times 10^{+01}$	$0.198 \times 10^{+01}$
^{169}Yb	49.77 (53)	$0.504 \times 10^{+01}$	$0.525 \times 10^{+00}$	$0.102 \times 10^{+00}$	$0.567 \times 10^{+01}$	$0.481 \times 10^{+01}$
	50.74 (93.8)	$0.478 \times 10^{+01}$	$0.509 \times 10^{+00}$	$0.102 \times 10^{+00}$	$0.539 \times 10^{+01}$	$0.457 \times 10^{+01}$
	57.50 (38.5)	$0.338 \times 10^{+01}$	$0.417 \times 10^{+00}$	$0.104 \times 10^{+00}$	$0.390 \times 10^{+01}$	$0.324 \times 10^{+01}$
	63.12 (44.2)	$0.260 \times 10^{+01}$	$0.360 \times 10^{+00}$	$0.104 \times 10^{+00}$	$0.307 \times 10^{+01}$	$0.251 \times 10^{+01}$
	109.78 (17.5)	$0.294 \times 10^{+01}$	$0.143 \times 10^{+00}$	$0.104 \times 10^{+00}$	$0.320 \times 10^{+01}$	$0.160 \times 10^{+01}$
	177.98 (35.8)	$0.799 \times 10^{+00}$	0.604×10^{-01}	0.962×10^{-01}	$0.960 \times 10^{+00}$	$0.584 \times 10^{+00}$
	307.74 (10.05)	$0.184 \times 10^{+00}$	0.221×10^{-01}	0.831×10^{-01}	$0.291 \times 10^{+00}$	$0.173 \times 10^{+00}$
^{198}Au	411.80 (95.58)	0.881×10^{-01}	0.127×10^{-01}	0.750×10^{-1}	$0.180 \times 10^{+00}$	0.999×10^{-01}
^{192}Ir	295.96 (28.72)	$0.204 \times 10^{+00}$	0.238×10^{-01}	0.841×10^{-01}	$0.313 \times 10^{+00}$	$0.188 \times 10^{+00}$
	308.46 (29.68)	$0.183 \times 10^{+00}$	0.220×10^{-01}	0.830×10^{-01}	$0.289 \times 10^{+00}$	$0.172 \times 10^{+00}$
	316.51 (82.71)	$0.172 \times 10^{+00}$	0.210×10^{-01}	0.823×10^{-01}	$0.277 \times 10^{+00}$	$0.164 \times 10^{+00}$
	468.07 (47.81)	0.637×10^{-01}	0.100×10^{-01}	0.717×10^{-01}	$0.146 \times 10^{+00}$	0.785×10^{-01}
	588.58 (4.52)	0.368×10^{-01}	0.643×10^{-02}	0.655×10^{-01}	$0.109 \times 10^{+00}$	0.556×10^{-01}
	604.41 (8.20)	0.346×10^{-01}	0.611×10^{-02}	0.648×10^{-01}	$0.105 \times 10^{+00}$	0.537×10^{-01}
	612.46 (5.34)	0.336×10^{-01}	0.596×10^{-02}	0.644×10^{-01}	$0.104 \times 10^{+00}$	0.529×10^{-01}
^{137}Cs	662 (85.1)	0.283×10^{-01}	0.512×10^{-02}	0.623×10^{-01}	0.961×10^{-01}	0.484×10^{-01}
^{60}Co	1173 (100)	0.852×10^{-02}	0.167×10^{-02}	0.478×10^{-01}	0.582×10^{-01}	0.287×10^{-01}
	1332 (100)	0.668×10^{-02}	0.130×10^{-02}	0.448×10^{-01}	0.536×10^{-01}	0.264×10^{-01}

TABLE 4.17

Total and Partial (Photoabsorption, Coherent, and Incoherent) Mass Attenuation Coefficients, as Well as Mass Energy Absorption Coefficients, for the Principal Energies of ^{103}Pd, ^{125}I, ^{170}Tm, ^{169}Yb, ^{198}Au, ^{192}Ir, ^{137}Cs, and ^{60}Co, for Lead

Nuclide	E (keV) Intensity (%)	μ^{ph}/ρ (cm²/g)	μ^{coh}/ρ (cm²/g)	μ^{incoh}/ρ (cm²/g)	μ/ρ (cm²/g)	μ_{en}/ρ (cm²/g)
^{103}Pd	20.07 (22.06)	$0.832 \times 10^{+02}$	$0.233 \times 10^{+01}$	0.691×10^{-01}	$0.856 \times 10^{+02}$	$0.684 \times 10^{+02}$
	20.22 (41.93)	$0.816 \times 10^{+02}$	$0.231 \times 10^{+01}$	0.693×10^{-01}	$0.840 \times 10^{+02}$	$0.672 \times 10^{+02}$
	22.70 (13.05)	$0.602 \times 10^{+02}$	$0.198 \times 10^{+01}$	0.729×10^{-01}	$0.623 \times 10^{+02}$	$0.505 \times 10^{+02}$
^{125}I	27.20 (39.90)	$0.374 \times 10^{+02}$	$0.157 \times 10^{+01}$	0.789×10^{-01}	$0.390 \times 10^{+02}$	$0.323 \times 10^{+02}$
	27.47 (74.50)	$0.364 \times 10^{+02}$	$0.155 \times 10^{+01}$	0.792×10^{-01}	$0.380 \times 10^{+02}$	$0.316 \times 10^{+02}$
	31.00 (25.90)	$0.265 \times 10^{+02}$	$0.132 \times 10^{+01}$	0.832×10^{-01}	$0.278 \times 10^{+02}$	$0.233 \times 10^{+02}$
	35.50 (6.68)	$0.184 \times 10^{+02}$	$0.109 \times 10^{+01}$	0.868×10^{-01}	$0.196 \times 10^{+02}$	$0.165 \times 10^{+02}$
^{170}Tm	51.35 (0.97)	$0.678 \times 10^{+01}$	$0.628 \times 10^{+00}$	0.952×10^{-01}	$0.750 \times 10^{+01}$	$0.628 \times 10^{+01}$
	52.39 (1.69)	$0.642 \times 10^{+01}$	$0.608 \times 10^{+00}$	0.955×10^{-01}	$0.712 \times 10^{+01}$	$0.595 \times 10^{+01}$
	59.38 (0.36)	$0.456 \times 10^{+01}$	$0.498 \times 10^{+00}$	0.972×10^{-01}	$0.515 \times 10^{+01}$	$0.426 \times 10^{+01}$
	84.25 (2.48)	$0.174 \times 10^{+01}$	$0.283 \times 10^{+00}$	0.993×10^{-01}	$0.213 \times 10^{+01}$	$0.245 \times 10^{+01}$
^{169}Yb	49.77 (53)	$0.738 \times 10^{+01}$	$0.660 \times 10^{+00}$	0.947×10^{-01}	$0.814 \times 10^{+01}$	$0.682 \times 10^{+01}$
	50.74 (93.8)	$0.700 \times 10^{+01}$	$0.640 \times 10^{+00}$	0.950×10^{-01}	$0.774 \times 10^{+01}$	$0.648 \times 10^{+01}$
	57.50 (38.5)	$0.498 \times 10^{+01}$	$0.524 \times 10^{+00}$	0.967×10^{-01}	$0.560 \times 10^{+01}$	$0.465 \times 10^{+01}$
	63.12 (44.2)	$0.385 \times 10^{+01}$	$0.452 \times 10^{+00}$	0.976×10^{-01}	$0.441 \times 10^{+01}$	$0.362 \times 10^{+01}$
	109.78 (17.5)	$0.410 \times 10^{+01}$	$0.181 \times 10^{+00}$	0.979×10^{-01}	$0.439 \times 10^{+01}$	$0.171 \times 10^{+01}$
	177.98 (35.8)	$0.115 \times 10^{+01}$	0.772×10^{-01}	0.917×10^{-01}	$0.133 \times 10^{+01}$	$0.746 \times 10^{+00}$
	307.74 (10.05)	$0.275 \times 10^{+00}$	0.285×10^{-01}	0.797×10^{-01}	$0.384 \times 10^{+00}$	$0.234 \times 10^{+00}$
^{198}Au	411.80 (95.58)	$0.133 \times 10^{+00}$	0.164×10^{-01}	0.721×10^{-01}	$0.219 \times 10^{+00}$	$0.132 \times 10^{+00}$
^{192}Ir	295.96 (28.72)	$0.304 \times 10^{+00}$	0.306×10^{-01}	0.807×10^{-01}	$0.415 \times 10^{+00}$	$0.253 \times 10^{+00}$
	308.46 (29.68)	$0.273 \times 10^{+00}$	0.284×10^{-01}	0.797×10^{-01}	$0.382 \times 10^{+00}$	$0.232 \times 10^{+00}$
	316.51 (82.71)	$0.256 \times 10^{+00}$	0.271×10^{-01}	0.790×10^{-01}	$0.364 \times 10^{+00}$	$0.221 \times 10^{+00}$
	468.07 (47.81)	0.970×10^{-01}	0.129×10^{-01}	0.690×10^{-01}	$0.179 \times 10^{+00}$	$0.103 \times 10^{+00}$
	588.58 (4.52)	0.566×10^{-01}	0.836×10^{-02}	0.631×10^{-01}	$0.128 \times 10^{+00}$	0.703×10^{-01}
	604.41 (8.20)	0.532×10^{-01}	0.795×10^{-02}	0.624×10^{-01}	$0.124 \times 10^{+00}$	0.675×10^{-01}
	612.46 (5.34)	0.517×10^{-01}	0.775×10^{-02}	0.621×10^{-01}	$0.122 \times 10^{+00}$	0.663×10^{-01}
^{137}Cs	662 (85.1)	0.436×10^{-01}	0.666×10^{-02}	0.600×10^{-01}	$0.111 \times 10^{+00}$	0.598×10^{-01}
^{60}Co	1173 (100)	0.132×10^{-01}	0.218×10^{-02}	0.462×10^{-01}	0.620×10^{-01}	0.316×10^{-01}
	1332 (100)	0.104×10^{-01}	0.170×10^{-02}	0.433×10^{-01}	0.564×10^{-01}	0.286×10^{-01}

TABLE 4.18

Total and Partial (Photoabsorption, Coherent, and Incoherent) Mass Attenuation Coefficients, as Well as Mass Energy Absorption Coefficients, for the Principal Energies of ^{103}Pd, ^{125}I, ^{170}Tm, ^{169}Yb, ^{198}Au, ^{192}Ir, ^{137}Cs, and ^{60}Co, for Glass Lead

	E (keV) Intensity (%)	μ^{ph}/ρ (cm²/g)	μ^{coh}/ρ (cm²/g)	μ^{incoh}/ρ (cm²/g)	μ/ρ (cm²/g)	μ_{en}/ρ (cm²/g)
^{103}Pd	20.07 (22.06)	$0.632 \times 10^{+02}$	$0.179 \times 10^{+01}$	0.887×10^{-01}	$0.651 \times 10^{+02}$	$0.515 \times 10^{+02}$
	20.22 (41.93)	$0.620 \times 10^{+02}$	$0.177 \times 10^{+01}$	0.889×10^{-01}	$0.639 \times 10^{+02}$	$0.497 \times 10^{+02}$
	22.70 (13.05)	$0.457 \times 10^{+02}$	$0.153 \times 10^{+01}$	0.925×10^{-01}	$0.473 \times 10^{+02}$	$0.320 \times 10^{+02}$
^{125}I	27.20 (39.90)	$0.284 \times 10^{+02}$	$0.121 \times 10^{+01}$	0.979×10^{-01}	$0.297 \times 10^{+02}$	$0.226 \times 10^{+02}$
	27.47 (74.50)	$0.276 \times 10^{+02}$	$0.119 \times 10^{+01}$	0.982×10^{-01}	$0.289 \times 10^{+02}$	$0.223 \times 10^{+02}$
	31.00 (25.90)	$0.201 \times 10^{+02}$	$0.101 \times 10^{+01}$	$0.102 \times 10^{+00}$	$0.212 \times 10^{+02}$	$0.181 \times 10^{+02}$
	35.50 (6.68)	$0.140 \times 10^{+02}$	$0.838 \times 10^{+00}$	$0.105 \times 10^{+00}$	$0.149 \times 10^{+02}$	$0.129 \times 10^{+02}$
^{170}Tm	51.35 (0.97)	$0.513 \times 10^{+01}$	$0.480 \times 10^{+00}$	$0.111 \times 10^{+00}$	$0.573 \times 10^{+01}$	$0.479 \times 10^{+01}$
	52.39 (1.69)	$0.486 \times 10^{+01}$	$0.466 \times 10^{+00}$	$0.111 \times 10^{+00}$	$0.544 \times 10^{+01}$	$0.456 \times 10^{+01}$
	59.38 (0.36)	$0.345 \times 10^{+01}$	$0.381 \times 10^{+00}$	$0.112 \times 10^{+00}$	$0.394 \times 10^{+01}$	$0.319 \times 10^{+01}$
	84.25 (2.48)	$0.132 \times 10^{+01}$	$0.216 \times 10^{+00}$	$0.112 \times 10^{+00}$	$0.165 \times 10^{+01}$	$0.183 \times 10^{+01}$
^{169}Yb	49.77 (53)	$0.559 \times 10^{+01}$	$0.505 \times 10^{+00}$	$0.110 \times 10^{+00}$	$0.621 \times 10^{+01}$	$0.518 \times 10^{+01}$
	50.74 (93.8)	$0.531 \times 10^{+01}$	$0.490 \times 10^{+00}$	$0.111 \times 10^{+00}$	$0.591 \times 10^{+01}$	$0.494 \times 10^{+01}$
	57.50 (38.5)	$0.377 \times 10^{+01}$	$0.401 \times 10^{+00}$	$0.112 \times 10^{+00}$	$0.428 \times 10^{+01}$	$0.360 \times 10^{+01}$
	63.12 (44.2)	$0.292 \times 10^{+01}$	$0.346 \times 10^{+00}$	$0.112 \times 10^{+00}$	$0.338 \times 10^{+01}$	$0.263 \times 10^{+01}$
	109.78 (17.5)	$0.309 \times 10^{+01}$	$0.139 \times 10^{+00}$	$0.109 \times 10^{+00}$	$0.334 \times 10^{+01}$	$0.131 \times 10^{+01}$
	177.98 (35.8)	$0.878 \times 10^{+00}$	0.594×10^{-01}	$0.100 \times 10^{+00}$	$0.104 \times 10^{+00}$	$0.582 \times 10^{+00}$
^{198}Au	307.74 (10.05)	$0.207 \times 10^{+00}$	0.217×10^{-01}	0.859×10^{-01}	$0.314 \times 10^{+00}$	$0.185 \times 10^{+00}$
	411.80 (95.58)	0.993×10^{-01}	0.126×10^{-01}	0.776×10^{-01}	$0.190 \times 10^{+00}$	$0.104 \times 10^{+00}$
^{192}Ir	295.96 (28.72)	$0.228 \times 10^{+00}$	0.234×10^{-01}	0.870×10^{-01}	$0.339 \times 10^{+00}$	$0.198 \times 10^{+00}$
	308.46 (29.68)	$0.205 \times 10^{+00}$	0.216×10^{-01}	0.858×10^{-01}	$0.313 \times 10^{+00}$	$0.184 \times 10^{+00}$
	316.51 (82.71)	$0.192 \times 10^{+00}$	0.206×10^{-01}	0.851×10^{-01}	$0.298 \times 10^{+00}$	$0.175 \times 10^{+00}$
	468.07 (47.81)	0.727×10^{-01}	0.988×10^{-02}	0.740×10^{-01}	$0.157 \times 10^{+00}$	0.800×10^{-01}
	588.58 (4.52)	0.425×10^{-01}	0.637×10^{-02}	0.675×10^{-01}	$0.116 \times 10^{+00}$	0.666×10^{-01}
	604.41 (8.20)	0.400×10^{-01}	0.605×10^{-02}	0.668×10^{-01}	$0.113 \times 10^{+00}$	0.560×10^{-01}
	612.46 (5.34)	0.388×10^{-01}	0.590×10^{-02}	0.664×10^{-01}	$0.111 \times 10^{+00}$	0.501×10^{-01}
^{137}Cs	662 (85.1)	0.326×10^{-01}	0.508×10^{-02}	0.643×10^{-01}	$0.102 \times 10^{+00}$	0.226×10^{-01}
^{60}Co	1173 (100)	0.993×10^{-02}	0.166×10^{-02}	0.493×10^{-01}	0.610×10^{-01}	0.230×10^{-01}
	1332 (100)	0.779×10^{-02}	0.130×10^{-02}	0.462×10^{-01}	0.558×10^{-01}	0.140×10^{-01}

TABLE 4.19

Total and Partial (Photoabsorption, Coherent, and Incoherent) Mass Attenuation Coefficients, as Well as Mass Energy Absorption Coefficients, for the Principal Energies of ^{103}Pd, ^{125}I, ^{170}Tm, ^{169}Yb, ^{198}Au, ^{192}Ir, ^{137}Cs, and ^{60}Co, for Ordinary Concrete

	E(keV) Intensity (%)	μ^{ph}/ρ (cm²/g)	μ^{coh}/ρ (cm²/g)	μ^{incoh}/ρ (cm²/g)	μ/ρ (cm²/g)	μ_{en}/ρ (cm²/g)
^{103}Pd	20.07 (22.06)	$0.247 \times 10^{+01}$	$0.158 \times 10^{+00}$	$0.153 \times 10^{+00}$	$0.278 \times 10^{+01}$	$0.244 \times 10^{+01}$
	20.22 (41.93)	$0.241 \times 10^{+01}$	$0.156 \times 10^{+00}$	$0.153 \times 10^{+00}$	$0.272 \times 10^{+01}$	$0.238 \times 10^{+01}$
	22.70 (13.05)	$0.169 \times 10^{+01}$	$0.131 \times 10^{+00}$	$0.156 \times 10^{+00}$	$0.198 \times 10^{+01}$	$0.168 \times 10^{+01}$
^{125}I	27.20 (39.90)	$0.969 \times 10^{+00}$	0.990×10^{-01}	$0.159 \times 10^{+00}$	$0.123 \times 10^{+01}$	$0.965 \times 10^{+00}$
	27.47 (74.50)	$0.940 \times 10^{+00}$	0.975×10^{-01}	$0.159 \times 10^{+00}$	$0.120 \times 10^{+01}$	$0.936 \times 10^{+00}$
	31.00 (25.90)	$0.646 \times 10^{+00}$	0.804×10^{-01}	$0.161 \times 10^{+00}$	$0.887 \times 10^{+00}$	$0.648 \times 10^{+00}$
	35.50 (6.68)	$0.423 \times 10^{+00}$	0.644×10^{-01}	$0.162 \times 10^{+00}$	$0.649 \times 10^{+00}$	$0.429 \times 10^{+00}$
^{170}Tm	51.35 (0.97)	$0.141 \times 10^{+00}$	0.349×10^{-01}	$0.161 \times 10^{+00}$	$0.337 \times 10^{+00}$	$0.145 \times 10^{+00}$
	52.39 (1.69)	$0.133 \times 10^{+00}$	0.337×10^{-01}	$0.161 \times 10^{+00}$	$0.327 \times 10^{+00}$	$0.138 \times 10^{+00}$
	59.38 (0.36)	0.829×10^{-01}	0.265×10^{-01}	$0.159 \times 10^{+00}$	$0.269 \times 10^{+00}$	0.988×10^{-01}
	84.25 (2.48)	0.294×10^{-01}	0.143×10^{-01}	$0.152 \times 10^{+00}$	$0.196 \times 10^{+00}$	0.520×10^{-01}
^{169}Yb	49.77 (53)	$0.146 \times 10^{+00}$	0.363×10^{-01}	$0.161 \times 10^{+00}$	$0.344 \times 10^{+00}$	$0.158 \times 10^{+00}$
	50.74 (93.8)	$0.137 \times 10^{+00}$	0.351×10^{-01}	$0.161 \times 10^{+00}$	$0.334 \times 10^{+00}$	$0.150 \times 10^{+00}$
	57.50 (38.5)	0.923×10^{-01}	0.281×10^{-01}	$0.160 \times 10^{+00}$	$0.280 \times 10^{+00}$	$0.107 \times 10^{+00}$
	63.12 (44.2)	0.687×10^{-01}	0.238×10^{-01}	$0.159 \times 10^{+00}$	$0.251 \times 10^{+00}$	0.842×10^{-01}
	109.78 (17.5)	0.117×10^{-01}	0.856×10^{-02}	$0.145 \times 10^{+00}$	$0.165 \times 10^{+00}$	0.334×10^{-01}
	177.98 (35.8)	0.257×10^{-02}	0.342×10^{-02}	$0.128 \times 10^{+00}$	$0.134 \times 10^{+00}$	0.286×10^{-01}
	307.74 (10.05)	0.472×10^{-03}	0.116×10^{-02}	$0.107 \times 10^{+00}$	$0.109 \times 10^{+00}$	0.298×10^{-01}
^{198}Au	411.80 (95.58)	0.204×10^{-03}	0.649×10^{-03}	0.958×10^{-01}	0.967×10^{-01}	0.303×10^{-01}
^{192}Ir	295.96 (28.72)	0.530×10^{-03}	0.125×10^{-02}	$0.109 \times 10^{+00}$	$0.110 \times 10^{+00}$	0.297×10^{-01}
	308.46 (29.68)	0.469×10^{-03}	0.115×10^{-02}	$0.107 \times 10^{+00}$	$0.109 \times 10^{+00}$	0.298×10^{-01}
	316.51 (82.71)	0.435×10^{-03}	0.109×10^{-02}	$0.106 \times 10^{+00}$	$0.108 \times 10^{+00}$	0.298×10^{-01}
	468.07 (47.81)	0.143×10^{-03}	0.504×10^{-03}	0.910×10^{-01}	0.917×10^{-01}	0.303×10^{-01}
	588.58 (4.52)	0.790×10^{-04}	0.319×10^{-03}	0.827×10^{-01}	0.831×10^{-01}	0.302×10^{-01}
	604.41 (8.20)	0.739×10^{-04}	0.303×10^{-03}	0.817×10^{-01}	0.821×10^{-01}	0.301×10^{-01}
	612.46 (5.34)	0.715×10^{-04}	0.295×10^{-03}	0.813×10^{-01}	0.816×10^{-01}	0.301×10^{-01}
^{137}Cs	662 (85.1)	0.589×10^{-04}	0.253×10^{-03}	0.785×10^{-01}	0.789×10^{-01}	0.300×10^{-01}
^{60}Co	1173 (100)	0.169×10^{-04}	0.805×10^{-04}	0.599×10^{-01}	0.600×10^{-01}	0.276×10^{-01}
	1332 (100)	0.136×10^{-04}	0.624×10^{-04}	0.561×10^{-01}	0.562×10^{-01}	0.268×10^{-01}

4.2.2   Coherent (Rayleigh) Scattering

Coherent (Rayleigh) scattering is the elastic scattering from atoms in which only the incident photon is scattered and the atom is neither ionized nor excited. It is a two-step process, in that the photon is initially absorbed by a bound electron which is raised to a higher energy state and then the photon is reemitted with the electron returning to its original state. The recoil momentum is taken up by the entire atom, resulting in negligible recoil energy owing to its large mass. Hence, the energy of the scattered photon is essentially the same as that of the incident photon and no energy is deposited.

The elastic scattering of a photon from a single free electron at low energies is described by Thomson's classical formula:

$$\frac{d_e \sigma^{Th}}{d\Omega} = \frac{1}{2} r_e^2 (1 + \cos^2 \theta) \qquad (4.18)$$

$d_e \sigma^{Th}/d\Omega$ is the Thomson electronic differential cross-section which gives the angular distribution of scattered photons. r_e is the classical electron radius ($= 2.818 \ 10^{-15}$ m) and θ is the scattering angle of the photon, defined as the angle between the direction of flight before and that after the interaction. Integrating the differential cross-section over all solid angles gives the total electronic cross-section $_e\sigma^{Th}$ for Thomson scattering:

$$_e\sigma^{Th} = \int_\Omega \frac{d_e \sigma^{Th}}{d\Omega} d\Omega = \int_\Omega \frac{d}{d\Omega} \frac{1}{2} r_e^2 (1 + \cos^2 \theta) 2\pi \sin \theta \, d\theta = \frac{8\pi}{3} r_e^2 \qquad (4.19)$$

$$= 0.662 \ \text{barn}$$

As seen in Equation 4.18 and Equation 4.19, Thomson scattering is independent of the incident photon energy and is symmetrical about the plane at $\theta = 90°$.

Elastic scattering from a bound electron implies that no change occurs in the energy state of the electron and the atom as a whole remains in its ground state. The atomic form factor F is used to correct Thomson scattering for electron binding effects, by modifying the differential cross-section to be

$$\frac{d_a \sigma^{coh}}{d\Omega} = F^2 \frac{d_e \sigma^{Th}}{d\Omega} \qquad (4.20)$$

and the atomic cross-section as

$$_a\sigma^{coh} = \int_\Omega F^2 \frac{d_e \sigma^{Th}}{d\Omega} d\Omega \qquad (4.21)$$

where $_a\sigma^{coh}$ is the atomic cross-section for coherent scattering. For pure elements, where F^2 has values approximately equal to the square of atomic number, Z^2, for small scattering angles θ, but is several orders of magnitude lower for large scattering angles θ. Therefore, coherent scattering is highly anisotropic and forward peaked. Also, since there is no energy transfer involved, coherent scattering is sometimes neglected in photon transport calculations.

Molecular cross-sections for coherent scattering, $_m\sigma^{coh}$, can be calculated based on the Bragg mixture rule of Equation 4.11, which for water gives

$$_m\sigma_{H_2O}^{coh} = 2\,_a\sigma_H^{coh} +\,_a\sigma_O^{coh} \qquad (4.22)$$

where $_m\sigma_{H_2O}^{coh}$ is the molecular coherent cross-section for water and $_a\sigma_H^{coh}$, $_a\sigma_O^{coh}$ are the atomic cross-sections for coherent scattering for hydrogen and oxygen, respectively. Accordingly, the molecular form factor for water, F_{H_2O}, is given by

$$F_{H_2O}^2 = 2F_H^2 + F_O^2 \qquad (4.23)$$

where F_H and F_O are the atomic form factors of hydrogen and oxygen. Consequently, mass attenuation coefficients due to coherent scattering, μ^{coh}/ρ, for water can be calculated based on Equation 4.7.

Figure 4.5 shows the mass attenuation coefficient due to coherent scattering, μ^{coh}/ρ, for selected elements, compounds, and mixtures as a function of the photon energy E. A first impression is that coherent scattering is of some significance at low photon energies and high-Z materials. The observed energy- and Z-dependence, however, is not as strong as in the photoelectric effect. For small momentum transfer $_a\sigma^{coh} \sim Z^2$,

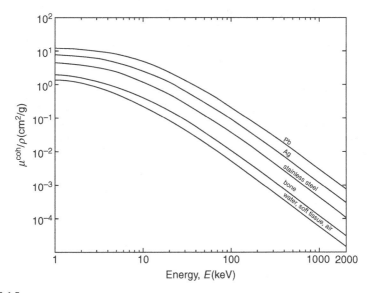

FIGURE 4.5

Mass attenuation coefficients due to coherent scattering, μ^{coh}/ρ, for selected elements, compounds, and mixtures, plotted as a function of incident photon energy, E.

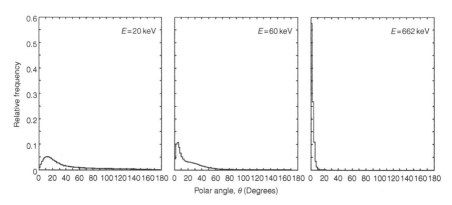

FIGURE 4.6
Angular distribution of coherent scattered photons in water medium for selected photon energies of 20, 60, and 662 keV. Each distribution is normalized to unit area.

according to Equation 4.7 results in

$$\mu^{coh}/\rho \propto Z \qquad (4.24)$$

since the Z/A ratio is approximately equal to $1/2$ for most elements. Consequently, in the low energy region, coherent scattering for high-Z materials is of reduced importance, since coherent attenuation coefficients are two to three orders of magnitude lower than the corresponding values for the photoelectric effect. Relative to photoabsorption, coherent scattering is of some significance only for low-Z materials. Specifically, for water as a medium, coherent scattering becomes greater than photoabsorption for energies above approximately 60 keV.

Figure 4.6 shows the angular distribution of coherent scattered photons for water and for selected photon energies of 20, 60, and 662 keV. Each distribution is normalized to unit area. It is observed that they are highly anisotropic and forward peaked especially at higher energies. Molecular bonds modify slightly these distributions for lower photon energies[14] but it is unlikely that these corrections have any detectable effect in the dosimetry of low energy emitters.[15]

The XCOM database[2] incorporates the atomic coherent scattering cross-section, $_a\sigma^{coh}$, values of Hubbell and Øverbø,[16] obtained from numerical integration of the Thomson[17] formula weighted by the F^2 factor. The coherent scattering attenuation data given in Table 4.2 through Table 4.19 were derived using the mixture rule of Equation 4.13 and the elemental composition of Table 4.1, by log–log interpolation at the principal photon energies on the values of Hubbell and Øverbø.[16]

4.2.3 Incoherent (Compton) Scattering

Incoherent (Compton) scattering, is the inelastic scattering of photons by atoms, in which the incident photon transfers a part of its energy to an atomic electron (which is then ejected from the atom), the photon being deflected with reduced energy to an angle θ relative to its original direction. Assuming that the electron is free and at rest, the relationship between the energy E_{sc} of the scattered photon and the scattering angle θ, is

$$E_{sc} = E \frac{1}{1 + (E/m_e c^2)(1 - \cos \theta)} \tag{4.25}$$

where E is the incident photon energy and $m_e c^2$ ($= 0.511$ MeV) is the rest energy of the electron. The kinetic energy T_e transferred to the Compton recoil electron is

$$T_e = E - E_{sc} = E \frac{(E/m_e c^2)(1 - \cos \theta)}{1 + (E/m_e c^2)(1 - \cos \theta)} \tag{4.26}$$

Equation 4.25 and Equation 4.26 show that the energies of both the scattered photon E_{sc} and the recoil electron T_e, depend not only on the scattering angle θ but also on the incident photon energy E. The maximum energy transfer from a photon to an electron occurs when the photon is back-scattered at an angle $\theta = 180°$, i.e., $\cos \theta = -1$. In this case, Equation 4.26 shows that the energy transferred to the electron is $T_e = 796.5$ keV for 1 MeV incident photon energy (80% of its energy) and $T_e = 11.4$ keV for a photon of 60 keV (19% of its energy). These calculations show that a high energy photon may suffer a large energy loss, in contrast to the situation with a low energy photon.

The probability that a photon of energy E is scattered by a single free electron at an angle θ is given by the quantum mechanically derived, relativistic Klein–Nishina (KN) formula:

$$\frac{d(_e\sigma^{KN})}{d\Omega} = \frac{r_e^2}{2} \left(\frac{E_{sc}}{E} \right)^2 \left(\frac{E}{E_{sc}} + \frac{E_{sc}}{E} - \sin^2 \theta \right) \tag{4.27}$$

where $_e\sigma^{KN}$, is the KN electronic cross-section and r_e is the classical electron radius. Substituting into Equation 4.27, the scatter photon energy E_{sc} in terms of scattering angle θ and incident photon energy E from Equation 4.25, an equivalent explicit expression for the differential KN cross-section is obtained:

$$\frac{d_e\sigma^{KN}}{d\Omega} = \frac{d_e\sigma^{Th}}{d\Omega}$$
$$\times \left(\frac{1}{1 + \alpha(1 - \cos \theta)} \right)^2 \left(1 + \frac{\alpha^2(1 - \cos \theta)^2}{(1 + \cos^2 \theta)(1 + \alpha(1 - \cos \theta))} \right) \tag{4.28}$$

where $_e\sigma^{Th}$ is the Thomson cross-section ($= 0.662$ b/e) given in Equation 4.19 and the dimensionless quantity $\alpha = E/m_e c^2$ gives the incident photon energy in terms of the rest energy of the electron. The angular distribution of scattered photons is shown in Figure 4.7 for selected values of the incident

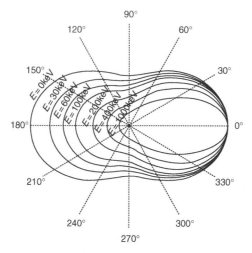

FIGURE 4.7
Angular distributions of scattered photons by a free electron for incident photon energies up to $E = 1$ MeV. The distribution for photon energy of $E = 0$ keV corresponds to the Thomson scattering.

photon energy up to ~ 1 MeV. For the low energy limit of $E = 0$, i.e., $\alpha = 0$, the angular distribution is that of Equation 4.18 for Thomson scattering. It can be observed that with the increase of photon energy E, the scattering tends forward.

Equation 4.25 and Equation 4.28 give the energy E_{sc} of the scattered photon as a function of the scattering angle θ and the probability that a photon is scattered at an angle θ, respectively. Therefore the average energy of the scattered photon $\langle E_{sc} \rangle$ is calculated by

$$\langle E_{sc} \rangle = \int_{\Omega} E_{sc} d_e \sigma^{KN} \tag{4.29}$$

Results from these calculations are given in Figure 4.8, where the fraction $\langle E_{sc} \rangle / E$ of the incident photon energy E which is scattered on average, and the fraction $\langle T_e \rangle / E$ of the incident photon energy E which is transferred to the electron on average, are given as a function of photon energy E. It is observed that at high energies a significant fraction of the incident photon energy is transferred on average to an electron, and consequently, the photon energy is noticeable reduced. At low photon energies only an insignificant energy fraction is transferred, with the photon maintaining its energy, and consequently, many such interactions are needed for a low energy photon to be finally absorbed (see also Chapter 9). Combining the data given in Figure 4.7 and Figure 4.8, an interesting conclusion can be drawn. High energy photons suffer on average a large energy change but only a small deflection. In contrast, low energy photons could even be deflected backwards, losing on average only a small fraction of their energies.

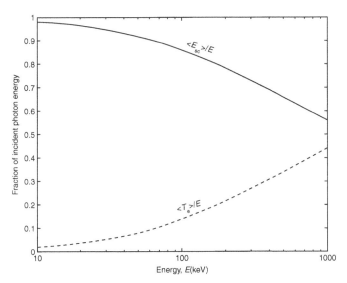

FIGURE 4.8
Mean scattered photon energy to incident photon energy ratio, $\langle E_{sc}\rangle/E$, and mean scattered electron kinetic energy to incident photon energy ratio, $\langle T_e\rangle/E$, plotted as a function of incident photon's energy, E.

The KN electronic cross-section $_e\sigma^{KN}$, describing the elastic scattering of a photon of energy E from a single free electron, is obtained by integration of Equation 4.26 over the total solid angle:

$$
\begin{aligned}
_e\sigma^{KN} &= \int_0^\pi \frac{d_e\sigma^{KN}}{d\Omega} 2\pi \sin\theta d\theta \\
&= \frac{3}{4}\,_e\sigma^{Th}\left\{\frac{1+\alpha}{\alpha^2}\left[\frac{2(1+\alpha)}{1+2\alpha}-\frac{\ln(1+2\alpha)}{\alpha}\right]+\frac{\ln(1+2\alpha)}{2\alpha}-\frac{1+3\alpha}{(1+2\alpha)^2}\right\}
\end{aligned}
$$

$$(4.30)$$

Assuming that all electrons in an atom or a molecule participate equivalently in the elastic scattering, atomic cross-sections are obtained by simply multiplying the KN electronic cross-section $_e\sigma^{KN}$ by the number of all electrons. That is,

$$_a\sigma^{KN} = Z_e\sigma^{KN} \tag{4.31}$$

and consequently,

$$\mu^{KN}/\rho = (Z/A)_e\sigma^{KN} \tag{4.32}$$

Correspondingly, molecular cross-sections are obtained by multiplying $_e\sigma^{KN}$ by the total number of electrons in the molecule. Figure 4.9 shows the energy dependence of the mass attenuation coefficient μ^{KN}/ρ based on Equation 4.32 for lead ($Z = 82$) and calculations for water (there are ten electrons in the water molecule and therefore $Z = 10$). It is observed that the mass attenuation coefficient for water is larger than that for the high-Z lead.

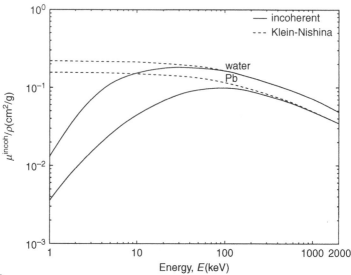

FIGURE 4.9
Mass attenuation coefficients due to incoherent scattering, μ^{incoh}/ρ, and corresponding ones based on the KN formula for water and lead, plotted as a function of incident photon energy, E.

This is due to Z/A ratio in Equation 4.30. This ratio is approximately $1/2$ for low-Z materials (for water ~ 0.55) but tends to lower values for high-Z materials (for lead ~ 0.40).

The KN formula is based on the assumption that the electron is free and at rest, which is a good approximation for incident photon energies much larger than the binding energies of electrons. At low incident photon energies E however, Equation 4.26 reveals that for small scattering angles θ the recoil energy T_e of the electron could be comparable to or smaller than the electron binding energy, especially for high-Z materials. Thus, the assumption of free electron may not apply. Deviations from the free electron KN scattering are greatest for inner-shell electrons but become negligible for loosely bound valence electrons. The incoherent scattering function S is used to correct the KN cross-section $_e\sigma^{\text{KN}}$, for electron binding effects and the atomic incoherent cross-section $_a\sigma^{\text{incoh}}$, is given by

$$_a\sigma^{\text{incoh}} = \int S d_e\sigma^{\text{KN}} \tag{4.33}$$

where S is the incoherent scattering function whose deviation from Z is a measure of the electron binding.

For homogeneous mixtures and chemical compounds the mixture rule of Equation 4.11 suggests that the incoherent scattering cross-sections of individual atoms combine independently. That is, for water:

$$_m\sigma^{\text{incoh}}_{\text{H}_2\text{O}} = 2_a\sigma^{\text{incoh}}_{\text{H}} + _a\sigma^{\text{incoh}}_{\text{O}} \tag{4.34}$$

where $_m\sigma_{H_2O}^{incoh}$, $_a\sigma_H^{incoh}$, and $_a\sigma_O^{incoh}$ are the incoherent cross-sections for water, hydrogen, and oxygen. The incoherent scattering function S is given by

$$S_{H_2O} = 2S_H + S_O \qquad (4.35)$$

where S_{H_2O}, S_H, and S_O are the incoherent scattering functions of water, hydrogen, and oxygen. Consequently, mass attenuation coefficients due to incoherent scattering, μ^{incoh}/ρ, for water can be calculated based on Equation 4.7.

Mass attenuation coefficient values due to incoherent scattering, μ^{incoh}/ρ, for water and lead are also shown in Figure 4.9 to allow for comparison with data based on free electron KN scattering. Noticeable deviations between the two data sets are observed for high-Z lead for photon energies lower than about 100 keV. For the low-Z water the free electron approximation is a safe assumption even for energies as low as 10 keV.

Figure 4.10 presents the mass attenuation coefficients due to incoherent scattering, μ^{incoh}/ρ, for selected elements, compounds, and mixtures, shown as a function of the photon energy E. A weak energy dependence, and especially, material dependence is observed. The material dependence is due to the Z/A ratio, which as explained before is lower for high-Z materials than for low-Z materials.

Figure 4.11a–Figure 4.11c shows differential cross-sections for water i.e., the possibility for incoherent scattering at specific θ, for selected incident

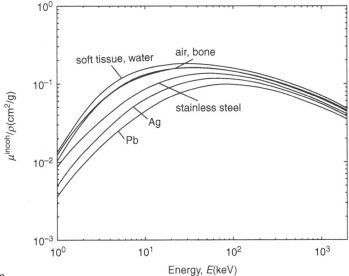

FIGURE 4.10

Mass attenuation coefficients due to incoherent scattering, μ^{incoh}/ρ, for selected elements, compounds, and mixtures, plotted as a function of incident photon energy, E.

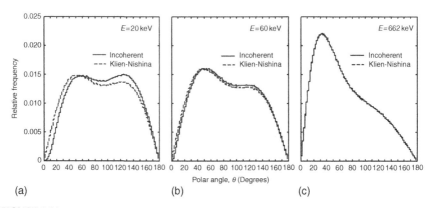

FIGURE 4.11
Angular distributions of incoherent scattered photons and corresponding ones based on the KN formula in water medium for selected incident photon energies of 20, 60, and 662 keV. Each distribution is normalized to unit area.

photon energies of 20, 60, and 662 keV, superimposed on calculations based on the KN formula. All angular distributions are normalized to unit area. It is seen that these distributions are forward peaked at high energies, but tend to be isotropic at the lower photon energies, where deviations from the free electron KN scattering are noticeable.

The XCOM database[2] incorporates the atomic incoherent scattering cross-section, $_a\sigma^{incoh}$, values of Hubbell et al.,[18] obtained from numerical integration of the KN[19] differential formula weighted by the incoherent scattering function S. The incoherent scattering attenuation data shown in Table 4.2 through Table 4.19 were derived based on the mixture rule of Equation 4.13 and the elemental composition of Table 4.1, using log–log interpolation at the principal photon energies on the values of Hubbell et al.[18]

4.3 Mass Attenuation Coefficient

Photoelectric absorption, coherent scattering, and incoherent scattering are independent interactions whose probabilities of occurrence are described by cross-sections and/or attenuation coefficients. The mass attenuation coefficient μ/ρ, is a measure of the total interaction probability and is obtained by summation over all individual processes, as in Equation 4.10. Total, partial, and differential cross-sections (or attenuation coefficients) are all key factors in photon transport computations, describing respectively the interaction probability, the type of interaction, and the angular distribution of scattered photon (see also Chapter 9). Energy fluence calculations rely heavily on these values for media subjected to photon irradiation.

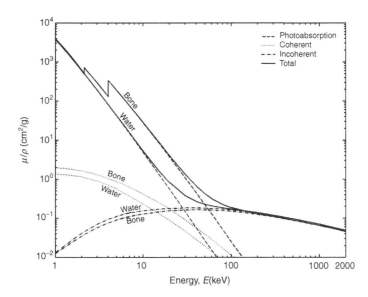

FIGURE 4.12
Total and partial mass attenuation coefficients, μ/ρ, due to photoabsorption, coherent and incoherent scattering, for water and bone, plotted as a function of the incident photon energy, E.

Figure 4.12 gives total mass attenuation coefficient μ/ρ values for water and bone, along with partial ones due to photoeffect, coherent and incoherent scattering, as a function of the incident photon energy. It is seen that the total attenuation curves for the two tissue materials are close to each other in the energy range above approximately 100 keV where incoherent scattering for these media predominates. As discussed in Section 4.2.3, the mass attenuation coefficient due to incoherent scattering shows a weak material dependence which is proportional to the ratio Z/A, and close to $1/2$ for almost all elements. At photon energies lower than about 100 keV, the attenuation curve for bone diverges from the water curve, by almost an order of magnitude towards higher μ/ρ values. In this energy region the predominant interaction is photoabsorption, which has strong Z dependence as described in Equation 4.16. Therefore the difference between water and bone is attributable mainly to $_{20}$Ca, which is a medium-Z element abundant in bone. Linear attenuation coefficient values for water and bone are also given in Figure 4.1, where the dependence on material density is obvious.

Figure 4.13 gives data corresponding to those presented in Figure 4.12, but for stainless steel and lead. Data for lead reveal that in almost the entire energy range relevant to brachytherapy, the photoelectric effect is the dominant interaction for high-Z materials. Hence these materials are most appropriate for shielding purposes for the high energy emitters in brachytherapy applications. High-density but medium-Z materials, such as stainless steel are not effective shielding materials for high energy emitters.[8,9]

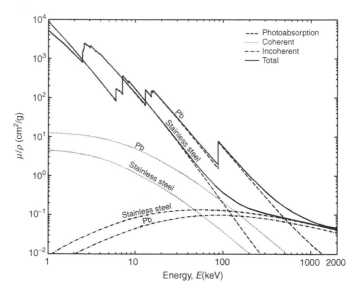

FIGURE 4.13
Total and partial mass attenuation coefficients, μ/ρ, due to photoabsorption, coherent and incoherent scattering, for stainless steel and lead, plotted as a function of the incident photon energy, E.

The XCOM database[2] incorporates total cross-section, σ, values using the combination of photoelectric cross-sections of Scofield,[13] coherent and incoherent scattering cross-sections of Hubbell and Øverbø[16] and Hubbell et al.[18] The total attenuation data seen in Table 4.2 through Table 4.19 were derived based on the mixture rule of Equation 4.13 and the elemental composition of Table 4.1, using log–log interpolation at the principal photon energies for the total cross-section values.

4.4   Mass Energy Absorption Coefficients

The mass energy absorption coefficient μ_{en}/ρ, is the major parameter in computations for energy deposited in media subjected to photon irradiation. As seen in Equation 4.4, the photon energy fluence $\Psi(E,r)$ when multiplied by the mass energy absorption coefficient μ_{en}/ρ, gives the absorbed dose. The prerequisites involved in these calculations are as follows.

Energy deposition is a two-step process involving first, the energy transfer from incident photons to kinetic energy of secondary electrons, and second, the dissipation of this kinetic energy in the absorbing medium. The energy transferred to the entire atom, as recoil energy, is negligible due to its large mass, as discussed in Section 4.2.1, and therefore is omitted in these calculations. Energy transfer to secondary electrons is described by the mass energy transfer coefficient μ_{tr}/ρ, defined as

$$\frac{\mu_{tr}}{\rho} = f^{ph}\frac{\mu^{ph}}{\rho} + f^{coh}\frac{\mu^{coh}}{\rho} + f^{incoh}\frac{\mu^{incoh}}{\rho} \qquad (4.36)$$

The dimensionless factors f^{ph}, f^{coh}, and f^{incoh} represent the energy fractions of incident photon energy E considered to be transferred to secondary electrons from each type of photoelectric, coherent and incoherent interaction.

In the photoelectric interaction, all the incident photon energy E, minus the energy E_{shell} binding the electron to the atom is carried away by the ejected electron, as described by Equation 4.14. Hence, a first estimate of the factor f^{ph} is

$$f^{ph} \approx 1 - \frac{E_{shell}}{E} \qquad (4.37)$$

which implies that the binding energy E_{shell} is lost and it is not locally available for deposition. Following photoabsorption the excited atom returns to its ground state and the binding energy is emitted promptly as characteristic x-ray fluorescence radiation or Auger electrons. Therefore, this energy is assumed to be available for deposit locally when an Auger electron is emitted, but it is lost to the volume of interest in the form of secondary photon in the case of fluorescence radiation, and the factor f^{ph} thus becomes

$$f^{ph} = 1 - \frac{X}{E} \qquad (4.38)$$

X is the average energy of fluorescence radiation emitted per photoabsorption interaction. The evaluation of fluorescence radiation energy X involves the calculations of the fluorescence yield. As discussed in Section 4.2.1 this is defined as the fraction of shell vacancies that are filled by photon emission rather than by Auger electron ejection. It should be noted that X in Equation 4.38 takes into account the contribution of all shells.

There is no energy transfer in coherent scattering and therefore the factor f^{coh} is implicitly set as

$$f^{coh} = 0 \qquad (4.39)$$

In the incoherent interaction, besides the fluorescence radiation, the energy of the Compton scattered photon is dissipated also from the volume of interest and hence:

$$f^{incoh} = 1 - \frac{\langle E_{sc}\rangle + X}{E} \qquad (4.40)$$

X is the average energy of fluorescence radiation emitted per incoherent interaction and $\langle E_{sc}\rangle$ is the average energy of the scattered photon calculated as described in Section 4.2.3.

The mass energy transfer coefficient μ_{tr}/ρ is written as

$$\frac{\mu_{tr}}{\rho} = f\frac{\mu}{\rho} \tag{4.41}$$

where the factor f is a weighting average over the individual processes (see Equation 4.36). The mass energy transfer coefficient μ_{tr}/ρ has the same dimensions as the mass attenuation coefficient μ/ρ, but it is reduced by the factor f, which represents the fraction of photon energy that on average is transferred to an electron. Hence, it is assumed that all secondary photons from the primary event, which are fluorescence, coherent, or incoherent scattered photons, are lost to the volume of interest.

For homogeneous mixtures and compounds, the mass energy transfer coefficient μ_{tr}/ρ, can be obtained as in the mixture rule of Equation 4.13:

$$\frac{\mu_{tr}}{\rho} = \sum_i w_i\left(\frac{\mu_{tr}}{\rho}\right)_i \tag{4.42}$$

where w_i is the fraction by weight of the ith atomic constituent.

The factor f in Equation 4.41 represents the fraction of the incident photon energy E which does not leave the site of the interaction in the form of secondary photon radiation (x-ray fluorescence radiation or Compton scattered photons) but goes into kinetic energy of secondary electrons for dissipation locally in the medium via collision losses as ionization and excitation. Hence, the ratio μ_{tr}/μ gives the mean fraction of incident photon energy that is changed into kinetic energy of electrons:

$$\frac{\mu_{tr}}{\mu} = \frac{\langle T_e \rangle}{E} \tag{4.43}$$

The photon energy fluence $\Psi(E,r)$ at a specific point in a medium, when multiplied by the energy transfer coefficient μ_{tr}/ρ, gives the dosimetric quantity kerma, i.e., the kinetic energy released by photons per unit mass at the specific point.

The mass energy absorption coefficient μ_{en}/ρ, is defined as

$$\frac{\mu_{en}}{\rho} = (1 - g)\frac{\mu_{tr}}{\rho} \tag{4.44}$$

which implies that relative to the energy transfer coefficient μ_{tr}/ρ, is further reduced by the dimensional factor g which represents the average fraction of the kinetic energy of all liberated secondary electrons that are subsequently lost in radiative energy loss processes as electrons slow down to rest in the medium. In the energy range relevant to brachytherapy these two coefficients, i.e., the energy transfer and the energy absorption coefficients are equal to within 1%, since radiation losses of the secondary electrons are insignificant for these photon energies. This is discussed in Section 4.5, where additionally it is emphasized that the ranges of the secondary electrons do not exceed 1 mm for the photon energies relevant to brachytherapy, ensuring electronic equilibrium. This allows the dose to be well approximated by kerma. Hence, the photon energy fluence $\Psi(E,r)$ when multiplied by the energy absorption coefficient, μ_{en}/ρ, or equivalently by the energy transfer coefficient, μ_{tr}/ρ, gives the absorbed dose.

Figure 4.1 shows the linear energy absorption coefficients for soft tissue and bone, and corresponding attenuation values. It is observed, that for photon energies less than 20 keV the energy absorption coefficient is approximately equal to the attenuation coefficient. This is due to the predominance of photoabsorption. For higher photon energies these two coefficients diverge owing to incoherent scattering.

In experimental source dosimetry, discussed in Chapter 10, ionization chambers, TLD dosimeters, and polymer gels are used. For these three detector materials, the ratio of the mass energy absorption coefficient of water to that of air, TLD, and VIPAR gel is given in Figure 4.14. These ratios represent response values relative to water response for these detectors (see also Section 4.6). The curves in Figure 4.14 thus show the energy dependence of this response. A strong energy dependence is observed for TLD, necessitating corrections to be applied when these dosimeters are used for source dosimetry of low energy emitters, such as ^{125}I and [103]Pd.[4,10] VIPAR polymer gel as a detector material presents the lowest energy dependence, being additionally a high water equivalent phantom material.[11]

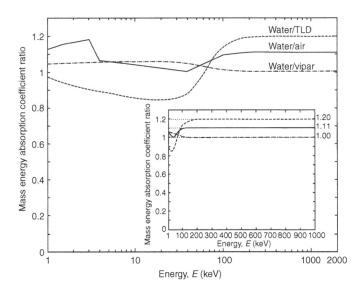

FIGURE 4.14
Mass energy absorption coefficient, μ_{en}/ρ, ratio values of water to TLD, water to air and water to VIPAR gel, plotted as a function of energy, E. The same ratios are plotted in the inset using a linear energy scale.

4.5 Electron Interaction Processes

Photon interactions in matter, which were discussed in previous sections, revealed a significant energy transfer from photons to atomic electrons as a consequence of successful photoabsorption or incoherent scattering events. This section considers the dissipation of these kinetic energies in the surrounding medium, and comments on the actual energy deposition from photon irradiation.

Ionization is the main cause of energy loss of electrons in the brachytherapy energy range. As an electron transverses matter it exerts electromagnetic forces on atomic electrons and imparts energy to them, leaving a trail of excitations and ionizations along its path. Since electrons can lose a large fraction of their energy in a single collision and undergo large deflections with atomic electrons, they do not move in straight lines but in complicated nonlinear paths. Another reason for energy loss is the rapid deceleration of an electron in the electric field of an atomic nucleus. This results in the emission of photon radiation termed bremsstrahlung. A well-known example of radiation losses is the continuous x-ray spectrum which is created when electrons are stopped by the anode in an x-ray tube.

Cross-sections as used to describe photon interactions are not appropriate to describe electron interactions. Unlike photons, for which a relatively small number of successful interactions is sufficient before their complete absorption (see Chapter 9), electrons interact approximately 10^5 times for the same energy loss. It is convenient to describe the extent to which electrons, as well as other charged particles, interact with matter in terms of the stopping power S, which is the mean energy loss per unit path length:

$$S = -\frac{dT_e}{dr} \qquad (4.45)$$

where $-dT_e$ is the energy loss (the minus sign is due to the continuous decrease in the kinetic energy T_e of electron) and dr is the unit path length transversed by the electron. For the energy range up to approximately 1 MeV, the stopping power of electrons can be written as

$$S = \left(-\frac{dT_e}{dr}\right)^{\text{ion}} + \left(-\frac{dT_e}{dr}\right)^{\text{rad}} \qquad (4.46)$$

where the first term on the right side of the equation is the energy loss per unit path length due to collisions giving rise to excitations and ionizations of the atoms. The second term gives the energy loss by bremsstrahlung, also termed braking radiation. The ionization term is also called the linear energy transfer (LET) (see also Chapter 2).

Figure 4.15 shows total and partial stopping powers in units of keV/cm, for water, bone, and lead, as a function of the electron kinetic energy T_e. It is seen that the contribution of radiation loss to the total energy loss is of

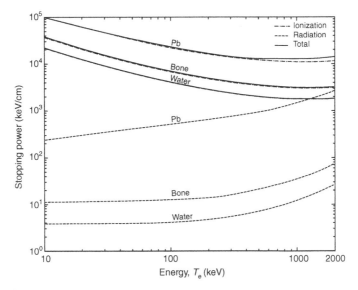

FIGURE 4.15
Total and partial stopping power values due to collision and radiation, in units of keV/cm for water, bone, and lead plotted as a function of electron energy, T_e.

significance only for high energy electrons and only for the high-Z lead. For water and bone this contribution is negligible even for 1 MeV electrons. The energy loss by radiation has a Z^2 dependence. This is because radiation losses are proportional to the square of acceleration and the force of Coulomb interaction for electrons is also proportional to the atomic number Z of the nucleus.

Radiation yield is the average fraction of its kinetic energy that an electron radiates as bremsstrahlung when completely slowing down. Figure 4.16 gives radiation yields for water, bone, and lead as a function of the electron kinetic energy, T_e. These data show that even for 1 MeV electrons, the fraction of electron kinetic energy that leaves the interaction site in the form of bremsstrahlung radiation, and therefore is not locally deposited, is lower than 1% for water and bone, rising to about 5% for lead. It is noted, however, that 1 MeV electrons are seldom liberated in brachytherapy by photon interactions even for the higher energy emitters, since (1) the photoabsorption, where the photon energy is transferred to a bound electron is relatively insignificant in the high energy range and (2) the incoherent interaction, dominant at high energy, results in energy transfer that on average does not exceed 40% of the photon energy (see also Figure 4.8). Indeed, the highest secondary electron energy in the brachytherapy photon energy range of up to approximately 1 MeV, does not exceed a few hundred keV. For these electrons radiation losses are totally unimportant for tissue materials.

The range of an electron is the distance it travels before coming to rest. Since stopping power S, is the mean energy loss per unit path length,

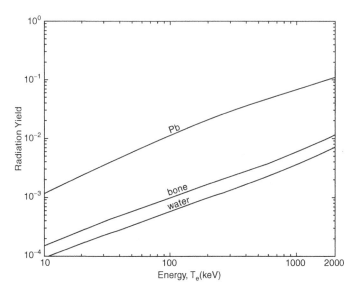

FIGURE 4.16
Radiation yield for water, bone, and lead plotted as function of electron energy, T_e.

the reciprocal of the stopping power gives the distance traveled per unit energy loss. Therefore, assuming that the kinetic energy of the electron changes as a continuous function as it slows down and comes to rest, the electron range $R(T_e)$, under the so-called continuously slowing down approximation (csda) is calculated by

$$\text{csda range}: \ R(T_e) = \int_0^{T_e} S^{-1}\mathrm{d}r \qquad (4.47)$$

Electrons however do not move in straight lines but follow very complicated paths, due to possible large deflections in a single collision. Therefore, the csda-range calculated by Equation 4.47, is only approximately equal to the average path length an electron travels.

Figure 4.17 shows csda ranges in units of centimeters for water, bone, and lead, as a function of the electron kinetic energy, T_e, in the energy range 10 keV to 1 MeV. It is observed that even for the highest electron energies that could be liberated by brachytherapy photons, the electron range does not exceed 1 mm.

Data shown in Figure 4.16 and Figure 4.17 indicate that in the photon energy range relevant to brachytherapy, radiation losses are totally insignificant for tissue materials and electronic disequilibrium may be present only in the proximity of a brachytherapy source. Therefore the kerma approximation is a safe assumption for brachytherapy dose computations.

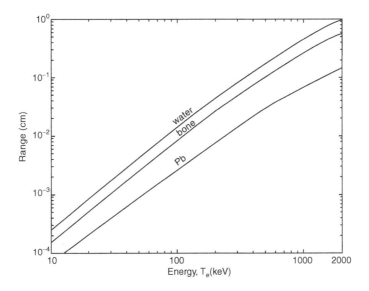

FIGURE 4.17

CSDA range of electrons, in centimeters, through water, bone, and lead media plotted as a function of electron energy, T_e.

4.6   Analytical Dose Rate Calculations

The radiation field in a medium surrounding a brachytherapy source includes both primary and scattered radiation. The contribution of primary radiation can be explicitly predicted based on the attenuation and energy absorption coefficients that describe the attenuation and scattering properties of the medium for the photon energies emitted by the source. Analytical calculations of scatter contribution are more complicated and Monte Carlo–based dosimetry, discussed in Chapter 9, has been acknowledged as a valuable tool in brachytherapy.

The absorbed dose rate $\dot{D}(E,r)$ in water medium at a distance r from a point source emitting photons of energy E, is written as

$$\dot{D}(E, r) = \dot{D}(E, r)_{\text{prim}} + \dot{D}(E, r)_{\text{scat}} = \dot{D}(E, r)_{\text{prim}}[1 + SPR(E, r)] \qquad (4.48)$$

where $\dot{D}(E,r)_{\text{prim}}$ and $\dot{D}(E,r)_{\text{scat}}$ are the absorbed doses at r due to primary and scattered photons, respectively, and $SPR(E,r)$ is the scatter to primary dose rate ratio, defined as

$$SPR(E, r) = \frac{\dot{D}_{\text{scat}}(E, r)}{\dot{D}_{\text{prim}}(E, r)} \qquad (4.49)$$

The absorbed dose rate $\dot{D}(E,r)$, is the product of the photon energy fluence $\Psi(E,r)$ at the distance r and the mass energy absorption coefficient $\mu_{en}(E)/\rho$, divided by time. Hence, assuming a monoenergetic photon point source of activity A emitting n photons of energy E per disintegration the dose rate, $\dot{D}(E,r)$, can be calculated as follows:[9,20]

$$\dot{D}(E,r) = AnE\frac{e^{-\mu(E)r}}{4\pi r^2}\left[\frac{\mu_{en}(E)}{\rho}\right]_{water}[1 + SPR(E,R)] \qquad (4.50)$$

where $\mu(E)$ and $[\mu_{en}(E)/\rho]_{water}$ are the linear attenuation and mass energy absorption coefficients for photons of energy E in water.

Activity is not an appropriate quantity to specify source strength in brachytherapy for reasons explained below. Instead, the source air kerma strength, S_k, is used, which is defined as the product of *air kerma rate*, $\dot{K}(r_c)_{air}$, measured at a calibration distance r_c (recommended to be 1 m) from the source *in free space* multiplied by the square of the distance r_c:[21]

$$S_k = \dot{K}(r_c)_{air}r_c \qquad (4.51)$$

In the presence of electronic equilibrium, i.e., $\dot{K}(r_c)_{air} = \dot{D}(r_c)_{air}$, S_k can be calculated by an equation similar to 4.50:

$$S_k = AnE\frac{1}{4\pi}\left[\frac{\mu_{en}(E)}{\rho}\right]_{air} \qquad (4.52)$$

where $[\mu_{en}(E)/\rho]_{air}$ is the mass energy absorption coefficient for photons of energy E in air. Since air kerma strength S_k, is defined in free space, there is the lack of an exponential term in Equation 4.52.

Dividing Equation 4.50 and Equation 4.52, gives

$$\frac{\dot{D}(E,r)}{S_k} = e^{-\mu(E)r}\left[\frac{\mu_{en}(E)}{\rho}\right]_{air}^{water}\frac{1}{r^2}[1 + SPR(E,r)] \qquad (4.53)$$

where $[\mu_{en}(E)/\rho]_{air}^{water}$ is the water to air mass energy absorption coefficient ratio, whose energy dependence is given in Figure 4.14.

The source activity does not participate in Equation 4.53 because it is present in both Equation 4.50 and Equation 4.52 and therefore is canceled out. This is the advantage of source calibration in terms of S_k according to the widely used AAPM TG-43 dosimetric formalism.[21] The recommended unit to express S_k is U ($= 1\,\mu Gy\,m^2\,h^{-1} = 1\,cGy\,cm^2\,h^{-1}$). The dimensions on the right side of Equation 4.53 are determined by the *geometry factor* r^{-2} and therefore dose rate arises in units of $cGy\,h^{-1}/U$, when distances are expressed in units of cm.

The quantity $(1 + SPR)$ in Equation 4.53 is the dose rate due to the scattered radiation. Ignoring for the interim this contribution, i.e. assuming that $SPR = 0$, the dose rate due to primary radiation per unit source strength as a function of the distance r is given in Figure 4.18 for selected photon energies of 20, 30, 60, 100, 400 keV and 1 MeV. Dose rate values were multiplied by radial distance

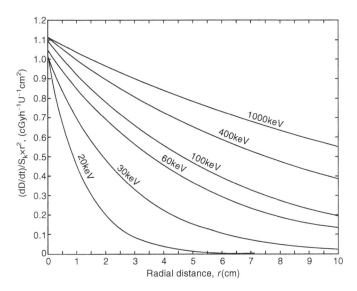

FIGURE 4.18
Dose rate per unit air kerma strength multiplied by radial distance squared, due to primary photons in water medium for selected photon energies of 20, 30, 60, 100, 400 keV and 1 MeV, plotted as a function of radial distance, r.

squared to remove the strong distance dependence. It is observed that primary dose rate distributions at low photon energies show a sharp fall-off, since in this energy range photoelectric absorption predominates.

As discussed in detail in Chapter 9, scatter dose rate gradually builds up with distance and reaches its maximum at about one photon mean free path. Therefore the contribution of scatter radiation for low energy emitters as ^{125}I and [103]Pd is limited close to the source. For intermediate energy emitters as [169]Yb, where incoherent scattering predominates, scatter overcompensates photon attenuation. At the high end of the brachytherapy energy range, scattering compensates for attenuation.

Equation 4.53 for a poly-energetic point source with emitted energy spectrum (n_i, E_i), i.e. emitting n_i photons of energy E_i per disintegration, becomes[9,20]

$$\frac{\dot{D}(r)}{S_k} = \frac{\sum n_i E_i (\mu_{en}(E_i)/\rho)_{water} e^{-\mu_{water}(E_i)r}[1 + SPR(E_i,r)]}{\sum n_i E_i (\mu_{en}(E_i)/\rho)_{air}} \frac{1}{r^2} \qquad (4.54)$$

where $\mu(E_i)$ and $\mu_{en}(E_i)/\rho$ are the linear attenuation and mass energy absorption coefficients and $SPR(E_i,r)$ are the scatter to primary dose rate ratios for the photon energy E_i. This expression is difficult to apply for a polyenergetic source such as [192]Ir. Instead, Equation 4.53 for mono-energetic sources can be applied for polyenergetic high energy emitters using effective attenuation coefficients and effective water to air mass energy absorption coefficient ratios weighted over the emitted energy spectrum according to (see also Section 9.4):

$$\langle \mu \rangle = \frac{\sum\limits_{i} n_i E_i \mu_i}{\sum\limits_{i} n_i E_i} \qquad (4.55)$$

$$\langle (\mu_{en}/\rho)_{air}^{water} \rangle = \frac{\sum\limits_{i} n_i E_i (\mu_{en}/\rho)_{air}^{water}}{\sum\limits_{i} n_i E_i} \qquad (4.56)$$

Specifically,[9] for the ^{192}Ir source the effective linear attenuation coefficient $\langle \mu \rangle$ and the water to air effective mass energy absorption coefficient ratio $\langle (\mu_{en}/\rho)_{air}^{water} \rangle$ have values of 0.109 cm^{-1} and 1.111, respectively, and the scatter to primary dose rate ratio is expressed by the polynomial:

$$SPR_{water}(r) = 0.123r + 0.005r^2 \qquad (4.57)$$

which is valid for radial distances of up to 10 cm.[9] This procedure has proved capable for calculating the dose distributions around point ^{192}Ir sources with clinical acceptable accuracy of within 1%. For low energy emitters, such as ^{125}I and ^{103}Pd, the accuracy of the proposed model is limited, resulting in dose errors as large as 6%.

References

1.   Hubbell, J.H. Review of photon cross interaction section data in the medical and biological context, *Phys. Med. Biol.*, 44, R1, 1999.

2.   Hubbell, J.H. and Seltzer, S.M., Tables of x-ray mass attenuation coefficients and mass energy-absorption coefficients (v.1.03, available online: http://physics.nist.gov/xaamdi), National Institute of Standards and Technology, Gaithersburg, MD 20899. Originally published as NISTIR 5632, National Institute of Standards and Technology, Gaithersburg, MD, 1995.

3.   ICRU, Tissue substitutes in radiation dosimetry and measurement, Report 44 of the International Commission on Radiation Units and Measurements, Bethesda, MD, 1989.

4.   Anagnostopoulos, G., Baltas, D., Karaiskos, P., Sandilos, P., Papagiannis, P., and Sakelliou, L. Thermoluminescent dosimetry of the selectSeed ^{125}I interstitial brachytherapy seed, *Med. Phys.*, 29, 709, 2002.

5.   Kipouros, P., Papagiannis, P., Sakelliou, L., Karaiskos, P., Sandilos, P., Baras, P., Seimenis, I., Kozicki, M., Anagnostopoulos, G., and Baltas, D. 3D dose verification in ^{192}Ir HDR prostate monotherapy using polymer gels and MRI, *Med. Phys.*, 30, 2031, 2003.

6.   Williamson, J.F. and Li, Z. Monte Carlo aided dosimetry of the microSelectron pulsed and high dose rate ^{192}Ir sources, *Med. Phys.*, 22, 809, 1995.

7.   Williamson, J.F., Perera, H., Li, Z., and Lutz, W.R. Comparison of calculated and measured heterogeneity correction factors for ^{125}I, ^{137}Cs, and ^{192}Ir brachytherapy sources near localized heterogeneities, *Med. Phys.*, 20, 209, 1993.

8.   Kirov, A.S., Williamson, J.F., Meigooni, A.S., and Zhu, Y. Measurement and calculation of the heterogeneity correction factors for an Ir-192 high dose-rate brachytherapy source behind tungsten alloy and steel shields, *Med. Phys.*, 23, 911, 1996.

9.   Anagnostopoulos, G., Baltas, D., Karaiskos, P., Pantelis, E., Papagiannis, P., and Sakelliou, L. An analytical dosimetry model as a step towards accounting for inhomogeneities and bounded geometries in ^{192}Ir brachytherapy treatment planning, *Phys. Med. Biol.*, 48, 1625, 2003.

10. Chiu-Tsao, S.T., Duckworth, T.L., Hsiung, C.Y., Li, Z., Williamson, J.F., Patel, N.S., and Harrison, L.B. Thermoluminescent dosimetry of the SourceTech Medical model STM1251 ^{125}I seed, *Med. Phys.*, 30, 1732, 2003.

11. Pantelis, E., Karlis, A.K., Kozicki, M., Papagiannis, P., Sakelliou, L., and Rosiak, J.M. Polymer gel water equivalence and relative energy response with emphasis on low photon energy dosimetry in brachytherapy, *Phys. Med. Biol.*, 49, 3495, 2004.

12. Williamson, J.F., Coursey, B.M., DeWerd, L.A., Hanson, W.F., Nath, R., and Ibbott, G. Guidance to users of Nycomed Amersham and North American Scientific, Inc., I-125 Interstitial Sources: Dosimetry and calibration changes: Recommendations of the American Association of Physicists in Medicine Radiation Therapy Committee Ad Hoc Subcommittee on Low-Energy Seed Dosimetry, *Med. Phys.*, 26, 570, 1999.

13. Scofield, J.H., Theoretical photoionization cross-sections from 1 to 1500 keV, Lawrence Livermore Laboratory Report UCRL-51326, 1973.

14. Morin, L.R.M. Molecular form factors and photon coherent scattering cross-sections of water, *J. Phys. Chem. Ref. Data*, 11, 1091–1098, 1982.

15. Karaiskos, P., Papagiannis, P., Sakelliou, L., Anagnostopoulos, G., and Baltas, D. Monte Carlo dosimetry of the selectSeed ^{125}I interstitial brachytherapy seed, *Med. Phys.*, 28, 1753, 2001.

16. Hubbell, J.H. and Øverbø, I. Relativistic atomic form factors and photon coherent scattering cross-sections, *J. Phys. Chem. Ref. Data*, 8, 69, 1979.

17. Thomson, J.J. *Conduction of Electricity Through Gases*, Cambridge University Press, London, p. 325, 1906.

18. Hubbell, J.H., Veigele, Wm.J., Briggs, E.A., Brown, R.T., Cromer, D.T., and Howerton, R.J. Atomic form factors, incoherent scattering functions, and photon scattering cross-sections, *J. Phys. Chem. Ref. Data*, 4, 471, 1975, [Erratum: *J. Phys. Chem. Ref. Data*, 6, 615, 1975.].

19. Klein, O. and Nishina, Y. Über die Streuung von Strahlung durch freie Elektronen nach der neuen relativistischen Quantendynamik von Dirac, *Z. Phy.*, 52, 853, 1929.

20. Chen, Z. and Nath, R. Dose rate constant and energy spectrum of interstitial brachytherapy sources, *Med. Phys.*, 28, 86–96, 2001.

21. Nath, R., Anderson, L., Luxton, G., Weaver, K., Williamson, J.F., and Meigooni, A.S. Dosimetry of interstitial brachytherapy sources: Recommendations of the AAPM Radiation Therapy Committee Task Group 43, *Med. Phys.*, 22, 209–234, 1995.

5

Brachytherapy Radionuclides and Their Properties

5.1 Introduction

Since Henri Becquerel's discovery of radioactivity in 1896 and Pierre and Marie Curie's discovery of radium in 1898, more than a century has elapsed and ^{226}Ra and its daughter product ^{222}Rn are no longer used for the treatment of cancer. They have now been totally replaced by artificially produced radionuclides, such as ^{192}Ir, ^{60}Co, ^{137}Cs, and ^{125}I. After the Manhattan Project, such replacements were inevitable for reasons of radiation safety because of the high energies of ^{226}Ra gamma-photons, its extremely long half-life of 1620 years, and the fact that ^{222}Rn is gaseous. In addition, the maximum obtainable specific activity of ^{226}Ra was such that it was impossible for thin wire sources to be manufactured and ^{226}Ra source geometries were only tubes and needles with a practical minimum length of about 2 cm. This limited the clinical applications of ^{226}Ra sources as they could not be used to treat all cancer sites, for example, the bronchus or bile duct.

5.1.1 ^{226}Radium Source Production

^{226}Ra sources were initially a problem for the worldwide medical community to purchase, from two points of view. Firstly, the enormous amount of work required for processing the uranium mineral ores and the fractional crystallization process involved limiting the supply of ^{226}Ra sources. Secondly, they were almost prohibitively expensive, see Table 5.1.

The Curies worked with the Joachimsthal uranium mineral pitchblende which was obtained free of charge from Austro-Hungarian Government since it was regarded as a useless residue material from silver mining. When it was realized that it was a source of ^{226}Ra, the Austro-Hungarians stopped the delivery of the ore to the Curies and alternative sources had to be found. In the U.S.A., the uranium mineral carnotite was much less rich in uranium

TABLE 5.1

Prices for 1 g of ^{226}Ra at the Beginning of the Twentieth Century

Year	U.S. Dollars
1902	3,000.00
1904	18,600.00
April 1906	50,000.00
August 1906	70,000.00
December 1906	90,000.00
1911	120,000.00
1914	150,000.00

Source: From Towpik, E. and Mould, R.F., *J. Oncol.*, Nowotwory, Warsaw, 1998.

than pitchblende but nevertheless formed the basis of the American radium industry from before World War I to the end of the 1920s.

By that time, uranium rich pitchblende had been discovered in the then Belgian Congo (now Zaire) in Katanga province and the Belgian company, the Union Miniere du Haut Katanga, with its processing and manufacturing plant at Oolen near Antwerp, effectively closed down the American radium companies as it was no longer economical to manufacture sources using carnotite.

Pitchblende was also discovered in Canada in the 1930s and by 1940 the price per gram of ^{226}Ra had significantly fallen and with then current large stockpiles of pitchblende there was effectively no further requirement for the continuation of uranium mining in order to produce clinical ^{226}Ra sources. Later, there was a short resurgence in uranium mining, but this was only for nuclear weapon and nuclear power applications.

5.1.2 Artificially Produced Radionuclides

In 1919, Ernest Rutherford reported the first artificial nuclear reaction. In an experiment he used gaseous nitrogen, ^{14}N, as a target that was bombarded by alpha particles (helium nuclei). He observed that very fast protons (hydrogen nuclei) were emitted and that the stable isotope ^{17}O was created.

$$^{14}_{7}N + {}^{4}_{2}He \Rightarrow {}^{17}_{8}O + {}^{1}_{1}H$$

In 1934, Iréne Joliot-Curie and Fréderic Joliot-Curie then proved that in such a type of reaction new radioactive isotopes (radionuclides) could be produced and not only stable isotopes. This discovery of artificial radioactivity opened the future to the production of the new radionuclides now used in radiation oncology. In the Joliot-Curie's experiment an aluminum target, ^{27}Al, was bombarded with alpha particles. The aluminum nucleus absorbed the alpha particles and emitted a neutron and a new radionuclide, the phosphorus isotope ^{30}P, was created.

$$^{27}_{13}\text{Al} + ^{4}_{2}\text{He} \Rightarrow ^{30}_{15}\text{P} + ^{1}_{0}\text{n}$$

Since the development of the nuclear reactor and of particle accelerators such as the cyclotron, a large number of artificially produced radionuclides have been, and are still being, produced for research and for industrial and medical applications.

5.1.3 Properties

There are several physical properties of radionuclides which define their relevance to brachytherapy, for both temporary and permanent implants.

5.1.3.1 Radioactive Decay Scheme

The radioactive decay scheme of a radionuclide determines the type of radiation: photons, beta particles, or neutrons that are emitted. This influences the possible form and general source design required for brachytherapy.

5.1.3.2 Half-Life $T_{1/2}$

The half-life of a radionuclide determines, in a large part, whether the radionuclide is going to be used in permanent or temporary brachytherapy implants. The half-life must be long enough to allow shipping of sources and subsequent preparation for an implant but also short enough, when required, for permanent implantation. It should be noted that the maximum specific activity is inversely proportional to the half-life of a radionuclide, see also Equation 2.60 in Chapter 2. In addition, the half-life influences the timeline of source renewal and therefore has economic implications for clinical practice.

5.1.3.3 Specific Activity

The specific activity determines the limits of miniaturization of the radio-nuclide source and of maximum possible dose rate. These factors are extremely important for remote controlled afterloading machine sources as well as for manual afterloaded ^{192}Ir wire sources such as those used in the Paris system.

5.1.3.4 Energy

The energy of the emitted radiation determines its penetration within tissue and the amount of shielding required for radiation protection. The half value layer (HVL) for lead is the thickness of lead required to reduce the exposure rate by one-half, usually in a narrow beam geometry situation. Hence, energy has a direct influence on the investment costs for radiation protection. The energy also determines the LET value and thus the biological effectiveness of the (particular type of) radiation. Usually, photon-emitting radionuclides have complex emission spectra and the emission weighted mean energy E_{mean} is defined as

$$E_{mean} = \frac{\sum_i f_i \cdot E_i}{\sum_i f_i} \tag{5.1}$$

Here, f_i is the intensity (or emission frequency) of the specific energy line E_i and the sum is taken over all photon ray energies E_i of the radionuclide spectrum.

For dosimetric purposes it is more appropriate to consider the effective energy E_{eff}, defined as the air-kerma weighted mean energy, see also Equation 2.35. This is given by

$$E_{eff} = \frac{\sum_i f_i \cdot E_i^2 \left(\frac{\mu_{tr}}{\rho}\right)_{\alpha, E_i}}{\sum_i f_i \cdot E_i \left(\frac{\mu_{tr}}{\rho}\right)_{\alpha, E_i}} \tag{5.2}$$

$(\mu_{tr}/\rho)_{\alpha, E_i}$ is the mass energy transfer coefficient for photons of energy E_i in air.

In a similar way to that described for E_{eff}, the effective mass energy transfer coefficient for air, $(\mu_{tr}/\rho)_{\alpha, E_i}$ is defined as the energy spectrum weighted mean value. It can be calculated using Equation 5.3:

$$\left(\frac{\mu_{tr}}{\rho}\right)_{\alpha, eff} = \frac{\sum_i f_i \cdot E_i \left(\frac{\mu_{tr}}{\rho}\right)_{\alpha, E_i}}{\sum_i f_i \cdot E_i} \tag{5.3}$$

For photon energies and radionuclides of relevance for brachytherapy, the fraction of the energy of the electrons liberated by photons in air that is lost to radiative processes (mostly bremsstrahlung) g_α is practically zero (see Table 7.9). Because of that, the mass energy absorption coefficient $(\mu_{en}/\rho)_\alpha$ instead of the mass energy transfer coefficient $(\mu_{tr}/\rho)_\alpha$ can be used in Equation 5.2 and Equation 5.3.

Since photons of energies up to a cut-off value of 10 keV are assumed to be not penetrating, the summations in Equation 5.1 through Equation 5.3 are carried out for photons with $E_i > 10$ keV (cut-off value $\delta = 10$ keV).

Figure 5.1 demonstrates the energy dependence of the air kerma-rate constant Γ_δ for mono-energetic point sources as in Equation 2.63 in the range 0 to 1400 keV. In the same figure, the Γ_δ values ($\delta = 10$ keV) for the seven

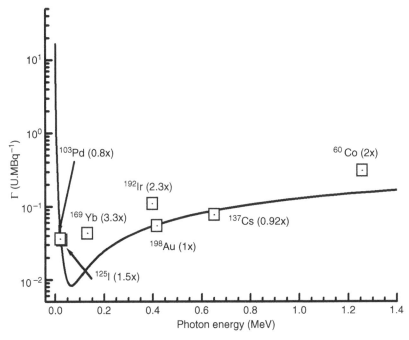

FIGURE 5.1
Energy dependence of the air kerma-rate constant Γ_δ for mono-energetic point sources calculated using Equation 2.63. Mass energy absorption coefficients for dry air are according to Hubbell and Seltzer,[17] where a log–log interpolation has been applied to these tables. Γ_δ values plotted against photon energy. The Γ_δ values ($\delta = 10$ keV) for the seven radionuclides of interest in brachytherapy are also plotted according to their effective energy, E_{eff} as in Equation 5.2. The numbers in parenthesis are the mean number of photons emitted per decay for each radionuclide considering only photons emitted with energy above the cut-off value of $\delta = 10$ keV (Table 5.2).

radionuclides of interest in brachytherapy are also plotted against their effective energy, E_{eff} as in Equation 5.2 and Table 5.2.

In Figure 5.2 the HVL values for lead listed in Table 5.2 and the tenth value layer (TVL) values for lead,[18,19] that is the thickness of lead required to reduce the exposure rate to one-tenth, are plotted vs. the mean photon energy for the different radionuclides as in Equation 5.1 and Table 5.2.

5.1.3.5 Density and Atomic Number

Both the density and the atomic number of the radionuclide determine its attenuation properties for x-ray radiography. This defines how easily the source can be localized using conventional radiographic techniques. In addition, the density and the atomic number of the radionuclide determine the amount of attenuation and absorption of the emitted radiation within

TABLE 5.2

Physical Characteristics of Radionuclides Used in Brachytherapy

Name	Symbol	Z	A	Density[a] ρ (g cm³)	Decay	Half-Life[a] $T_{1/2}$	Photon Energy Range (MeV)	Mean Photon Energy E_{mean} (MeV)	Effective Energy E_{eff} (MeV)	Mean γ-Energy (MeV)	HVL[b] (mm Pb)	Maximum Specific Activity (GBq mg⁻¹)	Γ_δ[c] (μGy h⁻¹ MBq⁻¹ m²)	Type of Implant
Cesium	Cs	55	137	1.873	β^-	30.07 a	0.032–0.662	0.615	0.652	0.662	7.0	3.202	0.0771	Temporary
Cobalt	Co	27	60	8.90	β^-	5.27 a	0.347–2.159	1.253	1.257	1.253	12.0	41.91	0.3059	Temporary
Gold	Au	79	198	19.32	β^-	2.695 d	0.069–1.088	0.406	0.417	0.415	2.8	9055.12	0.0545	Permanent
Iodine	I	53	125	4.93	EC	59.49 d	0.027–0.035	0.028	0.028	0.035	0.025	650.15	0.0348	Permanent or temporary
Iridium	Ir	77	192	22.42	β^-, EC	73.81 d	0.061–1.378	0.355	0.398	0.372	3.0	340.98	0.1091	Temporary
Palladium	Pd	46	103	12.02	EC	16.991 d[d]	0.020–0.497	0.021	0.021	0.137	0.008	2763.13	0.0361	Permanent
Radium	Ra	88	226	5.00	α	1600 a[d]	0.047–2.45	0.830[e]	—	0.830[e]	13.0	0.0366	—	Temporary
Thulium	Tm	69	170	9.321	β^-, EC	128.6 d[d]	0.048–0.084	0.066	0.067	0.084	0.17[f]	221.07	0.00053	Temporary
Ytterbium	Yb	70	169	6.73	EC	32.015 d	0.021–0.773	0.093	0.131	0.143	0.23[f]	893.29	0.0431	Permanent or temporary

Photon energy ranges, mean and effective photon energies are according to Table 5.4 through Table 5.12, where only penetrating energies above 10 keV have been considered. The maximum specific activity is calculated for ideal (point) pure radionuclide sources using Equation 2.60.

[a] According to NIST Physical Reference Data, on line version, 2004.[14,15]

[b] Half Value Layer (first) values most taken from ICRP 21 Figures 50 and 51.[16]

[c] Values calculated for pure radionuclides and ideal point sources using Equation 2.64 and the spectra given in Table 5.4 through Table 5.12 using a cut-off value δ = 10 keV.

[d] According to NUDAT vs. 2.0 on line tables, 2004.[3]

[e] Here all series decays of $^{222}_{86}$Rn to stable lead $^{206}_{82}$Pb have been considered.

[f] According to Granero et al.[18] and Lymperopoulou et al.[21]

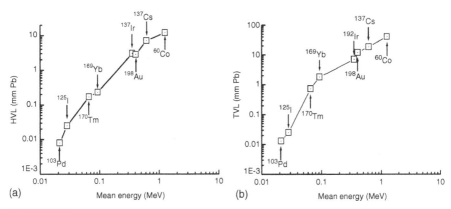

FIGURE 5.2
Thickness of lead required (a) for 50% transmission, HVL, and (b) for 10% transmission, TVL for the different radionuclides handled in this chapter. HVL values are according to the data presented in Table 5.2, where the tenth value layer data are according to Williamson.[19] The TVL values for [169]Yb and [170]Tm according to Granero et al.[18] (for [169]Yb and ^{192}Ir see also Lymperopoulou et al.[21]).

TABLE 5.3

Radionuclide Physical Properties

Physical Property	Relevance to Brachytherapy
Radiation emitted	Source geometry and structure
Half-life $T_{1/2}$	Determines if permanent or temporary implant or both are practical
Specific activity	Source size, dose rate
Energy of emitted radiation	Dose distribution within tissue, radiation protection requirements
Density and atomic number	Radiographic visibility/localization, isotropy/anisotropy of dose distribution

the source core and thus determines the isotropy or anisotropy of the dose distribution around the source, see Table 5.3.

5.2 Notation

The notation given below is used for the products of internal conversion (IC) processes of any γ-ray emitted as a result of gamma decay. These are the conversion electrons (CE) and characteristic x-rays. Characteristic x-rays are also produced directly by the electron capture (EC) process and, in addition, characteristic x-rays can result in the production of Auger electrons.

CE Y: CE emitted from the Y shell as a result of IC.

Auger Y: Auger electron emitted as a result of any transition of atomic electrons to a vacancy existing in the Y shell. Examples of such transitions for the K shell are given below.

> Auger KLL: Auger electron emitted from the L shell as a result of the transition of another electron of the L shell to a vacancy in the K shell.

> Auger KLX: Auger electron emitted from the X shell as a result of the transition of another electron of the L shell to a vacancy in the K shell. X is a shell higher than the L shell.

> Auger KXY: Auger electron emitted from the Y shell as a result of the transition of an electron of the X shell to a vacancy in the K shell. X and Y are shells higher than the K shell.

Spectroscopic notation for characteristic x-rays

$X\,K\alpha1$: X-Rays emitted due to an electron transition from the 3rd subshell of the L shell to the K shell: $K-L_{III}$ transition.

$X\,K\alpha2$: X-Rays emitted due to an electron transition from the 2nd subshell of the L shell to the K shell: $K-L_{II}$ transition.

$X\,K\beta1$: X-Rays emitted due to an electron transition from the 3rd subshell of the M shell to the K shell: $K-M_{III}$ transition.

$X\,K\beta2$: X-Rays emitted due to an electron transition from the 2nd subshell of the N shell to the K shell: $K-N_{II}$ transition.

$X\,K\beta3$: X-Rays emitted due to an electron transition from the 2nd subshell of the M shell to the K shell: $K-M_{II}$ transition.

$X\,L$: X-Rays emitted due to electrons transitions to a vacancy in the L shell independently of their original shell as a summarized process.

5.3 ^{60}Cobalt

In 1938, ^{60}Co was discovered by John Livingood and Glenn Seaborg at the University of California-Berkley; Livingood activated samples in a cyclotron and Seaborg separated the elements/isotopes of interest.

After World War II and the atomic bombs at Hiroshima and Nagasaki, scientists at the Canadian National Research Council identified ^{60}Co as a possible useful radioactive source for radiation therapy. This was due to the fact that in 1947 a heavy water reactor facility was available at Chalk River that made possible the production of large quantities of ^{60}Co.

5.3.1 Teletherapy

^{60}Co as a teletherapy source was considered as an alternative to both deep x-ray therapy tubes and sources of some 5 to 10 gm ^{226}Ra for treating deep-seated tumors. The economics for the new ^{60}Co was a major consideration. A ^{60}Co teletherapy machine could be constructed with costs for the required activity of the ^{60}Co source of 50,000 U.S.$. For an equivalent ^{226}Ra source the costs were a factor of 1000 higher.

In 1949, the Canadians started two independent ^{60}Co teletherapy projects: in Ottawa and in Saskatoon. The first patient was treated on October 27, 1951 by the Eldorado machine in Ottawa and on November 8 of the same year, a patient was treated at the University of Saskatchewan in Saskatoon.

5.3.2 Brachytherapy

In a letter to the journal *Science* in June 1948 on "Radioactive Needles Containing Cobalt 60", William Myers from the Department of Medicine, Ohio State University,[2] described studies with ^{60}Co started in October 1947.

For the production of ^{60}Co radioactive needles, an alloy wire composed of 45% cobalt and 55% nickel, the so-called cobanic, was used in order to overcome the machining difficulties related to pure cobalt. In this work, needles of a very small diameter, 1.0 mm, are described. The half-value thickness of the gamma irradiation of this type of source was 0.41″ (10.41 mm) Pb.

Myers also stated that the expected benefit when using ^{60}Co sources when compared to ^{226}Ra should be that the ^{60}Co radioactive wires can be bent to conform to the shape of tumors which are in a difficult anatomical environment (e.g., bones) and that there is no danger of loss by leaks or breakages. This was the worldwide first use of ^{60}Co radioactive needles for cancer therapy.

The initial design of the ^{60}Co sources was in the form of needles and were essentially a copy of existing ^{226}Ra needles. Due to its high specific activity, ^{60}Co (see Table 5.2) is appropriate for fabrication of small high-activity sources and has been mainly used to replace ^{226}Ra for gynecological brachytherapy. It is more expensive to produce than ^{137}Cs but is still in use in some centers in the developing world. However, because of the ^{60}Co half-life of 5.27 years it has important cost advantages over, for example, ^{192}Ir sources, which because of their much shorter half-life of 73.8 days have to be replaced far more often. In addition, all the associated bureaucratic problems of crossing customs boundaries and ensuring a rapid delivery of ^{192}Ir sources is not relevant for the longer lived ^{60}Co.

^{60}Co undergoes β^- decay to excited states of ^{60}Ni (see Figure 5.3). The de-excitation to the ground sate of ^{60}Ni occurs mainly via emission of γ-rays where there are two dominant energy lines of 1.1732 and 1.332 MeV,

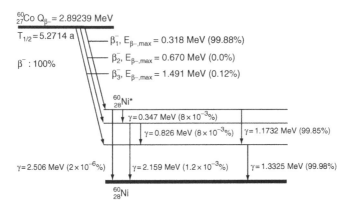

FIGURE 5.3

Schematic representation of β^- decay of $^{60}_{27}$Co which decays almost totally (99.92%) to the third excited state of $^{60}_{28}$Ni. The disintegration energy for the decay is $Q_{\beta^-} = 2.8239$ MeV. The two main γ-rays are 1.1732 and 1.3325 MeV, both with an intensity of approximately 100%.[3] On average, two photons are emitted per disintegration with energy above 10 keV (see Table 5.4c). The half-life for $^{60}_{27}$Co decay is 5.2714 years.[14]

each with an absolute intensity of approximately 100%. On average, 2.0 photons are emitted per disintegration.

The main beta ray has a maximum energy of 0.318 MeV, a mean energy of 0.096 MeV and an absolute intensity of 99.88%. The highest energy beta ray has a maximum energy of 1.491 MeV and a mean energy of 0.626 MeV with an emission probability of 0.12%. Due to the relative low energies of electrons emitted (see also Table 5.4a and Table 5.4b), their filtering can be easily achieved due to the cobalt material itself, as well as by using thin encapsulation layers. Figure 5.3 illustrates the decay scheme and the energy spectrum of the emitted photons for ^{60}Co (Table 5.4c).

TABLE 5.4a

Energies and Intensities of β^- Particles Emitted by β^- Decay of ^{60}Co as Shown in Figure 5.3

Mean Energy (MeV)	Maximum Energy (MeV)	Absolute Intensity (%)
0.0958	0.3182	99.88
0.6259	1.4914	0.12
Mean β^- energy	0.0964	100.00

Source: From National Nuclear Data Center (NNDC), Brookhaven National Laboratory NUDAT 2.0. Electronic Version available online at NNDC: www.nndc.bnl.gov/nudat2/, July 2005.

TABLE 5.4b

Energies and Intensities of Electrons Emitted by β^-
Decay of ^{60}Co as CEs and Auger Electrons

Electrons	Mean Energy (MeV)	Absolute Intensity (%)
Auger L	0.0008	0.0366
Auger K	0.0065	0.0154
CE K	0.3388	3.77×10^{-5}
CE L	0.3461	3.81×10^{-6}
CE K	0.8178	2.4×10^{-6}
CE L	0.8251	2.2×10^{-7}
CE K	1.1649	0.0151
CE K	1.3242	0.0115
CE K	2.1502	5.4×10^{-8}
CE L	2.1576	5.2×10^{-9}
CE K	2.4974	1.6×10^{-10}
CE L	2.5047	1.5×10^{-11}

Source: From National Nuclear Data Center (NNDC), Brookhaven National Laboratory NUDAT 2.0. Electronic Version available online at NNDC: www.nndc.bnl.gov/nudat2/, July 2005.

TABLE 5.4c

Energies and Intensities of γ-Rays as Result of γ-Decay and of Characteristic X-Rays as a Result of IC, by β^- Decay of ^{60}Co as Shown in Figure 5.3

Photon Rays	Energy (MeV)	Absolute Intensity (%)	$(\mu_{en}/\rho)_\alpha$ $(\text{cm}^2\,\text{g}^{-1})$	Γ_δ (μGy h^{-1} MBq^{-1} m^2 = U MBq^{-1}) (Equation 2.63)
X L	0.0009	0.00031	5.455×10^1	6.598×10^{-7}
X Kα2	0.0075	0.00325	1.170×10^1	1.302×10^{-5}
X Kα1	0.0075	0.0064	1.162×10^1	2.552×10^{-5}
X Kβ3	0.0083	0.000394	8.545	1.277×10^{-6}
X Kβ1	0.0083	0.00077	8.545	2.496×10^{-6}
γ	0.3471	0.0075	2.919×10^{-2}	3.488×10^{-6}
γ	0.8261	0.0076	2.871×10^{-2}	8.272×10^{-6}
γ	1.1732	99.85	2.704×10^{-2}	1.454×10^{-1}
γ	1.3325	99.9826	2.625×10^{-2}	1.605×10^{-1}
γ	2.1586	0.0012	2.288×10^{-2}	2.720×10^{-6}
γ	2.5057	2.0×10^{-6}	2.180×10^{-2}	5.015×10^{-9}
Mean energy E_{mean} (Equation 5.1)			1.2529 MeV	
Effective energy E_{eff} (Equation 5.2)			1.2568 MeV	
Effective mass energy absorption coefficient, $(\mu_{en}/\rho)_{\alpha,eff}$ (Equation 5.3)			2.662×10^{-2} cm^2 g^{-1}	

(continued)

TABLE 5.4c—*Continued*

Photon Rays	Energy (MeV)	Absolute Intensity (%)	$(\mu_{en}/\rho)_\alpha$ $(cm^2\,g^{-1})$	Γ_δ ($\mu Gy\,h^{-1}\,MBq^{-1}\,m^2$ = U MBq^{-1}) (Equation 2.63)
Air kerma-rate constant Γ_δ (Equation 2.64)				0.3059 $\mu Gy\,h^{-1}\,MBq^{-1}\,m^2$
Mean number of photons emitted per decay				2.0 (1.9985)

The mean energy E_{mean}, the effective energy E_{eff}, the air kerma-rate constant Γ_δ, as well as the number of photons emitted per decay, considers only rays above 10 keV (cut-off value $\delta = 10$ keV). Mass energy absorption coefficients for dry air are according to Hubbell and Seltzer,[17] where a log–log interpolation has been applied to these tables.

Source: From National Nuclear Data Center (NNDC), Brookhaven National Laboratory NUDAT 2.0. Electronic Version available online at NNDC: www.nndc.bnl.gov/nudat2/, July 2005.

5.4 ^{137}Cesium

^{137}Cs was discovered in the late 1930s by Glenn Seaborg and Margaret Melhase. It was first introduced as a ^{226}Ra substitute firstly for intracavitary brachytherapy for gynecological tumors and secondly for interstitial treatments. The principal advantages of ^{137}Cs over ^{226}Ra are the reduced amount of shielding required (see Table 5.2 for half-value layer values) and the absence of a gaseous daughter product.

The longer half-life of 30.07 years for ^{137}Cs, when compared to 5.27 years for ^{60}Co, enables the clinical use of ^{137}Cs sources over a long period of time before replacement is necessary: at least for 10 years. This feature, and the lower production costs of ^{137}Cs sources, as well as the smaller amount of shielding required when compared to ^{60}Co (see Table 5.2 for half-value layer values), made ^{137}Cs the main ^{226}Ra substitute in brachytherapy in the early 1960s.

^{137}Cs decays purely via β^- mainly (94.4%) to the second excited state of ^{137}Ba (see Figure 5.4 and Table 5.5a–Table 5.5c), where the de-excitation to the ground state of ^{137}Ba occurs by 90% via gamma decay and emission of practically a single γ-ray (γ-ray energy of 0.662 MeV with an absolute intensity of 85.1%, see Table 5.5c), and by 10% via IC. On average, 0.9 photons are emitted per disintegration of ^{137}Cs.

From the decay scheme of ^{137}Cs we observe higher energies for the emitted β^- particles (see Table 5.5a) when compared to ^{60}Co. This means that a greater encapsulation thickness is required to filter out these electrons. Also, this results in an increased anisotropic behavior of the dose distribution around ^{137}Cs sources, due to increased differences in attenuation and absorption in the source encapsulation materials.

Finally, the low specific activity of ^{137}Cs, which is the lowest of all radionuclides used as alternatives to ^{226}Ra, does not allow the production

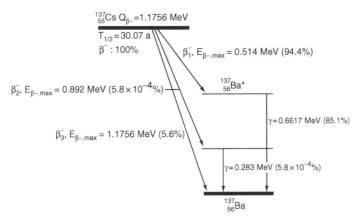

FIGURE 5.4

Schematic representation of β^- decay of $^{137}_{55}$Cs which decays mainly (94.4%) to the second excited state of $^{137}_{56}$Ba. The disintegration energy for the decay is $Q_{\beta^-} = 1.1756$ MeV. There is practically a single γ-ray emitted of 0.6617 MeV with an absolute intensity of 85.1%.[3] On average, 0.9 photons are emitted per disintegration with energy above 10 keV (see Table 5.5c). The half-life for $^{137}_{55}$Cs decay is 30.07 years.[14]

TABLE 5.5a

Energies and Intensities of β^- Particles Emitted by β^- Decay of ^{137}Cs as shown in Figure 5.4

Mean Energy (MeV)	Maximum Energy (MeV)	Absolute Intensity (%)
0.1743	0.5140	94.4
0.3006	0.8921	0.00058
0.4163	1.1756	5.6
Mean β^- energy	0.1879	100.00

Source: From National Nuclear Data Center (NNDC), Brookhaven National Laboratory NUDAT 2.0. Electronic Version available online at NNDC: www.nndc.bnl.gov/nudat2/, July 2005.

TABLE 5.5b

Energies and Intensities of Electrons Emitted by β^- Decay of ^{137}Cs as CEs and Auger Electrons

Electrons	Mean Energy (MeV)	Absolute Intensity (%)
Auger L	0.0037	7.29
Auger K	0.0264	0.77
CE K	0.6242	7.659
CE L	0.6557	1.387

Source: From National Nuclear Data Center (NNDC), Brookhaven National Laboratory NUDAT 2.0. Electronic Version available online at NNDC: www.nndc.bnl.gov/nudat2/, July 2005.

TABLE 5.5c

Energies and Intensities of γ-Rays as Result of γ-Decay and Characteristic
X-Rays as Result of IC, by β⁻ Decay of ^{137}Cs Shown in Figure 5.4

Photon Rays	Energy (MeV)	Absolute Intensity (%)	$(\mu_{en}/\rho)_\alpha$ $(cm^2 g^{-1})$	Γ_δ (μGy h^{-1} MBq^{-1} m^2 = U MBq^{-1}) (Equation 2.63)
X L	0.0045	0.9	5.521×10^1	1.019×10^{-2}
X Kα2	0.0318	1.96	1.291×10^{-1}	3.695×10^{-4}
X Kα1	0.0322	3.58	1.247×10^{-1}	6.597×10^{-4}
X Kβ3	0.0363	0.342	8.855×10^{-2}	5.046×10^{-5}
X Kβ1	0.0364	0.661	8.805×10^{-2}	9.718×10^{-5}
X Kβ2	0.0373	0.209	8.249×10^{-2}	2.948×10^{-5}
γ	0.2835	0.00058	2.850×10^{-2}	2.151×10^{-7}
γ	0.6617	85.1	2.936×10^{-2}	7.587×10^{-2}
Mean energy E_{mean} (Equation 5.1)			0.6154 MeV	
Effective energy E_{eff} (Equation 5.2)			0.6518 MeV	
Effective mass energy absorption coefficient, $(\mu_{en}/\rho)_{\alpha,eff}$ (Equation 5.3)			2.971×10^{-2} cm^2 g^{-1}	
Air kerma-rate constant Γ_δ (Equation 2.64)			0.0771 μGy h^{-1} MBq^{-1} m^2	
Mean number of photons emitted per decay			0.92 (0.9185)	

The mean energy E_{mean}, the effective energy E_{eff}, the air kerma-rate constant Γ_δ as well as the number of photons emitted per decay considers only rays above 10 keV (cut-off value $\delta = 10$ keV). mass energy absorption coefficients for dry air are according to Hubbell and Seltzer,[17] where a log–log interpolation has been applied to these tables.
Source: From National Nuclear Data Center (NNDC), Brookhaven National Laboratory NUDAT 2.0. Electronic Version available online at NNDC: www.nndc.bnl.gov/nudat2/, July 2005.

of miniature sources of very high activity (e.g., 370 GBq) for high dose rate (HDR) remote-controlled afterloading brachytherapy machines of the twenty-first century. ^{137}Cs is thus only appropriate for the production of low dose rate (LDR) sources.

5.5 ^{198}Gold

In 1952, ^{198}Au seeds[4] replaced ^{222}Rn seeds which had been used since approximately 1910. The original design of ^{222}Rn seed was a small glass capillary tube containing the radioactive gas, but in the 1920s the design was improved by using gold instead of glass. This prevented accidents due to breakages. The ^{222}Rn seeds had to be implanted into the tumor singly but

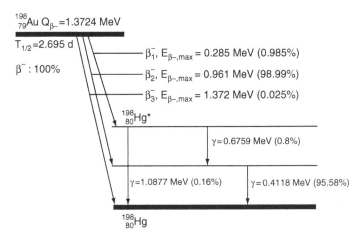

FIGURE 5.5

Schematic representation of β^- decay of $^{198}_{79}$Au which decays mainly (98.99%) to the second excited state of $^{198}_{80}$Hg. The disintegration energy for the decay is $Q_\beta{}_- = 1.3724$ MeV. There is mainly a γ-ray emitted of 0.412 MeV with an intensity of 95.58%.[3] On average, 1.0 photon is emitted per disintegration with energy above 10 keV (see Table 5.6c). The half-life for $^{198}_{79}$Au decay is 2.695 days.[14]

an advantage of using ^{198}Au seeds, or grains as they were sometimes termed, was that a special gun was designed to contain a cartridge of 14 ^{198}Au grains. This improved the probability of obtaining the required spatial distribution of the sources.

^{198}Au decays via β^- mainly (98.99%) to the first excited state of ^{198}Hg (see Figure 5.5 and Table 5.6a–Table 5.6c), where the de-excitation to the ground state of ^{198}Hg occurs by almost 97% via gamma decay and emission of practically a single γ-ray; γ-ray energy of approximately 0.412 MeV and an absolute intensity of 95.85% (see Table 5.6c). The other 3% of the de-excitation occurs via IC. On average, one photon is emitted per disintegration of ^{198}Au.

TABLE 5.6a

Energies and Intensities of β^- Particles Emitted by β^- Decay of ^{198}Au as Shown in Figure 5.5

Mean Energy (MeV)	Maximum Energy (MeV)	Absolute Intensity (%)
0.0794	0.2847	0.985
0.3147	0.9606	98.99
0.4673	1.3724	0.025
Mean β^- energy	0.3124	100.00

Source: From National Nuclear Data Center (NNDC), Brookhaven National Laboratory NUDAT 2.0. Electronic Version available online at NNDC: www.nndc.bnl.gov/nudat2/, July 2005.

TABLE 5.6b

Energies and Intensities of Electrons Emitted by β^- Decay of ^{198}Au as CEs and Auger Electrons

Electrons	Mean Energy (MeV)	Absolute Intensity (%)
Auger L	0.0076	2.18
Auger K	0.0538	0.11
CE K	0.3287	2.877
CE L	0.3970	1.0208
CE M	0.4082	0.2542
CE NP	0.4110	0.07933
CE K	0.5928	0.0179
CE L	0.6610	0.00323
CE K	1.0046	0.000663
CE L	1.0728	0.0001209

Source: From National Nuclear Data Center (NNDC), Brookhaven National Laboratory NUDAT 2.0. Electronic Version available online at NNDC: www.nndc.bnl.gov/nudat2/, July 2005.

TABLE 5.6c

Energies and Intensities of γ-Rays as Result of γ-Decay and Characteristic X-Rays as Result of IC, by β^- Decay of ^{198}Au as Shown in Figure 5.5

Photons	Energy (MeV)	Absolute Intensity (%)	$(\mu_{en}/\rho)_\alpha$ (cm^2 g^{-1})	Γ_δ (μGy h^{-1} MBq^{-1} m^2 = U MBq^{-1}) (Equation 2.63)
X L	0.0100	1.2	4.757	2.617×10^{-3}
X Kα2	0.0689	0.796	2.630×10^{-2}	6.620×10^{-5}
X Kα1	0.0708	1.34	2.575×10^{-2}	1.121×10^{-4}
X Kβ3	0.0798	0.162	2.409×10^{-2}	1.430×10^{-5}
X Kβ1	0.0802	0.31	2.404×10^{-2}	2.745×10^{-5}
X Kβ2	0.0825	0.1121	2.382×10^{-2}	1.011×10^{-5}
γ	0.4118	95.58	2.953×10^{-2}	5.336×10^{-2}
γ	0.6759	0.804	2.931×10^{-2}	7.310×10^{-4}
γ	1.0877	0.159	2.746×10^{-2}	2.180×10^{-4}
Mean energy E_{mean} (Equation 5.1)			0.4057 MeV	
Effective energy E_{eff} (Equation 5.2)			0.4166 MeV	
Effective mass energy absorption coefficient, $(\mu_{en}/\rho)_{\alpha,eff}$ (Equation 5.3)			2.950×10^{-2} cm^2 g^{-1}	
Air kerma-rate constant Γ_δ (Equation 2.64)			0.0545 μGy h^{-1} MBq^{-1} m^2	
Mean number of photons emitted per decay			1.0 (0.9926)	

The mean energy E_{mean}, the effective energy E_{eff}, the air kerma-rate constant Γ_δ as well as the number of photons emitted per decay considers only rays above 10 keV (cut-off value $\delta = 10$ keV). Mass energy absorption coefficients for dry air are according to Hubbell and Seltzer,[17] where a log–log interpolation has been applied to these tables.

Source: From National Nuclear Data Center (NNDC), Brookhaven National Laboratory NUDAT 2.0. Electronic Version available online at NNDC: www.nndc.bnl.gov/nudat2/, July 2005.

The average photon energy of gold is 0.406 MeV (see Table 5.6c and Table 5.2) making the radiation protection requirements much easier and cheaper to implement than those for ^{226}Ra, ^{222}Rn, ^{60}Co, or ^{137}Cs.

The main β^- rays have a maximum energy of 0.961 MeV and a mean energy of 0.315 MeV with an absolute emission probability of 99% (see Table 5.6a). These are easily filtered out by the gold itself and by thin layers of encapsulation materials.

Due to the very short half-life of approximately 2.7 days, ^{198}Au is only appropriate for permanent application techniques. Its specific activity is also the highest among the radionuclides considered in this chapter (see Table 5.2). However, with the advent of ^{125}I small sources for interstitial brachytherapy, ^{198}Au has now largely been replaced by ^{125}I.

5.6 ^{192}Iridium

Iridium has been used in brachytherapy since 1958, firstly as seeds by Ulrich Henschke and then, from the early 1960s, mainly as ^{192}Ir wires, forming the basis of the Paris System developed by Bernard Pierquin, Daniel Chassagne, Andree Dutreix, and others, mainly from the Institut Gustave Roussy.

Without doubt, ^{192}Ir has now become the most popular radionuclide as a replacement for ^{226}Ra.[5] Its half-life of 73.81 days allows ^{192}Ir to be easily used for temporary implants by making decay corrections of approximately 1% per day. The high specific activity of ^{192}Ir (see Table 5.2) makes it practical to supply sources of a very wide spectrum of activities to hundreds of GBqs. This makes very short treatment durations possible.

^{192}Ir decays mainly via β^-, almost 95% of the time, and in the majority to the third and fourth excited states of ^{192}Pt (see Figure 5.6a and Table 5.7a– Table 5.7c). The de-excitation to the ground state of ^{192}Pt occurs by almost 94% via gamma decay with emission of several γ-rays. The most frequent γ-rays are 0.296, 0.309, 0.317, and 0.468 MeV, emitted with absolute intensities of approximately 29, 30, 83, and 48% respectively (see Table 5.7c). The remaining 6% of the de-excitation occurs via IC. On average, 2.2 photons are emitted per β^- decay event with an average energy of 0.361 MeV.

There are six β^- rays emitted with a maximal energy of 0.675 MeV. The mean energy of all six emitted β^- rays is approximately 0.181 MeV (see Table 5.7a). In addition, many electrons are emitted as result of IC as CEs and as Auger electrons resulting from characteristic x-rays (see Table 5.7b). Although their energies are as high as 1.377 MeV, their emission probabilities are very low.

The second decay scheme of ^{192}Ir is that via EC, which is shown in Figure 5.6b (see also Table 5.8a and Table 5.8b) and occurs almost 5% of the time. ^{192}Ir decays via EC mainly to the fourth excited state of ^{192}Os. Here, the de-excitation to the ground state of ^{192}Os occurs almost 60% via gamma

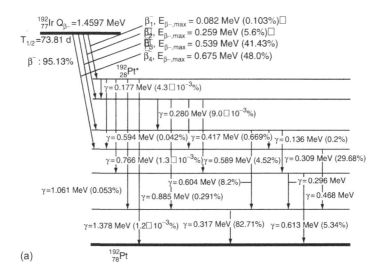

(a)

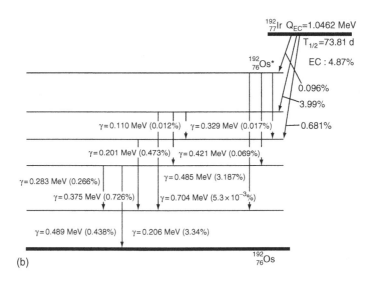

(b)

FIGURE 5.6
Schematic representations of the two different decay schemes of $^{192}_{77}$Ir. (a) β^- decay of $^{192}_{77}$Ir. This is the main disintegration process for this nuclide occurring in almost 95% of decays. $^{192}_{77}$Ir decays mainly to the third and fourth excited states of $^{192}_{78}$Pt. The disintegration energy for that decay is $Q_{\beta^-} = 1.4597$ MeV. On average, 2.2 photons are emitted with energy above 10 keV mainly with energies 0.296, 0.309, and 0.317 MeV (1.4 photons out of 2.2, see also Table 5.7c).[3] (b) EC decay of $^{192}_{77}$Ir. $^{192}_{77}$Ir decays mainly to the fourth excited state of $^{192}_{76}$Os. The disintegration energy for the decay is $Q_{EC} = 1.0462$ MeV. On average, 0.1 photons are emitted with energy above 10 keV mainly with energies 0.206 and 0.485 MeV (see also Table 5.7c).[3] In both decay schemes, several characteristic x-rays are emitted (see also Table 5.7c and Table 5.8b). The half-life for $^{192}_{77}$Ir decay is 73.81 days.[14]

TABLE 5.7a

Energies and Intensities of β^- Particles Emitted by β^- Decay of ^{192}Ir as Shown in Figure 5.6a

Mean Energy (MeV)	Maximum Energy (MeV)	Absolute Intensity (%)
0.0136	0.0535	0.0035
0.0195	0.0757	0.0039
0.0211	0.0817	0.103
0.0716	0.2587	5.6
0.1621	0.5388	41.43
0.2099	0.6751	48.0
Mean β^- energy	0.1807	95.13

Source: From National Nuclear Data Center (NNDC), Brookhaven National Laboratory NUDAT 2.0. Electronic Version available online at NNDC: www.nndc.bnl.gov/nudat2/, July 2005.

TABLE 5.7b

Energies and Intensities of Electrons Emitted by β^- Decay of ^{192}Ir as CEs and Auger Electrons

Electrons	Mean Energy (MeV)	Absolute Intensity (%)
Auger L	0.0072	8.0
Auger K	0.0510	0.39
CE K	0.0579	0.113
CE K	0.0986	0.00034
CE L	0.1225	0.149
CE M	0.1330	0.038
CE NP	0.1356	0.0112
CE L	0.1631	5.9×10^{-5}
CE M	0.1737	9.0×10^{-6}
CE NP	0.1763	1.2×10^{-5}
CE K	0.2019	0.001
CE K	0.2176	1.878
CE K	0.2301	1.805
CE K	0.2381	4.442
CE L	0.2664	0.00035
CE M	0.2770	9.0×10^{-5}
CE NP	0.2796	2.6×10^{-5}
CE L	0.2821	0.876
CE M	0.2927	0.2203
CE L	0.2946	0.778
CE NP	0.2952	0.0652
CE L	0.3026	1.952
CE M	0.3052	0.195
CE NP	0.3077	0.0579

(continued)

TABLE 5.7b—*Continued*

Electrons	Mean Energy (MeV)	Absolute Intensity (%)
CE M	0.3132	0.4888
CE NP	0.3158	0.1447
CE K	0.3381	0.025
CE K	0.3897	1.018
CE L	0.4026	0.0066
CE M	0.4132	0.00162
CE NP	0.4157	0.00049
CE L	0.4542	0.295
CE M	0.4648	0.0722
CE NP	0.4673	0.02161
CE K	0.5102	0.0583
CE K	0.5151	0.000232
CE K	0.5210	1.9×10^{-5}
CE K	0.5260	0.1747
CE K	0.5341	0.0635
CE L	0.5747	0.01405
CE L	0.5796	3.7×10^{-5}
CE M	0.5853	0.003392
CE L	0.5855	2.9×10^{-6}
CE NP	0.5879	0.001021
CE M	0.5902	1.38×10^{-5}
CE L	0.5905	0.03305
CE NP	0.5928	2.48×10^{-6}
CE M	0.5961	1.1×10^{-6}
CE L	0.5986	0.01485
CE NP	0.5987	2.0×10^{-7}
CE M	0.6011	0.00777
CE NP	0.6037	0.00237
CE M	0.6092	0.00357
CE NP	0.6117	0.001079
CE K	0.6874	6.0×10^{-6}
CE L	0.7519	9.0×10^{-7}
CE M	0.7625	1.8×10^{-7}
CE NP	0.7651	6.0×10^{-8}
CE K	0.8061	0.00165
CE L	0.8707	0.000314
CE M	0.8812	7.39×10^{-5}
CE NP	0.8838	2.25×10^{-5}
CE K	0.9831	8.8×10^{-5}
CE K	1.0115	6.2×10^{-6}
CE L	1.0476	1.29×10^{-5}
CE M	1.0582	3.07×10^{-6}
CE NP	1.0608	8.9×10^{-7}
CE L	1.0760	1.04×10^{-6}
CE M	1.0866	2.4×10^{-7}
CE NP	1.0892	7.4×10^{-8}

(continued)

TABLE 5.7b—*Continued*

Electrons	Mean Energy (MeV)	Absolute Intensity (%)
CE K	1.2998	5.9×10^{-6}
CE L	1.3643	1.2×10^{-6}
CE M	1.3749	2.8×10^{-7}
CE NP	1.3775	8.5×10^{-8}

Source: From National Nuclear Data Center (NNDC), Brookhaven National Laboratory NUDAT 2.0. Electronic Version available online at NNDC: www.nndc. bnl.gov/nudat2/, July 2005.

TABLE 5.7c

Energies and Intensities of γ-Rays as Result of γ-Decay and Characteristic X-Rays as Result of IC, by β$^-$ Decay of ^{192}Ir as Shown in Figure 5.6a

Photons	Energy (MeV)	Absolute Intensity (%)	$(\mu_{en}/\rho)_\alpha$ (cm^2 g^{-1})	Γ_δ (μGy h^{-1} MBq^{-1} m^2 = U MBq^{-1}) (Equation 2.63)
X L	0.0094	3.96	5.669	9.728×10^{-3}
X Kα2	0.0651	2.63	2.768×10^{-2}	2.176×10^{-4}
X Kα1	0.0668	4.46	2.700×10^{-2}	3.694×10^{-4}
X Kβ3	0.0754	0.533	2.475×10^{-2}	4.563×10^{-5}
X Kβ1	0.0757	1.025	2.468×10^{-2}	8.795×10^{-5}
X Kβ2	0.0778	0.365	2.435×10^{-2}	3.175×10^{-5}
γ	0.1363	0.2	2.438×10^{-2}	3.052×10^{-5}
γ	0.1770	0.0043	2.598×10^{-2}	9.076×10^{-7}
γ	0.2803	0.009	2.845×10^{-2}	3.294×10^{-6}
γ	0.2960	28.72	2.867×10^{-2}	1.118×10^{-2}
γ	0.3085	29.68	2.882×10^{-2}	1.211×10^{-2}
γ	0.3165	82.71	2.891×10^{-2}	3.474×10^{-2}
γ	0.4165	0.669	2.955×10^{-2}	3.779×10^{-4}
γ	0.4681	47.81	2.965×10^{-2}	3.045×10^{-2}
γ	0.4853	0.0023	2.966×10^{-2}	1.519×10^{-6}
γ	0.5886	4.517	2.956×10^{-2}	3.607×10^{-3}
γ	0.5935	0.0421	2.954×10^{-2}	3.388×10^{-5}
γ	0.5994	0.0039	2.953×10^{-2}	3.169×10^{-6}
γ	0.6044	8.2	2.952×10^{-2}	6.715×10^{-3}
γ	0.6125	5.34	2.950×10^{-2}	4.428×10^{-3}
γ	0.7658	0.0013	2.897×10^{-2}	1.324×10^{-6}
γ	0.8845	0.291	2.844×10^{-2}	3.360×10^{-4}
γ	1.0615	0.053	2.759×10^{-2}	7.125×10^{-5}
γ	1.0899	0.0012	2.745×10^{-2}	1.648×10^{-6}
γ	1.3782	0.0012	2.602×10^{-2}	1.976×10^{-6}
Mean energy E_{mean} (Equation 5.1)		0.3607 MeV		
Effective energy E_{eff} (Equation 5.2)		0.3987 MeV		

(continued)

TABLE 5.7c—*Continued*

Photons	Energy (MeV)	Absolute Intensity (%)	$(\mu_{en}/\rho)_\alpha$ (cm² g⁻¹)	Γ_δ (µGy h⁻¹ MBq⁻¹ m² = U MBq⁻¹) (Equation 2.63)
Effective mass energy absorption coefficient, $(\mu_{en}/\rho)_{\alpha,eff}$ (Equation 5.3)			2.915×10^{-2} cm² g⁻¹	
Air kerma-rate constant Γ_δ (Equation 2.64)			0.1049 µGy h⁻¹ MBq⁻¹ m²	
Mean number of photons emitted per decay			2.2 (2.1727)	

The mean energy E_{mean}, the effective energy E_{eff}, the air kerma-rate constant Γ_δ as well as the number of photons emitted per decay considers only rays above 10 keV (cut-off value $\delta = 10$ keV). Mass energy absorption coefficients for dry air are according to Hubbell and Seltzer,[17] where a log–log interpolation has been applied to these tables.
Source: From National Nuclear Data Center (NNDC), Brookhaven National Laboratory NUDAT 2.0. Electronic Version available online at NNDC: www.nndc.bnl.gov/nudat2/, July 2005.

decay. There are γ-rays emitted with energies in the range of 0.110 to 0.704 MeV where the 0.206 and 0.485 MeV energy lines have the highest absolute emission probabilities of approximately 3.3% each (see Table 5.8b).

On average, 0.1 photons are emitted with a mean energy of 0.252 MeV. There are also several characteristic x-rays and many conversion electrons, as well as Auger electrons, emitted (see Table 5.8a and Table 5.8b).

TABLE 5.8a

Energies and Intensities of Electrons Emitted by EC Decay of ^{192}Ir as CEs and Auger Electrons

Electrons	Mean Energy (MeV)	Absolute Intensity (%)
Auger L	0.0069	3.43
CE K	0.0365	0.035
Auger K	0.0483	0.189
CE L	0.0974	0.0107
CE M	0.1074	0.0026
CE NP	0.1097	0.00078
CE K	0.1274	0.108
CE K	0.1319	0.524
CE L	0.1883	0.0568
CE L	0.1928	0.371
CE M	0.1983	0.0142
CE NP	0.2007	0.00412
CE M	0.2027	0.0939
CE NP	0.2051	0.0271
CE K	0.2094	0.0215
CE K	0.2553	0.0015

(continued)

TABLE 5.8a—*Continued*

Electrons	Mean Energy (MeV)	Absolute Intensity (%)
CE L	0.2703	0.00846
CE M	0.2802	0.00209
CE NP	0.2826	0.000606
CE K	0.3006	0.02476
CE L	0.3162	0.00036
CE M	0.3261	8.7×10^{-5}
CE NP	0.3285	2.6×10^{-5}
CE K	0.3467	0.00178
CE L	0.3615	0.0082
CE M	0.3714	0.002004
CE NP	0.3738	0.000586
CE L	0.4076	0.00053
CE K	0.4107	0.0631
CE K	0.4152	0.008
CE M	0.4175	0.000128
CE NP	0.4199	3.8×10^{-5}
CE L	0.4716	0.0159
CE L	0.4761	0.00206
CE M	0.4815	0.00382
CE NP	0.4839	0.001128
CE M	0.4860	0.000499
CE NP	0.4884	0.000147
CE K	0.6300	4.4×10^{-5}
CE L	0.6909	8.9×10^{-6}
CE M	0.7008	2.1×10^{-6}
CE NP	0.7032	6.3×10^{-7}

Source: From National Nuclear Data Center (NNDC), Brookhaven National Laboratory NUDAT 2.0. Electronic Version available online at NNDC: www.nndc.bnl.gov/nudat2/, July 2005.

TABLE 5.8b

Energies and Intensities of γ-Rays as Result of γ-Decay and Characteristic X-Rays by EC Decay of ^{192}Ir as Shown in Figure 5.6b

Photons	Energy (MeV)	Absolute Intensity (%)	$(\mu_{en}/\rho)_\alpha$ $(cm^2 g^{-1})$	Γ_δ (μGy h^{-1} MBq^{-1} m^2 = U MBq^{-1}) (Equation 2.63)
X L	0.0089	1.53	6.778	4.241×10^{-3}
X Kα2	0.0615	1.2	2.949×10^{-2}	9.988×10^{-5}
X Kα1	0.0630	2.05	2.867×10^{-2}	1.699×10^{-4}
X Kβ3	0.0711	0.241	2.568×10^{-2}	2.019×10^{-5}
X Kβ1	0.0714	0.466	2.559×10^{-2}	3.909×10^{-5}
X Kβ2	0.0734	0.163	2.514×10^{-2}	1.380×10^{-5}

(continued)

TABLE 5.8b—*Continued*

Photons	Energy (MeV)	Absolute Intensity (%)	$(\mu_{en}/\rho)_\alpha$ $(cm^2\,g^{-1})$	Γ_δ (μGy h^{-1} MBq^{-1} m^2 = U MBq^{-1}) (Equation 2.63)
γ	0.1104	0.0122	2.342×10^{-2}	1.448×10^{-6}
γ	0.2013	0.473	2.676×10^{-2}	1.169×10^{-4}
γ	0.2058	3.34	2.689×10^{-2}	8.482×10^{-4}
γ	0.2833	0.266	2.849×10^{-2}	9.855×10^{-5}
γ	0.3292	0.0174	2.904×10^{-2}	7.633×10^{-6}
γ	0.3745	0.726	2.937×10^{-2}	3.665×10^{-4}
γ	0.4205	0.069	2.956×10^{-2}	3.937×10^{-5}
γ	0.4846	3.187	2.966×10^{-2}	2.102×10^{-3}
γ	0.4891	0.438	2.966×10^{-2}	2.916×10^{-4}
γ	0.7039	0.0053	2.921×10^{-2}	5.002×10^{-6}
Mean energy E_{mean} (Equation 5.1)			0.2524 MeV	
Effective energy E_{eff} (Equation 5.2)			0.3719 MeV	
Effective mass energy absorption coefficient, $(\mu_{en}/\rho)_{\alpha,eff}$ (Equation 5.3)			$2.879 \times 10^{-2}\,cm^2\,g^{-1}$	
Air kerma-rate constant Γ_δ (Equation 2.64)			$0.0042\,\mu$Gy h^{-1} MBq^{-1} m^2	
Mean number of photons emitted per decay			0.1 (0.1265)	

The mean energy E_{mean}, the effective energy E_{eff}, the air kerma-rate constant Γ_δ as well as the number of photons emitted per decay considers only rays above 10 keV (cut-off value $\delta = 10$ keV). Mass energy absorption coefficients for dry air are according to Hubbell and Seltzer,[17] where a log–log interpolation has been applied to these tables.
Source: From National Nuclear Data Center (NNDC), Brookhaven National Laboratory NUDAT 2.0. Electronic Version available online at NNDC: www.nndc.bnl.gov/nudat2/, July 2005.

Also, on average, 2.3 photons are emitted by both decay processes resulting in an average energy of the emitted photons of 0.355 MeV (Table 5.9) where the average energy of only the emitted γ-rays is 0.372 MeV (Table 5.2).

Finally, it is noted that ^{192}Ir is now fully established as the preferred radionuclide for temporary brachytherapy applications for virtually all tumor sites The low costs associated with its production and the half-life of 73.81 days make storage of sources practical. ^{192}Ir is also available as high-activity miniaturized sources.

5.7 ^{125}Iodine

Allen Reid and Albert Keston discovered the ^{125}I radionuclide in 1946,[6] but it took almost 20 years until ^{125}I sources became clinically available for interstitial brachytherapy.

TABLE 5.9

Mean Energy and Mean Number of Emitted Photons of γ-Rays as Result of
γ-Decay and Characteristic X-Rays, by Both β^- and EC Decay Processes of ^{192}Ir

Mean energy E_{mean} (Equation 5.1)	0.3547 MeV
Effective energy E_{eff} (Equation 5.2)	0.3977 MeV
Effective mass energy absorption coefficient, $(\mu_{en}/\rho)_{\alpha,eff}$ (Equation 5.3)	2.914×10^{-2} cm^2 g^{-1}
Air kerma-rate constant Γ_δ (Equation 2.64)	0.1091 μGy h^{-1} MBq^{-1} m^2
Mean number of photons emitted per decay	2.3 (2.299)

The mean energy E_{mean}, the effective energy E_{eff}, the air kerma-rate constant Γ_δ as well as the number of photons emitted per decay considers only rays above 10 keV (cut-off value $\delta = 10$ keV). Mass energy absorption coefficients for dry air are according to Hubbell and Seltzer,[17] where a log–log interpolation has been applied to these tables.

^{125}I decays via electron capture to the first excited state of ^{125}Te (see Figure 5.7 and Table 5.10a and Table 5.10b). The de-excitation to the ground state of ^{125}Te occurs almost 7% via emission of γ-rays (gamma decay) with the majority of almost 93% via IC which results in emission of characteristic x-rays. On average, 1.4 photons are emitted per disintegration of ^{125}I. Table 5.10b provides the photon emission spectrum for ^{125}I.

The low energy electrons emitted (maximum energy of 0.035 MeV with an intensity of approximately 2%, Table 5.10a) can be easily filtered by the iodine as well as by very thin layers of encapsulation material. The low energy of the emitted photons (mean energy of 0.028 MeV, Table 5.10b)

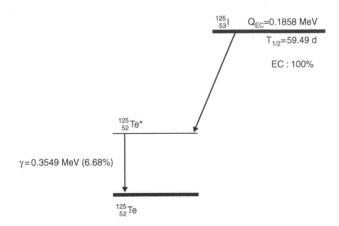

FIGURE 5.7
Schematic representation of EC decay of $^{125}_{53}$I to the first excited state of $^{125}_{52}$Te. The disintegration energy for the decay is $Q_{EC} = 0.1858$ MeV. There is a single γ-ray of 0.035 MeV with an intensity of 6.68% emitted, whereas there are several characteristic x-rays emitted in the range 0.027 to 0.032 MeV as a result of IC processes (see also Table 5.10b).[3] The average number of photons emitted per disintegration with energy above 10 keV is 1.4. The half-life for $^{125}_{53}$I decay is 59.49 days.[14]

TABLE 5.10a

Energies and Intensities of Electrons Emitted by EC Decay of ^{125}I as CEs and Auger Electrons

Electrons	Mean Energy (MeV)	Absolute Intensity (%)
Auger L	0.0032	158.4
CE K	0.0037	80.2
Auger K	0.0227	20.0
CE L	0.0306	10.75
CE M	0.0345	2.15

Source: From National Nuclear Data Center (NNDC), Brookhaven National Laboratory NUDAT 2.0. Electronic Version available online at NNDC: www.nndc.bnl.gov/nudat2/, July 2005.

TABLE 5.10b

Energies and Intensities of γ-Rays as Result of γ-Decay and Characteristic X-Rays by EC Decay of ^{125}I as Shown in Figure 5.7

Photons	Energy (MeV)	Absolute Intensity (%)	$(\mu_{en}/\rho)_{\alpha}$ (cm^2 g^{-1})	Γ_δ (μGy h^{-1} MBq^{-1} m^2 = U MBq^{-1}) (Equation 2.63)
X L	0.0038	14.9	$8.978 \times 10^{+1}$	2.315×10^{-1}
X Kα2	0.0272	40.1	2.067×10^{-1}	1.035×10^{-2}
X Kα1	0.0275	74.0	2.005×10^{-1}	1.871×10^{-2}
X Kβ3	0.0305	6.83	1.464×10^{-1}	1.399×10^{-3}
X Kβ1	0.0310	13.2	1.395×10^{-1}	2.619×10^{-3}
X Kβ2	0.0317	3.8	1.305×10^{-1}	7.214×10^{-4}
γ	0.0355	6.68	9.429×10^{-2}	1.026×10^{-3}
Mean energy E_{mean} (Equation 5.1)			0.0283 MeV	
Effective energy E_{eff} (Equation 5.2)			0.0281 MeV	
Effective mass energy absorption coefficient, $(\mu_{en}/\rho)_{\alpha,eff}$ (Equation 5.3)			1.851×10^{-1} cm^2 g^{-1}	
Air kerma-rate constant Γ_δ (Equation 2.64)			0.0348 μGy h^{-1} MBq^{-1} m^2	
Mean number of photons emitted per decay			1.5 (1.446)	

The mean energy E_{mean}, the effective energy E_{eff}, the air kerma-rate constant Γ_δ as well as the number of photons emitted per decay considers only rays above 10 keV (cut-off value $\delta = 10$ keV). Mass energy absorption coefficients for dry air are according to Hubbell and Seltzer,[17] where a log–log interpolation has been applied to these tables.
Source: From National Nuclear Data Center (NNDC), Brookhaven National Laboratory NUDAT 2.0. Electronic Version available online at NNDC: www.nndc.bnl.gov/nudat2/, July 2005.

allows minimum efforts regarding shielding (HVL of lead is approximately 0.03 mm, Table 5.2) and enables easy and safe handling of the sources.

The relatively long half-life of 59.49 days, compared for example to ^{222}Rn and ^{198}Au which ^{125}I has made obsolete, makes ^{125}I sources convenient for storage and enables their use for permanent as well as for temporary implants. However, the complex energy spectrum of the emitted photons (γ-rays and x-rays) makes the dosimetry of ^{125}I sources much more complicated than the dosimetry for ^{137}Cs or ^{60}Co sources.

The high specific activity of ^{125}I enables the production of miniature sources with sufficient activity for use in both permanent and temporary implants. Because of this factor, ^{125}I seed sources have been used for temporary and permanent interstitial implants for brain tumors; and permanent interstitial implants for lung, pancreas, breast, and prostate cancer, as well as for temporary implants for pediatric tumors.[7] The main current application field of ^{125}I radionuclide brachytherapy is permanent interstitial implants for prostate cancer.

5.8 ^{103}Palladium

Brachytherapy seed sources containing ^{103}Pd have been introduced as replacements for ^{125}I sources for permanent implants. Its very short half-life of approximately 17 days makes ^{103}Pd only appropriate for permanent implant treatments. Its very high specific activity (see Table 5.2) in combination with its short half-life enables dose delivery at an initial dose rate higher than that obtained when using ^{125}I. This feature should be of benefit when it is used for interstitial implantation of rapidly proliferating tumors, since the accumulation of dose with ^{103}Pd occurs in a shorter time period than for ^{125}I.

^{103}Pd decays via EC mainly to the first excited state of ^{103}Rh (99.9%, see Figure 5.8 and Table 5.11a and Table 5.11b). The de-excitation to the ground state of ^{103}Rh occurs in the majority (77%) via IC with the emission of characteristic x-rays and only in a small amount via emission of γ-rays (gamma decay). The mean energy of the emitted photons of approximately 0.021 MeV (see Table 5.11b) is lower than that of ^{125}I. On average, 0.8 photons are emitted per disintegration of ^{103}Pd.

Table 5.11a gives the energies and absolute intensities of electrons, CEs emitted as result of IC, and Auger electrons induced by x-rays. Table 5.11b provides the photon emission spectrum, γ-rays as result of γ-decay, and characteristic s-rays by EC decay for ^{103}Pd.

^{103}Pd seeds are currently used mainly for permanent implants of prostate cancer.

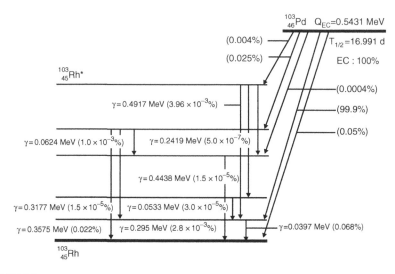

FIGURE 5.8

Schematic representation of EC decay of $^{103}_{46}$Pd to the different excited states and the ground state of $^{103}_{45}$Rh. The disintegration energy for the decay is $Q_{EC} = 0.5431$ MeV. The γ-rays emitted have very low, almost negligible intensities. There are several characteristic x-rays emitted in the range 0.020 to 0.023 MeV (see also Table 5.11b).[3] The average number of photons emitted per disintegration with energy above 10 keV is 0.8. The half-life for $^{103}_{46}$Pd decay is 16.991 days.[3]

TABLE 5.11a

Energies and Intensities of Electrons Emitted by EC Decay of ^{103}Pd as CEs and Auger Electrons

Electrons	Mean Energy (MeV)	Absolute Intensity (%)
Auger L	0.0024	168.0
CE K	0.0165	9.52
Auger K	0.0170	18.2
CE L	0.0363	71.2
CE M	0.0391	14.38
CE K	0.0392	0.00119
CE L	0.0590	0.000147
CE M	0.0618	2.73×10^{-5}
CE NP	0.0623	5.33×10^{-6}
CE K	0.3342	0.000302
CE L	0.3540	3.97×10^{-5}
CE M	0.3568	7.5×10^{-6}

Source: From National Nuclear Data Center (NNDC), Brookhaven National Laboratory NUDAT 2.0. Electronic Version available online at NNDC: www.nndc.bnl.gov/nudat2/, July 2005.

TABLE 5.11b

Energies and Intensities of γ-Rays as Result of γ-Decay and Characteristic X-Rays by EC Decay of ^{103}Pd as Shown in Figure 5.8

Photons	Energy (MeV)	Absolute Intensity (%)	$(\mu_{en}/\rho)_\alpha$ $(cm^2\,g^{-1})$	Γ_δ ($\mu Gy\,h^{-1}\,MBq^{-1}\,m^2$ = U MBq^{-1}) (Equation 2.63)
X L	0.0027	8.73	3.658E + 02	3.957E − 01
X Kα2	0.0201	22.4	5.327E − 01	1.099E − 02
X Kα1	0.0202	42.5	5.210E − 01	2.055E − 02
X Kβ3	0.0227	3.54	3.620E − 01	1.335E − 03
X Kβ1	0.0227	6.85	3.608E − 01	2.578E − 03
X Kβ2	0.0232	1.64	3.394E − 01	5.920E − 04
γ	0.0397	0.0683	6.945E − 02	8.655E − 06
γ	0.0533	3.0x10^{-5}	3.643E − 02	2.673E − 09
γ	0.0624	0.00104	2.898E − 02	8.633E − 08
γ	0.2419	5.0 × 10^{-7}	2.776E − 02	1.541E − 10
γ	0.2950	0.0028	2.866E − 02	1.086E − 06
γ	0.3177	1.5 × 10^{-5}	2.892E − 02	6.327E − 09
γ	0.3575	0.0221	2.926E − 02	1.061E − 05
γ	0.4438	1.5 × 10^{-5}	2.961E − 02	9.049E − 09
γ	0.4971	0.00396	2.966E − 02	2.680E − 06
Mean energy E_{mean} (Equation 5.1)		0.0207 MeV		
Effective energy E_{eff} (Equation 5.2)		0.0206 MeV		
Effective mass energy absorption coefficient, $(\mu_{en}/\rho)_{\alpha,eff}$ (Equation 5.3)		$4.922 \times 10^{-1}\,cm^2\,g^{-1}$		
Air kerma-rate constant Γ_δ (Equation 2.64)		0.0361 $\mu Gy\,h^{-1}\,MBq^{-1}\,m^2$		
Mean number of photons emitted per decay		0.8 (0.77)		

The mean energy E_{mean}, the effective energy E_{eff}, the air kerma-rate constant Γ_δ as well as the number of photons emitted per decay considers only rays above 10 keV (cut-off value $\delta = 10$ keV). Mass energy absorption coefficients for dry air are according to Hubbell and Seltzer,[17] where a log–log interpolation has been applied to these tables.
Source: From National Nuclear Data Center (NNDC), Brookhaven National Laboratory NUDAT 2.0. Electronic Version available online at NNDC: www.nndc.bnl.gov/nudat2/, July 2005.

5.9 ^{169}Ytterbium

^{169}Yb was discovered using artificial activation techniques in 1946, but its initial applications were industrial. It was not until 1990 that the first clinical application of ^{169}Yb sources took place in London, Ontario, Canada. This was followed in 1993 by a detailed report on the clinical application of the new ^{169}Yb brachytherapy seed source.[8] These seeds have been manually loaded in preimplanted catheters (for a temporary perineal implant) and also, using a total activity of approximately 3700 MBq, for treatment of a recurrence of rectal adenocarcinoma.

^{169}Yb is potentially applicable for permanent or temporary implants due to its intermediate half-life of 32 days and its relative high specific activity which is higher than that of ^{192}Ir and ^{125}I but lower than that of ^{103}Pd.

Considering the half-life and the specific activity of ^{169}Yb, it can be at least theoretically resumed that small sources of adequate activity can be produced for either permanent or temporary implants. Thus, a faster accumulation of dose over time can be achieved for permanent implants than that when ^{192}Ir is used.

Alternatively, considering this property, especially for rapidly proliferating tumors, ^{169}Yb is not advantageous when compared to ^{103}Pd which has an even shorter half-life of almost 17 days. In addition, the higher energy of ^{169}Yb, a mean energy of 0.093 MeV (see Table 5.12a), makes radiation protection for permanent implants with ^{169}Yb more problematic when compared to ^{125}I and ^{103}Pd.

It is likely that ^{169}Yb sources could be suitable for temporary implants using sources of high activity and could offer a potential substitute for ^{192}Ir.[8–13,20,22] The low first HVL in lead of ytterbium (almost a factor 10 lower than that for ^{192}Ir, see Table 5.2) clearly promises lower costs for radiation protection for the treatments rooms and environment when compared to iridium.

TABLE 5.12a

Energies and Intensities of γ-Rays as Result of γ-Decay and Characteristic X-Rays by EC Decay of ^{169}Yb as Shown in Figure 5.9

Photons	Energy (MeV)	Absolute Intensity (%)	$(\mu_{en}/\rho)_\alpha$ (cm^2 g^{-1})	Γ_δ (μGy h^{-1} MBq^{-1} m^2 = U MBq^{-1}) (Equation 2.63)
X L	0.0072	48.1	$1.316 \times 10^{+1}$	2.086×10^{-1}
γ	0.0084	0.333	8.099	1.041×10^{-3}
γ	0.0208	0.19	4.798×10^{-1}	8.683×10^{-5}
γ	0.0428	0.13	5.783×10^{-2}	1.475×10^{-5}
γ	0.0459	0.005	4.894×10^{-2}	5.160×10^{-7}
X Kα2	0.0498	53.2	4.135×10^{-2}	5.025×10^{-3}
γ	0.0506	0.3	4.003×10^{-2}	2.790×10^{-5}
X Kα1	0.0507	92.7	3.983×10^{-2}	8.599×10^{-3}
γ	0.0509	0.3	3.966×10^{-2}	2.777×10^{-5}
γ	0.0515	0.009	3.872×10^{-2}	8.239×10^{-7}
X Kβ3	0.0573	9.99	3.242×10^{-2}	8.517×10^{-4}
X Kβ1	0.0575	19.3	3.225×10^{-2}	1.643×10^{-3}

(continued)

TABLE 5.12a—*Continued*

Photons	Energy (MeV)	Absolute Intensity (%)	$(\mu_{en}/\rho)_\alpha$ $(\text{cm}^2\,\text{g}^{-1})$	$\Gamma_\delta\ (\mu\text{Gy h}^{-1}\,\text{MBq}^{-1}\,\text{m}^2$ $= \text{U MBq}^{-1})$ (Equation 2.63)
X Kβ2	0.0590	6.49	3.108×10^{-2}	5.465×10^{-4}
γ	0.0630	1.1	2.866×10^{-2}	9.119×10^{-5}
γ	0.0631	44.2	2.861×10^{-2}	3.663×10^{-3}
γ	0.0659	0.005	2.737×10^{-2}	4.137×10^{-7}
γ	0.0720	0.0018	2.544×10^{-2}	1.514×10^{-7}
γ	0.0851	0.0014	2.361×10^{-2}	1.291×10^{-7}
γ	0.0936	2.61	2.328×10^{-2}	2.611×10^{-4}
γ	0.0957	0.0011	2.325×10^{-2}	1.124×10^{-7}
γ	0.0959	0.0011	2.325×10^{-2}	1.125×10^{-7}
γ	0.0980	0.0009	2.324×10^{-2}	9.411×10^{-8}
γ	0.1014	0.004	2.326×10^{-2}	4.331×10^{-7}
γ	0.1052	0.0026	2.331×10^{-2}	2.926×10^{-7}
γ	0.1098	17.5	2.341×10^{-2}	2.064×10^{-3}
γ	0.1136	0.005	2.351×10^{-2}	6.131×10^{-7}
γ	0.1140	0.004	2.352×10^{-2}	4.923×10^{-7}
γ	0.1174	0.0397	2.363×10^{-2}	5.055×10^{-6}
γ	0.1182	1.87	2.366×10^{-2}	2.400×10^{-4}
γ	0.1299	0.3	2.411×10^{-2}	4.315×10^{-5}
γ	0.1305	11.31	2.414×10^{-2}	1.636×10^{-3}
γ	0.1567	0.01	2.523×10^{-2}	1.815×10^{-6}
γ	0.1739	0.0014	2.588×10^{-2}	2.891×10^{-7}
γ	0.1772	22.2	2.599×10^{-2}	4.694×10^{-3}
γ	0.1932	0.0074	2.651×10^{-2}	1.739×10^{-6}
γ	0.1980	35.8	2.666×10^{-2}	8.672×10^{-3}
γ	0.1998	0.016	2.671×10^{-2}	3.919×10^{-6}
γ	0.2060	0.00408	2.689×10^{-2}	1.037×10^{-6}
γ	0.2139	0.0029	2.711×10^{-2}	7.719×10^{-7}
γ	0.2263	0.00025	2.742×10^{-2}	7.119×10^{-8}
γ	0.2287	0.0002	2.747×10^{-2}	5.768×10^{-8}
γ	0.2403	0.1138	2.773×10^{-2}	3.481×10^{-5}
γ	0.2611	1.71	2.813×10^{-2}	5.765×10^{-4}
γ	0.2912	0.0043	2.861×10^{-2}	1.644×10^{-6}
γ	0.2945	0.001	2.865×10^{-2}	3.873×10^{-7}
γ	0.3017	0.0023	2.874×10^{-2}	9.155×10^{-7}
γ	0.3068	0.09	2.880×10^{-2}	3.651×10^{-5}
γ	0.3075	0.3	2.881×10^{-2}	1.220×10^{-4}
γ	0.3077	10.05	2.881×10^{-2}	4.090×10^{-3}
γ	0.3340	0.00179	2.908×10^{-2}	7.979×10^{-7}
γ	0.3366	0.00909	2.910×10^{-2}	4.087×10^{-6}
γ	0.3567	0.00014	2.926×10^{-2}	6.707×10^{-8}
γ	0.3709	0.0072	2.935×10^{-2}	3.597×10^{-6}
γ	0.3793	0.00122	2.939×10^{-2}	6.243×10^{-7}
γ	0.3867	0.00034	2.943×10^{-2}	1.776×10^{-7}
γ	0.4526	1.60×10^{-5}	2.963×10^{-2}	9.849×10^{-9}

(continued)

TABLE 5.12a—*Continued*

Photons	Energy (MeV)	Absolute Intensity (%)	$(\mu_{en}/\rho)_\alpha$ (cm² g⁻¹)	Γ_δ (μGy h⁻¹ MBq⁻¹ m² = U MBq⁻¹) (Equation 2.63)
γ	0.4647	3.60×10^{-6}	2.964×10^{-2}	2.276×10^{-9}
γ	0.4657	0.00019	2.965×10^{-2}	1.204×10^{-7}
γ	0.4666	1.93×10^{-5}	2.965×10^{-2}	1.225×10^{-8}
γ	0.4750	0.000193	2.965×10^{-2}	1.248×10^{-7}
γ	0.4944	0.00149	2.966×10^{-2}	1.003×10^{-6}
γ	0.5004	8.80×10^{-6}	2.966×10^{-2}	5.994×10^{-9}
γ	0.5078	1.50×10^{-6}	2.966×10^{-2}	1.037×10^{-9}
γ	0.5151	0.00414	2.966×10^{-2}	2.903×10^{-6}
γ	0.5286	0.000175	2.965×10^{-2}	1.259×10^{-7}
γ	0.5462	1.50×10^{-6}	2.963×10^{-2}	1.114×10^{-9}
γ	0.5624	0.000118	2.960×10^{-2}	9.018×10^{-8}
γ	0.5709	0.000125	2.959×10^{-2}	9.692×10^{-8}
γ	0.5799	0.00192	2.957×10^{-2}	1.511×10^{-6}
γ	0.6006	0.00113	2.953×10^{-2}	9.199×10^{-7}
γ	0.6249	0.0049	2.947×10^{-2}	4.141×10^{-6}
γ	0.6333	6.90×10^{-6}	2.944×10^{-2}	5.906×10^{-9}
γ	0.6429	7.59×10^{-5}	2.941×10^{-2}	6.588×10^{-8}
γ	0.6636	0.000189	2.935×10^{-2}	1.690×10^{-7}
γ	0.6935	8.70×10^{-6}	2.925×10^{-2}	8.099×10^{-9}
γ	0.7104	3.11×10^{-5}	2.919×10^{-2}	2.959×10^{-8}
γ	0.7394	1.83×10^{-6}	2.907×10^{-2}	1.806×10^{-9}
γ	0.7602	8.20×10^{-7}	2.899×10^{-2}	8.295×10^{-10}
γ	0.7734	0.000207	2.893×10^{-2}	2.126×10^{-7}
γ	0.7816	3.00×10^{-6}	2.890×10^{-2}	3.110×10^{-9}
Mean energy E_{mean} (Equation 5.1)			0.0933 MeV	
Effective energy E_{eff} (Equation 5.2)			0.1307 MeV	
Effective mass energy absorption coefficient, $(\mu_{en}/\rho)_{\alpha,eff}$ (Equation 5.3)			3.032×10^{-2} cm² g⁻¹	
Air kerma-rate constant Γ_δ (Equation 2.64)			0.0431 μGy h⁻¹ MBq⁻¹ m²	
Mean number of photons emitted per decay			3.3 (3.319)	

The mean energy E_{mean}, the effective energy E_{eff}, the air kerma-rate constant Γ_δ as well as the number of photons emitted per decay considers only rays above 10 keV (cut-off value $\delta = 10$ keV). Mass energy absorption coefficients for dry air are according to Hubbell and Seltzer,[17] where a log–log interpolation has been applied to these tables.

Source: From National Nuclear Data Center (NNDC), Brookhaven National Laboratory NUDAT 2.0. Electronic Version available online at NNDC: www.nndc.bnl.gov/nudat2/, July 2005.

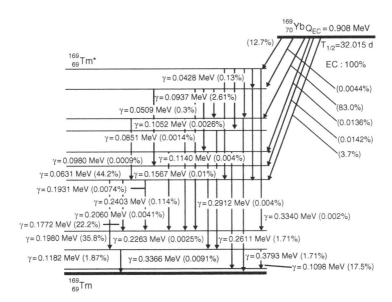

FIGURE 5.9

Schematic representation of EC decay of $^{169}_{70}$Yb to $^{169}_{69}$Tm. Here, only the disintegration to those excited states of $^{169}_{69}$Tm with the highest occurrence probability are shown. These are the tenth, ninth, eighth, sixth, fifth, and fourth excited levels. In 83% of the time, the decay goes through the eighth level.[3] The disintegration energy for that decay is $Q_{EC} = 0.908$ MeV. In total, 74 γ-rays are emitted (see Table 5.12a) and over 107 different electron energies are emitted (see Table 5.12b) as Auger electrons or as CEs resulting from IC. On average, 3.3 photons are emitted per disintegration with a mean energy of 0.0933 MeV. The half-life for EC decay of $^{169}_{70}$Yb is 32.015 days.[14]

^{169}Yb decays via electron capture to several excited states of ^{169}Tm (see Figure 5.9). The de-excitation to the ground state of ^{169}Tm occurs almost by 45% via emission of γ-rays (gamma decay) and by almost 55% via IC, which results in emission of several characteristic x-rays (see Table 5.12a). On average, 3.3 photons are emitted per disintegration of ^{169}Yb.

The emitted electrons are mainly of low energy (see Table 5.12b) so that they are easily filtered out either by the ytterbium itself or by thin layers of encapsulation materials commonly used. The complexity of the energy spectrum of the emitted photons (γ-rays and x-rays) makes the dosimetry of ^{169}Yb sources more complicated than the dosimetry for ^{137}Cs or ^{60}Co sources.

A possible further radioactive isotope produced in the reactor from the naturally occuring isotope ^{174}Yb is ^{175}Yb (see Chapter 6). The decay scheme of ^{175}Yb is shown in Figure 5.10, where the energies and intensities of the emitted electrons and photons are listed in Table 5.13a through Table 5.13c.

TABLE 5.12b

Energies and Intensities of Electrons Emitted by EC Decay of ^{169}Yb as CEs and Auger Electrons

Electrons	Mean Energy (MeV)	Absolute Intensity (%)
CE K	0.0037	40.4
Auger L	0.0057	164
CE M	0.0061	72
CE NP	0.0079	23.3
CE L	0.0106	8.3
CE M	0.0184	1.86
CE NP	0.0203	0.61
CE K	0.0342	8.24
Auger K	0.0409	10.8
CE K	0.0458	0.00063
CE K	0.0504	34.9
CE L	0.0530	7.21
CE K	0.0580	0.048
CE K	0.0588	1.32
CE M	0.0608	1.6
CE NP	0.0626	0.433
CE K	0.0711	6.19
CE L	0.0835	1.4
CE M	0.0913	0.315
CE NP	0.0931	0.091
CE L	0.0951	0.0001
CE K	0.0973	0.00085
CE L	0.0997	5.7
CE M	0.1029	2.2×10^{-5}
CE NP	0.1047	6.3×10^{-6}
CE L	0.1073	0.020
CE M	0.1075	1.28
CE L	0.1081	1.36
CE NP	0.1093	0.372
CE M	0.1151	0.0048
CE M	0.1159	0.329
CE NP	0.1169	0.0014
CE NP	0.1177	0.0921
CE K	0.1178	10.61
CE L	0.1204	5.28
CE M	0.1282	1.278
CE NP	0.1301	0.355
CE K	0.1338	0.003
CE K	0.1386	13.03
CE K	0.1466	0.000171
CE L	0.1466	0.000131
CE M	0.1544	2.9×10^{-5}
CE K	0.1545	0.00039
CE NP	0.1563	8.1×10^{-6}

(continued)

TABLE 5.12b—*Continued*

Electrons	Mean Energy (MeV)	Absolute Intensity (%)
CE L	0.1671	1.91
CE K	0.1693	2.2×10^{-5}
CE M	0.1749	0.43
CE NP	0.1767	0.123
CE K	0.1809	0.013
CE L	0.1830	0.00046
CE L	0.1878	2.15
CE M	0.1908	0.000102
CE NP	0.1927	2.9×10^{-5}
CE M	0.1957	0.483
CE L	0.1959	2.56×10^{-5}
CE NP	0.1975	0.137
CE K	0.2017	0.046
CE M	0.2037	5.67×10^{-6}
CE L	0.2038	0.000165
CE NP	0.2055	1.58×10^{-6}
CE M	0.2116	3.9×10^{-5}
CE NP	0.2135	1.08×10^{-5}
CE L	0.2186	9.0×10^{-6}
CE M	0.2264	2.0×10^{-6}
CE NP	0.2282	6.0×10^{-7}
CE L	0.2302	0.002
CE K	0.2318	7.5×10^{-5}
CE M	0.2380	0.0006
CE NP	0.2399	0.00016
CE K	0.2483	0.487
CE L	0.2510	0.007
CE M	0.2588	0.0016
CE NP	0.2606	0.00045
CE L	0.2811	1.1×10^{-5}
CE M	0.2889	2.42×10^{-6}
CE NP	0.2907	6.7×10^{-7}
CE L	0.2976	0.141
CE M	0.3054	0.0330
CE NP	0.3073	0.00908
CE K	0.3115	0.00178
CE K	0.3199	9.3×10^{-5}
CE K	0.3273	1.5×10^{-5}
CE L	0.3607	0.00033
CE M	0.3685	7.5×10^{-5}
CE L	0.3692	4.8×10^{-5}
CE NP	0.3704	2.12×10^{-5}
CE L	0.3766	2.7×10^{-5}
CE M	0.3770	1.18×10^{-5}
CE NP	0.3788	3.2×10^{-6}
CE M	0.3844	6.1×10^{-7}
CE NP	0.3862	1.7×10^{-7}
CE K	0.3932	4.1×10^{-7}

(continued)

TABLE 5.12b—*Continued*

Electrons	Mean Energy (MeV)	Absolute Intensity (%)
CE L	0.4425	7.4×10^{-8}
CE M	0.4503	1.7×10^{-8}
CE NP	0.4522	4.6×10^{-9}
CE K	0.4557	0.000125
CE K	0.5030	2.1×10^{-6}
CE L	0.5050	1.83×10^{-5}
CE K	0.5115	2.3×10^{-6}
CE K	0.5205	4.28×10^{-5}
CE K	0.5412	2.31×10^{-5}
CE L	0.5523	3.3×10^{-7}
CE L	0.5608	3.5×10^{-7}
CE K	0.5655	9.1×10^{-5}
CE L	0.5697	6.24×10^{-6}
CE L	0.5905	3.36×10^{-6}
CE L	0.6148	1.31×10^{-5}

Source: From National Nuclear Data Center (NNDC), Brookhaven National Laboratory NUDAT 2.0. Electronic Version available online at NNDC: www.nndc.bnl.gov/nudat2/, July 2005.

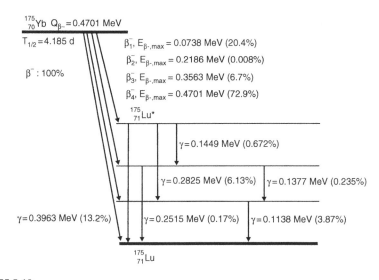

FIGURE 5.10
Schematic representation of β^- decay of $^{175}_{70}$Yb to $^{175}_{71}$Lu. The disintegration energy for that decay is $Q_{EC} = 0.4701$ MeV. In total, six γ-rays and six characteristic x-rays are emitted (see Table 5.13c). On average, 0.32 photons are emitted per disintegration with a mean energy of 0.2522 MeV. The half-life for β^- decay of $^{175}_{70}$Yb is 4.185 days.[3]

TABLE 5.13a

Energies and Intensities of β^- Particles Emitted by β^- Decay of ^{175}Yb as Shown in Figure 5.10

Mean Energy (MeV)	Maximum Energy (MeV)	Absolute Intensity (%)
0.0190	0.0738	20.40
0.0689	0.2186	0.008
0.1024	0.3563	6.70
0.1399	0.4701	72.90
Mean β^- energy	0.1127	100.00

Source: From National Nuclear Data Center (NNDC), Brookhaven National Laboratory NUDAT 2.0. Electronic Version available online at NNDC: www.nndc.bnl.gov/nudat2/, July 2005.

TABLE 5.13b

Energies and Intensities of Electrons Emitted by β^- Decay of ^{175}Yb as CEs and Auger Electrons

Electrons	Mean Energy (MeV)	Absolute Intensity (%)
Auger L	0.0060	6.34
Auger K	0.0435	0.40
CE K	0.0505	7.42
CE K	0.0743	0.261
CE K	0.0815	0.0746
CE L	0.1029	1.759
CE M	0.1113	0.41
CE NP	0.1133	0.116
CE L	0.1268	0.056
CE L	0.1340	0.01203
CE M	0.1352	0.013
CE NP	0.1372	0.00353
CE M	0.1424	0.00269
CE NP	0.1444	0.000706
CE K	0.1882	0.0152
CE K	0.2192	0.1409
CE L	0.2406	0.0059
CE M	0.2490	0.00142
CE NP	0.2510	0.00041
CE L	0.2717	0.0214
CE M	0.2800	0.0049
CE NP	0.2820	0.001471

Source: From National Nuclear Data Center (NNDC), Brookhaven National Laboratory NUDAT 2.0. Electronic Version available online at NNDC: www.nndc.bnl.gov/nudat2/, July 2005.

TABLE 5.13c

Energies and Intensities of γ-Rays as Result of γ-Decay and Characteristic X-Rays as a Result of IC, by β^- Decay of ^{175}Yb as Shown in Figure 5.10

Photons	Energy (MeV)	Absolute Intensity (%)	$(\mu_{en}/\rho)_\alpha$ $(cm^2\,g^{-1})$	Γ_δ (μGy h^{-1} MBq^{-1} m^2 = U MBq^{-1}) (Equation 2.63)
XR L	0.0077	2.09	10.79	7.931×10^{-3}
XR Kα2	0.0530	2.16	3.682×10^{-2}	1.933×10^{-4}
XR Kα1	0.0541	3.74	3.553×10^{-2}	3.298×10^{-4}
XR Kβ3	0.0611	0.412	2.975×10^{-2}	3.434×10^{-5}
XR Kβ1	0.0613	0.799	2.961×10^{-2}	6.655×10^{-5}
XR Kβ2	0.0629	0.271	2.870×10^{-2}	2.247×10^{-5}
γ	0.1138	3.87	2.352×10^{-2}	4.754×10^{-4}
γ	0.1377	0.235	2.444×10^{-2}	3.629×10^{-5}
γ	0.1449	0.672	2.475×10^{-2}	1.106×10^{-4}
γ	0.2515	0.17	2.796×10^{-2}	5.486×10^{-5}
γ	0.2825	6.13	2.848×10^{-2}	2.264×10^{-3}
γ	0.3963	13.2	2.947×10^{-2}	7.078×10^{-3}
Mean energy E_{mean} (Equation 5.1)			0.2522 MeV	
Effective energy E_{eff} (Equation 5.2)			0.3347 MeV	
Effective mass energy absorption coefficient, $(\mu_{en}/\rho)_{\alpha,eff}$ (Equation 5.3)			2.910×10^{-2} cm^2 g^{-1}	
Air kerma-rate constant Γ_δ (Equation 2.64)			0.0107 μGy h^{-1} MBq^{-1} m^2	
Mean number of photons emitted per decay			0.32 (0.3166)	

The mean energy E_{mean}, the effective energy E_{eff}, the air kerma-rate constant Γ_δ as well as the number of photons emitted per decay considers only rays above 10 keV (cut-off value $\delta = 10$ keV). Mass energy absorption coefficients for dry air are according to Hubbell and Seltzer,[17] where a log–log interpolation has been applied to these tables.
Source: From National Nuclear Data Center (NNDC), Brookhaven National Laboratory NUDAT 2.0. Electronic Version available online at NNDC: www.nndc.bnl.gov/nudat2/, July 2005.

5.10 ^{170}Thullium

^{170}Tm was originally considered to be used as a radiation source in portable x-ray devices for industrial applications. Its high production costs have prevented its commercial use. ^{170}Tm is produced by neutron irradiation of ^{169}Tm. Naturally occurring thulium is composed of one stable isotope, ^{169}Tm (100% natural abundance, see Appendix 1). The thulium element is never found in nature in its pure form, but it is found in small quantities in minerals with other rare earths. It is principally extracted from monazite ores ($\approx 0.007\%$ thulium) found in river sands through ion-exchange. Newer ion-exchange and solvent extraction techniques have led to easier separation of the rare earths, which has led to much lower thulium production costs. Nevertheless, high purity thulium is very expensive.

There is no published data on clinical or experimental use of ^{170}Tm sources. Recently, Granero et al.[18] published radiation protection related data for this radionuclide, indicating that ^{170}Tm could be of interest for substituting ^{137}Cs and ^{192}Ir in temporary implants.

There are mainly three physical properties of ^{170}Tm, causing interest in this radionuclide: (a) its half-life of 128.6 days, (b) its relatively high specific activity of 221 GBq mg^{-1} and (c) the energy of emitted photons with a mean energy of 0.066 MeV (see Table 5.2).

^{170}Tm decays mainly *via* β^-, almost 99.9% of the time, and in the majority (81.6%) to the ground state of ^{170}Yb (see Figure 5.11a). In only 18.3% of the time ^{170}Tm decays to the first excited state of ^{170}Yb, where the de-excitation to the ground state of ^{170}Yb occurs by 2.48% *via* gamma decay and emission of 0.0843 MeV γ-ray. The remaining 15.8% of the de-excitation occurs *via* internal conversion (IC), resulting in a total of six characteristic x-rays with energies in the range 0.007–0.061 MeV and total emission probability of 6.24%. There are on average 0.058 photons emitted per β^- decay event with an average energy of 0.067 MeV.

There are two β^--rays emitted with a maximal energy of 0.968 MeV. The mean energy of these two emitted β^--rays is about 0.317 MeV (see Table 5.14a). Additionally, six electrons are emitted as a result of internal conversion (IC) as conversion electrons (CE, 4x) and as Auger electrons

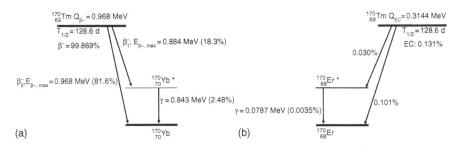

FIGURE 5.11
Schematic representation of the two different decay schemes of $^{170}_{69}$Tm. (a) β^- decay of $^{170}_{69}$Tm. This is the main disintegration process for this nuclide occurring in almost 99.9% of decays. $^{170}_{69}$Tm decays mainly to the ground state of $^{170}_{70}$Yb (with a probability of 81.6%). The disintegration energy for that decay is $Q_{\beta^-} = 0.968$ MeV. There are on average 0.058 photons emitted with energy above 10 keV, mainly with an energy of 0.084 MeV (0.025 photons out of 0.058, see also Table 5.14c).[3] (b) EC decay of $^{170}_{69}$Tm. $^{170}_{69}$Tm decays mainly to the ground state of $^{170}_{68}$Er. The disintegration energy for the decay is $Q_{EC} = 0.314$ MeV. There are on average 0.001 photons emitted with energy above 10 keV (see also Table 5.15b).[3] In each of these two decay schemes there are six characteristic x-rays emitted (see also Table 5.14c and Table 5.15b). The half-life for $^{170}_{69}$Tm decay is 128.6 days. (From National Nuclear Data Center (NNDC), Brookhaven National Laboratory NUDAT 2.0. Electronic Version available online at NNDC: www.nndc.bnl.gov/nudat2/, July 2005.)

TABLE 5.14a

Energies and Intensities of β^- Particles Emitted by β^- Decay of ^{170}Tm as Shown in Figure 5.11a

Mean Energy (MeV)	Maximum Energy (MeV)	Absolute Intensity (%)
0.2905	0.8837	18.3
0.3231	0.9680	81.6
Mean β^- energy	0.3170	99.9

Source: From National Nuclear Data Center (NNDC), Brookhaven National Laboratory NUDAT 2.0. Electronic Version available online at NNDC: www.nndc.bnl.gov/nudat2/, July 2005.

(Auger L and Auger K) resulting from characteristic x-rays (see Table 5.14b). Although their total emission probability is relatively high (25.34%), their energies are very low (maximum 0.084 MeV).

The second decay scheme of ^{170}Tm is *via* electron capture (EC) to ^{170}Er, which is shown in Figure 5.11b (see also Table 5.15a and Table 5.15b) and occurs only in 0.131% of the time. ^{170}Tm decays *via* EC mainly to the ground state of ^{170}Er (0.101%). In 0.030% of the time ^{170}Tm decays to the first excited state of ^{170}Er, and the de-excitation to its ground state occurs in 0.0035% *via* gamma decay. There is a single γ-ray emitted with an energy of 0.079 MeV (see Table 5.15b). Due to de-excitation *via* internal conversion (IC) there is a total of six characteristic x-rays emitted with energies ranging from 0.007 MeV to 0.057 MeV and total emission probability of 0.131%.

There are six electrons emitted as result of internal conversion (IC) as conversion electrons (CE, 4 ×) and as Auger electrons (Auger L and Auger K) resulting from characteristic x-rays (see Table 5.15a). On average, there are 0.001 photons emitted with a mean energy of 0.051 MeV.

Summarizing, there are 0.06 photons emitted by both decay processes with an average energy of 0.066 MeV (Table 5.16), where the average energy of only the emitted γ-rays amounts 0.084 MeV (Table 5.2).

TABLE 5.14b

Energies and Intensities of Electrons Emitted by β^- Decay of ^{170}Tm as Conversion Electrons (CE) and Auger Electrons

Electrons	Mean Energy (MeV)	Absolute Intensity (%)
Auger L	0.0058	9.360
CE K	0.0229	3.570
Auger K	0.0422	0.189
CE L	0.0738	9.310
CE M	0.0819	2.280
CE NP	0.0838	0.632

Source: From National Nuclear Data Center (NNDC), Brookhaven National Laboratory NUDAT 2.0. Electronic Version available online at NNDC: www.nndc.bnl.gov/nudat2/, July 2005.

TABLE 5.14c

Energies and Intensities of γ-Rays as Result of γ-Decay and Characteristic X-Rays as Result of Internal Conversion (IC), by β^- Decay of ^{170}Tm as Shown in Figure 5.11a

Photons	Energy (MeV)	Absolute Intensity (%)	$(\mu_{en}/\rho)_\alpha$ $(cm^2\,g^{-1})$	Γ_δ (μGy h^{-1} MBq^{-1} m^2 = U MBq^{-1}) (Equation 2.63)
X L	0.0074	2.92	11.9	1.183×10^{-2}
X Kα2	0.0514	0.97	3.894×10^{-2}	8.903×10^{-5}
X Kα1	0.0524	1.69	3.754×10^{-2}	1.526×10^{-4}
X Kβ3	0.0592	0.185	3.099×10^{-2}	1.557×10^{-5}
X Kβ1	0.0594	0.356	3.083×10^{-2}	2.991×10^{-5}
X Kβ2	0.0610	0.1205	2.980×10^{-2}	1.005×10^{-5}
γ	0.0843	2.48	2.367×10^{-2}	2.270×10^{-4}
Mean energy E_{mean} (Equation 5.1)			0.0667 MeV	
Effective energy E_{eff} (Equation 5.2)			0.0668 MeV	
Effective mass energy absorption coefficient, $(\mu_{en}/\rho)_{\alpha,eff}$ (Equation 5.3)			2.953×10^{-2} cm^2 g^{-1}	
Air kerma-rate constant Γ_δ (Equation 2.64)			0.00052 μGy h^{-1} MBq^{-1} m^2	
Mean number of photons emitted per decay			0.06 (0.058)	

The mean energy E_{mean}, the effective energy E_{eff}, the air kerma-rate constant Γ_δ as well as the number of photons emitted per decay considers only rays above 10 keV (cut-off value $\delta = 10$ keV). Mass energy absorption coefficients for dry air are according to Hubbell and Seltzer,[17] where a log–log interpolation has been applied to these tables.

Source: From National Nuclear Data Center (NNDC), Brookhaven National Laboratory NUDAT 2.0. Electronic Version available online at NNDC: www.nndc.bnl.gov/nudat2/, July 2005.

TABLE 5.15a

Energies and Intensities of Electrons Emitted by EC Decay of ^{170}Tm as Conversion Electrons (CE) and Auger Electrons (see Figure 5.11b)

Electrons	Mean Energy (MeV)	Absolute Intensity (%)
Auger L	0.0055	5.51×10^{-6}
CE K	0.0212	1.3×10^{-6}
Auger K	0.0397	2.49×10^{-6}
CE L	0.0689	1.06×10^{-5}
CE M	0.0765	2.8×10^{-6}
CE NP	0.0783	8.0×10^{-7}

Source: From National Nuclear Data Center (NNDC), Brookhaven National Laboratory NUDAT 2.0. Electronic Version available online at NNDC: www.nndc.bnl.gov/nudat2/, July 2005.

TABLE 5.15b

Energies and Intensities of γ-Rays as Result of γ-Decay and Characteristic X-Rays by EC Decay of ^{170}Tm as Shown in Figure 5.11b

Photons	Energy (MeV)	Absolute Intensity (%)	$(\mu_{en}/\rho)_\alpha$ $(cm^2\,g^{-1})$	Γ_δ ($\mu Gy\,h^{-1}\,MBq^{-1}\,m^2$ = U MBq^{-1}) (Equation 2.63)
X L	0.0070	0.0276	14.534	1.280×10^{-4}
X Kα2	0.0482	0.0291	4.409×10^{-2}	2.840×10^{-6}
X Kα1	0.0491	0.0513	4.244×10^{-2}	4.909×10^{-6}
X Kβ3	0.0555	0.00546	3.406×10^{-2}	4.736×10^{-7}
X Kβ1	0.0557	0.01055	3.387×10^{-2}	9.132×10^{-7}
X Kβ2	0.0571	0.00354	3.255×10^{-2}	3.022×10^{-7}
Γ	0.0787	0.0035	2.423×10^{-2}	3.064×10^{-7}
Mean energy E_{mean} (Equation 5.1)			0.0511 MeV	
Effective energy E_{eff} (Equation 5.2)			0.0510 MeV	
Effective mass energy absorption coefficient, $(\mu_{en}/\rho)_{\alpha,eff}$ (Equation 5.3)			$4.012 \times 10^{-2}\ cm^2\,g^{-1}$	
Air kerma-rate constant Γ_δ (Equation 2.64)			0.00001 $\mu Gy\,h^{-1}\,MBq^{-1}\,m^2$	
Mean number of photons emitted per decay			0.001	

The mean energy E_{mean}, the effective energy E_{eff}, the air kerma-rate constant Γ_δ as well as the number of photons emitted per decay considers only rays above 10 keV (cut-off value $\delta = 10$ keV). Mass energy absorption coefficients for dry air are according to Hubbell and Seltzer,[17] where a log–log interpolation has been applied to these tables.

Source: From National Nuclear Data Center (NNDC), Brookhaven National Laboratory NUDAT 2.0. Electronic Version available online at NNDC: www.nndc.bnl.gov/nudat2/, July 2005.

The relative high specific activity of ^{170}Tm probably enables the production of sources with active core volume in the range of 1.0–2.0 mm^3 and thus with total contained activity of potentially above 2000 GBq. Latter fact in combination with the long half-life of 128.6 days lets thulium appear as a possible candidate for afterloading technology using a single stepping source. By contrast, there are only 0.06 photons emitted per disintegration event resulting thus to a very low value for the air kerma-rate constant Γ_δ (see Chapter 2, Equation 2.64 and Table 5.2), which is approximately by a factor of 80 lower than for ^{169}Yb or by a factor of 200 lower than for ^{192}Ir. This implies that treatment times using high activity ^{170}Tm sources will be by a factor of 10–30 longer than for ^{192}Ir and ^{169}Yb HDR sources. The corresponding dose rates (e.g., at 1.0 cm radial distance from the source) will be significantly bellow the high dose rate limit of 2.0 Gy min^{-1} (see Section 1.8). The most probable application for high activity ^{170}Tm sources will be the pulsed dose rate technology (PDR, see also Chapter 8), where the low energies of

TABLE 5.16

Mean Energy and Mean Number of Emitted Photons of γ-Rays as Result of γ-Decay and Characteristic X-Rays, by Both β$^-$ and EC Decay Processes of ^{170}Tm

Mean energy E_{mean} (Equation 5.1)	0.0664 MeV
Effective energy E_{eff} (Equation 5.2)	0.0665 MeV
Effective mass energy absorption coefficient, $(\mu_{en}/\rho)_{\alpha,eff}$ (Equation 5.3)	2.967×10^{-2} cm^2 g^{-1}
Air kerma-rate constant Γ_δ (Equation 2.64)	0.00053 μGy h^{-1} MBq^{-1} m^2
Mean number of photons emitted per decay	0.06 (0.059)

The mean energy E_{mean}, the effective energy E_{eff}, the air kerma-rate constant Γ_δ as well as the number of photons emitted per decay considers only rays above 10 keV (cut-off value $\delta = 10$ keV). Mass energy absorption coefficients for dry air are according to Hubbell and Seltzer,[17] where a log–log interpolation has been applied to these tables.

the emitted photons for thulium lead to significantly smaller required thicknesses of the materials used in radiation protection compared to ^{192}Ir, which is the sole radionuclide currently used in PDR brachytherapy; the half-value layer in lead for ^{170}Tm is 0.17 mm compared to 3.0 mm for ^{192}Ir (Table 5.2) or 20.0 mm [18] *vs.* 65.0 mm [21] for ^{192}Ir in concrete.

References

1. Towpik, E. and Mould, R.F. Maria Sklodowska-Curie Memorial Issue, *J. Oncol.*, 1998, Nowotwory, Warsaw.

2. Myers, W.G. Radioactive needles containing cobalt-60, *Science*, 107, 621, 1948.

3. National Nuclear Data Center (NNDC). Brookhaven National Laboratory NUDAT 2.0 Electronic Version, available online at NNDC: www.nndc.bnl.gov/nudat2/, July 2005.

4. Sinclair, W.K. Artificial radioactive sources for interstitial therapy, *Br. J. Radiol.*, 25, 417–419, 1952.

5. Simon, N. Iridium-192 as a radium substitute, *Am. J. Roentgenol.*, 93, 170–178, 1965.

6. Reid, A.F. and Keston, A.S. Long-life radio-iodine, *Phys. Rev.*, 70, 987–988, 1946.

7. Hilaris, B., Nori, D., and Anderson, L.L. *Atlas of Brachytherapy*, New York, Macmillan, 1988.

8. Fisher, J.F., Porter, A.T., Barnett, R.B., Mason, D.L., Papiez, E., and Battista, J.J. First clinical application of a new brachytherapy source Yb-169, *Endocuriether-apy Hypertherm. Oncol.*, 9, 195–199, 1993.

9. Lazarescu, G.R. and Battista, J.J. Analysis of the radiobiology of ytterbium-169 and iodine-125 permanent brachytherapy implants, *Phys. Med. Biol.*, 42, 1727–1736, 1997.

10. Loft, S.M., Coles, I.P., and Dale, R.G. The potential of ytterbium-169 in brachytherapy: A brief physical and radiobiological assessment, *Br. J. Radiol.*, 65, 252–257, 1992.

11. Mason, D.L., Battista, J.J., Barnett, R.B., and Porter, A.T. Ytterbium-169: Calculated physical properties of a new radiation source for Brachytherapy, *Med. Phys.*, 19, 695–703, 1992.

12. Piermattei, A., Azario, L., and Montemaggi, P. Implantation guidelines for [169]Yb seed interstitial treatments, *Phys. Med. Biol.*, 40, 1331–1338, 1995.

13. Perera, H., Williamson, J.F., Li, Z., Mishra, V., and Meigooni, A. Dosimetric characteristics, air-kerma strength calibration and verification of Monte Carlo simulation for a new Ytterbium-169 brachytherapy source, *Int. J. Radiat. Oncol. Biol. Phys.*, 28:4, 953–970, 1994.

14. National Institute of Standards and Technology Physics Laboratory Physical Reference Data Electronic Version, National Institute of Standards and Technology, Gaithersburg, available online at http://physics.nist.gov/PhysRefData, August 2004.

15. Unterweger, M.P., Hoppes, D.D., and Schima, F.J. New and revised half-life measurement results, *Nucl. Instrum. Meth. Phys. Res.*, A312, 349–352, 1992.

16. International Commission on Radiological Protection, Data for protection against ionizing radiation from external sources, ICRP Publication 21 Supplement to ICRP Publication 15, Pergamon Press, New York, 1971.

17. Hubbell, J.H. and Seltzer, S.M. Tables of x-ray mass attenuation coefficients and mass energy-absorption coefficients, *National Institute of Standards and Technology*, Gaithersburg. Version 1.4 on http://physics.nist.gov/xaamdi. Originally published as NISTIR, 5632, 1995.

18. Granero, D., Pérez-Calatayud, J., Ballester, F., Bos, A.J.J., and Venselaar, J. Broad-beam transmission data for new brachytherapy sources Tm-170 and Yb-169, *Radiat. Prot. Dosim.*, 1–5, 2005.

19. Williamson, J.F. Recent developments in basic brachytherapy physics in radiation therapy physics, *Medical Radiology Diagnostic Imaging and Radiation Oncology, (A.L. Baert, L.W. Brady, H.P. Heilmann, M. Molls and K. Sartor, Series eds.)*, A.R. Smith, ed., Springer, Berlin, pp. 247–302, 1995.

20. Lymperopoulou, G., Papagiannis, P., Sakelliou, L., Milickovic, N., Giannouli, S., and Baltas, D. A dosimetric comparison of [169]Yb versus ^{192}Ir for HDR prostate brachytherapy, *Med. Phys.*, 32, 3832–3842, 2005.

21. Lymperopoulou, G., Papagiannis, P., Sakelliou, L., Georgiou, E., Chourdakis, C.I. and Baltas, D. A comparison of radiation shielding requirements for Brachytherapy using [169]Yb and ^{192}Ir sources, accepted for publication, *Med. Phys.*, 2006.

22. Medich, C.D., Tries, A.M., and Munro, J.J. Monte Carlo characterization of an ytterbium-169 high dose rate Brachytherapy source with analysis of statistical uncertainty, *Med. Phys.*, 33, 163–172, 2006.

6

Production and Construction of Sealed Sources

6.1 Introduction

Traditionally, brachytherapy was performed as a low-dose rate (LDR) therapy using radium (^{226}Ra) or its daughter element radon (^{222}Rn). Radium has the advantage of a very long half-life, but it also has the disadvantage of producing the alpha-emitting gaseous daughter product radon. By modern radiation safety standards, radium, and radon sources are considered to be unsafe and are no longer used. Today, the vast majority of interstitial brachytherapy treatments are being performed as a high-dose rate (HDR) temporary implant brachytherapy with ^{192}Ir or ^{60}Co, or LDR permanent implant brachytherapy with ^{125}I, or ^{103}Pd sources. The choice of radionuclides that can be used for brachytherapy, however, is limited because only a few have all the desirable properties of the ideal brachytherapy source (see also Table 5.3). These ideal properties are as follows:

1. The optimum gamma-ray emission energy should be high enough to avoid increased energy deposition in bone by the photoelectric effect and also high enough to minimize scatter. At the same time it must be low enough to minimize protection requirements.
2. The half-life should be such that correction for decay during treatment is minimal. In addition, a very long half-life is desirable to keep a permanent stock of sources. Then, radioactive decay within the lifetime of the source and its container is negligible and the stock can be easily used and stored.
3. Charged-particle emission should be absent or easily screened.
4. There should be no gaseous disintegration product.
5. The nuclide should have a high specific activity.

6. The material should be available in insoluble and nontoxic form.
7. The material should not powder or be otherwise dispersed if the source is damaged or incinerated.
8. It should be possible to manufacture the nuclide in different shapes and sizes including rigid tubes and needles, small spheres, and flexible wires. In the last case it should be possible to cut the wires into the required lengths without danger or contamination.
9. Damage during sterilization should not be possible.

Ever since radium was discovered in 1898, and first used therapeutically in 1900, many radionuclides have been used for brachytherapy over the years, but not one of them has satisfied all of the above criteria for the ideal source.

A radionuclide is either obtained from a natural source (as was radium from the uranium mineral pitchblende) or is artificially made (Table 6.1). Radionuclides can be produced in a nuclear reactor either as a fission product within a spent uranium fuel rod or as the result of a neutron capture reaction of a stable nuclide. In a radioactive neutron capture reaction (indicated by (n,γ)) a target nucleus $^A_Z X$ captures a neutron and converts to a product nucleus $^{A+1}_Z X^*$ in its excited state (see Section 3.3.3). The product nucleus immediately de-excites to its ground state by emitting prompt γ-rays. This reaction is schematically represented by

$$^A_Z X(n, \gamma)^{A+1}_Z X$$

The target and product nuclei in this capture reaction represent different isotopes of the same chemical element. The prompt γ-rays must not be confused with the decay γ-rays, which the product nucleus may eventually radiate.

Neutron capture reactions can also take place through the proton knock-out (n,p) reaction. With these reactions the target nucleus X captures a neutron and instantly emits a proton, forming another chemical element Y with the same mass number, but with a one unit lower atomic number. This reaction is represented schematically by

$$^A_Z X(n, p)^{\ A}_{Z-1} Y$$

The product nuclei $^{A+1}_Z X$ and $^{\ A}_{Z-1} Y$ are usually radioactive elements.

The activity of the radionuclide produced depends on the number of parent nuclei being bombarded, the probability of a nucleus capturing a neutron (the cross-section for the nuclear reaction which depends on the neutron energy), the neutron flux, the transformation constant of the isotope produced, and the length of time the material stays in the neutron flux. These parameters

TABLE 6.1

Physical Properties of Radionuclides Used for Brachytherapy and Their Mode of Production (See Also Table 5.2)

Nuclide		Half-Life	Type of Disintegration	Maximum Beta-Ray Energy (MeV)	Main Photon Energies (MeV)	Cross-Section for Thermal Neutrons[a] (b)	Method of Production
Cesium-131	^{131}Cs	9.689 d	EC (100%)	—	0.030	11.3	^{130}Ba$(n,\gamma)^{131}$Ba; ^{131}Ba $\rightarrow$ ^{131}Cs
Cesium-137	^{137}Cs	30.07 yr	β^-, γ	0.514	0.662 (^{137m}Ba)	—	Fission product
Cobalt-60	^{60}Co	5.27 yr	β^-, γ	0.318	1.17, 1.33	36.9	^{59}Co$(n,\gamma)^{60}$Co
Gold-198	^{198}Au	2.695 d	β^-, γ	0.962	0.412, 0.676	98.8	^{197}Au$(n,\gamma)^{198}$Au
Holmium-166	^{166}Ho	26.8 h	β^-	1.90	0.081	61	^{165}Ho$(n,\gamma)^{166}$Ho
Iodine-125	^{125}I	59.49 d	EC with x-ray	—	0.035	165	^{124}Xe$(n,\gamma)^{125}$Xe; ^{125}Xe $\rightarrow$ ^{125}I
Iridium-192	^{192}Ir	73.81 d	β^- (95%), EC(5%), γ	0.675	0.316	930	^{191}Ir$(n,\gamma)^{192}$Ir
Palladium-103	^{103}Pd	16.991 d	EC with x-ray	—	0.020	3.36	^{102}Pd$(n,\gamma)^{103}$Pd
Phosphorus-32	^{32}P	14.262 d	β^-	1.71	—	0.070	^{32}S$(n,p)^{32}$P
Radium-226	^{226}Ra[b]	1600 yr	α, β^-, γ	3.27	0.609	—	Naturally occurring
Radon-222	^{222}Rn[b]	3.8235 d	α, β^-, γ	3.27 (^{214}Bi)	0.609	—	Naturally occurring
Strontium-90	^{90}Sr[c]	28.79 yr	β^-	0.546	—	—	Fission product
Tantalum-182	^{182}Ta	114.43 d	β^-, γ	0.59	0.068	22.0	^{181}Ta$(n,\gamma)^{182}$Ta
Thulium-170	^{170}Tm	128.6 d	β^-, EC with x-ray	0.968	0.052	105	^{169}Tm$(n,\gamma)^{170}$Tm
Ytterbium-169	^{169}Yb	32.015 d	EC (100%)	—	0.063	2400	^{168}Yb$(n,\gamma)^{169}$Yb
Yttrium-90	^{90}Y	64.0 h	β^-	2.27	—	1.3	^{89}Y$(n,\gamma)^{90}$Y or ^{90}Sr $\rightarrow$ ^{90}Y

[a] For parent isotope.

[b] Gamma photons are produced by daughter elements in equilibrium with radium and radon.

[c] Used when in equilibrium with ^{90}Y.

not only affect the final activity of the isotope produced but also determine the activity per unit mass, the specific activity (see Equation 2.59 and Equation 2.60). It can be shown (see Equation 3.53 through Equation 3.56) that the activity of an isotope after being bombarded in a neutron field for a time, t, is given by

$$A(t) = N_t \sigma \dot{\Phi}[1 - \exp(-\lambda t)] \qquad (6.1)$$

where N_t is the number of nuclei being bombarded, σ is the cross-section for the nuclear reaction in cm^2 per atom or barns per atom and $\dot{\Phi}$ is the neutron fluence rate expressed by neutrons cm^{-2} sec^{-1}. The decay constant λ of the product nucleus defines the speed with which it can be formed.

Equation 6.1 indicates that the actual growth in activity is not linear with time but reaches a maximum, the saturation activity, after several half-lives. The product $N_t \sigma \dot{\Phi}$ is equivalent to the saturation activity of the radionuclide and occurs when the rate of production of the active atoms equals the rate at which they decay. Johns and Cunningham[1] have given a more detailed account of this method of isotope production.

Generally reactor-produced radionuclides have the following properties.

The product nuclei have one extra neutron compared with the stable target nucleus and this surplus of neutrons is usually solved by β^- emission.

With the (n,γ) reaction it is not possible to produce carrier- (or target-) free radionuclides, as the products belong to the same chemical element as the target material. Carrier-free products can be formed by using the (n,p) process (e.g., ^{32}P is formed in the process: ^{32}S(n,p)^{32}P) or by a multistep procedure through a short-lived intermediate state. As an example and as shown in Section 3.3.3 and in Equation 3.57 through Equation 3.61, ^{125}I is formed by activation of ^{124}Xe: ^{124}Xe(n,γ), ^{125}Xe, and subsequently ^{125}Xe decays, $T_{1/2} = 16.9$ h via electron capture to ^{125}I (see Table A.1.1 of Appendix 1).

Neutron activation is actually a rather ineffective method of removing target material; typically one in 10^6 to 10^9 target nuclei are activated, depending on the neutron flux. This causes the (n,γ) produced radionuclides to be of very low-specific activity because of the presence of large quantities of unactivated stable carrier material.

To ensure absolute encapsulation of the radioactive source the housing should be a nontoxic material that will not interact physically or chemically with body fluids. An additional constraint to the housing material is that it will not significantly attenuate the emitted radiation from the source. The housing, however, must be strong enough and also small enough to allow implantation with hypodermic needles or similar equipment for implanting the sources. An ideal size is a long and thin cylindrical shape of approximately 0.5 to 1 mm outer diameter and of 4 to 5 mm length. Low-atomic number metals such as titanium and stainless steel combine the nontoxicity, mechanical strength, and low-attenuation properties required.

6.2　^{192}Iridium Sources

6.2.1　HDR Remote Afterloading Machine Sources

^{192}Ir ($T_{1/2} = 73.81$ d, see Section 5.5) is produced from natural ^{191}Ir in a reactor using the (n,γ) reaction. However, the capture cross-section of the product ^{192}Ir is rather high (~ 1200 b) so that stable ^{193}Ir arises easily from it, and also some ^{194}Ir ($T_{1/2} = 19.28$ h). Owing to its short half-lifetime ^{194}Ir is not a problem, but the forming of the stable ^{193}Ir from (the original) ^{191}Ir is counterproductive. The optimum irradiation time amounts approximately to half the lifetime of ^{192}Ir, i.e., 5 to 6 weeks (see Figure 6.1).

The maximum achievable activity follows from the coupled differential equations for double neutron capture (in which the effect of an isomer must also be taken into account).

$$\frac{dN_{191}}{dt} = -(\sigma_{191} + \sigma_{191'})\phi N_{191}$$

$$\frac{dN_{192}}{dt} = (\sigma_{191} + \sigma_{191'})\phi N_{191} - \lambda_{192}N_{192} - \sigma_{192}\phi N_{192}$$

(6.2)

In the first part of the equation the neutron capture rate for the ^{191}Ir nuclei is given, leading to ^{192}Ir. The second part shows that production of ^{192}Ir is

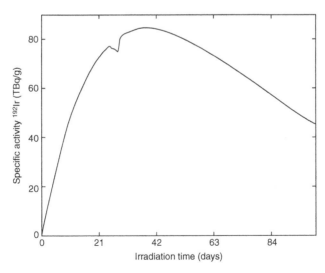

FIGURE 6.1
Specific activity of ^{192}Ir as a function of irradiation time in the Petten high-flux reactor at a thermal neutron fluence rate of 2.4×10^{14} n cm^{-2} sec^{-1}, using natural iridium targets. Decay during the reactor stop at 21 d is visible in the curve, subsequent reactor stops do not visibly influence the curve. Maximum specific activity is reached after 38 d at 200 TBq/g.

a balance between production by capture in ^{191}Ir and loss by both decay with decay constant λ_{192} ($= \ln(2)/T_{1/2}(^{192}\text{Ir})$) and neutron capture reactions in the formed ^{192}Ir nuclei. These equations can be solved by using the integrating factor $\exp([\lambda_{192} + \sigma_{192}\phi]t)$:

$$N_{191}(t) = N_{191}(0)e^{-(\sigma_{191}+\sigma_{191'})\phi t}$$

$$N_{192}(t) = \frac{N_{191}(0)(\sigma_{191} + \sigma_{191'})\phi}{\lambda_{192} + (\sigma_{192} - \sigma_{191} - \sigma_{191'})\phi} \left\{ e^{-(\sigma_{191}+\sigma_{191'})\phi t} - e^{-(\lambda_{192}+\sigma_{192}\phi)t} \right\} \quad (6.3)$$

The activity of ^{192}Ir $A_{192}(t)$ is obtained by multiplying the number of atoms $N_{192}(t)$ with its decay constant λ_{192}:

$$A_{192}(t) = \frac{\lambda_{192}N_{191}(0)(\sigma_{191} + \sigma_{191'})\phi}{\lambda_{192} + (\sigma_{192} - \sigma_{191} - \sigma_{191'})\phi} \left\{ e^{-(\sigma_{191}+\sigma_{191'})\phi t} - e^{-(\lambda_{192}+\sigma_{192}\phi)t} \right\} \quad (6.4)$$

The maximum achievable activity as seen in Equation 6.2, is actually less by the neutron self-shielding effect in the iridium target due to its high cross-section. Semiempirical formulae have been devised to give a quick indication of the neutron self-shielding f_n (e.g., Nisle's formula: $f_n = 1/2\xi(1 - e^{-2\xi})$, with $\xi = 2\sigma_{tot}$(volume/surface area) and σ_{tot} the total macroscopic cross-section for the target). Monte Carlo calculations provide a more specific way of estimating the self-shielding for any target geometry in combination with the irradiation rig.

Iridium is a very brittle and dense material, which will powder when the source is hassled. An iridium–platinum alloy is therefore used, as platinum is a much softer metal (a Brinnell hardness of 392 compared with 1670 for iridium). The iridium–platinum proportion is rather critical. On the one hand, the alloy must meet the mechanical requirements for a source (the more platinum the better) and, on the other hand, the source must have a HDR (the more iridium is being irradiated, the better, because longer irradiation works counterproductively).

Iridium sources are encapsulated in a thin titanium or stainless steel capsule to allow smoother handling in HDR afterloading equipment and to reduce absolutely the chance of chipping.

To reach higher specific activities targets with enriched ^{191}Ir abundances can be used. Thus natural iridium, consisting of 37.3% ^{191}Ir and 62.7% ^{193}Ir (Table A.1.1 of Appendix 1), can be enriched in ^{191}Ir by depleting its ^{193}Ir content through ultracentrifuge technology. The costs of enrichment, however, compel a search for a more economical use of this material, e.g., by using hollow cylinders. Enrichment in ^{191}Ir, on one hand, will create the formation of more ^{192}Ir but, on the other hand, the neutron self-shielding will increase with the higher cross-section and thus work counterproductively.

Most reactors that are suitable for the production of radioisotopes have an irradiation cycle of a month. Then after a month, minor maintenance is performed and spent fuel is replaced. The fresh and the partly spent rods/plates are redistributed so that the desired neutron distribution is obtained. The reactor shutdown usually lasts a week. In practice, the optimum irradiation time of 5 to 6 weeks is not reached, because it is not effective to wait for the next cycle to continue the irradiation.

When special circumstances require the supply of ^{192}Ir at moments other than after a complete cycle, it is practical to use enriched iridium. At 85% enrichment in ^{191}Ir the specific activity can be 2.1 times higher when irradiated in the same position and time. This means that the saving in irradiation time is almost directly proportional with the enrichment. Figure 6.2 shows the preparation of iridium for its activation in the high-flux reactor at Petten (Figure 6.3).

The HDR iridium sources, which are later mounted on wires, are usually cylindrical slices of some tenths mm thickness and with diameters of 2 to 4 mm, depending on their application purpose (see Figure 6.4). For irradiation, the slices are placed in a container of pure carbon, at a distance from each other because the neutron absorption of iridium is high. When the container is full, it is stored in an aluminum tube which is then sealed. To undergo constant irradiation this tube is rotated slowly in the reactor core position. Thus, iridium sources are created with a typical activity of 100 Ci (3700 GBq). After neutron activation during one reactor cycle the sources are placed in a stainless steel capsule and laser welded to the end of a flexible steel wire.

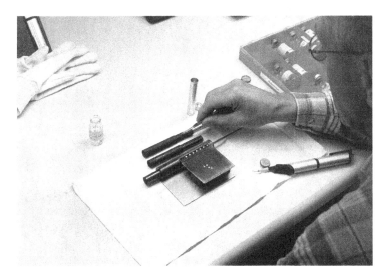

FIGURE 6.2
Slices of iridium are placed in a pure carbon container for irradiation in the Petten high-flux reactor.

FIGURE 6.3
A reactor's availability and reliability determine its market value. However, the reactor personnel's management of the logistics is equally important, because any waste of time means a loss of radioactivity and thus a loss of money. The high-flux reactor at Petten has an availability of 280 days per year. This makes the reactor one of the best performers in the world. It produces 70% of all the radioisotopes that are used in European hospitals.

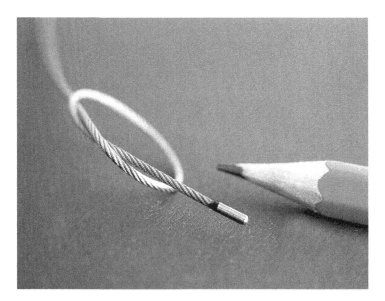

FIGURE 6.4
The well-known ^{192}Ir source, placed in a stainless steel capsule and laser welded to the end of a flexible steel wire.

6.2.2 LDR Seed and Hairpin Sources

For the production of LDR ^{192}Ir seeds, the iridium–platinum alloy to be irradiated is first packed in hollow cylindrical nitinol capsules. Nitinol is an alloy of nickel, titanium, and some other elements. It is also referred to as medical steel and hardly absorbs any neutrons. The seeds are sealed with a YAG laser. This is performed by the seed manufacturer.

As with the wire slices, the seeds are placed in pure carbon containers and irradiated. They usually remain in the reactor for a complete irradiation cycle, but because they contain less iridium they also contain less activity, typically 10 Ci (370 GBq). The seeds have typically a length of 3.5 mm and a diameter of 0.65 mm.

In addition, there is also a demand for so-called iridium hairpins (originally devised in the 1950s for treatment of bladder cancer using ^{182}Ta, which has now been superseded by ^{192}Ir; the hairpins were placed within the bladder wall). These hairpins have an activity of about 0.1 Ci (3.7 GBq) and require an irradiation time of only 15 min.

For some applications the wires, seeds, or hairpins require a relatively LDR. No separate products are manufactured for this small market and the sources are simply stored for a time. Within a year the dose rate will have dropped to 3% (five half-lives).

6.3 ^{125}Iodine LDR Seeds

^{125}I is a daughter of ^{125}Xe which is usually created in a nuclear reactor from ^{124}Xe (Equation 3.57 through Equation 3.61, see also Figure 3.9 and Table A.1.1, Appendix 1).

$$^{124}\text{Xe}(n, \gamma)^{125}\text{Xe} \text{ and } ^{125}\text{Xe}(\text{EC}, 16.9 \text{ h}) \rightarrow {}^{125}\text{I}$$

^{125}I ($T_{1/2} = 59.49$ d) emits low-energy gammas (27 to 35 keV, Table 5.10 and Figure 5.7).

Most manufacturers produce cylindrical seeds with a spherical front and back. Only InterSource and also Amersham have chosen an essentially different design (Figure 6.5): a hollow cylinder consisting of only a double-walled titanium tube. The iodine is inserted into the tube in ring-shaped segments, a configuration which permits a higher isotropy of the irradiation field around the source. The hollow cylinders can be pushed onto a biologically degradable wire. The hollow design reduces the chance of seed migration.

As the natural abundance of ^{124}Xe is only 0.1% (Table A.1.1 of Appendix 1) the benefit of isotopic enrichment will not easily exceed the costs. Enrichment can be achieved by centrifuge techniques to more than 99.9%. The thermal neutron capture cross-sections are 165 and 28 b to the ground state and the meta-stable state of ^{125}Xe, respectively. As ^{125}I formed by the

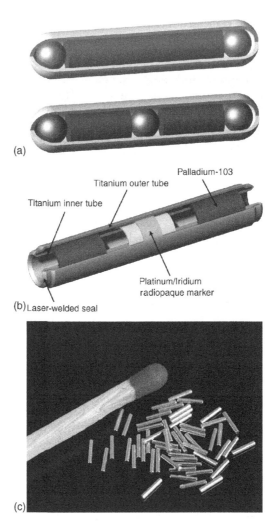

FIGURE 6.5
Conventionally shaped seeds and a hollow seed. The latter concept is chosen by InterSource and by GE Healthcare. (a) The geometry of hollow ^{125}I seed. (b) The geometry of hollow ^{103}Pd seeds (Section 6.4). (c) The photograph with the match demonstrates the small size of the seeds.

electron capture decay of ^{125}Xe has a fivefold higher thermal neutron capture cross-section of 894 b, as with ^{192}Ir, double neutron capture processes may diminish the ^{125}I production in long neutron irradiations. The advantages of a gas target as well as the carrier-free production, however, make it possible to extract ^{125}I shortly after production in a closed loop gas circulation target, thereby avoiding ^{126}I formation.

Several models of ^{125}I seed sources suitable for interstitial brachytherapy implants are commercially available. On the outside most models have an identical 4 to 5 mm length $\times$ 0.8 mm diameter encasing of very thin

(40 to approximately 60 μm) titanium sealed at each end. Usually, the welds have a thickness of 0.5 mm, but some sources have very thin welds of 0.1 to 0.15 mm. These sources display a flatter anisotropy function due to the lower attenuation at the ends. Inside, however, there is great variety in the design of internal cores.

Spheres or rods made of ion exchange resin or silver are used to absorb the ^{125}I into or on its surface. The hollow design sources by International Brachy Therapy (IBT s.a., Seneffe, Belgium) use an organic matrix to fix the ^{125}I to the titanium capsule by electrostatic charging comparable to laser printer technology. Very thin bands of 9 to 15 μm are achieved in this manner. A high-density metal radiopaque marker makes the source visible in x-ray fluoroscopy or CT imaging. The advantage of the hollow design is not only the lower probability of source migration, but also the smoothness of the insertion procedure by aligning the seeds on a (biodegradable) strand through the inner tubes. Most seeds today are designed to minimize the dosimetry anisotropy (butterfly effect) seen in earlier models.

6.4 ^{103}Palladium LDR Seeds

With ^{103}Pd ($T_{1/2} = 16.991$ d, see Table 5.11 and Figure 5.8) it is possible to reduce the dose to surrounding organs by the lower energy (mean energy of 21 keV, Table 5.2) of its photons in comparison to ^{125}I or ^{192}Ir. Also, the initial dose rate with which this dose is delivered is higher due to the shorter half-life. ^{103}Pd is produced by neutron capture in the stable ^{102}Pd nucleus:

$$^{102}\text{Pd}(n, \gamma)^{103}\text{Pd}$$

^{102}Pd has a thermal neutron capture cross-section of 3.4 b and has an isotopic abundance of 1.02% in natural palladium. The only other palladium isotope with a high cross-section is ^{108}Pd with $\sigma(n,\gamma) = 7.2$ b and an abundance of 26.46% (Table A.1.1 of Appendix 1). As the formed ^{109}Pd has a half-life of 13.7 h, a 3 d cooling period is long enough to remove this activity, although the emitted 361 keV β-particles will not seriously interfere with the ^{103}Pd dosimetry characteristics. Unfortunately, it is not possible to produce palladium in a gas-phase molecule to make it suitable for gas centrifuge enrichment. Laser isotope separation may be the process to produce enriched ^{102}Pd. Currently, palladium is enriched by the costly calutron (see Section 6.11.3) production method to $>85\%$ ^{102}Pd.

The geometry of the ^{103}Pd sources are very comparable to the ^{125}I sources. In fact, apart from the radionuclide, the ^{103}Pd source from IBT is exactly identical to their ^{125}I source. Silver rods, however, are not used in ^{103}Pd sources due to the lack of absorption of palladium in silver as well as to the larger attenuation of the lower energy photons in the silver.

6.5 ^{169}Ytterbium Sources

In the request for lower energy and shorter half-life photon emitters, ^{169}Yb ($T_{1/2} = 32.015$ d, see Table 5.12 and Figure 5.9) has been identified as a good alternative to ^{192}Ir. The isotope is produced in a nuclear reactor:

$$^{168}\text{Yb}(n, \gamma)^{169}\text{Yb}$$

^{168}Yb has a very high-thermal neutron cross-section of 2300 ± 170 b but a low-natural abundance of 0.13%. It is suitable for isotopic enrichment by either laser separation or the calutron method. With the latter method it is economically feasible to reach high-enrichment levels. An enrichment of 85% is common. From the other ytterbium isotopes only ^{174}Yb (31.83% abundance, Table A.1.1 of Appendix 1) produces a longer-lasting radioactive isotope. This is ^{175}Yb with a half-life of 4.185 d (see Table 5.13 and Figure 5.10).

6.6 ^{60}Cobalt HDR Sources

High-dose rate brachytherapy treatments using ^{60}Co sources have the economic advantage over lesser replacements of HDR sources due to the long half-life of 5.27 yr (Table 5.2). The higher photon energy (mean energy of 1.253 MeV, see Table 5.4 and Figure 5.3) of ^{60}Co in comparison to ^{192}Ir makes it more difficult to shield. ^{60}Co is produced by neutron irradiation:

$$^{59}\text{Co}(n, \gamma)^{60}\text{Co}$$

The cross-section of 36.9 b for ^{59}Co (which is the only stable isotope of cobalt; abundance: 100%) permits a good production rate in a high-flux reactor, but the long half-life of the ^{60}Co produced requires long irradiation times. The maximum activity is reached after 4.9 yr. Since ^{60}Co itself has a cross-section for thermal neutron capture of 2.0 b, longer irradiation times will lead to increased loss of ^{60}Co activity due to neutron capture reactions. The sources are cubes of the radioactive cobalt metal lined with a spherical titanium capsule. The typical overall diameter is about 3 mm with a 1.3 mm cobalt cube inside that has an activity of maximally 18.5 GBq. Treatment-specific radiation profiles can be composed by combining several sources and spacers in the catheter tube of the afterloader.

6.7 ^{137}Cesium LDR Sources

^{137}Cs ($T_{1/2} = 30.07$ yr, mean photon energy of 615 keV, see Table 5.2) is extracted from uranium-235 fission products; the accumulated fission yield to the $N = 137$ isobar is 6.2%. The production decay chain is (see also Figure 3.9):

$$^{235}U(n, f.p.)^{137}Te, \, ^{137}Te(\beta^-, 2.5 \text{ sec}) \rightarrow \, ^{137}I(\beta^-, 24 \text{ sec})$$
$$\rightarrow \, ^{137}Xe(\beta^-, 3.8 \text{ min}) \rightarrow \, ^{137}Cs$$

The cesium formed is trapped in an inert matrix material, such as gold, ceramic, or borosilicate glass, to prevent leaching. The sources are doubly encapsulated in totally 0.5-mm thick stainless steel. Spherical sources, such as the ^{60}Co sources, are made for use in intracavitary brachytherapy catheters. Several cylindrical models are also manufactured with 3 mm diameter and external lengths up to 21 mm. Nominal activities of these sources are between 0.5 and 4.7 GBq.

6.8 ^{198}Gold LDR Seeds

^{198}Au ($T_{1/2} = 2.695$ d, mean photon energy of 406 keV, see Table 5.2) is produced from natural ^{197}Au in a reactor using the (n,γ) reaction. Owing to its short half-lifetime ^{198}Au is merely of interest as a LDR source.

Better said: "^{198}Au used *to be* of interest as a permanent implant source. With the arrival of ^{103}Pd, ^{198}Au shifted to the background, particularly because of the high penetration in tissue and materials in general of the 406 keV gamma-rays and the accompanying shielding problems."

6.9 ^{170}Thulium High Activity Seeds

High activity sources with ^{170}Tm will be easier to shield, as thulium-170 emits mainly beta-particles, a gamma-ray at 84.3 keV (only 2.48% abundance) and several x-rays at 7.4 keV (2.92%) and between 51 and 61 keV (3.3%, see also Table 5.14, Table 5.15, and Figure 5.11). Production of high activity sources should enable therapy treatments inside the operating theatre without additional shielding requirements in the future. Due to the very low number of emitted photons per disintagration (0.06, Table 5.16) the air kerma-rate constant Γ_δ for ^{170}Tm is very low, nearly 200 times lower than of that of ^{192}Ir or 80 times lower than of ^{169}Yb sources (see Equation 2.64 and Table 5.2). Even though, if thulium sources of double active core volume of that of ^{192}Ir HDR sources would be constructed, ≈ 2.0 mm^3, the resulted treatment times would be by a factor 10–30 times longer, than for iridium HDR treatments. Thulium sources of such core volumes will result to dose rates lower than 1.0 Gy min^{-1} at 1.0 cm radial distance from the source center. This dose rate is lower than the 2.0 Gy min^{-1}-limit for the HDR treatments indicating that real ^{170}Tm high activity sources will be equal to currently used ^{192}Ir PDR sources, regarding the resulting dose rate and treatment duration.

The production route of ^{170}Tm by neutron irradiation is

$$^{169}\text{Tm}(n, \gamma)^{170}\text{Tm}$$

The mono-isotopic (100% abundance) ^{169}Tm has a thermal neutron cross-section of 105 b. Again there is a risk of neutron burn-up by the neutron capture in the produced ^{170}Tm with a cross-section of 100 b and the need for longer irradiation times by the half-life of 128.6 days. Thulium-171, however, is a low-energy beta-emitter with a half-life of 1.9 years and a cross-section of 160 b, so it will reduce the ^{170}Tm source strength and if irradiated long enough produce ^{172}Tm with high-energy gamma-rays. Thulium is the least abundant of the rare-earth metals (approximately 0.007% Tm in monazite ores) and high-purity thulium is therefore expensive.

6.10 ^{131}Cesium LDR Seeds

As a pure electron capture isotope cesium-131 is a promising option in producing sources with pure low energy photons. Especially its K_α x-rays at 29.5 and 29.8 keV with the high abundances of 21 and 39% and the K_β x-rays at 33.6–34.4 keV (12.8%) are very interesting, though necessitating thin encapsulation to reduce attenuation. The short half-life of 9.689 days enables production of higher dose rate sources, shortening the dose delivery time. Cesium-131 is produced by neutron capture in ^{130}Ba:

$$^{130}\text{Ba}(n, \gamma)^{131}\text{Ba and } ^{131}\text{Ba} \rightarrow ^{131}\text{Cs(EC, 11.5 d)}$$

The natural abundance of ^{130}Ba is just 0.106%, but can be enhanced by ultracentrifuge and is necessary as ^{132}Ba will produce the long-living ^{133}Ba ($T_{1/2} = 10.5$ yr). Another possible production method is by proton bombardment of a cesium-133 target in a cyclotron by

$$^{133}\text{Cs}(p, 3n)^{131}\text{Ba and } ^{131}\text{Ba} \rightarrow ^{131}\text{Cs(EC, 11.5 d)}$$

The proton energy must be high enough to jump over the ^{133}Cs$(p,n)^{133}$Ba reaction.

6.11 Enrichment Methods

6.11.1 Ultracentrifuge

The centrifuge consists of an ultralight, thin-walled tube made from advanced materials, containing a cylindrical rotor which rotates at high velocity in a vacuum on an almost frictionless bearing. The gaseous isotopic mixture, usually a metal hexafluoride such as UF6, is fed into the centrifuge where it adopts a rotational motion. The electric motor of the centrifuge produces heat at the base of the machine, causing a temperature profile along the length of the centrifuge and assisting the separation process. The centrifugal forces push the heavier isotope closer to the wall of the rotor than the lighter isotope in the mixture. With natural uranium, for example, consisting of 99.3% ^{238}U and 0.7% ^{235}U, the gas closer to the wall becomes depleted in ^{235}U whereas the gas nearer the rotor axis is slightly enriched in ^{235}U.

Figure 6.6 illustrates the basic centrifuge concept. The gaseous UF6 is fed through a pipe from the top of the centrifuge into the center of the cylinder, where it takes up the rotational motion and also flows along the temperature

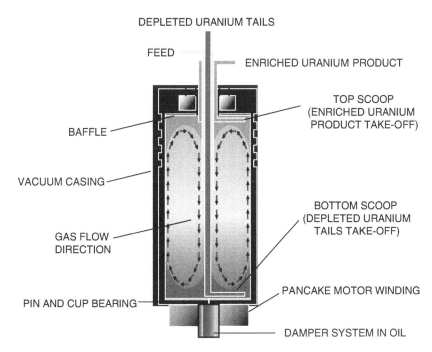

FIGURE 6.6
Enrichment with an ultracentrifuge.

gradient. The two streams of UF6, one enriched and one depleted in ^{235}U are removed from the centrifuge by two pipes (see Figure 6.6).

The enrichment level achieved by a single centrifuge is insufficient to obtain the desired concentration of the isotope of interest. It is therefore necessary to connect a number of centrifuges together in a series and in parallel. This arrangement of centrifuges is known as a cascade. Passing through the successive centrifuges of the cascade, it is possible to enrich isotopes to a high level >90%.

6.11.2   Laser Enrichment

Laser isotope separation (LIS) is based on the fact that different isotopes of the same element, while chemically identical, have different electronic energies and therefore absorb different colors of laser light. The isotopes of most elements can be separated by a laser-based process if they can be efficiently vaporized as atoms. For those elements to which it can effectively be applied, LIS is likely to be the least expensive enrichment technology. In LIS enrichment, the metal is first vaporized in a separator unit contained in a vacuum chamber. The vapor stream is then illuminated with laser light tuned precisely to a color at which the sought isotope of that metal absorbs energy (Figure 6.7).

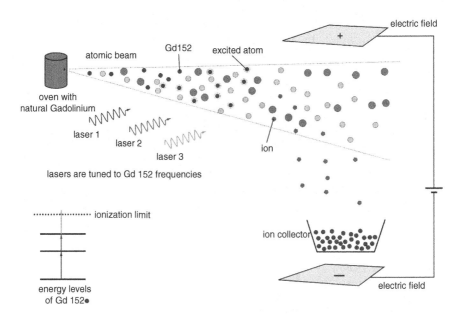

FIGURE 6.7
Laser isotope separation. Natural gadolinium consists of seven isotopes. With lasers tuned to excite and ionize only ^{152}Gd, this isotope can be electrically separated from the other six. In a nuclear reactor ^{152}Gd is converted into ^{153}Gd. This radionuclide serves as a line source for the SPECT camera. The attenuation of its 103 keV gamma by the patient's body is determined. The result is used to correct the SPECT image measured with the 140 keV gamma of Tc-99 m.

The generation of laser light starts with diode-pumped, solid-state lasers providing short, high-intensity pulses at high-repetition rates. This green light from the solid-state lasers travels via fiberoptic cable to energize high-power dye lasers. The dye laser absorbs green light and reemits it at a color that can be tuned to the isotope of interest. Each color selectively adds sufficient energy to ionize or remove an electron from the desired isotopic atoms, leaving other isotopes unaffected.

Since the ionized atoms are now tagged with a positive charge, they are easily collected on negatively charged surfaces inside the separator unit. The product material is condensed as liquid on these surfaces and then flows to a caster where it solidifies as metal nuggets. The unwanted isotopes, which are unaffected by the laser beam, pass through the product collector, condense on the tailings collector, and are removed.

6.11.3   The Calutron

In principle, the technique of isotopic separation by the calutron is simple and similar to the principle of mass spectrometry (Figure 6.8). When passing between the poles of a magnet, a monoenergetic beam of ions of a naturally occurring atom splits into several streams according to their momentum, one per isotope, each characterized by a particular radius of curvature. Collecting cups at the ends of the semicircular trajectories catch the homogeneous streams.

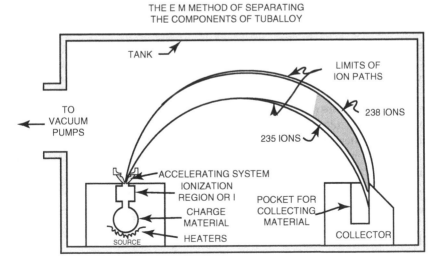

FIGURE 6.8
Isotopic separation with a calutron.

6.12   β-Ray Emitting Microparticles and Nanoparticles

Although in contradiction with having ideal property number 3 in the introduction (charged-particle emission should be absent or easily screened), a new class of unsealed brachytherapy sources is emerging. These are β-ray emitting microparticles and nanoparticles. β-rays have the advantage of delivering high doses in the vicinity of the emitting source and they are easily shielded by a few millimeters of plastic or tissue. However, this is also a main disadvantage as they cannot be used in a sealed source because of the attenuation of most of the β-rays in the metal casing.

Treatment is implemented using microspheres made from glass, resin, or biodegradable material labeled with the β-ray emitting nuclide. These are infused into the vascular supply of the target organ or of the tumor. Most tumors rely on the hepatic artery for their blood supply, and the principle of targeting the spheres in the tumor is based on the increased blood flow in the tumor neovasculature.

Microspheres depend on their size (10 to 30 μm in diameter) for mechanical trapping in the arterial branches leading into the tumor. Therapies for hepatic cancer are being developed using ^{90}Y glass or ^{166}Ho polylactic acid microspheres. Both types of microspheres are formed by neutron irradiation of, respectively, ^{89}Y glass or ^{165}Ho polylactic acid microspheres in the reactor. The production of these microspheres is technically very demanding, the energy production in the capture reaction may easily deform the glass spheres or even disintegrate and burn the polylactic acids. The advantage of the ^{166}Ho polylactic spheres over the ^{90}Y-glass spheres is that ^{166}Ho emits a gamma ray, making visualization of the spatial distribution possible. Also, the polylactic acid is biodegradable, thus enabling rapid clearance of untargeted spheres.

Selective targeting of the tumor neovasculature is achieved using nanoparticles coated with integrins or antibodies associated with angiogenesis. The perfluorocarbon spheres of nanometer size form a blood-like fluid. This much smaller size enables these nanoparticles to penetrate much deeper into the tumor microvasculature, thereby increasing their therapeutic potential. The nanoparticles are not only coated with targeting compounds, such as $\alpha_v\beta_3$ integrins and vascular endothelial growth factor receptor but are also equipped with binding sites for (radioactive) metals, such as ^{90}Y for therapy or Gd for what was originally termed nuclear magnetic resonance (NMR) imaging, but which is now known as magnetic resonance imaging (MRI).

Acknowledgments

Dr. Albert S. Keverling Buisman, Chief Health Physics Officer at NRG (Nuclear Research and Consultancy Group), Petten, The Netherlands, and

Chairman of the Netherlands' Health Physics Society, kindly reviewed the text and checked the physical data. Both the expertise and the experience of Mark Konijnenberg have been indispensable for the realization of this chapter.

(Dr. Mark W. Konijnenberg, Scientific Manager Research and Development, Mallinckrodt Medical B.V., P.O. Box 3, 1755 ZG Petten, The Netherlands, Mark.Konijnenberg@emea.tycohealthcare.com).

Physical data for the radionuclides considered in this chapter were extracted from the Online Tables: see http://atom.kaeri.re.kr/ton/nuc7.html (cross-sections from the Karlsruher Nuklidkarte, 6th Revised Edition, 1998). Urenco Nederland B.V. has provided the information for Section 6.11.

Reference

1. Johns, H.E. and Cunningham, J.R. *The Physics of Radiology,* 4th ed., C.C Thomas Publisher Ltd, Springfield, IL, 1983.

7

Source Specification and Source Calibration

7.1 Source Specification

7.1.1 Introduction

It is absolutely essential that source strength can be uniquely defined. This is because otherwise an accurate dose calculation cannot be made and calculations for radiation shielding cannot be performed. In the early years of the twentieth century only radium and radon were used for brachytherapy and dosimetry was essentially only milligram-hours (mgh) for radium or millicuries-destroyed (mcd) for radon. These measurements units were clearly inadequate.

Since the discovery of radium, there have been many recommendation and standards proposed for specifying the strength of sources. These can be classified by time periods which correlate with the times when different radionuclides dominated brachytherapy practice.

7.1.1.1 Mass of Radium

From the start of the radium era to the early 1950s, it was usual to specify the strength of radium sources available in form of tubes and needles in terms of the mass in mg of the radium, i.e., mass of radium. This period when the milligram radium (mgRa) unit was used included the year 1911 when Marie Curie prepared an international standard using a known mass of radium chloride in a Thüringian glass tube.

7.1.1.2 Activity

With the introduction of artificial radionuclides a few years after the end of World War II, the strength of sources was defined in terms of their activity content expressed in mCi.

However, since we are dealing exclusively with sealed brachytherapy sources and the activity of the contained radionuclide, the contained activity, was of limited practical interest. The exposure and exposure rate around the sources depend strongly on the form and dimensions of the source and its encapsulation due to self absorption and attenuation of the radiation emitted. To account for these factors the concept of apparent activity, A_{app}, has been introduced.

Apparent activity of a source was defined as the activity of a hypothetical point source, which is considered to be an ideal point source without the effects of self absorption and attenuation and having the same radionuclide delivering the same exposure rate in air at a specified distance equal to that for the actual source under consideration.

That distance should be large enough in comparison to the actual source dimensions such that the actual source could be considered as a point source. This distance was selected to be 1 m. The definition of apparent activity considers the effects of self absorption, attenuation, and filtration, as well as the production of Bremsstrahlung (x-rays) in the source and its encapsulation.

7.1.1.3   Equivalent Mass of Radium

Since the majority of experience in brachytherapy worldwide was gained using radium sources, it was convenient to retain the existing available dosage and dosimetry charts, i.e., the existing dosimetry systems, when changing from radium to a new radionuclide. Thus, for sources containing radionuclides considered as radium substitutes, their specification was given in milligram-radium equivalent (mg-Ra equivalent) enabling direct comparison with radium. At this point in time, to the early 1970s, tables used for radium sources could still be used for the new radionuclide sources, such as ^{60}Co and ^{137}Cs.

The radium equivalent mass of a source is defined as the mass in mg of radium filtered by 0.5 mm of platinum that leads to the same exposure rate as that resulting from the source under consideration at the same distance of 1 m in air.

7.1.1.4   Emission Properties

When activity or equivalent mass of radium was used for defining the source strength, the calculation of exposure rates and dose rates in tissue required the use of quantities relating the radiation output to the activity.

Such quantities were the specific gamma-ray constant and the exposure rate constant which are both sensitive to the method of calculation (e.g., point source assumption, photon energy spectrum etc.). Their uncertainties were then embedded into the dose calculation results.

Since these constants are defined for idealized point sources consisting of a pure radionuclide, they are characteristic of the radionuclide but not of the source itself. Furthermore, with these kinds of methodologies, the uncertainties in the corrections required for the actual sources, such as for taking into account absorption and filtration of radiation in the actual source and its encapsulation, further influence the accuracy of the final dosimetric calculation.

Last but not least, activity, apparent or contained, and equivalent mass of radium are not directly measurable quantities. They have to be extracted from measurements of radiation intensity at some distance from the source by applying conversion constants such as the exposure rate constant Γ_X:

$$\Gamma_X = \frac{r^2 \dot{X}}{A}$$

with $\dot{X}$ being the exposure rate at a distance r from a point source of activity A.

Because of these restrictions, output based quantities (i.e., measurable quantities) have been introduced and established at different time periods for specifying the strength of brachytherapy sources.

7.1.1.5 Reference Exposure Rate

The reference exposure rate has been defined as the exposure rate at 1 m distance from the source (see also Section 2.5.2.1) expressed in mR h^{-1} (milliroentgens per hour at 1 m).[1,2] At that period of time, in the mid-1970s, a conversion factor from exposure to air kerma of 0.873 cGy R^{-1} was used.

This was based on a value for the mean energy expended in air per ion pair formed (see Section 2.4.8) W of 33.85.[3] A reevaluation of this estimation by Boutillon and Perroche-Roux in 1987[4] resulted in a value of W of 33.97 eV and thus to a value of the ionization constant W/e of 33.97 C J^{-1} (see also Table A.2.2 in Appendix 2). Using this value for the ionization constant, the conversion factor from exposure to air kerma should be 0.876 cGy R^{-1}.

7.1.2 Reference Air Kerma Rate and Air Kerma Strength

In 1985, the ICRU Report 38[5] defined the reference air kerma rate as an emission specification quantity for γ-rays emitting brachytherapy sources. This was first recommended by the French Committee for the Measurement of Ionising Radiation in 1983[6] then by the British Committee on Radiation Units and Measurements in 1984[7] and has been widely established in Europe since the end of the 1980s.

The reference air kerma rate of a source is defined as the air kerma rate in air at a reference distance of 1 m from the center of the source, corrected for attenuation and scattering in air. This is the air kerma rate *in vacuo*. For rigid sources usually considered in the practice of brachytherapy (as distinct from malleable wire sources), the direction from the source center to the reference point at 1 m has to be at right angles to the long axis of the source.

For the reference air kerma rate, the unit μGy h^{-1} at 1 m was originally proposed in the ICRU Report 38.[5] Then, the ICRU Report 58[8] proposed that the reference air kerma rate should be expressed either in mGy h^{-1} at 1 m or μGy h^{-1} at 1 m. The SI unit for the reference air kerma rate is Gy s^{-1} but it is not practical for the sources used in brachytherapy.

The μGy h^{-1} at 1 m unit will be used when referring to the reference air kerma rate in the following. The notation RAKR has been proposed for this quantity but has not been widely used.[9] DIN-6809-2[10] and DIN-6814-3[11] use the notation $\dot{K}_{\alpha,100}$ for the reference air kerma rate, where 100 signifies the 100 cm/1 m reference distance. The notation $\dot{K}_R$ considered by IAEA TECDOC 1079[12] will be used here for the reference air kerma rate to indicate both the air and the air kerma rate, as well as for the reference for the distance of 1 m.

Further national and international recommendations were published in the 1980s and 1990s[8–10,12,13] and kept the reference air kerma rate as the specification quantity.

Recently, the BRACHYQS group of ESTRO, in their Booklet No 8 published in 2004, recommends the use of reference air kerma rate for source specification where the recommended unit is μGy h^{-1} for low dose rate (LDR) sources and mGy h^{-1} or μGy s^{-1} for sources of high dose rate (HDR). It is of interest to mention that in this recommendation the unit for the reference air kerma rate is given without the wording "at 1 m" as originally defined in the ICRU Report 38 and repeated in the ICRU Report 58.

In 1987, the American Association of Physicists in Medicine, in their Report No. 21 by the AAPM Task Group No 32,[14] introduced the idea for the specification of brachytherapy sources in terms of strength. That is, the air kerma strength, S_K, defined as the product of the air kerma rate in free space at a measurement distance r from the source center along the perpendicular bisector, $\dot{K}_{\alpha}(r)$, and the square of the distance r (see Figure 7.1):

$$S_K = \dot{K}_{\alpha}(r)r^2 \qquad (7.1)$$

The distance r must be chosen to be large enough so that the source can be

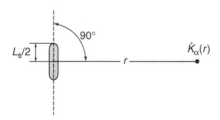

FIGURE 7.1
Schematic representation of the geometry as defined in AAPM Report 21 of the AAPM task group 32.[14] This report is for the source strength specification using the air kerma strength, S_K, concept for a cylindrical source. L_s is the length of the active part of the sealed source (active core) and r is the used radial distance for the measurement of air kerma rate in free space $\dot{K}_{\alpha}(r)$.

treated as a point source and so that the finite dimensions of the detector used for the measurement have no influence on the result.

The recommended unit for air kerma strength is μGy m^2 h^{-1} and has been denoted by the symbol U:[15]

$$1\,U = 1\,\mu\text{Gy m}^2\,\text{h}^{-1} = 1\,\text{cGy cm}^2\,\text{h}^{-1} \tag{7.2}$$

Although the air kerma strength S_K and the reference air kerma rate $\dot{K}_R$ are dimensionally different, their numerical values should be equal within the achievable dosimetric accuracy.

The S_K definition allows measurements of the source strength based on air kerma rate at distances other (shorter) than 1 m and accounts for this by multiplying the result by the square of the distance, thus making the resulting distance independent. In addition, the corresponding unit U is much simpler than previously proposed units, and avoids any words having to be included, unlike the case of the ICRU recommendation. The air kerma strength quantity is required in the U.S.A. as source strength descriptor.

The air kerma rate based source strength specifications, reference air rate kerma $\dot{K}_R$ and air kerma strength S_K, are a measure of the radiation field intensity in free space around brachytherapy sources, considering *a priori* the effects of source and source encapsulation, and making no use of intermediate constants.

Reference air rate kerma $\dot{K}_R$ and air kerma strength S_K are the currently recommended source strength specifications. Table 7.1 summarizes the units conversion factors for source strength defined as reference air kerma rate or air kerma strength.

It must be mentioned that the specification described and recommended is valid for cylindrical shaped γ-rays emitting sources, which are the majority of sources currently used for permanent as well as for temporary implants worldwide. In the case of wire sources such as ^{192}Ir wires, which are used for low dose rate implants, the source strength expressed either as $\dot{K}_R$ or S_K per unit length of 1 cm, is recommended.

7.1.3 Apparent Activity

Since there are still some requirements to estimate the source activity, apparent activity A_{app}, for radiation protection and regulatory issues, one can estimate this using the air kerma rate based strengths according to the following:[16]

$$A_{\text{app}} = 10^{-6}\left(\frac{1}{3600}\right)\frac{\dot{K}_R r_{\text{ref}}^2}{\Gamma_\delta} \tag{7.3}$$

TABLE 7.1

Units and Unit Conversion Factors for Reference Air Kerma Rate and
Air Kerma Strength

Source Strength Quantity	Symbol	SI	Current Unit	Conversion Factor
Reference air kerma rate	$\dot{K}_R$	Gy s^{-1}	$\mu\text{Gy h}^{-1}$	$(1/3600) \times 10^{-6} \text{ Gy s}^{-1}$ $= 2.778 \times 10^{-10} \text{ Gy s}^{-1}$
			mGy h^{-1}	$(1/3600) \times 10^{-3} \text{ Gy s}^{-1}$ $= 2.778 \times 10^{-7} \text{ Gy s}^{-1}$
			mGy h^{-1}	$1.0 \times 10^3 \ \mu\text{Gy h}^{-1}$
Air kerma strength	S_K	$\text{Gy m}^2 \text{s}^{-1}$	$\text{U}, 1 \text{ U} = 1$ $\mu\text{Gy m}^2 \text{h}^{-1}$ $= 1 \text{ cGy}$ $\text{cm}^2 \text{h}^{-1}$	$(1/3600) \times 10^{-6} \text{ Gy m}^2 \text{s}^{-1}$ $= 2.778 \times 10^{-10} \text{ Gy m}^2 \text{s}^{-1}$
Air kerma rate constant	Γ_δ	$\text{J kg}^{-1} \text{m}^2$ or Gy s^{-1} $\text{Bq}^{-1} \text{m}^2$	$\mu\text{Gy h}^{-1}$ $\text{MBq}^{-1} \text{m}^2$	$(1/3600) \times 10^{-12} \text{ Gy s}^{-1} \text{Bq}^{-1} \text{m}^2$ $= 2.778 \times 10^{-16} \text{ Gy s}^{-1} \text{Bq}^{-1} \text{m}^2$
			$\mu\text{Gy h}^{-1}$ $\text{MBq}^{-1} \text{m}^2$	1.0 U MBq^{-1}

where $\dot{K}_R$ is the reference air kerma rate expressed in $\mu\text{Gy h}^{-1}$ and r_{ref} is the reference distance of 1 m.

Alternatively,

$$A_{app} = 10^{-6} \left(\frac{1}{3600} \right) \frac{S_K}{\Gamma_\delta} \tag{7.4}$$

where the air kerma strength S_K is given in U, $1 \text{ U} = 1 \ \mu\text{Gy m}^2 \text{h}^{-1}$.

Γ_δ is the air kerma rate constant, see Equation 2.61 in Section 2.6.3, and is expressed in SI in $\text{Gy s}^{-1} \text{Bq}^{-1} \text{m}^2$. In brachytherapy practice, Γ_δ is expressed in $\mu\text{Gy h}^{-1} \text{MBq}^{-1} \text{m}^2$, or equivalently in U MBq^{-1}. Here, the notation Γ_δ for the air kerma rate constant introduced in ICRU Report 58[8] will be used instead of the $(\Gamma_\delta)_K$ which is used in some other publications. Thus, the apparent activity A_{app} given by Equation 7.3 and Equation 7.4 is in Bq or in MBq.

Γ_δ is based on the air kerma rate due to photons of energy greater than a certain cut-off value δ (see Section 2.6.3). The ICRU Report 58[8] generally recommends a cut-off value $\delta = 20$ keV. The selection of this value depends on the energy spectrum of emitted photons and on the cladding of the specific source. For the common source designs and radionuclides, δ values of 10 keV or even 5 keV are actually used in the literature. This is dependent on the assumption that photons with energies up to this cut-off value are not tissue penetrating.

The factor 10^{-6} is required to account for the conversion of μGy in the $\dot{K}_{R}$ and S_{K} units to Gy where the factor 1/3600 is to account for the conversion of hours to seconds for the same units.

Due to the fact that Γ_{δ} is defined for an ideal point source without considering any attenuation and scattering effect that exists for real sources (i.e. source material and source encapsulation), and if there is a need to consider this for the estimation of apparent activity A_{app}, then a correction factor for this attenuation has to be considered in the above equations such that it will adjust the source strength values for $\dot{K}_{R}$ and S_{K} for the case of missing attenuation (i.e., point source assumption).

Attention must to be paid to the situation when activity values are required for treatment planning systems. Since the measured and given values of source strength are either in reference air kerma rate or air kerma strength, the activity values for entries in planning systems have to be calculated using the planning system vendors applied values for the conversion factor and attenuation factor (if any is used). In this manner, and independently if these both values are accurate, the dose calculation results of the planning system will be correct due to the consistency of the factors used at both sites.

It must be underlined that apparent activity is not a recommended source specification quantity. It is referred to here because of practical brachytherapy needs when dealing with planning systems and calculation algorithms.

Table 7.2 summarizes the values published in the literature of the air kerma rate constant Γ_{δ} for the common brachytherapy radionuclides and for ideal point sources. In the same table, Γ_{δ} published values are listed for real source designs. The latest consider the attenuation due to source design for the source strength (i.e., dosimetry of real sources) and thus they are lower than those values calculated for ideal point sources of a pure radionuclide. This is most pronounced for the very low energy radionuclides ^{125}I and ^{103}Pa.

In the currently available literature on Monte Carlo–based dosimetry for brachytherapy sources, the activity contained in the source, A_{con}, is used to normalize the parameters calculated (taking into account the decay). A_{con} is the true actual activity of the specific radionuclide contained in the source model which is to be analyzed or simulated. The air kerma rate constant Γ_{δ} is thus defined per unit A_{con} and is usually expressed in UMBq. When the air kerma rate constant Γ_{δ} calculated in this manner is used, then by applying Equation 7.3 and Equation 7.4, A_{con} for the source can be calculated instead of the apparent activity A_{app}.

From a physics point of view, it is recommended that the contained activity A_{con} is considered when the activity of a source is to be described, e.g., for comparison purposes. Alternatively, when activity is required as an input for treatment planning systems, the user must very carefully use the planning system specific conversion factors for calculating the required input data from the measured source strength, the reference air kerma rate or the air kerma strength.

TABLE 7.2

Summary of Recommended Values of the Air Kerma Rate Constant Γ_δ for the Brachytherapy Radionuclides and for Ideal Point Sources

	Air Kerma Rate Constant Γ_δ (μGy h^{-1} MBq^{-1} m^2 = U MBq^{-1})			
Radionuclide	ICRU Report 58[8]	AAPM Report 51[15]	Table 5.2 ($\delta = 10$ keV)	Real Sources Literature
$^{137}_{55}$Cs	0.0772	—	0.0771	
$^{60}_{27}$Co	0.306	—	0.3059	
$^{198}_{79}$Au	0.0559	—	0.0545	
$^{125}_{53}$I	0.0337	0.0358/ 0.0355[a]	0.0348	0.0206 − 0.0219[b], 0.0217[c]
$^{192}_{77}$Ir	0.108 [0.100, 0.116]	0.111	0.1091	0.0993 − 0.975[d], 0.1027[e], 0.110[f], 0.0973[g], 0.0979[h], 0.1025[i], 0.107[j], 0.0992[k]
$^{103}_{46}$Pd	0.0343	0.035/ 0.0361[a]	0.0361	0.0173[l]
$^{170}_{69}$Tm	—	—	0.00053	
$^{169}_{70}$Yb	—	—	0.0431	0.0427[25] ($\delta = 10$ keV), 0.0359[m], 0.0297[n] ($\delta = 10$ keV),

Γ_δ values which are given in the literature for commercially available sources are also listed.
[a] Revised values using a cut-off value of $\delta = 5$ keV in AAPM Report TG-43 update.[17]
[b] STM1251-^{125}I-seed.[21]
[c] selectSeed-^{125}I.[23]
[d] PDR source designs for the microSelectron-PDR afterloader system from Nucletron B.V., The Netherlands considering $\delta = 11.3$ keV.[18]
[e] HDR source design (5 mm active length) for the VariSource afterloader from Varian Oncology Systems, USA.[19]
[f] HDR and PDR source designs for the microSelectron afterloaders from Nucletron B.V., The Netherlands.[20]
[g] New design for the microSelectron-HDR afterloader considering $\delta = 11.3$ keV.[26]
[h] Old design for the microSelectron-HDR afterloader considering $\delta = 11.3$ keV.[26]
[i] Old HDR source design (10 mm active length) for the VariSource afterloader considering $\delta = 11.3$ keV.[26]
[j] Seed source from Best Industries Inc., U.S.A. considering $\delta = 11.3$ keV.[26]
[k] Seed source from Alpha-Omega Services, Inc., U.S.A., considering $\delta = 11.3$ keV.[26]
[l] Model 200-^{103}Pd-seed.[22]
[m] X1267-^{169}Yb.[24]
[n] Model HDR 4140 ^{169}Yb by Implant Sciences Corporation, U.S.A., considering $\delta = 10$ keV.[79]

7.2 Source Calibration

The description "source calibration" refers to the estimation of the source strength expressed in one of the internationally recommended quantities: either reference air kerma rate or air kerma strength. Both are based on the measurement of a representative quantity of the radiation field surrounding the source. In practice, this means the measurement of ionization in a specified geometry and environment.

The air kerma rate at the reference distance of 1 m, or the product of the air kerma rate and the square of the measurement distance, is estimated. This estimation is based on the ionization measured, current or charge within the detector volume, and also after applying a number of corrections required. This is the so-called dosimetry protocol.

Although source manufacturers are providing calibration protocols for the sources they deliver, it is strongly recommended that source strength is validated by one's own measurements before their first application for patients' treatments.[8–10,12–14,27–36]

According to the 2002 guidelines for radiation protection in medicine, published by the German Ministry for Environment, Nature Conservation and Reactor Security,[37] it is a requirement for sources used in afterloading systems for temporary brachytherapy applications, that the source strength is in addition after a period of 2 weeks from the first calibration, measured to check the purity of the radionuclide. This is mainly focused on the use of ^{192}Ir afterloading systems where there exists the problem of possible contamination with [194]Ir (see Chapter 6). The requirement is also applicable for potential future usage of [169]Yb sealed sources in afterloading machines where the possibility of contamination with [175]Yb has to be considered[25] (see Table 5.13 and Figure 5.10 as well as Chapter 6).

There are generally three different methods of measuring the air kerma rate based source strength: (1) Measurement of air kerma rate in an in air set-up using an adequately calibrated ionization chamber and then extraction of the source strength; (2) Measurement using an in reference air kerma rate $\dot{K}_R$ or equivalently air kerma strength S_K calibrated well-type ionization chamber; and (3) Measurement using a solid phantom of well-defined geometry which is supplied with correction factors for extracting the source strength from the measured charge or current by using an ionization chamber.

7.2.1 Calibration Using an In-Air Set-Up

Here, an in air set-up for measurement of air kerma rate is considered (see Figure 7.2). The source is usually inserted in a tube, needle, or catheter that is normally used in brachytherapy implants. In this manner, the measured source strength includes the effect of attenuation of radiation resulting from the wall of the tube/needle/catheter. For sources used in permanent interstitial implants such as ^{192}Ir or [198]Au seeds, the effect of this wall thickness must be considered and the measurement result must be corrected for this factor.

This method is mainly appropriate for sources with high mean photon energies such as [198]Au, ^{192}Ir, ^{137}Cs, and for ^{60}Co with mean photon energies above 0.3 MeV. It is also appropriate for [169]Yb nuclide sources with a mean photon energy of 0.093 MeV. However, it is problematic for the very low energy radionuclide sources such as ^{125}I and [103]Pd which have a mean

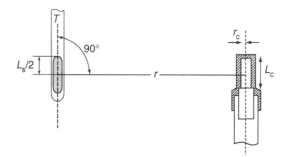

FIGURE 7.2

Geometry of an in-air calibration set-up at a distance r of a source using an ionization chamber with a build-up cap. The source is positioned within an suitable tubular structure (applicator/catheter) defined here by T. L_s is the length of the active part of the sealed source (active core) and r is the radial distance used for the measurement of air kerma rate in free space $\dot{K}_a(r)$. L_c is the internal length and r_c the internal radius of the cylindrical volume of the ionization chamber used for this example.

photon energy lower than 30 keV and which are available as low activity sources for permanent implants.

The ESTRO Guidelines Report[13] advises against using the in-air set-up calibration method for these low energy radionuclide sources, at least as a routine method. The IAEA-TECDOC-1079[12] considers this set-up explicitly for low and high dose rate sources of ^{192}Ir, ^{137}Cs, and ^{60}Co radionuclides. Alternatively, DIN 6809-2[10] also considers these sources to be able to be calibrated using in-air set-up measurements.

The real problems with the low energy, low activity seed sources, ^{125}I and ^{103}Pa, are related to the very low measurement signals which can be achieved with the usual cavity chambers which are available in a brachytherapy clinic and to the uncertainties of their calibration factors at these low energies. The use of large volume spherical chambers (with as large as 1000 cm^3 measuring volume) can solve these problems, but this is only practical in Accredited Dosimetry Calibration Laboratories (ADCLs) and not in the clinical routine.

Although the source strength when expressed as reference air kerma rate is defined at the reference distance of 1 m, this is not practical for the majority of cases. Generally, the measurements are taking place at distance r (in cm) free in-air, where r is defined along the transverse bisector of the source and is the distance between the center of the active core of the source and the reference point of the ionization chamber (see Figure 7.2).

Figure 7.3 demonstrates some available standard geometry calibration jigs for in-air measurement set-ups.

7.2.1.1 *Geometrical Conditions*

7.2.1.1.1 *Measurement Distance*

The measurement distance r must be selected in such a way that the source can be considered as a point source and that the dimensions of the collecting

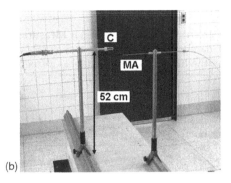

(a)　　　　　　　　　　　　　　　　(b)

FIGURE 7.3
Examples of in-air calibration set-ups for HDR and PDR sources. (a) The Nucletron source calibration jig. A fixed source to chamber distance of 10.0 cm is used, where two source positions are used which are symmetrical relative to the chamber. Plastic source catheters are used to minimize attenuation effects. The distance from the base plate where source and chamber are located is 30 cm. Adapters for any common thimble ionization chamber are provided. The outer dimensions of the jig are: height 32.5 cm, width and depth are each 52.0 cm. (b) Calibration jig constructed and available in the Offenbach Radiotherapy Clinic which enables variable source to chamber distances in the range 1.0 to 100.0 cm. This is a calibration set-up for an ^{192}Ir source of a microSelectron-HDR afterloader. The PTW type 23332 0.3 cm^3 ionization chamber with the standard ^{60}Co build-up cap (C) and the metallic dosimetry applicator (MA) are clearly shown.

volume of the chamber will not significantly influence the calibration result. This can be achieved if the distance r is at least three times the active length of the source (i.e., the length of source active core) L_s and at least five times that of the largest extension of the chamber, which is the chamber volume length L_c in Figure 7.2.

Furthermore, the distance r should be large enough so that positioning errors will not significantly influence the result. For example, let us consider an assumed positioning error of ± 0.5 mm for the in-air set-up. When a measurement distance of $r = 5.0$ cm is selected, the resultant inaccuracy can be as high as $\pm 2\%$ (see also Table 7.3). For large measurement distances, the positioning errors of the ionization chamber and source have a negligible influence on the measurement result: but the dose rate at such distances is probably not within the acceptable range for accuracy in measurements.

It is generally recommended that distances greater than 5.0 cm are considered for in-air measurement set-ups.[10] The selection of an appropriate distance further depends on the activity of the source and on the volume of the ionization chamber under consideration. For high activity (HDR) ^{192}Ir sources, when using typical chambers available for the dosimetry of high energy photon beams (0.3 to 1.0 cm^3 volumes), distances in the range of 5.0 to 40.0 cm can be considered as appropriate.

Another point of interest regarding the influence of the selected measurement distance on the result for in-air set-up is the attenuation and scattering of photons that can take place in air.

TABLE 7.3

Percent Uncertainty in In-Air Measurement Set-Up for Different Source to Chamber Distances r Due to Several Levels of Positional Error δr as a Result of the Inverse-Square Distance Correction

Uncertainty in Positioning δr (mm)	Uncertainty (%)						
	$r =$ 5.0 cm	$r =$ 10.0 cm	$r =$ 15.0 cm	$r =$ 20.0 cm	$r =$ 30.0 cm	$r =$ 40.0 cm	$r =$ 50.0 cm
0.1	0.40	0.20	0.13	0.10	0.07	0.05	0.04
0.2	0.80	0.40	0.27	0.20	0.13	0.10	0.08
0.3	1.20	0.60	0.40	0.30	0.20	0.15	0.12
0.4	1.60	0.80	0.53	0.40	0.27	0.20	0.16
0.5	2.00	1.00	0.67	0.50	0.33	0.25	0.20
0.6	2.40	1.20	0.80	0.60	0.40	0.30	0.24
0.8	3.20	1.60	1.07	0.80	0.53	0.40	0.32
1.0	4.00	2.00	1.33	1.00	0.67	0.50	0.40

Figure 7.4 demonstrates the dependence of mass interaction coefficients for attenuation μ/ρ, absorption μ_{en}/ρ, and scattering μ_{sc}/ρ on the photon energy in dry air (dry air according to ICRU Report 37[38]). As can be seen from this figure, for very low energies in the region 0.020 to 0.030 MeV

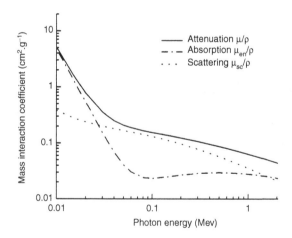

Photon energy (Mev)

FIGURE 7.4
Dependence of the mass interaction coefficients, μ/ρ, μ_{en}/ρ, and μ_{sc}/ρ for attenuation, absorption, and scattering, respectively, in dry air on the photon energy for the brachytherapy relevant energy range 0.01 to 2.0 MeV. Data from Hubell and Seltzer.[40] The scattering coefficient is defined as the difference between the corresponding attenuation and absorption coefficient. For very low energies in the region 0.020 to 0.030 MeV (mean energies of [103]Pd and ^{125}I) 30 to 60% of the total interaction coefficient, respectively, is due to scattering (Compton incoherent scattering). For energies in the range 0.050 to 0.150 MeV this rises to 80 to 85%. This means that most of the energy of the photons is being scattered in air instead of being absorbed. For an energy of 0.080 MeV the maximum ratio of the probability of scattering to the probability of absorption is ∼5.9.

(mean energies of ^{103}Pd and ^{125}I), 30 to 60% of the total interaction coefficient respectively is due to scattering (Compton incoherent scattering). For energies in the range 0.050 to 0.150 MeV this rises to 80 to 85%: thus most of the energy of photons is being scattered in air rather than being absorbed. For the energy of 0.080 MeV, the maximum observed ratio of probability of scattering to probability of absorption is approximately 5.9.

The dependence of the radionuclide and of the distance selected adequate corrections must be taken into account to include the effects of scattering and attenuation in air.

7.2.1.1.2 Source Positioning

The accuracy of source positioning within the catheter for the in-air set-up will also influence the accuracy of the measurement. This is mainly relevant to the accuracy in the source position definition along the catheter axis. Fortunately, due to the geometry considered for brachytherapy practice, this effect is very low. A source positioning error of ± 2.0 mm causes an uncertainty of 0.16% at a measuring distance r of 5.0 cm, whereas for source to chamber distances of greater than 10.0 cm, this has practically no influence on the result as far as the inverse-square distance correction is concerned (see Table 7.4).

The catheter used for the placement of the source must be selected such that its lumen diameter (inner diameter of the catheter), when compared to the outer diameter of the source, will not introduce a significant uncertainty in the estimation of the source to chamber distance (see Table 7.3).

TABLE 7.4

Percent Uncertainty in In-Air Measurement Set-Up for Different Source to Chamber Distances r Due to Several Levels of Source Positioning Uncertainty Along the Axis of the Applicator Used for the In-Air Set-Up Shown in Figure 7.2 as a Result of the Inverse-Square Distance Correction

Uncertainty in Source Positioning (mm)	Uncertainty (%)						
	$r =$ 5.0 cm	$r =$ 10.0 cm	$r =$ 15.0 cm	$r =$ 20.0 cm	$r =$ 30.0 cm	$r =$ 40.0 cm	$r =$ 50.0 cm
1.0	0.04	0.01	0.00	0.00	0.00	0.00	0.00
1.5	0.09	0.02	0.01	0.01	0.00	0.00	0.00
2.0	0.16	0.04	0.02	0.01	0.00	0.00	0.00
2.5	0.25	0.06	0.03	0.02	0.01	0.00	0.00
3.0	0.36	0.09	0.04	0.02	0.01	0.01	0.00
4.0	0.64	0.16	0.07	0.04	0.02	0.01	0.01
5.0	1.00	0.25	0.11	0.06	0.03	0.02	0.01
6.0	1.43	0.36	0.16	0.09	0.04	0.02	0.01
8.0	2.54	0.64	0.28	0.16	0.07	0.04	0.03
10.0	3.96	1.00	0.44	0.25	0.11	0.06	0.04

7.2.1.1.3 Room Dimensions

The room dimensions should be such that the minimum distance between the chamber and the possible scattering sources is at least 1 m. If this cannot be achieved, the influence of the room scattering to the measurement must be estimated and the result must be corrected for this factor. In general, the larger the source to detector distance the greater the influence of the room scattering to the measurement result.

7.2.1.2 Ionization Chamber

The selection of an ionization chamber depends on the source which is being considered. For sources of high activity (as in HDR brachytherapy applications), which are usually either ^{192}Ir or ^{60}Co or possibly ^{169}Yb sources for temporary implants, ionization chambers with volumes in the range 0.3 to 1.0 cm^3 can be considered adequate. The larger the volume, the greater the collecting signal (current or charge) and thus the greater the accuracy for a specific measurement distance. Alternatively, the larger the volume of the chamber the greater the role of radiation field gradients existing within that volume and thus the greater the uncertainty of the required (nonuniformity) correction.

For the example of the PTW Type 23332 0.3 cm^3 and Type 23332 1.0 cm^3 rigid chambers, a typical value of the leakage current is approximately 4×10^{-15} A and for their sensitivity 1×10^{-8} and 3.3×10^{-8} C Gy^{-1}, respectively. The resultant leakage for dose rate is estimated to be 0.14 and 0.04 cGy h^{-1} for the 0.3 and 1.0 cm^3 rigid chambers, respectively. This demonstrates the measurement limitations which are possible using these chambers. These are for the maximum distances for given source activities.

For very low activity sources, chambers with volumes much higher than 1 cm^3 and up to 1000 cm^3 may be required to achieve an adequate collecting signal and thus a large measurement distance between source and chamber up to 1 m is required.

7.2.1.2.1 Calibration Factor of the Chamber

The ionization chamber should have a calibration factor for air kerma for a photon energy near to that of the radionuclide source. In practice, the common calibration factors are for ^{60}Co external beams provided by all the Primary Standard Dosimetry Laboratories (PSDLs) and Secondary Standard Dosimetry Laboratories (SSDCLs).

In some of the laboratories in the U.S.A. and Germany it is also possible to have calibration factors for ^{137}Cs external beams. For sources of these two radionuclides it is possible to have direct calibration factors. For sources of radionuclides of lower energy, such as ^{198}Au and ^{192}Ir with mean photon energies of 0.406 and 0.355 MeV, respectively, (see Table 5.2) calibration factors supplied for an x-ray beam of some hundred keV are required. Typically, x-rays of 250 kVcp with additional Al and Cu filtration resulting in effective photon energies of 130 to 150 keV are available in the different

PSDLs and SSDCLs. These can be considered for extracting adequate air kerma calibration factors for ^{198}Au and ^{192}Ir sources.

For radionuclides of even lower energy, such as ^{169}Yb with a mean photon energy of 0.093 MeV, calibration at even lower x-rays, e.g., 140 to 200 kVcp, are needed for obtaining the necessary calibration factors.

This has the consequence that chambers with more than one calibration factor, for ^{60}Co and if possible ^{137}Cs, and for at least one for a suitable x-ray beams energy are needed.

For ^{192}Ir sources, simplified concepts exist for the estimation of an appropriate calibration factor for that photon energy spectrum. Thus, the BIR/IPSM[9] recommends that the calibration factor for the highest available kilovoltage quality, heavily filtered 280 kVcp x-rays, is used for ^{192}Ir and, in addition, a general correction of 3% per g cm^{-2} of low atomic number material (plastic or graphite) and for small thicknesses is to be considered.

DIN 6809-2[10] refers to the results of Pychlau[39] who recommended a simple (linear) interpolation between the available air kerma calibration factors for the PTW-chamber. In his analysis, chambers with volumes of 0.125 to 30.0 cm^3 were considered. He concluded, based on his own experimental results for ^{192}Ir and using a PTW Type 23331 1.0 cm^3 ionization chamber with a wall thickness of 0.5 mm PMMA, that the use of the ^{60}Co calibration for a build-up cap of 3.0 mm requires an attenuation correction of 0.5%. When the cap was not used, an effect due to secondary electron contamination of 1.8% at a measurement distance of 5.0 cm was demonstrated. The use of this build-up cap and the corresponding correction is then recommended for all sources except for the very low energy ones of ^{103}Pd and ^{125}I.

The DGMP Report 13[31] proposes the use of the calibration factor for ^{60}Co and of the corresponding build-up cap for ^{192}Ir HDR sources and for 0.3 to 1.0 cm^3 volume ionization chambers. In this manner, a correction for the attenuation of the build-up cap of 0.5% has to be considered, as proposed by Pychlau.[39]

All these values for the correction of attenuation of build-up caps for ^{192}Ir are in agreement with the corresponding mean mass absorption coefficient, for example for polymethyl methacrylate $C_5H_8O_2$ (PMMA)[40] $\mu_{en}/\rho = 0.03185$ cm^2 g^{-1} (see also Table 7.5) which results in a correction of approximately 3.2% for every g cm^{-2} of PMMA or of approximately 1.3% for a 3.0 mm PMMA build-up cap and 0.5 mm chamber wall thickness (PMMA) when using the Attix[41] approximation for this attenuation effect:

$$\left(\frac{\mu_{en}}{\rho}\right)\rho t 100 \text{ in\%} \tag{7.5}$$

where μ_{en}/ρ is the mass absorption coefficient for the chamber wall and cap material (here PMMA) for the γ-energy (for ^{192}Ir a rounded mean energy of 0.4 MeV is considered), ρ is the density of that material ($\rho_{PMMA} = 1.19$ g cm^{-3}) and t is the total thickness of the chamber wall and cap. This is valid for broad-beam conditions, i.e., it is always the case for in-air calibration of

TABLE 7.5

Composition and Material Constants According to ICRU Report 37[38] Which Is Relevant for Dosimetry Issues in Brachytherapy

Material	Mean Ratio of Atomic Number to Atomic Mass $\langle Z/A \rangle$	Mean Excitation Energy I (eV)	Density ρ (g cm^{-3})	Electron Density[a] ($\times 10^{23}$ e g^{-1})	Composition (Fraction by Weight)
A150 (Tissue equivalent plastic)	0.549031	65.1	1.127	3.306	H ($Z = 1$): 0.101327, C ($Z = 6$): 0.775501, N ($Z = 7$): 0.035057, O ($Z = 8$): 0.052316, F ($Z = 9$): 0.017422, Ca ($Z = 20$): 0.018378
C552 (Air equivalent plastic)	0.499687	86.8	1.760	3.009	H ($Z = 1$): 0.024680, C ($Z = 6$): 0.501610, O ($Z = 8$): 0.004527, F ($Z = 9$): 0.465209, Si ($Z = 14$): 0.003973
Dry air (near sea level)	0.499190	85.7	1.205×10^{-3}	3.006	C ($Z = 6$): 0.000124, N ($Z = 7$): 0.755268, O ($Z = 8$): 0.231781, Ar ($Z = 18$): 0.012827
Gafchromic Sensor[40]	0.543850	67.2	1.300	3.275	H ($Z = 1$): 0.089700, C ($Z = 6$): 0.605800, N ($Z = 7$): 0.112200, O ($Z = 8$): 0.192300
Graphite (Carbon)	0.499540	78.0	1.700	3.008	C ($Z = 6$): 1.000
Lithium fluoride LiF	0.462617	94.0	2.635	2.786	Li ($Z = 3$): 0.267585, F ($Z = 9$): 0.732415
Nylon type 6 and 6/6 C$_6$H$_{11}$ON	0.547902	63.9	1.140	3.299	H ($Z = 1$): 0.097976, C ($Z = 6$): 0.636856, N ($Z = 7$): 0.123779, O ($Z = 8$): 0.141389
Plastic water[b] (PW or PW2030)	0.540468		1.022	3.255	H ($Z = 1$): 0.086, C ($Z = 6$): 0.6126, N ($Z = 7$): 0.0157, O ($Z = 8$): 0.199, Al ($Z = 13$): 0.0845, Cl ($Z = 17$): 0.0022
Plastic water[b] (PW-LR)	0.538155		1.029	3.241	H ($Z = 1$): 0.0793, C ($Z = 6$): 0.5360, N ($Z = 7$): 0.0174, O ($Z = 8$): 0.2720, Mg ($Z = 12$): 0.00929, Cl ($Z = 17$): 0.0024
Polyoxymethylene[73,74] CH$_2$O (Delrin, POM)	0.532869	77.4	1.425	3.209	H ($Z = 1$): 0.067137, C ($Z = 6$): 0.400016, O ($Z = 8$): 0.532847
Polymethyl methacrylate C$_5$H$_8$O$_2$ (PMMA, Lucite, Plexiglas or Perspex)	0.539369	74.0	1.190	3.248	H ($Z = 1$): 0.080538, C ($Z = 6$): 0.599848, O ($Z = 8$): 0.319614

(*continued*)

TABLE 7.5—*Continued*

Material	Mean Ratio of Atomic Number to Atomic Mass $\langle Z/A \rangle$	Mean Excitation Energy I (eV)	Density ρ (g cm^{-3})	Electron Density[a] ($\times 10^{23}$ e g^{-1})	Composition (Fraction by Weight)
Polystyrene C$_8$H$_8$	0.537680	68.7	1.060	3.238	H ($Z = 1$): 0.077418, C ($Z = 6$): 0.922582
Polytetrafluoro-ethylene C$_2$F$_4$ (Teflon)	0.479925	99.1	2.200	2.890	C ($Z = 6$): 0.240183, F ($Z = 9$): 0.759817
Radiochromic dye filmnylon base[40]	0.549882	64.5	1.080	3.311	H ($Z = 1$): 0.101996, C ($Z = 6$): 0.654396, N ($Z = 7$): 0.098915, O ($Z = 8$): 0.144693
RW-3[76] (Polysty-rene + 2% TiO$_2$	0.536452		1.045	3.231	H ($Z = 1$): 0.075873, C ($Z = 6$): 0.904127, O ($Z = 8$): 0.008012, Ti ($Z = 22$): 0.011988
Silicon dioxide SiO$_2$	0.499298	139.2	2.320	3.007	O ($Z = 8$): 0.532565, Si ($Z = 14$): 0.467435
Solid water[75,77,78]	0.539470	73.3	1.017	3.249	H ($Z = 1$): 0.080901, C ($Z = 6$): 0.671710, N ($Z = 7$): 0.024149, O ($Z = 8$): 0.198734, Cl ($Z = 17$): 0.001306, Ca ($Z = 20$): 0.023200
Water liquid H$_2$O	0.555087	75.0	1.000	3.343	H ($Z = 1$): 0.111894, O ($Z = 8$): 0.888106

[a] Calculated using N_A. $\langle Z/A \rangle \times 10^{-23}$, where N_A is Avogadro's number ($N_A = 6.0221367 \times 10^{23}$ mol^{-1}, see also Appendix 2) and $\langle Z/A \rangle$ is the mean ratio of atomic number to atomic mass listed in the table.
[b] Material composition and density provided by Computerized Imaging Reference Systems CIRS Inc., Norfolk, Virginia, USA.

brachytherapy sources and for low Z materials such as PMMA, graphite, polystyrene, or other plastics which are used for the construction of chamber walls and build-up caps.

DIN 6809-2[10] also allows the use of ionization chambers which have been calibrated to give absorbed dose to water. The DGMP[29,31] recommendation for reference air kerma calibration of HDR ^{192}Ir sources also considers such calibration factors.

7.2.1.2.2 Use of Build-Up Caps

Calibration factors for commonly used ionization chambers at ^{137}Cs and ^{60}Co energies use build-up caps in order to achieve charged particle equilibrium and for filtering out possible contamination from electrons emitted by interaction processes of primary photons with the environment. Thus, when calibration set-ups for in-air measurements are considered for

^{137}Cs or ^{60}Co sources, the corresponding build-up cap considered in the calibration protocol of the chamber has to be used.

The required total thickness of the chamber wall and build-up cap for ^{137}Cs and ^{60}Co radiation is of 0.36 and 0.55 g cm^{-2}, respectively. For sources with radionuclides of lower mean energies, such as ^{169}Yb, ^{192}Ir and ^{198}Au, where interpolation exists between available calibration factors for x-ray beam(s) and for ^{137}Cs or ^{60}Co, there is a special problem when using or not using a build-up cap. The last option will influence the accuracy of the interpolation method used as well as the required correction for possible attenuation of the radiation due to the build-up cap material.

7.2.1.3 Formalism

Considering the in-air calibration set-up as is described in Figure 7.2, and assuming a source to detector distance r which is usually much less than the specification distance for reference air kerma rate of 1 m, the reference air kerma rate, K_R, or equivalently the air kerma strength, S_K, are determined from the detector measurement M using the following equations:

$$\dot{K}_R = N_K M k_u k_\tau k_\rho k_T k_p k_{ion} k_V k_{wall} k_{as,a} k_{scatt} k_{appl} k_r \qquad (7.6)$$

or

$$S_K = N_K M k_u k_\tau k_\rho k_T k_p k_{ion} k_V k_{wall} k_{as,a} k_{scatt} k_{appl} k_r' \qquad (7.7)$$

where:

$\dot{K}_R$:   is the reference air kerma rate expressed in μGy h^{-1} at 1 m.

S_K: is the air kerma strength in U, $1\,U = 1\,\mu Gy\,m^2\,h^{-1} = 1\,cGy\,cm^2\,h^{-1}$.

N_K: is the air kerma calibration factor of the chamber for the γ-energy of the radionuclide considered, usually expressed in Gy C^{-1}.

M: is the measured charge collected during a time interval of τ minutes.

k_u: is a unit conversion factor, to convert the product $N_K M$ to μGy as it is needed for $\dot{K}_R$ and S_K. In the case when the calibration factor N_K is in Gy C^{-1} and the measured charge M in C, then $k_u = 1.0 \times 10^6$.

$k_\tau = 60/\tau$:   is a factor to extrapolate the electrometer readout (charge) during the time interval τ (min) for 60 min duration to count for the air kerma rate per hour for the source strength definition. For high dose rate sources, such as ^{192}Ir and distances r around 20 cm, τ is normally 1 min.

k_ρ: is the correction factor for the current air pressure conditions P, expressed in hPa, other than that referred in the calibration protocol of the chamber, that is the normal pressure of 1013 hPa:

$$k_\rho = \left(\frac{1013 \text{ hPa}}{P} \right)$$

k_T: is the correction factor for the current air temperature conditions T in °C other than that referred in the calibration protocol of the chamber, that is the normal temperature T_0 in °C:

$$k_T = \left(\frac{273.15 + T \, (\degree C)}{273.15 + T_0 \, (\degree C)} \right)$$

Usually, $T_0 = 20\degree C$.

k_p: is the correction factor for the polarity effect of the bias voltage for the photon energy of the radionuclide. When M_1 and M_2 are the measured signals, charge or current, of the chamber for the standard and inverse voltage polarity respectively, then the polarity effect correction factor k_p is given by[42]

$$k_p = \frac{|M_1| + |M_2|}{2|M_1|}$$

This effect should not be higher than 1%.

k_{ion}: is the correction factor that counts for the unsaturated ion collection efficiency and thus for the charge lost to recombination for the specific radionuclide photon energy and the applied nominal voltage V. For continuous irradiation, such as that from radioactive decay, there is a linear dependence of $1/M$ vs. $1/V^2$, where M is the measured signal of the chamber (usually charge) at the voltage V and at the region of saturation.[42,43] Thus, the ion collection efficiency A_{ion}[43] for the nominal voltage V is given by[42]

$$A_{ion} = \frac{(V/V_2)^2 - M/M_2}{(V/V_2)^2 - 1}$$

where M_2 is the measured chamber signal at the voltage V_2, with $V_2 \ll V$. When V_2 is selected to be half of the nominal voltage V, $V_2 = V/2$, then[42]

$$A_{ion} = \frac{4}{3} - \frac{M}{3M_2}$$

k_{ion} is the reciprocal of the ion collection efficiency A_{ion}:

$$k_{ion} = \frac{1}{A_{ion}} = \frac{3}{4 - M/M_2}$$

with M_2 being the selected chamber signal at the half of the nominal voltage V and M the measured chamber signal (charge) at the nominal voltage V.

k_V: is a correction factor to take into account the effect of the chamber's finite size (volume) when the center of the chamber air cavity volume is considered as the point of reference for the positioning of the chamber and thus for referring the measurement result.

k_{wall}:   is the correction factor for accounting for attenuation and scattering effects of the build-up cap and chamber wall as far as this is not considered in the air kerma calibration factor N_K.

$k_{as,a}$:   is the correction factor for accounting for scattering and attenuation of the primary photons in air for the used measurement set-up (radial distance r, see also Figure 7.2).

k_{scatt}:   is the correction factor for correcting for the contribution of scattered radiation from room walls, floor and measurement set-up.

k_{appl}:   is the correction factor for accounting for the attenuation in the applicator/catheter wall used for positioning or fixating the source in the in-air setup.

$k_r = (r/r_0)^2$:   is the correction factor for the used radial distance r expressed in meters between source and ionization chamber (see Figure 7.2) with r_0 being the reference distance for the reference air kerma rate source specification, $r_0 = 1.0$ m.

$k_r' = r^2$:   for the air kerma strength S_K, calibration r is expressed either in cm or in m resulting in S_K expressed in U, $1\,U = 1\,\mu Gy\,m^2\,h^{-1} = 1\,cGy\,cm^2\,h^{-1}$.

7.2.1.3.1 *The Air Kerma Calibration Factor N_K*

As previously discussed in the general section on ionization chambers of this chapter, in practice, the common calibration factors are for ^{60}Co external beams provided by all PSDLs and SSDCLs. In some cases, as in the U.S.A. and Germany, calibration of the chambers in ^{137}Cs external beams is also possible.

Thus, for sources containing ^{60}Co or ^{137}Cs radionuclides, air kerma calibration factors can be provided directly. For sources of other radio-nuclides of lower energies, calibration factors supplied for x-ray beams for effective energies in the range 20 to 150 keV are needed. Then, the calibration factor N_K for the average energy of the source radionuclide under consideration can be estimated by interpolating between neighboring calibration points.

The method of interpolation depends on the mean energy of the radionuclide's γ-rays and on the specific type of ionization chamber.

For ^{192}Ir sources (mainly HDR sources), there have been several methods published for interpolating or averaging over the chamber response function (calibration factor as a function of photon energy).[12,29–31,43–47]

The published results and comparisons demonstrate that in extreme cases maximal differences of 1% will occur, depending on the method of extraction of the calibration factor. The method described by Goetsch et al.[44] has found wide application.

The average energy of ^{192}Ir spectrum varies according to the assumed spectrum, source design and method of calculation. For an ideal point source, a value of 0.355 MeV is calculated in Table 5.2 using Equation 5.1. For an air kerma weighted average energy, the effective energy E_{eff}, values of 0.397 MeV,[44] 0.390 MeV[49] and 0.370 MeV[45,48] for different source designs have been published. Although Goetsch et al. assumed a mean energy value (effective energy E_{eff}) of 0.397 MeV for their calculations, their method is applicable independently of which E_{eff} is selected or calculated to be the mean energy of the ^{192}Ir photons. In practice, a value in the range 0.350 to 0.400 MeV is acceptable. In Table 5.9 and Table 5.2, which are based on the photon spectra in Table 5.7 and Table 5.8, an effective energy of $E_{eff} = 0.398$ MeV for an ideal point source of ^{192}Ir is calculated using Equation 5.2. The result is virtually identical to that used by Goetsch et al.

The mean energy of ^{192}Ir falls approximately halfway between the mean energy of ^{137}Cs of 0.662 MeV and the effective energy of 0.146 MeV of a medium filtration 250 kVcp x-ray beam with an HVL of 3.2 mmCu. Thus, Goetsch et al. suggested that a simple averaging, practically a linear interpolation, of the calibration factors for these two photon energies (calibration points) is a rational basis for deriving an appropriate calibration factor (exposure or air kerma) for ^{192}Ir γ-rays.

Since for the ^{137}Cs calibration point an additional build-up cap is used, and since the attenuation of the chamber wall material and of the additional build-up cap is considered in the calibration factors provided for these two points, this attenuation effect has to be considered in the interpolation procedure. Finally, Goetsch et al. proposed the following interpolation/averaging formula:

$$N_{K,^{192}Ir} = \frac{A_{w,250kV}N_{K,250kV} + A_{w,^{137}Cs}N_{K,^{137}Cs}}{2A_{w,^{192}Ir}} \quad (7.8)$$

where $N_{K,^{192}Ir}$, $N_{K,^{137}Cs}$ and $N_{K,250kV}$ are the air kerma calibration factors for ^{192}Ir and ^{137}Cs γ-rays and for the 250 kVcp x-rays, respectively.

For this method, the wall attenuation factors A_w for these three energies need to be known. A_w is defined as the ratio of the ionization current or charge with the wall present and the ionization current or charge with a wall thickness of 0.0 cm or 0.0 g cm^{-2} ("wall" meaning the total material of the chamber wall itself and of the used build-up cap). Goetsch et al. estimated the wall factors A_w for all three ^{192}Ir and ^{137}Cs γ-rays and for the 250 kVcp x-rays experimentally. When the known calibration factors for ^{137}Cs and 250 kVcp qualities do not differ by more than 10% Equation 7.8 can be written as

$$N_{K,^{192}Ir} = (1 + x)\frac{N_{K,250kV} + N_{K,^{137}Cs}}{2} \quad (7.9)$$

where $x = 0.037 \times (t/9.3 \times 10^{22})$ for a total wall thickness of t electrons per cm^2. This simplified calculation formula assumes that for all three points, ^{192}Ir and ^{137}Cs γ-rays and for the 250 kVcp x-rays, the same build-up cap and thus the same total wall thickness is used, resulting in the same density t of electrons per cm^2.

Monte Carlo–based calculations for the wall attenuation factors A_w have been published[50] for several ionization chambers and for these three energy points, ^{192}Ir and ^{137}Cs γ-rays and for the 250 kVcp x-rays. These values are listed in Table 7.6 together with all other geometrical and material relevant data of the chambers with volumes above 0.1 cm^3.

Figure 7.5 summarizes published energy dependence figures of the air kerma calibration factors of several chambers and for photons energies in the range 0.2 to 1.25 MeV (^{60}Co).

According to Figure 7.5, the differences in the calibration factors of the considered chambers and for photon energies above 100 keV are a maximum of 3% and thus any interpolation method, practically an averaging or linear interpolation, will result in differences of significantly lower than 1% when compared to Equation 7.8 or Equation 7.9.

When the calibration factor for ^{60}Co is known instead of ^{137}Cs, then the ^{192}Ir photon rays calibration factor can be calculated using the following weighted interpolation:[12]

$$N_{K,^{192}Ir} = \frac{0.8A_{w,250kV}N_{K,250kV} + 0.2A_{w,^{60}Co}N_{K,^{60}Co}}{A_{w,^{192}Ir}} \quad (7.10)$$

The 0.8 and 0.2 weighting factors have been calculated using the air kerma weighted average energies for these three qualities.[12] In the case where the wall attenuation factors A_w for the specific chamber used are not known, Equation 7.10 can be simplified as follows:

$$N_{K,^{192}Ir} = 0.8N_{K,250kV} + 0.2N_{K,^{60}Co} \quad (7.11)$$

Due to this simplification, the uncertainty of the source calibration increases by approximately 0.5%.

TABLE 7.6

Geometrical and Material Parameters of the Most Common Compact Ionization Chambers with Cavity Volumes above 0.1 cm^3 and of Their Corresponding Build-Up Caps

Parameter	Ionization Chamber											
	Capintec PR-05 (Mini)	Capintec PR-05P (AAPM)	Capintec PR-06C (Farmer)	Capintec PR-06C (Farmer)	Capintec PR-06C (Farmer)	Exradin A2 (2 mm Cap)	Exradin A2 (4 mm Cap)	Exradin P2 (4 mm Cap)	Exradin T2 (4 mm Cap)	Exradin A12 (Farmer)	FZK TK 01 (Water-Proof)	NE 2515 (Farmer)
Air cavity volume (cm^3)	0.14	0.6	0.65	0.65	0.65	0.5	0.5	0.5	0.5	0.65	0.4	0.2
Air cavity length (cm)	1.15	2.38	2.23	2.23	2.23	1.14	1.14	1.14	1.14	2.42	1.2	0.7
Air cavity radius (cm)	0.2	0.33	0.32	0.32	0.32	0.48	0.48	0.48	0.48	0.31	0.35	0.3
Wall material	C552	Graphite	C552	C552	C552	C552	C552	Polystyrene	A150	C552	Delrin	Tufnol
Wall thickness t_{wall} (g cm^{-2})	0.220	0.046	0.050	0.050	0.050	0.176	0.176	0.105	0.113	0.088	0.071	0.074
Build-up cap material	Polystyrene	PMMA	C552	Polystyrene	PMMA	C552	C552	Polystyrene	A150	C552	Delrin	PMMA
Build-up cap thickness t_{cap} (g cm^{-2})	0.598	0.625	0.924	0.537	0.547	0.352	0.712	0.420	0.455	0.493	0.430	0.543
$A_{w,250kV}$	0.988	0.995	0.998	0.997	0.992	0.986	0.989	0.986	0.983	0.999	0.988	0.993
$A_{w,Ir-192}$	0.983	0.986	0.980	0.986	0.984	0.978	0.973	0.982	0.979	0.988	0.982	0.980
$A_{w,Co-60}$	0.989	0.986	0.984	0.990	0.989	0.984	0.976	0.988	0.985	0.991	0.989	0.987

(continued)

TABLE 7.6—Continued

Parameter	NE 2515/3 (Farmer)	NE 2577 (Farmer)	NE 2561	NE 2505 (Farmer)	NE 2505/A (Farmer)	NE 2505/3A (Farmer)	NE 2505/3B (Farmer)	NE 2571 (Farmer)	NE 2571 (Farmer)	NE 2581 (Farmer)	NE 2581 (Farmer)	PTW 31002 (Water-Proof)	PTW 31003 (Water-Proof)	PTW 23332 (Rigid)
Air cavity volume (cm^3)	0.2	0.2	0.325	0.6	0.6	0.6	0.6	0.6	0.6	0.6	0.6	00.125	0.3	0.3
Air cavity length (cm)	0.7	0.83	0.92	2.4	2.4	2.4	2.4	2.4	2.4	2.4	2.4	0.65	1.63	1.8
Air cavity radius (cm)	0.32	0.32	0.37	0.3	0.3	0.32	0.32	0.315	0.315	0.32	0.32	0.275	0.275	0.25
Wall material	Graphite	Graphite	Graphite	Tufnol	Nylon 66	Graphite	Nylon 66	Graphite	Graphite	A150	A150	PMMA	PMMA	PMMA
Wall thickness t_{wall} ($g\,cm^{-2}$)	0.066	0.066	0.09	0.075	0.063	0.065	0.041	0.065	0.065	0.040	0.041	0.079	0.079	0.054
Build-up cap material	PMMA	Delrin	Delrin	PMMA	PMMA	PMMA	PMMA	Delrin	PMMA	PMMA	Polystyrene	PMMA	PMMA	PMMA
Build-up cap thickness t_{cap} ($g\,cm^{-2}$)	0.543	0.552	0.600	0.545	0.545	0.551	0.551	0.551	0.550	0.584	0.584	0.357	0.357	0.357
$A_{w,250kV}$	0.994	0.988	0.987	0.997	0.996	0.998	0.995	0.999	0.996	0.986	0.991	0.990	1.000	1.000
$A_{w,Ir-192}$	0.982	0.981	0.984	0.989	0.984	0.989	0.990	0.989	0.983	0.988	0.990	0.992	0.993	0.993
$A_{w,Co-60}$	0.986	0.986	0.984	0.990	0.989	0.989	0.989	0.988	0.981	0.987	0.991	0.992	0.993	0.994

(continued)

TABLE 7.6—*Continued*

Parameter	Ionization Chamber												
	PTW 30010[a] (Farmer)	PTW 30011[b] (Farmer)	PTW 30012[c] (Farmer)	PTW 23331 (Rigid)	Victoreen Radocon III 550	Victoreen 30–348	Victoreen 30–361	Victoreen 30–351	Victoreen 30–349	Wellhöfer IC–15	Wellhöfer IC 28	Wellhöfer IC 69 (Farmer)	Wellhöfer IIC 70 (Farmer)
Air cavity volume (cm^3)	0.6	0.6	0.6	1.0	0.3	0.3	0.4	0.6	1.0	0.13	0.3	0.6	0.6
Air cavity length (cm)	2.3	2.3	2.3	2.2	2.3	1.8	2.23	2.3	2.2	0.58	0.9	2.3	2.3
Air cavity radius (cm)	0.305	0.305	0.305	0.395	0.24	0.25	0.24	0.31	0.4	0.3	0.31	0.31	0.31
Wall material	PMMA	Graphite	Graphite	PMMA	Polysty-rene	PMMA	PMMA	PMMA	PMMA	C552	C552	Delrin	Graphite
Wall thickness t_{wall}(g cm^{-2})	0.054	0.079	0.079	0.060	0.117	0.06	0.144	0.06	0.06	0.068	0.070	0.070	0.068
Build-up cap material	PMMA	PMMA	PMMA	PMMA	PMMA	PMMA	PMMA	PMMA	PMMA	PMMA	POM	POM	POM
Build-up cap thickness													
t_{cap} (g cm^{-2})	0.541	0.541	0.541	0.345	0.481	0.360	0.360	0.360	0.360	0.354	0.560	0.560	0.560
$A_{w,250kV}$	0.998	0.993	0.997	0.997	0.997	0.994	1.000	0.995	0.996	0.993	0.993	1.000	1.000
$A_{w,Ir-192}$	0.990	0.989	0.990	0.992	0.991	0.993	0.992	0.993	0.992	0.990	0.988	0.990	0.990
$A_{w,Co-60}$	0.991	0.989	0.990	0.993	0.991	0.994	0.992	0.994	0.992	0.990	0.988	0.990	0.990

The Monte Carlo calculated wall correction factors $A_{w,250 kV}$, $A_{w,Ir-192}$, and $A_{w,Co-60}$ for 250 kV x-rays as well as for the ^{192}Ir and ^{60}Co γ-rays according to Ferreira et al.[50] are also listed.

[a] Originally type 30001, PMMA made wall and aluminum made central electrode.
[b] Originally type 30002, graphite made wall and graphite made central electrode.
[c] Originally type 30004, graphite made wall and aluminum made central electrode.

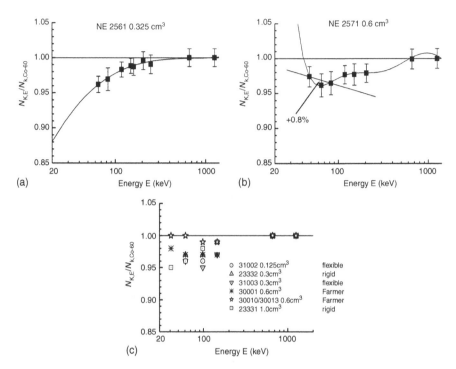

FIGURE 7.5

Photon energy dependence of the calibration factor $N_{K,E}$ for the air kerma rate for different x-ray and γ-ray energies. The calibration factors for each chamber have been normalized to the corresponding calibration factor for the reference mean γ-ray energy of ^{60}Co of 1.25 MeV. (a) Data for a NE 2561 0.325 cm^3 chamber. (b) Data for an NE 2571 0.6 cm^3 Farmer-type chamber. Both data sets are taken from Grimbergen and van Dijk,[46] Table 2 and Table 3, respectively. The calibration factors for the medium filtered x-ray beam of 250 kVcp with a HVL of 2.94 mm Cu and an effective energy of ~150 keV are also considered. For both chambers a polynomial fit was used for interpolating between the experimental values. A third order polynomial was used for the NE 2561 chamber but a sixth order polynomial was required for the NE 2571 chamber. When a linear interpolation method is used for estimating the calibration factor, a maximum error of +0.8% is observed when calculating N_K for 65 keV from the N_K values for 48 and 83 keV for the NE 2571 chamber. All measurements described by the authors used build-up caps for both chambers. (c) Typical relative calibrations factors for the five PTW ionization chambers with volumes 0.125 to 1.0 cm^3 as described in Table 7.6. The values given for ^{60}Co and ^{137}Cs are with the use of the corresponding standard build-up cap. The effective energies of the x-ray beams used by PTW are taken from Pychlau[39] where the values shown are typical values reported by the manufacturer.

Due to the very similar average energy of ^{198}Au (which has a mean effective energy of ~0.4 MeV, see also Table 5.14), the data and methods discussed for ^{192}Ir can also be applied for ^{198}Au.

For radionuclides of even lower energy, such as ^{169}Yb which has a mean photon energy of 0.093 MeV and an effective energy of 0.131 MeV for an ideal point source (see Table 5.2 this value of the effective energy was published by Chen and Nath[49]), calibration factors at even lower energy x-rays, e.g., 140 to 200 kVcp, with effective energies of ~0.1 MeV for the example of ^{169}Yb are needed to obtain the required calibration factors.

When several calibration points for an ionization chamber are available, such as for the cases shown in Figure 7.5, analytical formulae (usually polynomials of varying degrees) can be fitted to these data and thus accurate estimation of the calibration factor at any energy can be made.

For a nuclide such as ^{169}Yb, with a known photon emission spectrum (see Table 5.12), the calibration factor $N_{K,^{169}Yb}$ can be calculated using the energy spectrum weighted method:[24,48]

$$N_{K,^{169}Yb} = \frac{\sum\limits_{i=1}^{n} \dot{K}_i}{\sum\limits_{i=1}^{n} \dot{K}_i (N_{K,i})^{-1}} \tag{7.12}$$

$(N_{K,i})^{-1}$ is the response of the ionization chamber for photon energy E_i (inverse of calibration factor) and $\dot{K}_i$ is the air kerma contribution per decay for photons of energy E_i. The summation in Equation 7.12 is calculated over all n photon energies of the spectrum of the radionuclide.

$\dot{K}_i$ can be calculated according to

$$\dot{K}_i = E_i \left(\frac{\mu_{tr}}{\rho} \right)_{\alpha, E_i} \Phi(E_i) \tag{7.13}$$

where $(\mu_{tr}/\rho)_{\alpha, E_i}$ is the mass energy transfer coefficient for photons of energy E_i in air.

$\Phi(E_i)$ is the photon fluence per decay for photons of energy E_i in air; and $\Phi(E_i)$ can be approximated by the relative number of photons of energy E_i emitted per decay η_i.

If the spectrum of the radionuclide for the specific source design where the effects of self absorption and attenuation in the source core and in the encapsulation material influence $\Phi(E_i)$ is not known, then these effects have to be approximated. This is achieved by using $\eta_i f_i$, with f_i being the correction factor to account for the attenuation of the photons of energy E_i of the primary radionuclide spectrum in the active core and encapsulation material of the source.

Analyzing the data presented in Figure 7.5a and Figure 7.5b for the two commercially available ionization chambers NE 2561 (0.325 cm^3) and NE 2571 (0.6 cm^3), a change of N_K of 2.7 and 1.5% for photon energies in the range 60 to 150 keV for NE 2561 and NE 2571, respectively, is observed. Even simple interpolations among these values can result in maximal errors of 1.5%.

For the case of the PTW chambers shown in Figure 7.5c, and for effective energies in the range 60 to 150 keV, there are changes in the calibration factor N_K of 3% for the 1.0 cm^3 type 23331, 2% for the 0.3 cm^3 type 23332 (rigid), and of maximal 1% for all other three ionization chambers.

The above methods permit an acceptable estimation of the value of the air kerma calibration factor N_K for a specific source design and specific radionuclide. In some cases, simple linear interpolation can achieve accuracies of better than 1% and of as high as 0.5%. It is assumed that calibration factors for additional points other than those for ^{60}Co and ^{137}Cs are available which can adequately cover the energy region of the sources used in brachytherapy.

7.2.1.3.2 Chamber Finite Size Effect Correction Factor k_V

In the dosimetry around brachytherapy sources, we must consider the basic restriction of using dosimeters of finite size and thus of finite volume. Hence, we have to answer the question "What is their spatial resolution and at which location is the measurement taken?"

The dominating inverse square law for the particle fluence, and consequently also for the energy fluence, around a source results in remarkable gradients within the dosimeter volume. These are all the more remarkable the shorter the distance of the measurement. When considering the center of the chamber volume cavity as the reference point of measurement, a correction must be applied to correct for these gradient effects.

In 1960, Kondo and Randolph,[51] commencing from the ideas published by Spiers[52] in 1941, introduced a quantitative method for estimating a correction factor for the effect of the chamber's finite size, assuming point photon sources. Their method[51] is based on the assumption that the ionization, and thus the charge collecting by the chamber, is mainly generated by secondary electrons originating in the internal surface layers of the chamber wall.

For the latest, formulae for both the sideward irradiation as described in Figure 7.6 and for the endward irradiation (the source is positioned at the

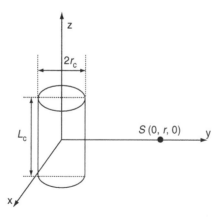

FIGURE 7.6
Sideward geometry for the calculation of the chamber finite size correction factor k_V, as in Equation 7.14, of an in-air calibration set-up for a point source S using a cylindrical ionization chamber. The point source S is positioned at a distance r on the perpendicular bisector of the chamber cylinder axis. Thus, S has $(0, r, 0)$ coordinates to the chamber coordinate system shown in the figure. L_c is the length and r_c the radius of the cylindrical volume (cavity) of the ionization chamber.

extension of the longitudinal axis of the chamber) situations have been provided.

The correction factor k_V for the sideward irradiation geometry considered here for the source calibration (Figure 7.2 and Figure 7.6) is given according to Kondo and Randolph by

$$k_V^{-1} = \int_{-1}^{1} G_1(x, \alpha, \sigma) f_1(x, \sigma) dx + \int_{0}^{1} G_2(x, \alpha, \sigma) f_2(x, \sigma) dx \qquad (7.14)$$

with

$$G_1(x, \alpha, \sigma) = [(1 + \alpha^2 + \alpha^2 x^2 / \sigma^2)^2 - 4\alpha^2]^{-\frac{1}{2}}$$

$$G_2(x, \alpha, \sigma) = [(1 + \alpha^2 / \sigma^2 + \alpha^2 x^2)^2 - 4\alpha^2 x^2]^{-\frac{1}{2}}$$

where $\alpha = r_c/r$ is the distance factor; and $\sigma = 2r_c/L_c$ is the shape factor of the cylindrical chamber for the sideward geometry, with r_c and L_c being the radius and the length of the chamber cavity respectively (see Figure 7.2 and Figure 7.6).

The functions $f_1(x, \sigma)$ and $f_2(x, \sigma)$ depend only on the shape factors and are given by

$$f_1(x, \sigma) = \left(\frac{1+x}{2\sigma\kappa}\right)[F_0(\kappa) - E_0(\kappa)]$$

$$f_2(x, \sigma) = x\sigma\left\{\sqrt{x}\left[\frac{F_0(l) - E_0(l)}{l}\right] + F_0(m)\left(1 - \frac{\sqrt{x}}{m}\right) + \frac{\sqrt{x}}{m}E_0(m) - \frac{l}{m}F_0(l)\right\}$$

$$+ x[2 - \Lambda_0(l, p_+) - \Lambda_0(l, p_-)]$$

and

$$\kappa^2 = \frac{4\sigma^2}{4\sigma^2 + (1+x)^2}, \qquad l^2 = \frac{4\sigma^2 x}{4 + \sigma^2(1+x)^2}, \qquad m^2 = \frac{4x}{(1+x)^2}$$

$$p_\pm = \frac{\pm 2\sigma x}{\sqrt{4 + \sigma^2 x^2} \pm \sigma x}$$

$$F_0(\kappa) = \frac{2}{\pi}\int_{0}^{\pi/2}\frac{d\phi}{\sqrt{1 - \kappa^2 \sin^2 \phi}}, \qquad E_0(\kappa) = \frac{2}{\pi}\int_{0}^{\pi/2}\sqrt{1 - \kappa^2 \sin^2 \phi}\,d\phi$$

$$\Lambda_0(l, p_\pm) = \frac{2}{\pi}\sqrt{(1 - p_\pm)\left(1 - \frac{l^2}{p_\pm}\right)}\int_{0}^{\pi/2}\frac{d\phi}{(1 - p_\pm \sin^2 \phi)\sqrt{1 - l^2 \sin^2 \phi}}$$

The values of correction factor k_V according to Equation 7.14 have been calculated for the different chambers listed in Table 7.6 using Monte Carlo integration methods.[53–56] The k_V values are tabulated for the absolute distances r for simplicity instead of using the distance factors α as originally proposed by Kondo and Randolph.[51] As demonstrated in Table 7.6, k_V is practically 1.0 (a maximal error of 0.4%) for measurement distances above 10.0 cm for all the chambers considered here.

The Kondo and Randolph theory is also based on the assumption that the angular distribution of the secondary electrons entering the chamber cavity (originating in the internal surface layers of the chamber wall) is isotropic. This is actually only valid at very low photon energies. That is, where photoelectric interaction and a high degree of multiple scattering result in secondary electron distributions that can be considered to be nearly isotropic. Bielajew[57,58] proposed an analytical theory to take into account this effect by expanding the Kondo and Randolph theory described above. For the distances and chambers considered in Table 7.7, this correction is of the order of a few tenths of 1%.

For shorter distances r between chamber and source, the finite size of source (spatial distribution of activity, instead of point source) will also have an effect. This is not considered by Equation 7.14.

7.2.1.3.3 Build-Up Cap and Chamber Wall Correction Factor k_{wall}

Thicknesses of chamber walls and of build-up caps must be selected in such a way that two conditions are satisfied: (1) the total resultant wall thickness is sufficient to provide charged particle equilibrium (CPE): here secondary electrons which are either photoelectrons and/or Compton recoil electrons depending on the energy/energy spectrum of the radionuclide; and (2) the total resultant wall thickness is sufficient to filter out any secondary electrons, photo- or Compton-electrons, generated in the source encapsulation, catheter/applicator used in the in-air setup and in air between source and chamber. In other words, to avoid secondary electron contamination.

Table 7.6 summarizes the materials and thicknesses of the chamber wall and standard build-up caps for the most commonly used ionization chambers with volumes greater than 0.1 cm^3 and which can be considered suitable for in-air calibration procedures.

The effect of using build-up caps has been extensively discussed in the literature and sufficient data have been published concerning ^{60}Co, ^{137}Cs and ^{192}Ir.[9,12,13,31,39,44,47,50,59–61]

The results of Goetsch et al.[44] indicate a necessity for wall thicknesses of at least 0.3 g cm^{-2} for PMMA or graphite in order to avoid secondary electron contamination for ^{192}Ir measurements. Their results are based on the use of several graphite build-caps and a graphite walled Farmer-type ionization chamber, thus resulting in different total graphite wall thicknesses. They demonstrated a 2.7 and 2.8% attenuation (reduction of ionization measurement) per g cm^{-2} of graphite for ^{192}Ir and ^{137}Cs, respectively.

This is again in very good agreement with a mass energy absorption coefficient value $\mu_{en}/\rho = (0.0293 - 0.0295)$ cm^2 g^{-1} for the mean photon energies of ^{192}Ir and ^{137}Cs for graphite[40] which gives, according to Attix's[41]

TABLE 7.7

Values of the Chamber Volume Corrections Factor k_V According to Kondo and Randolph,[51] Equation 7.14, for the Different Cylinder or Almost Cylinder Shaped Ionization Chambers Listed in Table 7.6

K_V

Distance r (cm)	Capintec PR-05 0.14 cm³	Capintec PR-05P 0.6 cm³	Capintec PR-06C 0.65 cm³	Exradin A2/P2/T2 0.5 cm³	Exradin A12 0.65 cm³	FZK TK 01 0.4 cm³	NE 2515 0.2 cm³	NE 2515/3 0.2 cm³	NE 2577 0.2 cm³	NE 2561 0.325 cm³	NE 2505 2505/A 0.6 cm³	NE 2505/3A 2505/3B 0.6 cm³	NE 2571 0.6 cm³	NE 2581 0.6 cm³
5.0	1.004	1.017	1.015	0.999	1.018	1.002	0.999	0.999	1.000	0.999	1.018	1.017	1.018	1.018
8.0	1.001	1.007	1.006	0.999	1.007	1.001	1.000	1.000	1.000	1.000	1.007	1.007	1.007	1.007
10.0	1.001	1.004	1.004	1.000	1.004	1.001	1.000	1.000	1.000	1.000	1.004	1.004	1.004	1.004
12.0	1.001	1.003	1.003	1.000	1.003	1.000	1.000	1.000	1.000	1.000	1.003	1.003	1.003	1.003
15.0	1.000	1.002	1.002	1.000	1.002	1.000	1.000	1.000	1.000	1.000	1.002	1.002	1.002	1.002
18.0	1.000	1.001	1.001	1.000	1.001	1.000	1.000	1.000	1.000	1.000	1.001	1.001	1.001	1.001
20.0	1.000	1.001	1.001	1.000	1.001	1.000	1.000	1.000	1.000	1.000	1.001	1.001	1.001	1.001
25.0	1.000	1.001	1.001	1.000	1.001	1.000	1.000	1.000	1.000	1.000	1.001	1.001	1.001	1.001
30.0	1.000	1.000	1.000	1.000	1.000	1.000	1.000	1.000	1.000	1.000	1.000	1.000	1.000	1.000
35.0	1.000	1.000	1.000	1.000	1.000	1.000	1.000	1.000	1.000	1.000	1.000	1.000	1.000	1.000
40.0	1.000	1.000	1.000	1.000	1.000	1.000	1.000	1.000	1.000	1.000	1.000	1.000	1.000	1.000
45.0	1.000	1.000	1.000	1.000	1.000	1.000	1.000	1.000	1.000	1.000	1.000	1.000	1.000	1.000
50.0	1.000	1.000	1.000	1.000	1.000	1.000	1.000	1.000	1.000	1.000	1.000	1.000	1.000	1.000

(continued)

TABLE 7.7—*Continued*

						K_V							
Distance r (cm)	PTW 31002 0.125 cm³	PTW 31003 0.3 cm³	PTW 23332 0.3 cm³	PTW 3001/ 3002/3004 0.6 cm³	PTW 23331 1.0 cm³	Victoreen Radocon III 550 0.3 cm³	Victoreen 30-348 0.3 cm³	Victoreen 30-361 0.4 cm³	Victoreen 30-351 0.6 cm³	Victoreen 30-349 1.0 cm³	Wellhöfer IC-15 0.13 cm³	Wellhöfer IC 28 0.3 cm³	Wellhöfer IC 69/IC 70 0.6 cm³
5.0	1.000	1.007	1.010	1.016	1.013	1.017	1.010	1.016	1.016	1.013	0.999	1.000	1.016
8.0	1.000	1.003	1.004	1.006	1.005	1.006	1.004	1.006	1.006	1.005	1.000	1.000	1.006
10.0	1.000	1.002	1.002	1.004	1.003	1.004	1.002	1.004	1.004	1.003	1.000	1.000	1.004
12.0	1.000	1.001	1.002	1.003	1.002	1.003	1.002	1.003	1.003	1.002	1.000	1.000	1.003
15.0	1.000	1.001	1.001	1.002	1.001	1.002	1.001	1.002	1.002	1.001	1.000	1.000	1.002
18.0	1.000	1.001	1.001	1.001	1.001	1.001	1.001	1.001	1.001	1.001	1.000	1.000	1.001
20.0	1.000	1.000	1.001	1.001	1.001	1.001	1.001	1.001	1.001	1.001	1.000	1.000	1.001
25.0	1.000	1.000	1.000	1.001	1.001	1.001	1.000	1.001	1.001	1.001	1.000	1.000	1.001
30.0	1.000	1.000	1.000	1.000	1.000	1.000	1.000	1.000	1.000	1.000	1.000	1.000	1.000
35.0	1.000	1.000	1.000	1.000	1.000	1.000	1.000	1.000	1.000	1.000	1.000	1.000	1.000
40.0	1.000	1.000	1.000	1.000	1.000	1.000	1.000	1.000	1.000	1.000	1.000	1.000	1.000
45.0	1.000	1.000	1.000	1.000	1.000	1.000	1.000	1.000	1.000	1.000	1.000	1.000	1.000
50.0	1.000	1.000	1.000	1.000	1.000	1.000	1.000	1.000	1.000	1.000	1.000	1.000	1.000

Here the measurement set-up is as described in Figure 7.6, thus for the sideward irradiation geometry, of the chamber at a specific distance r from an ideal point like source is considered. The k_V values at absolute distances r have been calculated and are listed here for simplicity instead of using the distance factors α as was originally proposed by Kondo and Randolph.[51]

approximation assuming broad-beam geometry and straight-ahead scattering (see also Equation 7.5), approximately 2.9% attenuation of the photons per g cm^{-2} of graphite for these two radionuclides. The graphite chamber wall of 0.065 g cm^{-2} alone results in a secondary electron contamination of 2.5% at 10 cm and 1.2% at 20 cm source to chamber distance.

In 1991, Pychlau[39] published similar experimental work to that of Goetsch et al. He used a PTW type 23331 1.0 cm^3 ionization chamber in an in-air setup positioned 1 m from the floor and having a fixed 5.0 cm source to chamber distance for a HDR ^{192}Ir source. Three different PMMA build-up caps were used in addition to the PMMA chamber wall at a thickness of 0.05 cm. This resulted in total wall thicknesses in the range of 0.05 to 0.4 cm, or equivalently 0.06 to 0.48 g cm^{-2} of PMMA.

Pychlau's results are shown in Figure 7.7. The three measurements with additional build-up caps were used to estimate the attenuation of the PMMA caps. For this, a linear fit to the results is obtained when using additional caps and a total wall thickness in the range 0.15 to 0.4 cm. This results in an attenuation of 0.9% per cm PMMA total wall thickness, or equivalently 0.8% per g cm^{-2} of PMMA total material. This value is significantly lower than the 2.7% per g cm^{-2} obtained by Goetsch et al. for graphite as well as that expected using Attix's method, Equation 7.5, of approximately 3.2% for every g cm^{-2} of PMMA.

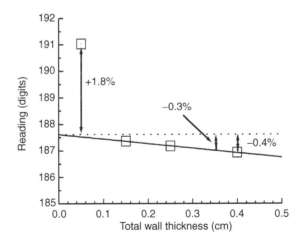

FIGURE 7.7
Experimental data by Pychlau[39] using a PTW type 23331 1.0 cm^3 ionization chamber in an in-air set-up at 5.0 cm distance from an HDR ^{192}Ir source at 1.0 m height above the floor. Three different build-up caps were used in addition to the PMMA wall thickness of the chamber of 0.05 cm. This resulted in a total wall thicknesses in the range 0.050 to 0.4 cm or equivalently 0.06 of 0.48 g cm^{-2} of PMMA. The linear fit to the results of using additional caps, with a total wall thickness of 0.15 to 0.4 cm results in an attenuation of 0.9% per cm PMMA total wall thickness, or equivalently 0.8% per g cm^{-2} of PMMA total material. The use of this ionization chamber with only its 0.05 cm PMMA wall thickness results in a contribution of 1.8% of secondary electrons due to scattering from the applicator and air and thus to an overestimation of 1.8% for the source strength at the measurement distance of 5.0 cm.

The use of an ionization chamber with only 0.05 cm PMMA wall thickness results in a contribution of 1.8% of secondary electrons due to scattering from the applicator and air and thus to an overestimation by 1.8% of the source strength at the measurement distance of 5.0 cm. Although this is an experimental set-up, which is chamber and room dimensions dependent, Pychlau's result is of the same order as that by Goetsch et al., namely 2.5% secondary electron contamination at 10 cm.

It is possible, at least theoretically, that due to the angular distribution and finite range of the secondary electrons in air, the effect of secondary electron contamination will decrease with increasing source to chamber distance.

In the DGMP Report 13,[31] data for experimental work by Krieger were published. Krieger used an in-air calibration set-up with a PTW type 23332 0.3 cm^3 ionization chamber at different distances from an HDR ^{192}Ir source. For each distance measurement, with, M_w, and without $M_{w/o}$, the standard ^{60}Co build-up cap of 0.3 cm PMMA provided by the manufacturer was used. The ratio $M_{w/o}/M_w$ is given in Figure 7.8 plotted vs. the source to chamber distance.

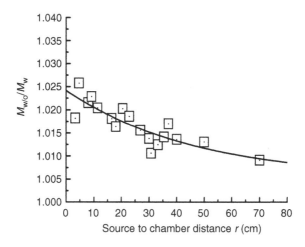

FIGURE 7.8

Experimental data by Krieger published in DGMP Report 13[31] for an in-air calibration set-up using a PTW type 23332 0.3 cm^3 ionization chamber at different distances from an HDR ^{192}Ir source. Results were obtained for each measurement distance with, M_w, and without M_w, the standard ^{60}Co build-up cap of 0.3 cm PMMA. The ratio $M_{w/o}/M_w$ has been plotted vs. the source to chamber distance. These data demonstrate that for very large distances the use of this build-up cap reduces the chamber reading by 0.5%. The absence of the build-up cap results in an overestimation of the source strength due to increased chamber reading because of secondary electron contribution. This increases with decreasing distance from the source and achieves a level of ~2.2% at a distance of 5.0 cm from the source. This agrees very well with the results of Pychlau[39] which are shown in Figure 7.7, where there is an overestimation of 1.8% for the 1.0 cm^3 ionization chamber at a distance of 5.0 cm.

Krieger's data demonstrates that for very large distances the use of this build-up cap reduces the chamber reading by 0.5% (progressive extension of the fitted curve to very large distances). This observation is in good agreement with the expected attenuation of 0.2% resulting from the use of the 0.05 cm PMMA chamber wall calculated using the theoretical value of 3.2% attenuation per g cm^{-2} PPMA, as seen in Equation 7.5, and $\rho_{PMMA} = 1.19$ g cm^{-3} (see Table 7.5).

When the build-up cap is omitted, an overestimation of the source strength due to an increased chamber reading because of secondary electron contribution is observed. This increases with decreasing distance from the source as theoretically expected. This contamination effect is estimated to be 2.2, 2.1, and 1.8% at a distance of 5.0, 10.0, and 20 cm from the source, respectively. All three values are compatible with the values estimated by Goetsch et al. and Pychlau.

The Monte Carlo calculated total wall attenuation factors, A_w, for the γ-rays of ^{192}Ir and ^{60}Co, as well as for x-rays of 250 kV with an HVL of 2.5 mm Cu for different ionization chambers with volumes above 0.1 cm^3 according to Ferreira et al.,[50] are listed in Table 7.6 together with the geometrical and material characteristics of these chambers. The A_w values listed include the effect of the chamber wall itself and of the standard build-up cap provided by the manufacturer for ^{60}Co γ-rays.

The build-up cap and chamber wall correction factor k_{wall} in Equation 7.6 and Equation 7.7 is defined as the inverse of the wall factor A_w:

$$k_{wall} = \frac{1}{A_w}$$

Considering these values and the experimental results described, it can be concluded that for ^{192}Ir and ^{137}Cs γ-rays, experimental and Monte Carlo published data show good agreement with theoretical values based on the Attix[41] approximation assuming broad-beam geometry and straight-ahead scattering, as seen in Equation 7.5.

For clinical source calibration procedures, Table 7.6 demonstrates that for the majority of the chambers a k_{wall} correction factor of approximately 1.01 (1%) for ^{192}Ir and ^{60}Co γ-rays is applicable, whereas for the 250 kV x-rays it can be assumed that $k_{wall} = 1.00$ for all wall and build-up cap materials listed in Table 7.6.

For chambers or radionuclides not listed, the wall correction factor k_{wall} can be estimated experimentally based on ionization measurements in-air using at least two different build-up cap thicknesses or using the approximation formula in Equation 7.5. Due to the similar average photon energy for ^{198}Au, it can be safely assumed that the values given for ^{192}Ir can also be applied for this radionuclide.

For radionuclides of very low photon energy, experimental or Monte Carlo simulation methods values for k_{wall} must be estimated. Due to the larger angle Compton scattering it can be expected that Equation 7.5 overestimates the effect of the build-up cap and chamber wall thickness.

7.2.1.3.4   In-Air Scatter and Attenuation Correction Factor $k_{as,a}$

The determination of source strength in terms of reference air kerma rate, K_R, or air kerma strength, S_K, requires that in-air made measurements are corrected for the attenuation of the primary photons in-air for the source to chamber distance r, see Equation 7.6 and Equation 7.7. Assuming broad beam geometry and dry air conditions (see Table 7.5), values for the in-air scatter and attenuation correction factor $k_{as,a}$ have been calculated for the seven main brachytherapy radionuclides ^{198}Au, ^{60}Co, ^{137}Cs, ^{125}I, ^{192}Ir, ^{103}Pd, and ^{169}Yb and for several source to chamber distances r in the range of 5.0 to 100.0 cm.

These results are summarized in Table 7.8. The $k_{as,a}$ values were calculated using the air-effective mass energy absorption coefficient values $(\mu_{en}/\rho)_{\alpha,eff}$, see also Equation 5.3, which are given in Table 5.4 through Table 5.12 for different radionuclides; and using the dry air material constants which are given in Table 7.5 for ideal point sources and broad beam conditions.

With the exception of the low energy radionuclides, ^{125}I and ^{103}Pd, and for source to chamber distances up to 50.0 cm, $k_{as,a}$ can be assumed to be 1.00 with an error of maximal 0.2% ($k_{as,a} \leq 1.002$).

7.2.1.3.5   Room Scatter Correction Factor k_{scatt}

Depending on the room dimensions, measurement set-up and radionuclide, scattering of radiation from room walls, floor and measurement set-up itself have a potential influence on the measurement result.

TABLE 7.8

Correction Factor $k_{as,a}$ for Attenuation in Air for Primary Photons for Brachytherapy Radionulides for Different Measurement Distances

Distance (cm)	^{137}Cs	^{60}Co	^{198}Au	^{125}I	^{192}Ir	^{103}Pd	^{170}Tm	^{169}Yb
5	1.000	1.000	1.000	1.001	1.000	1.003	1.000	1.000
10	1.000	1.000	1.000	1.002	1.000	1.006	1.000	1.000
20	1.001	1.001	1.001	1.004	1.001	1.012	1.001	1.001
30	1.001	1.001	1.001	1.007	1.001	1.018	1.001	1.001
40	1.001	1.001	1.001	1.009	1.001	1.024	1.001	1.001
50	1.002	1.002	1.002	1.011	1.002	1.030	1.002	1.002
60	1.002	1.002	1.002	1.013	1.002	1.036	1.002	1.002
70	1.003	1.002	1.002	1.016	1.002	1.042	1.003	1.003
80	1.003	1.003	1.003	1.018	1.003	1.049	1.003	1.003
90	1.003	1.003	1.003	1.020	1.003	1.055	1.003	1.003
100	1.004	1.003	1.004	1.023	1.004	1.061	1.004	1.004

Values calculated using the effective mass energy absorption coefficient values, $(\mu_{en}/\rho)_{\alpha,eff}$, given in Table 5.4 through Table 5.16, and dry air material constants given in Table 7.5 for an ideal point source and broad-beam conditions.

In 1991, Goetsch et al.[44] published a method to estimate such types of effects focusing on ^{192}Ir high dose rate sources. This method can be generally considered for in-air set-ups, such as those shown in Figure 7.3, and also for any source radionuclide and activity appropriate for accurate in-air calibration set-ups.

Assuming that measurements M for several distances r have been made, usually in the range of 5.0 to 50.0 cm, and assuming that there is a systematic set-up error (offset) of δr in the definition of the source to chamber distance r (see Figure 7.2), and scattered radiation from room walls, floor and measurement set-up can be assumed to contribute with a constant amount of M_{scatt} to each of the measurement M at any of the distances r (constant scatter irradiation background), then we have

$$(Mk_V k_{as,a} - M_{scatt}) = \frac{f}{(r + \delta r)^2}$$

or equivalently

$$Mk_V k_{as,a} = \frac{f}{(r + \delta r)^2} + M_{scatt} \tag{7.15}$$

where f is a constant which is independent of the distance of measurement r.

The dosimeter reading M has to be previously corrected for the chamber finite size effect, correction factor k_V, and for the attenuation and scattering of the primary photons in air, correction factor $k_{as,a}$, for each distance r. Having measurements at a minimum of three different sources to chamber distances r, a robust estimation of the three parameters f, δr, and M_{scatt} can be obtained by applying nonlinear curve fitting of Equation 7.15 to the experimental data pairs $(r, M' = Mk_V k_{as,a})$.

Because of dependence on the half-time of the radionuclide and on the duration of the measurement procedure, corrections of the values M for the radioactive decay can be necessary.

Thus, instead of the product

$$Mk_V k_{as,a} k_{scatt} k_r = f \tag{7.16}$$

the estimated parameter f can be used according to Goetsch et al. This offers a more stable and accurate procedure by averaging over several distances and thus filtering errors.

Alternatively, once the multi-distance measurement procedure has been achieved, the correction factor:

$$k_{scatt} = 1 - \frac{M_{scatt}}{Mk_V k_{as,a}} \tag{7.17}$$

can be estimated for a specific distance r and can be used in the future when the same set-up and the same source type are used.

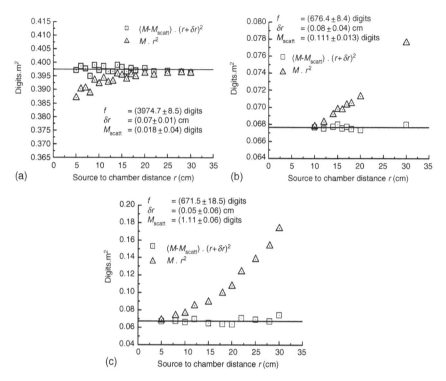

FIGURE 7.9

Results of our Offenbach Radiotherapy Clinic in-air calibration set-ups for three different sources. Goetsch's method,[44] as described by Equation 7.15, has been applied for estimating the room scatter correction factor k_{scatt}. The dosimeter reading M shown in the figures was previously corrected for the chamber's finite size effect, correction factor k_V, and for the attenuation and scattering of the primary photons in air. The correction factor $k_{as,a}$, was used for each distance r; $M = M \cdot k_V \cdot k_{as,a}$. For all three procedures the Offenbach calibration jig shown in Figure 7.3 was used. (a) ^{192}Ir source of a microSelectron-HDR afterloader (classic type). Measurements were made for 19 different distances in the range of 5.0 to 30.0 cm using a PTW type 23331 1.0 cm³ ionization chamber with a total wall (chamber wall and standard ^{60}Co build-up cap) thickness of 3.5 mm PMMA or 0.42 g cm⁻². (b) Single ^{60}Co pellet source of a Selectron-HDR afterloading unit. The same 1.0 cm³ ionization chamber and build-up cap as for the ^{192}Ir procedure have been used, where nine different source to chamber distances in the range of 10.0 to 30.0 cm have been considered. (c) ^{137}Cs source of a Selectron-MDR afterloader. A PTW type 23332 0.3 cm³ ionization chamber with a total wall thickness of 3.5 mm PMMA or 0.42 g cm⁻² was used. Measurements at 11 different distances in the range 5.0 to 30.0 cm were used. These results demonstrate that an offset of 0.05 to 0.08 cm exists in the fixation mechanism or scale reading for the source to chamber distance r. There exists a systematic positioning error δr of ~0.07 cm (mean value over all three estimated values δr).

Figure 7.9a shows the results of fitting Equation 7.15 to our own experimental results for an ^{192}Ir source (microSelectron-HDR classic source) at 19 different distances in the range of 5.0 to 30.0 cm from the source using our Offenbach Radiotherapy Clinic in-house calibration jig, shown in Figure 7.3.

Measurements were made using a PTW Type 23331 1.0 cm^3 ionization chamber including the ^{60}Co build-up cap, resulting in a total wall thickness of 3.5 mm PMMA or 0.42 g cm^{-2} (see also Table 7.6 and Table 7.7).

Figure 7.9b summarizes our results when using the Goetsch et al. method for calibrating a single ^{60}Co pellet of a Selectron-HDR afterloading unit and nine different distances, whereas Figure 7.9c shows the results for a ^{137}Cs source of a Selectron-MDR afterloader for 11 different distances. The same in-air jig has been used in both cases as in the procedure shown in Figure 7.9a for ^{192}Ir (see also Figure 7.3).

For the ^{60}Co pellet calibration, the same 1.0 cm^3 ionization chamber and build-up cap as for the ^{192}Ir procedure have been used. For the calibration of the ^{137}Cs source, a PTW type 23332 0.3 cm^3 ionization chamber with a total wall thickness of 3.5 mm PMMA or 0.42 g cm^{-2} was used (see also Table 7.6 and Table 7.7).

For all these three calibration procedures, correction factors for scattering and attenuation in air $k_{as,a}$ for ^{192}Ir, ^{60}Co, and ^{137}Cs, which are listed in Table 7.8, have been considered in Equation 7.15. We found that for the distances r in the range of 5.0 to 30.0 cm, this correction is negligible (i.e., a maximum of 0.1%). Thus, without any hesitation it can be assumed that $k_{as,a} = 1.000$.

All these results demonstrate that an offset of 0.05 to 0.08 cm exists in the fixation mechanism for defining the source to chamber distance r and thus there exists a systematic positioning error of approximately 0.07 cm (calculated as a mean value of all three estimated values).

Ezzell[61] has published a method similar to that of Goetsch et al. for the estimation of an assumed constant room scattering contribution by making measurements at several source to chamber distances r. The Ezzell formulation is given in Equation 7.15, when the positioning systematic error δr is ignored ($\delta r = 0.0$ cm). As demonstrated by the results in Figure 7.9, Goetsch's method is more robust and more general than that of Ezzell.

7.2.1.3.6 *Applicator/Catheter Wall Attenuation Correction Factor k_{appl}*

The source calibration provides an estimation of the source strength in terms of either reference air kerma rate or air kerma strength. When sources are placed within catheters/applicators previously inserted into the patient anatomy, as is the case for temporary implants, then the effect of the catheter/applicator on the source strength should be considered.

One method should be that the same catheter/applicator type is considered for source positioning and fixation in the calibration procedure. Hence, this effect is then included in the derived source strength. If this is not practical because there are several type of catheter/applicator with different wall thicknesses and materials, then the effect of the standard catheter/applicator used for the calibration should be considered and the measured ionization should be corrected using the corresponding correction factor k_{appl} in Equation 7.6 and Equation 7.7.

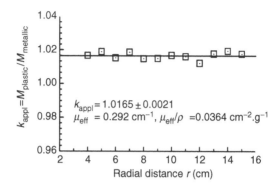

FIGURE 7.10
Experimental results from Offenbach Radiotherapy Clinic for estimating the applicator attenuation correction factor k_{appl} for the stainless steel made (ANSI 303/304) metallic dosimetry applicator used for the calibration of the ^{192}Ir source of a microSelectron-HDR afterloader. The measurements were in a water phantom of dimensions $48 \times 48 \times 41$ cm^3 for 12 different source to chamber distance in the range of 4.0 to 15.0 cm using the metallic dosimetry applicator and a thin plastic applicator (6F) in an alternating manner. The ratio of chamber reading corrected for the chamber finite size effect, correction factor k_V, and the decay of iridium, when using the plastic to those when using the metallic applicator, $k_{appl} = M_{plastic}/M_{metallic}$ was calculated for each of the 12 distances. The overall distances mean value of k_{appl} was found to be $k_{appl} = 1.0165 \pm 0.0021$ resulting in an effective attenuation coefficient of $\mu_{eff} = 0.292$ cm^{-1} or $\mu_{eff}/\rho = 0.0364$ cm^2 g^{-1} for stainless steel and for ^{192}Ir γ-rays.

Figure 7.10 shows our own experimental results for the estimation of the attenuation effect on the dosimetry for a metallic applicator typically used for all the calibration procedures in the Offenbach Radiotherapy Clinic for the ^{192}Ir sources of the microSelectron-HDR afterloader. This applicator is the so-called dosimetry applicator, Part-No 084.119 manufactured by Nucletron B.V., and constructed of stainless steel (ANSI 303/304 with a weight composition of 1% $_{14}$Si, 19% $_{24}$Cr, 2% $_{25}$Cr, 68% $_{26}$Fe, and 10% $_{28}$Ni, $\rho = 8.02$ g cm^{-3}, see also Table 4.1b) and having an outer diameter of 0.3175 cm and an inner diameter of 0.2055 cm. This results in a wall thickness of 0.056 cm of stainless steel.

Measurements were obtained in a water phantom of dimensions $48 \times 48 \times 41$ cm^3 for 12 different source to chamber distances in the range of 4.0 to 15.0 cm using the metallic dosimetry applicator and a thin (6F) plastic applicator in an alternating manner. The ratio of chamber reading corrected for the chamber finite size effect, correction factor k_V, and the decay of iridium, when using the plastic to those when using the metallic applicator, $k_{appl} = M_{plastic}/M_{metallic}$ has been calculated for each of the 12 distances. A mean value of $k_{appl} = 1.0165 \pm 0.0021$ was found.

Considering the metallic applicator wall thickness of 0.056 cm of stainless steel, an effective attenuation coefficient of $\mu_{eff} = (0.292 \pm 0.038)$ cm^{-1} or $\mu_{eff}/\rho = (0.0364 \pm 0.0047)$ cm^2 g^{-1} can be calculated. This value is in good agreement ($+2.8\%$) with the theoretical value of the absorption coefficient,

$\mu_{en} = 0.284$ cm^{-1} or $\mu_{en}/\rho = 0.0354$ cm^2 g^{-1}, which was calculated for stainless steel and for a 0.4 MeV effective photon energy, using the interaction coefficient data by Hubell and Seltzer.[40]

7.2.1.3.7 Use of Other Calibration Factors

DIN 6809-2[10] also allows the use of ionization chambers which have been calibrated to give exposure dose to water. The DGMP[29,31] recommendation for reference air kerma calibration of HDR iridium sources also considers such calibration factors. Considering Equation 7.6 and Equation 7.7, the air kerma calibration factor N_K of the chamber for the γ-energy of the radionuclide considered can be calculated from the corresponding known exposure calibration factor N_X according to (see also Section 2.5.1.3, Equation 2.41):

$$N_K = N_X \left(\frac{W}{e} \right) \left(\frac{1}{1 - g_\alpha} \right) k_{RC} \tag{7.18}$$

where W/e is the ionization constant, $W/e = 33.97$ J C^{-1} (energy-independent for energies above 1 keV, see Table A.2.2 in Appendix 2) with e being the elementary charge and W the mean energy expended in air per ion pair formed.[4]

g_α is the fraction of the energy of the electrons liberated by photons in air that is lost to radiative processes (mostly bremsstrahlung). g_α is for the photon energies relevant to brachytherapy practically 0.0, where for the highest energy of ^{60}Co radionuclide $g_\alpha = 0.00315$ (see Table 7.9).

TABLE 7.9

Fraction of the Energy of the Electrons Liberated by Photons in Air That Lost to Radiative Processes (Mostly Bremsstrahlung) g_α for Different Photon Energies According to DIN 6814-3[11]

Photon Energy (MeV)	g_α
0.01	0.00011
0.02	0.00019
0.03	0.00025
0.04	0.00028
0.05	0.00028
0.06	0.00027
0.08	0.00024
0.10	0.00024
0.15	0.00033
0.20	0.00045
0.30	0.00071
0.40	0.00097
0.60	0.00148
0.80	0.00198
1.00	0.00249
^{60}Co	0.00315

Since usually exposure and, thus N_X, are expressed in terms of roentgen (R) and $R\,C^{-1}$ respectively, k_{RC} the roentgen to $C\,kg^{-1}$ conversion factor (see Table A.2.1 in Appendix 2) has to be considered:

$$k_{RC} = 2.58 \times 10^{-4}\,C\,kg^{-1}\,R^{-1} \qquad (7.19)$$

7.2.1.4 Summary

Table 7.10 summarizes the source/radionuclide, ionization chamber, and geometrical environment dependent correction factors which are required for Equation 7.6 and Equation 7.7 and which must be carefully considered for an in-air calibration procedure in clinical environment.

The set-up independent factors N_K, k_P, k_{ion}, and k_{wall} must be initially estimated and then can be applied for any measurement which uses the same radionuclide source type and ionization chamber. If the geometrical environmental conditions are reproducible (i.e., same room, jig, applicator, measurement distances, etc.) then the environment dependent correction factors k_V, $k_{as,a}$, k_{scatt}, and k_{appl}, once initially estimated, can also be used for other calibration procedures where these k factors are constant.

For high energy radionuclides such as ^{60}Co and ^{137}Cs, as well as for the intermediate energy radionuclides ^{192}Ir and ^{198}Au, the in-air calibration procedure can be regarded as standardized. There is extensive experience published and this aids the user in selecting optimum chamber volumes and measurement distances according to the activity/strength of the source.

TABLE 7.10

Summary of Correction Factors to Be Handled Carefully When Applying In-Air Calibration Set-Ups and Their Dependency on Source (Radionuclide), Chamber, and Set-Up Environment Conditions

Correction factor		Parameter		
Name	Symbol	Source/ Radionuclide	Chamber	Set-Up Environment
Air kerma calibration factor of the chamber for the photon energy spectrum of the radionuclide	N_K	+	+	−
Polarity effect	k_P	+	+	−
Unsaturated ion collection efficiency	k_{ion}	+	+	−
Chamber finite size	k_V	−	+	+
Build-up cap and chamber wall	k_{wall}	+	+	−
Scattering and attenuation of the primary photons in air	$k_{as,a}$	+	−	+
Scattered radiation from room walls, floor, and measurement set-up	k_{scatt}	+	+	+
Attenuation in the applicator/ catheter wall	k_{appl}	+	−	+

These will vary depending on the brachytherapy dose rate to be used: HDR, PDR, or LDR.

For [169]Yb there is no extensive literature and therefore the user should consider only ionization chambers which are largely energy independent in terms of their response within the energy range relevant for [169]Yb. This is 60 to 150 keV. Build-up cap and chamber wall correction factors k_{wall}, must also be carefully considered since in most cases the calibration of ionization chambers for x-ray energies are made without the use of a build-up cap.

The maximum photon energy with a significant emission intensity in the spectrum of [169]Yb is 0.308 MeV and has an absolute intensity of $\sim 10\%$ (see Table 5.12a). Thus, the expected maximum energy of the emitted Compton recoil electrons is ~ 0.160 MeV, seen in Equation 4.26, with a range of ~ 0.04 g cm^{-2} in PMMA.

According to Table 7.6, for all the chambers listed the wall thickness is at least ~ 0.04 g cm^{-2} in PMMA. Thus, CPE conditions are fulfilled without the use of any additional build-up cap. Furthermore, it can also be assumed that contamination electrons from the source core and encapsulation reaching the chamber wall will be totally filtered out. Based on these findings, the chambers considered in Table 7.6 can be used without additional build-up material for in-air calibration set-ups of [169]Yb sources.

For the low energy radionuclides ^{125}I and [103]Pd, and because the clinically available sources of these radionuclides are of very low strength (activities of usually less than 37 MBq or equivalently 1 mCi), chambers of much larger volumes are required. This means that in-air calibration set-ups are not practical for clinically routine calibration procedures with ^{125}I and [103]Pd.

The equipment required for an in-air calibration set-up is as follows:

- *Calibration jig*
  With adapter and fixation mechanism for the source and ionization chamber enabling variation of the source to chamber distance in the range 5.0 to 50.0 cm. The jig should be made of low Z and low density material in order to avoid scattered radiation from the jig itself.

- *Ionization chamber*
  Compact 0.3 to 1.0 cm^3 for HDR sources or large volume ionization chambers of up to 1000 cm^3 or more for LDR sources of low energy such as ^{125}I and [103]Pd seeds. Ensure that appropriate air kerma calibration factors are used.

- *Build-up cap*
  Use an appropriate build-up cap for ^{60}Co, ^{137}Cs, ^{192}Ir, and [198]Au sources.

- *Electrometer*
  High resolution electrometer: dependent on the type of source and radionuclide.

- *Thermometer and barometer or alternatively a check source*

- *Applicator/catheter*
  Of known attenuation for the specific radionuclide source.

- *Treatment room*
  Room and geometrical conditions must be adequate to enable positioning of the in-air set-up such that there is a minimum of 1 m distance from room floor and walls to the ionization chamber position.

7.2.2 Calibration Using a Well-Type Ionization Chamber

Due to the complexity of the in-air set-up and the number and complexity of the parameters and effects required to be considered, the in-air method is currently limited to relatively few centers worldwide.

Currently, the most widely used method for calibration of sources of any radionuclide is the use of a well-type ionization chamber.[12,13,30,31,44,64,65] Such chambers can be provided with calibration factors for reference air kerma rate or air kerma strength for any commonly used radionuclide such as ^{192}Ir, ^{125}I, or ^{103}Pd. The calibration of the well-type chamber in a SSDL is made either (a) with a source previously calibrated using an in-air set-up (see Section 7.2.1) or (b) with a source of the same radionuclide calibrated by a PSDL. (a) is commonly used for ^{192}Ir sources whereas (b) is common practice for ^{125}I and ^{103}Pd sources.

Any type of source, ^{192}Ir LDR wires, ^{192}Ir HDR sources, or LDR seeds of any radionuclide (see Table 5.2), can be calibrated with a well-type ionization chamber.

Figure 7.11 through Figure 7.13 give three examples of commercially available well-type chambers. Table 7.11 summarizes the technical specifications of these chambers.

7.2.2.1 Formalism

The reference air kerma rate or air kerma strength of a brachytherapy source when using a well-type chamber calibration procedure is determined from

$$\dot{K}_R = N_{\dot{K}_R} M_{\text{max}} k_\rho k_T k_\text{p} k_{\text{ion}} k_{\text{appl}} \tag{7.20}$$

or

$$S_K = N_{S_K} M_{\text{max}} k_\rho k_T k_\text{p} k_{\text{ion}} k_{\text{appl}} \tag{7.21}$$

$\dot{K}_R$:   is the reference air kerma rate expressed in μGy h^{-1} at 1 m.

S_K: is the air kerma strength in U, $1\,U = 1\,\mu$Gy m^2 h^{-1} = 1 cGy cm^2 h^{-1}.

$N_{\dot{K}_R}$:   is the reference air kerma rate calibration factor of the chamber for the radionuclide studied. It is usually expressed in Gy m^2 h^{-1} A^{-1}, or μGy m^2 h^{-1} A^{-1}.

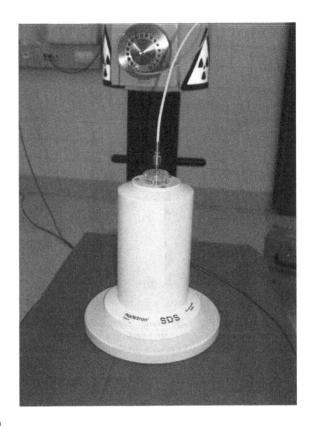

FIGURE 7.11
The SDS Nucletron (Nucletron B.V., Veenendaal, the Netherlands) well-type chamber in a
calibration set-up with a microSelectron-HDR afterloader.

N_{S_K}:   is the air kerma strength calibration factor of the chamber
            for the radionuclide studied. It is expressed in $U A^{-1} = 1$
            $\mu Gy\ m^2\ h^{-1}\ A^{-1} = 1\ cGy\ cm^2\ h^{-1}\ A^{-1}$.

M_{max}:   is the maximum measured ionization current value with
            the well-type chamber and is usually expressed in A. If the
            charge Q_{max} in C is collected during a time interval of
            minutes, then $M_{max} = Q_{max}/(60\tau)$.

k_ρ: is the correction factor for the current air pressure
            conditions P, expressed in hPa, other than the referred in
            the calibration protocol of the chamber, i.e., the normal
            pressure of 1013 hPa:

$$k_\rho = \left(\frac{1013\ hPa}{P} \right)$$

k_T: is the correction factor for the current air temperature
            conditions T in °C other than that referred to in the

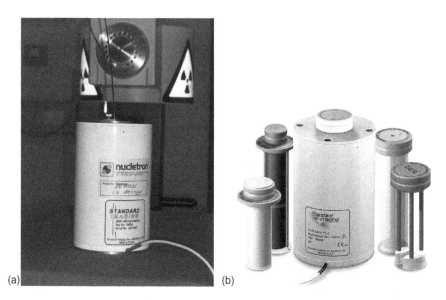

(a)   (b)

FIGURE 7.12
The HDR-1000 Standard Imaging (Standard Imaging Inc., Middleton, Wisconsin, USA) well-type chamber. This is the commercially available version of a chamber originally designed by Attix.[65] (a) HDR-1000 in a calibration set-up with a microSelectron HDR afterloader. (b) The current HDR-1000 Plus model with several inserts/adapters for different sources/radio-nuclides.

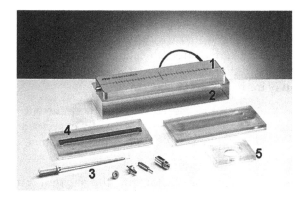

FIGURE 7.13
SourceCheck Type 34051 chamber by PTW is a vented ionization chamber with a volume of $55\ cm^3$ and is appropriate to measure the sources strength of sealed sources either by positioning the sources inside the chamber or on the flat side of the chamber. Typical response for ^{125}I seeds given by the manufacturer is 28.3 fA MBq^{-1} (1.05 pA mCi^{-1}). The SourceCheck chamber (1) backscatter phantom (2) accessories for measurement of single seeds (3) or seeds strands (4) and the adapter for positioning of a check source (5) for testing the long-term stability of the chamber. This chamber is appropriate for the measurement of LDR seeds as well as of seed strands and HDR and PDR sources.

TABLE 7.11

Technical Specifications of Three Examples of Commercially Available Well-Type Ionization Chambers

Parameter	Well-Type Chambers			Compact Chamber
	SDS	HDR-1000/HDR-1000 Plus[a]	SourceCheck	Type 23332
Manufacturer	Nucletron (Nucletron B.V., Veenendaal, The Netherlands)	Standard Imaging (Standard Imaging Inc., Middleton, Wisconsin, U.S.A.)	PTW (PTW, Freiburg, Germany)	PTW (PTW, Freiburg, Germany)
Type	Vented to the atmosphere	Vented to the atmosphere	Vented to the atmosphere	Vented to the atmosphere
Active volume (cm^3)	200	245	55	0.30
Dimensions				
Height/length (cm)	17.80	15.60	22.00	1.85
Diameter/width (cm)	20.05	10.20	1.40	0.60
Depth (cm)	—	—	6.00	0.60
Weight (kg)	2.00	2.70	0.200	0.11
Outer wall material/thickness (cm)	Polyamide/0.02 + Aluminum/0.30	Aluminum/2.0, Aluminum/1.91[a]	Polycarbonate/0.25	PMMA/0.05
Sensitivity	1.0[b] nA (mGy h^{-1})$^{-1}$ or 1.0[b] pA U^{-1}	1.8[b] nA (mGy h^{-1})$^{-1}$/ 2.1 nA (mGy h^{-1})$^{-1}$ or 1.8[b] pA U^{-1}/2.1 pA U^{-1}	1.5[c] nA (mGy h^{-1})$^{-1}$ or 1.5 pA U^{-1}	10^{-8} C Gy^{-1}
Bias voltage (V)	300	300	400	500
A_{ion}[d]	>0.996[d]	0.9996[d]	—	>0.995
Maximum response position/reference point	5.7–5.9 cm[e]	5.1–5.3 cm[e]	11.0 cm[f]	0.95 cm[g]

In addition a PTW type 23332 0.3 cm^3 compact ionization is also listed that has been considered in comparison studies.

a HDR-1000 Plus has a changed inner wall construction for enabling calibrations of ^{125}I and ^{103}Pd sources.

b Values derived from those quoted in nA Ci^{-1} by the manufacturers and are valid for ^{192}Ir sources.

c Typical value given by the manufacturer for ^{192}Ir sources.

d Values valid for ^{192}Ir HDR sources.

e Measured from the bottom of the chamber.

f Middle of chamber length.

g Measured from the top of the chamber.

calibration protocol of the chamber, i.e., the normal temperature T_0 in °C:

$$k_T = \left(\frac{273.15 + T\,(°C)}{273.15 + T_0\,(°C)} \right)$$

usually $T_0 = 20$°C.

Both correction factors k_ρ and k_T must be considered when the chamber is vented to ambient atmospheric conditions: otherwise both factors are set to 1.00.

k_p: is the correction factor for the polarity effect of the bias voltage for the photon energy of the radionuclide, as seen in Equation 7.6 and Equation 7.7.

k_{ion}:   is the correction factor that accounts for the unsaturated ion collection efficiency and thus for the charge lost to recombination for the specific radionuclide photon energy and the applied nominal voltage V. As discussed for the in-air calibration set-ups, k_{ion} is the reciprocal of the ion collection efficiency A_{ion}. It is given by

$$k_{ion} = \frac{1}{A_{ion}} = \frac{3}{4 - M/M_2}$$

where M_2 is the selected chamber signal at the half the nominal voltage V and M is the measured chamber signal (current or charge) at the nominal voltage V.

k_{appl}:   is the correction factor accounting for the attenuation in the applicator/catheter wall used for positioning or fixing the source within the well-type chamber. When the same applicator is used as in the chamber calibration procedure then $k_{appl} = 1.00$.

7.2.2.1.1   *The Calibration Factor N_{K_R} or N_{S_K}*

For radionuclide sources such as ^{60}Co, ^{137}Cs, ^{192}Ir, and [198]Au, which emit high energy or intermediate energy photons, it is to be expected that the chamber calibration factor will depend to a certain extent on the source design. However, for the low energy photon sources of ^{125}I and [103]Pd, which are used clinically as seed sources, this is much more evident due to the significant influence of the source encapsulation and core design on the emitted photon spectrum (e.g., emission of characteristic x-rays and fluorescent radiation from the material surrounding the radionuclide source) and also on the filtration effect which occurs in the well chamber walls (see Table 7.11). Since well-type chamber walls are much thicker than those of compact ionization chambers (Table 7.6) the energy and source type dependence of the response of well-type chambers is much more pronounced than for compact chambers.

Mitch et al.[64] studied the response of two different commercially available well-type chambers for different [125]I (nine types) and [103]Pa (four types) seed designs from nine different manufacturers. For one of the chambers they found a variation of approximately 6% in the calibration factor among the [103]Pa seeds and up to 25% for the ^{125}I seeds.

The variation of the second well chamber was 5% for [103]Pa and up to 35% for [125]I (Figure 2 and Figure 1 after Mitch et al.). The higher variations for ^{125}I are mainly due to the silver material used in the seed designs and thus to the resultant differences in the emission characteristic x-rays from silver. These findings emphasize the necessity for having a calibration factor available for the specific seed design that is to be measured for these low energy radionuclides.

Figure 7.14 summarizes a calibration procedure carried out at the Offenbach Radiotherapy Clinic for the [125]I SelectSeeds which are manufactured by Isotron, Isotopentechnik GmbH in Berlin, for Nucletron B.V. of Veenendaal in The Netherlands.

For this purpose, SelectSeeds calibrated by NIST were provided by the manufacturer for the chamber calibration procedure within the clinic. The results shown were based on using the No. 3 NIST calibrated seed (seed No. 3, Ref. No. 130002-00, Batch No. 13911) that had an air kerma strength of 2.220 U (apparent activity of 64.676 MBq equivalent to 1.748 mCi) at 19/10/2001, 00:00:01 EST) and a UNIDOS electrometer operated at $+400$ V.

The air kerma strength calibration factor was found to be $N_{S_K} = 1.122$ U pA^{-1} equivalent to 0.883 mCi pA^{-1} for the apparent activity. Both calibration factors account for the attenuation effect of the metallic seed adapter which was used. (This is component 3 in Figure 7.13 and in Figure 7.14b).

The calibration follow-up of the same chamber over a period of 25 days with a total of eight measurements using the same NIST calibrated seed is summarized in Figure 7.14c. With the exception of measurement number 6 (-1.5%), the remaining 7/8 follow-up results are all within 1% agreement.

For the estimation of the collection efficiency, current measurements were performed using a SelectSeed source and varying the collecting voltage. The results shown in Figure 7.15 indicate that the chamber collection efficiency is not strongly dependent on collection voltage since current measurements at voltages of 50 to 400 V are within 0.75%. The ion collection efficiency A_{ion} and, thus k_{ion} in Equation 7.20 and Equation 7.21, was found to be 1.000. This is explained by the low dose rate of the ^{125}I seeds combined with the chamber geometry which results in insignificant ion recombination within the chamber volume. The polarity effect correction factor of the chamber, see also explanations for Equation 7.6 and Equation 7.7, was $k_p = 0.995$.

Figure 7.16 summarizes the results of our clinic measurements using the SDS and HDR-1000 well-type chambers with collecting voltages from 50 to

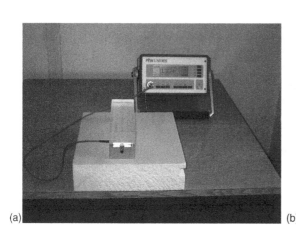

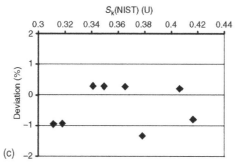

FIGURE 7.14
Calibration of a SourceCheck chamber (TM 34051-0013) for ^{125}I SelectSeeds at the Offenbach radiotherapy clinic. (a) Set-up for the calibration using a NIST calibrated ^{125}I SelectSeed (seed No. 3, Ref. No.130002-00, Batch No. 13911) and an UNIDOS electrometer operated at $+400$ V collection voltage. The calibration factor was found to be $N_{S_K} = 1.122$ U.pA^{-1} or the for apparent activity 0.883 mCi pA^{-1} which already count for the attenuation effect of the seed adapter used (component 3 in Figure 7.13). (b) Detailed view of the adapter used for fixing the SelectSeed and inserting this into the SourceCheck chamber. (c) Calibration follow-up of the chamber over a period of 25 days with a total of eight measurements using the NIST calibrated seed. Here the deviation of measured air kerma strength from the NIST calibration is shown.

400 V^{30} and with a microSelectron-HDR afterloader ^{192}Ir source. For both chambers there is a very small increase of the measured signal (current or charge) of less than 0.5% for a corresponding increase of the voltage from 100 to 500 V. The resulting ion collection efficiency A_{ion} was 0.9996 and 0.9997 for the HDR-1000 and SDS chambers, respectively. This resulted in a k_{ion} of 1.0004 and 1.0003, respectively, for the standard operating voltage of 300 V for both chambers.

The polarity effect, expressed as the percentage variation of the electrometer reading by changing the polarity of the bias voltage, was 0.09%

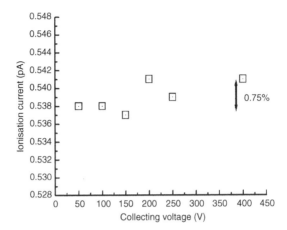

FIGURE 7.15

Ionization current measurements for the estimation of the collection efficiency performed using a SelectSeed source and by varying the collecting voltage in the range of 50 to 400 V for the SourceCheck chamber shown in Figure 7.14. The maximum observed change in the measured current was 0.75% resulting in an ion collection efficiency $A_{ion} = 1.00$ and hence $k_{ion} = 1.00$.

for the HDR-1000 and 0.45% for the SDS well-type chamber, resulting in a polarity correction factor k_p of 0.999 and 0.998, respectively. All measurements were made with the source at the position of maximum chamber response.

Figure 7.17 provides an overview of the radionuclide/source dependence of the air kerma strength calibration factor N_{S_K} for the HDR-1000 and SDS well-type chambers and is based on typical values provided by the manufacturers. The original chamber response data values published by the manufacturers have been converted to N_{S_K}.

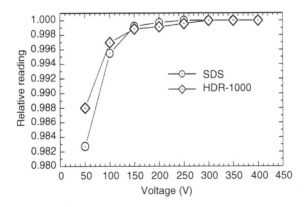

FIGURE 7.16

Saturation curves for the SDS and HDR-1000 well-type ionization chambers measured using an ^{192}Ir source of a microSelectron-HDR afterloader. The electrometer readings are normalized at the maximum observed value for each chamber.

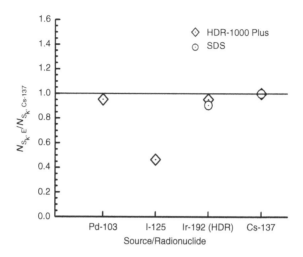

FIGURE 7.17

Source/radionuclide dependence of the calibration factor N_{S_K} for air kerma strength for [103]Pd, ^{125}I, ^{192}Ir, and ^{137}Cs sources for the two well-type chambers, HDR-1000 Plus and SDS. Values are typical values given by the manufacturers and are normalized to the calibration factor for ^{137}Cs source. The calibration factors for ^{192}Ir are for HDR sources. The original values published by the manufacturers have been converted to N_{S_K}. The ^{137}Cs calibration point for the SDS chamber is based on Offenbach Radiotherapy Clinic data using an in-air calibration set-up as the primary standard to calibrate this chamber.

For ^{192}Ir calibration points, only the results for HDR afterloader sources are considered and not those for LDR wire sources. The ^{137}Cs calibration point for the SDS chamber is based on our own data using an in-air calibration set-up to calibrate this chamber. There is a significant change in the calibration factor which is demonstrated when studying low energy sources such as ^{125}I and [103]Pd.

7.2.2.2   Geometrical Conditions

7.2.2.2.1   Chamber Sensitivity to Source Position

Figure 7.18 shows, in a simplified manner, a cross-sectional view of a standard well-type chamber. The ionization current or charge depends on the position of the source along the longitudinal axis of the adapter or source insertion tool. This is shown as distance h in Figure 7.18. The sensitivity of the chamber response to the source position also depends on the individual design of the well chamber.

Figure 7.19 summarizes our own measurements[30] of the sensitivity of the HDR-1000 and SDS well-type chambers using a microSelectron-HDR ^{192}Ir stepping source. The maximum ionization current was found to be at a distance of ~5.2 cm from the bottom of the HDR 1000 well-type chamber. This is exactly the median value of the range 5.1 to 5.3 cm given by the manufacturer, see Table 7.11.

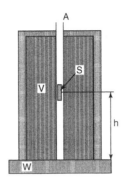

FIGURE 7.18
Simplified cross-sectional schematic of a well-type chamber. There is an outer wall (W), an opening or adapter (A) for inserting and positioning the radioactive source (S) that is at a distance h from the base of the chamber. The collecting volume V of well-type chambers in clinical use is usually >100 cm^3 and in the range 200 to 300 cm^3, to enable an adequate signal (current or charge) to be produced even for very low activity sources.

For the SDS chamber this was observed at $\sim$5.8 cm, which again is at the median value of the range 5.7 to 5.9 cm stated by Nucletron, see Table 7.11. The plateau defined as the region around the position of maximum response which shows a maximum of 0.1% variation in the ionization current in the sensitivity curves is shown in Figure 7.19. It is estimated to have a width of 0.75 cm for both the HDR-1000 and the SDS chambers.

7.2.2.2.2 Room Scatter

Unlike the majority of other radiation detectors, well-type chambers are expected to show only a weak sensitivity dependence on environmental

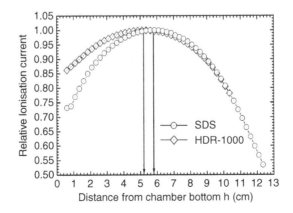

FIGURE 7.19
Experimentally measured sensitivity curves for the HDR-1000 and SDS well-type chambers. The measured ionization current is normalized for each curve at the maximum observed value. Both curves have been measured with the same HDR ^{192}Ir source of a microSelectron-HDR afterloader. The source distance from the chamber base h is seen in Figure 7.18.

scattering conditions, i.e. on scattering from walls and floor. This is mainly due to the fact that the radionuclide sources are positioned very close to the sensitive volume and also because the outer wall of these chambers are made of metallic materials of much greater thickness than those of conventional compact ionization chambers or of large volume ionization chambers used in the in-air calibration set-ups. Even so, it is important to investigate the possible influence of the well-chamber response and the radioactive source on the specific measurement set-up in the laboratory or clinical environment.

The sensitivity of the HDR-1000 and SDS chambers on scattering from room walls and floor was investigated in our own clinic[30] for ^{192}Ir HDR source calibrations: keeping the distance from the room floor at 1 m. The results for both systems are shown in Figure 7.20. The distance of the outer lateral surface of the chamber to room wall was varied in the range 0 to 100 cm.

For the SDS chamber a minimum distance of 5.0 cm could be achieved due to the specific design of the base plate of that chamber (see Figure 7.11) where a contribution of scattered radiation to chamber response of ~1% was measured. The maximum lateral scattered radiation effect for the HDR-1000 chamber was measured to be 1% for direct contact to the room wall: $d = 0$ cm in Figure 7.20.

These results show that a minimum lateral distance from the walls of 15 and 25 cm is needed for the HDR-1000 and SDS well-type chambers,

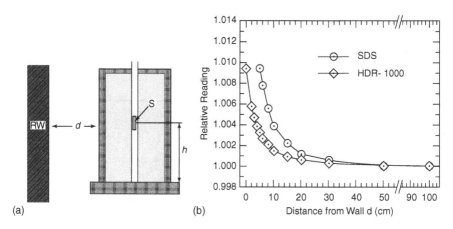

FIGURE 7.20
Influence of lateral wall scattering on the measurements for the two well chambers using a ^{192}Ir source of microSelectron-HDR afterloader. (a) Simplified drawing of the experimental set-up. The radioactive source (S) has been positioned at distance h from the base of the chamber, which is always the location of the maximum chamber response (see also Figure 7.19). Both chambers have been placed 1 m above the floor to guarantee negligible influence of the floor scattering. The distances d given in this figure are measured from the outer surface of the chamber to the room wall surface (RW). (b) Chamber readings have been normalized to their response for $d = 100$ cm. The minimum chamber to room wall distance for the SDS system was 5.0 cm due to the special shape of the base plate of this chamber (see Figure 7.11).

respectively, in order to keep the scattering contribution to below 0.1% of the readout (current or charge).

As was expected the two well-type chambers were independent of their proximity to a wall surface above the 25.0 cm. This is in agreement with the measurements of Podgorsak et al.[66] The experimental results in Figure 7.20 show a greater effect of wall scatter on measurements for the SDS chamber. This is due to the much thinner outer wall of that chamber and thus a much smaller wall scatter irradiation filtration effect when compared to the HDR-1000. This was 0.3 cm aluminum compared to 2.0 cm aluminum outer wall thickness for the HDR-1000 well chamber: see Table 7.11.

7.2.2.2.3   Other Effects

Podgorsak et al.[66] published results showing a thermal effect on the response of the originally designed Attix chamber. They measured a gradual decrease in sensitivity of this chamber of 0.15% over 30 min of radiation exposure. This was observed when no insulator material was in place and thus the air temperature within the ion collecting volume was increased during exposure. This was caused by heat transfer from the radiation emitted by the ^{192}Ir HDR source of ~ 300 GBq.

When a styrofoam insulator is used, this thermal effect disappears and transfer of heat from source to the chamber collecting volume is prevented. Hence these authors recommended that the insulator should always be kept in place when calibrating or operating these chambers. The HDR-1000 is always provided with a styrofoam insulating sleeve around the central source tube. This is taken into account in the calibration protocol of this chamber. There are no results published for similar observations with the SDS or other open type of well-type chambers.

In any event, since practical measurements with well-type chambers using high strength ^{192}Ir sources of ~ 370 GBq take less than 30 min, the expected effect for thermally noninsulated collecting volumes can be expected to be negligible. For low activity seed sources used for permanent implants this has no measurable effect.

7.2.2.3   Calibration of ^{192}Ir LDR Wires

When using ^{192}Ir line sources cut from a coil of wire, or single pins or hairpins, the well-type chambers should be provided by the SSDL with a calibration factor for a 1.0 cm length of wire or single pin, or for the hairpin type of source. Due to the sensitivity of a well chamber response to the source position (see Figure 7.18 and Figure 7.19), it is to be expected that the chamber response per unit length of wire will depend on the total length of the source.

In general, a wire length correction factor k_L, must be considered in Equation 7.20 and Equation 7.21,[12] where the reference air kerma rate and air kerma strength chamber calibration factors N_{K_R} and N_{S_K} respectively are stated with reference to a 1.0 cm wire length.

TABLE 7.12

Wire Length Correction Factors k_L for the HDR-1000 and SDS
Well-Type Chambers According to TECDOC-1079[12] for ^{192}Ir
Wire Lengths in the Range of 1.0 to 9.0 cm

	Well-Type Chamber	
Wire Length (cm)	HDR-1000	SDS
1.0	1.000	1.000
3.0	1.005	1.012
5.0	1.012	1.017
7.0	1.029	1.038
9.0	1.050	1.070

Values are normalized to that for 1.0 cm. The center of the wire source is
each time positioned at the reference point of the well-type chamber (peak
of the response curves in Figure 7.19).

Table 7.12 provides wire length correction factors k_L for the HDR-1000 and
SDS chambers from the IAEA TECDOC-1079[12] for ^{192}Ir wire lengths in the
range 1.0 to 9.0 cm. Values are normalized to that for 1.0 cm wire and are
valid when the center of the wire source is positioned at the reference
point of the well-type chamber. This is at the peak of the response curves
in Figure 7.19. The higher correction factors for the SDS chamber can be
explained by the significantly lower response of that chamber for source
positions at distances h from chamber base (see Figure 7.18) which are
less than the peak response when compared to the HDR-1000 chamber
in Figure 7.19. The chamber response curves for the two chambers and
for distances h above the peak are almost identical: see Figure 7.19. Thus, for
wires of longer length than 2.0 cm, according to this figure the SDS response
is the lower and thus the correction factor k_L value is the higher when
compared to that of the HDR-1000 for the longer the wire source.

When a similar table for another well chamber is not available, the user
can estimate this correction factor by measuring the response curve of the
chamber in a manner similar to that described for the data in Figure 7.18
and Figure 7.19. In these figures the center of a 1.0 cm long wire source is
either positioned at different distances around the reference point of the
chamber or by using a measured response curve for an ^{192}Ir HDR source.

Based on a known response curve of a well chamber, the length correction
factor k_L for a wire source of length L_s can be calculated using the following
formula:[62]

$$k_L = \frac{L_s}{\displaystyle\int_{-L_s/2}^{+L_s/2} R(x)\, dx} \tag{7.22}$$

where $R(x)$ is the measured chamber current or charge normalized to that at
the reference point of the chamber. It is the value from the response curve,

such as those in Figure 7.19, for a distance x from the chamber reference point which is the point/position of the chamber peak response (not the distance from the chamber base h as defined in Figure 7.18). This presupposes that the x-axis in Figure 7.19 is, for each of the chambers, individually shifted to the position of peak response such that at this position $x = 0.0$ cm.

Considering now the HDR-1000 curve in Figure 7.19, and a source length of 5.0 cm, calculating the integral of that curve for x in the interval -2.5 to 2.5 cm we obtain a value of 4.933 and thus a length correction factor $k_L = 5.0/4.933 = 1.014$. This fits very well to the value of 1.012 given in Table 7.12. Similarly for the SDS chamber the value of integral in Equation 7.22 is 4.905 and thus $k_L = 5.0/4.905 = 1.019$ which again is in very good agreement with the value of 1.017 given in Table 7.12 according to the IAEA TECDOC-1079.[12]

7.2.2.4 Stability of Well-Type Chamber Response

In order to identify drifts in the response of a well-type chamber, which can be expected when the chamber is used over a prolonged period of time, it is recommended that the response is regularly checked. This can be achieved by investigating the stability of the corrected chamber signal, current or charge, under reproducible irradiation conditions and at regular time intervals. There are three possible methods for such an investigation.

7.2.2.4.1 Use of a Long-Lived Source

Since a ^{137}Cs source has a half-life $T_{1/2}$ of 30.2 years, this radionuclide is suitable for such a stability check procedure. However, it is emphasized that the same source insert/adapter, the same position of the source (same distance from chamber bottom), and the standard voltage and polarity according to the chamber protocol must always be used. In addition, to achieve a charge measurement with as high as possible accuracy over a long time, the same time period always should be used.

Firstly, the reference chamber measurement, charge M_{ref}, has to be defined, corrected for ambient atmospheric conditions, temperature, and pressure when we consider chambers vented to the atmosphere. This should be performed immediately after delivery of the chamber. Then, the chamber reading M_{act} is obtained under the same conditions and using the same source and set-up and corrected again if necessary for air pressure and temperature conditions.

The deviation

$$\left(1 - \frac{M_{act} \cdot e^{(\ln 2/T_{1/2})\Delta t}}{M_{ref}}\right)100(\%) \tag{7.23}$$

is then calculated. In Equation 7.23, the actual chamber reading M_{act} is, in addition, corrected for source decay during the time interval Δt between reference and current measurement times. This deviation should be in the range $\pm 0.5\%$.[12,13]

PTW supports the SourceCheck chamber with an appropriate adapter and a ^{90}Sr (half-life $T_{1/2} = 28.79$ years[67]) check source for the stability test procedure (see also Figure 7.13).

7.2.2.4.2   *Use of a ^{192}Ir HDR or PDR Source*

When well-type chambers are considered for calibrating HDR or PDR ^{192}Ir sources, a clearly defined procedure must be established. In HDR and PDR brachytherapy the ^{192}Ir sources are regularly exchanged ~ 4 times per year.

The ^{192}Ir source prior to the exchange can be used for checking the well chamber response stability using a similar method to that described when using a long-lived ^{137}Cs or ^{90}Sr source. The calibration of the ^{192}Ir source is then repeated and the result is compared to the first calibration after correcting for the decay of ^{192}Ir. The deviation is then estimated using Equation 7.23. Again, the response of the well chamber can be considered to be unchanged when the measured deviation is only in the range $\pm 0.5\%$.[12,13]

7.2.2.4.3   *Use of an External Radiation Source*

An alternative method is to use a standardized set-up with an external radiation source such as a ^{60}Co external beam radiotherapy unit or a linear accelerator. Figure 7.21 shows such a set-up. This was proposed by IAEA[12] originally for SSDLs, but was also later adopted by ESTRO.[13] The source to chamber surface distance SSD should be the standard distance used in external beam dosimetry, i.e., SSD = 100 cm. The field size should be selected to be larger than the outer diameter of the well chamber. This is to ensure that the entire chamber volume is definitely covered by the radiation field. The irradiation parameters should be selected so as to deliver 1.0 Gy at the surface of the chamber.[12,13] The chamber reading, i.e., the charge, is then

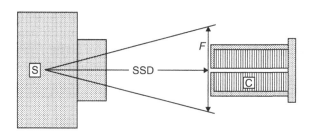

FIGURE 7.21
Set-up for checking the stability over a time period of the response of well-type chambers. (C) the well chamber, (S) the focus of the external radiation source (^{60}Co or accelerator). The source to chamber surface distance SSD should be standardized and preferably 100 cm, whereas the field size F should be selected to be larger than the diameter of the well-type chamber.

corrected for ambient atmospheric conditions, temperature, and pressure when the chamber is vented to the atmosphere.

In the case when a ^{60}Co external beam unit is used, the chamber reading must be corrected in addition to the source decay, such as when small long-lived sources are used. Then, Equation 7.23 can be applied for estimating the deviation of the chamber response. The reference value M_{ref} must be fixed in exactly the same set-up prior to the start of any constancy check procedure.

Currently, the majority of radiotherapy centers worldwide have linear accelerators at their disposal and it is much more practical to use an x-ray beam from an accelerator for this procedure than use a ^{60}Co external beam unit. No decay correction factor (the exponential part of Equation 7.23) is required when an accelerator is used for this procedure. The chamber response measured in the way described should remain unchanged, as in Equation 7.23, to within $\pm 1.0\%$.[12,13]

7.2.2.5 Summary

Table 7.13 summarizes the source/radionuclide, chamber, and geometrical environment dependent correction factors considered in Equation 7.20 and

TABLE 7.13

Summary of Correction Factors to Be Handled Carefully When Applying Well-Type Chamber Calibration Set-Ups and Their Dependency on Source (Radionuclide), Chamber and Set-Up Environment Conditions

Correction Factor		Parameter		
Name	Symbol	Source/ Radionuclide	Chamber	Set-Up Environment[a]
Air kerma calibration factor of the chamber for the photon energy spectrum of the radionuclide	N_{K_R}, N_{S_K}	+	+	$-/+$[b]
Maximum chamber response	M_{max}	+	+	+
Polarity effect	k_p	+	+	−
Unsaturated ion collection efficiency	k_{ion}	+	+	−
Scattered radiation from room walls and floor	k_{scatt}	+	+	+
Attenuation in the applicator/ catheter wall	k_{appl}	+	$-/+$[c]	+
Wire length correction factor	k_L	+	+	+

[a] Includes also the well-type chamber inserts.
[b] The calibration factor refers to a specific insert/adapter for the source and for the case of stepping sources (^{192}Ir HDR) to a specific applicator.
[c] Depending on the photon spectrum emitted by the specific source design and on the chamber wall material and thickness, the attenuation factor of a specific applicator to be considered in the calibration procedure can be also chamber dependent.

Equation 7.21, which must be carefully considered for a well-type chamber based calibration procedure in a clinical environment.

The set-up independent factors k_P and k_{ion} only have to be estimated once and then can be applied for any other measurement using the same source type and well ionization chamber. If the same applicator type, or even the same applicator, is always used then k_{appl} has to be estimated only once and can be applied for future set-ups.

For well-type chambers, the applicator attenuation factor for the case which is not the same as that used for the calibration of the chamber can be easily estimated by comparing measurements using this specific applicator and a plastic applicator, which can be assumed to have a negligible attenuation effect for its material and wall thickness. Hence, the applicator correction factor k_{appl} to be used in Equation 7.20 or Equation 7.21 can be estimated.

If the geometrical environmental conditions are reproducible (i.e., the same room and the same distances from floor and walls) then, once estimated, the environment dependent correction factor k_{scatt} can also be used for repeated calibration procedures.

Care must be taken with the position of the maximum response (Figure 7.19) when plastic tubes and plastic connecting cables/transfer tubes to the afterloading device are used. This is because these cables/tubes can have different physical lengths and their length is temperature dependent. The differences caused can be outside the ± 0.38 cm range defined in Figure 7.19 although the chamber response remains unchanged within 0.1%.

Since a well-type chamber response generally exhibits a clear dependence on the source radionuclide and the source design, it is important that the user ensures that the chamber has been adequately calibrated. That is, a calibration factor is known for all the radionuclides and source designs under consideration. In addition, all conditions described in the chamber calibration protocol must be strictly fulfilled.

For LDR ^{192}Ir wire sources the calibration factor per 1.0 cm wire reference length must be provided and a table for the wire length correction factor k_L should be available or calculated using Equation 7.22 and the corresponding response curve of the chamber.

Finally, when storage of the chamber is at a different place than that used for the calibration procedure, enough time should be allocated so that the open chamber reaches an equilibrium with the ambient conditions. Otherwise, the correction factor for the current air temperature conditions k_T which is based on the measured ambient temperature is not representative for the chamber. This will introduce an additional error in the calibration procedure.

Although measurement of the air temperature in the collecting volume of the chamber is practically impossible, the temperature of the chamber wall itself is indicative for such an equilibrium not existing. In practice, a period of several hours may be required for achieving this independence on the initial temperature difference. To avoid such deviations the chamber can be left overnight in the calibration room to ensure that enough time is available for reaching temperature equilibrium with the ambient conditions.

The most feasible method in clinical practice for checking the stability of the well chamber response is that of using a long-lived radionuclide source such ^{137}Cs or ^{90}Sr. Manufacturers are requested to provide standardized set-ups with such check sources in order to simplify calibration in clinical practice. The required equipment for source calibration with well type chamber is as follows:

- *Well-type chamber*
  With an appropriate adapter for the insertion and positioning of the sources in use

- *Electrometer*
  High resolution electrometer according to the type of source and radionuclide considered

- *Thermometer and barometer*

- *Applicator/catheter*
  Of known attenuation for the specific source

- *Treatment room*
  Room and geometrical conditions that will enable a positioning of the in-air set-up with a minimum of 1 m distance from the room floor and at least 25 cm from room walls

7.2.3 Calibration Using Solid Phantoms

Solid phantoms have been considered by several physics groups for calibration procedures, especially for ^{192}Ir HDR[28–32,42,61,63,68,69] sources and other types of high energy radionuclide sources such as ^{137}Cs and ^{60}Co and for both LDR or HDR brachytherapy.[36,42,63,70] There are two main arguments for using solid phantoms for this category of brachytherapy source.

Firstly, solid phantoms permit a high degree of accuracy and thus in the reproducibility of measurements. Phantom calibration procedures are superior to the in-air set-up and equivalent to the well-type chamber calibration.

Secondly, commonly available ionization chambers in radiotherapy departments can be used without any need for additional accessories, chamber calibration, and QA procedures. Therefore, again, phantom calibration procedures are superior to the in-air set-up methods and equivalent to the calibration using the well-type chamber: at least for non-LDR sources of low energy such as ^{125}I and ^{103}Pd. For the in-air calibration set-up methods as stated previously, ionization chambers of a large volume (>1.0 cm^3) are required and these are not commonly available in radiotherapy clinics.

The principle of the phantom method for reference air kerma rate or air kerma strength is to convert the ionization current or charge measured via a specific formalism defined for the radionuclide to be used and for the phantom's specific geometry and material. All factors are therefore phantom and radionuclide/source dependent. As a first approximation, due to the limited energy dependence of the calibration factors for the commonly used ionization chambers (see Figure 7.5), such a calibration system might also be appropriate for the low energy radionuclides ^{125}I and ^{103}Pd.

However, such sources are available for brachytherapy in very low activities. Thus, in addition, because of the very high attenuation and absorption of the low energy photons within the phantom material at the distance of measurement, the corresponding dose rate will be too low. This means that the measuring signal of chambers with volumes in the range 0.3 to 1.0 cm^3 is outside an acceptable range. Therefore, solid phantom set-ups are not appropriate for the low energy and low activity ^{125}I and ^{103}Pd sources.

Finally, it is noted that the phantom material PMMA is now established, at least for commercially available phantoms, as the standard construction material, see Table 7.14.

7.2.3.1 Geometrical Conditions

7.2.3.1.1 Measurement Distance

In similarity with the in-air calibration set-up, the measurement distance r must be selected in such a way that the source can be seen as a point source and that the dimensions of the collecting volume of the chamber will not significantly influence the calibration result. In addition, the distance r should be large enough such that positioning errors and commonly encountered manufacturing inaccuracies will not significantly influence the result. From Table 7.3 it is seen that measurement distances greater than 5.0 cm can be considered to be appropriate. As an example, a total uncertainty in the measurement distance of $\delta r = \pm 0.2$ mm results in a measurement uncertainty of less than $\pm 1\%$.

7.2.3.1.2 Source Positioning

Due to the fact that for in-air set-ups several distances are used, systematic errors in defining the source to chamber distance r can be filtered out, as in Equation 7.15. Also, the distances considered for in-air set-ups are, in general, larger than those used in solid phantom calibration procedures. Thus, for the in-phantom calibrations, as opposed to in-air calibrations, two types of uncertainties must be considered for the source positioning.

Firstly, it is important that the applicator used has to enable accurate centering positioning of the source with reference to the applicator's longitudinal axis. Depending on the distance r used for the specific phantom and the outer diameter of the source, the maximum inner diameter of the applicator can be estimated using the uncertainties listed in Table 7.3.

TABLE 7.14

Geometrical Characteristics of the Three PMMA Solid Phantoms

Phantom	Manufacturer	Height (cm)	Diameter/ Width (cm)	Length (cm)	Weight (kg)	Measurement Distance (cm)	Typical Ionization Chamber
Krieger cylindrical phantom	PTW (PTW, Freiburg, Germany)	12.0	20.0	—	4.3	4 holes[a] at 8 cm	PTW type 23332 0.3 cm^3
Meertens cylindrical phantom	—	15.0	20.0	—	5.4	3 holes for sources[b] at 5 cm	NE 2571 (Farmer-type) 0.6 cm^3
Baltas plate phantom	GfM (GfM-Medizintechnik, Weiterstadt, Germany)	4.0	30.0	38.0	5.2	8	PTW type 23332 0.3 cm^3

[a] Equally spaced every 90° (see also Figure 7.23).
[b] Equally spaced every 120° (see also Figure 7.24).

An effective method to reduce the influence of uncertainties in the axial source positioning, as well the manufacturing uncertainties, is to use a specific source to chamber distance r, either: (a) multiple angular chamber positions around the source, or (b) multiple angular source positions around the chamber (see Figure 7.22). The Krieger phantom design is (a), Figure 7.23, whereas (b) is the Meertens phantom design, Figure 7.24.

Secondly, the accuracy in the source position definition along the catheter axis fortunately has a very low influence, as has been shown previously in the discussion on the in-air set-up (see Table 7.4) and for the measuring geometries shown in Figure 7.2. This is also the situation for all available solid phantoms. Thus a positioning error of several millimeters (1 to 5 mm) causes a maximum uncertainty of 1.0% at a measuring distance of 5.0 cm.

7.2.3.1.3 Room Dimensions

Solid phantoms have limited dimensions (see Table 7.14) and depending on the individual design, the scattering originating from the room walls and floor can influence the measurement result. However, since there is always a certain amount of phantom material between the chamber and phantom surface (Figure 7.22) this effect is much lower than for in-air calibration set-ups.

7.2.3.2 Ionization Chambers

Compact cylindrical or thimble ionization chambers with collecting volumes in the range 0.3 to 1.0 cm^3, such as for in-air calibration set-ups, can be used for in-phantom measurements. Chambers calibrated either for air kerma or absorbed dose in water or for exposure can be considered, depending on the requirement for radionuclide calibration factors at various energies.

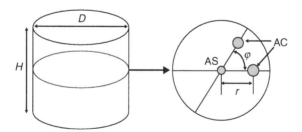

FIGURE 7.22
Simplified schematic of a cylindrical solid phantom of height H and diameter D. The transaxial cut through the geometrical center of the phantom, the reference plane that contains the center of the source and the chamber reference point, is shown in more detail at the right of the figure. In the center is the adapter for inserting the catheter for source positioning AS. At a radial distance r (center to center) is the adapter for the ionization chamber AC. To reduce the effect of positioning uncertainties, averaging over several equally spaced positions of the ionization chamber at distance r from the source can be seen at angular intervals φ. Usually angles $\varphi = 90°$ or 120° are considered. It is also possible to have a single chamber position AC at the center and several source positions (AS) at distance r and angle φ intervals.

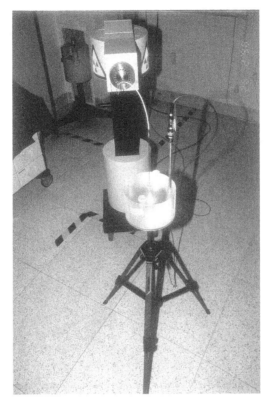

FIGURE 7.23
The PMMA cylindrical phantom of Krieger[29,31,68] shown in a measurement set-up with a ^{192}Ir source of a microSelectron HDR afterloader.

7.2.3.3 Formalism

The only detailed protocol for source calibration using solid phantoms is that published by DGMP.[29–31] This focuses on the Krieger cylindrical phantom which is manufactured of PMMA and on ^{192}Ir HDR sources. The DGMP formalism can be generalized to also include other solid phantom designs as well as other radionuclides of higher or similar photon energies such as ^{60}Co, ^{137}Cs, and [198]Au or intermediate energies such as [169]Yb.

Considering an in-phantom calibration set-up such as that described in Figure 7.22 (see also Figure 7.23 through Figure 7.25), and assuming a source to detector distance r, the reference air kerma rate, $\dot{K}_R$, or equivalently the air kerma strength, S_K, are determined from the detector measurement M using the following equations.

Chamber calibrated in air kerma

$$\dot{K}_R = N_K M k_u k_r k_\rho k_T k_p k_{ion} k_V k_{wall} k_{appl} k_r k_{ap} k_{ph} \tag{7.24}$$

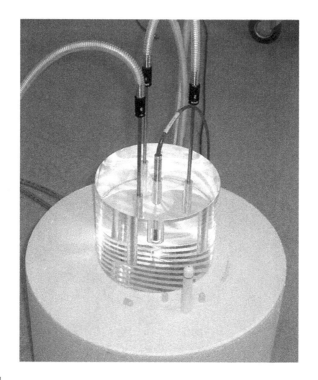

FIGURE 7.24
The PMMA cylindrical phantom of Meertens[70] shown in a measurement set-up placed on a foam block with an ionization chamber positioned centrally and three stainless steel catheters at 120° for Selectron-LDR ^{137}Cs sources.

or

$$S_K = N_K M k_u k_\tau k_\rho k_T k_p k_{\text{ion}} k_V k_{\text{wall}} k_{\text{appl}} k'_r k_{\text{ap}} k_{\text{ph}} \tag{7.25}$$

Chamber calibrated in absorbed dose to water

$$\dot{K}_R = N_W \left(\frac{1}{1 - g_\alpha} \right) \left(\frac{\mu_{\text{en}}}{\rho} \right)_W^\alpha M k_u k_\tau k_\rho k_T k_p k_{\text{ion}} k_V k_{\text{wall}} k_{\text{appl}} k_r k_{\text{wp}} k_{\text{ph}} \tag{7.26}$$

or

$$S_K = N_W \left(\frac{1}{1 - g_\alpha} \right) \left(\frac{\mu_{\text{en}}}{\rho} \right)_W^\alpha M k_u k_\tau k_\rho k_T k_p k_{\text{ion}} k_V k_{\text{wall}} k_{\text{appl}} k'_r k_{\text{wp}} k_{\text{ph}} \tag{7.27}$$

Chamber calibrated in exposure

$$\dot{K}_R = N_X \left(\frac{1}{1 - g_\alpha} \right) \left(\frac{W}{e} \right) k_{\text{RC}} M k_u k_\tau k_\rho k_T k_p k_{\text{ion}} k_V k_{\text{wall}} k_{\text{appl}} k_r k_{\text{ap}} k_{\text{ph}} \tag{7.28}$$

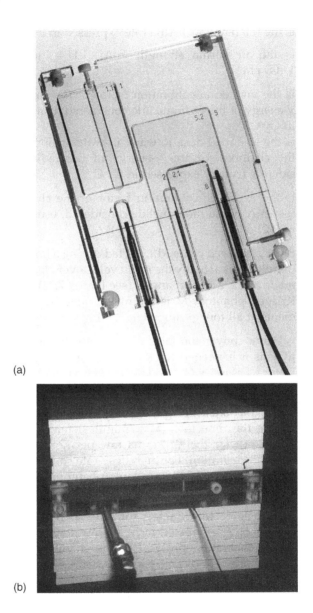

(a)

(b)

FIGURE 7.25
The PMMA plate phantom of Baltas.[28,30,69] (a) Details of the phantom where the chamber adapter is clearly seen. (b) The phantom embedded in RW-3 (Polystyrol + 2% TiO_2) plates (30 × 30 cm).

or

$$S_K = N_X \left(\frac{1}{1 - g_\alpha} \right) \left(\frac{W}{e} \right) k_{RC} M k_u k_T k_\rho k_T k_p k_{ion} k_V k_{wall} k_{appl} k'_r k_{\alpha p} k_{ph} \quad (7.29)$$

$\dot{K}_R$:   is the reference air kerma rate expressed in μGy h^{-1} at 1 m.

S_K: is the air kerma strength in U, $1\,U = 1\,\mu$Gy m^2 h^{-1} = 1 cGy cm^2 h^{-1}.

N_K: is the air kerma calibration factor of the chamber for the γ-energy of the radionuclide considered, usually expressed in Gy C^{-1}.

N_W: is the absorbed dose to water in-water calibration factor of the chamber for the γ-energy of the radionuclide considered, usually expressed in Gy C^{-1}.

N_X: is the exposure calibration factor of the chamber for the γ-energy of the radionuclide considered, usually expressed in R C^{-1}.

M: is the measured charged collected during a time interval of τ minutes. It is actually the mean value over all measurements made at the different angles (see Figure 7.22). In the case of Krieger, phantom M is the mean value over all measurements at all four positions ($\phi = 90°$).

k_u: is a unit conversion factor, to convert the product $N_K M$ to μGy as it is needed for $\dot{K}_R$ and S_K. In case the calibration factor N_K is in Gy C^{-1} and the measured charge M in C, then $k_u = 1.0 \times 10^6$.

$k_\tau = 60/\tau$: is a factor to extrapolate the electrometer readout (charge) during the time interval τ (minutes) for 60 min duration to account for the air kerma rate per hour for the source strength definition. For high dose rate sources such as ^{192}Ir and distances r of the available phantoms (5.0 to 8.0 cm), τ is normally 1 min.

k_p: is the correction factor for the current air pressure conditions P, expressed in hPa, other than that referred to in the calibration protocol of the chamber, which is the normal pressure of 1013 hPa:

$$k_p = \left(\frac{1013\ \text{hPa}}{P} \right)$$

k_T: is the correction factor for the current air temperature conditions T in °C other than that referred to in the calibration protocol of the chamber, i.e., the normal temperature T_0 in °C:

$$k_T = \left(\frac{273.15 + T\ (°C)}{273.15 + T_0\ (°C)} \right)$$

This is usually $T_0 = 20°C$. Alternatively to air pressure and temperature correction factors, a low activity check radioactive source (^{90}Sr) can been used to determine $k_p k_T$ according to DGMP recommendations.[10,29,31] This is the same source as used in the calibration protocol.

k_p: is the correction factor for the polarity effect of the bias voltage for the photon energy of the radionuclide, see also Equation 7.6 and Equation 7.7.

k_{ion}: is the correction factor that accounts for the unsaturated ion collection efficiency and thus for the charge lost to recombination for the specific radionuclide photon energy and the applied nominal voltage V. As has been discussed for the other two calibration set-ups, k_{ion} is the reciprocal of the ion collection efficiency A_{ion}. For details see Equation 7.6 and Equation 7.7.

k_V: is a correction factor to account for the effect of the chamber's finite size (i.e., volume) when the center of the chamber air cavity volume is considered as the point of reference for the positioning of the chamber and thus a reference for the measurement result.

k_{wall}: is the correction factor accounting for attenuation and scattering effects of the chamber wall, so far as this is not taken into account in the corresponding calibration factor. This must be considered as described in the section on in-air calibration set-up.

k_{appl}: is the correction factor to account for the attenuation in the applicator/catheter wall, used for positioning or fixing the source in the in-phantom set-up.

$k_r = (r/r_0)^2$: is the correction factor for the used radial distance r expressed in meters between source and ionization chamber (see Figure 7.22) with r_0 being the reference distance for the reference air kerma rate source specification, $r_0 = 1.0$ m.

$k'_r = r^2$: for the air kerma strength S_K calibration r is expressed either in cm or in m, resulting in S_K expressed in U, where $1\ U = 1\ \mu Gy\ m^2\ h^{-1} = 1\ cGy\ cm^2\ h^{-1}$.

$k_{\alpha p}$: is the perturbation correction factor accounting for differences when changing from a surrounding medium of air to one of a phantom material. This is chamber type dependent. For ^{192}Ir and PMMA material a good approximation is $k_{\alpha p} = 1.0$.[10,29,31]

k_{wp}: is the perturbation correction factor accounting for differences when changing from a surrounding medium of water to one of a phantom material. This is chamber type dependent. For phantom materials with density and effective atomic number which closely approximate that of water, $k_{wp} = 1.0$.[10,42]

g_α: is the energy fraction of the electrons liberated by photons in air, that is lost to radiative processes (mostly bremsstrahlung). g_α for different photon energies according to DIN 6814-3[11] is given in Table 7.9. g_α is for the photon energies relevant to brachytherapy very low and is virtually zero. For the highest energy, which is that of ^{60}Co, $g_\alpha = 0.00315$, and for ^{192}Ir which has an effective energy of approximately 0.4 MeV, it is $g_\alpha = 0.00097$.

W/e: is the ionization constant, $W/e = 33.97$ J C^{-1} and is energy independent for energies above 1 keV, (see Table A.2.2 in Appendix 2) with e being the elementary charge and W the mean energy expended in air per ion pair formed.[4]

k_{RC}: is the roentgen to C kg^{-1} conversion factor (see Table A.2.1 in Appendix 2 and Equation 7.19), with

$$k_{RC} = 2.58 \times 10^{-4} \text{ C kg}^{-1} \text{ R}^{-1}$$

$(\mu_{en}/\rho)^\alpha_w$: is the ratio of mass energy absorption coefficient of air to that of water. Table 7.15 summarizes values for the different brachytherapy radionuclides based on the corresponding effective energies listed in Table 5.2.

k_{ph}: is the correction factor to account for the existence of the phantom material and hence, some absorption and scattering effects, when the material is compared to air.

TABLE 7.15

Values of the Ratio of Mass Energy Absorption Coefficient of Air to That of Water $(\mu_{en}/\rho)^\alpha_w$

Parameter	Radionuclide							
	^{137}Cs	^{60}Co	^{198}Au	^{125}I	^{192}Ir	^{103}Pd	^{170}Tm	^{169}Yb
Effective energy E_{eff} (MeV)	0.652	1.257	0.417	0.028	0.398	0.021	0.067	0.131
$\left(\dfrac{\mu_{en}}{\rho}\right)^\alpha_w$	0.898	0.898	0.899	0.987	0.899	0.981	0.942	0.905

For the different brachytherapy radionuclides using the effective energies of Table 5.2 and mass energy absorption coefficients for dry air and liquid water according to Hubbell and Seltzer,[40] where a log-log interpolation has been applied to these tables.

7.2.3.3.1 Calibration Factors N_K, N_W and N_X

For the calibration factors, the discussion and methods described for the in-air calibration set-up are also valid here. The only exception is that, effectively, the chamber wall replaces the build-up cap. The chambers are in direct contact with the phantom material as in the case of external beam radiotherapy calibration procedures in which solid phantoms are also used for dosimetry.

However, it is emphasized that, in general, additional calibration points (i.e., calibration factors at specific photon energies) to those usually given for ^{60}Co or ^{137}Cs are required, depending on the energy of the radionuclide being considered.

7.2.3.3.2 Chamber Finite Size Effect Correction Factor k_V

The correction factor according to Kondo and Randolph[51] for sideward irradiation geometry (Figure 7.22 through Figure 7.25), as described by Equation 7.14 and listed in Table 7.7, for different ionization chambers is considered to account for the field inhomogeneity within the chamber volume. This correction was derived for in-air set-ups and assumes an isotropic distribution of secondary electrons in the chamber wall and medium. This formulation can be used with reasonable accuracy.[29,31]

Steggerda and Mijnheer[63] recommend the use a replacement correction factor, originally denoted as p_r to account for the finite size of the chamber air volume. In particular, they state that the use of the Kondo and Randolph correction is not appropriate for this purpose in solid phantoms. They calculated values experimentally for p_r and for a Farmer-type ionization chamber NE2571 (see Table 7.6 for further details).

Steggerda and Mijnheer[63] used ^{137}Cs spherical sources from a Selectron-LDR afterloader and the Meertens phantom (Figure 7.24) with a source to chamber distance of $r = 5.0$ cm. In each of the three catheters at a 120° angle, two sets of five ^{137}Cs spheres were used so as to reduce gradients at the location of the chamber volume. Measurements with ^{192}Ir seeds, with ^{192}Ir wires of up to a length of 7.0 cm, and with a ^{192}Ir HDR source (source length of 4.0 mm) were also made. They found p_r values of 1.007 and 0.997 for the ^{192}Ir HDR source and for the ^{137}Cs sphere configuration, respectively. The K_V value at 5.0 cm from a point source for the NE2571 chamber is, according to Table 7.7, $K_V = 1.018$.

When short measurement distances are used and sources with lengths that cannot be considered as point sources, Equation 7.14 should be expanded to account for the source active volume. The simplest way is to subdivide the source into subelements which are small enough to be considered to be point sources. The procedure is then to integrate Equation 7.14 over the source volume. By doing so, one observes that for an increasing source length the correction factor K_V decreases. In the case of the PTW 23332 chamber with a 0.3 cm^3 volume, four K_V values of 1.010, 1.009, 1.009, and 1.008 were obtained. They were calculated in the manner described for the four sources: point source; source of 3.5 mm length and 0.06 cm diameter core; source 1.15 cm length and 0.15 cm diameter core; and source 2.4 cm length and 0.15 cm diameter core. All calculations were for a distance from source center to chamber center of 5.0 cm and sideward irradiation geometry. Taking these into account, the p_r value for an ^{192}Ir HDR source given by Steggerda and Mijnheer agrees to within 1% of the K_V value given in Table 7.7.

Elfrink et al.[33,36] published p_r values of 1.016 and 0.997 for the ^{192}Ir HDR source and the ^{137}Cs sphere sources, respectively, for the Meertens phantom. The value for the ^{192}Ir sources, which can be considered as point sources, is in very good agreement (+0.2%) with the K_V value of 1.018 given in Table 7.7. This is for a measurement at 5.0 cm and for the NE2571 chamber which was used with that phantom.

In addition, since for the ^{137}Cs measurements, source length in two parts of 0.75 cm which are separated by 3.5 cm were used, it is to be expected that source length dependent K_V values are necessary for measurements at the larger distances of 8.0 to 10.0 cm.

7.2.3.3.3 Applicator/Catheter Wall Attenuation Correction Factor k_{appl}

When applicators are considered which introduce additional attenuation, e.g., metallic applicators, the applicator attenuation factor can be easily estimated by comparing measurements using this specific applicator and a plastic applicator with a very thin wall so that it can be assumed that it has a negligible attenuation effect (material and wall thickness). For this purpose a suitable phantom adapter for the relevant type of catheter must be available.

Since k_{appl} depends only on the catheter material and the emitted photon energy, this factor can be accurately estimated using the effective energy of the source photon energy spectrum and the corresponding absorption coefficient of the catheter material.

7.2.3.3.4 Phantom Correction Factor K_{ph}

This correction factor takes into account the attenuation and scattering effects which occur within the phantom at the measurement distance. Its value depends on the phantom material, measurement geometry, and source photon energy spectrum. This factor can be estimated theoretically or by using a Monte Carlo simulation for the scattering and attenuation effects within the phantom.

Alternatively, a calibrated (PSDL or SSDL) source of the same type can be used in the phantom measurement set-up and then the appropriate equation from the series, Equation 7.24 through Equation 7.29, can be solved for the phantom correction factor K_{ph} based on the known source strength at the time of measurement. The same procedure is also applicable if, firstly, an in-air calibration or a well-type chamber calibration is performed and the same source is then measured with the solid phantom. For both cases the source strength at the time of the in-phantom measurement must be calculated using the half-life of the radionuclide.

Krieger published a correction factor $K_{ph} = 1.187 \pm 0.012$[29,31,71] for his PMMA phantom for the measurement distance of $r = 8.0$ cm and a ^{192}Ir miniaturized source (HDR or PDR). This value was estimated comparing the in-phantom measurement to in-air calibration for the same source according to the method described here. This value includes the chamber's finite size effect correction factor k_V for the PTW 23332 type ionization chamber.[31] According to Table 7.7, $k_V = 1.004$ for that chamber and a 8.0 cm measurement distance. Thus, the chamber independent phantom correction factor as used in Equation 7.24 through Equation 7.29 is $K_{ph} = 1.182$. This value can also be derived theoretically.

When scaled according to the PMMA density, the 8.0 cm PMMA material, $\rho = 1.19$ g cm^{-3} (see Table 7.5), corresponds to a 9.52 cm path in water. Using the scatter and absorption correction factor for the ^{192}Ir HDR source of the microSelectron-HDR afterloader, a correction of ~ 1.121 has to be applied for the distance of 8.0 cm in PMMA. In addition, the missing scattering effect (the cylindrical phantom has limited outer dimensions so that backscatter is missing) is experimentally measured and results in an additional correction of 1.058. Measurements were made with the standard set-up of the phantom and also with the cylindrical phantom placed within a $40 \times 40 \times 30$ cm^3 water filled plastic tank. Such a water tank is assumed to result in providing full scattering conditions at the location of the chamber. The ratio of the two measurements is the missing scattering correction factor. Thus the phantom correction factor should be

$$K_{ph} = 1.121 \times 1.058 = 1.186$$

This is only 0.3% higher than the experimentally estimated value of 1.182.

For the plate phantom (Figure 7.25), Baltas et al.[30] have published a phantom correction factor $K_{ph} = 1.048$ for ^{192}Ir HDR sources. This factor was determined in an initial calibration set-up taking a previous calibration of the same source with an HDR-1000 well-type chamber as a secondary standard. However, it is emphasized that this phantom correction factor is only valid with the plate phantom embedded within 30×30 cm^2 PMMA plates which provide a total phantom thickness of 24 cm. This is in order to achieve saturation of phantom scattering effects (see also Figure 7.26). For smaller thicknesses, an additional correction is required for missing scattering according to Figure 7.26.

Meertens[70] published a K_{ph} value of 1.041 and of 1.045 estimated experimentally and theoretically, respectively, for ^{137}Cs LDR sources. Both values already take into account the attenuation effect of the stainless steel catheters used with the ^{137}Cs LDR sources. The experimental value of 1.041 was referred to in Report No. 13 of The Netherlands Commission on Radiation Dosimetry.[33,36] The phantom correction published for ^{192}Ir sources for the Meertens phantom is $K_{ph} = 1.033$.[33,36] Table 7.16 summarizes all these phantom correction factors.

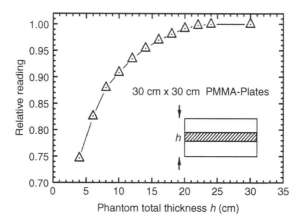

FIGURE 7.26

Influence of the total plate phantom thickness h on the measurement results for the Baltas phantom for an ^{192}Ir source of a microSelectron-HDR afterloader. As seen in this figure there is phantom scattering saturation for total thicknesses above 24 cm.

7.2.3.3.5 Room Scatter Effects

Attention must be paid to the influence of scattered radiation originating at the room wall or floor on the measurement. Due to the fact that in the Meertens phantom the chamber is at the center of the phantom, and due to the large height of 15.0 cm of that phantom, it is expected that these effects are very small for this design of phantom.

Experimental data on the influence of the room wall scattered radiation are available for the Krieger and Baltas phantoms for ^{192}Ir sources.[30] Figure 7.27 summarizes these data. According to this figure, a minimum lateral distance from the walls of 15 and 75 cm is required, respectively, for the Baltas and Krieger phantoms in order to keep the scattering contribution below 0.1% of the chamber readout. Both phantoms were placed 1.0 m above the floor.

TABLE 7.16

Phantom Correction Factors k_{ph} for the Three PMMA Phantoms

	Solid Phantom		
Source/Radionuclide	Krieger Cylindrical Phantom	Baltas Plate Phantom	Meertens Cylindrical Phantom
^{192}Ir (miniaturized sources, HDR, PDR)	1.182[29,31,68 a]	1.048[30]	1.033[33]
^{137}Cs (LDR spherical sources, 10 spheres)	—	—	1.041[33,70]

[a] Corrected to exclude the chamber finite size effect for the PTW 23332 type ionization chamber from the originally published value of 1.187.[31]

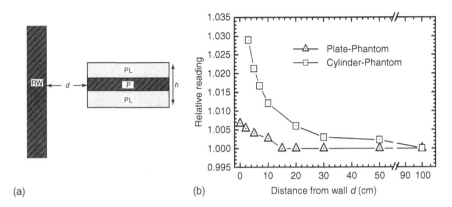

(a) (b) Distance from wall d (cm)

FIGURE 7.27
Influence of lateral wall scattering on the measurements for the two solid phantoms using a ^{192}Ir source of a microSelectron-HDR afterloader. (a) Simplified drawing of the experimental set-up for the case of the plate phantom (P) embedded in PMMA plates (PL) resulting in a total phantom thickness $h = 24$ cm which is required to achieve phantom scattering saturation (see Figure 7.26). Both phantoms have been placed 1 m above the floor to guarantee negligible influence of the floor scattering. The distances d given in this figure are measured from the outer lateral surface of the phantom to the room wall surface (RW). (b) Chamber readings have been normalized to a response for $d = 100$ cm. The minimum chamber to room wall distance for the cylindrical phantom system was limited to only 3.0 cm due to the photo tripod used routinely to position that phantom at 1 m above the floor (see Figure 7.23).

The plate phantom was embedded within 30×30 cm^2 PMMA plates resulting in a total phantom thickness of 24 cm and thus achieving saturation of phantom scattering (see Figure 7.26). The minimum chamber to room wall distance for the cylindrical phantom system was limited to only 3.0 cm due to the tripod routinely used to position that phantom (see Figure 7.23). The maximum wall scatter contribution was 0.7% for direct contact to the wall ($d = 0$ cm in Figure 7.27) for the plate phantom and 2.9% at the minimum distance of 3.0 cm for the cylindrical phantom. The significantly higher effect for the Krieger phantom can be easily explained due to the peripheral placement of the ionization chamber at a distance of 8.0 cm from the phantom center and source, thus resulting in a minimum of 2.0 cm backscattering material (PMMA).

The effect of room scatter depends on the energy spectrum of the source and the scatter is lower the lower the energy. Generally, distances above 75 cm from room scattering regions of floor and walls for in-phantom calibration set-ups are recommended.

7.2.3.3.6 Other Effects

As with well-type chambers, it is important that the phantom temperature is in equilibrium with the ambient/air temperature. Furthermore, if a radioactive check source is used to obtain the $k_\rho k_T$ correction, then it must be ensured that this source has also achieved room air temperature. To do so, the optimum method is to leave the phantom, chamber, and check source overnight in the room where the calibration will take place. In any other case, correction factors must be implemented to account for these temperature differences.

7.2.3.4 Summary

Solid phantoms offer a stable and easy method of implementing source calibration alternative to in-air methods, except for low energy and low activity sources such as ^{125}I and ^{103}Pd. Phantoms also provide a high degree of geometrical accuracy and reproducibility.

Ionization chambers used for the external beam radiotherapy dosimetry procedures and volumes 0.3 to 1.0 cm^3 have been discussed. As in the case for the in-air calibration method, depending on radionuclide energy, multiple calibration points can be required for interpolating chamber calibration factors, especially for ^{198}Au ^{192}Ir, and ^{169}Yb sources.

The set-up independent factors k_P and k_{ion} only have to be estimated once and then can be applied for any other measurement using the same source type and ionization chamber. If the same applicator type, or even the same applicator, is always used then k_{appl} also only has to be estimated once and can then be applied for future set-ups.

Attention must be paid in providing or estimating the phantom correction factor that is dependent on the material, on the phantom geometry, and on the radionuclide/source. Special care should be taken when sources of extended length are used, since these are not covered by the K_V correction for a specific chamber, as described by Kondo and Randolph[51] (see Equation 7.14). In these cases the finite source volume has also to be considered.

If room wall or floor scattering influences the measurement (although if at all possible this must be avoided), the effect of this scattering must be experimentally estimated and a room scatter correction factor k_{scatt} as for set-ups using well-type chambers has to be applied.

Table 7.17 provides a summary of the source/radionuclide, chamber, and geometrical environment including phantom dependent correction factors which are considered in Equation 7.24 through Equation 7.29.

The required equipment for the in-phantom set-up, where the phantom is a solid phantom, is as follows:

- *Solid phantom*
  With adapters for the insertion of the ionization chamber and applicator/catheter, as well as dummy adapters for filling empty spaces (multiple positions). The phantom related correction factor for the specific radionuclide/source has to be provided.

- *Build-up material or support*
  Depending on the phantom type, build-up material or a support (tripod) for positioning the phantom and achieving the required geometrical conditions.

TABLE 7.17

Summary of Correction Factors to Be Handled Carefully When Applying in Solid Phantom Calibration Set-Ups and Their Dependency on Source (Radionuclide), Chamber and Set-Up Environment Conditions

Correction Factor		Parameter		
Name	Symbol	Source/ Radionuclide	Chamber	Set-Up Environment[a]
Calibration factor of the chamber for the photon energy spectrum of the radionuclide	N_K, N_W, N_X	+	+	−
Polarity effect	k_p	+	+	−
Unsaturated ion collection efficiency	k_{ion}	+	+	−
Chamber finite size	k_V	−	+	+
Chamber wall	k_{wall}	+	+	−
Scattered radiation from room walls and floor	k_{scatt}	+	+	+
Attenuation in the applicator/catheter wall	k_{appl}	+	−	+
Phantom correction factor	k_{ph}	+	−	+

[a] Also includes the solid phantom itself.

- *Ionization chamber*
  Compact/thimble 0.3 to 1.0 cm^3 ionization chamber with appropriate calibration factors for at least two energy points which encompass the effective energy of the source photon spectrum.

- *Electrometer*

- *Thermometer and barometer or alternatively a check source*

- *Applicator/catheter*
  Of known attenuation for the specific source.

- *Adequate treatment room*
  Room and geometrical conditions that will enable the positioning of the phantom under negligible room walls and floor scatter influences. This can be achieved generally, keeping a minimum distance of 1 m from all possible scattering sources.

7.2.4 Comparison of Methods

Although different in the ionization chambers used, set-ups and correction factors, all three calibration methods, in-air, using a well chamber, and in-solid phantom, should result in the same source strength results for the same source. However, a precondition for this is that all correction factors are correctly known and all recommendations are correctly implemented.

Until the end of the 1980s the calibration of ^{192}Ir brachytherapy sources was a major problem for clinical physicists, mainly due to missing calibration factors for the polyenergetic spectrum of this radionuclide and no existing national or international standards. However, the establishment of national and international protocols during the 1990s, as described in detail in this chapter, introduced standardization for calibration procedures, especially for HDR and PDR ^{192}Ir brachytherapy sources.

In 1999, Baltas et al. published results on a comparison of well-type chamber and in-solid phantoms based calibration protocols for ^{192}Ir sources for microSelectron-HDR afterloaders. The HDR-1000 and SDS well chambers, as well as the PMMA phantoms of Krieger (cylindrical) and of Baltas (plate phantom), were considered in this study. The ^{192}Ir source strength used in these measurements was 35.48 mGy h^{-1} or equivalently 35.48×10^3 U. The plate phantom was used in a set-up with PMMA build-up plates with a total thickness of 24 cm as recommended (see Figure 7.26). For the measurements in both phantoms, a PTW type 23332 ionization chamber was used which was calibrated in absorbed dose in water. The measured collection efficiency at 8.0 cm distance was $A_{ion} = 0.9997$ resulting in a correction factor of $k_{ion} = 1.000$ and a polarity effect of $k_p = 0.997$.

The metallic dosimetry applicator relevant to Figure 7.10 was used for all measurements in the phantoms. The ^{60}Co calibration factor was used since a low energy dependence (1%) is observed for this chamber type (Figure 7.5c) in the range ^{60}Co to 145 keV photon effective energies. The corrections factors $k_V = 1.004$ (Table 7.7), $k_{wp} = 1.0$, and $k_{appl} = 1.017$ for the metallic applicator (see Figure 7.10) were used for both PMMA phantoms. The phantom correction factors k_{ph} are listed in Table 7.16.

To minimize source positioning dependent uncertainties, a plastic 5F catheter was always inserted into the dosimetry applicator and thus the maximum error was reduced to $\pm 0.7\%$ for the plate phantom. For the longitudinal direction of the applicator, this error remained within $\pm 0.1\%$ for a longitudinal shift of ± 2.5 mm, i.e., one source step for the micro-Selectron-HDR afterloader.

Within a period of 5.5 years (1992–1997) a total of 20 ^{192}Ir HDR sources have been used in two institutions. The total number of contributions of each separate system and the calibration results are given in Table 7.18. The HDR-1000 well chamber was considered as a reference.

The residual mean square (RMS), was used for the analysis of the results. RMS is calculated according to

$$\text{RMS} = \frac{1}{N} \sum_{i=1}^{N} \left[\left(\frac{\dot{K}_{R,system}}{\dot{K}_{R,HDR-1000}} \right)_i - 1 \right]^2 \qquad (7.30)$$

where $\dot{K}_{R,system}$ is the reference air kerma rate measured with any of the systems and $\dot{K}_{R,HDR-1000}$ is the reference air kerma rate measured with

TABLE 7.18

Overall Percentage Root Mean Square Error Values (RMS, see Equation 7.30)
for Each Calibration System Obtained by Taking the Results of the HDR-1000
Well-Type Chamber as a Reference

Calibration System	RMS	Number of Measurements
HDR-1000	—	108
SDS	0.11% [−0.0028, 0.0036]	100
Cylindrical phantom	0.44% [−0.0096, 0.0094]	103
Plate phantom	0.60% [−0.0113, 0.0115]	54

The values in brackets are the ranges of the fractional errors. A total of 20 ^{192}Ir sources were used
for these measurements.

the HDR-1000 well chamber during the same ith comparison set-up. N is
the total number of comparison measurements.

A maximum RMS of 0.6% was observed for the plate phantom system.
This is explained by the higher expected geometrical fluctuation of the
source position within the catheter resulting in a maximum inaccuracy of
$\pm 0.7\%$. The cylindrical phantom showed an RMS value of 0.44%. The best
results, as expected, were observed for the SDS well-type chamber with an
RMS value of 0.11%.

These results underline the equivalence of these calibration methods
under the extreme conditions of involving seven dosimetrists in two
institutions with two afterloaders, a total of 20 sources and up to a total of
108 individual calibration set-ups (see Table 7.18).

As part of this investigation, one source with the production number
29578-106 (source number 664) after it has been calibrated in the
Radiotherapy Clinic and before returning it to the manufacturer (Mallinckrodt Medical B.V., Petten, The Netherlands), was sent to PTB for a
comparison calibration according to the PTB protocol.[47]

PTB used a secondary standard ionization chamber LS01 1000 cm^3
spherical ionization chamber (Austrian Research Centre, Seibersdorf,
Austria) calibrated against the PTB's primary standard and an in-air
experimental set-up at 1 m distance from the source. The chamber was
calibrated for air kerma for ^{137}Cs γ-radiation. An average correction factor
corresponding to the radiation field of the ^{192}Ir source was used. This was
derived from the energy dependence of the air kerma response of this
chamber in the range 0.030 of 1.25 MeV and an emission probability and
photon fluence weighted chamber response averaging method.[47]

Table 7.19 summarizes the results of that comparison where both well-
type chambers and only the cylindrical phantom were considered. All values
have been corrected for decay and have been referred to at the time of the
first calibration measurement. All calibrations are in very good agreement
with the PTB calibration, i.e., within 0.4%. This again proves the equivalence

TABLE 7.19

Comparison of the Calibration Results for Both Well-Type Chambers and the Cylindrical Phantom with a PTB[47a] Reference Calibration (Source No. 664, Production No. 29578-106)

Calibration System	Reference Air Kerma Rate (mGy h^{-1})	Deviation from PTB
PTB	19.52	0.0%
HDR-1000	19.57	+0.3%
SDS	19.60	+0.4%
Cylindrical phantom	19.56	+0.2%

[a] Physikalisch-Technische Bundesanstalt, Braunschweig, Germany.

of all three calibration methods and the compatibility and correctness of the calibration and the correction factors.

In 1994, Venselaar et al.[72] published the results of an intercomparison study of calibration for ^{192}Ir HDR sources in centers in The Netherlands and Belgium. A total of 13 radiotherapy departments in both countries participated, where 12/13 of those used a microSelectron HDR afterloader and only 1/13 a GammaMed 12i (Isotopen-Technik Dr. Sauerwein GmbH, now Varian Medical Systems Inc., U.S.A.). All institutions used in-air calibration set-ups whereas 3/13 of these claimed to use in-PMMA phantom measurements in addition. No institution made use of a well-type chamber calibration set-up.

As a reference, the Nucletron calibration jig (Figure 7.3) and an NE 2505/3 Farmer-type ionization chamber calibrated to ^{60}Co, ^{137}Cs, and 250-kvp

TABLE 7.20

Comparison of the Three Source Calibration Methods Regarding Their Complexity, Time Costs, and Application Field

Calibration Method	Parameter				
	Complexity	Duration	Permanent or Temporary Implants Brachytherapy Sources[a]	Sources	Nuclide Energy
In-air set-up	+++++	+++++	Both	All	All
Well-type chamber	+	+	Both	All	All
In-solid phantom	++	++	Temporary	Limited according to phantom design	Intermediate and high (^{169}Yb, ^{192}Ir, ^{137}Cs, ^{60}Co)

[a] For permanent implants sources of low energy (^{125}I, ^{103}Pd) and very low source strength are used, where for temporary implants (afterloading technology) sources of intermediate to high energy and of low to high source strength (LDR, PDR or HDR) are used.

x-rays with the standard build-up cap were used. The source reference air kerma rate measured by the hospital physicists was compared to that with the reference set-up and calibration made by the authors. The air kerma calibration factor N_K for ^{192}I of the reference ionization chamber was estimated according to the method of Goetsch et al.[44] (see Section 7.2.1 and Equation 7.8 through Equation 7.10).

Based on 19 comparisons including 16 sources, the mean deviation of the in-house calibrations from the reference calibration system was $+1.3\%$ with a range of -0.4 to $+3.0\%$. Thus, in addition to the most complicated in-air calibration set-up, a high degree of agreement was demonstrated over several institutions and several sources.

A comparison of complexity, time requirements, and applicability (source design, source strength/activity, and nuclide energy) of all three calibration methods is finally presented in Table 7.20.

References

1. Wambersie, A., Prignot, A., and Gueulette, J. A propos du remplacement du radium par le caesium 137 en curiéthérapie, *J. Radiol. D'Electrol. Méd. Nucl*, 54, 261–270, 1973.
2. Dutreix, A. and Wambersie, A. Specification of gamma-ray brachytherapy sources, *Br. J. Radiol*, 48, 1034, 1975.
3. International Commission on Radiation Units and Measurements, *Average Energy Required to Produce an Ion Pair, ICRU Report 31*. ICRU, Bethesda, 1979.
4. Boutillon, M. and Perroche-Roux, A.M. Re-evaluation of the W value for electrons in dry air, *Phys. Med. Biol.*, 32:2, 213–219, 1987.
5. International Commission on Radiation Units and Measurements, *Dose and Volume Specification for Reporting Intracavitary Therapy in Gynecology, ICRU Report 38*. ICRU, Bethesda, 1985.
6. Comite Francais Measure des Rayonments Ionizants, *Recommendations pour la détermination des doses absorbées en curiéthérapie, CFMRI Report No 1*. Bureau National de Métrologie, Paris, 1983.
7. BCRU, Specification of brachytherapy sources: Memorandum from the British Committee on Radiation Units and Measurements, *Br. J. Radiol.*, 57, 941–942, 1984.
8. International Commission on Radiation Units and Measurements, *Dose and Volume Specification for Reporting Interstitial Therapy, ICRU Report 58*. ICRU, Bethesda, 1997.
9. British Institute of Radiology and Institute of Physical Sciences in Medicine, *Recommendations for Brachytherapy Dosimetry Report of a Joint BIR/IPSM Working Party*, BIR, London, 1993.
10. Deutsches Institut für Normung e.V. [DIN], *Clinical Dosimetry. Part 2. Brachytherapy with Sealed Gamma Sources (Klinische Dosimetrie. Teil 2. Brachytherapie mit umschlossenen gammastrahlenden radioaktiven Stoffen)*, DIN 6809. DIN, Berlin, 1993.
11. Deutsches Institut für Normung e.V. [DIN], *Terms in the Field of Radiological Technique - Part 3: Dose Quantities and Units (Begriffe in der radiologischen Technik. Teil 3: Dosisgrößen und Dosiseinheiten)*, DIN 6814-3. DIN, Berlin, 2001.

12. (a) International Atomic Energy Agency, *Calibration of Brachytherapy Sources IAEA-TECDOC-1079, Guidelines on Standardized Procedures for the Calibration of Brachytherapy Sources at SSDLs and Hospitals.* IAEA, Vienna, 1999. (b) International Atomic Energy Agency, *Calibration of Photon and Bety Ray Sources Used in Brachytherapy IAEA-TECDOC-1079, Guidelines on Standardized Procedures at Secondary Standards Dosimetry Laboratories (SSDLs) and Hospitals.* IAEA, Vienna, 2002.

13. European Guidelines for Quality Assurance in Radiotherapy, *A Practical Guide to Quality Control of Brachytherapy Equipment Booklet No 8*, J. Venselaar and J. Pérez-Calatayud, eds., ESTRO, Belgium, 2004.

14. American Association of Physicists in Medicine, *Specification of Brachytherapy Sources Strength, AAPM Report no. 21.* American Institute of Physics, New York, 1987.

15. Nath, R., Anderson, L.L., Luxton, G., Weaver, K.A., Williamson, J.F., and Meigooni, A.S. Dosimetry of interstitial brachytherapy sources: recommendations of the AAPM Radiation Therapy Committee Task Group No 43 AAPM report no. 51, *Med. Phys.*, 22:2, 209–234, 1995.

16. Williamson, J.F., Coursey, B.M., DeWerd, L.A., Hanson, W.F., Nath, R., Rivard, M.J., and Ibbott, G. On the use of apparent activity (A_{app}) for treatment planning of ^{125}I and ^{103}Pd interstitial sources: Recommendations of the American Association of Physicists in Medicine Radiation Therapy Committee Subcommittee on Low-Energy Brachytherapy Source Dosimetry, *Med. Phys.*, 26:12, 2529–2530, 1999.

17. Rivard, M., Coursey, B., Hanson, W., Hug, S., Ibbot, G., Mitch, M., Nath, R., and Williamson, J. Update of AAPM Task Group No 43 Report: A revised AAPM protocol for brachytherapy dose calculations, *Med. Phys.*, 31:3, 633–674, 2004.

18. Karaiskos, P., Angelopoulos, A., Pantelis, E., Papagiannis, P., Sakelliou, L., Kouwenhoven, E., and Baltas, D. Monte Carlo dosimetry of a new ^{192}Ir pulsed dose rate brachytherapy source, *Med. Phys.*, 30, 9–16, 2003.

19. Angelopoulos, A., Baras, P., Sakelliou, L., Karaiskos, P., and Sandilos, P. Monte Carlo dosimetry of a new ^{192}Ir high dose rate brachytherapy source, *Med. Phys.*, 27, 2521–2527, 2000.

20. Williamson, J.F. and Li, Z. Monte Carlo dosimetry of the microSelectron pulsed and high dose rate ^{192}Ir sources, *Med. Phys.*, 22, 809–819, 1995.

21. Kirov, A.S. and Williamson, J.F. Monte Carlo-aided dosimetry of the Source Tech Medical Model STM1251 I-125 interstitial brachytherapy source, *Med. Phys.*, 28, 764–772, 2001.

22. Williamson, J.F. Monte Carlo modelling of the transverse-axis dose distribution of the Model 200 ^{103}Pd interstitial brachytherapy source, *Med. Phys.*, 27, 643–654, 2000.

23. Karaiskos, P., Papagiannis, P., Sakelliou, L., Anagnostopoulos, A., and Baltas, D. Monte Carlo dosimetry of the selectSeed ^{125}I interstitial brachytherapy seed, *Med. Phys.*, 28, 1753–1760, 2001.

24. Piermattei, A., Azario, L., Rossi, G., Soriani, A., Arcovito, G., Ragona, R., Galelli, M., and Taccini, G. Dosimetry of ^{169}Yb seed model X1267, *Phys. Med. Biol.*, 40, 1317–1330, 1995.

25. Mason, D.L., Battista, J.J., Barnett, R.B., and Porter, A.T. Ytterbium-169: calculated physical properties of a new radiation source for Brachytherapy, *Med. Phys.*, 19, 695–703, 1992.

26. Borg, J. and Roggers, D.W.O. Spectra and air kerma strength for encapsulated ^{192}Ir sources, *Med. Phys.*, 26, 2441–2444, 1999.

27. American Association of Physicists in Medicine, *Remote Afterloading Technology, AAPM Report no 41*. American Institute of Physics, New York, 1993.

28. Baltas, D. Quality Assurance in Brachytherapy with special Reference to the MicroSelectron-HDR, *Activity, Int. Selectron Brachytherapy Journal*, Special Report No. 2, 1993.

29. Krieger, H. Messung der Kenndosisleistung punkt- und linienförmiger HDR-^{192}Ir-Afterloadingstrahler mit einem PMMA-Zylinderphantom, *Z. f. Med. Phys.*, 1, 38–41, 1991.

30. Baltas, D., Geramani, K., Ioannidis, G., Hierholz, K., Rogge, B., Kolotas, C., Müller-Sievers, K., Milickovic, N., Kober, B., and Zamboglou, N. Comparison of Calibration Procedures for ^{192}Ir HDR Brachytherapy Sources, *Int. J. Rad. Oncol. Biol. Phys.*, 3:43, 653–661, 1999.

31. H. Krieger and D. Baltas, eds., Deutsche Gesellschaft für Medizinische Physik, Praktische Dosimetrie in der HDR, *Brachytherapie DGMP, Bericht Nr. 13*, 1999.

32. Balta, S., Ertan, E., Kober, B., and Müller-Sievers, K. Apparative Voraussetzungne für eine Kurzzeit-Afterloading-Therapie, *Z. Med. Phys.*, 1, 100–102, 1991.

33. Elfrink, R.J.M., Kolkman-Deurloo, I.K.K., van Kleffens, H.J., Rijnders, A., Schaeken, B., Aalbers, T.H.L., Dries, W.J.F., and Venselaar, J.L.M. Determination of the accuracy of implant reconstruction and dose delivery in brachytherapy in The Netherlands and Belgium, *Radiother. Oncol.*, 59, 297–306, 2001.

34. Yu, Y., Anderson, L.L., Li, Z., Mellenberg, D.E., Nath, R., Schell, M.C., Waterman, F.M., Wu, A., and Blasko, J.C. Permanent prostate seed implant brachytherapy: Report of the American Association of Physicists in Medicine Task Group No. 64, *Med. Phys.*, 26:10, 2054–2076, 1999.

35. Interstitial Collaborative Working Group, *Interstitial Brachytherapy: Physical Biological and Clinical Considerations*, Raven Press, New York, 1990.

36. Netherlands Commission on Radiation Dosimetry (NCS), *Quality Control in Brachytherapy: Current Practice and Minimum Requirements Task Group Quality Control in Brachytherapy Report No. 13, Med Phys*. NCS, Delft, 2000.

37. Bundesministerium für Umwelt, Naturschutz und Reaktorsicherheit, Richtlinie Strahlenschutz in der Medizin: Richtlinie nach der Verordnung über den Schutz von Schäden durch ionisierende Strahlen (Strahlenschutzverordnung - StrlSchV) (Berlin: Hoffmann Verlag), 2002.

38. International Commission on Radiation Units and Measurements, *Stopping Powers for Electrons and Positrons, ICRU report 37*. ICRU, Bethesda, 1984.

39. Pychlau, P. Messung kleiner Aktivitäten mit Teletherapie-Ionisationskammern, *Ermittlung des Kalibrierfaktors N_e, Z. med. Phys*, 1, 194–198, 1991.

40. Hubbell, J.H. and Seltzer, S.M. *Tables of x-Ray Mass Attenuation Coefficients and Mass Energy-Absorption Coefficients Version 1.4 on http://physics.nist.gov/xaamdi*, National Institute of Standards and Technology, Gaithersburg, Originally published as NISTIR 5632, 1995.

41. Attix, H.F. *Introduction to Radiological Physics and Radiation Dosimetry*, Wiley, New York, USA, 1986.

42. Deutsches Institut für Normung e.V. [DIN], *Clinical Procedures of Dosimetry With Probe-Type Detectors for Photon and Electron Radiation. Part 2. Ionization dosimetry (Dosimessverfahren nach der Sondenmethode für Photonen- und Elektronenstrahlung. Teil 2. Ionizationsdosimetrie), DIN 6800-2*. DIN, Berlin, 1997.

43. Attix, H.F. Determination of A_{ion} and P_{ion} in the new AAPM radiotherapy dosimetry protocol, *Med. Phys.*, 11:5, 714–716, 1984.
44. Goetsch, S.J., Attix, F.H., Pearson, D.W., and Thomadsen, B.R. Calibration of ^{192}Ir high-dose-rate afterloading systems, *Med. Phys.*, 18, 462–467, 1991.
45. van Dijk, E., Kolkman-Deurloo, I.K.K., and Damen, P.M.G. Determination of the reference air kerma rate for ^{192}Ir brachytherapy sources and the related uncertainty, *Med. Phys.*, 31, 2286–2833, 2004.
46. Grimbergen, T.W.M. and van Dijk, E. Comparison of Methods for Derivation of Iridium-192 Calibration Factors for the NE 2561 & NE 2571 Ionization Chambers, *Nucletron-Oldelft Activity Report*, 7, 52–56, 1995.
47. Büermann, L., Kramer, H.M., Schrader, H., and Selbach, H.J. Activity determination of 192Ir solid sources by ionization chamber measurements using calculated corrections for self absorption, *Nucl. Instr. and Meth. in Phys. Res., A*, 339, 369–376, 1994.
48. Verhaegen, F., van Dijk, E., Thierens, H., Aalberts, A., and Seuntjens, J. Calibration of low activity ^{192}Ir brachytherapy sources in terms of reference air kerma rate with large volume spherical ionization chambers, *Phys. Med. Biol.*, 37, 2071–2082, 1992.
49. Chen, Z. and Nath, R. Dose rate constant and energy spectrum of interstitial brachytherapy sources, *Med. Phys.*, 28, 86–96, 2001.
50. Ferreira, I.H., de Almeida, C.E., Marre, D., Marechal, M.H., Bridier, A., and Chavaudra, J. Monte Carlo calculations of the ionization chamber wall correction factors for ^{192}Ir and ^{60}Co gamma rays and 250 kV x-rays for use in calibration of ^{192}Ir HDR brachytherapy sources, *Phys. Med. Biol.*, 44, 1897–1904, 1999.
51. Kondo, S. and Randolph, M.L. Effect of finite size of ionization chambers on measurement of small photon sources, *Rad. Res.*, 13, 37–60, 1960.
52. Spiers, F.W. Inverse square law errors in gamma-ray dose measurements, *Brit. J. Radiol.*, 14, 147–156, 1941.
53. James, F. *Monte Carlo Theory and Practice, Data Handling Division (CERN)*. CERN, Geneva, 1980.
54. Koonin, S. and Meredith, D. *Computational Physics Fortran Version*, Addison-Wesley, Redwood City, 1990.
55. Press, W., Flannery, B., Teukolsky, S., and Vetterling, W. *Numerical Recipes in FORTRAN 77: The Art of Scientific Computing*, Cambridge University Press, Cambridge, 1992.
56. Baltas, D., Giannouli, S., Garbi, A., Diakonos, F., Geramani, K., Ioannidis, G.T., Tsalpatouros, A., Uzunoglu, N., Kolotas, C., and Zamboglou, N. Application of the Monte Carlo integration (MCI) method for calculation of the anisotropy of ^{192}Ir brachytherapy sources, *Phys. Med. Biol.*, 43, 1783–1801, 1998.
57. Bielajew, A.F. Correction factors for thick-walled ionization chambers in point-source photon beams, *Phys. Med. Biol.*, 35, 501–516, 1990.
58. Bielajew, A.F. An analytic theory of the point-source non-uniformity correction factor for thick-walled ionization chambers in photon beams, *Phys. Med. Biol.*, 35, 517–538, 1990.
59. Maréchal, M.H., de Almeida, C.E., Ferreira, I.H., and Sibata, C.H. Experimental derivation of wall correction factors for ionization chambers used in high dose rate ^{192}Ir source calibration, *Med. Phys.*, 29, 1–5, 2002.
60. Poynter, A.J. Direct measurement of air kerma rate in air from CDCS J-type caesium-137 therapy sources using a Farmer ionization chamber, *Brit. J. Radiol.*, 73, 425–428, 2000.

61. Ezzel, G. Evaluation of calibration techniques for the MicroSelectron-HDR Activity, *The Selectron User's Newsletter*, 1, 10–14, 1989.
62. Brezovich, I.A., Popple, R.A., Duan, J., Shen, S., and Pareek, P.N. Assaying ^{192}Ir line sources using a standard length well chamber, *Med. Phys.*, 29, 2692–2697, 2002.
63. Steggerda, M.J. and Mijnheer, B.J. Replacement corrections of Farmer-type ionization chamber for the calibration of Cs-137 and Ir-192 sources in a solid phantom, *Radiother. Oncol.*, 31, 76–84, 1994.
64. Mitch, M.G., Zimmermann, B.E., Lamperti, P.J., and Seltzer, S.M. Well-ionization chamber response to NIST air kerma strength standard for prostate brachytherapy seeds, *Med. Phys.*, 27, 2293–2296, 2000.
65. Goetsch, S.J., Attix, F.H., DeWerd, L.A., and Thomadsen, B.R. A new re-entrant ionization chamber for the calibration of iridium-192 high dose rate sources, *Int. J. Rad. Oncol. Biol. Phys.*, 24, 167–170, 1992.
66. Podgorsak, M.B., DeWerd, L.A., and Thomadsen, B.R. Thermal and scatter effects on the radiation sensitivity of well chambers used for high dose rate Ir-192 calibrations, *Med. Phys.*, 19, 1311–1314, 1992.
67. National Nuclear Data Center (NNDC), *Brookhaven National Laboratory NUDAT 2.0 Electronic Version*, 2004, Available online at NNDC: www.nndc.bnl.gov/nndc/nudat/
68. Krieger, H. A new Solid State Phantom for the Measurement of the Characteristic Dose Rate of HDR Afterloading Proceedings of the International Meeting on Remote controlled Afterloading in Cancer Treatment, Sept. 6–9, 187–195, 1988.
69. Baltas, D. *Quality Assurance in Brachytherapy German Brachytherapy Conference 1992*, 1993, June 24–26, Cologne, Nucletron B.V. Veenendaal, 57–83.
70. Meertens, H. In-phantom calibration of Selectron-LDR sources, *Radiother. Oncol.*, 17, 369–378, 1990.
71. Krieger, H. Fundamental Investigations on the Dosimetry with High-Dose-Rate Sources for Afterloading Devices, *Proceedings of the International Meeting on Remote Controlled Afterloading in Cancer Treatment*, Sept. 6–9, 30–51, 1988.
72. Venselaar, J.L.M., Brower, W.F.M., van Straaten, B.H.M., and Aalbers, A.H.L. Intercomparison of calibration procedures for Ir-192 HDR sources in The Netherlands and Belgium, *Radiother. Oncol.*, 30, 155–161, 1994.
73. Gruber, P. *Material Constants for Muon Cooling Simulations European Organisation for Nuclear Research CERN/NUFACT Note 023, Geneva, Switzerland*, 2000.
74. Lide, D.R. *Handbook of Chemistry and Physics, 2004–2005*, 85th ed., CRC Press, Boca Raton, FL, 2004.
75. Hiraokat, T., Kawashimat, K., Hoshinojand, K., and Bichselt, H. Energy loss of 70 MeV protons in tissue-substitute materials, *Phys Med Biol*, 39, 983–991, 1994.
76. Deutsches Institut fur Normunge.V. [DIN], *Clinical dosimetry—Part 5: Application of x-Rays With Peak Voltages Between 100 and 400 kV in Radiotherapy*, DIN 6809-5. DIN, Berlin, 1996.
77. Constantinou, C., Attix, F.H., and Paliwal, B.R. A solid water phantom material for radiotherapy x-ray and gamma-ray beam calibrations, *Med. Phys.*, 9, 436–441, 1982.
78. Luxton, G. Comparison of radiation dosimetry in water and in solid water phantom materials for I-125 and Pd-103 brachytherapy sources: EGS4 Monte Carlo study, *Med. Phys.*, 21, 631–641, 1994.
79. Medich, D.C., Tries, M.A., and Munro, J.J. Monte Carlo characterization of an ytterbium-169 high dose rate Brachytherapy source with analysis of statistical uncertainty, *Med. Phys.*, 33, 163–172, 2006.

8

Source Dosimetry

8.1 Introduction

The term dosimetry, as used here, means the methodology of calculating the dose rate value at a specific point from a source in a given medium, which is usually water. This enables us to calculate the dose distribution in the three-dimensional (3D) space around the source. In fact, all available sources applied in modern brachytherapy have a cylindrical geometry, and are constructed (core and encapsulation) in such a way that one can assume a cylindrical symmetry of the dose distribution with respect to their longitudinal axis.

The accuracy of the dose rate or dose calculations around the brachytherapy sources is limited by the accuracy of the dosimetric data and the parameters of the sources used. Furthermore, the accuracy of such calculations depends on the assumptions and limitations of the model/formalism considered. Thus, for the case of a cylindrical source, there is an anisotropy in the dose distribution around the source caused by the attenuation and absorption of the radiation through the source itself, and its encapsulation. This is pronounced mainly along the longitudinal axis of the source. The degree of anisotropy is reduced with increasing distance from a source, since the contribution of scattering at longer distances gains in importance and compensates partly for the attenuation and absorption of the primary radiation field. Therefore, methods or calculation models that do not consider the effect of anisotropy have limited accuracy.

8.2 Coordinate Systems and Geometry Definition

In order to be able to calculate the dose rate around a source at a specific point in space, an adequate coordinate system has to be utilized. The coordinate system used for such calculations is source-based. Figure 8.1 demonstrates the source-based coordinate systems and the related geometries for an ideal

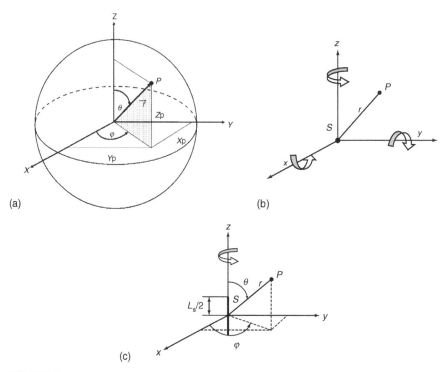

FIGURE 8.1

Source based coordinate system and the related geometries. (a) Schematic representation of the relationship between spherical coordinates and Cartesian orthogonal coordinates. If P is a point in space, then the position of P is described in the Cartesian orthogonal coordinates by the vector $\vec{r} = (x_P, y_P, z_P)$. The radial distance r of P is given by the magnitude of the position vector $\vec{r}: r = |\vec{r}| = \sqrt{x_P^2 + y_P^2 + z_P^2}$. In spherical coordinates the position vector of P is $\vec{r} = (r, \theta, \varphi)$, with θ the polar angle, φ the azimuthal angle and r the radial distance of P, where $x = r \sin \theta \cos \varphi$, $y = r \sin \theta \sin \varphi$ and $z = r \cos \theta$. (b) The case of an ideal point source S. There is a spherical symmetry (rotational symmetry around any of the three axes) and thus dose and dose rate at P depend simply on the radial distance r from the point source. This is the simplest case of 1D geometry, where for the source dosimetry any point P in space is sufficiently described by its radial distance from the source, $P = P(r)$. (c) The case of an ideal line source S of length L_s. The origin of the coordinate system is at the midpoint of the source, at the midpoint of length L_s, and the source is aligned along the z-axis. Here there is a cylindrical symmetry around the z-axis and thus dose and dose rate at any point P depend on the radial distance r from the center of the source and on the angle under which the source length L_s is viewed by that point, which is the polar angle θ. Consequently, the geometry here is a 2D geometry and any point P around the source is adequately described by its radial distance r and the polar angle θ, $P = P(r, \theta)$. Thus, for the case of an ideal line source, it is sufficient to know the dose distribution on the principal $y - z$-plane.

point and an ideal line source. In the first case there is a spherical symmetry and thus dose and dose rate at a point in space depend simply on the radial distance r of that point from the point source. In the second case of an ideal line source, source is simply extended on the z-axis over a length of L_s, where the midpoint of the source is positioned at the origin of the coordinate system. Here, there is a cylindrical symmetry around the z-axis and thus dose and dose rate at a point in space depend on the radial distance r of that point from the center of the source, midpoint of source line, and on the angle under which the source length L_s is viewed by that point (the polar angle θ).

Figure 8.2a shows the case of a cylindrical source consisting only of an active core of length L_s and diameter D_s, positioned in such a way that the longitudinal axis of the source lies on the z-axis and the center of the source is positioned at the origin of the coordinate system.

Here, assuming a homogeneous distribution of the activity over the total volume of the source cylinder V_s,

$$V_s = \pi (D_s^2/4) L_s$$

there is, as in the case of an ideal line source ($D_s = 0$), a cylindrical symmetry around the z-axis, and the dose or dose rate at any point depends on the radial distance from the source center r and on the polar angle θ.

In Figure 8.2b, a realistic source design consisting of a radioactive source core of length L_s and diameter D_s, and an encapsulation with a simplified cylindrical shape with an outer length of L_o and outer diameter of D_o, is shown. Once more, the position of the source radioactive core is as described above (Figure 8.2a): the longitudinal axis of the source cylinder lies on the

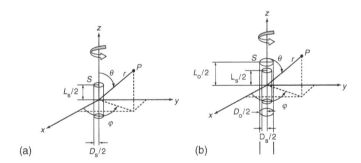

(a)   (b)

FIGURE 8.2
Coordinate system and geometry of realistic source designs. (a) The spherical coordinate system and geometry for a cylindrically shaped source S consisting only of a cylindrical radioactive core of length L_s and diameter D_s. The origin is positioned at the geometrical center of the source cylinder (active core) and the z-axis is positioned along the longitudinal axis of the source. Due to the assumed cylindrical symmetry of radioactivity distribution within the source volume this is again a 2D geometry and the dose rate at any point $P = P(r,\theta)$ depends simply on the radial distance and polar angle θ with reference to the source coordinate system. (b) The case of a cylindrical source design consisting of a cylindrical radioactive core of length L_s and diameter D_s encapsulated in a cylindrical-shaped outer clad resulting in a total outer length L_o and outer diameter D_o. The longitudinal axis of the source cylinder lies on the z-axis and the center of the source core is positioned at the origin of the coordinate system. Also, here the dose and dose rate at P depends on the radial distance r from the center of the source core and on the polar angle θ (2D geometry).

z-axis and the center of the source core is positioned at the origin of the coordinate system. Also in this case, the dose and dose rate at a point depends on the radial distance r from the center of the source core and on the polar angle θ. For high activity sources, mainly ^{192}Ir sources, used in afterloader systems with stepping source technology, the source, core, and encapsulation are fixed on a driving cable. Here, the tip of the source is at the $+z$ direction where the cable lies along the $-z$ direction.

8.3   Models of Dose Rate and Dose Calculation

The only individually measured quantity of a brachytherapy source is its strength, expressed either as reference air kerma rate $\dot{K}_R$ or air kerma strength S_K.

This gives the kerma rate in a small volume dV filled with air at a reference distance of 1 m from the center of the source in air filled space, corrected for attenuation and scattering in air (thus, *in vacuo*, see definition of kerma and kerma rate in Chapter 2, Equation 2.34 through Equation 2.38 and Section 7.1.2).

The aim of dosimetry is to accurately calculate the dose or dose rate distribution in a medium surrounding a single or several brachytherapy sources, an implant. Generally, in brachytherapy the absorbed dose to water in water filled unbounded space (water medium) around the source is considered.

The question arising is, how can the dose rate in water at a specific point around a source positioned in a water-filled space be calculated. The simplest example that we could consider to demonstrate this is the case of an ideal point source with a spherical symmetry of the radiation field around it. In this case the dose rate depends only on the radial distance r from the source (Figure 8.1a).

8.3.1   Ideal Point Sources

Based on the definitions of kerma and dose (see Section 2.5) a four-step procedure can be demonstrated to result in the desired calculation.

Step 1: At the place of point of interest P, at a distance r (Figure 8.3a) from the source *in vacuo*, a small air volume containing an air mass dm_α is considered.

The air kerma rate in dm_α can then be calculated based on the reference air kerma rate $\dot{K}_R$, and the inverse square law of change of the particle fluence, and thus of air kerma rate with the distance from the source according to

$$\dot{K}_\alpha(r) = \dot{K}_R \left(\frac{r_0}{r} \right)^2 \tag{8.1}$$

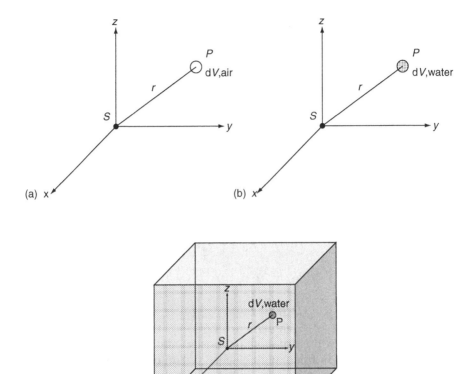

FIGURE 8.3
Schematic representation of the four different steps involved in the calculation of the absorbed dose rate to water in water at a point $P = P(r)$ around an ideal point source S. (a) Here a small air volume dV containing an air mass dm_α and cantered at P *in vacuo* (empty space) is considered. *Step 1*: Based on the reference air kerma rate $\dot{K}_R$ the air kerma rate $\dot{K}_\alpha(r)$ *in vacuo* can be calculated, Equation 8.1 and Equation 8.2. *Step 2*: The absorbed dose rate to air in the air volume dV, $\dot{D}_\alpha(r)$, is given by the collision kerma rate $\dot{K}_{col,\alpha}(r)$ in that air volume, Equation 8.3 and Equation 8.4. (b) At P, a small water filled volume dV containing a water mass dm_w is considered. *Step 3*: The absorbed dose rate to water $\dot{D}_w(r)$ at the point P *in vacuo* can be calculated from the absorbed dose to air $\dot{D}_\alpha(r)$ (Step 2) at the same place by correcting for the differences in the mass absorption coefficients between the two media assuming secondary particle (electron) equilibrium, CPE, conditions,[1] Equation 8.5 and Equation 8.6. (c) Finally, the space around point source S and point of interest P is filled with water. *Step 4*: In contrast to the situation *in vacuo* ((a) and (b) above) there are now attenuation and scatter affects that influence the radiation field at the distance r from the source that have to be considered in addition, Equation 8.7 and Equation 8.8.

or based on the air kerma strength S_K:

$$\dot{K}_\alpha(r) = S_K\left(\frac{1}{r}\right)^2 \qquad (8.2)$$

r_0 is the reference distance for $\dot{K}_R$, $r_0 = 1.0$ m. It is important to note that both r_0 and r are expressed in the same unit, commonly in cm.

Step 2: The absorbed dose rate to air in the air volume dV, $\dot{D}_\alpha(r)$, is given by the collision kerma rate (see Equation 2.42 and Section 2.5) $\dot{K}_{col,\alpha}(r)$ in that air volume

$$\dot{D}_\alpha(r) = \dot{K}_{col,\alpha}(r) = \dot{K}_\alpha(r)(1 - g_\alpha)$$

g_α is the fraction of the energy of the electrons liberated by photons in air that is lost due to radiative processes (bremsstrahlung and fluorescence). Values of g_α for different photon energies are given in Table 7.9, where for almost all radionuclides g_α is practically 0.

Considering Equation 8.1 and Equation 8.2, the absorbed dose rate to air *in vacuo* is

$$\dot{D}_\alpha(r) = \dot{K}_R(1 - g_\alpha)\left(\frac{r_0}{r}\right)^2 \tag{8.3}$$

or

$$\dot{D}_\alpha(r) = S_K(1 - g_\alpha)\left(\frac{1}{r}\right)^2 \tag{8.4}$$

Step 3: Consider now the situation demonstrated in Figure 8.3b, where the volume element dV is filled with water containing a water mass dm_w. Assuming secondary particle (electron) equilibrium, CPE, conditions,[1] the absorbed dose rate to water $\dot{D}_w(r)$ at the point P *in vacuo* can be calculated from the absorbed dose to air $\dot{D}_\alpha(r)$ at the same place, by correcting for the differences in the mass absorption coefficients between the two media:[1]

$$\dot{D}_w(r) = \dot{D}_\alpha(r)\left(\frac{\mu_{en}}{\rho}\right)_\alpha^w$$

$(\mu_{en}/\rho)_\alpha^w$ is the ratio of mass energy absorption coefficient of water to that of air. In Table 7.15, the inverse value of this ratio is shown for the radionuclides of interest in brachytherapy.

Using Equation 8.3 and Equation 8.4 the absorbed dose rate to water at a distance r, P(r), from the source in empty space, *in vacuo*, is given by

$$\dot{D}_w(r) = \dot{K}_R(1 - g_\alpha)\left(\frac{\mu_{en}}{\rho}\right)_\alpha^w\left(\frac{r_0}{r}\right)^2 \tag{8.5}$$

or

$$\dot{D}_w(r) = S_K(1 - g_\alpha)\left(\frac{\mu_{en}}{\rho}\right)_\alpha^w\left(\frac{1}{r}\right)^2 \tag{8.6}$$

Step 4: Finally, the situation is considered where the space around source and point of interest P(r) is also filled with water (Figure 8.3c). In contrast to the situation *in vacuo*, in water there are attenuation and scatter effects that influence the radiation field at the distance r from the source. Let $f_{as,w}(r)$ be the correction for the absorption and scattering effects of the photons emitted from the source at the distance r in water, when compared to the same point *in vacuo*. The absorbed dose to water in water-filled space and distance r from the ideal point source is

$$\dot{D}_w(r) = \dot{K}_R(1 - g_\alpha)\left(\frac{\mu_{en}}{\rho}\right)_\alpha^w \left(\frac{r_0}{r}\right)^2 f_{as,w}(r) \tag{8.7}$$

or

$$\dot{D}_w(r) = S_K(1 - g_\alpha)\left(\frac{\mu_{en}}{\rho}\right)_\alpha^w \left(\frac{1}{r}\right)^2 f_{as,w}(r) \tag{8.8}$$

Concerning the above equations, there are factors depending on the emitted photon energy spectrum and thus radionuclide/source and on the distance, $f_{as,w}(r)$, and factors that depend only on the source radionuclide/ source, g_α and $(\mu_{en}/\rho)_\alpha^w$.

g_α is practically 0 for all radionuclides where for the highest energies of ^{60}Co $g_\alpha = 0.003$ (see Table 7.9) and is thus insensitive to individual source designs.

$(\mu_{en}/\rho)_\alpha^w$ for high-energy radionuclides, ^{60}Co, ^{137}Cs, ^{198}Au, and ^{192}Ir, is insensitive to source designs, and thus to photon spectra changes due to the individual source construction (see also Figure 8.4). For lower energy

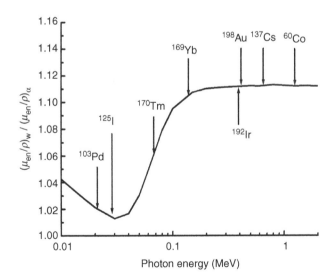

FIGURE 8.4
Energy dependence of the ratio of mass energy absorption coefficients of water to that of air, $(\mu_{en}/\rho)_\alpha^w$, in the brachytherapy related energy range 0.01 to 2.0 MeV. The mass energy absorption coefficients for dry air and liquid water considered are according to Hubbell and Seltzer.[19] The corresponding values for the different brachytherapy radionuclides calculated using their effective energies as listed in Table 5.2, and a log–log interpolation in the tables of Hubbell and Seltzer,[19] are also indicated.

sources such as ^{169}Yb, changes in the effective energy in the range of 0.080 to 0.150 MeV (intermediate energies) will result in a more than 2% change in the value of the ratio of mass energy absorption coefficient of water to air. Concerning the very low energy radionuclides ^{125}I and ^{103}Pd, with effective energies in the range 0.020 to 0.030 MeV, the individual construction of the sources will significantly influence the spectrum of emitted photons, and thus the value of $(\mu_{en}/\rho)_{\alpha}^{W}$. The absorption and scatter correction $f_{as,w}(r)$ for water is distance and source type (photon spectrum) dependent.

8.3.2 Real Sources

Let us consider now a real source of cylindrical shape, as shown in Figure 8.2. Here, the radiation field is assumed to be cylindrically symmetric (independent of azimuthal angle φ) and thus the dose rate at a point P in space will depend on both distance r, and polar angle θ.

Regarding Equation 8.7 and Equation 8.8, there are two additional factors that have to be considered for a real source.

8.3.2.1 Finite Source Core Dimensions

For the case of an ideal point source the inverse square law was derived to describe the dependence of dose rate on the distance *in vacuo*, thus when no absorption and scattering occurs. For an activity having a finite spatial distribution, this generally will not be the case. The influence of the finite sources active core dimensions on the distance dependence of the radiation field *in vacuo* has to be considered and estimated.

This can be done by subdividing the source active core in small elements that can be considered as point sources, and then apply an inverse square law for each one of these elements. If $\rho(\vec{r}')$ is the density of radioactivity at the point $\vec{r}'$ within the source active core volume (see Figure 8.5 and Figure 8.2a), then the air kerma rate in empty space at a point $\vec{r} = (r,\theta,\phi)$ resulting from that small core element at $\vec{r}'$ is

$$dK_{\alpha}(\vec{r}) = \dot{K}_R \left[\frac{\rho(\vec{r}')}{\int_{V_s} \rho(\vec{r}')dV'} \right] \left(\frac{r_0}{|\vec{r} - \vec{r}'|} \right)^2$$

or

$$d\dot{K}_{\alpha}(\vec{r}) = S_K \left[\frac{\rho(\vec{r}')}{\int_{V_s} \rho(\vec{r}')dV'} \right] \left(\frac{1}{|\vec{r} - \vec{r}'|} \right)^2$$

with V_s denoting integration over the active source volume (core) and dV' referring to the source core volume element located at $\vec{r}'$ (Figure 8.5).

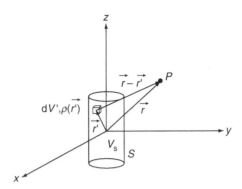

FIGURE 8.5
A cylindrically shaped source active core S of volume V_s and the point of interest P *in vacuo* (empty space). The active core is subdivided into small elements dV' that can be considered as point sources and, thus, then inverse square law for each one of these elements can be applied. $\rho(\vec{r}')$ is the density of radioactivity at the point $\vec{r}'$ centered in the volume element dV' which is positioned within the source active core volume. The air kerma rate *in vacuo* at the point $P = P(\vec{r})$ with $\vec{r} = (r,\theta,\phi)$ (see Figure 8.2a) can be calculated by integrating over the source core volume V_s the air kerma rate contribution of such elements dV', Equation 8.9 through Equation 8.11. An isotropic radioactivity distribution within V_s is usually assumed so that $\rho(\vec{r}') = \rho \; \forall \; \vec{r}'$ within V_s.

The integral in the denominator is used to normalize the radioactivity distribution within the source core.

By integrating the above equations with the source core volume V_s, the following equations result instead of Equation 8.1 and Equation 8.2 that are, as pointed out, valid only for ideal point sources, giving the air kerma rate at $\vec{r}$ which results from a real source with an active core volume V_s as shown in Figure 8.5:

$$\dot{K}_\alpha(\vec{r}) = \int_{V_s} d\dot{K}_\alpha(\vec{r}) dV' = \dot{K}_R \left[\frac{\displaystyle\int_{V_s} \frac{\rho(\vec{r}') dV'}{|\vec{r} - \vec{r}'|^2}}{\displaystyle\int_{V_s} \rho(\vec{r}') dV'} \right] r_0^2 \tag{8.9}$$

or based on the air kerma strength S_K:

$$\dot{K}_\alpha(\vec{r}) = \int_{V_s} d\dot{K}_\alpha(\vec{r}) dV' = S_K \left[\frac{\displaystyle\int_{V_s} \frac{\rho(\vec{r}') dV'}{|\vec{r} - \vec{r}'|^2}}{\displaystyle\int_{V_s} \rho(\vec{r}') dV'} \right] \tag{8.10}$$

It has been established to name the fraction in the brackets in the above two equations as geometry function, $G(\vec{r})$:

$$G(\vec{r}) = \frac{\displaystyle\int_{V_s} \frac{\rho(\vec{r}') dV'}{|\vec{r} - \vec{r}'|^2}}{\displaystyle\int_{V_s} \rho(\vec{r}') dV'} \tag{8.11}$$

From a physics point of view, the geometry function neglects scattering and attenuation, and simply provides an effective inverse square-law correction based upon the spatial distribution of radioactivity within the source core, thus depending only on the 3D volume shape of the active core of the source and not on the radionuclide, nor on the encapsulation.

Thus, for a real source of finite size active core, the term $1/r^2$ in Equation 8.7 and Equation 8.8, has to be replaced by the geometry function $G(\vec{r})$.

8.3.2.2 Self Absorption and Attenuation

The finite size of the active core of a real source causes an anisotropy in the angular distribution of particle fluence, and thus of the air kerma rate and absorbed dose rate, due to self absorption and attenuation of the emitted photons within the source core itself and in the encapsulation material(s) of the sealed sources.

By taking this into account, the function $f_{as,w}$, which corrects for the absorption and scattering effects in water environment, has no more spherical symmetry with regard to the source, but depends in addition for the case of a cylindrical source on the polar angle θ (cylindrical symmetry, see also Figure 8.2b) in order to also consider the attenuation and absorption within the core and encapsulation of the source; $f_{as,w}(\vec{r}) = f_{as,w}(r,\theta)$. For a source of any shape, in general this correction is a function of radial distance r, polar angle θ, and azimuthal angle φ, $f_{as,w}(\vec{r}) = f_{as,w}(r,\theta,\varphi)$.

$f_{as,w}(r,\theta)$ thus depends not only on the radionuclide and consequently on the energy spectrum of the emitted photons, but also on the source core, encapsulation materials, and geometry.

Considering now Equation 8.9 through Equation 8.11 into Equation 8.7 and Equation 8.8, the following general equations are derived for the absorbed dose rate to water in water at a point $P(\vec{r}) = P(r,\theta,\varphi)$ in the neighborhood of a real source with an active core volume V and encapsulation

$$\dot{D}_w(\vec{r}) = \dot{K}_R(1 - g_\alpha)\left(\frac{\mu_{en}}{\rho}\right)^w_\alpha r_0^2 G(\vec{r}) f_{as,w}(\vec{r}) \tag{8.12}$$

or

$$\dot{D}_w(\vec{r}) = S_K(1 - g_\alpha)\left(\frac{\mu_{en}}{\rho}\right)^w_\alpha G(\vec{r}) f_{as,w}(\vec{r}) \tag{8.13}$$

For cylindrical shaped sources, cylindrical shaped source core, and encapsulation construction cylindrically symmetric around the z-axis that is the longitudinal axis of the source cylinder (Figure 8.2b), geometry function $G(\vec{r})$ and absorption and scattering correction function $f_{as,w}(\vec{r})$, depend only on radial distance r and polar angle θ. Thus, for cylindrical source designs the following equations are derived:

$$\dot{D}_w(r, \theta) = \dot{K}_R(1 - g_\alpha)\left(\frac{\mu_{en}}{\rho}\right)^w_\alpha r_0^2 G(r, \theta) f_{as,w}(r, \theta) \tag{8.14}$$

or

$$\dot{D}_{w}(r, \theta) = S_{K}(1 - g_{\alpha})\left(\frac{\mu_{en}}{\rho}\right)_{\alpha}^{w} G(r, \theta)f_{as,w}(r, \theta) \qquad (8.15)$$

$f_{as,w}(r,\theta)$ corrects in this case not only for the in-water medium absorption and scattering, but also for the absorption and attenuation of the emitted photons within the core itself and the encapsulation of the source.

According to how the geometry function $G(r,\theta)$ and the in-water absorption and scattering function $f_{as,w}(r,\theta)$ are handled, and to the terminology followed for the product $(1 - g_{\alpha})(\mu_{en}/\rho)_{\alpha}^{w}$ in the above two general equations, a dosimetry model—a protocol—is defined.

8.3.3 The TG-43 Dosimetry Formalism

Based on the calculation model proposed by the Interstitial Collaborative Working Group (ICWG),[2] Task Group No 43 of AAPM Radiation Therapy Committee,[3] published recommendations on dosimetry formalism and dosimetry parameters for interstitial brachytherapy sources in 1995. Even though this was primarily only considering low dose rate (LDR) sources (in the original publication it was explicitly mentioned that high activity sources and iridium wires were beyond the scope of that report), the TG-43 formalism has been widely used, and is also virtually internationally accepted for high dose rate (HDR) and pulsed dose rate (PDR) iridium sources used in remote afterloading systems.[4]

Previous calculation formalisms were based upon apparent activity (A_{app}), equivalent mass of radium, exposure-rate constants, and tissue-attenuation coefficients. These older formalisms did not account for source-to-source differences in active core construction and encapsulation design. With the exception of radium, the exposure-rate constants and other input parameters to these algorithms depended only on the radionuclide.[5]

In contrast to these methods, the TG-43 formalism introduced and incorporated dose rate constants and several other dosimetric parameters that all depend on the specific source design. TG-43 is a consistent formalism, simple to implement, and based on a small number of parameters/quantities that can be easily extracted from Monte Carlo (MC) calculated dose rate distributions around the sources in a water equivalent medium, or from measurements in such a medium. This increases the accuracy of calculations that are carried out in the clinic, which are always for water medium and not in free space.

8.3.3.1 The Basic Concept

The basic concept of the TG-43 dosimetry protocol is to define a clear formalism expressed in mathematical formulas, and incorporate parameters and quantities that enable users to accurately calculate dose and dose rate distributions around common designs of radioactive sources in the clinical routine. This enables the use of common and consistent data and databases for commercially available source designs.

8.3.3.2 Reference Medium

The TG-43 defines water as the reference dosimetry medium, where water equivalent materials can also be considered for the measurement of dose distributions and thus of TG-43 related parameters.

8.3.3.3 Reference of Data

MC simulation results are verified by experimental measurements and published in peer-reviewed journals and are defined as reference data.

8.3.3.4 Source Geometry

The TG-43 protocol generally addresses cylindrically shaped sources having an active core that allows the assumption of a cylindrical symmetry in the distribution of activity. The geometry of such sources and the corresponding polar coordinate system used is shown in Figure 8.2. Figure 8.6 demonstrates the cross-section on the $y - z$ plane which is considered as the basic plane for any description according to TG-43. Due to the cylindrical symmetry of the handled sources, this is equivalent to any plane containing the z-axis (any azimuthal angle φ other than 90°). Consequently, a cylindrical symmetry of the dose distribution is assumed.

As in the case of Figure 8.2b, and in the cross-sectional view of Figure 8.6b, the source construction includes a driving cable attached at the one end of the source capsule at the $-z$ direction in order to also cover the case of single stepping high activity sources, mainly ^{192}Ir sources, used in modern afterloader systems. The tip of the source is at the $+z$ direction.

As will be shown later, TG-43 offers analytical expressions regarding the effect of finite size of source active core, (geometry function, Equation 8.11) when these sources are approximated either with ideal line sources (diameter $D_s = 0$ cm in Figure 8.2 and Figure 8.6) or by ideal point sources (diameter $D_s = $ length $L_s = 0$ cm, see also Figure 8.1b).

8.3.3.5 Reference Point of Dose Calculations (r_0, θ_0)

The reference point for the TG-43 formalism is chosen to be a point lying on the transverse bisector of the source at a distance of 1 cm from its center. Expressed in polar coordinates, that is $(r_0, \theta_0) = (1 \text{ cm}, 90°)$.

8.3.3.6 Formalism

According to the TG-43 protocol, the dose rate to water in water medium $\dot{D}(r, \theta)$ at a point $P(r, \theta)$ from a source is given by (see also Figure 8.6 for

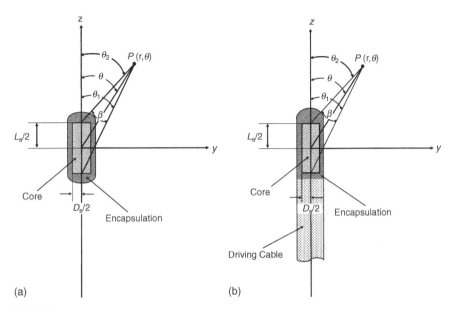

FIGURE 8.6
Geometry and parameters as used by the TG-43 protocol for cylindrically shaped sources having an active core that allows the assumption of a cylindrical symmetry in the distribution of activity. Due to the cylindrical symmetry of the dose rate distribution, the polar coordinate system as shown in Figure 8.2 is adequate for defining the needed geometry. Here, the cross-section on the principal $y - z$ plane, which is considered as the basic plane for any description according to TG-43, is shown. The point of interest P is described by its radial distance r and the polar angle θ, $P = P(r,\theta)$. (a) Typical case of a source geometry for permanent implant brachytherapy (seeds). The cylindrical active core of length L_s and diameter D_s is symmetrically encapsulated by a specific material. All TG-43 geometry related parameters are shown. (b) Corresponding geometry and parameters for source construction including a driving cable attached at the one end of the source capsule at the $-z$-direction as it is the case of single stepping high activity source (HDR source), mainly ^{192}Ir sources, used in modern afterloader systems. The tip of the source is at the $+z$ direction.

coordinates and coordinate system):[3]

$$\dot{D}(r, \theta) = S_K \Lambda \frac{G(r, \theta)}{G(r_0, \theta_0)} g(r) F(r, \theta) \tag{8.16}$$

where r is the radial distance from the source center and θ is the polar angle.

In the following discussion involving the absorbed dose or absorbed dose rate to water, the index "w" for water will be omitted.

S_K: The air kerma strength in U, $1\,U = 1\,\mu Gy\,m^2\,h^{-1} = 1\,cGy\,cm^2\,h^{-1}$, as seen in Equation 7.1 and Equation 7.2.

Λ: The dose rate constant in water. Λ is expressed in units of $cGy\,h^{-1}\,U^{-1}$.

$G(r,\theta)$: The geometry function at radial distance r and polar angle θ as defined already in Equation 8.11 (see also Equation 8.15). $G(r,\theta)$ is a dimensionless quantity.

$G(r_0,\theta_0)$: The geometry function at the reference point $(r_0,\ \theta_0)$ with $r_0 1.0$ cm and $\theta_0 90°$. Attention has to be paid to distinguish between the r_0 radial distance coordinate of the TG-43 reference point and the reference distance r_0 for the definition of reference air kerma rate $\dot{K}_R$, as seen in Equation 8.1.

$g(r)$: The radial dose function that considers the distance dependence of absorption and scatter of the photon rays in the water medium along the transversal axis, that is the y-axis or equivalently for $\theta = 90°$. $g(r)$ is a dimensionless quantity.

$F(r,\theta)$: The anisotropy function that considers the effect of absorption and scatter of the photons within the source active core and encapsulation material as well as part of the driving cable if any. $F(r,\theta)$ is a dimensionless quantity.

According to the above, the dose rate to water in water medium $\dot{D}(r,\theta)$ as obtained by Equation 8.16 is in cGy h^{-1}.

It is essential to note that the TG-43 formulation is based on the air kerma strength S_K for the determination of the source strength.

The TG-43 formula Equation 8.16 can be derived based on the general formulation demonstrated previously for the air kerma strength based source calibration given by Equation 8.15, now avoiding the index "w" for water.

For the reference point of the TG-43 formalism $(r_0,\theta_0) = (1\ \text{cm},\ 90°)$ Equation 8.15 becomes

$$\dot{D}(r_0,\ \theta_0) = S_K(1 - g_\alpha)\left(\frac{\mu_{en}}{\rho}\right)_\alpha^w G(r_0,\ \theta_0)f_{as,w}(r_0,\ \theta_0) \qquad (8.17)$$

By dividing Equation 8.15 with Equation 8.17, the following expression for the dose rate to water in water medium $\dot{D}(r,\theta)$, with respect to that at the TG-43 reference point (r_0,θ_0), is derived:

$$\frac{\dot{D}(r,\ \theta)}{\dot{D}(r_0,\ \theta_0)} = \left(\frac{G(r,\ \theta)}{G(r_0,\ \theta_0)}\right)\left(\frac{f_{as,w}(r,\ \theta)}{f_{as,w}(r_0,\ \theta_0)}\right)$$

or equivalently:

$$\dot{D}(r,\ \theta) = S_K\left(\frac{\dot{D}(r_0,\ \theta_0)}{S_K}\right)\left(\frac{G(r,\ \theta)}{G(r_0,\ \theta_0)}\right)\left(\frac{f_{as,w}(r,\ \theta)}{f_{as,w}(r_0,\ \theta_0)}\right) \qquad (8.18)$$

Thus, the term in the first brackets corresponds to the dose rate constant in water Λ shown in the TG-43 formula Equation 8.16. The ratio of the value of the geometry function for the actual point of calculation (r,θ) to that for the TG-43 reference point (r_0,θ_0) is identical in both expressions. The ratio of the value of the absorption and scattering function $f_{as,w}(r,\theta)$ at the point of interest to that at the reference point $f_{as,w}(r_0,\theta_0)$ in Equation 8.18 corresponds to the product, $g(r)F(r,\theta)$, in TG-43 formula Equation 8.16.

8.3.3.6.1 The Dose Rate Constant Λ

The dose rate constant, Λ, is defined as the dose rate to water in water at the reference point, namely at a distance of $r_0 = 1$ cm on the transverse axis $(\theta = 90°)$, $\dot{D}(r_0, \theta_0)$, per unit air kerma strength, S_K, as seen in Equation 8.18:

$$\Lambda = \frac{\dot{D}(r_0, \theta_0)}{S_K} \tag{8.19}$$

The dose rate constant depends on both the radionuclide and source model, and is influenced by both the source design (radioactive core and encapsulation) and the methodology used to determine S_K.

Based on the above equation and considering Equation 8.17, the dose rate constant $\Lambda(E)$ for a monoenergetic photon ideal point source (pure radionuclide source with $L_s = D_s = 0$ cm) can be generally estimated using the following expression:

$$\Lambda(E) = (1 - g_\alpha)\left(\frac{\mu_{en}(E)}{\rho}\right)_\alpha^w G(r_0, \theta_0)f_{as,w}(r_0, \theta_0) \tag{8.20}$$

with $G(r_0, \theta_0) = 1/r_0^2$ and $f_{as,w}(r_0, \theta_0) = f_{as,w}(r_0)$ for the case of an ideal point source with $r_0 = 1.0$ cm, as seen in Equation 8.8.

$(\mu_{en}(E)/\rho)_\alpha^w$ is the ratio of mass energy absorption coefficients of water to that of air for the photon energy E.

Figure 8.7 shows the energy dependence of the dose rate constant for monoenergetic ideal point sources in the energy range of 0.020 to 0.700 MeV, where Λ values corresponding to the effective energy values, E_{eff}, of the

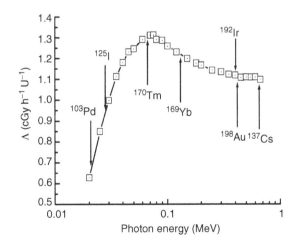

FIGURE 8.7
Energy dependence of the dose rate constant Λ expressed in cGy h^{-1} U^{-1} for monoenergetic point sources in the energy range of 0.020 to 0.700 MeV. The Λ values for the effective energies E_{eff}, see Equation 5.2, of the brachytherapy relevant radionuclides listed in Table 5.2 are indicated by arrows. The dose rate constant values shown have been calculated using MC simulations.

radionuclides relevant to brachytherapy as listed in Table 5.2 are marked (see also Figure 9.10).

For an ideal point source with a known energy spectrum, the corresponding dose rate constant Λ can be calculated according to[6]

$$\Lambda = \frac{\sum_i \left(f_i E_i \left(\frac{\mu_{en}}{\rho} \right)_{\alpha, E_i} \Lambda(E_i) \right)}{\sum_i \left(f_i E_i \left(\frac{\mu_{en}}{\rho} \right)_{\alpha, E_i} \right)} \qquad (8.21)$$

where f_i is the intensity (or emission frequency) of the specific energy line E_i.

$(\mu_{en}/\rho)_{\alpha, E_i}$ is the mass energy absorption coefficient of air for the energy E_i. $\Lambda(E_i)$ is the dose rate constant for a monoenergetic photon point source of energy E_i, calculated according to Equation 8.20. The sum in the nominator and denominator in Equation 8.21 is extended over all photon ray energies E_i of the source spectrum.

The energy dependence of $(\mu_{en}/\rho)_{\alpha, E_i}$ and of the product $E_i(\mu_{en}/\rho)_{\alpha, E_i}$ is demonstrated in Figure 8.8.

Due to the fact that for the photon energies of the radionuclide of interest the energy absorption coefficient can be set equal to the energy transfer coefficient, the fraction of the energy of the electrons liberated by photons in air that lost to radiative processes $g_\alpha \approx 0$ (see Table 7.9), Equation 8.21

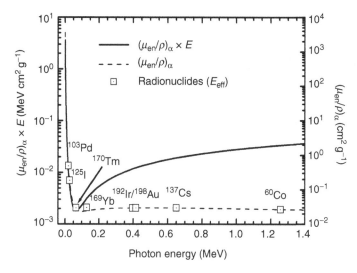

FIGURE 8.8
Mass energy absorption coefficient values for dry air, $(\mu_{en}/\rho)_\alpha$, and product $(\mu_{en}/\rho)_\alpha E$, in dependence on the photon energy. The effective mass energy absorption coefficient values for dry air, $(\mu_{en}/\rho)_{\alpha,eff}$, according to Equation 5.3 listed in Table 5.4 through Table 5.15 for the different radionuclides in dependence on the corresponding effective energy E_{eff} of the radionuclides, are also shown as squares.

defines the dose rate constant Λ of a polyenergetic photon point source to be the effective dose rate constant. That is the air kerma weighted mean dose rate constant of the individual $\Lambda(E_i)$ values calculated in a way analogous to that of defining the effective energy E_{eff} (see Equation 5.2).

For real sources, the dose rate constant Λ can be estimated either experimentally by measuring the dose rate to water $\dot{D}(r_0,\theta_0)$ in water or water equivalent medium using a calibrated source (with a given source strength S_K at the time of measurement), or alternatively it can be calculated using MC simulation techniques of the real source geometry in an adequately dimensioned water medium (see also Chapter 9). With the latter, assumptions and thus possible inaccuracies when using the scatter and attenuation correction $f_{\text{as,w}}(r_0,\theta_0)$ of water in Equation 8.20 are avoided.

Furthermore, the use of the analytic expression Equation 8.21 assumes that the photon spectrum emitted from the real source, as well as the corresponding correction, $f_{\text{as,w}}(r_0,\theta_0)$, are accurately known.

8.3.3.6.2 The Radial Dose Function g(r)

As demonstrated in the process of deriving Equation 8.18, the ratio of the value of the absorption and scattering function $f_{\text{as,w}}(r,\theta)$ at the point of interest to that at the reference point $f_{\text{as,w}}(r_0,\theta_0)$ corresponds to the product, $g(r)F(r,\theta)$, in the TG-43 formulation given in Equation 8.16.

According to this observation, the TG-43 protocol incorporates the following additional concept. Considering the dose rate at the reference polar angle $\theta_0 = 90°$ for a radial distance r, $\dot{D}(r,\theta_0)$, and applying Equation 8.18, we obtain

$$\dot{D}(r, \theta_0) = S_K \left(\frac{\dot{D}(r_0, \theta_0)}{S_K} \right) \left(\frac{G(r, \theta_0)}{G(r_0, \theta_0)} \right) \left(\frac{f_{\text{as,w}}(r, \theta_0)}{f_{\text{as,w}}(r_0, \theta_0)} \right) \tag{8.22}$$

This equation describes the dose rate distribution along the transverse axis of the source. By dividing this equation with the dose rate at the reference point (r_0,θ_0), $\dot{D}(r_0,\theta_0)$, the normalized dose rate distribution along the transverse y-axis ($\theta = \theta_0 = 90°$, see Figure 8.6) is derived:

$$\frac{\dot{D}(r, \theta_0)}{\dot{D}(r_0, \theta_0)} = \left(\frac{G(r, \theta_0)}{G(r_0, \theta_0)} \right) \left(\frac{f_{\text{as,w}}(r, \theta_0)}{f_{\text{as,w}}(r_0, \theta_0)} \right)$$

or

$$\frac{f_{\text{as,w}}(r, \theta_0)}{f_{\text{as,w}}(r_0, \theta_0)} = \left(\frac{G(r_0, \theta_0)}{G(r, \theta_0)} \right) \left(\frac{\dot{D}(r, \theta_0)}{\dot{D}(r_0, \theta_0)} \right) \tag{8.23}$$

The ratio in the left part of this equation describes the absorption and attenuation correction in water at a radial distance r to that at the reference distance r_0 at polar angle $\theta = \theta_0$. As mentioned previously, $f_{\text{as,w}}(r,\theta)$ also includes the correction for the self absorption and attenuation of photons within the source core and encapsulation.

TG-43 protocol defines this expression as the radial dose function $g(r)$:

$$g(r) = \frac{f_{as,w}(r, \theta_0)}{f_{as,w}(r_0, \theta_0)} = \left(\frac{G(r_0, \theta_0)}{G(r, \theta_0)} \right) \left(\frac{\dot{D}(r, \theta_0)}{\dot{D}(r_0, \theta_0)} \right) \tag{8.24}$$

which is the radial dependence of the dose rate value at the reference polar angle $\theta = \theta_0$, or equivalently along the transversal y-axis, corrected for the distance related effect by using the effective inverse square-law correction, resulting from the source core finite volume as described by the geometry function $G(r,\theta)$. According to this, the radial dose function value at the reference distance $r_0 = 1.0$ cm is for any source/radionuclide per definition 1.0; $g(r_0) = 1.0$.

Having in mind the above, and in order to derive the values of $g(r)$ for the different radial distances r, the dose rate at the corresponding r along the transverse y-axis (at $\theta = \theta_0$), the dose rate at the TG-43 reference point (r_0, θ_0) and the geometry function values at these points are needed. These dose rate values can be calculated using MC simulations or, alternatively, they can be measured in water or in water equivalent material.

Due to the normalization realized in deriving Equation 8.23, the values of $g(r)$ can thus be calculated without knowledge of the values of the function $f_{as,w}(r,\theta)$ at the corresponding distances r along the y-axis.

8.3.3.6.3   The Anisotropy Function F(r,θ)

Consider now the angular dose rate distribution for a given radial distance r given by Equation 8.18 and normalize this relative to the dose rate value at the (r,θ_0):

$$\frac{\dot{D}(r, \theta)}{\dot{D}(r, \theta_0)} = \left(\frac{G(r, \theta)}{G(r, \theta_0)} \right) \left(\frac{f_{as,w}(r, \theta)}{f_{as,w}(r, \theta_0)} \right)$$

or

$$\frac{f_{as,w}(r, \theta)}{f_{as,w}(r, \theta_0)} = \left(\frac{G(r, \theta_0)}{G(r, \theta)} \right) \left(\frac{\dot{D}(r, \theta)}{\dot{D}(r, \theta_0)} \right) \tag{8.25}$$

The ratio in the left part of this equation describes the absorption and attenuation correction in water at a radial distance r and at a polar angle θ to that at the same radial distance but at polar angle $\theta = \theta_0$. As mentioned previously, $f_{as,w}(r,\theta)$ also considers the self absorption and attenuation of photons within the source core and encapsulation, and thus the dominating effect for the angular dependence (polar angle θ) of this ratio for a fixed r, is the self absorption and attenuation in the source.

TG-43 protocol defines this expression as the anisotropy function $F(r,\theta)$:

$$F(r, \theta) = \frac{f_{as,w}(r, \theta)}{f_{as,w}(r, \theta_0)} = \left(\frac{G(r, \theta_0)}{G(r, \theta)} \right) \left(\frac{\dot{D}(r, \theta)}{\dot{D}(r, \theta_0)} \right) \tag{8.26}$$

which is the angular dependence of the dose rate value for a given radial distance *r*, corrected for the distance related effect using the effective inverse square-law correction resulting from the source core finite volume as described by the geometry function $G(r,\theta)$. According to the definition of the anisotropy function, its value at the reference angle $\theta_0 = 90°$ and at any radial distance *r* is always 1.0; $F(r,\theta_0) = 1.0, \forall \ r$.

Similar to the case of the radial dose function, the values of the anisotropy function can be derived from MC calculated or measured dose rate values for different radial distances and different polar angles, and the corresponding values of the geometry function. Once more there is no need to know the values of the function $f_{as,w}(r,\theta)$ for calculating the anisotropy function values $F(r,\theta)$.

Using Equation 8.24 and Equation 8.26, the product $g(r)F(r,\theta)$ in the TG-43 formula Equation 8.16 is

$$g(r)F(r, \theta) = \left(\frac{f_{as,w}(r, \theta_0)}{f_{as,w}(r_0, \theta_0)} \right) \left(\frac{f_{as,w}(r, \theta)}{f_{as,w}(r, \theta_0)} \right) = \frac{f_{as,w}(r, \theta)}{f_{as,w}(r_0, \theta_0)}$$

as was mentioned in the discussion for deriving Equation 8.18.

Summarizing the dose rate calculation at a point $P(r,\theta)$ around a cylindrically symmetric source (Figure 8.6), according to the TG-43 formalism described in Equation 8.16, can be explained as a two-step procedure (see Figure 8.9).

Step 1: Based on the known dose rate at the reference point (r_0,θ_0), $\dot{D}(r_0,\theta_0) = S_K \Lambda$, the dose rate along the transversal *y*-axis at the distance *r*, $\dot{D}(r,\theta_0)$, is calculated based on the radial dose function $g(r)$, and corrected using the effective inverse square-law correction resulting from the source core finite volume as described by the geometry function $G(r,\theta)$ for changing from r_0 to *r* at $\theta_0 = 90°$.

Step 2: The dose rate at the point of interest $P(r,\theta)$, $\dot{D}(r,\theta)$, is calculated from the dose rate at (r,θ_0), $\dot{D}(r,\theta_0)$ using the anisotropy function $F(r,\theta)$ and correcting again using the geometry function $G(r,\theta)$ for changing from $\theta_0 = 90°$ to θ at radial distance *r*.

8.3.3.7 The 2D TG-43 Formulation

At the introduction of the TG-43 formalism, Equation 8.16, it was mentioned that it is actually valid for cylindrically symmetric sources. For such sources, the dose rate to water in water at any point around the source depends only on the radial distance *r* of that point form the center of the active core (Figure 8.6) and on the polar angle θ. This means that the dose rate calculation is carried out on the $y - z$-plane that contains the longitudinal axis of the source, as well as the point of interest. In other words, this is a two-dimensional (2D) calculation formalism.

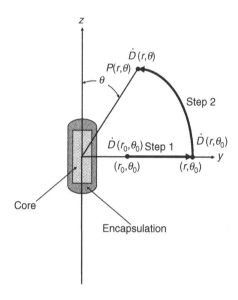

FIGURE 8.9

Schematic representation of the two steps procedure of calculating the dose rate $\dot{D}(r,\theta)$ at a point $P(r,\theta)$ around a cylindrically symmetric source according to the TG-43 formalism obtained by Equation 8.16. The starting point is the known dose rate at the reference point (r_0,θ_0), $\dot{D}(r_0,\theta_0) = S_K \Lambda$. *Step 1*: Firstly, the dose rate along the transversal y-axis ($\theta = \theta_0 = 90°$) at the distance of interest r, $\dot{D}(r,\theta_0)$, is calculated from $\dot{D}(r_0,\theta_0)$ by applying the radial dose function $g(r)$ and correcting using the effective inverse square-law correction resulting from the source core finite volume, as described by the geometry function $G(r,\theta)$, for moving from r_0 to r at the reference polar angle $\theta_0 = 90°$. *Step 2*: Finally, the dose rate at the point of interest $\dot{D}(r,\theta)$ is calculated from the dose rate at (r,θ_0), $\dot{D}(r,\theta_0)$, by applying the anisotropy function $F(r,\theta)$ and correcting again using the geometry function $G(r,\theta)$ for moving from the polar angle $\theta_0 = 90°$ to the polar angle of interest θ at the radial distance r.

Regarding the geometry function $G(r,\theta)$, defined in Equation 8.11, it can be assumed that it is a 3D based calculation of an effective inverse square-law correction considering the finite size of the source core.

TG-43 considers two simplified cases for analytical calculation of the geometry function: (a) the line source approximation, where the source core is considered to be an ideal line (see Figure 8.1b), and (b) the point source approximation, where the source core is considered to be an ideal point source (see Figure 8.1a).

8.3.3.7.1 Line Source Approximation

Figure 8.10 summarizes the geometry that is to be considered for deriving the analytical expression of the geometry function for the case when an active core is considered to be an ideal line segment, extended from $z = -L_s/2$ to $z = +L_s/2$.

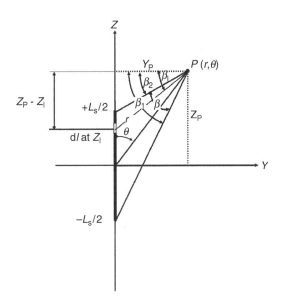

FIGURE 8.10

Geometry and parameters used for deriving the analytical expression of the geometry function for the line source approximation, the active core of the source is considered to be an ideal line segment, extended from $z = -L_s/2$ to $z = +L_s/2$ (see also Figure 8.1c). A longitudinal source element dl is centered at z_l. The volume integrals in Equation 8.11 are now replaced by line integrals from $z = -L_s/2$ to $z = +L_s/2$, Equation 8.27 and Equation 8.28, where a constant radioactivity per unit length ρ_l along L_s is assumed so that $\rho_l(z_l) = \rho_l$ for $-L_s/2 \leq z_l \leq +L_s/2$.

Assuming a uniformly distributed radioactivity along L_s, constant radioactivity per unit length ρ_l, the integral in the denominator in Equation 8.11 is

$$\int_{V_s} \rho(\vec{r}')dV' = \int_{-L_s/2}^{+L_s/2} \rho_l(z_l)dl = \rho_l L_s \tag{8.27}$$

Before calculating the integral in the nominator of the definition of geometry function, the following can be written based on the geometry shown in Figure 8.10:

$$|\vec{r} - \vec{r}'|^2 = |\vec{r} - \vec{r}_l|^2 = y_P^2 + (z_P - z_l)^2$$

we thus have

$$\int_{V_s} \frac{\rho(\vec{r}')dV'}{|\vec{r} - \vec{r}'|^2} = \rho_l \int_{-L_s/2}^{+L_s/2} \frac{dl}{y_P^2 + (z_P - z_l)^2} = \rho_l \int_{-L_s/2}^{+L_s/2} \frac{dz_l}{y_P^2 + (z_P - z_l)^2}$$

$$= \rho_l \left(\frac{1}{y_P}\right)\left[\text{Arc} \tan\left(\frac{z_P - z_l}{y_P}\right)\right]_{z_l=+L_s/2}^{z_l=-L_s/2}$$

with $y_P = r \sin \theta$ and $z_P = r \cos \theta$.

Based on Figure 8.10:

$$\text{Arc tan}\left(\frac{z_P - z_l}{y_P}\right) = \beta_l$$

and thus combining the above equations with Equation 8.11, the geometry function $G(r,\theta)$ for the line source approximation is given by

$$G(r, \theta) = \frac{\rho_l\left(\dfrac{1}{y_P}\right)\left[\text{Arc tan}\left(\dfrac{z_P - z_l}{y_P}\right)\right]_{z_l=+L_s/2}^{z_l=-L_s/2}}{\rho_l L_s}$$

and finally:

$$G(r, \theta) = \frac{\beta_1 - \beta_2}{L_s r \sin\theta} = \frac{\beta}{L_s r \sin\theta} \tag{8.28}$$

Comparing Figure 8.6 and Figure 8.10 it can be found that

$$\beta = \beta_1 - \beta_2 = \theta_2 - \theta_1$$

and is the angle subtended by the point of interest $P(r,\theta)$ and the two ends of the active core, which is always in radians.

According to Equation 8.28, the line approximation geometry function $G(r,\theta)$ is symmetrical with reference to 90°. $G(r,\theta)$ angular distributions to either side of the reference polar angle of 90° are mirror images of each other:

$$G(r, \theta) = G(r, 180° - \theta)$$

Karaiskos et al.[7] compared the TG-43 proposed line source approximation of the geometry function to its explicit 3D version as described by Equation 8.11. The 3D geometry function has been calculated using MC integration methods for several commercially available source designs ($D_s > 0$ in Figure 8.2 and Figure 8.6).

It was demonstrated that the line source approximation may be safely applied for radial distances greater than half the source core length L_s. At shorter distances, errors above 3% were found due to the fact that the line source approximation ignores the radial dimension D_s of the source core. The deviations depend on the ratio D_s/L_s and they increase as this ratio increases. Similar results were also found by Rivard.[8]

Figure 8.11 demonstrates a comparison on the $y - z$-plane of the line source approximation of the geometry function obtained using Equation 8.28 to the 3D geometry function obtained using the original formulation for $G(\vec{r})$ given by Equation 8.11 for the example of a source with a cylindrical active core with a diameter of 0.06 cm and length of 0.35 cm (microSelectron HDR classic type ^{192}Ir source; see Table 8.2 and Figure 8.16 in Section 8.4.1.1.1.1 for more details). The 3D geometry function values have been calculated using the MC integration method[9–12] for the integral at the nominator of Equation 8.11. For radial distances above 0.3 cm, there is practically no difference between the 3D and the line source model for the total range of polar angle θ (0 to 90° for those source dimensions).

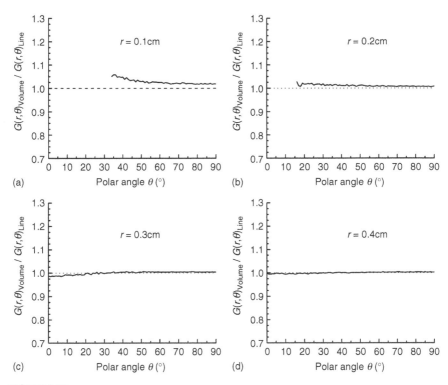

FIGURE 8.11

Graphical comparison on the $y - z$-plane of the line source approximation of the geometry function obtained using Equation 8.28 to the 3D geometry function obtained using the original formulation for $G(\vec{r})$ as described by Equation 8.11. A source with a cylindrical active core with a diameter of 0.06 cm and length of 0.35 cm (microSelectron HDR classic type ^{192}Ir source; see Table 8.2 and Figure 8.16 in Section 8.4.1.1.1.1 for more details) is considered. The 3D geometry function values have been calculated using the MC integration method[9–12] for the integral at the nominator of Equation 8.11. The angular dependence relative to the polar angle θ of the ratio of the $G(\vec{r})$ function calculated according to the 3D model to that according to the line source approximation $G(r,\theta)$ for radial distances $r = 0.1, 0.2, 0.3,$ and 0.4 cm is shown. Angles where the point of calculation falls within the source core have been omitted. For radial distances above 0.3 cm there is practically no difference between 3D and line source model for the total range of polar angle θ, 0 to 90° for these source dimensions.

If we incorporate the line source approximation of the geometry function as described by Equation 8.28 to the TG-43 formula given by Equation 8.16, we obtain the 2D TG-43 formulation.[3]

It is important to mention that for using the 2D formulation, the orientation of the individual source in space has to be known.

8.3.3.7.2 *Point Source Approximation*

When considering the source core to be an ideal point, point source approximation—whereby the radioactivity is assumed to subtend

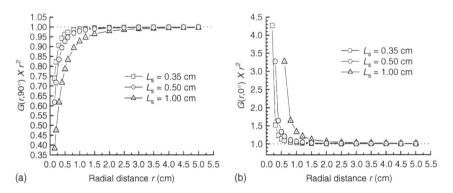

FIGURE 8.12
Dependence of the product $G(r,\theta)r^2$, where $G(r,\theta)$ is for the line source approximation as described by Equation 8.28, on the radial distance r for three typical source active core lengths, $L_s = 0.35$, 0.5, and 1.0 cm. (a) Along the y-axis, polar angle $\theta = \theta_0 = 90°$. (b) Along the z-axis, polar angle $\theta = 0°$. For radial distances r greater than three times the source core length L_s, the source core behaves practically as a point source and thus point source approximation obtained by Equation 8.29 can adequately describe the effective inverse square-law correction $G(r,\theta)$ resulting from the finite source core volume when compared to the line source approximation obtained by Equation 8.28.

a dimensionless point—then the geometry function is

$$G(r, \theta) = G(r) = \frac{1}{r^2} \tag{8.29}$$

In Figure 8.12 the product $G(r,\theta)r^2$, where $G(r,\theta)$ is the line source approximated geometry function according to Equation 8.28, is plotted against the radial distance r for three typical source core lengths, $L_s = 0.35$, 0.5, and 1.0 cm, and for two extreme values of the polar angle, $\theta = \theta_0 = 90°$ and $\theta = 0°$.

For radial distance r three times greater than the source core length L_s, the source core behaves as a point source, and thus the inverse square-law can appropriately describe the dependence of the air kerma and absorbed dose rate on distance:

$$G(r, \theta)r^2 \approx 1.0 \ \forall_r > 3L_s$$

The polar angular dependence of the line source approximation $G(r,\theta)$, Equation 8.28, for three representative radial distances $r = 0.5$, 1.0, and 2.0 cm, for the three source core lengths $L_s = 0.35$, 0.5, and 1.0 cm, are shown in Figure 8.13. This figure demonstrates the enormous increase of the geometry function for small radial distances r and small polar angles θ, where the longer the source the more pronounced the increase.

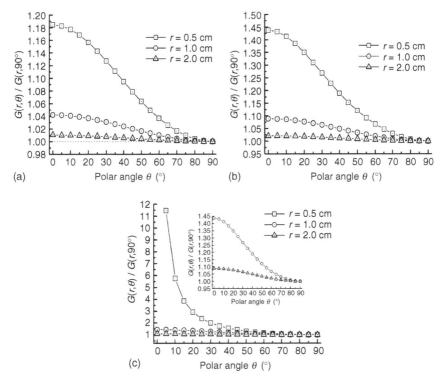

FIGURE 8.13
Polar angular dependence of the line source approximation of $G(\vec{r})$, $G(r,\theta)$ obtained according to Equation 8.28 for three radial distances, $r = 0.5$, 1.0, and 2.0 cm for the three source core lengths considered in Figure 8.12: (a) $L_s = 0.35$ cm, (b) $L_s = 0.5$ cm, and (c) $L_s = 1.0$ cm. The $G(r,\theta)$ values at each one of the radial distances r are normalized to the corresponding value at the reference polar angle $\theta_0 = 90°$, $G(r,90°)$. An enormous increase of the geometry function for small radial distances r and small polar angles θ is observed, where the longer the source the more pronounced the increase.

8.3.3.8 The 1D TG-43 Approximation

For some brachytherapy applications, as in the case for permanent prostate implants, it is not possible or meaningful to define the orientation of each individual implanted sources. Implanted seeds are often randomly orientated and, due to their miniaturization, it is very difficult and in some cases impossible to define the actual orientation of each individual seed using reconstruction methods. In these cases the 2D TG-43 formulation is not appropriate. A more practical approach would be that for each individual seed to average the dose rate distribution at a specific radial distance r over the total possible orientation range, which is over the entire sphere of radius r:

$$\dot{D}(r) = \left(\frac{1}{4\pi} \right) \int_0^{4\pi} \dot{D}(r, \theta, \varphi)\mathrm{d}\Omega \qquad (8.30)$$

with $d\Omega = \sin\theta\, d\theta\, d\varphi$, the solid angle (see Figure 2.1). Mathematically, Equation 8.30 obtains the solid angle-weighted average dose rate value (averaged over the entire 4π steradian space). Due to the cylindrical symmetry of the dose distribution for all sources addressed by the TG-43 protocol we have

$$\dot{D}(r, \theta, \varphi) = \dot{D}(r, \theta)$$

for any azimuthal angle φ. Thus the solid angle element $d\Omega$, after integrating for φ in the full range $[0, 2\pi]$, is $d\Omega = 2\pi\sin\theta\, d\theta$ and Equation 8.30 becomes

$$\dot{D}(r) = \left(\frac{1}{2}\right)\int_0^\pi \dot{D}(r,\theta)\sin\theta\, d\theta \tag{8.31}$$

or using the dose rate at the point (r,θ_0), $\dot{D}(r,\theta_0)$:

$$\dot{D}(r) = \dot{D}(r, \theta_0)\frac{\displaystyle\int_0^\pi \dot{D}(r,\theta)\sin\theta\, d\theta}{2\dot{D}(r, \theta_0)}$$

Considering the TG-43 formula, Equation 8.16, for the point (r,θ_0) and noting that $F(r,\theta_0) = 1.0 \forall r$, the following equation is obtained:

$$\dot{D}(r) = S_K\Lambda\frac{G(r, \theta_0)}{G(r_0, \theta_0)}g(r)\left[\frac{\displaystyle\int_0^\pi \dot{D}(r, \theta)\sin\theta\, d\theta}{2\dot{D}(r, \theta_0)}\right] \tag{8.32}$$

TG-43 terms the fraction shown in brackets at the right side of the above equation anisotropy factor $\varphi_{an}(r)$[3]:

$$\varphi_{an}(r) = \frac{\displaystyle\int_0^\pi \dot{D}(r, \theta)\sin\theta\, d\theta}{2\dot{D}(r, \theta_0)} \tag{8.33}$$

The anisotropy factor $\varphi_{an}(r)$ is thus defined as the ratio of the dose rate at distance r, averaged with respect to solid angle, to the dose rate on the transverse y-axis ($\theta = \theta_0 = 90°$) at the same distance.

The 1D TG-43 model results from Equation 8.32 using the anisotropy factor $\varphi_{an}(r)$:

$$\dot{D}(r) = S_K\Lambda\frac{G(r, \theta_0)}{G(r_0, \theta_0)}g(r)\varphi_{an}(r) \tag{8.34}$$

For distances greater than the source active length, TG-43 alternatively proposed the simplified expression of the above equation, using the point source approximation for the geometry function $G(r,\theta) = 1/r^2$, as seen in

Equation 8.29:

$$\dot{D}(r) = S_K \Lambda \left(\frac{r_0}{r} \right)^2 g(r) \varphi_{an}(r) \tag{8.35}$$

Finally, for the seed sources used in interstitial brachytherapy, the original TG-43 report stated that the anisotropy function $\varphi_{an}(r)$ can be approximated by a distance-independent overall anisotropy parameter, the anisotropy constant φ_{an}.

A common method of obtaining the value of the anisotropy constant φ_{an} is by using a $1/r^2$ weighted-average of anisotropy factors for radial distances $r > 1$ cm:

$$\varphi_{an} = \frac{\displaystyle\sum_i \frac{\varphi_{an}(r_i)}{r_i^2}}{\displaystyle\sum_i \frac{1}{r_i^2}} \tag{8.36}$$

where the sum in both nominator and denominator is over all radial distances r_i where anisotropy factors $\varphi_{an}(r_i)$ are available.

8.3.4 The Revised TG-43 Dosimetry Formalism

Overall, the TG-43 protocol has resulted in significant improvements in the standardization of both dose calculation methodologies as well as dose rate distributions used for clinical implementation of brachytherapy. Most treatment planning software vendors have implemented the TG-43 formalism and the recommended dosimetry parameters in their systems. Due to the fact that the TG-43 parameters and factors can easily be extracted from dose rate measurements, LiF–TLD dose measurements and MC dose calculations have largely replaced the semi-empirical dose calculation models used in the past. Since the publication of the TG-43 protocol 10 years ago, significant advances have taken place in the field of permanent source implantation and brachytherapy dosimetry.

The AAPM[13] deemed it necessary to update this original protocol for the following reasons:

1. To eliminate minor inconsistencies and omissions in the original TG-43 formalism and its implementation
2. To incorporate subsequent AAPM recommendations, addressing requirements for acquisition of dosimetry data as well as clinical implementation
3. To critically reassess published brachytherapy dosimetry data for the ^{125}I and ^{103}Pd source models introduced both priory and subsequent to publication of the TG-43 protocol in 1995,[3] and to recommend consensus datasets where appropriate

4. To develop guidelines for the determination of reference-quality dose distributions by both experimental and MC methods, and promote consistency in derivation of parameters used in TG-43 formalism

There are two important clarifications in the revised AAPM TG-43 protocol, called TG-43 U1.[13]

8.3.4.1 The Transverse-Plane

The transverse-plane of a cylindrically symmetric source is that plane which is perpendicular to the longitudinal axis of the source and bisects the radioactivity distribution. This is the $x - y$-plane shown in Figure 8.2. Regarding this, TG-43 U1 offers concrete instructions of how to handle this in the clinical treatment planning practice, as well as for specific source constructions.

8.3.4.2 Seed and Effective Length L_{eff}

A seed is defined to be a cylindrical brachytherapy source such as that shown in Figure 8.6a, with an active length L_s that is the less than or equal to 0.5 cm.

In the case where the radioactivity is distributed over a cylindrical volume or annulus, TG-43 U1 recommends using active length L_s, the length of this cylinder.

In the case where a brachytherapy source contains uniformly spaced multiple radioactive components, e.g., pellets or spheres as shown in Figure 8.14, the active source length L_s should be considered to be the

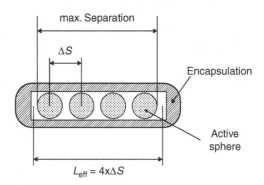

FIGURE 8.14
Graphical demonstration of the determination of the effective source length L_{eff} according to TG-43 U1[13] for the example of a seed source containing four uniformly spaced radioactive spheres. ΔS is the center-to-center spacing of the radioactive spheres. In this specific case the effective source length L_{eff} is longer than maximum separation of the radioactivity defined as the distance between the most distal edges of the radioactive spheres (1st and 4th spheres).

effective length L_{eff} defined as

$$L_{eff} = N\Delta S \qquad (8.37)$$

with N being the number of the discrete radioactive components contained in the source with a nominal component center-to-center spacing ΔS.

Calculated in this way, if the L_{eff} is greater than the physical length of the source capsule, the maximum separation of the activity distribution should be used as the effective length (Figure 8.14).

8.3.4.3 The Basic Consistency Principle

The derivation of the TG-43 function values, radial dose function $g(r)$, Equation 8.24, and anisotropy function $F(r,\theta)$, Equation 8.26, from 2D calculated or measured dose rate tables, is based on the considered geometry function $G(r,\theta)$.

Practically (see Figure 8.15) based on MC calculated or measured 2D dose rate tables, $\dot{D}(r,\theta)$ per unit source strength, per U, and applying a geometry function model, the $g(r)$ and $F(r,\theta)$ values/tables are derived.

Thereafter, and according to the original TG-43 formulation described by Equation 8.16, the product of the $g(r)$, $F(r,\theta)$, geometry function ratio, and dose rate constant Λ values is built which should result in the originally used, calculated, or measured dose rate values/table $\dot{D}(r,\theta)$.

This is true only in the case where the consistency criterion is fulfilled, namely in both steps the same geometry function model is used. This is the basic principle for the updated TG-43, the TG-43 U1 formulation.

The same is also valid for the case of the anisotropy factor $\varphi_{an}(r)$, which according to its definition, Equation 8.33, is also based on the $\dot{D}(r,\theta)$ values.

At that point a violation of the consistency criterion is identified when considering the original TG-43 formulation with regard to the point source approximation described by Equation 8.35. There radial dose function values $g(r)$ obtained using the line point approximation for the geometry function $G(r,\theta)$ are mixed in this case with the point source approximation of $G(r,\theta) = 1/r^2$.

The TG-43 U1 clearly fulfils the consistency criterion and recommends the use of corresponding methods of calculation.

8.3.4.4 The Geometry Function $G_X(r,\theta)$

For being able to clearly distinguish among the two source approximations for calculating the geometry function $G(r,\theta)$, as has been introduced by the TG-43 original protocol, the line source approximation obtained by Equation 8.28, and point source approximation obtained by Equation 8.29, TG-43 U1 introduced separate notations of the corresponding geometry function:

$$G_X(r, \theta) = \begin{cases} G_P(r, \theta) \text{ for the point source approximation} \\ G_L(r, \theta) \text{ for the line source approximation} \end{cases}$$

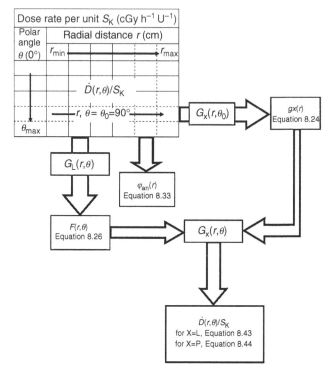

FIGURE 8.15

The basic consistency principle for derivation of parameter and function values of the TG-43 (revised) formalism. The starting point is an MC calculated or measured 2D dose rate values table, $\dot{D}(r,\theta)$ per unit source strength, per U (column is the polar angle dependence and row is the radial distance dependence). Applying a geometry function model $G_X(r,\theta)$ to the row for the reference polar angle value $\theta = \theta_0 = 90°$, the values of the corresponding radial dose function $g_X(r)$ are obtained using Equation 8.24. Applying the line source approximation of the geometry function $G_L(r,\theta)$ (X = L) according to Equation 8.38 to any of the columns of the dose rate matrix, the corresponding values of the 2D anisotropy function $F(r,\theta)$ are obtained using Equation 8.26. Alternatively, for the case of 1D TG-43 model, for each one of the radial distances (columns) the value of the anisotropy factor $\varphi_{an}(r)$ is obtained using the integration over the whole polar angle range 0 to 180° (column) as described in Equation 8.33. Thereafter, and using either the 2D model obtained using Equation 8.43, X = L, or the 1D model obtained using Equation 8.44, X = P, the corresponding values of the TG-43 parameters are multiplied together with the consistent geometry function ratio values and the dose rate constant Λ to result in the originally used MC calculated or measured dose rate values/table $\dot{D}(r,\theta)$.

with

$$G_P(r, \theta) = G_P(r) = \frac{1}{r^2}$$

$$G_L(r, \theta) = \begin{cases} \left(r^2 - \dfrac{L_s^2}{4}\right)^{-1} & \text{if } \theta = 0 \\[3mm] \dfrac{\beta}{L_s r \sin \theta} & \forall\, \theta \neq 0 \end{cases} \qquad (8.38)$$

where, for the definition of the corresponding parameters, the same geometry as for TG-43 shown in Figure 8.6 and Figure 8.8 is incorporated. The subscripts "L" and "P" have been added to denote the line and point source approximation, respectively, used for the calculation of the geometry function.

For the line source approximation $G_L(r,\theta)$, TG-43 U1 explicitly handles the case of $\theta = 0°$ by solving Equation 8.11 directly for that case. In the original formulation of 1995, this case was undefined since Equation 8.28 results for $\theta = 0°$ to the degenerative case of zero divided by zero.

Based on the geometry shown in Figure 8.10, and for the specific case $\theta = 0°$, it is $y_P = 0$, $z_P = r$ and thus

$$|\vec{r} - \vec{r}'|^2 = |\vec{r} - \vec{r}_l|^2 = (z_P - z_l)^2 = (r - z_l)^2$$

Using this, the integral in the nominator of Equation 8.11 takes the following form:

$$\int_{V_s} \frac{\rho(\vec{r}')dV'}{|\vec{r} - \vec{r}'|^2} = \rho_l \int_{-L_s/2}^{+L_s/2} \frac{dl}{(r - z_l)^2} = \rho_l \int_{-L_s/2}^{+L_s/2} \frac{dz_l}{(r - z_l)^2} = \rho_l \left[\frac{1}{r - z_l} \right]_{-L_s/2}^{+L_s/2}$$

and finally:

$$\int_{V_s} \frac{\rho(\vec{r}')dV'}{|\vec{r} - \vec{r}'|^2} = \rho_l \left[\frac{1}{r - \frac{L_s}{2}} - \frac{1}{r + \frac{L_s}{2}} \right] = \rho_l \left[\frac{L_s}{r^2 - \frac{L_s^2}{4}} \right] \quad (8.39)$$

Considering Equation 8.39 and Equation 8.27 into the general definition of the geometry function described by Equation 8.11, and for the specific case of $\theta = 0°$, it is

$$G_L(r, \theta = 0°) = \left[\frac{\int_{V_s} \frac{\rho(\vec{r}')dV'}{|\vec{r} - \vec{r}'|^2}}{\int_{V_s} \rho(\vec{r}')dV'} \right]_{\theta=0°} = \frac{1}{r^2 - \frac{L_s^2}{4}} \quad (8.40)$$

According to the source design, the source length L_s or the effective length L_{eff} as defined in Equation 8.37 has to be used in the line source approximation $G_L(r,\theta)$ obtained by Equation 8.38.

8.3.4.5 The Radial Dose Function $g_X(r)$

Regarding the definition of the radial dose function described by Equation 8.24, and considering the definitions of the line source and point source approximations for the geometry function, $G_L(r,\theta)$ and $G_P(r,\theta)$, respectively, the following two separate calculations of the radial dose function are obtained to fulfill the consistency criterion:

$$g_X(r) = \begin{cases} g_L(r) \text{ for the line source approximation} = \left(\dfrac{G_L(r_0, \theta_0)}{G_L(r, \theta_0)}\right)\left(\dfrac{\dot{D}(r, \theta_0)}{\dot{D}(r_0, \theta_0)}\right) \\[3mm] g_P(r) \text{ for the point source approximation} = \left(\dfrac{G_P(r_0, \theta_0)}{G_P(r, \theta_0)}\right)\left(\dfrac{\dot{D}(r, \theta_0)}{\dot{D}(r_0, \theta_0)}\right) \end{cases}$$

$$(8.41)$$

where $G_L(r,\theta)$ and $G_P(r,\theta)$ are obtained according to Equation 8.38. Once more the subscripts "L" and "P" have been added to the radial dose function to denote the line and point source approximation, respectively, used for the calculation of the geometry function.

Compared to the original TG-43 protocol this is essentially different, since now the radial dose function has to be calculated according to the approximation model considered for the geometry function, line, or point source approximation. The $g_L(r)$ radial dose function is identical to the $g(r)$ obtained by Equation 8.24 for the TG-43 protocol.

8.3.4.6   The Air Kerma Strength S_K

The air kerma strength S_K, established practically by the TG-43-report published in 1995 (see also Chapter 7, Equation 7.1), is defined as the product of the air kerma rate in free space, *vacuo*, at a measurement distance r from the source center along the perpendicular bisector, $\dot{K}_\alpha(r)$, and the square of the distance r. It is numerically identical to the reference air kerma rate, $\dot{K}_R$, recommended by ICRU report 38[14] and ICRU report 58.[15]

In order to be able to exclude low energy or contaminant photons such as characteristic x-rays originating in the outer layers of titanium or steel source cladding material, which are not tissue penetrating (they contribute insignificantly to absorbed dose at distances above 0.1 cm), TG-43 U1 uses a photon energy cut-off value δ for defining the air kerma strength S_K (similar to the definition of the air kerma rate constant Γ_δ Equation 2.61, see also Chapter 5):

$$S_K = \dot{K}_{\alpha,\delta}(r)r^2 \qquad\qquad (8.42)$$

where $\dot{K}_{\alpha,\delta}(r)$ is the air kerma rate *in vacuo* at a distance r from the source center along the perpendicular bisector, due to photons of energy greater than the cut-off value δ which is expressed in keV. The recommended unit for air kerma strength remains μGy m^2 h^{-1} denoted by the symbol U, as was introduced by the original TG-43 publication.[15]

According to the TG-43 U1 report, the distance r can be any distance that is large relative to the maximum linear dimension of the radioactivity distribution.

Considering the energy cut-off value δ in the source strength definition makes air kerma rate *in vacuo* consistent to the dose rate in water due to the fact that the photons with energy below this cut-off value will have no, or at least no significant, contribution to the dose rate in water and consequently their contribution to air kerma rate has also to be excluded. Thus consistent values for the dose rate constant Λ are also derived. Generally, when using the energy cut-off, the air kerma strength S_K is reduced and consequently the dose rate constant Λ, as seen in Equation 8.19, is increased.

For the low energy radionuclides ^{125}I and ^{103}Pd TG-43 U1 recommends the use of an energy cut-off value $\delta = 5$ keV. For the intermediate energy ^{169}Yb, and high-energy sources of ^{198}Au, ^{192}Ir, ^{137}Cs, and ^{60}Co, a cut-off value $\delta = 10$ keV is commonly used (see also Chapter 5 and Chapter 7). For ^{192}Ir a value $\delta = 11.3$ keV has also been proposed.[16,17] The ICRU report 58[15] recommends a cut-off value of 20 keV where it is noted that this may depend on cladding of the specific source.

For the energy spectra of pure radionuclides presented in Chapter 5, there is no difference expected for the intermediate and high energy sources when either $\delta = 10$ keV or $\delta = 11.3$ keV is used.

8.3.4.7 *The Revised TG-43 2D Dosimetry Formalism*

The general 2D equation for calculating the absorbed dose rate to water is the one from the original 1995 TG-43 protocol, Equation 8.16, where the line source approximation based geometry function, $G_L(r,\theta)$ from Equation 8.38 and radial dose function $g_L(r)$ from Equation 8.41 are included:

$$\dot{D}(r, \theta) = S_K \Lambda \frac{G_L(r, \theta)}{G_L(r_0, \theta_0)} g_L(r) F(r, \theta) \tag{8.43}$$

The corresponding geometry and coordinate system is the exact one defined in the TG-43 protocol (also shown in Figure 8.6).

The anisotropy function $F(r,\theta)$ is defined by Equation 8.26 and incorporates, per definition, the line source approximated geometry function $G_L(r,\theta)$. Thus for the anisotropy function the subscript "L" is not used.

8.3.4.8 *The Revised TG-43 1D Dosimetry Approximation*

Using the anisotropy factor $\varphi_{an}(r)$ introduced by TG-43 Equation 8.33 and defined as the ratio of the dose rate at the distance r, averaged with respect to solid angle, to the dose rate on the transverse y-axis ($\theta = 90°$) at the same distance, the dose rate around a source can be calculated without the need to determine the source orientation in space using the 1D approximation:

$$\dot{D}(r) = S_K \Lambda \frac{G_X(r, \theta_0)}{G_X(r_0, \theta_0)} g_X(r) \varphi_{an}(r) \tag{8.44}$$

Considering the two possible approximations for the calculation of the geometry function, line and point source approximation obtained in Equation 8.38, the following two possible 1D dose rate calculation formulas are derived.

When the line source approximation $G_L(r,\theta)$ is considered:

$$\dot{D}(r) = S_K \Lambda \frac{G_L(r, \theta_0)}{G_L(r_0, \theta_0)} g_L(r) \varphi_{an}(r) \qquad (8.45)$$

and in the case where the point source approximation $G_P(r,\theta)$ is considered:

$$\dot{D}(r) = S_K \Lambda \left(\frac{r_0}{r}\right)^2 g_P(r) \varphi_{an}(r) \qquad (8.46)$$

where the line and point source approximation based radial dose functions, $g_L(r)$ and $g_P(r)$, respectively, are according to Equation 8.41.

TG-43 U1 recommends the use of Equation 8.45 due to the improved accuracy at small distances. This simplifies the data handling as well, since there is no need to store and handle an additional table with $g_P(r)$ values.

Although not recommended, TG-43 U1 proposes that if the application of anisotropy constant φ_{an} (see Equation 8.35) cannot be avoided, e.g., is used by the treatment planning system, then the following equation should be considered to minimize errors at small distances of $r \le 1.0$ cm:

$$\dot{D}(r) = S_K \Lambda \left(\frac{r_0}{r}\right)^2 g_L(r) \varphi_{an} \qquad (8.47)$$

The anisotropy constant φ_{an} is obtained by Equation 8.36 where the sum in both nominator and denominator is overall radial distances, $r_i \ge 1.0$ cm.

8.3.4.9 *Implementation Recommendations*

The TG-43 U1 recommends rules and methodologies for measuring and calculating using MC simulation techniques, dose rate values, and dose rate distributions around brachytherapy sources, and thus for extracting the different quantities considered in the 2D and 1D formulations.

8.3.4.9.1 *Dose Rate Constant Λ*

Having available an experimentally determined dose rate constant value $_{EXP}\Lambda$ with a percentage uncertainty σ_{EXP}, and a MC obtained value $_{MC}\Lambda$ with a percentage uncertainty σ_{MC}, the following consensus, average, value for $_{CON}\Lambda$ is recommended:

$$_{CON}\Lambda = \frac{_{EXP}\Lambda + _{MC}\Lambda}{2} \qquad (8.48)$$

and the corresponding percentage uncertainty σ_Λ is

$$\sigma_{CON\Lambda} = \sqrt{a^2 \left(\frac{_{EXP}\Lambda}{_{CON}\Lambda}\right)^2 \sigma_{EXP}^2 + (1-a)^2 \left(\frac{_{MC}\Lambda}{_{CON}\Lambda}\right)^2 \sigma_{MC}^2 + \sigma_B^2} \qquad (8.49)$$

with $a = 0.5$. σ_B is an additional type B uncertainty (systematic), due to the possible bias in the average of the results of experimental and MC methods, and is obtained by[13]

$$\sigma_B = 100\left(\frac{|_{EXP}\Lambda - _{MC}\Lambda|}{2\sqrt{3}\Lambda}\right) \tag{8.50}$$

8.3.4.9.2 Radial Dose Function g(r)

The radial dose function $g(r)$ obtained according to Equation 8.40, which is either line source or point source approximated, is a relative quantity and its values are extracted from MC based dosimetry. Experimental values are only consider for validation of the MC simulation results. TG-43 U1 recommends a radial distance range of 0.5 to 7.0 cm for ^{125}I and 0.5 to 5.0 cm for ^{103}Pd, where values of the radial dose function have to be available. In the range of 0.5 to 1.5 cm, a more dense distribution of radial distances should be considered to increase the accuracy of calculations.

Although TG-43 U1 does not address the HDR sources used in afterloading devices, most commonly ^{192}Ir sources, for such intermediate energy, ^{169}Yb, or high energy sources, ^{192}I, ^{198}Au, ^{137}Cs, ^{60}Co, radial distances ranges of 0.5 to 15.0 cm are commonly used. Depending on the source encapsulation design and on the thickness of applied catheters for the temporary implants, distances as low as 0.2 to 0.3 cm or even 0.1 cm could also be needed, e.g., for ^{192}Ir HDR sources.

8.3.4.9.3 Anisotropy Function F(r,θ)

Similar to the radial dose function, the anisotropy function $F(r,\theta)$ is obtained using Equation 8.26, and incorporating the line source approximated geometry function $G_L(r,\theta)$ from MC calculated dose rate distributions.

For the low energy seeds ^{125}I and ^{103}Pd used in permanent implants which have a cylindrical symmetry as shown in Figure 8.6a, values of the polar angle θ in the range of 0 to 90° in increments of 10° are recommended. For ^{125}I seeds radial distances, $r = \{0.5, 1.0, 2.0, 3.0, 5.0, 7.0\}$ cm and for ^{103}Pd $r = \{0.5, 1.0, 2.0, 3.0, 5.0\}$ cm are recommended.

For source designs that are asymmetrical about the transverse plane $(x - y$-plane in Figure 8.2), such as the ^{192}Ir HDR sources in afterloading devices that are fixed to driving cables (Figure 8.6b), polar angles in the range of 0 to 180° in increments of 10° are advised. Radial distances should be selected to be in the range of 0.5 to 15.0 cm and, if possible, smaller distances up to 0.1 to 0.3 cm (usually 0.25 cm) should be considered if applicable for a given source design. If possible, and for high gradient regions, around $\theta = 0°$ and $\theta = 180°$, higher spatial resolution should be selected in the tabulation of the anisotropy values.

Table 8.1 summarizes the ranges of radial distance and polar angles recommended for the radial dose and anisotropy function and for the different types of sources, low, intermediate, and high energy.

TABLE 8.1

Recommended Values or Ranges of Values for the Radial Distance r and Polar Angle θ

Quantity	Source Type	Radial Distance r (cm)	Polar Angle θ (°)
Radial dose function $g_X(r)$	Low energy sources[a]	^{125}I: 0.5–7.0 ^{103}Pd: 0.5–5.0	—
	Intermediate and high energy sources	0.2–15.0 if possible 0.1–15.0	—
Anisotropy function $F(r,\theta)$	Low energy sources[a]	^{125}I: {0.5, 1, 2, 3, 5, 7} ^{103}Pd: {0.5, 1, 2, 3, 5}	0–90 in 10° steps
	Intermediate and high energy sources	In the range of 0.2–15.0, e.g., {0.2, 0.5, 1, 3, 5, 7, 10, 12, 15}	0–180 in 10° steps[b]

Tabulated values of the radial dose function $g_X(r)$ and anisotropy function $F(r,\theta)$ should be available.

[a] According to TG-43 U1.[13]
[b] Assumed to be asymmetric about the transverse plane.

8.3.4.9.4   Interpolation and Extrapolation Rules

Radial dose function and anisotropy function are usually available in tabular form. The corresponding function values for radial distance and polar angle values between tabulated values of $g(r)$ and $F(r,\theta)$, are obtained by interpolation.

Care must be taken in calculating and evaluating dose rate values for positions outside the tabulated data.

8.3.4.9.4.1   2D Dose Calculation

Equation 8.42 is utilized for dose calculation. The geometry function value at the point (r,θ) is calculated explicitly using Equation 8.38.

Radial Dose Function $g_L(r)$

For a radial distance lying between two tabulated distances r_1 and r_2, the radial dose function value at r can be calculated using linear interpolation:

$$g_L(r) = g_L(r_1) + \left(\frac{g_L(r_2) - g_L(r_1)}{r_2 - r_1} \right)(r - r_1) \qquad (8.51)$$

with $r_1 \leq r \leq r_2$.

For $r > r_{max}$, with r_{max} being the maximum radial distance of the tabulated values, or $r < r_{min}$, with r_{min} being the minimum radial distance of the tabulated values, linear extrapolation can be used considering the two previous tabulated $g(r)$ values.

In any case, for those calculation points with radial distances larger than the recommended maximum available in the table, the resulted dose rate will be extremely low and thus the contribution of the extrapolation error to the overall accuracy is extremely low.

Special attention should be paid when polynomial functions are considered for interpolating between table values and extrapolating outside the table. Depending on the order of the polynomial function, abnormal behavior can result, e.g., extrapolating to large distances can result in an increase in radial dose function value.

For both LDR seeds[18] and HDR[18,19] brachytherapy sources, more advanced analytical functions for interpolation and extrapolation have been considered in the literature with adequate accuracy results.

Anisotropy Function F(r,θ)

The anisotropy function $F(r,\theta)$ defined by Equation 8.26 incorporates, per definition, the line source approximated geometry function $G_L(r,\theta)$. $F(r,\theta)$ is available in the form of a 2D table with values in the radial distances range of $r \in [r_{min}, r_{max}]$ and in the polar angle range of $\theta \in [\theta_{min}, \theta_{max}]$. It is expected (as listed in Table 8.1) that for sources symmetrical about the transverse plane, the limits are $\theta_{min} = 0°$ and $\theta_{max} = 90°$, whereas for all other cases $\theta_{max} = 180°$.

For any point of calculation, $P(r,\theta)$ with (r,θ) lying between tabulated values, a bilinear interpolation can be used for obtaining the anisotropy function value at that point.

$$F(r, \theta) = F_1 + (r - r_1)\left[\frac{F_2 - F_1}{r_2 - r_1}\right] \tag{8.52}$$

with

$$F_1 = F(r_1, \theta) = F(r_1, \theta_1) + (\theta - \theta_1)\left[\frac{(F(r_1, \theta_2) - F(r_1, \theta_1))}{(\theta_2 - \theta_1)}\right]$$

$$F_2 = F(r_2, \theta) = F(r_2, \theta_1) + (\theta - \theta_1)\left[\frac{(F(r_2, \theta_2) - F(r_2, \theta_1))}{(\theta_2 - \theta_1)}\right]$$

where $r_1 \leq r \leq r_2$ and $\theta_1 \leq \theta \leq \theta_2$ with $F(r_1, \theta_1)$ and $F(r_2, \theta_2)$ being tabulated values of the anisotropy function.

For the case of radial distance value r outside the tabulated range, $r \notin [r_{min}, r_{max}]$, the following scheme is recommended:

$$F(r, \theta) = \begin{array}{ll} F(r_{min}, \theta) & \text{for } r < r_{min} \\ F(r_{max}, \theta) & \text{for } r > r_{max} \end{array} \tag{8.53}$$

where for the polar angle θ, $\theta_1 \leq \theta \leq \theta_2$, linear interpolation can be used to obtain the anisotropy function value at the resulting radial distance r, r_{min} or r_{max} according to Equation 8.53.

Important

In order that dose calculations based on either the 2D formulation described by Equation 8.43 or 1D formulation described by Equation 8.44 are within the recommended accuracy achievable using the TG-43 parameter tables and constant values, the consistency criterion has to be fulfilled. This requires that the same source length determination L_s (for seed sources see Figure 8.14), and consequently the same geometry function approximation, is used in the calculations as has been incorporated for the conversion of MC calculated or measured dose rate tables into TG-43 quantities (see also Figure 8.15).

8.4 TG-43 Data for Sources

Dozens of source designs of several radionuclides, of high, intermediate, and low energy, have been developed and many of these are currently in clinical use. A wide variety is particularly observed for the low energy and low source strength sources of ^{125}I and ^{103}Pd, used mainly for permanent implants in early stages of prostate cancer. The TG-43 dosimetry data for several source designs will be addressed in the following paragraphs, predominantly dealing with two main categories of implants, permanent and temporary, used in brachytherapy.

In general, sources containing the high energy and extremely long half-life radionuclides ^{60}Co and ^{137}Cs and the intermediate energy and intermediate half-life radionuclide ^{192}Ir are considered only for temporary implants using manually or automatic afterloading technologies. Even if ^{192}Ir seeds have been also used for permanent implants, this never gained the interest of the radiotherapy community.

The low energy, intermediate to short half-life radionuclides ^{125}I and ^{103}Pd are only considered for permanent implants, currently the majority for prostate cancer brachytherapy. ^{198}Au containing sources, being of intermediate energy and having a very short half-life, are also considered for permanent implants, but such sources are currently not widely applied.

Finally, ^{169}Yb sources, an intermediate energy and intermediate half-life radionuclide, have been considered in the past in a few clinical applications for permanent as well as for temporary brachytherapy implants. Recent improvements in source production technology have made ytterbium a promising candidate for high strength sources for temporary implants[42] using afterloading technology, being theoretically able to achieve an activity three to seven times of that of ^{192}Ir (in means of contained activity).

All figures and values given are generally based on the TG-43/TG-43 U1 source coordinate system as shown in Figure 8.6. In case there is any deviation, this will be clearly stated and explained.

8.4.1 Sources Used in HDR Brachytherapy

In HDR brachytherapy, ^{192}Ir and ^{60}Co sources are currently used in afterloading devices. Iridium sources with an intermediate energy have a great advantage over cobalt sources regarding radiation protection issues. On the other hand, the intermediate half-life of iridium makes three to four source exchanges per year mandatory in order to keep total treatment times short. HDR ^{60}Co sources can be used for several years, thus making the logistics for radioactive material transportation and regulation issues simpler. This is of great importance for developing countries and perhaps for centers with a limited number of brachytherapy applications where the three to four iridium source exchanges per year and the related costs are hindering the establishment of brachytherapy as a mode of therapy.

8.4.1.1 [192]Ir Sources

The commercially available source designs for ^{192}Ir containing HDR sources are listed in Table 8.2. It is common practice to name the sources according to the HDR afterloader system they are used for, and to the special version of design for that system. This means that the afterloader system manufacturer, and not the source manufacturer/producer, is mainly listed in the tables and figures in this chapter. Table 8.3 summarizes the published values of the dose rate constant Λ based either on MC simulations, $_{MC}\Lambda$, or experimentally measured $_{EXP}\Lambda$, for all HDR iridium sources. The MC code used for the simulations are also listed, as well as the phantom material and dosimeters used for the experimental dosimetry.

8.4.1.1.1 Sources for the MicroSelectron HDR Afterloader System

There are two different source designs used with the Version 1 and Version 2 microSelectron afterloaders by Nucletron (Nucletron B.V, Veenendaal, The Netherlands) named classic (old) with an active core of 0.35 cm length, and new with an active core of 0.36 cm in length, both shown in Figure 8.16.

8.4.1.1.1.1 The Classic Design
Williamson and Li[20] were the first to publish TG-43 parameter values for the classic design based on MC simulation studies, using their in-house developed MCPT code. The calculations have been realized with the source cantered in a liquid water sphere of 30 cm diameter. For the extraction of the TG-43 parameters, the line source approximation of the geometry function $G_L(r,\theta)$ according to Equation 8.38, has been considered using the length of the active core as source length. Authors performed the MC simulation study including a 3 mm long drive cable consisting of solid stainless steel. Although the original publication considered a source coordinate system with $+z$ direction towards the drive cable (polar angle θ defined relative to drive cable), all data have been transformed to follow the TG-43/TG-43 U1 convention of Figure 8.6.

TABLE 8.2

Geometrical and Material Characteristics of Common Commercial High Dose Rate (HDR) Sources all Containing ^{192}Ir

| Source Type | Active Core | | | Encapsulation | | | Cable | |
	Material	Length L_s (cm)	Diameter D_s (cm)	Material	Thickness (cm)	Outer Diameter (cm)	Material	Outer Diameter (cm)
Microselectron HDR (old design, classic)	Ir	0.35	0.060	Stainless steel AISI 316L	0.0250	0.110	Stainless steel AISI 304	0.110
MicroSelectron HDR (new design)	Ir	0.36	0.065	Stainless steel AISI 316	0.0125	0.090	Stainless steel AISI 316	0.070
VariSource HDR (old design)	Ir	1.00	0.034	Ti/Ni	0.0125	0.059	Ti/Ni	0.059
VariSource HDR (new design)	Ir	0.50	0.034	Ti/Ni	0.0125	0.059	Ti/Ni	0.059
Buchler HDR	Ir	0.13	0.100	Stainless steel AISI 321	0.0200	0.160	Stainless steel AISI 301	0.110
GammaMed HDR Model 12i	Ir	0.35	0.060	Stainless steel AISI 316L	0.0200	0.110	Stainless steel AISI 304	0.110
GammaMed HDR Model plus	Ir	0.35	0.060	Stainless steel AISI 316L	0.0100	0.090	Stainless steel AISI 304	0.090
BEBIG MultiSource HDR GI192M11	Ir	0.35	0.060	Stainless steel	0.0200	0.100	Stainless steel	0.100

All sources are used in computer controlled afterloading devices. The commercial names of the corresponding afterloading devices are commonly used to identify source types, and not the source manufacturer itself.

TABLE 8.3

Monte Carlo and Experimental Estimated Values of the Dose Rate Constant Λ for the Different ^{192}Ir High Dose Rate Brachytherapy Sources Listed in Table 8.2

Source Type	$_{MC}\Lambda$	Monte Carlo Code	δ (keV) (Equation 8.42)	$_{EXP}\Lambda$	Phantom	Detector	Author
MicroSelectron HDR (old design, *classic*)	1.115 ± 0.5%	MCPT, own	—				Williamson and Li[20]
	1.131 ± 1.0%	—	—	1.143 ± 5.0%	Solid water	LiF–TLDs	Kirov et al.[28]
	—	EGS4/PRESTA	—				Russel and Anhesjo[29]
	—	—	—	1.134 ± 2.9%	Polystyrene	LiF–TLDs	Anctil et al.[33]
	1.116 ± 0.5%	UoA, own	—				Karaiskos et al.[30]
	1.115 ± 0.5%	UoA, own	10.0				Papagiannis et al.[31]
MicroSelectron HDR (new design)	1.108 ± 0.13%	MCPT, own	—				Daskalov et al.[21]
VariSource HDR (old design)	1.109 ± 0.5%	UoA, own	10.0				Papagiannis et al.[31]
	1.044 ± 2.0%	EGS4/DOSRZ	1.0				Wang and Sloboda[32]
	1.043 ± 0.5%	UoA, own	10.0				Karaiskos et al.[22]
	1.043 ± 0.5%	UoA, own	10.0				Papagiannis et al.[31]
	—	—	—	1.069 ± 5%	Solid water	LiF–TLDs and RCF[a]	Meigooni et al.[34]
VariSource HDR (new design)	1.101 ± 0.5%	UoA, own	10.0				Angelopoulos et al.[23]
	1.101 ± 0.5%	UoA, own	10.0				Papagiannis et al.[31]

(continued)

TABLE 8.3—*Continued*

Source Type	$_{MC}\Lambda$	Monte Carlo Code	δ (keV) (Equation 8.42)	$_{EXP}\Lambda$	Phantom	Detector	Author
Buchler HDR	1.115 ± 0.3%	GEANT	10.0				Ballester et al.[24]
	1.115 ± 0.5%	UoA, own	10.0				Papagiannis et al.[31]
GammaMed HDR Model 12*i*	1.118 ± 0.3%	GEANT3	—				Ballester et al.[25]
GammaMed HDR Model Plus	1.118 ± 0.3%	GEANT3	—				Ballester et al.[25]
BEBIG Multi-Source HDR GI192M11	—	—	—	1.119 ± 4.5%	Water phantom	Pin-point-chamber	Selbach and Andrássy[71]
	1.108 ± 0.3%	GEANT4	10.0				Granero et al.[76]

All Λ values are expressed in cGy h^{-1} U^{-1}. Energy cut-off values δ referred in the original publications are also listed.

[a] Radiochromic film value is the mean of values with TLDs and RCF.

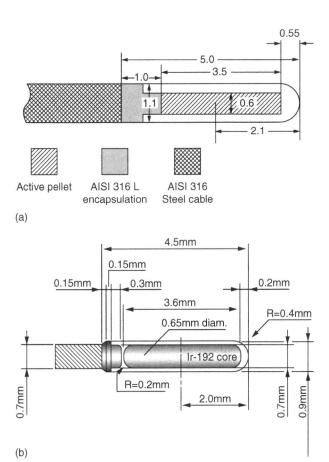

FIGURE 8.16
MicroSelectron HDR ^{192}Ir source type designs (Nucletron B.V., Veenendaal, The Netherlands). (a) The old "classic" design with 0.6 mm cylindrical core diameter and 3.5 mm core length having a total outer diameter (including encapsulation) of 1.1 mm.[20] (b) The new design with a cylindrical core diameter of 0.65 mm and core length of 3.6 mm. The total outer diameter of this source design is 0.9 mm (figure provided by Nucletron). All dimensions shown are in mm.

Russel and Anesjo[29] also published MC based data for the classic source design. They used the EGS4/PRESTA simulation code.[50] Their simulation geometry consisted of an 80 cm high water cylinder of 80 cm diameter, thus approximating an infinite-phantom geometry (unbounded geometry). They carried out simulations using several applicator materials. Only the dose rate constant value for water applicator and pure source is considered in the data presentation. In their MC calculations, the stainless steel cable attached at the distal end of the source (see Figure 8.16a) was also considered to extend up to the end of the water cylinder, being 40 cm long. A point source approximation of the geometry function was considered as given by Equation 8.29 and Equation 8.38.

Karaiskos et al.[30] published MC simulation-based data, as well as experimental data, measured using LiF TLD-100 rods in a polystyrene phantom with external dimensions of $30 \times 30 \times 30$ cm^3. The simulation study has been performed using the in-house developed MC code, noted here as UoA MC code, including a drive cable of sufficient length (at least 14 cm). The authors considered the line source approximation of the geometry function.

Papagiannis et al.[31] also published MC simulation results including dose rate constant and radial dose function values for the classic source design. The MC calculations were carried out with the source centered in a 30 cm diameter water sphere, considering the drive cable of the source to extend up to the end of the mathematical phantom. An energy cut-off value for photons of 10.0 keV was also used. In this paper the exact source geometry function according to Equation 8.11 has been considered.

Kirov et al.,[28] and Anctil et al.[33] published experimental parameter values using LiF–TLD dosimeters in solid water ($28 \times 28 \times 20$ cm^3) and poly-styrene phantom ($30 \times 30 \times 30$ cm^3), respectively. Anctil et al. also published radial dose function $g(r)$ and anisotropy function $F(r,\theta)$ values up to radial distances of 10.0 cm, also using LiF–TLDs and a $30 \times 30 \times 16.5$ cm^3 polystyrene phantom.

With the exception of Russel and Anesjo,[29] in all other MC studies the primary photon spectra for ^{192}Ir of Glasgow and Dillman[16] were used. Russel and Anesjo[29] considered the ^{192}Ir decay spectrum from MIRD.[46]

As demonstrated in Table 8.3, all MC-based published values of the dose rate constant $_{MC}\Lambda$ are in good agreement and practically within the standard error, 1.115 to 1.131 cGy h^{-1} U^{-1}. The experimentally estimated values $_{MEXP}\Lambda$ of 1.143 and 1.134 cGy h^{-1} U^{-1} are generally higher than the MC based values but, due to their higher experimental uncertainty, 5.0 and 2.9%, respectively, are in good agreement to the MC values.

The radial dose function values of all mentioned publications, MC based or experimentally measured, are shown in Figure 8.17. Very good agreement over all published data in the radial distance range of 0.1 to -15 cm can be observed. The radial dose function values shown in this figure are listed in Table A.3.1 of Appendix 3.

The anisotropy function values and anisotropy factor values according to the MC results by Williamson and Li[20] are listed in Table A.3.2 and Table A.3.3 of Appendix 3.

Figure 8.18 demonstrates a comparison of the anisotropy function values for representative radial distances in the range of 0.25 to 5.0 cm, according to the MC data by Williamson and Li.[20]

Figure 8.19 shows the comparison of the anisotropy function values published by several authors for the representative radial distance of 3.0 and 5.0 cm. Baltas et al.[43] has used a water phantom ($48 \times 48 \times 41$ cm^3) and a 0.1 cm^3 ionization chamber, whereas Muller-Runkel and Cho[44] made measurements using LiF–TLDs in a polystyrene phantom consisting of 30×30 cm^2 plates and thickness ensuring a minimum of 10 cm backscatter

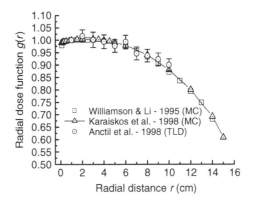

FIGURE 8.17
Radial dose function values, $g(r)$, plotted against radial distance, r, for the microSelectron HDR
^{192}Ir type old "classic" source design. The MC simulation based data by Williamson and Li[20] and
Karaiskos et al.[30] are shown together with the experimentally estimated values using LiF–TLD
dosimeters by Anctil et al.[33] (Table A.3.1 of Appendix 3).

material in each direction for each of the measuring points. Mishra et al.[45]
carried out anisotropy function measurements in a water phantom
$(48 \times 48 \times 48 \text{ cm}^3)$ using a 0.147 cm^3 ionization chamber. Experimentally
measured and MC calculated anisotropy values are in good agreement, as
demonstrated in Figure 8.19. The consistently higher values provided by
Russel and Anesjo[29] for polar angles below 50° and above 130° can be
explained by the practically unbounded simulation geometry, and thus
saturated scatter conditions used in their MC simulation study when
compared to the 30 cm diameter sphere geometry considered in the other
two MC simulation studies and the smaller phantoms used in the
experimental investigations.

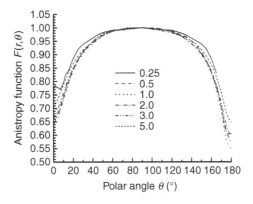

FIGURE 8.18
Anisotropy function, $F(r,\theta)$, polar angle profiles for the microSelectron HDR ^{192}Ir type old
"classic" source at six different radial distances in the range 0.25 to 5.0 cm according to the MC
data by Williamson and Li[20] (Table A.3.2, Appendix 3).

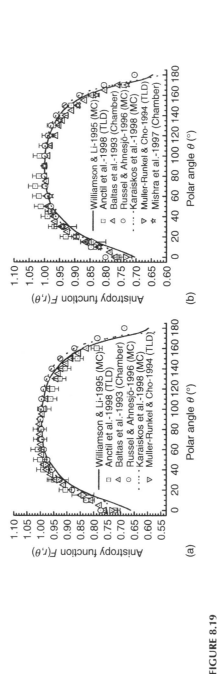

FIGURE 8.19

Comparison of the anisotropy function, $F(r,\theta)$, values published by several authors, estimated either experimentally or using MC simulations for the microSelectron HDR ^{192}Ir type old "classic" source at radial distance (a) $r = 3.0$ cm and (b) $r = 5.0$ cm.

8.4.1.1.1.2   The New Design

Daskalov et al.[21] published MC-based dosimetric data for the new source design with an active core of 0.36 cm in length for the microSelectron HDR system. The corresponding source material and geometrical data are listed in Table 8.2 and are shown in Figure 8.16b. The in-house MC simulation code MCPT was used and the simulation geometry was with the source cantered in a 30 cm outer diameter liquid water sphere. A 2.0 mm long drive cable was taken into account for the simulation geometry. The line source approximation for the geometry function $G_L(r,\theta)$, according to Equation 8.38, using as source length the length of the active core, 0.36 cm, has been considered. The dose rate constant was found to be $_{MC}\Lambda = (1.108 \pm 0.13\%)$ cGy h^{-1} U^{-1}. This value is in perfect agreement with that of $_{MC}\Lambda = (1.109 \pm 0.13\%)$ cGy h^{-1} U^{-1}, calculated by Papagiannis et al.,[31] considering the same simulation geometry (water sphere of 30 cm diameter) and using the in-house developed MC code but simulating a sufficient length (at least 14 cm) for the drive cable. Papagiannis et al. considered a photon energy cut-off value of 10.0 keV in their simulation study. Both dose rate constant values are very close (within 1%) to those for the classic source design (see Table 8.3). Both investigator groups used the primary photon spectra for ^{192}Ir reported by Glasgow and Dillman.[16]

Figure 8.20 shows a comparison of the radial dose functions as obtained by the two author groups, where the corresponding values according to Daskalov et al.[21] are listed in Table A.3.4 of Appendix 3.

Figure 8.21 shows a comparison of the radial dose function values of the classic and new source designs based on the MC calculated data by Williamson and Li[20] (classic) and Daskalov et al.[21] (new).

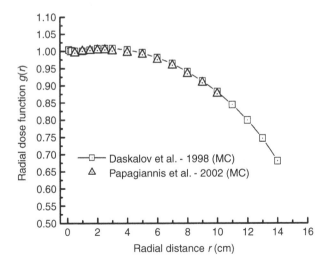

FIGURE 8.20
Radial dose function values, $g(r)$, plotted against radial distance, r, for the microSelectron HDR ^{192}Ir type new source design. The MC simulation based data by Daskalov et al.[21] are compared to the data by Papagiannis et al.[31] also estimated using MC simulation (Table A.3.2 of Appendix 3).

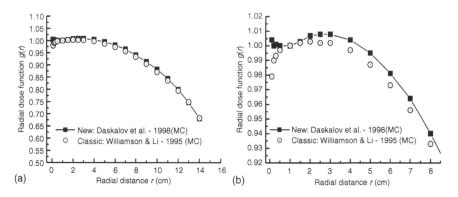

FIGURE 8.21
Comparison of MC simulation based radial dose function values, $g(r)$, for the old "classic" (data by Williamson and Li[20]) and new (data by Daskalov et al.[21]) microSelectron HDR ^{192}Ir source designs. $g(r)$ plotted against radial distance, r, (a) in the complete range up to $r = 14.0$ cm, and (b) for $r \leq 8.0$ cm (zoom).

The anisotropy function values and anisotropy factor values according to the MC results by Daskalov et al.[21] are listed in Table A.3.5 and Table A.3.6 of Appendix 3. Figure 8.22 demonstrates a comparison of the anisotropy function values for radial distances of 0.25, 0.5, and 1.0 cm for the classic[20] and new[21] microSelectron HDR source designs. The differences in design between these two source types result in obvious differences in the anisotropy function values purely for extremely short radial distances ($r = 0.25$ cm), as shown in this figure.

8.4.1.1.2   Sources for the VariSource HDR Afterloader System

8.4.1.1.2.1   A. The Old Design
This involves a 1.0 cm long iridium pellet core of 0.034 cm in diameter that is encapsulated in a titanium/nickel wire with an outer diameter of 0.059 cm and a total length of 150 cm (see Table 8.2 and Figure 8.23).

The first data on this source design were published by Meigooni et al.[34] based on experimental measurements using LiF–TLDs and MD-55-2 radiochromic film in a solid water phantom. The source strength was measured using a well-type chamber (HDR-1000) calibrated by the ADCL of University of Wisconsin, Madison, U.S.A. The authors published dose rate constant value, radial dose function values up to a distance of 10.0 cm, and anisotropy function values for 2.0 and 5.0 cm radial distances. The line source approximation was used for calculating the geometry function $G_L(r,\theta)$ as described by Equation 8.28 or Equation 8.38. Since the original publication considered a source coordinate system with $+z$ direction towards the drive cable (polar angle θ defined relative to drive cable), all data have been transformed to follow the TG-43/TG-43 U1 convention of Figure 8.6.

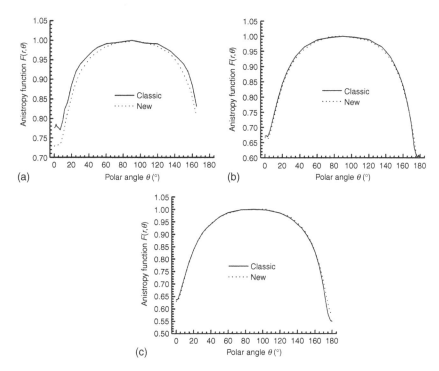

FIGURE 8.22

Comparison of MC simulation based anisotropy function, $F(r,\theta)$, values for the old "classic" (data by Williamson and Li[20]) and new (data by Daskalov et al.[21]) microSelectron HDR ^{192}Ir source designs (Table A.3.2 and Table A.3.5 of Appendix 3). $F(r,\theta)$ polar angle profiles at radial distance (a) $r = 0.25$ cm, (b) $r = 0.5$ cm, and (c) $r = 1.0$ cm.

Wang and Sloboda[32] published an MC dosimetry study for the VariSource old source design using the EGS4 MC code[50] (EGS4/DOSRZ). The simulations were carried out in a spherical water phantom of 30 cm diameter and an energy cut-off value for photon transport of 1 keV was considered. These data are also considered as the reference data for this source type.

Karaiskos et al.[22] also published an MC investigation for this source design using their own MC code. The analytical MC simulations have been carried out in a spherical water phantom with a diameter of 30 cm. Authors used an energy cut-off value of 1 keV up to radial distances of 2.0 cm, and 10 keV for distances greater than 2.0 cm.

Both of these studies[22,32] considered the line source approximation for the geometry function $G_L(r,\theta)$. Both author groups used a polar angle definition relative to the drive cable and thus all results have been converted to follow the coordinate system convention shown in Figure 8.6.

The most recent publication on this source type is by Papagiannis et al.[31] The authors published MC simulation results for this source design from

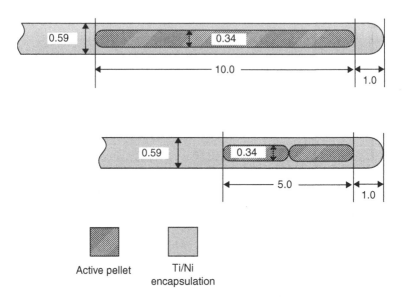

FIGURE 8.23
VariSource HDR ^{192}Ir source type designs (Varian Oncology Systems, Palo Alto, USA). (*Upper*)
The old design with 10.0 mm core length having a total outer diameter (including encapsulation)
of 0.59 mm.[22] (*Lower*) The new design with a core length of 5.0 mm and the same diameter as the
old design.[23] All dimensions shown are in mm.

simulations that were made with the source cantered in a 30 cm diameter
water sphere, considering the drive cable of the source to extend up to the
end of the mathematical phantom. Authors included an energy cut-off value
for photons of 10.0 keV and the exact geometry function $G(r,\theta)$ as given by
Equation 8.11.

All three MC investigations used the primary photon spectra for ^{192}Ir of
Glasgow and Dillman.[16]

Table 8.3 includes a summary of the dose rate constant, Λ, values
published by these four author groups. The MC calculated values $_{MC}\Lambda$ are
practically identical for all three studies, 1.043 to 1.044 cGy h^{-1} U^{-1}. The
experimentally estimated value by Meigooni et al.[34] of $_{EXP}\Lambda = (1.069 \pm 5\%)$
cGy h^{-1} U^{-1} represents the mean value of the TLD and radiochromic
film based dosimetry and is 2.5% higher than the mean $_{MC}\Lambda$ value of
1.043 cGy h^{-1} U^{-1}. Due to the significantly longer iridium core of this source
design, the dose rate constant value is 6 to 7% lower than those of the
microSelectron HDR, classic, and new source designs (see also Table 8.3).

Figure 8.24 shows the radial dose function values of these publications,
MC based or experimentally measured. Very good agreement over all these
data is observed. The radial dose function values shown in this figure
are listed in Table A.3.7, where the anisotropy function values for radial
distances in the range 0.25 to 10.0 cm according to Wang and Sloboda[32] are
listed in Table A.3.8 of Appendix 3.

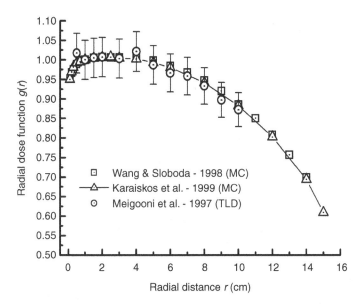

FIGURE 8.24

Radial dose function values, $g(r)$, plotted against radial distance, r, for the VariSource HDR ^{192}Ir old source design. The MC simulation based data by Wang and Sloboda[32] and Karaiskos et al.[22] are shown together with the experimentally estimated values using LiF–TLD dosimeters by Meigooni et al.[34] (Table A.3.7 of Appendix 3).

The change of the anisotropy function values over the whole range of polar angle θ with the radial distance and for radial distances in the range 0.25 to 5.0 cm is shown in Figure 8.25 based on the data in Table A.3.8. Figure 8.26 compares the $F(r,\theta)$ values of the two MC studies with the experimental data by Meigooni et al.[34] for the radial distances of 2.0 and 5.0.

8.4.1.1.2.2 The New Design

Angelopoulos et al.[23] published MC dosimetry data for the new source design for the VariSource afterloader system. The active core of that design consists of two 0.034 cm diameter, 0.25 cm-long cylinders with semispherical endings made of pure iridium metal, thus resulting in a total active core length of 0.5 cm. The source is encapsulated at the end of an approximately 150 cm long titanium/nickel wire of an outer diameter of 0.059 cm (see Figure 8.23 and Table 8.2). Simulations were made using the own MC code (UoA, University of Athens) in a 30 cm diameter spherical water phantom with the source positioned at its center. An energy cut-off of 10 keV was taken into account. Authors considered the primary photon spectra for ^{192}Ir of Glasgow and Dillman[16] in the simulation calculations.

For the extraction of the TG-43 parameter values, the line source approximation was used for the geometry function $G_L(r,\theta)$ as described by Equation 8.28 or Equation 8.37.

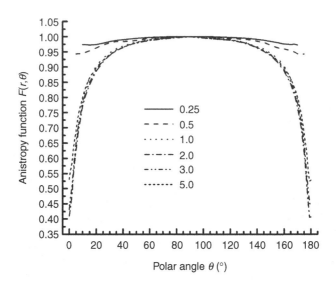

FIGURE 8.25
Anisotropy function, $F(r,\theta)$, polar angle profiles for the VariSource HDR ^{192}Ir old source design at six different radial distances in the range 0.25 to 5.0 cm according to the MC data by Wang and Sloboda[32] (Table A.3.8, Appendix 3).

The original publication considered a source coordinate system with $+z$ direction towards the drive cable, (polar angle θ defined relative to drive cable), and all data were transformed to follow the TG-43/TG-43 U1 convention of Figure 8.6.

Angelopoulos et al.[23] published a dose rate constant value of $_{MC}\Lambda = (1.101 \pm 0.5\%)$ cGy h^{-1} U^{-1} which is identical to the value published by Papagiannis et al.[31] using the same MC code and simulation geometry. This value is very close to those for the microSelectron HDR afterloader

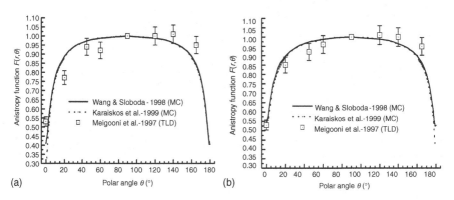

FIGURE 8.26
Comparison of the anisotropy function, $F(r,\theta)$, values published by several authors, estimated either experimentally or using MC simulations for the VariSource HDR ^{192}Ir old source design at radial distance (a) $r = 2.0$ cm and (b) $r = 5.0$ cm.

source designs (≈1%), explained by the similar geometrical design and similar dimensions. When compared to the old VariSource design, the dose rate constant value is 5.6% higher, due to the halved active core length of the new design (0.5 cm compared to 1.0 cm).

Table A.3.9 and Table A.3.10 of Appendix 3 summarize the radial dose function and anisotropy function values according to Angelopoulos et al.[23]

Figure 8.27 compares the radial dose function values of the two VariSource ^{192}Ir source designs, where differences are only present for radial distances below 1.0 cm.

The new design shows much higher anisotropy when compared to the old design for radial distances $r < 1.0$ cm and polar angles $\theta \leq 50°$ and $\theta \geq 130°$ (Figure 8.28).

8.4.1.1.3 Sources for the Buchler HDR Afterloader System

The ^{192}Ir Buchler HDR source consists of a cylindrical active core of 0.1 cm diameter and 0.13 cm length made of iridium metal. The active core is encapsulated in a stainless steel wire of an outer diameter of 0.16 cm which, after an initial length of 0.55 cm, ends in a driving cable of 0.11 cm outer diameter also made of stainless steel (see Figure 8.29 and Table 8.2).

Ballester et al.[24] was the first to publish MC dosimetry data for this source design. They used the GEANT MC simulation code and the decay scheme for ^{192}Ir was taken from the nuclear data sheet published by Shirley.[46] The simulations were performed in a water cylinder of 40 cm height and 40 cm diameter, where a cut-off energy for photons of 10 keV was considered. The proximal end of the stainless steel wire was modeled as a 6.0 cm long cylinder.

Ballester et al. estimated a dose rate constant value of $_{MC}\Lambda = (1.115 \pm 0.3\%)$ cGy h^{-1} U^{-1} which is identical to that published by

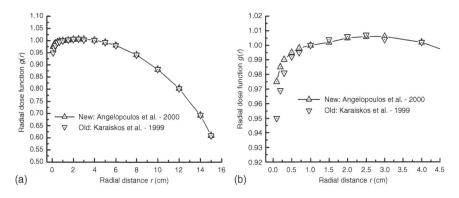

FIGURE 8.27
Comparison of MC simulation based radial dose function values, $g(r)$, for the old (data by Karaiskos et al.[22]) and new (data by Angelopoulos et al.[23]) VariSource HDR ^{192}Ir source designs. $g(r)$ plotted against radial distance, r, (a) in the complete range up to $r = 15.0$ cm and (b) for $r \leq 4.0$ cm (zoom).

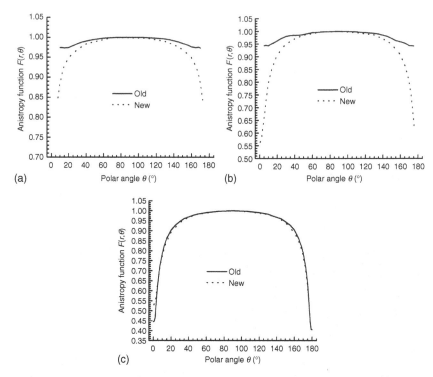

FIGURE 8.28
Comparison of MC simulation based anisotropy function, $F(r,\theta)$, values for the old (data by Wang and Sloboda[32]) and new (data by Angelopoulos et al.[23]) VariSource HDR ^{192}Ir source designs (Table A.3.8 and Table A.3.10 of Appendix 3). $F(r,\theta)$ polar angle profiles at radial distance (a) $r = 0.25$ cm, (b) $r = 0.5$ cm, and (c) $r = 1.0$ cm.

Papagiannis et al.[31] (Table 8.3). This value is extremely close to those values published for the microSelectron HDR source designs and for the VariSource new source design (see Table 8.3).

The radial dose function and the anisotropy function values according to these authors are listed in Table A.3.11 and Table A.3.12 of Appendix 3.

Figure 8.30 demonstrates a comparison of the radial dose functions calculated by Ballester et al.[24] to those by Papagiannis et al.[31] The values given by Ballester et al. are significantly greater, especially for radial distances $r > 5.0$ cm. This can be explained by the much larger simulation volume (356%) used by the first group, which approaches unbounded geometry (see also Chapter 9) and results in a much higher scatter dose contribution in this radial distance range. It is doubtful if such large simulation geometries are adequate for extracting parameter values to be used in clinical implants.

The anisotropy factor values $\varphi_{an}(r)$ have been fitted by the authors to the linear function $\varphi_{an}(r) = 0.981 + 0.00034r$ and are listed in Table A.3.13.

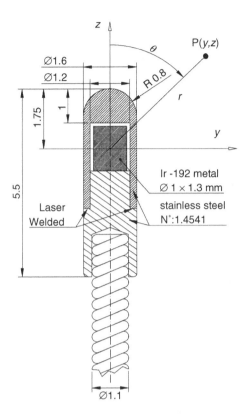

FIGURE 8.29

Mechanical design of the HDR ^{192}Ir source used with the Buchler afterloading device. This source contains a cylindrical core of 1.3 mm length and 1.0 mm diameter and is manufactured by Amersham (product code ICCB2113, source G0814). All dimensions shown are in mm. (From Ballester, F. et al. *Phys. Med. Biol.*, 46, N79–N90, 2001. Institute of Physics Publishing 2001. With permission.)

8.4.1.1.4   Sources for the GammaMed HDR Afterloader Systems 12i and Plus

There are two different source designs used for the 12*i* and Plus GammaMed afterloaders (Figure 8.31 and Table 8.2). In both designs the active core consists of a cylinder with a diameter of 0.06 cm and length of 0.35 cm, made of iridium metal. There are different encapsulation geometries resulting in an outer diameter of 0.11 and 0.09 cm for the 12*i* and Plus model, respectively. The encapsulation is made of stainless steel. Both sources are very similar to the classic and new source designs for the microSelectron HDR afterloader (see Table 8.2).

Ballester et al.[25] published MC dosimetry results for both GammaMed source designs. They used the GEANT3 MC simulation code[51] and simulations were performed in a water cylinder of 40 cm height and 40 cm diameter. A 6.0 cm-long stainless steel cylinder at the proximal end of the

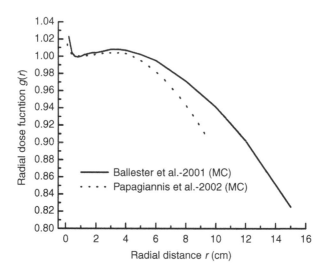

FIGURE 8.30
Radial dose function values, $g(r)$, plotted against radial distance, r, for the Buchler HDR ^{192}Ir source design. The MC simulation based data by Ballester et al.[24] and Papagiannis et al.[31] are shown (Table A.3.11 of Appendix 3). The significantly higher values by Ballester et al. especially for $r > 5.0$ cm can be explained by the practically unbounded simulation geometry used in their MC calculations. Papagiannis et al. performed MC simulations in a liquid water spherical phantom of 30 cm in diameter.

source was modeled to adequately consider the drive wire. The decay scheme for ^{192}Ir was taken from the nuclear data sheet published by Shirley[46] and an energy cut-off value of 10 keV was used for photons. The TG-43 dosimetry parameter values were extracted using the line source approximation for the geometry function $G_L(r,\theta)$.

Ballester et al. found identical dose rate constant values for both source designs, $_{MC}\Lambda = (1.118 \pm 0.3\%)$ cGy h^{-1} U^{-1}, which is very closed to the values for the microSelectron HDR sources, the VariSource new source design, and the Buchler HDR source (see Table 8.3).

The radial dose function values were found to be identical for both source models and adequately fitted to a third-order polynomial for radial distances in the range of 0.15 to 15.0 cm. $g(r)$ values for the two GammaMed sources are listed in Table A.3.14 of Appendix 3. Figure 8.32 shows a comparison of the radial dose functions for the Buchler HDR and GammaMed 12i and Plus sources. Differences are only observed at very short distances of $r < 0.5$ cm.

The anisotropy function values $F(r,\theta)$ for the 12i and Plus source models are listed in Table A.3.15 and Table A.3.16, respectively.

8.4.1.1.5   Sources for the BEBIG MultiSource HDR Afterloader System
The HDR ^{192}Ir source design for the MultiSource afterloader by BEBIG contains an active core of 0.35 cm length and 0.06 cm diameter that is made

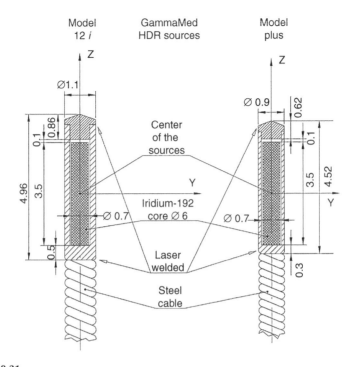

FIGURE 8.31

GammaMed HDR ^{192}Ir source type geometrical designs for the GammaMed 12*i* and GammaMed Plus afterloaders (Varian Oncology Systems, Palo Alto, USA). For both sources the active cores are cylindrical shaped. All dimensions shown are in mm. (From Ballester, Technical Note: Monte Carlo Dosimetry of the HDR 12*i* and Plus [192]Ir Sources, AAPM, 2001. With permission.)

of pure iridium metal. This is encapsulated in a stainless steel wire resulting in a total outer diameter of the source of 0.1 cm and in a total length including the stainless steel drive wire of 206.5 cm (see Figure 8.33).

Granero et al.[76] published MC dosimetry data for the BEBIG source design. In this MC study the GEANT4 MC code[74] was used, where the decay scheme for ^{192}Ir was taken from the NuDat Nuclear Data Database.[75] The simulations were performed with the source situated in the center of a spherical 40 cm in radius water phantom that can be considered to result in unbounded phantom conditions (full scattering) up to a radial distance of 20.0 cm. Liquid water was used with a density of 0.998 g cm^{-3} at 22°C as recommended by the TG-43 U1.[13] The authors considered a cut-off energy for photons history tracking of 10 keV.

The only data for this source design are provided by Selbach and Andrássy,[71] which are based on experimentally measured dose rate values in a $30 \times 30 \times 30$ cm^3 water phantom using a pin-point-ionization chamber constructed by PTB with collecting volume dimensions of 0.35 cm in diameter and 0.7 cm in length. The iridium source was positioned within a polyethylene catheter and scans have been performed in a Cartesian grid

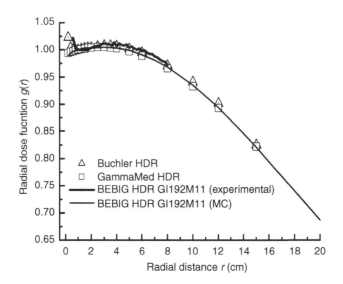

FIGURE 8.32
Radial dose function values, $g(r)$, plotted against radial distance, r, for the GammaMed HDR ^{192}Ir source designs according to the MC data by Ballester et al.[25] (Table A.3.14, Appendix 3). Authors found identical $g(r)$ values for both 12i and Plus source designs. The MC based $g(r)$ values for the Buchler HDR source (data by Ballester et al.,[24] Table A.3.11, Appendix 3), and the experimental values (data by Selbach and Andrássy[71]) for the BEBIG MultiSource HDR ^{192}Ir source model GI192M11 for the line source approximation, $g_L(r)$, together with the corresponding MC simulation based values by Granero et al.[76] (Table A.3.17, Appendix 3) are also plotted.

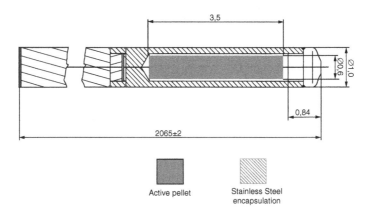

FIGURE 8.33
[192]Ir source type GI192M11 design by BEBIG used in the MultiSource HDR afterloader (figure provided by BEBIG Isotopen- und Medizintechnik GmbH, Berlin, Germany). The active core is made of iridium metal where all other parts including drive wire are made of stainless steel. The total length of source drive wire is 206.5 cm. All dimensions shown are in mm. The same source design is considered for the BEBIG HDR ^{60}Co source model GK60M21, where the active core is made by cobalt metal.

enabling TG-43 data extraction up to radial distances of $r = 8.0$ cm. The source was previously calibrated in PTB using the shadow-shield calibration set-up and an LS01 1000 cm^3 spherical ionization chamber as described by Büermann et al.[77]

In both studies the line source approximation $G(r,\theta)$, Equation 8.38, using $L_s = L_{eff} = 0.35$ cm, has been utilized for calculating the TG-43 U1 values for $g_L(r)$ and $F(r,\theta)$.

The experimentally measured dose rate constant value (Table 8.3) is $_{EXP}\Lambda = (1.119 \pm 4.5\%)$ cGy h^{-1} U^{-1} which is in very close agreement (1.0%) to the MC estimated value[76] of $_{MC}\Lambda = (1.108 \pm 0.3\%)$ cGy h^{-1} U^{-1}. Both values are very similar, 1.0 to 2.0%, to the values derived for all other HDR iridium source types with similar designs, such as those for the Micro-Selectron and GammaMed afterloader systems (see Table 8.3).

Table A.3.17 of Appendix 3 summarizes the radial dose values for the BEBIG source for both point sources, $g_P(r)$, and line source, $g_L(r)$, approximations of the geometry function (see Equation 8.41) as derived in both studies. In Figure 8.32, a close agreement between radial dose function values of the Buchler, GammaMed HDR, and BEBIG (line approximation $g_L(r)$) iridium source models is demonstrated. This is also the case for large radial distances up to 15 cm due to the very similar simulation geometries (practically unbounded phantom geometries) used in all these MC studies.[24,25,76]

The MC based anisotropy function values are listed in Table A.3.18 of Appendix 3 for radial distances of 0.25 to 20.0 cm. A graphical comparison of the MC calculated and experimental $F(r,\theta)$ profiles vs. polar angle θ for three representative radial distances of 1.0, 3.0, and 5.0 cm is demonstrated in Figure 8.34.

8.4.1.2 ^{60}Co Sources

8.4.1.2.1 Sources for the Ralstron HDR Afterloader System

There are three different source designs used with the Ralstron HDR afterloader which are specially designed for being used in gynecological brachytherapy. All three contain two cobalt active pellets each 0.1 mm long and 0.1 cm in diameter. The pellets are either in contact (Type 2) or are spaced 0.9 cm (Type 1) and 1.1 cm (Type 3) apart. The two active pellets in the Type 1 and Type 3 designs are fixed together using a 0.1 cm in diameter and 0.9 or 1.1 in length stainless steel cylinder (SUS 316). The total is then encapsulated into a 0.3 cm in diameter cylinder also made of SUS. Finally, there is a stainless steel drive wire adjusted to the distal end with an outer diameter for all three designs of 0.16 cm (see Table 8.4 and Figure 8.35).

Papagiannis et al.[36] published MC dosimetry data for all three source designs. The authors used their own MC simulation code and the source coordinate system was for all the three sources positioned at the geometrical center of the active core (Figure 8.33). Due to the rather unique configuration

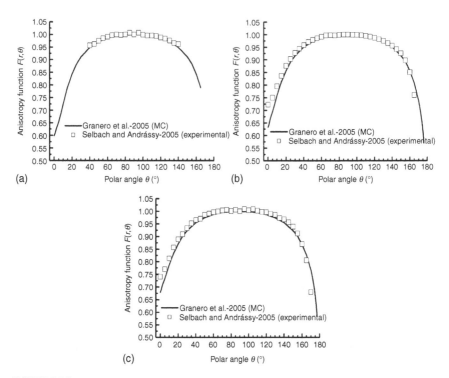

FIGURE 8.34

Comparison of MC simulation based (data by Granero et al.,[76] Table A.3.18, Appendix 3) and experimentally measured (in water using a pin-point-ionization chamber, data by Selbach and Andrássy[71]) anisotropy function, $F(r,\theta)$, polar angle profiles for the BEBIG MultiSource HDR ^{192}Ir source model GI192M11 at three different radial distances (a) $r = 1.0$ cm, (b) $r = 3.0$ cm, and (c) $r = 5.0$ cm.

of these source designs, they used the point source approximation of the geometry function for extracting all TG-43 related parameter values, $G(r,\theta) = G(r) = 1/r^2$. The consequence of the above is that the published tables always have to be used with the point source approximated geometry function in order to guarantee data consistency and keep the accuracy described by the authors. This is obviously not in agreement with the TG-43 U1 formulation (Equation 8.43 through Equation 8.45), but it guarantees being within the 0.5 to 15.0 cm radial distance range published by the authors, the achievement of the same accuracy of the MC originally derived dose rate values.

The MC simulations were made in a spherical water phantom of 15.0 cm in radius.

The calculated dose rate constants were $_{MC}\Lambda = (0.878 \pm 0.5\%)$ cGy h^{-1} U^{-1}, $_{MC}\Lambda = (1.101 \pm 0.5\%)$ cGy h^{-1} U^{-1}, and $_{MC}\Lambda = (0.800 \pm 0.6\%)$ cGy h^{-1} U^{-1} for the Type 1, Type 2, and Type 3 source designs, respectively (see Table 8.5).

TABLE 8.4

Geometrical and Material Characteristics of Common Commercial High Dose Rate (HDR) ^{60}Co Sources

Source Type	Active Core							Encapsulation			Cable	
	Material	No Pellets	Material between Pellets	Gap Length between Pellets (cm)	Effective Length L_{eff} (cm)	Pellet Length L_s (cm)	Pellet Diameter D_s (cm)	Material	Thickness (cm)	Outer Diameter (cm)	Material	Outer Diameter (cm)
Ralstron HDR[a] Type 1	Co	2	Stainless steel SUS 316	9.0	—	0.10	0.10	Stainless steel SUS 316	0.075	0.30	Stainless steel SUS 316	0.160
Ralstron HDR Type 2	Co	2	—	0.0	0.20	0.10	0.10	Stainless steel SUS 316	0.075	0.30	Stainless steel SUS 316	0.160
Ralstron HDR Type 3	Co	2	Stainless steel SUS 316	11.0	—	0.10	0.10	Stainless steel SUS 316	0.075	0.30	Stainless steel SUS 316	0.160
BEBIG MultiSource HDR, GK60M21	Co	1	—	0.0	0.35	0.35	0.060	Stainless steel	0.020	0.100	Stainless steel	0.100

All three sources are used in computer controlled afterloading device. The commercial names of the corresponding afterloading devices are commonly used to identify source types, and not the source manufacturer itself.

a Shimadzu Corporation, Japan.

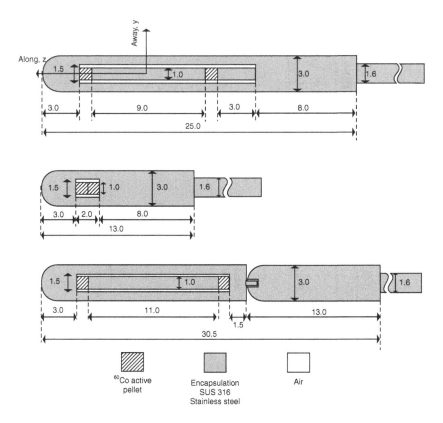

FIGURE 8.35
Schematic diagrams of the three ^{60}Co source types, type 1, type 2, and type 3, used with the Ralston remote afterloader (Shimadzu Corporation, Japan) showing geometries and materials. The coordinate system used for published data[36] coincides in each case with the corresponding center of symmetry of the two active pellets and it is also shown for the type 1 source design. All dimensions shown are in mm.

Table A.3.19 lists the values of the radial dose function for the three sources, where Figure 8.36 presents their graphical comparison.

The anisotropy function values $F(r,\theta)$ are summarized in Table A.3.20 through Table A.3.22 of Appendix 3. In Figure 8.37, $F(r,\theta)$ values are plotted vs. polar angle, θ, for three representative radial distances of $r = 0.5$, 1.0, and 3.0 cm. The anisotropy profiles for Type 1 and Type 3 sources show an untypical shape with $F(r,\theta)$ values much higher than unity. This is obviously an artifact owing to the utilization of the point source approximation of the geometry function in the extraction of the anisotropy values. For the Type 2 source, and due to the fact that both active cobalt pellets are fixed together, the curves show normal shapes indicating that the point source approximation is adequately describing the activity distribution within this source.

TABLE 8.5

Monte Carlo Calculated and Experimentally Estimated Values of the Dose Rate Constant Λ for the Different ^{60}Co High Dose Rate Brachytherapy Sources Listed in Table 8.4

Source Type	$_{MC}\Lambda$	Monte Carlo Code	δ (keV) (Equation 8.42)	$_{EXP}\Lambda$	Phantom	Detector	Author
Ralstron HDR Type 1	$0.878 \pm 0.5\%$	UoA, own	10.0				Papagiannis et al.[36]
Ralstron HDR Type 2	$1.101 \pm 0.5\%$	UoA, own	10.0				
Ralstron HDR Type 3	$0.800 \pm 0.6\%$	UoA, own	10.0				
BEBIG MultiSource HDR, GK60M21	—	—	—	$1.070 \pm 3.5\%$	Water phantom	Pin-point-chamber	Selbach and Andrássy[72]
	$1.084 \pm 0.5\%$	GEANT4	10.0				Ballester et al.[73]

All Λ values are expressed in cGy h^{-1} U^{-1}.

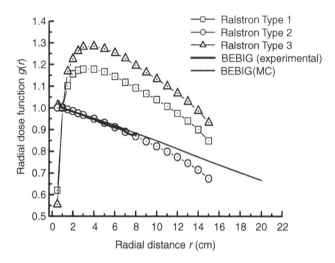

FIGURE 8.36
Radial dose function values, $g(r)$, plotted against radial distance, r, for the Ralstron remote HDR afterloader, type 1, 2, and 3 ^{60}Co source designs according to the MC data by Papagiannis et al.[36] (Table A.3.19, Appendix 3). The point source approximation of the geometry function has been considered for the extraction of all these values. The experimental $g(r)$ values by Selbach and Andrássy[72] for the line source approximation, $g_L(r)$, and the corresponding MC simulation based values by Ballester et al.[73] for the BEBIG MultiSource HDR ^{60}Co source model GK60M21 (Table A.3.24, Appendix 3) are also plotted. These are in close agreement with the data for the type 2 Ralstron source design. The higher values for the BEBIG MultiSource for distances above 10 cm are due to the fact that these MC simulation data are obtained within a spherical water phantom of 50.0 cm in radius (unbounded geometry), where the simulation study for the type 2 source used a phantom of 15.0 cm in radius. This reflects the lack of backscatter photons for those distances near the phantom boundaries for the 15.0 cm phantom.

8.4.1.2.2 Sources for the BEBIG MultiSource HDR Afterloader System

The design of this ^{60}Co source is identical to that for ^{192}Ir shown in Figure 8.33 (see also Table 8.4) with the exception that the active core is made of active cobalt material.

Experimental data for this source design are only available by Selbach and Andrássy[72] which are based on measured dose rate values in a $30 \times 30 \times 30$ cm^3 water phantom using a pin-point-ionization chamber constructed by PTB with collecting volume dimensions of 0.35 cm in diameter and 0.7 cm in length. The ^{60}Co source was positioned within a polyethylene catheter and scans have been performed in a Cartesian grid enabling TG-43 data extraction up to radial distances of $r = 8.0$ cm. The source was previously calibrated in PTB using the shadow-shield calibration set-up and an LS01 1000 cm^3 spherical ionization chamber as described by Büermann et al.[77]

Ballester et al.[73] published MC dosimetry data for this source design. In this MC study the GEANT4 MC code[74] was used, and the decay scheme for ^{60}Co was taken from the NuDat Nuclear Data Database.[75]

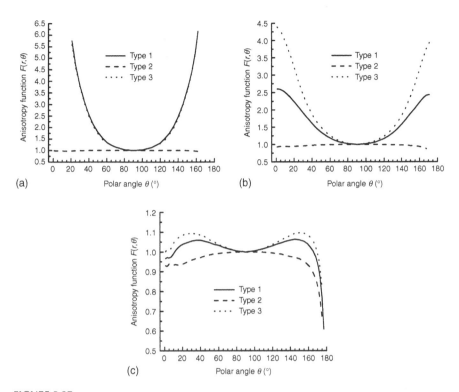

FIGURE 8.37

Comparison of MC simulation based anisotropy function, $F(r,\theta)$, values for the Ralstron remote HDR afterloader, type 1, 2, and 3 ^{60}Co source designs according to the MC data by Papagiannis et al.[36] (Table A.3.20 through Table A.3.22 of Appendix 3). $F(r,\theta)$ polar angle profiles at radial distance (a) $r = 0.5$ cm, (b) $r = 1.0$ cm, and (c) $r = 3.0$ cm.

The simulations were performed in a spherical water phantom of 50 cm in radius, for achieving unbounded phantom conditions (full scattering) up to a radial distance of 20.0 cm. Liquid water was used with a density of 0.998 g cm^{-3} at 22°C as recommended by the TG-43 U1.[13] The authors considered a cut-off energy for both photons and electrons history tracking of 10 keV.

In both studies the line source approximation $G_L(r,\theta)$, Equation 8.38, using $L_s = L_{eff} = 0.35$ cm has been utilized for calculating the TG-43 U1 values for $g_L(r)$ and $F(r,\theta)$.

The measured[72] dose rate constant value (Table 8.5) is $_{EXP}\Lambda = (1.070 \pm 3.5\%)$ cGy h^{-1} U^{-1} that is in close agreement (1.3%) to the MC estimated value[73] of $_{MC}\Lambda = (1.084 \pm 0.5\%)$ cGy h^{-1} U^{-1}. Both values are similar, 1.6 to 3.0%, to the MC calculated value for the Type 2 Ralstron HDR source despite the obvious differences in their geometry. The same is observed for the radial dose function values shown in Figure 8.36 together with those of the three Ralstron HDR source types. There is a remarkable agreement between the $g_L(r)$ values of the BEBIG design and the Type 2 design of Ralstron afterloader up to a radial distance of 8.0 cm. For larger distances the differences caused by the lack of backscatter of photons are obvious; the MC simulation based values for the BEBIG design by Ballester et al.[73] using a quasi unbounded phantom geometry are higher than those for the Type 2 design obtained by Papagiannis et al.[36] using a spherical phantom geometry of 15 cm in radius. The observed difference at $r = 15$ cm is 12%.

Table A.3.24 of Appendix 3 summarizes both the experimental and the MC simulation based radial dose function values $g(r)$ for the BEBIG source.

The MC estimated anisotropy function values are listed in Table A.3.25 for radial distances of 0.25 to 20.0 cm. Figure 8.38 graphically compares the experimental and MC calculated $F(r,\theta)$ values for three representative radial distances of 1.0, 3.0, and 5.0 cm.

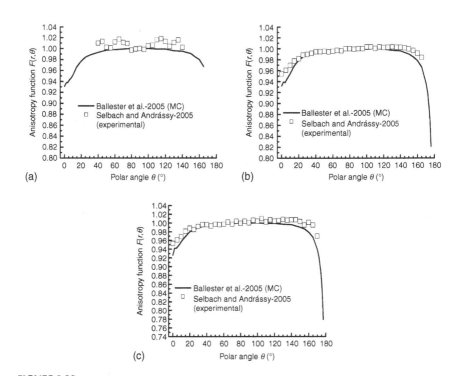

FIGURE 8.38
Comparison of MC simulation based (data by Ballester et al.,[73] Table A.3.25, Appendix 3) and experimentally measured (in water using a pin-point-ionization chamber, data by Selbach and Andrássy[72]) anisotropy function, $F(r,\theta)$, polar angle profiles for the BEBIG MultiSource HDR ^{60}Co source model GK60M21 at three different radial distances (a) $r = 1.0$ cm, (b) $r = 3.0$ cm, and (c) $r = 5.0$ cm.

8.4.1.3 ^{169}Yb Sources

The only commercially available HDR ^{169}Yb source is the HDR 4140 model manufactured by Implant Sciences Corporation (Wakefield, MA, USA). Medich et al.[78] published MC dosimetry data for this source type. The geometrical design of this model is shown in Figure 8.39. It contains a cylindrical active core made of ytterbium oxide (ytterbia, Yb_2O_3) with 0.073 cm diameter and 0.36 cm length. The physical density of Yb_2O_3 of $\rho = 6.9$ g cm^{-3} as given by the authors, deviates from the value (theoretical maximum value) referring to the pure form being in the range of 9.1 to 9.3 g cm^{-3}. Latter fact is due to the difficulties in achieving reproducible ytterbia material in solid form under laboratory conditions. Hereby, sintered ceramic has been experimentally found by the manufacturer to be a stable production method resulting in a measured Yb_2O_3 density of 6.9 g cm^{-3}. The source capsule is attached to a source cable made by stainless steel with a density of $\rho = 6.9$ g cm^{-3} and a total length of 2.1 m.

Medich et al. used the MCNP5 MC computer code[80] and ^{169}Yb photon energy spectrum as provided by the Lund/LBNL Nuclear Data Search vs. 2.0.[81] The simulation geometry used a source capsule (Figure 8.39) with a total drive cable length of 0.2 cm positioned at the center of liquid water filled phantom sphere with a diameter of 40.0 cm.

Only the penetrating part of the ytterbium spectrum was taken into account with photon energies above 10 keV. Simulations were performed with a total of 1×10^9 source photon histories processed. Dosimetric data were determined at radial distances ranging from 0.5 to 10.0 cm and over

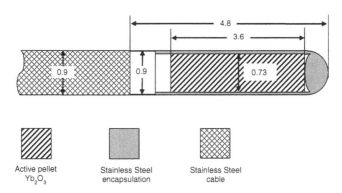

FIGURE 8.39
^{169}Yb model HDR 4140 source design (Implant Sciences Corporation, Wakefield, MA, USA) for HDR brachytherapy according to Medich et al.[78] The cylindrical core with 0.73 mm diameter and 3.6 mm length consists of ytterbium oxide (ytterbia, Yb_2O_3) with a density of $\rho = 6.9$ g cm^{-3}. The core is encapsulated in a 0.085 mm thick stainless steel shell with a density of $\rho = 7.8$ g cm^{-3} resulting thus to a source cylinder with a total outer diameter of 0.9 mm. There is a 0.5 mm thick hemispherical welded tip and a 0.75 mm thick solid plug. The source capsule is attached to a source cable made by stainless steel with a density of $\rho = 6.9$ g cm^{-3} and a total length of 2.1 m. All dimensions shown are in mm.

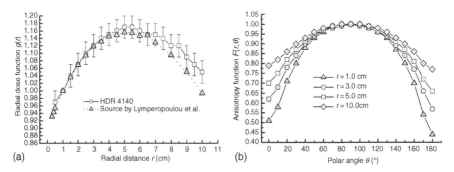

FIGURE 8.40
Monte Carlo simulation based dosimetry data for the ^{169}Yb model HDR 4140 source design
according to Medich et al.[78] (a) Radial dose function values for distances between 0.5 and
10.0 cm (Table A.3.26 of Appendix 3) with a demonstrated maximum for r in the interval of
5.0–5.5 cm. Error bars represent calculated uncertainties at $\pm 1 \times$ standard deviation. The
$g(r)$ values for a hypothetical ytterbium source having the same geometrical dimensions of
the microSelectron HDR ^{192}Ir new source design (see Figure 8.16) but with an active core
containing pure ytterbium according to Lymperopoulou et al.,[79] are also shown. The significant
lower values for that source type for radial distances above 7.0 cm can be explained mainly by
the smaller dimensions of the MC simulation phantom used by Lymperopoulou et al.
(spherical water phantom of 15.0 cm radius vs. 20.0 cm radius as used by Medich et al.) and
the increased role of scattered radiation at the energies emitted by ^{169}Yb. (b) Anisotropy
function, $F(r,\theta)$, values for radial distances $r = 1.0$, 3.0, 5.0, and 10.0 cm (Table A.3.27 of
Appendix 3).

polar angles in the range 0–180° in steps of 10° (Table A.3.26 and Table A.3.27
in Appendix 3 and Figure 8.40). The line source approximation for the
geometry function $G_L(r,\theta)$ according to Equation 8.38 using as source length
the length of the active core, 0.36 cm, has been considered. The dose rate
constant was found to be $_{MC}\Lambda = (1.19 \pm 2.5\%)$ cGy h^{-1} U^{-1}. This value is
only 3.7% higher than that of $_{MC}\Lambda = (1.147 \pm 0.3\%)$ cGy h^{-1} U^{-1}, calculated
by Lymperopoulou et al.[79] for a HDR ^{169}Yb source design with the same
geometrical configuration as the microSelectron HDR ^{192}Ir new source type
design (Figure 8.16b) but with a pure ytterbium active core with a density of
6.73 g cm^{-3}.

8.4.2 Sources Used in PDR Brachytherapy

In pulsed dose rate (PDR) brachytherapy, a continuous low dose rate
brachytherapy treatment typically delivered at a prescribed dose rate of
0.5 Gy h^{-1} is mimicked by delivering a small dose fraction, a pulse, at a
medium dose rate in the range of 1 to 3 Gy h^{-1}. This is carried out in a certain
time period (pulse duration of 10 to 15 min), and is repeated once per hour
over the same overall elapsed time.[47–49]

For PDR, ^{192}Ir sources have been used with strengths 10 to 20 times lower
than those used for HDR applications. There are currently two vendors
offering PDR afterloaders using ^{192}Ir sources, namely Nucletron B.V. and
Varian Oncology Systems.

8.4.2.1 Sources for the MicroSelectron PDR Afterloader System

There have been two different PDR source designs delivered with the microSelectron PDR afterloader system, the old and the new design (see Table 8.6 and Figure 8.41). The old design consists of two cylindrical iridium pellets of 0.060 ± 0.005 cm length and 0.060 ± 0.005 cm diameter each, the most distal being radioactive.

The new design consists of two cylindrical, radioactive, ^{192}Ir pellets of 0.05 ± 0.01 cm length and 0.050 ± 0.005 cm diameter, each thus presenting an increase in the active core length of 0.04 cm and a decrease in active core diameter of 0.01 cm when compared to the old design.

Both sources are encapsulated in an AISI 316 stainless steel cylinder that is 0.11 cm in the outer diameter and bears a cavity 0.12 cm long and 0.06 cm in diameter. In the new source design, an air gap exists between the active pellets and the encapsulation.

Williamson and Li[20] were the first to publish TG-43 parameter values for the old PDR source design together with classic HDR design based on MC simulation studies using their in-house developed MCPT code. As described previously, the calculations were realized with the source cantered in liquid water sphere of 30 cm diameter and the extraction of the TG-43 parameters was realized using the point source approximation of the geometry function $G_P(r,\theta) = G(r) = 1/r^2$ (see also Equation 8.38). Authors included in the MC simulation study a 0.3 cm-long drive cable consisting of AISI 304 solid stainless steel of 0.11 cm diameter. Although the original publication considered a source coordinate system with $+z$ direction towards the drive cable (polar angle θ defined relative to drive cable), all data have been transformed to follow the TG-43/TG-43 U1 convention as described in Figure 8.6.

Karaiskos et al.[26] published an MC simulation based study on both old and new PDR source designs for the microSelectron afterloader. The simulations have been performed with each source positioned in the center of a 30 cm diameter liquid water sphere using the UoA MC simulation code. The drive wire of each of the sources was simulated to the full extent of the mathematical phantom used. The point source approximation of the geometry function in this study was also considered for extracting TG-43 parameter values.

In both MC papers the ^{192}Ir primary photon spectrum was taken from Glasgow and Dillman,[16] whereas Karaiskos et al.[26] took only the penetrating part of this spectrum into account with photon energies above 11.3 keV. They also considered an energy cut-off of 2 keV for terminating the simulation of photon's history.

For the old PDR source design the only experimental study available is that by Valicenti et al.,[35] using TLD dosimeters placed in a solid water phantom of total outer dimensions of $20 \times 20 \times 28$ cm^3. In this study, a PDR iridium source previously used in clinical implants was adequately decayed (11 months after delivery), removed from the afterloader, and the drive-cable was trimmed to a length of 6.0 mm.

TABLE 8.6

Geometrical and Material Characteristics of Common Commercial Pulsed Dose Rate (PDR) Sources all Containing ^{192}Ir. All Sources Are Used in Computer Controlled Afterloading Devices

Source Type	Active Core			Encapsulation			Cable	
	Material	Length L_s (cm)	Diameter D_s (cm)	Material	Thickness (cm)	Outer Diameter (cm)	Material	Outer Diameter (cm)
MicroSelectron PDR (old design)	Ir	0.06	0.060	Stainless steel AISI 316	0.0250	0.110	Stainless steel AISI 316	0.110
MicroSelectron PDR (new design)	Ir	0.10	0.050	Stainless steel AISI 316	0.0250	0.110	Stainless steel AISI 316	0.110
GammaMed PDR Model 12i	Ir	0.14	0.060	Stainless steel AISI 316L	0.0200	0.110	Stainless steel AISI 304	0.110
GammaMed PDR Model plus	Ir	0.14	0.060	Stainless steel AISI 316L	0.0100	0.090	Stainless steel AISI 304	0.090

The commercial names of the corresponding afterloading devices are commonly used to identify source types, and not the source manufacturer itself.

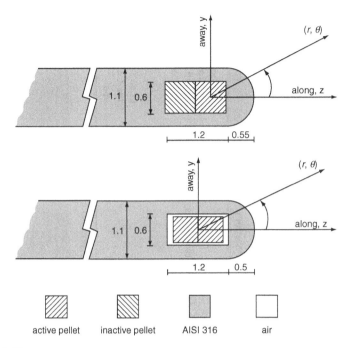

FIGURE 8.41
MicroSelectron PDR ^{192}Ir source type designs (Nucletron B.V., Veenendaal, The Netherlands). (*Upper*) The old design containing one active pellet at the tip and a second inactive one both cylindrical with a diameter of 0.6 mm and length of 0.6 mm.[20] (*Lower*) The new design with a core consisting of two cylindrical radioactive pellets of 0.5 mm length and 0.5 mm diameter each.[26] In each case the origin of the source coordinate system is aligned to the corresponding geometrical center of the activity (center of the single active pellet for the old design and center of both active pellets in the new design). All dimensions shown are in mm.

The MC calculated value for the dose rate constant was $_{MC}\Lambda = (1.128 \pm 0.5\%)$ cGy h^{-1} U^{-1} according to Williamson and Li[20] and $_{MC}\Lambda = (1.124 \pm 0.5\%)$ cGy h^{-1} U^{-1} according to Karaiskos et al.[26] (see Table 8.7). The experimental estimated value by Valicenti et al.[35] was $_{EXP}\Lambda = (1.122 \pm 4.0\%)$ cGy h^{-1} U^{-1}. All these values are in close agreement to each other within their stated statistical and experimental uncertainties.

The radial dose function values $g_p(r)$ for both PDR source designs are listed in Table A.3.28 of Appendix 3. Figure 8.42 compares the two $g_p(r)$ curves for radial distances in the range 0.1 to 14.0 cm, where differences are only present for very short distances, $r < 0.5$ cm.

The anisotropy function values $F(r,\theta)$ are listed in Table A.3.29 and Table A.3.30 of Appendix 3. Due to the increased active core length by 0.04 cm, the new design demonstrates lower $F(r,\theta)$ values and thus higher anisotropy for polar angles smaller than 50°, when compared to the old design (see Figure 8.43). On the other hand, since only the most distal pellet is radioactive in the old source design (see Figure 8.41), this source

TABLE 8.7

Monte Carlo and Experimental Estimated Values of the Dose Rate Constant Λ for the Different ^{192}Ir Pulsed Dose Rate Brachytherapy Sources Listed in Table 8.6

Source Type	$_{MC}\Lambda$	Monte Carlo Code	δ (keV) (Equation 8.42)	$_{EXP}\Lambda$	Phantom	Detector	Author
MicroSelectron PDR (old design)	1.128 ± 0.5%	MCPT, own	—				Williamson and Li[20]
	1.124 ± 0.5%	UoA, own	11.3				Karaiskos et al.[26]
	—	—		1.122 ± 4.0%	Solid water	LiF–TLDs	Valicenti et al.[35]
MicroSelectron PDR (new design)	1.121 ± 0.5%	UoA, own	11.3				Karaiskos et al.[26]
GammaMed PDR Model 12i	1.122 ± 0.3%	GEANT3	10.0				Perez-Calatayud et al.[27]
GammaMed PDR Model Plus	1.122 ± 0.3%	GEANT3	10.0				Perez-Calatayud et al.[27]

All Λ values are expressed in cGy h^{-1} U^{-1}. Energy cut-off values δ referred in the original publications are also listed.

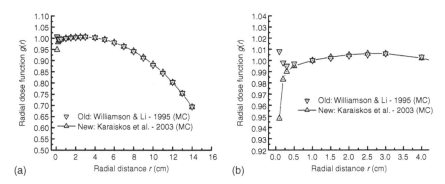

FIGURE 8.42
Comparison of MC simulation based radial dose function values, $g(r)$, for the old (data by Williamson and Li[20]) and new (data by Karaiskos et al.[26]) microSelectron PDR ^{192}Ir source type designs (Table A.3.28 of Appendix 3). $g(r)$ plotted against radial distance, r, (a) in the complete range up to $r = 14.0$ cm and (b) for $r \leq 4.0$ cm (zoom).

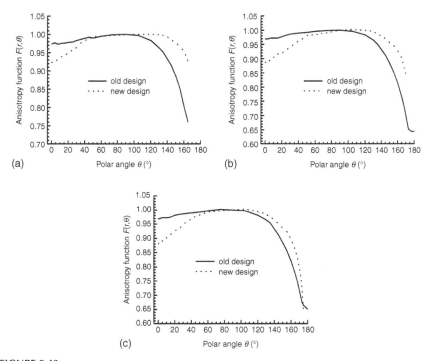

FIGURE 8.43
Comparison of MC simulation based anisotropy function, $F(r,\theta)$, values for the old (data by Williamson and Li[20]) and new (data by Karaiskos et al.[26]) microSelectron PDR ^{192}Ir source type designs (Table A.3.29 and Table A.3.30 of Appendix 3). $F(r,\theta)$ polar angle profiles at radial distance (a) $r = 0.25$ cm, (b) $r = 0.5$ cm, and (c) $r = 1.0$ cm.

demonstrates higher anisotropy for polar angles above 120° as compared to the new design. Finally, anisotropy factor, $\varphi_{an}(r)$, values for both designs are listed in Table A.3.31 with the new design showing higher $\varphi_{an}(r)$ values for all listed radial distances of 0.25 to 5.0 cm.

Due to the fact that TG-43 data sets for both source designs have been derived by using the point source approximation $G_P(r,\theta)$, the published TG-43 tables always have to be used with the point source approximated geometry function in order to guarantee data consistency and keep the accuracy described by the authors. Again, this is obviously not in agreement with the TG-43 U1 formulation (Equation 8.43 through Equation 8.45), but it guarantees being within the radial distance range published by the authors of 0.5 to 10.0 cm, the achievement of the same accuracy of the MC originally derived dose rate values.

8.4.2.2   Sources for the GammaMed PDR Afterloader Systems 12i and Plus

The mechanical designs of the two available ^{192}Ir source designs for the GammaMed 12*i* and Plus PDR afterloaders are shown in Figure 8.44, material details are also listed in Table 8.6.

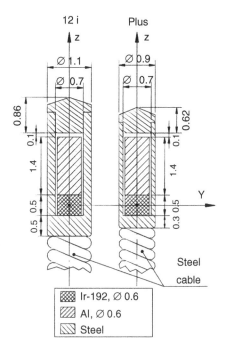

FIGURE 8.44
GammaMed PDR ^{192}Ir source type geometrical designs for the GamaMed PDR 12*i* and GammaMed PDR Plus afterloaders (Varian Oncology Systems, Palo Alto, U.S.A.) Both sources have identical active core designs, cylindrically shaped with a length of 1.4 mm and diameter of 0.6 mm. All dimensions shown are in mm. (From Perex-Calatayud and Ballester, Dosimetry Characteristics of the Plus and 12*i* PDR [192]Ir Sources, AAPM 2001. With permission.)

Both designs contain a cylindrical active iridium core of 0.06 cm in diameter and 0.05 cm in length, which is encapsulated in a stainless steel wire. The active core in both designs is filtered in the most distal end by a cylinder of aluminum and a stainless steel cap. The 12*i* PDR source design thus results in a total outer diameter of 0.11 cm, and the Plus PDR source to 0.09 cm. These are also the outer diameters of the drive cables for the two sources.

Pérez-Calatayud et al.[27] published MC based dosimetry data for both GammaMed PDR sources. They modeled a 6 cm-long stainless steel cable. The simulations have been performed using the GEANT3 MC code,[51] and the decay scheme for ^{192}Ir was taken from the nuclear data sheet published by Shirley.[46] The simulations were performed in a water cylinder of 40 cm height and 40 cm diameter, for achieving full scatter conditions. The authors considered a cut-off energy for photons of 10 keV. The proximal end of the stainless steel wire was modeled as a 6.0 cm-long cylinder of an effective density of 5.6 g cm^{-3}. For the extraction of the TG-43 parameter values, the line source approximation was used for the geometry function $G_L(r,\theta)$ as described by Equation 8.28 or Equation 8.37.

The dose rate constant was found to have the same value for both 12*i* and Plus PDR source designs of $_{MC}\Lambda = (1.122 \pm 0.3\%)$ cGy h^{-1} U^{-1}. This value is very close to the MC-based values for the old and new microSelectron PDR source designs listed in Table 8.7, 1.121 to 1.128 cGy h^{-1} U^{-1}.

The radial dose function $g(r)$ values for both sources are listed in Table A.3.32 of Appendix 3, where Figure 8.45 demonstrates a comparison of the $g(r)$ curves. The $g(r)$ values for the Plus source design are, on average, 0.3% higher than those for the 12*i* PDR source design.

The anisotropy function values are listed in Table A.3.33 and Table A.3.34 for the 12*i* and Plus source models, respectively. There are small differences in the $F(r,\theta)$ values, where the 12*i* model demonstrates generally higher anisotropy detectable mainly for radial distances bellow 1.0 cm (see also Figure 8.46).

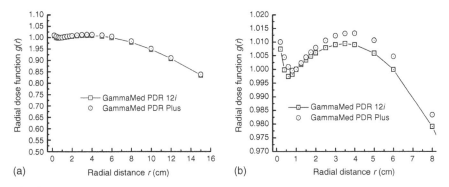

FIGURE 8.45
Comparison of MC simulation based radial dose function values, $g(r)$, for the 12*i* and Plus GammaMed PDR ^{192}Ir source type designs (data by Pérez-Calatayud et al.,[27] Table A.3.32 of Appendix 3). $g(r)$ plotted against radial distance, r, (a) in the complete range up to $r = 15.0$ cm, and (b) for $r \leq 8.0$ cm (zoom).

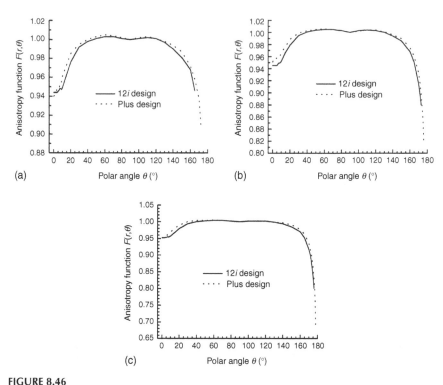

FIGURE 8.46
MC simulation based anisotropy function, $F(r,\theta)$, values for the 12i and Plus GammaMed PDR ^{192}Ir source type designs (data by Pérez-Calatayud et al.,[27] Table A.3.33 and Table A.3.34 of Appendix 3). $F(r,\theta)$ polar angle profiles at radial distance (a) $r = 0.4$ cm, (b) $r = 1.0$ cm, and (c) $r = 2.0$ cm.

Based on the linear fits published by Pérez-Calatayud et al.[27] for the anisotropy factor $\varphi_{an}(r)$, values for representative radial distances in the range of 0.25 to 12.0 cm have been calculated, and are listed in Table A.3.35 of Appendix 3. For both source designs $\varphi_{an}(r)$ remains constant within the accuracy of three digits and for the described range r.

8.4.3 Sources Used in LDR Brachytherapy

Several types of sources of different radionuclides, such as the high energy ^{137}Cs, the intermediate energy ^{192}Ir, and the very low energy ^{125}I and ^{103}Pd, have been developed and used in clinical applications. Iridium wires and iridium seed source configurations, although still in use in a few countries and centers, are of limited interest. Furthermore, for iridium wires (single pin or double length hairpin sources), the TG-43 and its update TG-43 U1, are not applicable. Analytical calculation methods or tabulated dose rate values for a specific source strength have been used for enabling dosimetric treatment planning when using these sources. When such types sources are used, detailed information on their dosimetric data and dose calculation models can be requested directly from the manufacturer and planning system vendor.

Regarding LDR sources of the intermediate energy radionuclide ^{169}Yb, there have been some data published on different experimental seed designs. Mason et al.[54] published dosimetric results using MC simulation for the prototype seed designs Type 4 and Type 5 manufactured by Amersham. They published only normalized dose rate tables, but no TG-43 related data. Perera et al.[55] investigated the dosimetric characteristics for Type 6 seed designs (Amersham). They published only dose rate constant values and a radial dose function table. Finally Piermattei et al.[56,57] published experimental and MC simulation data for the X1267 ytterbium seed (Amersham) where only dose rate constant values and radial dose rate graphs were presented. ^{169}Yb seeds of low strength originally thought to be used in LDR permanent implants of prostate cancer did not find adequate acceptance in the clinical environment, and thus have been used in some limited experimental studies.

8.4.3.1 ^{137}Cs Sources

Due to the very low specific activity of ^{137}Cs (3.202 GBq mg^{-1}, the lowest among all currently used radionuclides, see Table 5.2), practically, only sources for LDR brachytherapy applications can be manufactured. Several ^{137}Cs LDR source designs are used either in automatic remote afterloading systems or in manual procedures for intracavitary brachytherapy applications. Cesium sources are, in general, short (maximal lengths around 2.0 cm) with an outer diameter in the range of 0.2 to 0.3 cm. Complete TG-43 published data[4] only exist for some of these. There are six source types manufactured by Amersham (Amersham Health, Buckinghamshire, U.K.) and the CSM11 source design manufactured by BEBIG/CIS (Berlin Isotopen-und Medizintechnik GmbH, Berlin, Germany). For all of these seven source designs the encapsulation is made of stainless steel (see Figure 8.47 through Figure 8.51). A summary of the geometrical characteristics of all seven ^{137}Cs source designs is presented in Table 8.8. The effective source length values, L_{eff}, listed in this table are according to TG-43 U1 notation as described by Equation 8.37 and Figure 8.14.

The only pellet-shaped source is that for the Selectron-LDR afterloader system (Figure 8.47), consisting of an 0.15 cm in diameter spherical active core made of borosilicate glass and having an outer diameter of 0.25 cm thus resulting in a stainless steel encapsulation thickness of 0.05 cm. All other sources are cylindrically shaped, having either a single cylindrical active core (type CSM11), or containing a single or several active spherical pellets containing ^{137}Cs. All these cylindrical sources are used both in manual and automatic afterloading systems.

All published papers are MC dosimetry studies all using the GEANT MC simulation code. With the exception of the study by Pérez-Calatayud et al.[37]

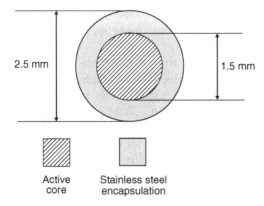

FIGURE 8.47
Schematic diagram of the low dose rate ^{137}Cs pellet used with the Selectron LDR afterloader device.

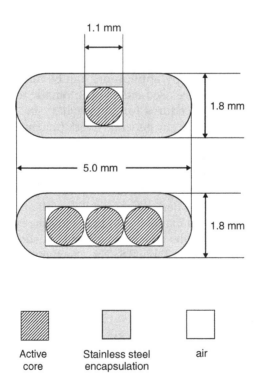

FIGURE 8.48
Geometry, dimensions, and materials of the two CDC type ^{137}Cs low dose rate sources.[38] (*Upper*) The CDC-1 design. The origin of the coordinate system is placed in the center of the single active sphere. (*Lower*) The CDC-3 design consisting of three active spheres. The origin of the coordinate system is placed in the center of the middle active sphere.

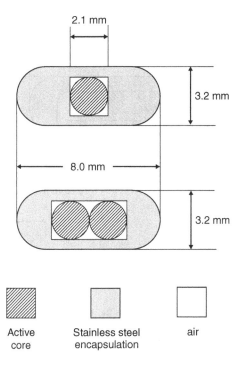

FIGURE 8.49
Geometry, dimensions and materials of the Walstam type ^{137}Cs low dose rate sources.[41] (*Upper*) The CDC.K1–K3 design. For K1 to K3 models the same geometry is valid with different source strengths. The origin of the coordinate system is placed in the center of the single active sphere. (*Lower*) The CDC.K4 design consisting of two active spheres. The origin of the coordinate system is placed in between the two active spheres (geometrical center of the two spheres).

for the Selectron-LDR pellet where the simulation geometry consisted of a 30 cm in diameter spherical water phantom, all other studies used a cylindrical water phantom of 40 cm high and 40 cm in diameter. In all these simulation geometries the source, sphere, or cylindrical, was positioned at the center of the water phantom.[37–41]

In all studies a 10 keV energy cut-off for photon and electron history tracking was considered. With the exception of the study by Ballester et al.[39] for the CSM11 source type and of the study by Casal et al.[40] for the CDCS-M source type, where the origin of the coordinate system was placed in the geometric center of the source itself and not of the cylindrical active core, all other studies followed the TG-43/TG-43 U1 convention of Figure 8.6.

The dose rate constant values $_{MC}\Lambda$ are summarized in Table 8.9. The calculated $_{MC}\Lambda$ values for all sources with the exception of that for the CDCS-M, are very close due to the very similar effective lengths (0.11 to 0.42 cm). For the case of the CDS-M design (Figure 8.51) the effective

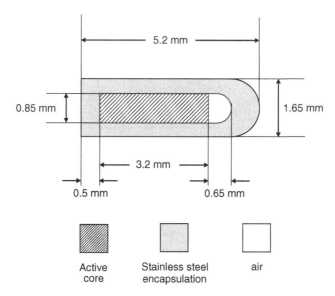

FIGURE 8.50
Schematic diagram of the CIS CSM11 type (CIS Bio International, France) low dose rate ^{137}Cs
cylindrically shaped source.[39] The origin of the coordinate system is in the center of the
cylindrical active core, where the $+z$ direction is aligned towards the round tip-end of the source
(right direction in the figure). The origin was assumed by Ballester et al.[39] at the center of the
source and not the center of the active core length (displacement of 0.05 cm).

length is 1.5 cm and this explains the significantly lower $_{MC}\Lambda$ value for
that source.

In Figure 8.52 the geometry-normalized dose rate constant values, $\Lambda_{L_{eff}}/$
$G_{L_{eff}}(1 \text{ cm}, 90°)$ have been analyzed for all seven source designs. $\Lambda_{L_{eff}}$ is the
dose rate constant value that corresponds to the L_{eff} (in cm) effective source
length and $G_{L_{eff}}(1 \text{ cm}, 90°)$ is the value of the corresponding geometry
function for L_{eff} source length and for the TG-43/TG-43 U1 reference point,
$r = 1.0$ cm and $\theta = 90°$, calculated using the line source approximation as
described in Equation 8.28 or Equation 8.38. It can be shown that the dose
rate constant, $\Lambda_{L_{eff}}$, for all these cesium sources expressed in cGy h^{-1} U^{-1} is
very accurately calculated according to

$$\Lambda_{L_{eff}} = (1.109 \pm 0.004)G_{L_{eff}}(1 \text{ cm}, 90°) \tag{8.54}$$

where the geometry function value $G_{L_{eff}}(1 \text{ cm}, 90°)$ is expressed in cm^{-2}.

Similar results were found for the dose rate constant for ^{192}Ir sources
and are presented in Chapter 9 on MC-based dosimetry (see also Figure 9.19).

As in the case of iridium,[31,52,53] the dose rate constant for the cesium
sources depends marginally on the source diameter. This behavior indicates
that the dose rate constant Λ is not affected by scattering and absorption
within source and encapsulation (see data in Table 8.8 and Table 8.9).
This is due to the fact that the TG-43/TG-43 U1 dosimetric formalism

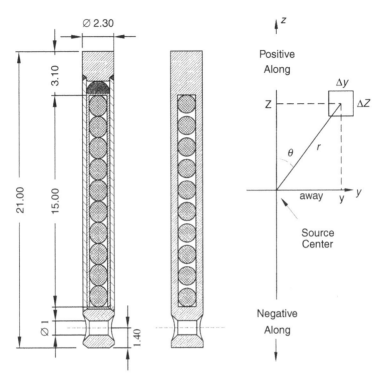

FIGURE 8.51
Geometry of the Amersham CDCS-M-type low dose rate ^{137}Cs cylindrically shaped source containing a total of ten active spheres encapsulated in stainless steel. The origin of the coordinate system is in the center of the 15 mm long cylindrical core volume containing the 10 active spheres. The +z direction is aligned towards the upper end of the source. (*Left*) Design according to the manufacturer. (*Middle*) Geometry used for the Monte Carlo simulation. (*Right*) The coordinate system. All dimensions shown are in mm. The origin was assumed by Casal et al.[40] at the center of the source and not at the center of the active core length (displacement of 0.1 mm). (From Casal and Ballester, Monte Carlo Calculation of Dose Role Distributions around the Amersham CDSC-M-type 137-Cs Source AAPM 2000. With permission.)

demands on the encapsulated sources being calibrated in terms of reference air kerma rate. Thus, for calculation of the dose rate constant, the effect of the encapsulated source geometry on the scattering and absorption of photons are cancelled out.

For the extraction of the TG-43 data, all publications with the exception of that by Pérez-Calatayud et al.[37] for the Selectron-LDR pellet considered the effective source length L_{eff} as described by Equation 8.37 and Figure 8.14 and listed in Table 8.8.

Due to the spherical shape of the active core and of its encapsulation, Pérez-Calatayud et al.[37] considered the point source approximation of the geometry function $G(r,\theta) = 1/r^2$. The radial dose function values for all these sources are listed in Table A.3.36 and Table A.3.37 of Appendix 3, whereas Figure 8.53

TABLE 8.8

Geometrical and Material Characteristics of Common Commercial Low Dose Rate (LDR) ^{137}Cs Sources

Source Type	Active Core Material	No Pellets	Material between Pellets	Gap Length between Pellets (cm)	Effective Length L_{eff} (cm)	Pellet Length L_s (cm)	Pellet Diameter D_s (cm)	Encapsulation Material	Thickness (cm)	Outer Diameter (cm)	Outer Length (cm)
Selectron LDR[a] pellet[b]	Pollucite/borosilicate glass[c]	1	—	0.0	0.15	0.15	0.15	Stainless steel AISI 316L	0.050	0.250	0.25
CDC-1	Pollucite/borosilicate glass[c]	1	—	0.0	0.11	0.11	0.11	Stainless steel AISI 316L	0.035	0.180	0.50
CDC-3	Pollucite/borosilicate glass[c]	3	Air	0.0	0.33	0.11	0.11	Stainless steel AISI 316L	0.035	0.180	0.50
CDC.K1-K3	Borosilicate glass[c]	1	—	0.0	0.21	0.21	0.21	Stainless steel AISI 316L	0.055	0.320	0.80
CDC.K4	Borosilicate glass[c]	2	Air	0.0	0.42	0.21	0.21	Stainless steel AISI 316L	0.055	0.320	0.80
CIS[d] CSM11	Pollucite[c]	1	—	0.0	0.32	0.32	0.085	Stainless steel AISI 316L	0.040	0.165	0.52
CDCS-M-type[b]	Borosilicate glass[c]	10	—	0.0	1.50	0.13	0.13	Stainless steel AISI 316L	0.050	0.230	2.10

All sources are used either in manual or automatic afterloading systems.

[a] Nucletron B.V., Veenendaal, The Netherlands.

[b] Produced by Amersham, U.K.

[c] Atomic composition and fractions by weight: Si(26.18%), Ti(3.00%), Al(1.59%), B(3.73%), Mg(1.21%), Ca(2.86%), Na(12.61%), Cs(0.94%), O(47.89%).

[d] CIS Bio International (France).

TABLE 8.9

Monte Carlo Calculated Values of the Dose Rate Constant Λ for the Different ^{137}Cs Low Dose Rate Brachytherapy Sources Listed in Table 8.8

Source Type	$_{MC}\Lambda$	Monte Carlo Code	δ (keV) (Equation 8.42)	Author
Selectron LDR pellet	1.107 ± 0.3%	GEANT4	10.0	Pérez-Calatayud et al.[37]
CDC-1	1.113 ± 0.3%	GEANT3	10.0	Pérez-Calatayud et al.[38]
CDC-3	1.103 ± 0.3%	GEANT3	10.0	Pérez-Calatayud et al.[38]
CDC.K1–K3	1.106 ± 0.1%	GEANT	10.0	Pérez-Calatayud et al.[41]
CDC.K4	1.092 ± 0.1%	GEANT	10.0	Pérez-Calatayud et al.[41]
CSM11	1.096 ± 0.2%	GEANT	10.0	Ballester et al.[39]
CDCS-M	0.946 ± 0.7%	GEANT	10.0	Casal et al.[40]

All Λ values are expressed in cGy h^{-1} U^{-1}. Energy cut-off values δ referred in the original publications are also listed.

presents a graphical comparison of the $g(r)$ curves. A very high degree of agreement between all sources can be demonstrated for radial distances above 0.8 cm. For smaller distances the differences in the active core configuration, and thus in the spatial distribution of activity among these source designs, results in pronounced deviations among the $g(r)$ curves.

It has to be mentioned that Ballester et al.,[39] in their MC study on the CSM11 source, used as origin of the source coordinate system the center of the source itself instead of the center of the active core (see Figure 8.6 and Figure 8.50). There is a difference of 0.05 cm between these two centers as can be seen in Figure 8.50. Similarly Casal et al.[40] used in their MC-study on the

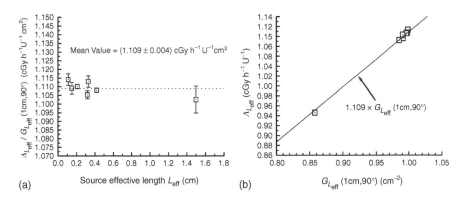

(a)   (b)

FIGURE 8.52

Dependence of the dose rate constant, $\Lambda_{L_{eff}}$, for the different ^{137}Cs LDR sources on the effective source length L_{eff} (Table 8.9). (a) The $\Lambda_{L_{eff}}$ values normalized to the corresponding line source approximated geometry function values $G_{L_{eff}}(1\,cm, 90°)$ as described by Equation 8.38 plotted against the source effective length L_{eff}. A mean value of (1.109 +/− 0.004) cGy h^{-1} U^{-1} cm^2 can be estimated (dotted line). (b) $\Lambda_{L_{eff}}$ plotted against $G_{L_{eff}}(1\,cm, 90°)$. The line $\Lambda_{L_{eff}} = 1.109 \times G_{L_{eff}}(1\,cm, 90°)$ is also presented.

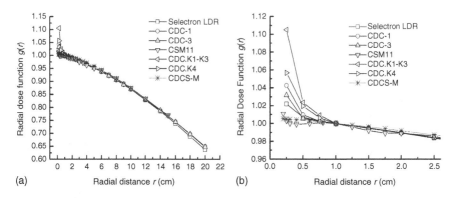

FIGURE 8.53
Comparison of MC simulation based radial dose function values, $g(r)$, for all seven ^{137}Cs LDR sources (Table A.3.36 and Table A.3.37 of Appendix 3). $g(r)$ plotted against radial distance, r, (a) in the complete range up to $r = 20.0$ cm and (b) for $r \leq 2.5$ cm (zoom).

CDCS-M source the center of the source, as the origin of the coordinate system resulting in a displacement of 0.01 cm from the center of the active core (see Figure 8.51).

The anisotropy function values for the seven cesium seed designs are listed in Table A.3.38 through Table A.3.43 of Appendix 3. For the spherical Selectron LDR pellet source anisotropy corrections based on the anisotropy function are irrelevant, $F(r,\theta) = 1.0$, $\forall\, r$ and $\forall\, \theta$.

Since the CSM11 and CDCS-M have an asymmetric encapsulation design regarding the two ends of the seed (asymmetric about the transverse plane), the corresponding anisotropy tables (Table A.3.42 and Table A.3.43) are extended up to 180° for the polar angle θ.

For the CDC-1 and CDC-3 designs,[38] as well as for the CDC.K1–K3 and CDC.K4 designs,[41] the original publications also offered values for the anisotropy factor $\varphi_{an}(r)$ which are listed in Table A.3.44.

8.4.3.2 ^{125}I Sources

Since the first clinical applications using ^{125}I seeds for the treatment of prostate cancer in the early 1970s, many seed designs have been manufactured for clinical use. New designs are developed every now and then, all intended for use in prostate brachytherapy.[4,13,58] This is supported by the rapid increase in the utilization of permanent brachytherapy implants for the treatment of early stages of prostate cancer. An increasing number of manufacturers are offering ^{125}I seeds for clinical applications.

The main reasons for the continuous efforts of developing new seeds are to improve (a) the anisotropy of dose distribution, (b) visualization of seeds using x-rays, (c) visualization of seeds in CT imaging by reducing artifacts, (d) enhancement of the visualization of seeds in ultrasound imaging by improving the ultrasound signature of these over a wider range of angular orientations and to reduce migration risk.

These can be achieved by altering the design of the internal seed core, changing the material(s) used for the core, adapting the distribution of radioactivity, and changing the encapsulation materials and design.

The 6702 and 6711 seed models by Amersham (Amersham Health, U.K.) have the longest history of clinical use and are therefore considered the benchmarks for the industry.

The revised TG-43 report, the TG-43 U1,[13] offers reference data sets for several ^{125}I-seeds and these data are electronically available on the web (www.rpc.mdanderson.org/rpc/html/Home_html/Low-energy.htm or www.uv.es.brachyqs).

The consensus reference data for the 6702 and 6711 Amersham seeds, BEBIG Symmetra I25.S06 seed as provided by TG-43 U1[13] together with the data for the Nucletron selectSeed model 130.002, are shortly considered.

Table 8.10 summarizes the geometrical characteristics of these seeds which almost have the same outer dimensions and titanium encapsulation material. Figure 8.54 shows the geometrical and material designs of the four iodine seeds. The consensus dose rate constant values, $_{CON}\Lambda$, are listed in Table 8.11. The selectSeed $_{CON}\Lambda$ of (0.962 ± 0.043) cGy h^{-1} U^{-1} is according to Papagiannis et al.[82] and has been calculated according to Equation 8.47. Papagiannis et al. considered a $_{MC}\Lambda$ value of (0.937 ± 0.014) cGy h^{-1} U^{-1} that is the average value of MC results using a detector resembling the solid angle subtended to the seed by the NIST WAFAC detector,[13] and a point detector.[59,82] The $_{EXP}\Lambda = (0.987 \pm 0.077)$ cGy h^{-1} U^{-1} value published[82] is based on measurements using TLD-100 rods in a solid water phantom[60] and three NIST calibrated selectSeeds.

The radial dose function values for point source approximation and line source approximation as defined in Equation 8.41 are listed in Table A.3.45 and Table A.3.46 of Appendix 3. The $g(r)$ values are plotted against radial distance r in Figure 8.55 for both the line and point source approximation. Values for the selectSeed are close to those for the model 6711, and values for the Symmetra I25.S06 are close to those for the 6702 seed model.

The consensus anisotropy function values[13] for the 6702, 6711 Amersham models, as well as for the BEBIG Symmetra I25.S06 seed, are listed in Table A.3.47 through Table A.3.49 of Appendix 3. The corresponding values based on MC dosimetry for the selectSeed[59] are listed in Table A.3.50. $F(r,\theta)$ values are plotted against polar angle θ in Figure 8.56 for radial distances $r = 0.5$ cm and $r = 1.0$ cm for all four seeds.

8.4.3.3 ^{103}Pd Sources

For almost 20 years, brachytherapy sources containing ^{103}Pd have been clinically introduced and are in use for performing mainly permanent implants of prostate cancer as an alternative to the ^{125}I seeds. They have also been employed in the treatment of other anatomical sites.[62-69] The first ^{103}Pd interstitial source was introduced by Theragenics Corporation in 1987, and was the Model 200 (TheraSeed®, Theragenics Co, Norcross, U.S.A.).

TABLE 8.10

Geometrical and Material Characteristics of Some Representative Commercial Low Dose Rate (LDR) ^{125}I Seed Sources

Source Type	Active Core							Encapsulation			
	Material	No Pellets	Material between Pellets	Gap Length between Pellets (cm)	Effective Length L_{eff} (cm)	Pellet Length L_s (cm)	Pellet Diameter D_s (cm)	Material	Thickness (cm)	Outer Diameter (cm)	Outer Length (cm)
Amersham 6702	Resin	3	—	—	0.30	0.06	0.06	Ti	0.006	0.08	0.46
Amersham 6711	Ag	1	—	—	0.30	0.30	0.05	Ti	0.006	0.08	0.46
BEBIG Symmetra I25.S06	Al$_2$O$_3$ + AgI	1	—	—	0.35	0.35	0.06	Ti	0.005	0.08	0.46
Nucletron selectSeed 130.002	Ag	1	—	—	0.34	0.34	0.05	Ti	0.005	0.08	0.45

All sources are used for permanent prostate implants.

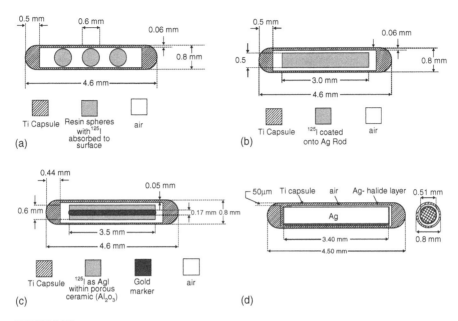

FIGURE 8.54

Cross-sectional drawings of the four ^{125}I LDR seed sources.[13] (a) Amersham model 6702,[13] (b) Amersham model 6711 OncoSeed,[13] (c) BEBIG Symmetra I25.S06,[13] and (d) Nucletron select Seed model 130.002.[82]

The Model 200 remained the sole palladium brachytherapy source available in the market until 1999.[62] In February 1999 the model MED3633 ^{103}Pd source manufactured by NASI (North American Scientific Inc., Chatsworth, California, U.S.A.) was introduced. The geometric and material design of both palladium seeds are graphically presented in Figure 8.57 and summarized in Table 8.12. Both seed designs have been considered in the TG-43 U1 publication.[13] The consensus values for the dose rate constant, $_{CON}\Lambda$, are listed in Table 8.13 demonstrating very similar values, 0.686 cGy h^{-1} U^{-1} for Model 200 vs. 0.688 cGy h^{-1} U^{-1} for the MED3633 seed. As for the case of the ^{125}I seeds, for these two ^{103}Pd seed designs, the $_{CON}\Lambda$ values of dose rate constant also have to be used in conjunction with

TABLE 8.11

Consensus Values of the Dose Rate Constant, $_{CON}\Lambda$ in Equation 8.48, for Four Representative ^{125}I Seeds Listed in Table 8.10

Source Type	$_{CON}\Lambda$ (cGy h^{-1} U^{-1})
Amersham 6702[13]	1.036
Amersham 6711[13]	0.965
BEBIG Symmetra I25.S06[13]	1.012
Nucletron SelectSeed 130.002[82]	0.962

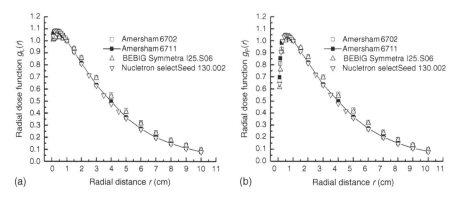

FIGURE 8.55
Comparison of the radial dose function values, $g(r)$, for all four ^{125}I LDR sources of Figure 8.54. These are the consensus values for the 6702, 6711, and Symmetra I25.S06 source designs[13] (Table A.3.45 of Appendix 3) and the MC dosimetry based values for the selectSeed model 130.002 according to Karaiskos et al.[60] (Table A.3.46 of Appendix 3). $g(r)$ plotted against radial distance, r, (a) line source approximated radial dose function $g_L(r)$ and (b) point source approximated radial dose function $g_P(r)$.

source strength measurements which are in accordance with, and traceable to, the NIST WAFAC 1999 standard as corrected in 2000.[13,61]

The radial dose function values for line $g_L(r)$ and point $g_P(r)$ source approximation (see Equation 8.41) are listed in Table A.3.51 of Appendix 3. In Figure 8.58 the radial dose function values for both approximations are plotted against radial distance r.

The consensus anisotropy function values $F(r,\theta)$ and anisotropy factor values $\varphi_{an}(r)$[13] for the Model 200 and MED3633 are listed in Table A.3.52 and A.3.53 of Appendix 3, respectively. In Figure 8.59 the anisotropy function

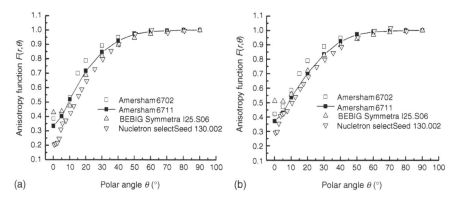

FIGURE 8.56
Anisotropy function, $F(r,\theta)$, values for the four ^{125}I LDR sources (Table A.3.47 through Table A.3.50 of Appendix 3). $F(r,\theta)$ polar angle profiles at radial distance (a) $r = 0.5$ cm and (b) $r = 1.0$ cm.

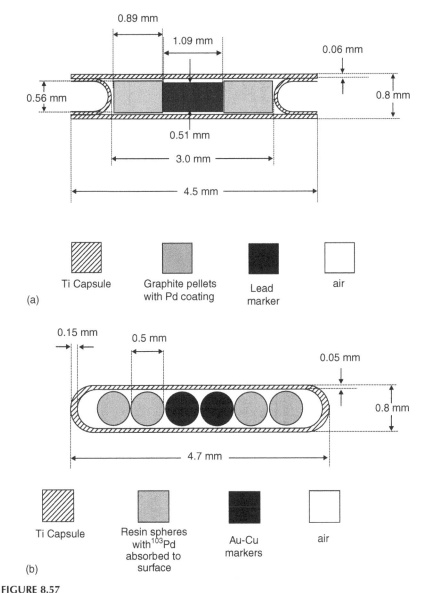

FIGURE 8.57
Cross-sectional drawings of the ^{103}Pd LDR seed sources considered in the TG-43 U1 report.[13] (a) Theragenics TheraSeed model 200, and (b) North American Scientific model MED3633.

values for both seeds and for radial distances $r = 0.25$ cm and $r = 1.0$ cm are plotted against polar angle θ.

When calculations are to be performed using these consensus TG-43 tables and data it is mandatory that the effective source lengths, L_{eff}, listed in Table 8.12 have to be utilized for calculating geometry function, $G(r,\theta)$, values using Equation 8.38.

TABLE 8.12

Geometrical and Material Characteristics of the Two Commercial Low Dose Rate (LDR) ^{103}Pd Seed Sources

Source Type	Active Core							Encapsulation		
	Material	No Pellets	Material between Pellets	Gap Length between Pellets (cm)	Effective Length L_{eff} (cm)	Pellet Length L_s (cm)	Pellet Diameter D_s (cm)	Material	Thick-ness (cm)	Outer Diameter (cm)	Outer Length (cm)
Theragenics Model 200	Graphite	2	Pb (cylindrical marker)	0.109	0.423[a]	0.089	0.056	Ti	0.006	0.08	0.45
NASI MED3633	Resin	4	Au-Cu (2 × spherical markers)	—	0.42[a]	0.05	0.05	Ti	0.005	0.08	0.47

[a] As used to derive consensus $g(r)$ and $F(r,\theta)$ values.[13]

TABLE 8.13

Consensus Values[13] of the Dose Rate Constant, $_{CON}\Lambda$ in Equation 8.48, for the Two ^{103}Pd Seeds, Model 200 and MED3633, Listed in Table 8.12

Source Type	$_{CON}\Lambda$ (cGy h^{-1} U^{-1})
Theragenics Model 200	0.686
NASI MED3633	0.688

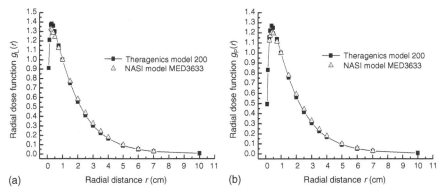

FIGURE 8.58
Comparison of the radial dose function values, $g(r)$, for the two ^{103}Pd LDR sources, TheraSeed model 200 and MED3633, shown in Figure 8.57. These are the consensus values according to TG-43 U1[13] (Table A.3.51 of Appendix 3). $g(r)$ plotted against radial distance, r, (a) line source approximated radial dose function $g_L(r)$, and (b) point source approximated radial dose function $g_P(r)$.

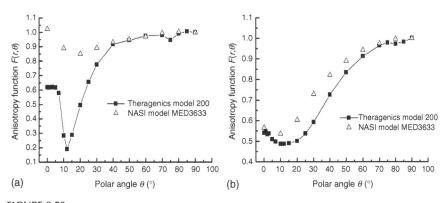

FIGURE 8.59
Anisotropy function, $F(r,\theta)$, values for TheraSeed model 200 and MED3633 ^{103}Pd LDR sources (Table A.3.52 and Table A.3.53 of Appendix 3). $F(r,\theta)$ polar angle profiles at radial distance (a) $r = 0.25$ cm and (b) $r = 1.0$ cm.

The consensus TG-43 data and tables for these two ^{103}Pd sources are electronically available at www.rpc.mdanderson.org/rpc/html/Home_html/Low-energy.htm or www.uv.es.brachyqs.

8.5 Dose Rate Look-Up Tables

As shown in Figure 8.15, the primary information used for deriving the TG-43/TG-43 U1 data and tables is a dose rate table per unit source strength, DRT, which is usually expressed in cGy h^{-1} U^{-1}. This is a "dissemble-and-assemble" factorization approach;[70] first derive from the DRT the TG-43 U1 parameters, and then apply these utilizing either Equation 8.43 or Equation 8.44 to calculate the dose rate value at the required place.

Although from a practical point of view DRTs could be much more useful if they are expressed as radial distance r and polar angle θ, dose rate tables (in polar coordinates), commonly published DRTs, are expressed as along-away tables (Cartesian coordinates). Along means the z- and away the y-axis of the TG-43 coordinate system shown in Figure 8.6. This is due to the fact that using polar coordinates a higher resolution at short distances, and respectively, lower resolution at large radial distances can be achieved in accordance to the variation of the dose rate gradient around the sources.

Although extrapolations outside an existing DRT can be easily handled (similarly to that described for the TG-43 parameters, Equation 8.51 and Equation 8.52), the accuracy of interpolation between DRT values is questionable.

Despite the physical content/meaning of the geometry function $G(\vec{r})$ described in Equation 8.11 and its line or point source approximations (Equation 8.38), this can be utilized to filter out, to a very high degree, the large dose rate gradients existing when moving along a ray, at a specific polar angle θ or moving along an arc, for a specific radial distance r. The same is valid when an along-away table is considered, and a bilinear interpolation has to be applied along the y- and the z-axis. In this way the accuracy of the bilinear interpolation is significantly improved.

In Figure 8.60 this effect is graphically demonstrated for the example of the new microSelectron HDR ^{192}Ir source by Nucletron using the DRT in Table A.4.2 of Appendix 4.

Look-up tables (LUT) in Cartesian (along-away) or polar coordinates (r, θ) of dose rate per unit source length multiplied by the corresponding geometry function value obtained using the line source approximation could be used either for computations or for verification reasons:

$$\mathrm{LUT}(y,z) = \left(\frac{\dot{D}(y,z)}{S_K} \right) \left(\frac{1}{G_L(y,z)} \right) \tag{8.55}$$

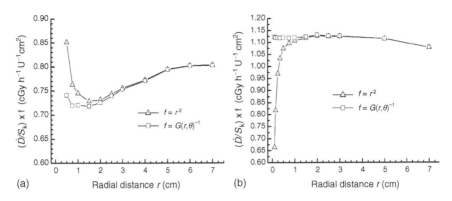

FIGURE 8.60
Demonstration of improving interpolation accuracy when using dose rate look-up tables (LUTs) instead of analytic expressions such as the TG-43 U1[13] dosimetry protocol. The dose rate per unit source strength, $\dot{D}/S_K$, table values for the microSelectron HDR new source design according to Daskalov et al.[20] (Table A.4.2 of Appendix 4) have been multiplied by the factor f, which is the inverse of the corresponding geometry function value either for the point source approximation, $f = r^2$, or the line source approximation, $f = G(r,\theta)^{-1}$. $(\dot{D}/S_K)f$, plotted against radial distance r for (a) polar angle $\theta = 0°$, which is the column with away distance $y = 0.0$ cm and (b) polar angle $\theta = 90°$, which is the row with along distance $z = 0.0$ cm in Table A.4.2. The homogenization effect when applying the line source approximation of the geometry function based factor f is clearly demonstrated, thus resulting in very accurate results when performing interpolations among tabulated values (LUT).

or

$$\text{LUT}(r, \theta) = \left(\frac{\dot{D}(r, \theta)}{S_K}\right)\left(\frac{1}{G_L(r, \theta)}\right)$$

with $G_L(y,z)$ and $G_L(r,\theta)$ being the value of the line approximation of the geometry function as given by Equation 8.38. When the Cartesian coordinate system is used (along-away LUT), then

$$r = \sqrt{y^2 + z^2} \text{ and } \theta = \text{Arc} \tan\left(\frac{y}{z}\right)$$

The LUT values are thus expressed in cGy h^{-1} U^{-1} cm^2. The use of the line source approximation for the geometry function improves interpolation for radial distances $r < 3L_s$.

After the interpolation among LUT values, the final value has to be multiplied by the corresponding geometry function value to obtain the dose rate per unit source strength at the point of interest.

This methodology, using of LUTs, simplifies the data handling (only a single LUT) and also enables the handling of new source types such as miniaturized x-ray sources or cylindrically asymmetric source designs, where 3D LUT can be easily considered.

For HDR or PDR single stepping source afterloaders, the orientation of the source at any point is easily estimated based on the catheter geometry.

Thus, only a single LUT is needed for computations. In the case of LDR seeds in permanent implants, where in the majority of the cases it is very difficult to identify the individual seed orientation (seed direction vector), a second LUT can be utilized where only the distance dependence is considered after averaging over the angular dependence effect.[70]

In Appendix 4, examples of DRTs for several source types are listed.

8.6 Dose from Dose Rate

All formalisms described previously enable the estimation of the dose rate value $\dot{D}(r,\theta)$ at a point of interest $P(r,\theta)$ around a single source aligned at the origin of the coordinate system, as shown in Figure 8.6. For calculating the absorbed dose value $D(r,\theta)$ at that point we assume that the source strength expressed as air kerma strength, and thus the dose rate, are obtained at the start point of consideration, $t_0 = 0$, $\dot{D}(r,\theta,t_0)$.

If λ is the decay constant of the source radionuclide, $\lambda = \ln(2)/T_{1/2}$ with $T_{1/2}$ the half-life of the radionuclide (see Section 2.6.1) the dose $dD(r,\theta)$ delivered in the time interval dt at the time t ($t \geq t_0$), with dt being small enough so that the dose rate during dt can be considered as unchanged (see Equation 2.51):

$$dD(r, \theta) = \dot{D}(r, \theta, t)dt = \dot{D}(r, \theta, t_0)e^{-\lambda(t-t_0)}dt$$

If the dwell time of the source is T then the total dose delivered at the point of interest due to that source and during the time period T is obtained by integrating the above equation in the range $[0, T]$:

$$D(r, \theta) = \int_{t=t_0=0}^{t=T} \dot{D}(r, \theta, t_0)e^{-\lambda(t-t_0)}dt = \dot{D}(r, \theta, t_0)\left(\frac{1}{\lambda}\right)(1 - e^{-\lambda T})$$

and finally:

$$D(r, \theta) = \dot{D}(r, \theta, t_0)\tau\left(1 - e^{-\frac{T}{\tau}}\right) \qquad (8.56)$$

with $t_0 = 0$ and τ the mean life of the radionuclide, $\tau = 1/\lambda$, see Equation 2.54.

Since all formalisms described in Equation 8.16, Equation 8.43, Equation 8.45, and Equation 8.46, result to dose rate values expressed in cGy h^{-1}, the mean life τ in Equation 8.56 has to be expressed in hours.

For irradiation duration $T \ll \tau$, the exponential part in Equation 8.56 is

$$e^{-\frac{T}{\tau}} \approx 1 - \frac{T}{\tau}$$

TABLE 8.14

Units Conversion Factor k_u Considered in Equation 8.57 and Equation 8.58 for Converting the Dose Value into the Desired Dose Unit, cGy or Gy

Dose Unit	Source Dwell Time T Unit (Equation 8.57)	Source Mean Life τ Unit (Equation 8.58)	Conversion Factor
cGy	sec	d	$1/3{,}600 = 2.778 \times 10^{-4}$ h sec^{-1}
	min		$1/60 = 1.667 \times 10^{-2}$ h min^{-1}
	h		1.000
	d		24 h d^{-1}
Gy	sec	d	$1/3600 \times 10^{-2} = 2.778 \times 10^{-6}$ h sec^{-1} Gy cGy^{-1}
	min		$1/60 \times 10^{-2} = 1.667 \times 10^{-4}$ h min^{-1} Gy cGy^{-1}
	h		1.0×10^{-2} Gy cGy^{-1}
	d	y	0.24 h d^{-1} Gy cGy^{-1}

Based on the calculated initial dose rate value expressed in cGy h^{-1} and the source dwell time T expressed either in seconds or minutes or hours or days, or source mean life τ expressed in days (see also Table 2.3).

and thus

$$D(r, \theta) = \dot{D}(r, \theta, t_0)Tk_u \tag{8.57}$$

k_u is a units conversion factor to convert the dose rate expressed in cGy h^{-1} and considering the time unit of the value T to the desired dose unit, Gy or cGy.

For the case of medium and high energy sources of high strength (HDR sources) the dwell time T is expressed in seconds (sec). Thus, the units conversion factor k_u is 1/3600 for expressing dose in cGy or 1/360,000 for expressing dose in Gy. Table 8.14 summarizes values for the units conversion factor.

Thus, for ^{192}Ir, ^{60}Co, ^{137}Cs, or ^{169}Yb sources used for temporary implants, $T \ll \tau$, the dose delivered at a point equals the initial dose rate at that point multiplied by the source dwell time and corrected for the units considered.

Figure 8.61 demonstrates the behavior of the time dependent part of Equation 8.56 for the temporary implant related radionuclides ^{192}Ir and ^{169}Yb, with the lowest half-life values for different irradiation duration times T. According to this figure the application of Equation 8.57 results in differences above 1% when compared to the original Equation 8.56, for times above 22 h and 2.1 days for the ^{169}Yb and ^{192}Ir radionuclides, respectively.

For permanent implants the situation can be approximated by considering the irradiation time T of the source much higher than the mean life of the radionuclide τ, $T \gg \tau$, and thus the exponential part in Equation 8.56 is

$$e^{-\frac{T}{\tau}} \approx 0$$

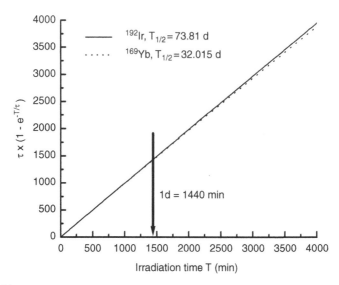

FIGURE 8.61
Irradiation time and decay dependent right part of Equation 8.56 for the temporary implant related radionuclides ^{192}Ir and ^{169}Yb with the lowest half-life values, plotted vs. the irradiation duration time T. It is demonstrated that the approximation described by Equation 8.57 results to deviations above 1% when compared to the original Equation 8.56, for irradiation times above 22 h and 2.1 days for the ^{169}Yb and ^{192}Ir radionuclides, respectively, which is approximately 3% of the corresponding half-life value.

and consequently

$$D(r, \theta) = \dot{D}(r, \theta, t_0)\tau k_u \qquad (8.58)$$

Thus, the dose delivered at a point in the case of a permanent stay of the source at its initial position, $T \gg \tau$, equals the initial dose rate at that point multiplied by the source mean life τ and corrected for the units considered.

References

1. Attix, H.F. *Introduction to Radiological Physics and Radiation Dosimetry*, Wiley, New York, 1986.
2. Interstitial Collaborative Working Group, *Interstitial Brachytherapy: Physical Biological and Clinical Considerations*, Raven Press, New York, 1990.
3. Nath, R., Anderson, L., Luxton, G., Weaver, K., Williamson, J.F., and Meigooni, A.S. Dosimetry of interstitial brachytherapy sources: Recommendations of the AAPM radiation therapy committee task group 43, *Med. Phys.*, 22, 209–234, 1995.
4. J. Venselaar and J. Pérez-Calatayud, eds., *European Guidelines for Quality Assurance in Radiotherapy, A Practical Guide to Quality Control of Brachytherapy Equipment Booklet No 8*, ESTRO, Belgium, 2004.

5. Williamson, J.F. *Physics of Brachytherapy in Principles and Practice of Radiation Oncology,* 3rd ed., C.A. Perez and L.W. Brady, eds., Lippincott-Raven Press, Philadelphia, USA, pp. 405–467, 1977.

6. Chen, Z. and Nath, R. Dose rate constant and energy spectrum of interstitial brachytherapy sources, *Med. Phys.,* 28, 86–96, 2001.

7. Karaiskos, P., Sakeliou, L., Sandilos, P., and Vlachos, L. Limitations of the point and line source approximations for the determination of geometry factors around brachytherapy sources, *Med. Phys.,* 27, 124–128, 2000.

8. Rivard, M.J. Refinements to the geometry factor used in the AAPM task group report NO. 43 necessary for brachytherapy dosimetry calculations, *Med. Phys.,* 26, pp. 2445–2450, 1999.

9. James, F. *Monte Carlo Theory and Practice Data Handling Division,* CERN, Geneva, 1980.

10. Koonin, S. and Meredith, D. *Computational Physics Fortran Version,* Addison-Wesley, Redwood City, 1990.

11. Press, W., Flannery, B., Teukolsky, S., and Vetterling, W. *Numerical Recipes in Fortran 77: The Art of Scientific Computing,* Cambridge University Press, Cambridge, 1992.

12. Baltas, D., Giannouli, S., Garbi, A., Diakonos, F., Geramani, K., Ioannidis, G.T., Tsalpatouros, A., Uzunoglu, N., Kolotas, C., and Zamboglou, N. Application of the Monte Carlo integration (MCI) method for calculation of the anisotropy of ^{192}Ir brachytherapy sources, *Phys. Med. Biol.,* 43, 1783–1801, 1998.

13. Rivard, M., Coursey, B., Hanson, W., Hug, S., Ibbot, G., Mitch, M., Nath, R., and Williamson, J. Update of AAPM task group no. 43 report: a revised AAPM protocol for brachytherapy dose calculations, *Med. Phys.,* 31:3, 633–674, 2004.

14. International Commission on Radiation Units and Measurements, Dose and Volume Specification for Reporting Intracavitary Therapy in Gynecology ICRU Report 38. ICRU, Bethesda, 1985.

15. International Commission on Radiation Units and Measurements, Dose and Volume Specification for Reporting Interstitial Therapy ICRU Report 58. ICRU, Bethesda, 1997.

16. Glasgow, G.P. and Dillman, L.T. Specific γ-ray constant and exposure rate constant of ^{192}Ir, *Med. Phys.,* 6:1, 49–52, 1979.

17. Borg, J. and Roggers, D.W.O. Spectra and air-kerma strength for encapsulated ^{192}Ir sources, *Med. Phys.,* 26, 2441–2444, 1999.

18. Moss, C.D. Improved analytical fir to the TG-43 radial dose function $g(r)$, *Med. Phys.,* 27, 659–661, 2000.

19. Hubbell, J.H. and Seltzer, S.M. Tables of x-ray mass attenuation coefficients and mass energy-absorption coefficients Version 1.4 on http://physics.nist.gov/xaamdi (Gaithersburg: National Institute of Standards and Technology) Originally published as NISTIR 5632, 1995.

20. Williamson, J.F. and Li, Z. Monte Carlo aided dosimetry of the microSelectron pulsed and high dose rate ^{192}Ir sources, *Med. Phys.,* 22, 809–819, 1995.

21. Daskalov, G.M., Löffler, E., and Williamson, J.F. Monte Carlo-aided dosimetry of a new high dose-rate brachytherapy source, *Med. Phys.,* 25, 2200–2208, 1998.

22. Karaiskos, P., Angelopoulos, A., Baras, P., Sakelliou, L., Sandilos, P., Dardoufas, K., and Vlachos, L. A Monte Carlo investigation of the dosimetric characteristics of the varisource ^{192}Ir high dose rate brachytherapy source, *Med. Phys.,* 26, 1498–1502, 1999.

23. Angelopoulos, A., Baras, P., Sakelliou, L., Karaiskos, P., and Sandilos, P. Monte Carlo dosimetry of a new ^{192}Ir high dose rate brachytherapy source, _Med. Phys._, 27, 2521–2527, 2000.

24. Ballester, F., Pérez-Calatayud, J., Puchades, V., Lluch, J.L., Serrano-Andrés, M.A., Limami, Y., Lliso, F., and Casal, E. Monte Carlo dosimetry of the Buchler high dose rate ^{192}Ir source, _Phys. Med. Biol._, 46, N79–N90, 2001.

25. Ballester, F., Puchades, V., Lluch, J.L., Serrano-Andrés, M.A., Limami, Y., Pérez-Calatayud, J., and Casal, E. Technical note: Monte Carlo dosimetry of the HDR 12i and Plus ^{192}Ir sources, _Med. Phys._, 28, 2586–2591, 2001.

26. Karaiskos, P., Angelopoulos, A., Pantelis, E., Papagiannis, P., Sakelliou, L., Kouwenhoven, E., and Baltas, D. Monte Carlo dosimetry of a new ^{192}Ir pulsed dose rate brachytherapy source, _Med. Phys._, 30, 9–16, 2003.

27. Pérez-Calatayud, J., Ballester, F., Serrano-Andrés, M.A., Puchades, V., Lluch, J.L., Limami, Y., and Casal, E. Dosimetry characteristics of the plus and 12i gammamed PDR ^{192}Ir sources, _Med. Phys._, 28, 2576–2585, 2001.

28. Kirov, A.S., Williamson, J.F., Meigooni, A.S., and Zhu, Y. TLD, diode and Monte Carlo dosimetry of an ^{192}Ir source for high dose-rate brachytherapy, _Phys. Med. Biol._, 40, 2015–2036, 1995.

29. Russel, K.R. and Ahnesjo, A. Dose calculation in brachytherapy for ^{192}Ir source using a primary and scatter dose separation technique, _Phys. Med. Biol._, 41, 1007–1024, 1996.

30. Karaiskos, P., Angelopoulos, A., Sakelliou, L., Sandilos, P., Antypas, C., Vlachos, L., and Koutsouveli, E. Monte Carlo and TLD dosimetry of an ^{192}Ir high dose-rate brachytherapy source, _Med. Phys._, 25, 1975–1984, 1998.

31. Papagiannis, P., Angelopoulos, A., Pantelis, E., Sakelliou, L., Baltas, D., Karaiskos, P., Sandilos, P., and Vlachos, L. Dosimetry comparison of ^{192}Ir sources, _Med. Phys._, 29, 2239–2246, 2002.

32. Wang, R. and Sloboda, R.S. Influence of source geometry and materials on the transverse axis dosimetry of ^{192}Ir brachytherapy sources, _Phys. Med. Biol._, 43, 37–48, 1998.

33. Anctil, J.C., Clark, B.G., and Arsenault, C.J. Experimental determination of dosimetry functions of Ir-192 sources, _Med. Phys._, 25, 2279–2287, 1998.

34. Meigooni, A.S., Kleiman, M.T., Johnson, L.J., Mazloomdoost, D., and Ibbott, G.S. Dosimetric characteristics of a new high-intensity ^{192}Ir source for remote afterloading, _Med. Phys._, 24, 2008–2013, 1997.

35. Valicenti, R.K., Kirov, A.S., Meigooni, A.S., Mishra, V., Das, R.K., and Williamson, J.F. Experimental validation of Monte Carlo dose calculations about a high-intensity Ir-192 source for pulsed dose-rate brachytherapy, _Med. Phys._, 22, 821–829, 1995.

36. Papagiannis, P., Angelopoulos, A., Pantelis, E., Sakelliou, L., Karaiskos, P., and Shimizu, Y. Monte Carlo dosimetry of ^{60}Co HDR brachytherapy sources, _Med. Phys._, 29, 712–721, 2003.

37. Pérez-Calatayud, J., Granero, D., Ballester, F., Puchades, V., and Casal, E. Monte Carlo dosimetric characterization of the Cs-137 selectron/LDR source: evaluation of applicator attenuation and superposition approximation effects, _Med. Phys._, 31, 493–499, 2004.

38. Pérez-Calatayud, J., Ballester, F., Serrano-Andrés, M.A., Llich, J.L., Puchades, V., Limami, Y., and Casal, E. Dosimetric characteristics of the CDC-type miniature cylindrical ^{137}Cs brachytherapy sources, _Med. Phys._, 29, 538–543, 2002.

39. Ballester, F., Lluch, J.L., Limami, Y., Serrano, M.A., Casal, E., Pérez-Calatayud, J., and Lliso, F. A Monte Carlo investigation of the dosimetric characteristics of the CSM11 ^{137}Cs source from CIS, *Med. Phys.*, 27, 2182–2189, 2000.

40. Casal, E., Ballester, F., Lluch, J.L., Pérez-Calatayud, J., and Lliso, F. Monte Carlo calculations of dose rate distributions around the Amersham CDCS-M-type ^{137}Cs source, *Med. Phys.*, 27, 132–140, 2000.

41. Pérez-Calatayud, J., Ballester, F., Lluch, J.L., Serrano-Andrés, M.A., Casal, E., Puchades, V., and Limami, Y.F. Monte Carlo calculation of dose rate distributions around the Walstam CDC.K-Type ^{137}Cs sources, *Phys. Med. Biol.*, 46, 2029–2040, 2001.

42. Perera, H., Williamson, J.F., Li, Z., Mishra, V., and Meigooni, A.S. Dosimetric characteristics, air-kerma strength calibration and verification of Monte Carlo simulation for a new Ytterbium-169 brachytherapy source, *Int. J. Rad. Oncol. Biol. Phys.*, 28:4, 953–970, 1994.

43. Baltas, D., Kramer, R., and Loeffler, E. Measurements of the anisotropy of the new Ir-192 source for the microSelectron-HDR, Activity, Special Report No. 3, Nucletron B.V., The Netherlands, 1993.

44. Muller-Runkel, R. and Cho, S.H. Anisotropy measurements of a high dose rate Ir-192 source in air and in polystyrene, *Med. Phys.*, 21:7, 1131–1134, 1994.

45. Mishra, V., Waterman, F.M., and Suntharalingam, N. Anisotropy of an ^{192}iridium high dose rate source measured with a miniature ionization chamber, *Med. Phys.*, 24:7, 751–755, 1997.

46. Shirley, V.S. Revised A-chains A=192. *Nucl. Data Sheets*, 64, 205, 395–401,1991.

47. Brenner, D.J. and Hall, E.J. Conditions for the equivalence of continuous to pulsed low dose rate brachytherapy, *Int. J. Rad. Oncol. Biol. Phys.*, 20, 181–190, 1991.

48. Armour, E., Wang, Z., Corry, P., and Martinez, A. Equivalence of continuous and pulse simulated low dose rate irradiation in 9L gliosarcoma cells at 37° and 41°, *Int. J. Rad. Oncol. Biol. Phys.*, 22, 109–114, 1992.

49. Fowler, J.F. and Mount, M. Pulsed brachytherapy: the conditions for no significant loss of therapeutic ration compared with traditional low dose rate brachytherapy, *Int. J. Rad. Oncol. Biol. Phys.*, 23, 661–669, 1992.

50. Nelson, W.R., Hirayama, H., and Rogers, D.W.O. The EGS4 code system Stanford Linear Accelerator Report SLAC-265, 1985.

51. Brun, R, Bruyant, F., Maire, M., McPherson, A.C., and Zanarini, P. GEANT3 CERN DD/EE/84-1, 1987.

52. Karaiskos, P., Papagiannis, P., and Angelopoulos, A. Dosimetry of ^{192}Ir wires for LDR interstitial brachytherapy following the AAPM TG-43 dosimetric formalism, *Med. Phys.*, 28, 156–166, 2001.

53. Pantelis, E., Baltas, D., Dardoufas, K., Karaiskos, P., Papagiannis, P., Rosaki-Mavrouli, H., and Sakelliou, L. On the dosimetric accuracy of a Sievert integration model in the proximity of ^{192}Ir HDR sources, *Int. J. Rad. Oncol. Biol. Phys.*, 53:4, 1071–1084, 2002.

54. Mason, D.L.D., Battista, J.J., Barnett, R.B., and Porter, A.T. Ytterbium-169: calculated physical properties of a new radiation source for brachytherapy, *Med. Phys.*, 19, 695–703, 1992.

55. Perera, H., Williamson, J.F., Li, Z., Mishra, V., and Meigooni, A. Dosimetric characteristics, air-kerma strength calibration and verification of Monte Carlo simulation for a new Ytterbium-169 brachytherapy source, *Int. J. Rad. Oncol. Biol. Phys.*, 28:4, 953–970, 1994.

56. Piermattei, A., Azario, L., and Montemaggi, P. Implantation guidelines for
     ^{169}Yb seed interstitial treatments, Phys. Med. Biol., 40, 1331–1338, 1995.
57. Piermattei, A., Azario, L., Rossi, G., Arcovito, G., Ragonat, R., Galelli, M., and
     Taccini, G. Dosimetry of ^{169}Yb seed model X1267, Phys. Med. Biol., 40,
     1317–1330, 1995.
58. Heintz, B.H., Wallace, R.E., and Hevezi, J.M. Comparison of I-125 sources
     used for permanent interstitial implants, Med. Phys., 28:4, 671–682, 2001.
59. Karaiskos, P., Papagiannis, P., Sakelliou, L., Anagnostopoulos, G., and Baltas,
     D. Monte Carlo dosimetry of the selectseed ^{125}I interstitial brachytherapy
     seed, Med. Phys., 28, 1753–1760, 2001.
60. Anagnostopoulos, G., Baltas, D., Karaiskos, P., Sandilos, P., Papagiannis, P.,
     and Sakelliou, L. Thermoluminescent dosimetry of the selectseed ^{125}I
     interstitial brachytherapy seed, Med. Phys., 29, 709–716, 2002.
61. DeWerd, L.A., Huq, M.S., Das, I.J., Ibbott, G.S., Hanson, W.F., Slowey, T.W.,
     Williamson, J.F., and Coursey, B.M. Procedures for establishing and main-
     taining consistent air-kerma strength standards for low-energy, photon-
     emitting brachytherapy sources: recommendations of the calibration labora-
     tory accreditation subcommittee of the American association of physicists in
     medicine, Med. Phys., 31:3, 675–681, 2004.
62. Williamson, J.F., Coursey, B.M., DeWerd, L.A., Hanson, W.F., Nath, R., Rivard,
     M.J., and Ibbott, G. Recommendations of the American association of
     physicists in medicine on ^{103}Pd interstitial source calibration and dosimetry:
     implications for dose specification and prescription, Med. Phys., 27:4,
     634–642, 2000.
63. Cha, C.M., Potters, L., Ashley, R., Freeman, K., Wang, X.H., Walbaum, R., and
     Leibel, S. Isotope selection for patients undergoing prostate brachytherapy,
     Int. J. Rad. Oncol. Biol. Phys., 45:2, 391–395, 1999.
64. Blasko, J.C., Grimm, P.D., Sylvester, J.E., Badiozamani, K.R., Hoak, D., and
     Cavanagh, W. Palladium-103 brachytherapy for prostate carcinoma, Int. J. Rad.
     Oncol. Biol. Phys., 46:4, 839–850, 2000.
65. Potters, L., Hunag, D., Fearn, P., and Kattan, M.W. The effect of isotope
     selection on the prostate-specific antigen response in patients treated with
     permanent prostate brachytherapy, Brachytherapy, 2, 26–31, 2003.
66. Nag, S., Beyer, D., Friedland, J., Grimm, P., and Nath, R. American
     brachytherapy society (ABS) recommendations for transperineal permanent
     brachytherapy of prostate cancer, Int. J. Rad. Oncol. Biol. Phys., 44:4, 789–799,
     1999.
67. Ash, D., Flynn, A., Battermann, J., de Reijke, T., Lavagnini, P., and Blank, L.
     IESTRO/EAU/EORTC recommendations on permanent seed implantation
     for localized prostate cancer, Radiother. Oncol., 57, 315–321, 2000.
68. Sharkey, J., Cantor, A., Solc, Z., Huff, W., Chovnivk, S.D., Behar, R.J., Perez, R.,
     Otheguy, J., and Rabinowitz, R. ^{103}Pd brachytherapy versus radical
     prostatectomy in patients with clinically localized prostate cancer: a 12-year
     experience from a single group practice, Brachytherapy, 4, 34–44, 2005.
69. Martínez-Monge, R., Garrán, C., Vivas, I., and López-Picazo, J.M. Percuta-
     neous CT-guided ^{103}Pd implantation for the medically inoperable patient with
     T1N0M0 non-small cell lung cancer: a case report, Brachytherapy, 3, 179–181,
     2004.
70. Song, H., Luxton, G., and Hendee, W.R. Calculation of brachytherapy doses
     does not need TG-43 factorization, Med. Phys., 30:6, 997–999, 2003.

71. Selbach, H.J. and Andrássy, M., Experimentally determined TG-43 data for the BEBIG MultiSource ^{192}Ir HDR source, unpublished data, personal communication, 2005.

72. Selbach, H.J., Andrássy, M., Experimentally determined TG-43 data for the BEBIG MultiSource ^{60}Co HDR source, unpublished data, personal communication, 2005.

73. Ballester, F., Granero, D., Pérez-Calatayud, J., Casal, E., Agramunt, A., and Cases, R. Monte Carlo dosimetric study of the BEBIG Co-60 HDR source, *Phys. Med. Biol.*, 50, N309–N316, 2005.

74. Agostinelli, S., Allison, J., and Amako, K. GEANT4—a simulation toolkit, *Nucl. Instr. Meth. Phys. Res.*, A506, 250–303, 2003.

75. National Nuclear Data Center (NNDC) Brookhaven National Laboratory NUDAT 2.0 Electronic Version, available online at NNDC: www.nndc.bnl. gov/nudat2/, July 2005.

76. Granero, D., Pérez-Calatayud, J., and Ballester, F. Monte Carlo calculation of the TG-43 dosimetric parameters of a new BEBIG Ir-192 HDR source, *Radiother. Oncol.*, 76, 79–85, 2005.

77. Büermann, L., Kramer, H.M., Schrader, H., and Selbach, H.J. Activity determination of ^{192}Ir solid sources by ionization chamber measurements using calculated corrections for self-absorption, *Nucl. Instr. Meth. Phys. Res.*, A339, 369–376, 1994.

78. Medich, C.D., Tries, A.M., and Munro, J.J. Monte Carlo characterization of an ytterbium-169 high dose rate brachytherapy source with analysis of statistical uncertainty, *Med. Phys.*, 33, 163–172, 2006.

79. Lymperopoulou, G., Papagiannis, P., Sakelliou, L., Milickovic, N., Giannouli, S., and Baltas, D. A dosimetric comparison of ^{169}Yb versus ^{192}Ir for HDR prostate brachytherapy, *Med. Phys.*, 32, 3832–3842, 2005.

80. Booth, T.E., Brown, F.B., Bull, J.S., Cox, L.J., Forster, R.A., Goorley, J.T., Hughes. H.G., Mosteller, R.D., Prael, R.E., Selcow, E.C., Sood, A., and Sweeney, J.E., Report No. MCNP-A General Monte Carlo N-Particle Transport Code, version 5, L-UR-03-1987, 2003.

81. Chu, S.Y.F., Ekström, L.P., and Firestone, R.B. The Lund/LBNL Nuclear Data Search, database Version 2 1999-02-28, http://nucleardata.nuclear.lu.se/nucleardata/toi/index.asp, 1999.

82. Papagiannis, P., Sakelliou, L., Anagnostopoulos, G., and Baltas, D. Letter to the Editor: On the dose rate constant of the selectSeed ^{125}I interstitial Brachytherapy seed, *Med. Phys.*, 33, 1522–1523, 2006.

9

Monte Carlo–Based Source Dosimetry

9.1 Introduction

Monte Carlo–based dosimetry has been acknowledged as a valuable tool in brachytherapy, constituting one of the dosimetric prerequisites for routine clinical use of new low energy photon interstitial sources.[1,2]

The Monte Carlo method is a means of statistical simulation of all processes associated with radiation emission and transport by using random numbers and appropriate probability distribution functions. The main characteristic of the method is its stochastic character, meaning that unlike analytical methods, each calculation for the same problem will, in general, provide different results. The accuracy of these results depends among other factors on the number of statistical simulations performed. Thus, Monte Carlo calculations are more rigorous and time-consuming than analytical methods. However, the availability of computational resources and the significant increase in computer speed over the last decade has significantly reduced calculation times. Monte Carlo simulation is still too CPU intensive to support commercial treatment planning systems.

In this chapter, the main principles of Monte Carlo photon transport simulations are presented. Dosimetry results are discussed in the energy range of 20 to 700 keV, that covers low energy ^{103}Pd and ^{125}I, medium energy ^{241}Am and ^{169}Yb, and higher energy ^{192}Ir and ^{137}Cs nuclides, but excludes the infrequently used ^{60}Co (see Section 5.3 and Section 8.4.1.2). In this energy range, electronic equilibrium may be safely assumed[3,4] and therefore collision kerma is used to approximate absorbed dose, omitting secondary electron tracking. Electronic nonequilibrium exists only at points in close proximity (i.e., less than 1 mm) to higher energy ^{192}Ir sources. Moreover, in this distance range a dose rate enhancement effect due to the beta spectrum emitted by ^{192}Ir has been reported.[5] These points will be discussed within the overview section on Monte Carlo–based dosimetry for the most commonly used brachytherapy source designs.

9.2 Monte Carlo Photon Transport Simulations

9.2.1 Random Numbers and Sampling

The basis of any Monte Carlo simulation code is a random number generator. It is imperative that this source of random numbers (R_N) provides a uniform distribution within the unitary interval [0, 1], meaning that every number within this interval is of equal probability. A uniform distribution can be used for sampling from any of the probability distributions describing the different physical phenomena involved in photon transport.

Figure 9.1 shows the relative frequency of occurrence distribution of $N = 10^2$, 10^3, and 10^6 random numbers, R_N. Results correspond to 0.1 intervals, and therefore, the expected frequency distribution is uniform with a value of 0.1. It can be seen that sampling of $N = 10^2$ R_N results in a frequency distribution with significant fluctuations: the standard deviation of the mean is $\sigma = 0.0236$. Sampling a different set of $N = 10^2$ R_N will provide, in general, a different distribution of comparable fluctuations. Sampling of $N = 10^3$ R_N improves results but fluctuations are still noticeable: the mean frequency of occurrence presents a standard deviation of $\sigma = 0.0124$. Finally, sampling of $N = 10^6$ R_N provides an accurate estimate of the expected frequency distribution with a mean relative frequency of 0.0999 ± 0.0003.

Results in Figure 9.1 illustrate the strong dependence of Monte Carlo results on the number of histories simulated. A relatively small number of simulated histories will not reproduce the frequency distributions of

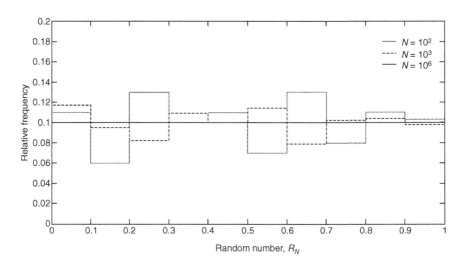

FIGURE 9.1
Relative frequency of occurrence distribution for $N = 10^2$, 10^3, and 10^6 random numbers (R_N) in the [0, 1] interval.

the underlying physical phenomena accurately and will therefore lead to increased statistical uncertainty. In general, statistical uncertainty is proportional to the inverse root square of the number of histories simulated and in order for the uncertainty to be reduced by 50%, the number of simulated photon histories must be quadrupled. It is emphasized that this will increase the computational time required.

9.2.2 Simulating Primary Radiation

In brachytherapy, Monte Carlo radiation dose calculation in a medium is a three-step procedure. First, the energy fluence of photons emitted by the active source core must be simulated. Second, these photons must be transported through the source core and source encapsulation materials. This involves the sampling of photon interaction points and the simulation of photon interactions in order to define the energy fluence of photons emerging from the source. This energy fluence may differ significantly from that of the initially emitted photon energy fluence, especially in the case of low energy brachytherapy sources such as ^{125}I and ^{103}Pd. Third, photons must be transported in the medium surrounding the source in order to estimate the energy deposited in predefined volume elements: i.e., scoring voxels.

In order to illustrate this procedure, we first assume a monoenergetic point source of radiation positioned in liquid water. Photons of given energy are emitted isotropically. In order to sample the directions of primary photons, the direction cosines $(u, v, w) = (\sin \theta \cos \varphi, \sin \theta \sin \varphi, \cos \theta)$ are used where φ is the azimuthal angle and θ is the polar angle of emitted photon directions in spherical coordinates. The point within a medium that a photon has reached after traveling a length, r, is given in Cartesian coordinates by $(x,y,z) = (ru,rv,rw)$ while direction cosines also facilitate the calculation of direction change after a scatter interaction of a photon in the medium.

Since isotropic emission means that the same number of photons is emitted per solid angle element $d\Omega$, that is given by $d\Omega = d\varphi \, d(\cos \theta)$, sampling of the azimuthal angle, φ, is a problem of sampling a number x from a uniform distribution in the interval $[a, b]$ using a uniform distribution of R_N in the $[0, 1]$ interval according to $x = a + (b - a)R_N$. Therefore, given that the azimuthal angle, φ, lies in the $[0, 2\pi]$ interval, it can be sampled by the azimuthal angle:

$$\varphi = 2\pi R_N \tag{9.1}$$

For the polar angle, θ, we cannot sample from a uniform distribution in the $[0, \pi]$ interval. The polar direction of the emitted photon has to be calculated by sampling $\cos \theta$ that varies within $[-1, 1]$ from a uniform R_N distribution in the $[0, 1]$ interval as above and for the polar angle:

$$\cos \theta = 2R_N - 1 \tag{9.2}$$

Having sampled the initial photon direction, the next step is to sample the traveling distance (path length, d) of these photons before their first interaction with the material surrounding the source. The probability density function for a photon of energy E, traveling a distance between r and $r + dr$ in a medium without interacting is given by

$$P(r) = \mu \exp(-\mu r) \tag{9.3}$$

where μ is the total linear energy attenuation coefficient that depends on photon energy, E, as well as the medium (see Chapter 4).

Following the probability density function of Equation 9.3, the probability of an emitted photon having a path length of d or less, $P(d)$, is given by integration:

$$P(d) = \int_0^d \mu \exp(-\mu r) dr \Rightarrow P(d) = 1 - \exp(-\mu d) \tag{9.4}$$

Since the value of probability $P(d)$, ranges from 0 to 1 it can be sampled by a uniform distribution of random numbers R_N so that the path length of a photon is given by

$$d = -(1/\mu) \ln R_N \tag{9.5}$$

It should be noted that it is not necessary to subtract R_N from unity when solving Equation 9.4 with respect to d since R_N and $(1 - R_N)$ are both uniformly distributed numbers ranging from 0 to 1 and thus, results for d would be the same. We now have the ability to simulate the primary radiation field of a monoenergetic point source.

Figure 9.2 gives results in Cartesian coordinates of Monte Carlo calculations for 10 and 10^3 photons emitted by a 50 keV point source. It can be seen that when an insufficient number of photon histories are simulated, the results fail to reproduce the isotropic point source primary radiation field.

Monte Carlo results may be evaluated by comparison with nonstochastic quantities relevant to the simulation. For example, the photon path length is a stochastic quantity that cannot be predicted since its value is determined by a probability density function. However, photon mean free path (mfp), defined as the average distance traversed by photons of given energy prior to interaction:

$$\mathrm{mfp} = \int_0^\infty r \exp(-\mu r) dr = 1/\mu \tag{9.6}$$

is the mean path length and therefore it constitutes a nonstochastic quantity (as do all mean values of stochastic quantities) that can be accurately predicted.

Figure 9.3 gives the partial and total linear attenuation coefficients in water in the energy range of interest to brachytherapy applications: 20 to 700 keV. In this figure, it can be seen that the total attenuation coefficient shows a dramatic increase with decreasing energy at lower photon energies due

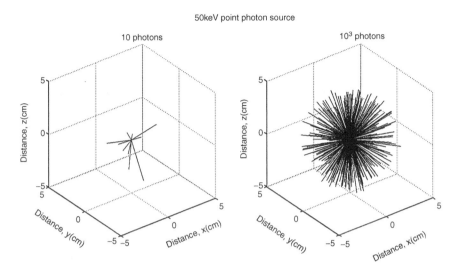

FIGURE 9.2
Monte Carlo results for primary radiation field simulation for 10 and 10^3 photons emitted uniformly by a 50 keV point source in liquid water.

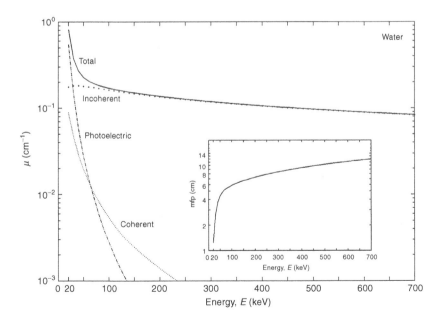

FIGURE 9.3
Partial and total linear attenuation coefficients in water in the energy range of interest to brachytherapy: 20 to 700 keV. In the inset, photon mean free path, mfp, is plotted vs. energy, E, in the same energy range.

to the predominance of the photoelectric effect in this energy range. The inverse of the total linear attenuation coefficient for a given photon energy (this equals the corresponding mfp according to Equation 9.5), is plotted in the inset of Figure 9.3 vs. energy.

In Figure 9.2 the average of the photon path lengths provides a statistical estimate of mfp for the particular photon energy (50 keV) with statistical uncertainty depending on the number of photon histories simulated. For example, compared with an expected mfp of 4.48 cm for 50 keV photons, results for 10 primary photons in Figure 9.2 yield a mfp of 5.41 cm which deviates by 21% from the expected value. Whereas results for 10^3 primary photons in Figure 9.2 yield a mfp of 4.23 cm deviating by less than 6% from the expected value. This emphasizes the importance of sampling a sufficient number of primary photons.

The results shown in Figure 9.4 are similar to those of Figure 9.2 but are presented for 10^3 primary photons emitted by different monoenergetic point sources in the energy range of interest to brachytherapy applications: 20 to 700 keV. The increase of the primary radiation field extent with increasing photon energy corresponds to the expected increase of mfp according to the inset in Figure 9.3.

Results of the simulations shown in Figure 9.4 may also be evaluated by comparison with the nonstochastic quantity of the percentage of photons that have not interacted with water, as a function of distance from the point

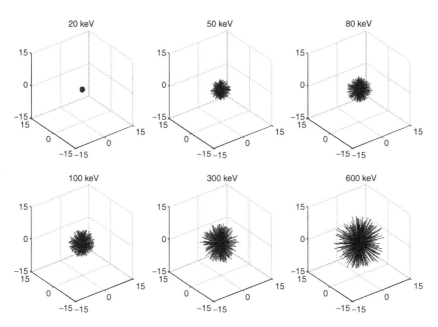

FIGURE 9.4
Monte Carlo results for primary radiation field simulation for 10^3 photons emitted uniformly by different monoenergetic point sources in liquid water.

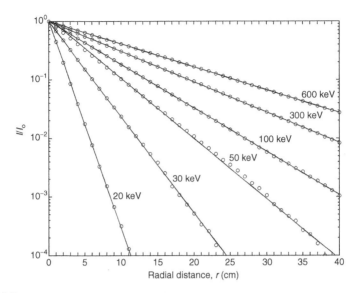

FIGURE 9.5
Ratio of photons that have not interacted with water (I) to primary photons (I_0), as a function of distance, r, from monoenergetic point sources positioned in liquid water.

source. This percentage, I/I_0, may be predicted by the exponential law of attenuation: $I = I_0 \exp(-\mu r)$.

In Figure 9.5, I/I_0 is plotted vs. distance, r, on a logarithmic scale to illustrate exponential attenuation using the data in Figure 9.4. A fit to the data of each of the curves in Figure 9.5 gives their slopes and these equate to statistical estimates of the linear attenuation coefficients for the corresponding energies. This procedure would provide an effective attenuation coefficient in the case of a polyenergetic source, which could be of use, as primary radiation is the main dose contributor close to brachytherapy sources (see Section 9.4).

9.2.3   Choosing Interaction Type

The next step towards fully simulating radiation photon transport around monoenergetic point sources is the sampling of the interaction type at the end of the path length of primary photons. For this purpose, a simplified rejection method of sampling is used which is often called the branching ratio method.

For the energy range relevant to brachytherapy sources, the interactions of interest are photoelectric absorption, coherent (Rayleigh) and incoherent (Compton) scattering. As reviewed in Chapter 4, the probability of each type of photon interaction per unit length traversed by photons is quantified by the corresponding interaction cross-section, or equivalently the linear attenuation coefficient, so that the properly normalized total attenuation coefficient is given by

$$\mu = \mu^{ph} + \mu^{coh} + \mu^{incoh} \qquad (9.7)$$

The branching ratio for each interaction type for a given photon energy corresponds to its relative probability. That is

$$\mu^{ph}/\mu + \mu^{coh}/\mu + \mu^{incoh}/\mu = 1 \qquad (9.8)$$

For example, total and partial linear attenuation coefficients in water are given in Figure 9.3 as a function of energy, and the normalized relative probabilities (i.e., branching ratios) for 50 keV photons are 0.120 for photoelectric absorption, 0.085 for coherent scattering, and 0.795 for incoherent scattering. By sampling a random number, R_N, we are able to determine the type of interaction that a photon undergoes by finding the branching ratio where the photon belongs. A 50 keV photon will undergo photoelectric absorption if $R_N < 0.120$, coherent scattering if $0.120 < R_N < 0.205$, and incoherent scattering if $R_N > 0.205$.

From the above analysis, it can be seen that the accuracy of Monte Carlo–based brachytherapy dosimetry depends on the interaction coefficients used in the simulations since they are used for the determination of photon path length and interaction type. These two variables determine energy deposition in a medium. Uncertainty in interaction coefficients and other input data, such as photon spectra, utilized in the simulations is added to the statistical uncertainty, in order to obtain the overall uncertainty in the results. It is emphasized that the use Monte Carlo codes with different cross-section libraries obscures comparison of corresponding dosimetry results.

9.2.4 Simulating Interactions

In incoherent scattering, photons interacting with atomic electrons undergo a direction change as well as an energy reduction. The method outlined by Kahn[6] is the usual method of treating incoherent scattering. It uses the Klein–Nishina differential cross-section equation for photon scattering with free electrons for sampling of the photon polar angle, θ, after the interaction and assumes a uniform distribution for the corresponding azimuthal angle, φ, thus ignoring polarization effects. Having sampled the photon polar scattering angle, θ, the ejected electron, T_e, and scattered photon, E_{sc}, energies are calculated by conservation of energy and momentum according to

$$T_e = (E^2/m_e c^2)(1 - \cos\theta)/(1 + E/m_e c^2(1 - \cos\theta)) \qquad (9.9)$$

and to

$$E_{sc} = E/(1 + E/m_e c^2(1 - \cos\theta)) \qquad (9.10)$$

where E is the incident photon energy.

A more detailed approach, which is beyond the scope of this discussion, includes polarization effects[7,8] and accounts for the electron binding energy by modifying the Klein–Nishina equation by use of appropriate incoherent scattering functions[9] following, for example, the technique outlined in Chan and Doi.[10] It should be noted here that, in contrast to the external radiotherapy photon beams of a few MeV energy where the incoherently scattered photon angular distribution is forward peaked, in the brachytherapy energy range this distribution tends to isotropy with decreasing photon energy (see also Figure 4.7). Therefore, a considerable contribution of backscattered radiation is expected in brachytherapy dosimetry.

When a photon undergoes coherent scattering with orbital electrons, only its direction is altered. The polar scattering angle, θ, is commonly sampled using the Thompson differential cross-section per target electron modified by appropriate form factors[11,12] to account for electron binding in the atom. Liquid water molecular form factors may also be incorporated in the simulations.[13] Polarization effects may be ignored and the azimuthal angle, φ, can be sampled from a uniform distribution. Overall, as shown in Figure 9.3, the probability of coherent scattering is small while, in any case, the coherently scattered photon angular distribution is highly anisotropic and forward peaked (see also Figure 4.6).

Following a photoelectric interaction of a photon with an atom, photons are absorbed and an inner shell orbital electron is ejected with kinetic energy equal to the photon energy minus the electron binding energy. When an outer orbital electron fills the shell vacancy, characteristic x-rays, also referred to as fluorescence radiation, are emitted. It is also possible that the atom absorbs characteristic x-rays internally, resulting in the emission of monoenergetic Auger electrons (see also Section 4.2.1).

Auger electron emission predominates in lower Z materials such as water and soft tissue and, in any case, Auger electrons may be considered to be absorbed locally due to their reduced range. Fluorescence radiation, however, is of importance in low energy sources such as ^{125}I and ^{103}Pd seeds. For these sources the probability of photoelectric absorption is significant and high Z materials are used as encapsulation and radioopaque markers, e.g., Ti, Au, and Ag. These materials give rise to fluorescence photons of energies sufficient to escape the interaction site and therefore needing to be simulated separately. For example, the calibration protocol for low energy brachytherapy seeds has been recently reviewed to exclude Ti characteristic x-rays of 4.5 keV which contribute to calibration measurements in air while their contribution to dose rate in water is negligible.[14,15]

Characteristic x-rays following photoelectric absorption are simulated using tabulated vacancy probabilities to decide whether a K- or an L-shell characteristic will occur, transition probabilities to decide which K or L characteristic will occur, and fluorescence yields to determine whether this characteristic x-ray will escape the atom.[16] Isotropic emission may be safely assumed for characteristic x-rays, and their tracks are, thereafter, separately simulated.

The above sampling procedures allow the tracking of secondary photons between interaction points that are determined by sampling consecutive track lengths as described earlier.

A simplified flow chart of Monte Carlo tracking of primary and secondary photons is given in Figure 9.6. Powerful, general-purpose simulation codes such as the EGS[17] and the MCNP[18] have been used for brachytherapy dosimetry. Also, an overview of experimentally verified Monte Carlo codes by independent investigators can be found in the publications of Daskalov et al.[19] and Angelopoulos et al.[20] for ^{192}Ir sources, and Kirov and Williamson[21] and Karaiskos et al.[4] for ^{125}I sources. The code described in Angelopoulos et al.[20] and Karaiskos et al.[4] was used for the calculation of some of the dosimetry results that follow.

In Figure 9.7, photon tracks are shown in Cartesian coordinates for 15 and 50 primary photons emitted by a 50 keV point source, up to the point of their absorption. In this figure it can be seen that a number of photon tracks corresponds to a sole interaction: photoelectric absorption. The majority of photon tracks, however, are the result of successive scattering interactions where, in some cases, significant direction changes and even backscattering occur. These results raise the question of predicting the mean number of interactions that a photon of given energy undergoes.

During successive interactions, photon energy is degraded and energy is transferred to secondary electrons. The deposition of this energy is assumed to take place on the exact site of the interaction, according to the water kerma

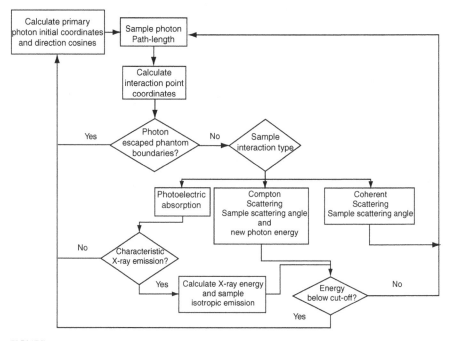

FIGURE 9.6
Simplified flow chart of the Monte Carlo method of sampling primary and secondary photons emitted by a radionuclide source.

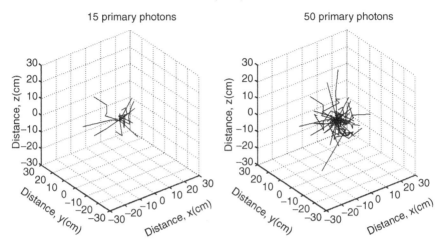

FIGURE 9.7

Primary and secondary photon tracks in Cartesian coordinates for 15 and 50 photons emitted by a 50 keV point source in the center of a liquid water spherical phantom of 15 cm radius.

dose approximation that can be safely applied in the energy range of interest to brachytherapy applications.

The secondary electron energy may be estimated by $T_e = (\mu_{en}/\mu)E_\gamma$ where the ratio of linear energy absorption to linear attenuation coefficients corresponds to the fraction of photon energy E that is transferred to an electron: in this energy range $\mu_{en} = \mu_{transfer}$. For example, let us assume a primary ^{137}Cs photon of energy equal to 662 keV in water. After the first interaction the photon energy will be reduced, on average, to

$$E_{sc} = (1 - \mu_{en}/\mu)E \tag{9.11}$$

which equals 410 keV. By calculating the new ratios of μ_{en}/μ and applying Equation 9.11, each time substituting E by the new E_{sc}, we can calculate the successive, mean, photon energies for all successive photon interactions.

Results of this are shown in Figure 9.8, where the ratio of μ_{en}/μ is plotted vs. photon energy, E, and photon energies prior to each interaction marked by arrows. In this figure it can be seen that, on average, a primary 662 keV ^{137}Cs photon undergoes 14 interactions in water before its final absorption. The majority of these interactions takes place at low photon energies, 50 to 80 keV, where μ_{en}/μ reaches its minimum, i.e., energy transfer to secondary electrons per interaction is minimal. These observations may be explained by the fact that in this photon energy region, μ starts increasing while μ_{en} reaches its minimum. This can be seen in the inset in Figure 9.8 where both coefficients are plotted vs. energy. The component of 50 to 80 keV scattered photons builds up gradually with increasing radial distance, r, from the source.

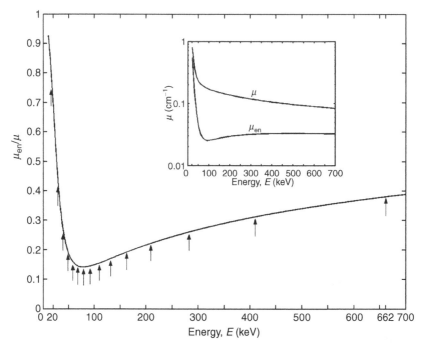

FIGURE 9.8

Ratio of linear energy absorption, to linear energy attenuation coefficients (μ_{en}/μ) plotted vs. photon energy (E). Arrows mark the successive photon energies for all 14 successive photon interactions that a primary ^{137}Cs photon undergoes on average in water.

However, it must be emphasized that the above calculations are only indicative of the photon energy degradation during successive interactions. In Figure 9.9, the mean number of interactions per photon emitted by a monoenergetic point source positioned in a liquid water phantom are plotted vs. primary photon energy, E. Figure 9.9a corresponds to simulations in a spherical phantom of radius equal to eight photon mfp that is practically equivalent to an infinite (unbounded) medium phantom. In this figure it can be seen that there is a rapid increase of the mean number of interactions with increasing energy in the low photon energy region. For photon energies greater than 100 keV only a slight increase of the mean number of interactions with increasing energy is observed. Furthermore, the majority of interactions corresponds to incoherent scattering while, on average, only one photoelectric interaction occurs at all energies corresponding to the end of the photon track where the photon is absorbed.

In general, the data for unbounded medium geometries in Figure 9.9a are in agreement with the results of Figure 9.8. Unbounded medium geometries, however, are not realistic compared with the usual cancer sites treated by brachytherapy, which is as finite patient dimensions resemble bounded phantom geometries.

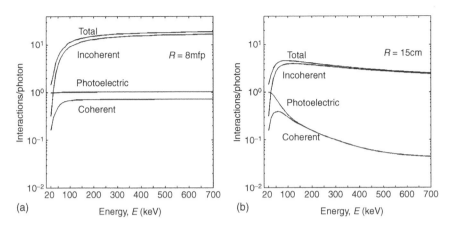

FIGURE 9.9
Mean number of interactions per photon plotted vs. primary photon energy (*E*) for photons emitted by a monoenergetic point source: (a) in a liquid water phantom of radius equal to eight photon mean free paths; and (b) in a spherical phantom of 15 cm radius.

In Figure 9.9b simulation results are shown for a spherical phantom of 15 cm in radius that is commonly used in Monte Carlo–based dosimetry studies. Comparison of Figure 9.9a and Figure 9.9b reveals that at lower primary photon energies the mean total number of interactions is almost the same. At higher photon energies though, the total number of interactions is reduced compared with the unbounded geometry of Figure 9.9a due to the fact that a number of photons escape the phantom boundaries.

9.3 Monte Carlo–Based Dosimetry of Monoenergetic Photon Point Sources

The results in this section correspond to Monte Carlo simulation for monoenergetic point sources positioned in the center of a spherical liquid water phantom of 15 cm radius. These results as well as those in subsequent sections are reported in the widely used dosimetric formalism of the AAPM Task Group 43 recommendations[22] detailed in Chapter 8.

In Figure 9.10, the dose rate constant, Λ, of the sources is plotted as a function of photon energy, *E*, in the energy range of 20 to 700 keV which is relevant to brachytherapy. It should be noted that for the calculation of the dose rate constant, Λ, of a source which constitutes the only absolute quantity in the AAPM TG-43 dosimetric formalism, two different Monte Carlo simulations are required. First, the dose rate at $r = 1$ cm along the transverse bisector of the source must be calculated in liquid water medium. Then, the source air kerma strength must be calculated by simulation either in free space

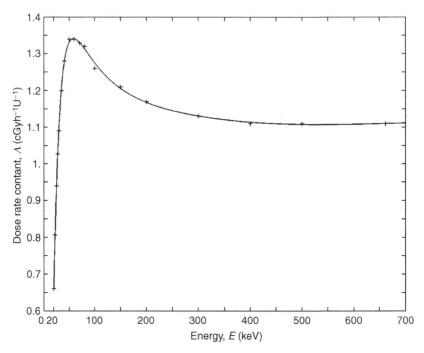

FIGURE 9.10
Monte Carlo calculated dose rate constant, Λ, of monoenergetic point sources plotted as a function of photon energy, E, in the energy range relevant to brachytherapy: 20 to 700 keV.

or in dry air. In the latter case, however, results should be corrected for photon build-up due to scattering of photons emerging from the source in air.[4,20]

In Figure 9.10 it can be seen that the dose rate constant, $\Lambda(E)$, peaks at about 60 keV (with a value of approximately 1.34 cGy h^{-1} U^{-1}) where the ratio of μ_{en}/μ, which quantifies the mean energy transfer to secondary electrons per photon interaction, is minimal according to data in Figure 9.8. The dose rate constant presents strong energy dependence in the low energy region where photoelectric absorption predominates. This suggests that accuracy in the photon energy spectrum and other input data such as cross-sections is of importance for accurate Monte Carlo–based dosimetry of low energy sources such as ^{125}I and ^{103}Pd. At energies greater than 100 keV, however, $\Lambda(E)$ presents a weak energy dependence, attaining a value of approximately 1.11 cGy h^{-1} U^{-1} at 700 keV. Therefore, details of the photon energy spectra used in Monte Carlo simulations are not significant for higher energy sources such as ^{192}Ir.

The results of Figure 9.10 are in agreement with the Monte Carlo calculated results of Luxton and Jozsef,[23] using the EGS4 code and full secondary electron tracking instead of using the water kerma approximation, as well as with the findings of Chen and Nath,[24] who used

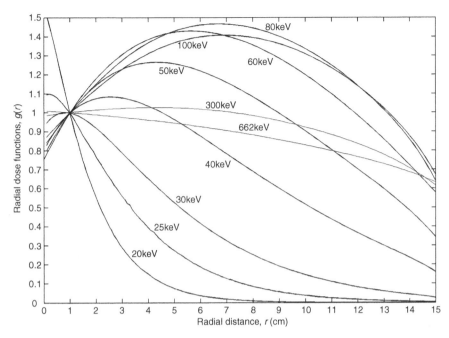

FIGURE 9.11
Monte Carlo calculated radial dose functions, $g(r)$, for different monoenergetic point sources in a liquid water phantom of 15 cm radius plotted vs. radial distance, r.

an analytical calculation method and provided a polynomial expression for the calculation of brachytherapy source dose rate constants.

Radial dose function, $g(r)$, results for different monoenergetic point sources in a liquid water phantom of 15 cm radius are plotted vs. radial distance, r, in Figure 9.11. It can be seen that $g(r)$ for photons in the low energy region shows a sharp fall-off, since in this energy range photoelectric absorption predominates. For photons in the intermediate energy range around 60 keV, incoherent scattering results in overcompensation of photon absorption as expected from the findings of Figure 9.8 and Figure 9.9.b. At the high end of the brachytherapy energy range scattering compensates for absorption, and $g(r)$ values present only a slight variation with varying radial distance, r.

9.4 Monte Carlo–Based Dosimetry of ^{103}Pd, ^{125}I, ^{169}Yb, and ^{192}Ir Point Sources

For the transition from monoenergetic sources to real nuclide point sources, the emitted energy spectra must be accurately known so that the energy of each primary emitted photon is properly sampled. This sampling procedure

TABLE 9.1

Photon Energy Spectra of ^{103}Pd, ^{125}I, ^{169}Yb, and ^{192}Ir Nuclides

^{103}Pd		^{125}I		^{125}I (Ag)		^{139}Yb		^{192}Ir	
E (keV)	I (%)	E (keV)	I (%)	E (keV)	I (%)	E (keV)	I (%)	E (keV)	I (%)
20.07	22.06	27.20	39.90	22.1	25	49.77	53.0	295.96	28.72
20.22	41.93	27.47	74.50	25.2	7	50.74	93.8	308.46	29.68
22.70	13.05	31.00	25.90	27.4	100	57.50	38.5	316.51	82.71
—	—	35.50	6.68	31.4	25	63.12	44.2	468.07	47.81
—	—	—	—	35.5	6	109.78	17.5	588.58	4.52
—	—	—	—	—	—	177.96	35.8	604.41	8.20
—	—	—	—	—	—	307.74	10.05	612.46	5.34

The spectrum of an 6711-type ^{125}I source (i.e., including the characteristic x-rays of Ag that is used as a core material in the 6711 ^{125}I source design) is given under the symbol ^{125}I(Ag).

may be performed using the branching ratios method described in Section 9.2.3 for choosing the photon interaction type. In the present case, branching ratios correspond to the normalized emission probabilities for each of the spectral energies. The main photon energies in terms of intensity for the ^{103}Pd, ^{125}I, ^{169}Yb, and ^{192}Ir nuclides have been taken from Nudat[25] and are shown in Table 9.1. These data are normalized to 100 disintegrations of the parent nuclide. A hypothetical ^{125}I point source is also given under the symbol ^{125}I(Ag). The relative spectrum of this hypothetical source corresponds to that emerging from a 6711-type ^{125}I source (i.e., including the characteristic x-rays of Ag that are used as a core material) taken by Chen and Nath[24] and normalized to the principle line of 27.4 keV. It should be noted that according to the currently used dosimetry protocols including those of AAPM TG-43[22] absolute photon spectra are not necessary for dosimetry of sources calibrated in terms of air kerma strength.

Transport of the photon energy flux in a medium surrounding a source provides the ability for scoring the photon energy distribution at any given distance from the source. Energy distributions of energy, E_i, vs. relative frequency, f_i, normalized so that

$$\sum_i E_i f_i = 1 \qquad (9.12)$$

are given in Figure 9.12 for low energy ^{103}Pd and ^{125}I point sources and in Figure 9.13 for higher energy ^{169}Yb and ^{192}Ir point sources. Simulations were performed with the point sources positioned at the center of a spherical liquid water phantom of 15 cm radius and results are given at selected radial distances, r, of 1, 5, and 10 cm distances from the sources.

In Figure 9.12 it can be seen that for the low energy sources where photoelectric absorption predominates, a significant attenuation of primary radiation occurs and only a small scatter radiation component builds up as the radial distance, r, increases. The primary spectral lines are attenuated

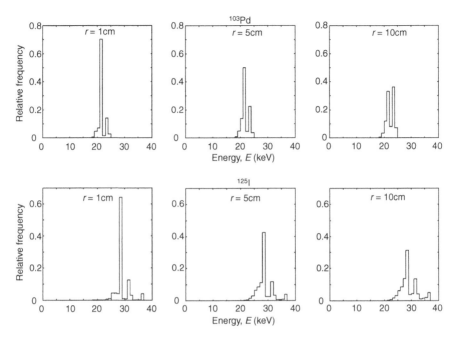

FIGURE 9.12

Monte Carlo-calculated, normalized photon energy distributions at selected radial distances, r, from ^{103}Pd and ^{125}I point sources. Results correspond to simulations in a spherical liquid water phantom of 15 cm radius.

exponentially as $\exp(-\mu r)$, and μ presents a dramatic change in this low energy region, as illustrated in Figure 9.3. For example, while the two ^{103}Pd principal lines that fall within the 20 to 21 keV and 22 to 23 keV energy bins present a relative frequency ratio of approximately 5 at $r = 1$ cm, they appear of almost equal relative frequency at $r = 10$ cm.

In Figure 9.13 it can be seen that for the presented, higher energy, ^{169}Yb and ^{192}Ir point sources scattering is the predominant interaction process. Thus, the photon energy distribution spreads over lower energies with increasing radial distance due to photon energy degradation, owing to successive incoherent interactions as illustrated previously in Figure 9.8. For example, in the ^{192}Ir energy distribution, for the radial distance $r = 10$ cm it can be seen that scattered radiation is the main contributor to absorbed dose in water.

Figure 9.14 and Figure 9.15 illustrate the effect of photon energy on the manner in which energy is deposited in a medium surrounding a brachytherapy source. Figure 9.14 shows the full tracks of 15 primary photons emitted by a ^{125}I point source positioned in the center of a 15 cm radius liquid water sphere. It can be seen that for the low photon energies of ^{125}I (and similarly for ^{103}Pd), the mean number of interactions is small with the photoelectric effect being significant. In contrast, corresponding

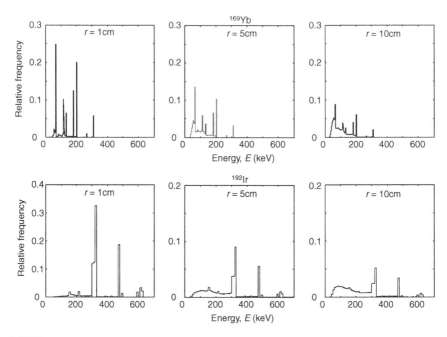

FIGURE 9.13
Monte Carlo-calculated, normalized photon energy distributions at selected radial distances, r, from ^{169}Yb and ^{192}Ir point sources. Results correspond to simulations in a spherical liquid water phantom of 15 cm radius.

results presented in Figure 9.15 illustrate that for the higher energy photons of ^{192}Ir (and similarly ^{169}Yb) incoherent scattering predominates resulting in irregular tracks with an increased mean number of interactions prior to photon absorption.

In Table 9.2, Monte Carlo-calculated dose rate constant values, Λ, of ^{103}Pd, ^{125}I, ^{125}I(Ag), ^{169}Yb, and ^{192}Ir point sources are given. These values are in agreement with the corresponding analytical calculations of Chen and Nath.[24] In the same table corresponding radial dose function $g(r)$ results are also shown. These $g(r)$ results are plotted vs. radial distance, r, in Figure 9.16 along with corresponding data for ^{241}Am (60 keV) and ^{137}Cs (662 keV) point sources. It can be seen that in accordance with data for monoenergetic point sources in Figure 9.11, $g(r)$ for the low photon energy sources ^{103}Pd and ^{125}I demonstrates a sharp fall-off with increasing r. As the mean photon energy of the sources increases to those of ^{241}Am and ^{169}Yb, scattering over-compensates for absorption. Finally, in the highest energy range considered, that of ^{192}Ir and ^{137}Cs, scattering compensates absorption and $g(r)$ shows only a slight decrease with increasing r.

Figure 9.17 illustrates the total, primary, and scatter radial dose rate distributions around a ^{192}Ir point source for different liquid water phantom dimensions. That is, results for spherical phantom radii of $R = 5, 10, 15,$ and

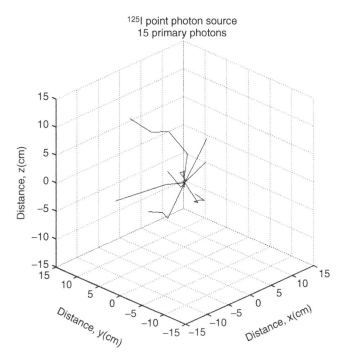

FIGURE 9.14
Photon tracks around an ^{125}I point source in a liquid water spherical phantom of 15 cm radius, shown in Cartesian coordinates for 15 primary photons.

20 cm are combined in Figure 9.17. Dose rates are multiplied by the square of the radial distance to remove the inverse square law dependence. In this figure it can be observed that the scatter dose rate component reaches its maximum at approximately one photon mfp. Most importantly, the scatter dose rates and consequently the total dose rates differ significantly between different phantom dimensions near the phantom boundaries.

This is due to the increasing lack of backscatter as the phantom dimensions decrease. For example, the total dose rate near the boundary of the $R = 10$ cm radius phantom is 17 and 20% lower than the corresponding dose rates in the $R = 15$ cm and $R = 20$ cm radii phantoms. This suggests that choice of appropriate phantom dimensions is of importance for accurate dosimetry of ^{192}Ir sources, and that caution is needed for dosimetry near the body surface in clinical applications due to lack of backscatter radiation.

In Figure 9.18, results similar to those in Figure 9.17, are shown for a ^{125}I point source where the effect of the reduced dose rate near the phantom boundaries is again evident, being significant near the boundaries of the $R = 5$ cm radius phantom where the dose rate is 20% lower than corresponding dose rates for the other phantom dimensions. For greater

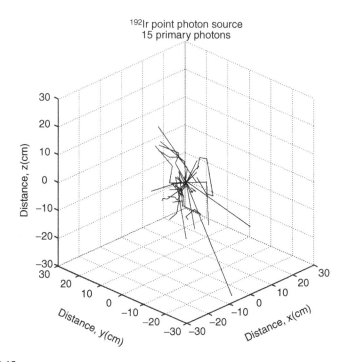

FIGURE 9.15
Photon tracks around an ^{192}Ir point source in a liquid water spherical phantom of 15 cm radius, shown in Cartesian coordinates for 15 primary photons.

phantom dimensions of $R = 10$ and 15 cm this effect is unimportant due to the sharp dose rate fall-off at ^{125}I energies.

In both Figure 9.17 and Figure 9.18, the primary dose rate component is described by exponential attenuation and is, of course, the same for all phantom dimensions. Fitting to this primary dose rate component yields an effective linear attenuation coefficient, μ_{eff}. Alternatively, μ_{eff} may be calculated by weighting the linear attenuation coefficients of the energies in a spectrum with the corresponding emission probabilities, f_i, according to

$$\mu_{eff} = \frac{\sum_i f_i E_i \mu_i}{\sum_i f_i E_i} \qquad (9.13)$$

This μ_{eff} has been successfully utilized in analytical dose rate calculation methods for ^{192}Ir sources.[26–29] Results of Monte Carlo-calculated μ_{eff} coefficients are given in Table 9.2 for ^{103}Pd, ^{125}I, ^{125}I(Ag), ^{169}Yb, and ^{192}Ir point sources, along with the corresponding effective energy, E_{eff}, defined as the photon energy with a linear attenuation coefficient equal to μ_{eff}.

In summary, Figure 9.17 and Figure 9.18 illustrate that at distances close to brachytherapy sources the primary dose component prevails. The scatter

TABLE 9.2

Radial Dose Functions, $g(r)$, Effective Attenuation Coefficients, μ_{eff}, Effective Energies, E_{eff}, and Dose Rate Constants, Λ, for ^{103}Pd, ^{125}I, ^{125}I (Ag), ^{169}Yb, and ^{192}Ir Point Sources

r (cm)	^{103}Pd	^{125}I	^{125}I(Ag)	^{169}Yb	^{192}Ir
0.1	1.443	0.996	1.058	0.926	0.990
0.2	1.408	1.008	1.065	0.934	0.991
0.3	1.364	1.019	1.065	0.944	0.992
0.5	1.264	1.023	1.058	0.959	0.995
0.7	1.155	1.020	1.041	0.975	0.997
1.0	1.000	1.000	1.000	1.000	1.000
1.5	0.767	0.939	0.916	1.035	1.004
2.0	0.577	0.865	0.822	1.068	1.007
2.5	0.432	0.786	0.729	1.097	1.009
3.0	0.320	0.706	0.645	1.117	1.010
4.0	0.175	0.556	0.489	1.157	1.006
5.0	0.094	0.430	0.368	1.179	0.999
6.0	0.050	0.327	0.274	1.185	0.987
8.0	0.014	0.184	0.150	1.158	0.949
10.0	0.005	0.101	0.080	1.073	0.891
μ_{eff} (cm^{-1})	—	0.713	0.417	0.45	0.157	0.108
E_{eff} (keV)	—	22	29	27	134	388
Λ (cGy h^{-1} U^{-1})	—	0.68	1.03	0.97	1.20	1.12

dose rate component that builds up with increasing distance from the source reaches its maximum at approximately one primary photon mfp. This can be determined from the inverse of the μ_{eff} coefficients quoted in Table 9.2.

Finally, it is emphasized that at distances close to the treatment site boundaries, an important effect, due to the lack of backscatter, may occur if dosimetry data corresponding to a geometry of different shape or dimensions are used.

9.5 Monte Carlo–Based Dosimetry of Commercially Available ^{192}Ir Source Designs

The final requirement for performing Monte Carlo–based dosimetry of real brachytherapy source designs is the geometric modeling of the sources and the random generation of primary photons within their active core. The cylindrical symmetry of brachytherapy source designs facilitates their mathematical simulation while most of the currently available codes, such as EGS, MCNP, and GEANT, include powerful geometric simulation packages for creating any geometrical object as a configuration of simple mathematical objects such as cylinders, spheres, or ellipses.

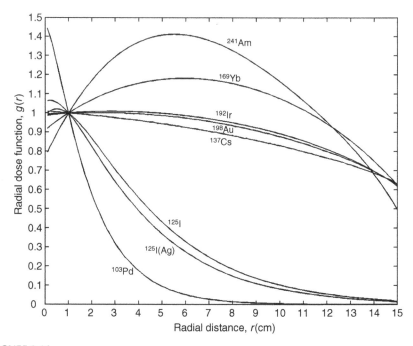

FIGURE 9.16
Monte Carlo calculated radial dose function results, $g(r)$, for ^{103}Pd, ^{125}I, ^{241}Am, ^{169}Yb, ^{192}Ir, and ^{137}Cs point sources, as well as for a hypothetical 6711-type ^{125}I (Ag) source vs. radial distance, r. Results correspond to simulations in a liquid water spherical phantom of 15 cm radius.

The most widely utilized nuclide in current brachytherapy applications is ^{192}Ir used in the form of HDR source designs for remote afterloaders as well as LDR seed and wire source designs. The difference lies in the source air kerma strength that ranges from 2.9×10^4 to 4.1×10^4 U for HDR sources, 1.4 to 7.2 U for seed sources, and 140 U for wire sources.[22]

The most widespread ^{192}Ir HDR source designs, to date, are the old[30,31] and new microSelectron,[19] the old[32,33] and new VariSource,[20] and the Buchler[34] source designs, presenting a variety of source and encapsulation structural as well as compositional details.

Common seed source designs are the stainless steel encapsulated seed manufactured by Best Industries (Springfield, VA) and the Pt encapsulated seed by Alpha Omega (Bellflower, CA).[22] Seed sources are supplied in nylon strands of varying length. Pt encapsulated wire sources are available in different diameters (0.3, 0.5, and 0.6 mm) and 14 cm standard length that is cut down to smaller lengths (mostly in the order of 4 to 6 cm) to facilitate specific applications.[27,35,36] A Ti/Ni encapsulated thin wire under the commercial name Angiorad™ has also become available to facilitate contemporary intravascular applications.[37] More information about the geometry and structure details of these sources may be found in the corresponding references and Chapter 8.

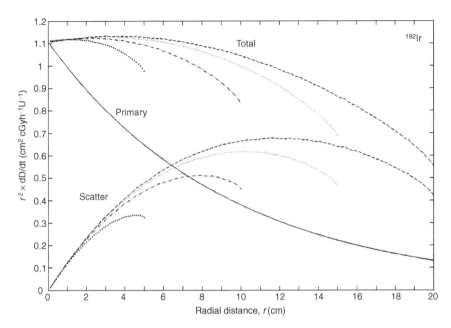

FIGURE 9.17
Monte Carlo calculated total, primary, and scatter radial dose rate distributions for an ^{192}Ir point source, multiplied by the square of the radial distance to remove the inverse square law dependence, for different spherical liquid water phantom dimensions of $R = 5, 10, 15$, and 20 cm radius.

Table 9.3 summarizes dose rate constant, Λ, results found in the literature for ^{192}Ir source designs. It can be seen that differences in the results of independent investigators for the same source design are negligible. This confirms that details in the photon energy spectrum and other input data used in Monte Carlo–based dosimetry are not significant for higher energy sources, such as ^{192}Ir due to the weak energy dependence of $\Lambda(E)$ in this energy region, as seen in Figure 9.10. In Table 9.3 the source geometry factors, $G(1\ \text{cm}, 90°)$, at the point where Λ is defined (see Chapter 8) are also given, and it can be seen that Λ values are almost proportional to $G(1\ \text{cm}, 90°)$.

Theoretically, the dose rate constant per unit source air kerma strength, Λ, of the sources includes the effect of filtration by the source core and encapsulation, in-water scattering, and spatial distribution of radioactivity within the source.[22] However, differences in the normalized energy spectrum emerging from the ^{192}Ir sources are not significant,[31] and additionally, Λ presents only a minor energy dependence in the range close to the effective ^{192}Ir energy. Moreover, it has already been shown that for the ^{192}Ir energies scattering compensates for absorption.[31,38,39] Therefore, the spatial distribution of radioactivity, addressed by the source geometry factor at the point where Λ is calculated, is the prevailing influence on the dose rate constant of an ^{192}Ir source.[40]

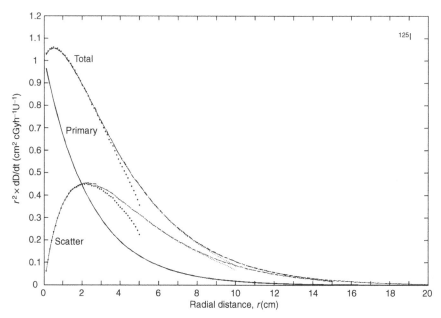

FIGURE 9.18
Monte Carlo calculated primary, scatter and total radial dose rate distributions for an ^{125}I point source, multiplied by the square of the radial distance to remove the inverse square law dependence, for different spherical liquid water phantom dimensions of $R = 5, 10, 15$, and 20 cm radius.

TABLE 9.3

Dose Rate Constant Values, Λ (cGy h^{-1} U^{-1}), of Commercially Available ^{192}Ir Sources

Source Type	Λ (cGy h^{-1} U^{-1})	G (1 cm, 90°)	Refs.
MicroSelectron (old design)	1.115	0.990	30
	1.116		31
MicroSelectron (new design)	1.108	0.989	19
VariSource (old design)	1.044	0.927	32
	1.043		33
VariSource (new design)	1.101	0.980	20
Buchler	1.115	0.999	34
Seed (Best Medical)	1.109	0.993	3
AngioRad™	0.716	0.655	37
Wire $L = 1$ cm	1.040	0.927	27
	1.047		35
Wire $L = 5$ cm	0.521	0.476	27
	0.521		35

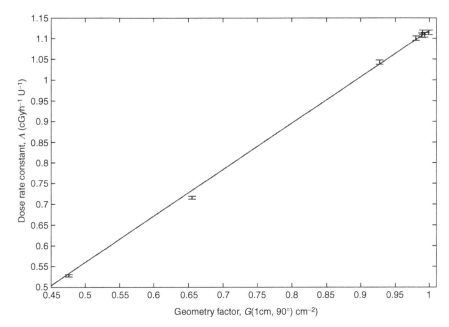

FIGURE 9.19
Monte Carlo calculated dose rate constant results, Λ, of ^{192}Ir source designs and a point ^{192}Ir source vs. the corresponding geometry factors, $G(1\ \text{cm}, 90°)$. A linear fit of the form: $\Lambda_{\text{ro}} = a\ G(r_0, 90°)$ is also drawn.

This is illustrated in Figure 9.19, where for the source designs summarized in Table 9.3, Λ as well as a point ^{192}Ir source are plotted vs. the corresponding geometry factors $G(1\ \text{cm}, 90°)$. A linear fit was performed for the data in Table 9.3 and is also shown in Figure 9.19. The resulting equation:

$$\Lambda_{\text{ro}}(\text{cGy h}^{-1}\ \text{U}^{-1}) = 1.12(\text{cGy h}^{-1}\ \text{U}^{-1}\ \text{cm}^{-2}) \times G(r_0, 90°)(\text{cm}^{-2}) \quad (9.14)$$

provides accurate dose rate constant results for any ^{192}Ir source design.[27,40] The only source that appears to deviate (>1 SD) from Equation 9.14 in Figure 9.19 is the Angiorad wire source. However, this deviation could be explained by the utilization of a conversion factor relating air kerma strength and source activity in the calculations of Patel et al.,[37] that contradicts the specifications of the TG-43 formalism for calibration of brachytherapy sources in air kerma strength units[22] and leads to an underestimation of approximately 1.5%.

The use of Equation 9.14 for dose rate constant calculation overcomes problems such as the use of custom-length ^{192}Ir wire sources in clinical practice where, alternatively, a separate simulation would be required for the dose rate constant calculation of each different wire source length.[27]

With regard to radial dose function, $g(r)$, caution is needed when comparing results calculated in different Monte Carlo simulations since

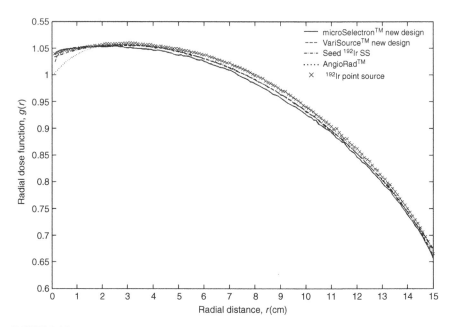

FIGURE 9.20
Monte Carlo-calculated radial dose functions, $g(r)$ of the new VariSource $(--)$, the new microSelectron (———), the stainless steel encapsulated ^{192}Ir seed $(\cdot\text{———}\cdot)$, the Angiorad^{TM} $(\cdots)$, and an ^{192}Ir point source $(\times)$, vs. radial distance, r.

these results are sensitive to the mathematical phantom shape and dimensions that may result from different backscatter conditions near the phantom boundaries, as shown earlier in Figure 9.17. In general, source and encapsulation geometry does not significantly affect radial dose function values of ^{192}Ir sources.[19,20,32] This is evident in Figure 9.20 where $g(r)$ results for four representative ^{192}Ir source designs (the new microSelectron, the new VariSource, the stainless steel encapsulated ^{192}Ir seed, and the Angiorad wire source) as well as a point ^{192}Ir source are plotted vs. radial distance, r. The results correspond to simulations in a 15 cm radius spherical liquid water phantom and it can be seen that $g(r)$ values of the sources agree to within 2% with that of the point ^{192}Ir source given in Table 9.2.

The anisotropy function of ^{192}Ir sources, however, is known to depend strongly on source geometry in terms of active core length and diameter and varies significantly with calculation point (r,θ). Therefore, a thorough set of anisotropy function data are needed for clinical dosimetry purposes around any particular ^{192}Ir source design; Monte Carlo simulation is an invaluable tool for generating such data sets since it does not suffer from the inherent limitations of experimental dosimetry methods, such as volume averaging, energy dependence, and limited number of experimental points.

In Figure 9.21, anisotropy function results, $F(r,\theta)$ for the same four sources as in Figure 9.20 are plotted vs. polar angle, θ, at the radial distance of

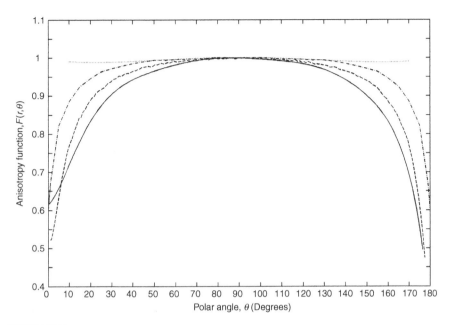

FIGURE 9.21
Monte Carlo calculated anisotropy function results, $F(1 \text{ cm}, \theta)$, of the new VariSource ($--$), the new microSelectron (——), the stainless steel encapsulated ^{192}Ir seed (-----), and the Angiorad™ (.....) vs. polar angle, θ at the radial distance of $r = 1$ cm.

$r = 1$ cm: $\theta = 180°$ refers to the drive wire side of the microSelectron and VariSource HDR designs. Comparison of anisotropy functions reveals that the stainless steel seed source is less anisotropic than the microSelectron and the VariSource designs due to its relatively reduced active core length and diameter: 0.3 and 0.01 cm, respectively. The new VariSource design produces a smaller anisotropy for polar angles close to the source midsection ($\theta = 90°$) compared with the new microSelectron source due to its smaller active core diameter: 0.034 cm for the new VariSource vs. 0.065 cm for the new microSelectron.

At polar angles very close to the longitudinal source axes, however, larger anisotropy is observed for the new VariSource design mainly due to the fact that it presents a slightly longer active core relative to the new microSelectron: 0.5 cm for the new VariSource vs. 0.36 cm for the new microSelectron. Anisotropy function results for the Angiorad source are almost equal to unity at all points lying outside the source structure at the selected radial distance of $r = 1$ cm where Figure 9.21 corresponds. This is also the case for the anisotropy function values of all ^{192}Ir source designs at radial distances close to the source. In general, anisotropy is of importance only at points lying at polar angles $\theta < 30°$ and radial distances at least greater than half the length of the sources since at shorter radial distances the source segment closer to a particular point is the main dose contributor,

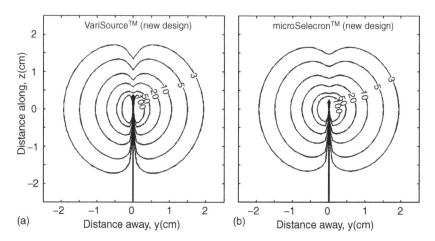

FIGURE 9.22
Monte Carlo calculated dose rate distributions (cGy h^{-1}) in Cartesian coordinates, in a plane containing the longitudinal source axis of (a) the new VariSource, and (b) the new microSelectron designs for hypothetical source air kerma strengths of 10 U.

thus compensating for the filtration of radiation emitted by other segments.[27,40] In other words, at radial distances in close proximity to ^{192}Ir sources it is the geometry factor and not the anisotropy function that defines the dose rate distribution of the source.

In summary, dosimetry differences between ^{192}Ir sources may be reviewed in Figure 9.22, where dose distributions in Cartesian coordinates are shown in a plane containing the longitudinal source axis of the new VariSource and the new microSelectron designs for hypothetical source air kerma strengths of 10 U. In Figure 9.22 it can be seen that considerable differences exist at points close to the sources due to differences in their geometry factors which depend strongly on source design. Significant differences are also observed at points close to the longitudinal source axes due to the significant dependence of anisotropy on source geometry.

It should be noted that for the ^{192}Ir energies, electronic equilibrium, and consequently, water kerma dose approximation, may be safely assumed only at distances greater than 1 mm from the sources.[3,5] At distances less than 1 mm the water kerma dose approximation introduces errors of up to 6%. Moreover, Baltas et al.,[5] owing to the beta spectrum emitted by ^{192}Ir, were the first to report a significant dose rate enhancement in this short distance range. This dose rate enhancement depends strongly on the geometrical as well as compositional details of the source active core and encapsulation, and was found equal to approximately 35% of the gamma dose rate for the Angiorad source,[37] 15% for both VariSource designs, and 5% for the new microSelectron source at 1 mm along the transverse source bisectors.[5]

^{169}Yb could be an alternative to ^{192}Ir in high dose rate applications since it presents a relatively high specific activity and reduced demands for radiation shielding and critical organ protection: see Table 9.1. A ^{169}Yb HDR source has recently become commercially available.[41] ^{169}Yb is also suitable for permanent implantation due to its short physical half-life (32.02 days). However, ^{169}Yb prototype seeds that have been produced have only found use in a limited number of clinical trials (see Section 8.4.3). ^{169}Yb sources present slightly increased anisotropy compared with ^{192}Ir source designs due to increased filtration of their relatively lower energy photons within the source structure.[42]

9.6 Monte Carlo–Based Dosimetry of ^{125}I and ^{103}Pd LDR Seeds

Currently, ^{125}I and ^{103}Pd radioactive seeds are widely used as permanent implants for the treatment of early stages of prostate cancer, and a great number of source designs have become commercially available (see for example the AAPM/RPC registry of low energy brachytherapy seeds at: http://rpc.mdanderson.org/rpc/htm/Home_htm/Low-energy.htm). ^{125}I and ^{103}Pd seeds are generally of small dimensions and are mostly of cylindrical shape (about 4 mm long and 1 mm in diameter) using Ti as the encapsulation material. These sources often utilize high Z materials, such as Ag, Pd, Mo or Au, as radioopaque markers.

Both ^{125}I and ^{103}Pd are low energy photon emitters (see Table 9.1), with an increased probability of x-ray emission following photoelectric absorption in the source core and encapsulation materials. This fluorescence radiation may be of significant impact on source dosimetry. For example, the x-ray characteristic radiation of approximately 4.5 keV emerging from the commonly used Ti encapsulation, although resulting in a significant contribution to air kerma strength, S_K, calculations, does not contribute to absorbed dose in water since 4.5 keV x-rays are readily absorbed within the first mm. Therefore, according to the current AAPM recommendations, and in order to be consistent with the new 1999 NIST calibration standard, these x-rays must be suppressed in S_K calculations.[14,15] Otherwise, Ti x-rays lead to an overestimation of S_K, which is of the order of 5% at radial distance $d = 30$ cm for simulations in air, and even greater for calculations in vacuum where they result in a 22% decrease of Λ.[4]

X-ray characteristic radiation emerging from high Z materials incorporated in the radioactive source core may also influence dose rate constant, Λ, and radial dose function, $g(r)$, values. For example, Ag and Pd K-shell characteristic x-ray energy is approximately 22 keV while Au emits only

TABLE 9.4

Dose Rate Constant Values, Λ (cGy h^{-1} U^{-1}), of ^{125}I Sources

Source Type	Λ (cGy h^{-1} U^{-1})	Refs.
6702	1.016	43
Symmetra	1.101	43
InterSource	1.020	46
Best	1.110	47
STM1251	0.980	21
6711	0.952	4
selectSeed	0.954	4
Pharma Seed	0.950	48

L-shell characteristic radiation for the ^{125}I energies ($E_K = 80.7$ keV for Au), that is of approximately 11 keV energy and therefore readily absorbed by the Ti encapsulation.

Table 9.4 summarizes dose rate constant, Λ, results for different ^{125}I source designs found in the literature, following the new 1999 NIST calibration standard. It can be seen that ^{125}I sources can be grouped in two major categories:

1. Sources not including Ag, Pd or other high Z materials in their active core that present Λ values similar to that of the ^{125}I point source presented in Table 9.2 (6702, Symmetra, InterSource, Best, STM1251).

2. Sources containing Ag, Pd or other high Z materials in their active core that present Λ values similar to that of the ^{125}I(Ag) point source presented in Table 9.2 (6711, selectSeed, Pharma-Seed).

It should also be noted that filtration and shielding in low energy brachytherapy sources of the second category with radioactivity distributed on each end their (typically) cylindrical, high Z marker results in photon fluence anisotropy close to their transverse bisector. This effect renders Monte Carlo calculated dose rate constant results for these sources sensitive to whether the Wide Angle Free Air Chamber (WAFAC) geometry of the NIST 1999 air kerma strength primary standard is employed in the simulation for air kerma strength calculation.[44,49,50]

In Figure 9.23 radial dose function, $g(r)$, values for the 6702[43] and 6711[4] ^{125}I source designs as well as the ^{125}I and ^{125}I(Ag) point sources presented in Table 9.2 are plotted vs. radial distance, r. It can be seen that the 6702 source design presents similar $g(r)$ values to the ^{125}I point source while the 6711 source design presents $g(r)$ values that are similar to the ^{125}I(Ag) point source, in agreement with the above grouping of ^{125}I sources. Moreover, $g(r)$ of the 6771-type sources is increased at radial distances under 1 cm and falls off more rapidly thereafter compared with $g(r)$ results for the

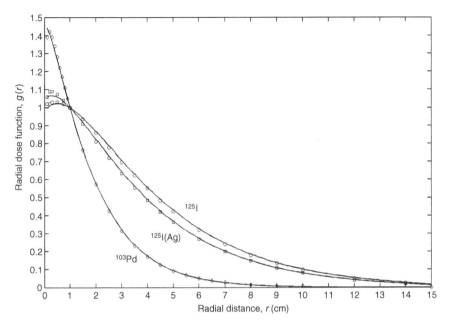

FIGURE 9.23

Monte Carlo-calculated radial dose function, $g(r)$, values vs. radial distance, r, for the 6702[43] (O) and 6711[4] (□) [125]I source designs, the Bebig IsoSeed[45] [103]Pd source design (◇) as well as the ^{125}I, [125]I (Ag), and [103]Pd point sources (data presented in Table 9.2).

6702-type sources, due to the influence of the low energy Ag x-ray characteristics.

Dose rate constant values of commonly used [103]Pd sources are in good agreement, to within 3%, with the corresponding value obtained for a [103]Pd point source presented in Table 9.2.[43,44] In general, dosimetry results at radial distances greater than 0.5 cm along the transverse bisector of [103]Pd sources agree to within 3% for all different source designs.[44] This is also evident in Figure 9.23 where the radial dose functions of the Bebig IsoSeed[45] [103]Pd source design as well as the [103]Pd point source presented in Table 9.2 are plotted vs. radial distance, r.

Dose anisotropy around low energy ^{125}I and [103]Pd sources is more significant compared with ^{192}Ir sources due to increased filtration within the source. This can be seen in Figure 9.24 where anisotropy function results, $F(r, \theta)$, for the 6702[43] and 6711[4] [125]I source designs as well as the Bebig Isoseed [103]Pd source design[45] are plotted vs. polar angle, θ, at the radial distance of $r = 1$ cm. Moreover, anisotropy function values depend significantly on active core and encapsulation structural details (end welds shape, radioactive coating thickness, presence of radioactive material on the source core ends, etc.), and a thorough set of anisotropy function data is needed for accurate dosimetry around a particular source.[4]

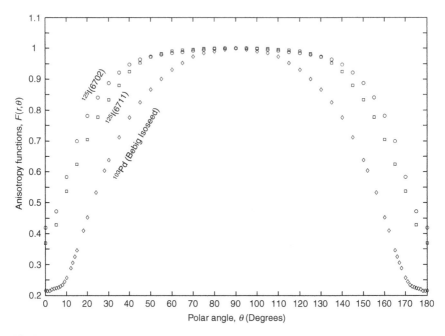

FIGURE 9.24

Monte Carlo-calculated anisotropy function results, $F(1 \text{ cm}, \theta)$, of the 6702^{43} (O) and 6711^4 ($\square$) ^{125}I source designs as well as the Bebig Isoseed[45] ^{103}Pd source design ($\Diamond$) vs. polar angle, θ at the radial distance of $r = 1$ cm.

References

1. Williamson, J.F., Coursey, B.M., DeWerd, L.A., Hanson, W.F., and Meigooni, A.S. Dosimetric prerequisites for routine clinical use of new low energy photon interstitial brachytherapy sources, *Med. Phys.*, 25, 2269, 1998.

2. Weaver, K. Anisotropy functions for ^{125}I and ^{103}Pd sources, *Med. Phys.*, 25, 2271, 1998.

3. Wang, R. and Li, X.A. A Monte Carlo calculation of dosimetric parameters of ^{90}Sr/^{90}Y and ^{192}Ir SS sources for intravascular brachytherapy, *Med. Phys.*, 27, 2528, 2000.

4. Karaiskos, P., Papagiannis, P., Sakelliou, L., Anagnostopoulos, A., and Baltas, D. Monte Carlo dosimetry of the selectSeed ^{125}I interstitial brachytherapy seed, *Med. Phys.*, 28, 1753, 2001.

5. Baltas, D., Karaiskos, P., Papagiannis, P., Sakelliou, L., Loeffler, E., and Zamboglou, N. Beta versus gamma dosimetry close to Ir-192 brachytherapy sources, *Med. Phys.*, 28, 1875, 2001.

6. Kahn, H. *Applications of Monte Carlo*, RAND Corp, Santa Monica, CA, Publication AECU-3259, 1954.

7. Hubbell, J.H. *Polarization Effects in Coherent and Incoherent Photon Scattering: Survey of Measurements and Theory Relevant to Radiation Transport Calculations,* National Institute of Standards and Technology, Gaithersburg, MD, 1992.

8. Fernàndez, J.E., Hubbell, J.H., Hanson, A.L., and Spencer, L.V. Polarization effects on multiple scattering gamma transport, *Radiat. Phys. Chem.,* 41, 579, 1993.

9. Hubbell, J.H., Veigele, W.J., Briggs, E.A., Brown, R.T., Cromer, D.T., and Howerton, R.J. Atomic form factors, incoherent scattering functions, and photon scattering cross-sections, *J. Phys. Chem. Ref. Data,* 4, 471, 1975 [Erratum: *J. Phys. Chem. Ref. Data,* 6, 615, 1977].

10. Chan, H.P. and Doi, K. The validity of Monte Carlo simulation in studies of scattered radiation in diagnostic radiology, *Phys. Med. Biol.,* 28, 109, 1983.

11. Hubbell, J.H. and Overdo, I. Relativistic atomic form factors and photon coherent scattering cross-sections, *J. Phys. Chem. Ref. Data,* 8, 69, 1979.

12. Schaupp, D., Schumacher, M., Smend, F., Rullhusen, P., and Hubbell, J.H. Small-angle Rayleigh scattering of photons at high energies: Tabulation of relativistic HFS modified atomic form factors, *J. Phys. Chem. Ref. Data,* 12, 467, 1983.

13. Morin, L.R.M. Molecular form factors and photon coherent scattering cross-sections of water, *J. Phys. Chem. Ref. Data,* 11, 1091, 1982.

14. Kubo, H.D., Coursey, B.M., Hanson, W.F., Kline, R.W., Seltzer, S.M., Shuping, R.E., and Williamson, J.F. Report of the ad hoc Committee of the AAPM Radiation Therapy Committee on ^{125}I sealed source dosimetry, *Int. J. Rad. Onc. Biol. Phys.,* 40, 697, 1998.

15. Williamson, J.F., Coursey, B.M., DeWerd, L.A., Hanson, W.F., Nath, R., and Ibbot, G. Dosimetry and calibration changes: Recommendations of the American Association of Physicists in Medicine Radiation Therapy Committee ad hoc Subcommittee on low-energy seed dosimetry, *Med. Phys.,* 26, 570, 1999.

16. Plechaty, E.F., Cullen, D.E., and Howerton, R.J. *Tables and Graphs of Photo-Interaction Cross-Sections from 0.1 keV to 100 MeV Derived from the LLL Evaluated Nuclear Data Library,* Vol. 6. Lawrence Livermore Laboratory Publications, UCRL-50400, Rev. 3, NISTIR 4881, 1981.

17. Nelson, W.R. and Rogers, D.W.O. *The EGS4 Code System, SLAC-265,* Stanford Linear Accelerator Center, Stanford, CA, 1996.

18. Briesmeister, J.F. *MCNP—A General Monte Carlo N-Particle Transport Code, Version 4C, LA-13709-M,* Los Alamos National Laboratory, 2000.

19. Daskalov, G.M., Löffler, E., and Williamson, J.F. Monte Carlo-aided dosimetry of a new high dose-rate brachytherapy source, *Med. Phys.,* 25, 2200, 1998.

20. Angelopoulos, A., Baras, P., Sakelliou, L., Karaiskos, P., and Sandilos, P. Monte Carlo dosimetry of a new ^{192}Ir high dose rate brachytherapy source, *Med. Phys.,* 27, 2521, 2000.

21. Kirov, A.S. and Williamson, J.F. Monte Carlo-aided dosimetry of the Source Tech Medical Model STM1251 I-125 interstitial brachytherapy source, *Med. Phys.,* 25, 764, 2001.

22. Nath, R., Anderson, L., Luxton, G., Weaver, K., Williamson, J.F., and Meigooni, A.S. Dosimetry of interstitial brachytherapy sources: Recommendations of the AAPM Radiation Therapy Committee Task Group 43, *Med. Phys.,* 22, 209, 1995.

23. Luxton, G. and Jozsef, G. Radial dose distribution, dose to water and dose rate constant for monoenergetic photon point sources from 10 keV to 2 MeV:EGS4 Monte Carlo model calculation, *Med. Phys.,* 26, 2531, 1999.

24.   Chen, Z. and Nath, R. Dose rate constant and energy spectrum of interstitial brachytherapy sources, *Med. Phys.*, 28, 86, 2001.

25.   Nudat available online at National Nuclear Data Centre (NNDC): http://www.nndc.bnl.gov/nudat2/.

26.   Karaiskos, P., Angelopoulos, A., and Baras, P. Dose rate calculations around ^{192}Ir brachytherapy sources using a Sievert integration model, *Phys. Med. Biol.*, 45, 383, 2000.

27.   Karaiskos, P., Papagiannis, P., and Angelopoulos, A. Dosimetry of ^{192}Ir wires for LDR interstitial brachytherapy following the AAPM TG-43 dosimetric formalism, *Med. Phys.*, 28, 156, 2001.

28.   Anagnostopoulos, G., Baltas, D., Karaiskos, P., Pantelis, E., Papagiannis, P., and Sakelliou, L. An analytical dosimetry model as a step towards accounting for inhomogeneities and bounded geometries in ^{192}Ir brachytherapy treatment planning, *Phys. Med. Biol.*, 48, 1625, 2003.

29.   Pantelis, E., Papagiannis, P., Anagnostopoulos, G., Baltas, D., Karaiskos, P., Sandilos, P., and Sakelliou, L. Evaluation of a TG-43 compliant analytical dosimetry model in clinical Ir-192 HDR brachytherapy treatment planning and assessment of the significance of source position and catheter reconstruction uncertainties, *Phys. Med. Biol.*, 49, 55, 2004.

30.   Williamson, J.F. and Li, Z. Monte Carlo dosimetry of the microSelectron pulsed and high dose rate ^{192}Ir sources, *Med. Phys.*, 22, 809, 1995.

31.   Karaiskos, P., Angelopoulos, A., Sakelliou, L., Sandilos, P., Antypas, C., Vlachos, L., and Koutsouveli, E. Monte Carlo and TLD dosimetry of an ^{192}Ir high dose-rate brachytherapy source, *Med. Phys.*, 25, 1975, 1998.

32.   Wang, R. and Sloboda, R.S. Influence of source geometry and materials on the transverse axis dosimetry of ^{192}Ir brachytherapy sources, *Phys. Med. Biol.*, 43, 37, 1998.

33.   Karaiskos, P., Angelopoulos, A., Baras, P., Sakelliou, L., Sandilos, P., Dardoufas, K., and Vlachos, L. A Monte Carlo investigation of the dosimetric characteristics of the VariSource ^{192}Ir high dose rate brachytherapy source, *Med. Phys.*, 26, 1498, 1999.

34.   Ballester, F., Pérez-Calatayud, J., Puchades, V., Lluch, J.L., Serrano-Andrés, M.A., Limami, Y., Lliso, F., and Casal, E. Monte Carlo dosimetry of the Buchler high dose rate ^{192}Ir source, *Phys. Med. Biol.*, 46, N79, 2001.

35.   Ballester, F., Hrenandez, C., Perez-Calatayud, J., and Lliso, F. Monte Carlo calculation of dose rate distributions around ^{192}Ir wires, *Med. Phys.*, 24, 1221, 1997.

36.   Perez-Catalayud, J., Lliso, F., Carmona, V., Balester, F., and Hernandez, C. Monte Carlo calculation of dose rate distributions around 0.5 and 0.6 mm in diameter ^{192}Ir wires, *Med. Phys.*, 26, 395, 1999.

37.   Patel, N.S., Chiu-Tsao, S.T., Fan, P., Tsao, H.S., Liprie, S.F., and Harrison, L.B. The use of cylindrical coordinates for treatment planning parameters of an elongated ^{192}Ir source, *Int. J. Rad. Oncol. Biol. Phys.*, 51, 1093, 2001.

38.   Angelopoulos, A., Perris, A., Sakellariou, K., Sakelliou, L., Sarigiannis, K., and Zarris, G. Accurate Monte Carlo calculations of the combined attenuation and build-up factors, for energies (20–1500 keV) and distances (0–10 cm) relevant in brachytherapy, *Phys. Med. Biol.*, 36, 763, 1991.

39.   Sakelliou, L., Sakellariou, K., Sarigiannis, K., Angelopoulos, A., Perris, A., and Zarris, G. Dose rate distributions around ^{60}Co ^{137}Cs, ^{198}Au, ^{192}Ir, ^{241}Am, ^{125}I

(models 6702 and 6711) brachytherapy sources and the nuclide ^{99}Tc^{m}, *Phys. Med. Biol.*, 37, 1859, 1992.

40. Papagiannis, P., Angelopoulos, A., Pantelis, E., Karaiskos, P., Sandilos, P., Sakelliou, L., and Baltas, D. Dosimetry comparison of ^{192}Ir sources, *Med. Phys.*, 29, 2239, 2002.

41. Medich, C.D., Tries, A.M., and Munro, J.J. Monte Carlo characterization of an ytterbium-169 high dose rate brachytherapy source with analysis of statistical uncertainty, *Med. Phys.*, 33, 163–172, 2006.

42. Lymperopoulou, G., Papagiannis, P., Sakelliou, L., Milickovic, N., Giannouli, S., and Baltas, D. A dosimetric comparison of ^{169}Yb versus ^{192}Ir for HDR prostate brachytherapy, *Med. Phys.*, 32, 3832–3842, 2005.

43. Hedtjärn, H., Carlsson, G.A., and Williamson, J.F. Monte Carlo-aided dosimetry of the symmetra model I25.S06 ^{125}I, interstitial brachytherapy seed, *Med. Phys.*, 27, 1076, 2000.

44. Williamson, J.F. Monte Carlo modelling of the transverse-axis dose distribution of the Model 200 ^{103}Pd interstitial brachytherapy source, *Med. Phys.*, 27, 643, 2000.

45. Daskalov, G.M. and Williamson, J.F. Monte Carlo-aided dosimetry of the new Bebig Isoseed® ^{103}Pd interstitial brachytherapy seed, *Med. Phys.*, 28, 2154, 2001.

46. Reniers, B., Vynckier, S., and Scalliet, P. Dosimetric study of the new InterSource125 iodine seed, *Med. Phys.*, 28, 2285, 2001.

47. Meigooni, A.S., Bharucha, Z., Yoe-Sein, M., and Sowards, K. Dosimetric characteristics of the Best® double-wall ^{103}Pd brachytherapy source, *Med. Phys.*, 28, 2568, 2001.

48. Popescu, C.C., Wise, J., Sowards, K., Meigooni, A.S., and Ibbott, G.S. Dosimetric characteristics of the Pharma Seed™ model BT-125-I source, *Med. Phys.*, 27, 2174, 2000.

49. Lymperopoulou, G., Papagiannis, P., Sakelliou, L., Karaiskos, P., Sandilos, P., Przykutta, A., and Baltas, D. Monte Carlo and Thermoluminescence dosimetry of the new IsoSeed model l25.S17 ^{125}I interstitial brachytherapy seed, *Med. Phys.*, 32, 3313, 2005.

50. Papagiannis, P., Sakelliou, L., Anagnostopoulos, G., and Baltas, D. On the dose rate constant of the selectSeed ^{125}I interstitial brachytherapy seed, *Med. Phys.*, 33, 1522, 2006.

10

Experimental Dosimetry

10.1 Introduction

The methodology for quantifying the strength of a brachytherapy source in terms of air kerma strength, S_K, measurement as well as the established dosimetry protocols for dose rate calculations around such a source based on its strength, have been extensively discussed in Chapter 7 and Chapter 8.

As mentioned in Chapter 8, for the determination of the different parameter values implemented in the internationally established TG-43[1] and TG-43 U1[2] dosimetry protocol experimental validation of MC simulation results is required. This applies to both the dose rate constant, Λ, that requires measurement of the absolute dose or dose rate, as well as the radial dose function $g(r)$ and anisotropy function $F(r,\theta)$ which are relative values as seen in Equation 8.24 and Equation 8.26.

Furthermore, the need to investigate and quantify the dose perturbation arising from simple or complex applicator geometries and applicator materials (plastic or metal), shielding materials, and tissue inhomogeneities, also necessitates the availability of appropriate experimental dosimetry systems in modern brachytherapy.

The radiation field around brachytherapy sources is characterized by (1) high dose gradients, (2) an extended dose rate range, and (3) photon energies typically lower than those in the standardized dosimetry of external beam fields, which spreads from those of high (^{60}Co and ^{137}Cs) and intermediate (^{198}Au, ^{192}Ir, and ^{169}Yb), to those of low energy radionuclides (^{125}I and ^{103}Pd). Given these three characteristics of the radiation field, experimental brachytherapy dosimetry does place severe demands on candidate detectors:[3]

- Wide dynamic range
- Flat energy response
- Small active volume

- High sensitivity
- Isotropic angular response

The size of a detector defines both the maximum spatial resolution that can be achieved as well as the minimum meaningful distance of measurement. Especially in intermediate and high energy radionuclides, a softening of the photon spectrum is observed with increased distance from the source due to the gradual build-up of scattered (Compton effect), lower energy photons and the attenuation of the primary emitted photons (see Figure 10.1 through Figure 10.3). Owing to this energy softening effect, the energy response of a detector becomes crucial with regards its suitability for experimental studies (see Figure 10.4). Finally, pronounced anisotropic angular response of detectors is expected to lead to underestimation of the dose with increased distance from the source since the fraction of backscattered photons increases with distance.

There is a variety of experimental dosimeters that have been used for dosimetry around brachytherapy sources, such as ionization chambers,[4-11] thermoluminescence dosimetry (TLD),[12-31] diodes,[32,33] plastic scintillators,[34-36] diamond detectors,[37-39] radiographic and radiochromic film,[40-44] polymer gels, and chemical dosimeters.[45-53]

Table 10.1 summarizes the sensitivity of several detectors based on the data published by Perera et al.[54,55] Based on these values, ionization chambers are not appropriate for measurements with low energy radionuclide sources such as ^{125}I and ^{103}Pd. Diodes (silicon) and plastic scintillators demonstrate higher sensitivity and very small dimensions thus allowing for an adequate spatial resolution. On the other hand, owing to the potential energy dependence of their response (Figure 10.4 and Figure 10.5) they have to be used carefully.[35] New types of polyvinyl toluene (PVT) mixtures with medium atomic number atoms very closely approximate the radiological properties of water (within 10%) in the energy range 0.020 to 0.662 MeV and yield improved energy response.[36] Unfortunately, there is no extended experience with such types of plastic detectors. Radiographic films such as Kodak X-V demonstrate a supralinearity of dose response for doses above some tens of cGy and a pronounced photon energy dependence for energies below 0.127 MeV (rising to a factor of 10 for photon

---

FIGURE 10.1

Photon energy spectra in liquid water at radial distances in the range 1.0 to 10.0 cm from ideal point sources of ^{103}Pd (a), ^{125}I (b), ^{170}Tm (c), ^{169}Yb (d), ^{192}Ir (e), ^{198}Au (f), ^{137}Cs (g), and ^{60}Co (h) radionuclides. These are the results of MC simulation calculations with the MCNP 4c Monte Carlo general code,[67] where the point sources were located in the center of a liquid water spherical phantom of an external radius of 80.0 cm. The energy spectra for these radionuclides used are according to the corresponding tables in Chapter 5. A maximum number of 2×10^8 particle histories were initiated for MC simulation study resulting in a maximum relative percentage deviation of 1σ less than 0.1%. In the inserts, the decrease of the mean energy, Equation 5.1, with the distance from the source for the corresponding radionuclide is shown.

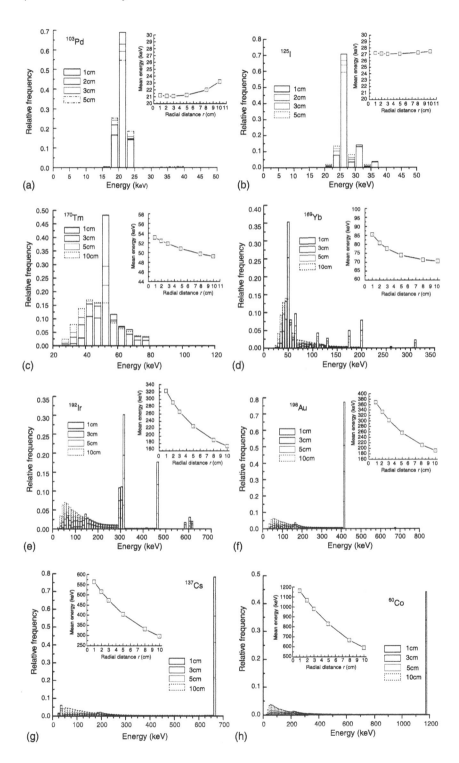

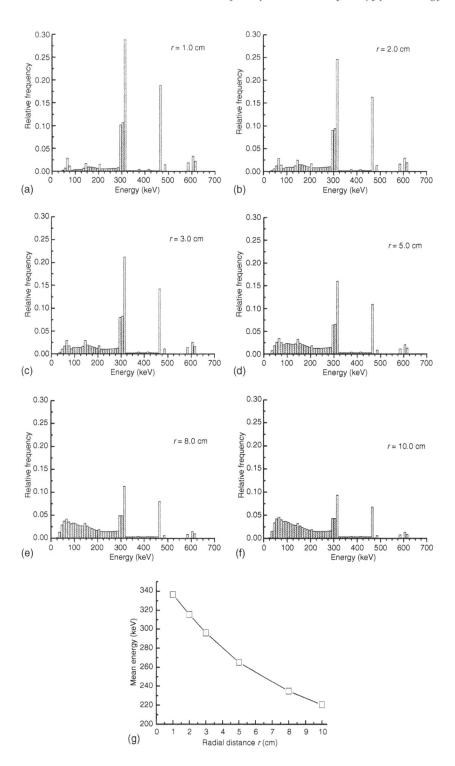

energies of 0.020 to 0.040 MeV).[40] Radiochromic films (GafChromic), on the other hand, present a linear dose response over an extended dose range (up to hundreds of Gy) and a much lower photon energy dependence with gaining sensitivity as photon energy rises in the low energy range below 0.127 MeV.[40] This gain is much lower than for that of radiographic films (for 0.020 to 0.040 MeV about 40 to 45% lower than for the photon energies of ^{137}Cs or ^{60}Co).[40,41,44] Despite these advantages, dosimetry using radiochromic films requires exposures at several Gy. For example, a dose above 100 Gy is required for low energy photons, such as those emitted by ^{125}I, to achieve an optical density of 1.0, even for the high-sensitivity type GafChromic films.[41] This severely limits the use of radiochromic films for dosimetry studies for low energy radionuclides.

Diamond detectors of small sensitive volume (1.8 to 1.9 mm^3) present promising properties for high-energy photon beams and for the intermediate energy of ^{192}Ir.[37-39] Their response has been shown to be nearly independent of incident photon energy for energies above 0.1 MeV since the ratio of carbon to water mass energy absorption coefficient is nearly constant in this energy region, as shown in Figure 10.5. This implies that for intermediate and high energy radionuclides (^{169}Yb, ^{192}Ir, ^{198}Au, ^{137}Cs, ^{60}Co) diamond detectors offer very low energy dependence on their response. For low energies, however, (0.020 to 0.030 MeV) a significant change in their energy response is expected, as shown in Figure 10.5 (50%). The angular dependence of diamond detector response is very low in both high energy photon beams (maximum 2%[37]) and the intermediate photon energies of ^{192}Ir (maximum 1.5%[39]). Furthermore, the temperature dependence of their response was found to be less than 2.5% in the range 14 to 40°C.[39] In addition, diamond detectors exhibit good spatial resolution due to their small sensitive volume and high sensitivity due to their high density. The combination of the properties discussed above makes diamond detectors appropriate for both relative and absolute dosimetry for intermediate and high energy brachytherapy sources. Unfortunately, there is only very limited experience with diamond detectors which is mainly due to their low availability and their high price. Finally, their validity for obtaining

---

FIGURE 10.2
Photon energy spectra in liquid water at 1.0, 2.0, 3.0, 5.0, 8.0, and 10.0 cm from the new design microSelectron HDR ^{192}Ir source (see Table 8.2 and Figure 8.16b). These are the results of MC calculations with the MCNP 4c Monte Carlo general code,[67] where the source was located in the center of a liquid water spherical phantom of an external radius of 15.0 cm. The energy spectrum used for ^{192}Ir is that listed in Table 5.7 and Table 5.8 in Chapter 5. A maximum number of 2×10^8 particle histories were initiated resulting in a maximum relative percentage deviation of 1σ less than 0.1%.[57] The spectra have been scored in spherical shells around the source at the specified distances. There is an increase of frequency for energies below 0.20 MeV and a decrease of frequency for energies above 0.40 MeV with distance. A significant decrease of the mean photon energy, Equation 5.1, with distance from the source in water can be observed (g).

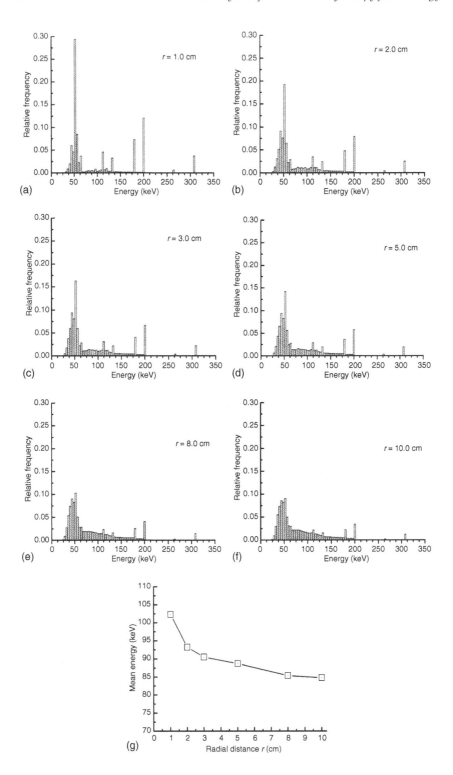

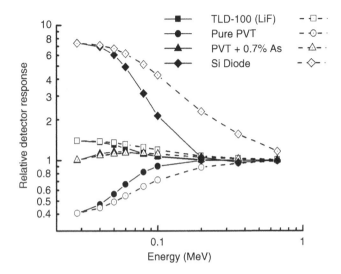

FIGURE 10.4

Theoretical detector response ratio defined as the ratio of the dose to detector material to the dose to water estimated using MC simulations for TLD-100 (LiF), pure PVT, arsenic-doped PVT, and Si diode detectors at 1.0 cm (closed symbols) and 10.0 cm (open symbols) distances in water from monoenergetic point sources with energies in the range 0.028 to 0.667 MeV. The detector response ratio for each detector type has been normalized to the corresponding value at 1.0 cm distance and for the ^{137}Cs photon energy (0.667 MeV). (These data are according to the work of Perera, H. et al., *Int. J. Radiat. Oncol. Biol. Phys.*, 23, 1059–1069, 1992, and have been kindly provided by Williamson, J.F., *Medical Radiology Diagonostic Imaging and Radiation Oncology*, Springer, Berlin, pp. 247–302, 1995.)

brachytherapy dosimetry parameters has not been convincingly demonstrated in the vicinity of low energy photon emitting brachytherapy sources.

Ionization chambers, diodes, plastic scintillators, and diamond detectors have the advantage of allowing direct measurement in water phantoms, thus obviating the need for solid-state phantoms and the influence of corresponding corrections and their associated uncertainties.

LiF thermoluminescence dosimetry is currently the method of choice for both absolute and relative dose rate measurement in the entire energy range of brachytherapy sources.

FIGURE 10.3

Photon energy spectra in liquid water at 1.0, 2.0, 3.0, 5.0, 8.0 and 10.0 cm from a hypothetical [169]Yb HDR source[57] with the same geometrical design as the new microSelectron HDR ^{192}Ir source (see Table 8.2 and Figure 8.16b), with the active core consisting of pure [169]Yb. These are the results of MC calculations with the MCNP 4c Monte Carlo general code,[67] where the same simulation geometry and methodology described in Figure 10.2 were applied. The energy spectrum used for [169]Yb is that listed in Table 5.12 in Chapter 5. A significant decrease of the mean photon energy, Equation 5.1, with the distance in water can be observed (g).

TABLE 10.1

Absolute Response Characteristics of 1.0 mm^3 Detector to 0.001 Gy Absorbed Dose

Detector	Energy Dissipated Per Created Quantum (eV)	No. Quanta Emitted	Typical Quantum Efficiency × Geometric Collection Efficiency	Practical Response Relative to Ionization Chamber
LiF-TLD-100	8400	1.8×10^6	0.20×0.08	0.12
Silicon diode	3.6	4.0×10^9	1.00×1.00	1.7×10^4
Ionization chamber (air)	33.8	2.4×10^5	1.00×1.00	1.00
Plastic scintillator	100	6.2×10^7	0.20×0.05	2.60

Source: Data are according to Perera, H. et al., *Int. J. Radiat. Oncol. Biol. Phys.*, 23, 1059–1069, 1992.

This chapter focuses on three experimental dosimetry methods: (1) the standardized ionization dosimetry and (2) TLD dosimetry, as well as the emerging technique of (3) polymer gel dosimetry, which is under investigation, and has the advantages of full three-dimensional dosimetry with fine spatial resolution and adaptive detector geometry and size.[53]

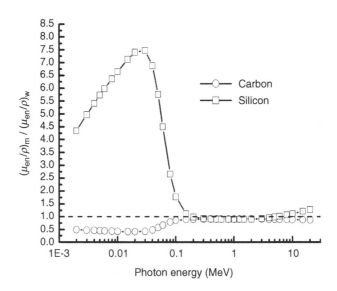

FIGURE 10.5

Ratio of mass energy absorption coefficients of carbon and silicon to that of water, $(\mu_{en}/\rho)_m/(\mu_{en}/\rho)_w$, depending on photon energy. The corresponding values of the mass absorption coefficients have been calculated using the elemental interaction coefficients by the XCOM database[56] available online at NIST.

10.2   Phantom Material

The reference medium for the dosimetry of brachytherapy sources is water.[2,58,59] Rivard et al.[2] defined pure, degassed water, composed of two parts hydrogen atoms and one part oxygen atoms, with a mass density of $\rho = 0.998 \, \text{g cm}^{-3}$ at 22°C as the recommended water composition for reference dosimetry. In the TG-43 U1 report,[2] the reference conditions for dry air as a dosimetry material are given as 22°C and 101.325 kPa (760 mmHg) with a mass density of $\rho = 0.001196 \, \text{g cm}^{-3}$. In this report a relative humidity of 40% and an average density of $\rho = 0.00120 \, \text{g cm}^{-3}$ for air are also recommended for MC calculations. Table 10.2 lists the composition of air depending on relative humidity in the range of 0 to 100% and for the reference pressure of 101.325 kPa based on the data published in the TG-43 U1.

LiF TLDs are currently considered as the dosimeter of choice for experimental dosimetry in the entire energy range of brachytherapy sources. Several solid state materials have been investigated as water substitutes for brachytherapy experimental dosimetry using TLD detectors.[2,58-66] Given that accuracy in source-detector positioning is crucial in dosimetry of brachytherapy sources due to the high dose gradients in their vicinity, solid phantoms that can be precisely machined have also been considered for use with other kinds of detectors, such as ionization chambers, for intermediate and high energy radionuclides.[58,59] Appropriate solid phantoms can also facilitate measurements using film dosimeters, which normally cannot be used in liquid water medium. The physical properties and the chemical composition of all common dosimetry materials, solid water, plastic water, PMMA, polystyrene, and RW3 (polystyrene plus 2% TiO_2) are listed in Table 7.5 in Chapter 7.

Experimental dosimetry in solid phantoms is more convenient than in water. However, the dosimetric characteristics of these materials depend on the energy spectrum of the emitted photons and thus on the specific source under consideration. Furthermore, owing to the energy softening effect with increasing distance from a source (see Figure 10.1 through Figure 10.3), the

TABLE 10.2

Composition Given as Percent Mass of Air as a Function of Relative Humidity at the Reference Pressure of 101.325 kPa

Relative Humidity (%)	H (Z = 1)	C (Z = 6)	N (Z = 7)	O (Z = 8)	Ar (Z = 18)
0	0.0000	0.0124	75.5268	23.1781	1.2827
10	0.0181	0.0124	75.4048	23.2841	1.2806
40	0.0732	0.0123	75.0325	23.6077	1.2743
60	0.1101	0.0123	74.7837	23.8238	1.2701
100	0.1842	0.0122	74.2835	24.2585	1.2616

Source: Data are according to TG-43 U1, Rivard, M. et al., *Med. Phys.*, 31, 633–674, 2004.

dosimetric characteristics of a specific phantom material are expected to be also distance dependent.

What is of practical interest is the correction required for absorption and scattering of the radiation at a specific radial distance r from a source in a phantom when compared with liquid water, $k_{m,w}$, that can be calculated using the following:

$$k_{m,w}(r) = \frac{\dot{D}_m(r)}{\dot{D}_w(r)} \qquad (10.1)$$

where $\dot{D}_m(r)$ is the dose rate to water in a medium (phantom material) per unit source strength at a distance r from the source and $\dot{D}_w(r)$ is the dose rate to water in water per unit source strength at the same distance. Although the energy shifting effect is generally distance and polar angle dependent for cylindrical sources, which is especially the case for the intermediate and high energy radionuclides, for reasons of simplification only the radial distance dependence is considered herein. This simplification, however, does not significantly influence or limit the results of the following discussion.

Figure 10.6 presents the partial and total mass attenuation, (μ/ρ), coefficients, the mass energy absorption, (μ_{en}/ρ), coefficients as well as selected ratios plotted vs. energy in the energy range of interest to brachytherapy (0.01 to 2.0 MeV). Plotted data were calculated using elemental interaction coefficients by the XCOM database,[56] available online at NIST, weighted by the atomic weight fractions given in Table 7.5 for the different materials according to the Bragg additivity rule as in Equation 4.13 and Equation 4.42. Figure 10.6 is presented to facilitate the following discussion. For example, it should be noted that when the dose rate to a medium in the same medium is considered, this also includes the differences in the mass energy absorption coefficients, μ_{en}/ρ, between medium and water. This difference is demonstrated in Figure 10.6e, where the ratio of the coefficient for medium to that of water, $(\mu_{en}/\rho)_m/(\mu_{en}/\rho)_w$, is plotted vs. photon energy in the range of 0.005 to 2.0 MeV.

The majority of the available literature on phantom materials is, of course, related to the low energy sources, ^{103}Pd and ^{125}I.[60–63,65] Some of the phantom materials considered herein have been altered in terms of their chemical composition over time, i.e., solid water and plastic water materials. There are only data, published by Luxton[64] on the correction factor described in Equation 10.1 for radial distances up to 10.0 cm for PMMA, solid water, and RW-1 material for the energy spectra of a point ^{103}Pd source and two point sources with the spectrum of ^{125}I model 6702 and 6711 seeds. Meigooni et al.,[63] published figures showing the dependence of the correction factor $k_{m,w}$ on both distance and energy for monoenergetic point sources in the range 0.010 to 1.0 MeV for plastic water and solid water phantom materials. The data published by Meli et al.[61] for a HDR ^{192}Ir source in PMMA, solid

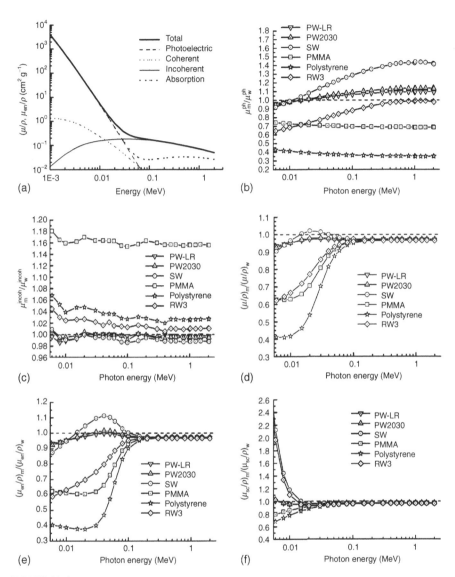

FIGURE 10.6

Interaction and absorption coefficients for the different phantom materials depending on photon energy. (a) Photoelectric, coherent, incoherent (Compton), total mass interaction, μ/ρ, and mass energy absorption, μ_{en}/ρ, coefficients for water. (b) and (c) Ratio of photoelectric, μ_m^{ph}/μ_w^{ph}, and incoherent, $\mu_m^{incoh}/\mu_w^{incoh}$, interaction coefficients of the phantom materials to that of water. (d)–(f) Ratio of total mass attenuation, $(\mu/\rho)_m/(\mu/\rho)_w$, energy absorption, $(\mu_{en}/\rho)_m/(\mu_{en}/\rho)_w$, and scattering, $(\mu_{sc}/\rho)_m/(\mu_{sc}/\rho)_w$, coefficients of the phantom materials to that of water. The scattering coefficient, μ_{sc}/ρ, is defined as the difference between the corresponding attenuation and absorption coefficient. All coefficient values have been calculated using the elemental interaction or absorption coefficients by the XCOM database[56] available online at NIST weighted by the atomic weight fractions given in Table 7.5 for the different materials according to the Bragg additivity rule in Equation 4.13 and Equation 4.42.

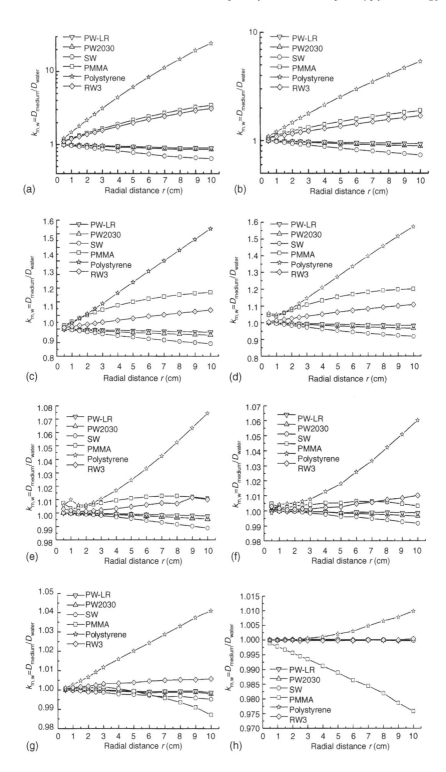

water, and polystyrene offer no specific information on the distance dependence of the correction factor $k_{m,w}$.

Based on the energy spectra listed in Chapter 5 for all seven common brachytherapy radionuclides and using the MCNP 4C Monte Carlo code,[67] simulations were carried out for candidate phantom materials for brachytherapy experimental dosimetry: the two plastic water types, PW2030 and PW-LR, solid water, PMMA, polystyrene, RW3, and liquid water. For each radionuclide an ideal point source was located at the center of a spherical phantom of an external radius of 80.0 cm to simulate an unbounded phantom geometry, consisting each time of the phantom material in question. The elemental composition as well as the mass density of the phantom materials were as listed in Table 7.5. The photon transport was simulated and the dose to water within each phantom material was assumed equal to the water kerma as scored by weighting the energy fluence crossing spherical shells situated at selected radial distances of $r = 0.5, 1, 1.5,$ 2, 2.5, 3, 4, 5, 6, 7, 8, 9, and 10 cm with the corresponding mass energy absorption coefficients of water. A maximum number of 2×10^8 particle histories were initiated resulting in a maximum relative percentage deviation of 1σ less than 0.1%.

The results of these simulations are summarized in Figure 10.7. In order to consider these figures, a short discussion on the role of the physical properties and elemental composition of the different materials in relation to photon energy is required. The differences in the phantom materials relative to water are mainly due to the elemental composition and physical density of these materials.

For low Z materials and in the energy range above 0.1 MeV the incoherent scattering (Compton scattering, see also Section 4.2) dominates the corresponding linear attenuation coefficient:

$$\mu^{\text{incoh}} \approx \mu^{\text{KN}} \propto \rho \left(\frac{Z}{A}\right)_e \sigma^{\text{KN}} \tag{10.2}$$

where $_e\sigma^{\text{KN}}$ is the electronic Klein–Nishima (KN) scattering, as in Equation 4.30, and Z/A is the ratio of atomic to mass number. In the above equation

---

FIGURE 10.7
Correction required for absorption and scattering of the radiation depending on the radial distance r from a point source in a phantom when compared with liquid water, $k_{m,w}$ according to Equation 10.1. All five phantom materials, plastic water (PW-LR and PW2030), solid water (SW), PMMA, Polystyrene and RW3, have been considered. The physical density and the elemental composition of the phantom materials are listed in Table 7.5 and Table 10.3. These are MC simulation calculation results for ^{103}Pd (a), ^{125}I (b), ^{170}Tm (c), ^{169}Yb (d), ^{192}Ir (e), ^{198}Au (f), ^{137}Cs (g), and ^{60}Co (h) radionuclides, where the energy spectra used are those listed in the tables in Chapter 5. The MCNP 4c Monte Carlo general code[67] has been used, where the point sources were located in the center of a liquid water spherical phantom of an external radius of 80.0 cm achieving practically an unbounded geometry. A maximum number of 2×10^8 particle histories were initiated for MC simulation study resulting in a maximum relative percentage deviation of 1σ less than 0.1%.

$\rho_e = \rho(Z/A)$ is the electronic density of the concrete element, which also quantifies the material dependence of the incoherent scattering. For a mixture or compound the mean ratio $\langle Z/A \rangle$ is used, calculated using the Bragg additivity rule:

$$\left\langle \frac{Z}{A} \right\rangle = \sum_i w_i \left(\frac{Z}{A} \right)_i \qquad (10.3)$$

where w_i is the fractional elemental composition for the ith element with atomic to mass number ratio $(Z/A)_i$. Since the ratio Z/A is nearly constant (with the exception of hydrogen) the incoherent interaction coefficient depends mainly on the density.

In the low energy region, the relative importance of the photoelectric absorption, which is the predominant interaction mode, depends on the elemental composition of the material.

The concept of the effective atomic number, Z_{eff}, which has no sound theoretical foundation,[53,68–74] even when calculated in the more general form, is a crude yet useful tool for lower photon energies where the photoelectric effect predominates, since for the corresponding linear attenuation coefficient, see also Equation 4.16:

$$\mu^{ph} \propto \rho \left(\frac{Z}{A} \right) Z^m \qquad (10.4)$$

ρ is the physical density, where $3 < m < 4$ due to which photoabsorption does not present the same energy dependence for all Z or the same Z dependence for all energies. Therefore, the photoelectric effective atomic number values, Z_{eff}^{ph}, for the materials considered herein were calculated using the Bragg additivity rule for a mixture or compound:

$$Z_{eff}^{ph} = \sqrt[m]{\sum_i \alpha_i Z_i^m} \qquad \text{where} \qquad \alpha_i = \frac{w_i \left(\frac{Z}{A} \right)_i}{\sum_i w_i \left(\frac{Z}{A} \right)_i} \qquad (10.5)$$

For $m = 3.5$[53,74] we have

$$Z_{eff}^{ph} = \sqrt[3.5]{\sum_i \alpha_i Z_i^{3.5}} \qquad (10.6)$$

Usually the effective atomic number, Z_{eff}, is expressed as

$$Z_{eff} = \sum_i w_i Z_i \qquad (10.7)$$

Both effective atomic numbers, Z_{eff} and Z_{eff}^{ph}, together with the physical density, mean ratio $\langle Z/A \rangle$, and the electron densities for the phantom materials considered herein are listed in Table 10.3, based on the elemental compositions given in Table 7.5.

TABLE 10.3

Physical Density, Electronic Density (Table 7.5) and Effective Atomic Numbers (Equation 10.4 and Equation 10.5) for the Different Phantom Materials

Material	Density ρ (g cm^{-3})	Mean Ratio of Atomic Number to Atomic Mass $\langle Z/A \rangle$	$\rho\langle Z/A \rangle$	Z_{eff} (Equation 10.7)	Z_{eff}^{ph} (Equation 10.6)	$\rho\langle Z/A \rangle(Z_{eff}^{ph})^{3.5}$
Plastic water (PW or PW2030)	1.022	0.540468	0.552	6.60	7.61	671.1
Plastic water (PW-LR)	1.029	0.538155	0.554	6.65	7.51	643.1
Solid water	1.017	0.539470	0.549	6.36	7.94	774.4
Polymethyl methacrylate (PMMA)	1.190	0.539369	0.642	6.24	6.56	464.2
Polystyrene	1.060	0.537680	0.570	5.61	5.74	258.3
RW3	1.045	0.536452	0.561	5.83	7.13	543.0
Water liquid H$_2$O	1.000	0.555087	0.555	7.22	7.51	644.2

Elemental composition of these materials is listed in Table 7.5.

In this table, solid water has, among all materials, the highest effective atomic number. In the low energy region, a higher attenuation in solid water is expected due to the higher density and higher effective atomic number which declares the dominance of the photoelectric absorption (see Figure 10.6b). Therefore, dose rate measurements in solid water underestimate the dose in water as shown by the MC results in Figure 10.7a and Figure 10.7b for ^{103}Pd and ^{125}I radionuclides. At a radial distance of 1.0 cm there is an underestimation of 4 and 2% relative to 23 and 15% at 5.0 cm from the ^{103}Pd and ^{125}I point sources, respectively. As energy increases, the solid water material is expected to approximate water better since the incoherent scattering gradually becomes the dominating interaction process (see Figure 10.6a and Figure 10.6c) and the electron density of solid water is comparable to that of water. The differences observed in Figure 10.7d–Figure 10.7h are mainly attributed to the slightly higher density of solid water. For example, measurements in solid water at a radial distance of 1.0 cm are equivalent to those in liquid water for both ^{169}Yb and ^{192}Ir radionuclides, whereas at a distance of 5.0 cm an underestimation of the dose in water by 5 and 0.4%, respectively, is observed for these two radionuclides.

In the case of the two types of plastic water, PW and PW-LR, the effective atomic number as well as the electron density are in good agreement with those in water (see Figure 10.6b). In the low energy region of ^{103}Pd and ^{125}I radionuclides, it is expected that these materials better approximate water than solid water, and the differences observed in Figure 10.7a and Figure 10.7b are due to their higher density. For PW there is an underestimation of the dose in water of only 2.0 and 1.3% at 1.0 cm from the ^{103}Pd and ^{125}I point sources, respectively, whereas for PW-LR the corresponding differences are reduced to 1.4 and 0.7%. At 5.0 cm, PW results in an underestimation of the dose in water by 9.0 and 6.0%, and PW-LR by 6.0 and 4.0% for the ^{103}Pd and ^{125}I radionuclides, respectively. Similarly to solid water, owing to the close agreement of electron density values of these materials with water, as energy increases dose rate measurements in these materials converge with that in water, and thus any differences observed in Figure 10.7d–Figure 10.7h are mainly attributed to their higher density. For the case of ^{169}Yb there is an underestimation by only 2.0% for PW and 1.0% for PW-LR at the radial distance of 5.0 cm. At 1.0 cm both materials show a perfect water equivalence (differences less than 0.3%). PW and PW-LR plastic water materials also demonstrate a perfect water equivalence for both 1.0 and 5.0 cm distances for the energy spectrum of ^{192}Ir (differences within the uncertainties of the MC simulation results).

As shown in Table 10.3, PMMA, polystyrene, and RW3 present lower effective Z and higher electron density compared with water. Hence, the attenuation in these materials is lower than in water medium in the low energy region besides the fact that their density is significantly higher than that of water (see also Figure 10.6b). In Figure 10.7a and Figure 10.7b it can be seen that dose rate measurements in PMMA, polystyrene, and RW3 for ^{103}Pd and ^{125}I radionuclides overestimate the dose rate in water. For ^{103}Pd there is an overestimation of the dose in water by 22.0, 48.0, and 20.0% at 1.0 cm in PMMA, polystyrene, and RW3, whereas at 5.0 cm these values increase dramatically to 117.0, 516.0, and 100.0%, respectively. For ^{125}I the overestimation of the dose in water at 1.0 cm is 12.0, 21.0, and 9.0% for PMMA, polystyrene, and RW3, and increases to 51.0, 154.0, and 37.0%, respectively, at 5.0 cm. The highest overestimation is observed, as expected, for polystyrene that demonstrates by far the lowest attenuation coefficient among all six materials at the low energy region (see Table 10.3 and Figure 10.6d). These differences are gradually reduced as energy increases (see Figure 10.7d–Figure 10.7h) and, in fact, use of PMMA as a phantom material results in a slight underestimation of the dose rate in water for the high energy radionuclides ^{137}Cs and ^{60}Co, which can be explained by its higher electron density $\rho(Z/A)$ (see also Figure 10.6c). At 5.0 cm there is an underestimation of the dose in water by 1.1% for PMMA and for ^{60}Co. Use of polystyrene as phantom material results in overestimation of the dose even for the high energy radionuclides ^{137}Cs and ^{60}Co, whereas for ^{60}Co this occurs only for radial distances above 3.0 cm (see Figure 10.7h). At 5.0 cm from the source and in the polystyrene phantom there is an overestimation of 2.0% for ^{137}Cs, whereas for ^{60}Co there is only a slight overestimation of 0.2%.

A summary of the evaluation of water equivalence of the different phantom materials for several radionuclides, based on the MC results of Figure 10.7, is shown in Table 10.4. Owing to the fact that the dosimetric differences of the materials when compared to liquid water are most pronounced for photon energies below 0.1 MeV and that the changes of the emitted photon spectrum due to the source construction (active core and encapsulation) are most evident at that low energy region exact distance correction factors $k_{m,w}$ have to be calculated for the specific design of a source with such photon energies.

Finally, according to the scope of the experimental measurements and to the involved distances, adequate sizes for the phantoms have to be selected in order to warrant full scatter conditions.[75] This depends on both the energy of the radionuclide and the phantom material and it is correlated with the mean free path (mfp) of photons in this material (see also Section 4.1). In the case that this condition is not met, additional corrections for the missing scatter have to be considered for the actual experimental set-up.

10.3 Ionization Dosimetry

Ionization chambers can be used in either water phantoms or solid material phantoms for the measurement of relative dose distributions as well as of absolute dose rate and dose values. In general, ionization chambers with small dimensions and thus of small collecting volumes, usually in the range 0.01 to 0.6 cm^3, should be considered in order to achieve appropriate spatial resolution[76–80] and keep dose gradient-related effects at an acceptable level. Thus, an optimum chamber volume has to be selected to achieve, on the one hand, adequate spatial resolution (the smaller the volume the better), and on

TABLE 10.4

Summary of the Water Equivalence of the Different Phantom Materials for the Seven Brachytherapy Radionuclides of Interest Subdivided According to the Energy of the Emitted Photons

Material	Low Energy Radionuclides ^{103}Pd, ^{125}I	Intermediate Energy Radionuclides ^{170}Tm, ^{169}Yb	Intermediate Energy Radionuclides ^{192}Ir, ^{198}Au	High Energy Radionuclides ^{137}Cs, ^{60}Co
Plastic water (PW or PW2030)	++++ distance dependent $k_{m,w}$	+++++ distance dependent $k_{m,w}$	+++++ slightly distance dependent $k_{m,w}$	+++++ distance independent $k_{m,w}$
Plastic water (PW-LR)	++++ distance dependent $k_{m,w}$	+++++ distance dependent $k_{m,w}$	+++++ slightly distance dependent $k_{m,w}$	+++++ distance independent $k_{m,w}$
Solid water	+++ distance dependent $k_{m,w}$	++++ distance dependent $k_{m,w}$	+++++ slightly distance dependent $k_{m,w}$	+++++ distance independent $k_{m,w}$
Polymethyl methacrylate (PMMA)	—	+++ distance dependent $k_{m,w}$	+++++ slightly distance dependent $k_{m,w}$	++++ distance dependent $k_{m,w}$
Polystyrene	—	—	+++ distance dependent $k_{m,w}$	++++ distance dependent $k_{m,w}$
RW3	—	++++ distance dependent $k_{m,w}$	+++++ slightly distance dependent $k_{m,w}$	+++++ distance independent $k_{m,w}$

$k_{m,w}$ is the correction factor required for absorption and scattering of the radiation at a specific radial distance r from a source in a phantom when compared to liquid water as defined in Equation 10.1.

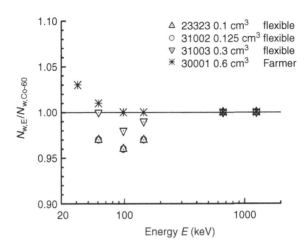

FIGURE 10.8

Photon energy dependence of the calibration factor $N_{w,E}$ for the dose/dose rate to water for different x-ray and γ-ray energies and for four PTW ionization chambers with volumes 0.1 to 0.6 cm^3 (geometrical and material characteristics are described in Table 7.6). The calibration factors for each chamber have been normalized to the corresponding calibration factor for the reference mean γ-ray energy of ^{60}Co of 1.25 MeV. These are typical relative calibrations factors reported by the manufacturer. (The effective energies of the x-ray beams used by PTW are taken from Pychlau, P. *Zeit. Med. Physik*, 1, 194–198, 1991.)

the other hand, a proper chamber signal (the larger the volume the better). Depending on the source strength and due to their small collecting volume (lower chamber response), certain restrictions, which will result in proper chamber readings, will be placed on the maximum possible measurement distance.

In addition, the chamber should demonstrate a flat angular response over the angle range relevant to the experimental set-up.

Chambers with the lowest possible energy dependence in their response at the energy region of interest should be considered, especially for intermediate energy radionuclides, such as ^{169}Yb and ^{192}Ir where a clear energy softening effect is expected (compare here Figure 10.1 to Figure 10.3). Figure 10.8 demonstrates typical energy dependence of some commercially available chambers with small collecting volumes in the range 0.1 to 0.6 cm^3.

10.3.1 Measurement of Dose or Dose Rate

The estimation of the required dose to water in water, D_w, from the ionization chamber reading depends on the available calibration factor N of the chamber used.

The most commonly used calibration protocol for ionization dosimetry in radiotherapy is that of dose to water in water.[58,59,82–88] Thus, for most ionization chambers available in radiotherapy departments, calibration factors in terms of dose to water in water, N_w, for a reference beam quality are available. Usually, ^{60}Co beams are considered as a reference quality.

In general, the dose to water in water medium, D_w, expressed in Gy can be determined from the detector measurement M in a medium m (phantom material) using the following equations.

Chamber calibrated in absorbed dose to water

$$D_w = N_w M k_\rho k_T k_p k_{ion} k_V k_{appl} k_{m,w} k_{wp} \tag{10.8}$$

Chamber calibrated in exposure

$$D_w = N_X \left(\frac{W}{e}\right)\left(\frac{\mu_{en}}{\rho}\right)_\alpha^w k_{RC} M k_\rho k_T k_p k_{ion} k_V k_{appl} k_{m,w} k_{\alpha\rho} \tag{10.9}$$

Chamber calibrated in air kerma

$$D_w = N_K (1 - g_\alpha)\left(\frac{\mu_{en}}{\rho}\right)_\alpha^w M k_\rho k_T k_p k_{ion} k_V k_{appl} k_{m,w} k_{\alpha\rho} \tag{10.10}$$

N_w: is the absorbed dose in-water calibration factor of the chamber for the -energy of the radionuclide considered, usually expressed in Gy C^1.

N_X: is the exposure calibration factor of the chamber for the γ-energy of the radionuclide considered, usually expressed in R C^{-1}.

N_K: is the air kerma calibration factor of the chamber for the -energy of the radionuclide considered, usually expressed in Gy C^1.

M: is the measured charge (C).

k_ρ: is the correction factor for the current air pressure conditions P, expressed in hPa, other than that referred in the calibration protocol of the chamber, that is the normal pressure of 1013 hPa:

$$k_\rho = \left(\frac{1013 \text{ hPa}}{P}\right)$$

k_T: is the correction factor for the current air temperature conditions T in °C other than that referred in the calibration protocol of the chamber, i.e., the normal temperature T_0 in °C.

$$k_T = \left(\frac{273.15 + T(°C)}{273.15 + T_0(°C)}\right)$$

This is usually $T_0 = 20$°C. Alternatively to air pressure and temperature correction factors, a low activity check radioactive source (^{90}Sr) can been used to determine $k_\rho k_T$ according to DGMP recommendations;[56,59,82] the same source as used in the calibration protocol.

k_p: is the correction factor for the polarity effect of the bias voltage for the photon energy of the radionuclide, see also Equation 7.6 and Equation 7.7 in Chapter 7.

k_{ion}:   is the correction factor that accounts for the unsaturated ion collection efficiency and thus for the charge lost to recombination for the specific radionuclide photon energy and the applied nominal voltage V. k_{ion} is the reciprocal of the ion collection efficiency A_{ion}. For details see Equation 7.6 and Equation 7.7.

k_V: is a correction factor to account for the effect of the chamber's finite size (i.e., volume) when the center of the chamber air cavity volume is considered as the point of reference for the positioning of the chamber and thus as a reference for the measurement result.

k_{appl}:   is the correction factor to account for the attenuation in the applicator/catheter wall used for positioning or fixing the source in the in-phantom set-up when the measured dose has to exclude this (see also Chapter 7).

$k_{m,w}$:   is the correction required for absorption and scattering of the radiation at the specific point of measurement in the phantom material when compared with liquid water (see also Equation 10.1). This is, in general, radionuclide and source design, especially for the low energy sources, distance and polar angle dependent. $k_{m,w}$ has to be estimated experimentally or using MC simulation calculations for the specific measurement set-up.

$k_{\alpha p}$:   is the perturbation correction factor accounting for differences when changing from a surrounding medium of air to one of a phantom material. This is chamber type dependent. For ^{192}Ir and PMMA material a good approximation is $k_{\alpha p} = 1.0$.[58,59]

k_{wp}:   is the perturbation correction factor accounting for differences when changing from a surrounding medium of water to one of a phantom material. This is chamber type dependent. For phantom materials with density and effective atomic number which closely approximate that of water, $k_{wp} = 1.0$.[58,59,82]

g_α: is the energy fraction of the electrons which are liberated by photons in air that are lost to radiative processes (bremsstrahlung). g for different photon energies is given in Table 7.9 and is, for the photon energies relevant to brachytherapy, practically zero.

$\dfrac{W}{e}$:   is the ionization constant, $W/e = 33.97 \, \mathrm{J \, C^{-1}}$ and is energy independent for energies above 1 keV (see Table A.2.2 in Appendix 2), with e being the elementary charge and W the mean energy expended in air per ion pair formed.

k_{RC}:   is the roentgen to $\mathrm{C \, kg^1}$ conversion factor (see Table A.2.1 in Appendix 2 and Chapter 7, Equation 7.19), with

$$k_{RC} = 2.58 \times 10^{-4} \, \mathrm{C \, kg^{-1} \, R^{-1}}$$

$\left(\frac{\mu_{en}}{\rho}\right)_{\alpha}^{w}$:   is the ratio of mass energy absorption coefficient of water to that of air. In Table 7.15 the reciprocal values of this ratio are given for the different brachytherapy radionuclides based on the corresponding effective energies listed in Table 5.2. The energy dependence of this ratio in the energy range 0.01 to 2.0 MeV is shown in Figure 8.4 in Chapter 8. The photon energy spectrum at the place of measurement has to be considered for an accurate estimation of $(\mu_{en}/\rho)_{\alpha}^{w}$, see also Equation 4.13 and Equation 4.42.

10.3.1.1 Calibration Factors N_K, N_W, and N_X

For calibration factors, the discussion and methods described in Chapter 7 for the source calibration, are also valid here. It is though, emphasized that, in general, additional calibration points (i.e., calibration factors at specific photon energies) to those usually given for ^{60}Co or ^{137}Cs are required, depending on the energy of the radionuclide being considered (see also Figure 10.8).

10.3.1.2 Chamber Finite Size Effect Correction Factor k_V

In Chapter 7 the correction factor regarding the radiation field inhomogeneity within the chamber volume has been extensively discussed. Table 7.7 summarizes such correction factors for the different cylinder or almost cylinder-shaped ionization chambers for the sideward irradiation geometry (Figure 7.6) based on Equation 7.14.

In the case of irradiation geometries other than that of the sideward one, detailed calculation of the corresponding gradient correction factor value k_V has to be realized.

10.3.1.3 Phantom Dimensions

Depending on the radionuclide, on the dimensions of the used phantom, and on the measurement geometry, additional correction for missing scattering at chamber positions lying at the periphery of the phantom (little backscatter material available) could become necessary for the individual experimental set-up.

10.3.1.4 Room Scatter Effects

When measurement geometries are utilized with the chamber positioned near the border of the phantom (water or solid phantom), attention must be paid to the potential influence of scattered radiation on the measurement originating on the wall or floor of the room. For small distances from, for example the wall of the room, an increase in the chamber reading of some percents can be the case (see Figure 7.27 for PMMA phantoms and ^{192}Ir sources). This can be generally avoided by keeping a minimum distance of 1.0 m from all possible scattering sources.

10.3.1.5 Other Effects

As has been discussed in Chapter 7 regarding the source calibration with well-type chambers or using solid phantoms, it is important that the phantom temperature is in equilibrium with the ambient/air temperature. Furthermore, if a radioactive check source is used to obtain the $k_p k_T$ correction, then it must be ensured that this source has also achieved room air temperature. In any other case correction factors must be implemented to account for these temperature differences.

10.3.1.6 Calculating Dose Rate from Dose

When the dose rate values, $\dot{D}_w$, are of interest, these can be calculated based on the estimated dose to water in water D_w, Equation 10.8 through Equation 10.10, using the following equation:

$$\dot{D}_w = \frac{D_w}{\tau} k_u \qquad (10.11)$$

τ :   is the time interval during which the electrometer reading (charge) was taken. For high dose rate sources such as [169]Yb, ^{192}Ir, and ^{60}Co and distances up to 10.0 cm, τ is normally in the order of some minutes.

k_u: is a unit conversion factor, to convert the result of Equation 10.11 to the required dose rate unit, usually cGy h^{-1}, cGy min^{-1}, or Gy min^{-1}, depending on the strength of the source.

10.4 TLD Dosimetry

10.4.1 Introduction

McKeever[89] has commented that it is difficult to pinpoint exactly when the word thermoluminescence (TL) was first used in the published literature, but it was certainly used in 1895 by Wiedemann and Schmidt.[90] Then, Marie Curie in her doctoral thesis of 1904[91] showed that radium can restore to their TL property to certain thermoluminescent materials. TL has, therefore, been studied for more than a century.

Among the various thermally stimulated processes, TL was probably the first to be observed: by Robert Boyle in 1663[89,92] and maybe even earlier by medieval alchemists.[89] Certainly, TL is now the most widely utilized of these processes. TL can be defined as the phenomenon of visible light emission caused by heating a dielectric material that has previously absorbed energy from an external source. It should not be confused with the spontaneous light emission resulting from heating to incandescence. TL heating is simply the triggering mechanism for the liberation of energy that has been previously stored in the material.

Given that the external source of energy absorbed by TL material can be ionizing radiation and that a large number of dielectric minerals which exhibit natural TL exist, the potential use of TL for dosimetry purposes was realized early. The first proper application of TLD, however, was in 1953[89] when LiF was used to measure radiation following an atomic weapon test. It was therefore in the late 1950s that research on materials and models for radiation-induced TL began to escalate and resulted in the commercial availability of TL dosimeters (TLDs) and the necessary associated instrumentation.

10.4.2 Summary of the Simplified Theory of TL

A simplified theory of TL is based on the energy band model in simple crystalline solids. When atoms are brought together to form a crystal their discrete electron energy levels split into energy bands: the valence band and the conductivity band. In the case of a semiconductor or an insulator in its ground state, the valence band is fully occupied by valence electrons while the conductivity band is empty. The two bands are separated by an energy gap of sufficient magnitude so that energy absorption is required for electrons to move from the valence to the conductivity band. When the crystal absorbs energy, part of it is spent on electron excitation. The process of electron deexcitation accompanied by light emission is known as luminescence. For the terminology of different luminescence phenomena a first term is added to describe the type of excitation energy. Examples are chemoluminescence, photoluminescence, triboluminescence, electroluminescence, and radioluminescence.

In an ideal perfect crystal excited electrons would promptly return to their ground state with light emission occurring after a characteristic time $\tau < 10^{-8}$ sec. This process is known as fluorescence. However, a perfect crystal does not exist and defects in the crystal lattice periodicity are relatively common. These defects may be due to thermal or inherent imperfections and crystal impurities or even be radiation induced. In the case of a defect causing a localized surplus of positive charge, an electron trap is said to be formed that gives rise to a permitted metastable energy level lying between the valence and conductivity energy bands. It is also possible that an electron transition occurs between the conductivity band and an electron trap. Since direct deexcitation from a metastable energy level is not permitted, the existence of these traps delays the effect of luminescence as an amount of energy, E, equal to the difference between the minimum conductivity band energy and the energy level of the trap, is required for the electron to be excited out of the trap. This delayed luminescence process is termed phosphorescence.

The probability of a trapped electron escaping is proportional to the exponent of $(-E/kT_0)$ where k is the Boltzmann constant and T_0 is the temperature during excitation. This is useful when the characteristic time at which luminescence is observed cannot definitively classify the luminescence as fluorescence or phosphorescence. While fluorescence is temperature independent, a rise of temperature increases the rate of phosphorescence. If the electron trap is deep enough, such that $E \gg kT_0$, the probability of electrons escaping the trap is very small. This means that practically, no luminescence will be observed at T_0. However, luminescence can be triggered by raising the temperature, thus the term thermoluminescence (TL). Following irradiation of a TL material (or excitation by any other means), which builds a population of trapped electrons, the rise in temperature (assumed to be linear) will allow a temperature value to be reached at which the probability of trapped electrons escaping is substantial. Luminescence will then be observed. Further temperature increase will raise the light intensity as more electrons are freed from the trap and allowed to deexcite. A decrease in light intensity will occur as the trap is gradually depleted. Thus, the common method of presenting TL data is a plot of light intensity vs. temperature. This is usually called a glow curve and will contain at least one peak. The number of peaks in a glow curve corresponds to the number of different types of traps existing in the TL material. The temperature at which the maximum of each peak occurs is correlated with its energy depth, E. The area under a glow curve is related to the number of electrons trapped, and consequently, is correlated with the amount of radiation received. This is useful information in TL dosimetry.

For further reading see existing textbooks.[89,93–96]

10.4.3 TL Brachytherapy Dosimetry

As detailed in Chapter 5, the radionuclides used for brachytherapy extend from low energy [103]Pd (22 keV) and [125]I (27 keV) seeds via intermediate energy [169]Yb (134 keV) and [192]Ir (388 keV) to higher energy [137]Cs (667 keV) and [60]Co (1250 keV) sources. A steep spatial dose gradient exists in all three dimensions around the sources that is $\sim 5\%$ per mm in the centimeter distance range from a source and up to $\sim 40\%$ per mm in the millimeter distance range. Dose rates around a single brachytherapy source vary from

~ 100 Gy h^{-1} at 1 cm to ~ 100 cGy h^{-1} at 10 cm along the transverse axis of an ^{192}Ir HDR source and from ~ 0.7 cGy h^{-1} at 1 cm to ~ 0.01 cGy h^{-1} at 5 cm along the transverse axis of an ^{125}I LDR seed.

For the purpose of experimental dosimetry in brachytherapy it is necessary to introduce a dosimeter in the radiation field of a source. This dosimeter should provide a measurable reading and present an adequate sensitivity, which is defined as the ratio of reading per unit absorbed dose per unit mass of its sensitive volume. This sensitive volume should be small so that it does not perturb the radiation field. The previously mentioned steep spatial dose gradient around brachytherapy sources also imposes a size limitation in order to avoid severe dose volume-averaging problems. Since dose in human tissue is the required outcome, the dosimeter should preferably approximate the radiation interaction and absorption properties of water. Also, given the wide dose rate range existing around brachytherapy sources, the dosimeter should exhibit a wide dose range with adequate sensitivity. Such sensitivity would ideally be constant with dose reflecting linearity of dosimeter reading with absorbed dose. However, nonlinear behavior is acceptable provided that the dynamic dose range (the dose range from the lowest detectable dose to the dose where saturation sets in) covers the investigated dose range. Finally, it is noted that the shift of photon spectrum to lower energies as the distance from a source increases requires a uniform dosimeter energy response. This is defined as the variation of signal-to-dose ratio in water with radiation quality.

The dosimetric system that provides the optimum compromise between all the above prerequisites is TL dosimetry. This has now been accepted as a standard method for experimental brachytherapy dosimetry owing to independent studies as well as the collaborative work of the Interstitial Brachytherapy Dosimetry Committee.[12,14,15,97] In fact, TLD has, in its turn, helped validate Monte Carlo-aided dosimetry as a valuable brachytherapy dosimetry tool.[17,18,98] Among the variety of commercially available TLDs, TLD-100 (LiF doped with Mg and Ti in small concentrations to enhance its physical radiation-induced TL) is still the most common material in experimental TL dosimetry in brachytherapy. Some studies, however, suggest that other materials such as TLD-700H (LiF doped with Mg, Cu, and P) are advantageous for clinical dosimetry purposes.[99-102]

TLD-100 presents an effective atomic number of 8.4 which is close to that of soft tissue (7.4), and therefore it may be considered almost tissue equivalent with an energy response that is uniform down to energies of about 200 keV. It then begins to increase with decreasing energy due to the increasing predominance of the photoelectric effect (see Figure 10.4). A correction is thus necessary. It is commercially available in various forms of small dimension so that it does not perturb the measured radiation field (cubes of $1 \times 1 \times 1$ mm^3, rods of $1 \times 1 \times 6$ mm^3, chips of 3.1×3.1 mm^2 and thickness 0.2 to 0.9 mm, discs of 4.2 mm in diameter and thickness 0.25 to 0.89 mm thickness and powder). Owing to its solid state it presents a satisfactory sensitivity and a signal-to-noise ratio that can be enhanced by prolonging its irradiation time since it is an integrating dosimeter. TLD-100 also presents a wide dynamic dose range[102] (linear from 0.01 cGy up to ~100 cGy for the ^{60}Co quality and supralinear up to the dose of ~20% of the saturation dose value D_s). In the following two sections, the practical considerations involved in brachytherapy TLD dosimetry will be reviewed using TLD-100 as the reference material of choice. This excludes the fields of contemporary intravascular brachytherapy applications[103] using beta emitters where TL dosimetry also finds a use[104] and brachytherapy using ^{252}Cf sources of mixed neutron–photon radiation.[105,106]

10.4.4 Annealing of TL Dosimeters

The term annealing is generally used to describe the thermal treatment of TLDs that allows for the reuse of dosimeters after an irradiation-readout cycle, ensuring at the same time a reproducible sensitivity.

Several annealing procedures have been utilized for TLD-100 in brachytherapy dosimetry applications[14,21,25] that differ both in the temperature at which the TL crystals are annealed as well as the corresponding duration. However, the standard preirradiation annealing treatment of TLD-100 (LiF:Mg,Ti) consists of heating at the high temperature of 400°C for 1 h followed by a slow cool down, and finally, heating at the low temperature of 80°C for 24 h.[19,107–109] Alternatively, heating at the low temperature of 100°C for 2 h instead of 24 h at 80°C has been proposed,[110] and is commonly used in TLD brachytherapy dosimetry.[21,22,24] In both cases, the high temperature annealing fully depletes the electron traps in the TL crystal while the low temperature annealing accounts for the redistribution of the traps to the desired stable glow curve peak of LiF and suppression of the unstable low temperature peaks, as explained in the following.

A TLD-100 glow curve may be deconvoluted in six different peaks corresponding to the number of different types of traps existing in the TL material and peaking at a temperature that correlates with the trap energy depth. Preirradiation annealing reestablishes the defect equilibrium in the crystal and enhances the intensity of peak 5 (which occurs at a temperature of 224°C) at the expense of lower temperature peaks 1 and 2 in the glow curve of TLD-100 acquired in the next readout (see Figure 10.9). This is desirable because peak 5 is the main dosimetric peak due to its favorable characteristics.[89] Low temperature peaks are more sensitive to loss of signal after irradiation due to thermal energy transfer to the trapped electrons even at room temperatures (thermal fading), exposure to ultraviolet or visible light (optical fading), and/or other causes (anomalous fading).

Total elimination of low temperature peaks prior to readout would further ensure reproducibility of results. This is the role of postirradiation annealing, which consists of heating of the dosimeters for 10 min at a temperature of 100°C prior to their readout, thus depleting low temperature glow curves.[111] It has been shown[19] that annealing of the TLDs to 400°C for 1 h prior to their irradiation, combined with the postirradiation annealing, yields the same

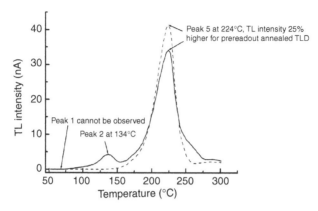

FIGURE 10.9
Typical glow curve of a TLD-100 dosimeter that has been preirradiation annealed (full line) and
both pre- and postirradiation annealed (broken line). The two TLDs were irradiated under the
same conditions using a 6 MV linear accelerator beam to deliver a dose of 0.8 Gy.

readout precision vs. dose as the common preirradiation procedure, and
thus shortens the duration of the annealing-readout cycle. Several
investigators[21,22,24] have combined the preirradiation and postirradiation
annealing techniques in order to eliminate completely the influence of the
low temperature glow curve peaks on the TL readout.

In Figure 10.9 the glow curves of two TLD-100 dosimeters irradiated
under the same conditions using a 6 MV linear accelerator beam to deliver a
dose of 0.8 Gy are presented. The first TLD was only preirradiation annealed
while the second one was annealed both pre- and postirradiation.
Comparison reveals that peak 1 cannot be observed in both glow curves
due to its very short half-life (10 min)[112] while peak 2 can only be observed
for the TLD that was not postirradiation annealed. Moreover, the maximum
intensity of peak number 5 (or to be precise, the convolved 3, 4, and 5 peaks)
is about 25% higher for the TLD that underwent postirradiation annealing
compared with the corresponding value for the TLD that was not
postirradiation annealed.

The reproducibility of TLD annealing procedures is of paramount
importance. For example, an increase in the duration of the cooldown from
400 to 100°C in preirradiation annealing by as much as 4 min could cause an
8% decrease in TLD sensitivity.[113] Reproducible thermal treatment of the TL
dosimeters is technically achieved through microprocessor-controlled TLD
annealing ovens (see Figure 10.10), which also utilize air fans to ensure
homogeneous heating. TLD ovens should, however, be checked for the
reproducible heating and cooldown rate and duration as well as for their
temperature stability. The dosimeters should be loaded on metallic trays
which exhibit good thermal conductivity and do not interact with the LiF
material at high temperatures. Trays made of aluminum used to accommo-
date as well as cover the TLDs fulfill this condition, and additionally, protect

FIGURE 10.10
TLD annealing oven (PTW-TLDO).

the dosimeters from chemical interactions with impurities, such as machine oil or cleaning solution vapors, which emerge from mechanical parts of the fan when air is left in the TLD oven during the cooldown phase.[113]

10.4.5   Readout of TL Dosimeters

The information stored in the TLD after irradiation in the form of electrons in the energy traps of the crystal lattice, is derived by heating up the dosimeter at a certain predefined temperature rate to a maximum temperature in the TLD reader (see Figure 10.11). As electrons recombine with holes in the recombination centers located in the energy gap visible photons are emitted and strike the cathode of a photomultiplier tube (PMT). The cathode usually consists of mixtures of alkali metals making it sensitive to the entire visible light spectrum. The ejected photoelectrons are accelerated to the anode generating along their way additional electrons at the dynodes of the PMT. The amplification factor depends on the acceleration voltage and the number of dynodes in the PMT. The amplified electrical signal in terms of electrical current, which is proportional to the TL intensity, is then registered in a number of channels associated with a specific temperature or time value which is reached at a specific time defined by the chosen heating rate. This results in the TL intensity curve over the temperature, known as the glow curve of the TLD.

The readout procedure of the dosimeters may be divided into preheat, acquisition, and annealing phases. It should be noted that preheat and annealing of the TL crystals may be performed in the TLD ovens as

FIGURE 10.11
TLD reader (Bicron Harshaw QS 5500).

described in the previous section. The heating rate from ambient temperature to the preheat temperature used for the elimination of the lower temperature peaks is invariant in most commercial readout systems, while the preheat temperature and duration may be varied. The annealing phase is intended to remove the residual TL signal of the dosimeters. In most commercially available TLD readers the heating rate from the maximum acquisition temperature to the anneal temperature is invariant (usually $50°C\ s^{-1}$).

In the acquisition phase, where TL light is emitted and translated to amplified electrical signal by the PMT, the maximum temperature (of about 300°C), the duration of signal acquisition, as well as the heating rate (usually of the order of 5 to $20°C\ s^{-1}$) can be set up by the user and significantly affects the shape of the glow curve. Usually, glow curve peaks 3, 4, and 5[24] of the LiF TLD are observed as a unique broader peak (see Figure 10.9). Increasing the heating rate during the acquisition phase causes a shift of the glow curve peaks to higher temperatures. For example, Figure 10.9 reveals that a heating rate of $15°C\ s^{-1}$ causes a temperature shift of both peaks 2 and 5 to higher temperatures than those expected for lower heating rates (peak 2 at 120°C and peak 5 at 210°C).[24] This shift has proved to be a logarithmic function of the heating rate.[114] Moreover, the acquisition heating rate increase results in an increase of signal for the same dose, or in other words to supralinearity of the TLD response.[114]

As described earlier, the useful information is the area under the glow curve since it is related to the number of electrons trapped and is consequently correlated with the amount of radiation received. Thus, the output of the TLD readout procedure is a charge value acquired by integration of the electrical current signal of the whole glow curve or within selected regions of interest. However, since according to the above analysis the position of a particular glow curve peak (i.e., the temperature or time channel at which its maximum appears) as well as its relative intensity depend not only on the thermal history of the TLD (as described in the previous subsection) but also on the acquisition heating rate, evaluation of TLD integrated charge in specific regions of interest calls for reproducibility of the individual peak positions.

TLD readers, designed to provide a reproducible linear heating rate to dosimeters, are either loaded with individual TLDs on a metallic plate, which is heated by a molded thermoelement, or in groups of 40 to 50 TLDs mounted on special trays and heated by hot inert gas (usually N_2). The use of nitrogen gas flow is currently considered a prerequisite especially for low dose measurements since it reduces the effects of thermally stimulated chemoluminescence at the dosimeters' surface.[115] It has been shown by measurements around a ^{137}Cs brachytherapy source[19] that, especially in the low dose region, <10 cGy, TLD readout without nitrogen flow leads to a profound increase in the mean TL response and a dispersion of responses by almost a factor of two. In TLD readers decay in the TL light output appears due to impurities that settle on the metallic plate on which the dosimeters are mounted during the readout. However, in automatic TLD reading devices, each TLD is held by a vacuum pinchette in the continuous nitrogen flow, and may be dislocated in such a way that both TLD ends do not lie in the hot inert gas flow. This results in thermal transfer effects along the TL crystal and a delayed emission of TL light from the dosimeter ends.[113] This, in turn, gives rise to delayed registration of the signal in channels corresponding to higher temperatures than those for which this signal would normally have been measured. Whenever this occurs, a shoulder-shaped extension of the tail of the high temperature peak 4 and 5 complex is observed.

Overall, it is important to evaluate and maintain the accuracy and efficiency of readout instruments over long periods of time and analyze trends of irregularities that would affect the TLD readouts. In order to accomplish this, TLD readers contain a test light source (a LED or a ^{14}C β^- source), which provides a stable signal used to test the consistency of the electronic circuits. Additionally, background noise, generated by PMT dark current, light leaks and/or contamination can be read and recorded.

10.4.6   Calibration of TL Dosimeters

Prior to using TLDs a batch should be generated. A batch is defined as a group of TLDs that experience the same thermal history in the sense that they undergo the same successive annealing and readout cycles and are even stored together at room temperature even if they are irradiated under totally different conditions. Before any new TLD batch is calibrated, the dosimeters have to undergo a stabilization phase consisting of several cycles (usually 20 to 30) of irradiation at a specific dose, readout, and annealing. This is necessary in order to ensure a stable response and a decreased background signal for each of the dosimeters in the batch. This is because it has been observed that the response of new TL crystals increases with consecutive irradiations: a phenomenon known as sensitization.[89,94]

Ideally, all TLDs belonging to the same batch would present the same dose response. However, even for TLDs in the same batch the dose response varies since each TLD is unique with respect to its physical properties such as mass and dopant concentration. Although commercial TLD suppliers guarantee the uniformity of TLD groups to within 5%, individual TLD calibration is considered necessary in brachytherapy dosimetry,[3,25,26,29] assigning a relative detector sensitivity (chip factor) to each dosimeter. This increases measurement precision.

First, each TLD within a batch is assigned a unique identification number related to its position in the storage tray. The relative detector sensitivity is then derived by normalizing the TL signal of each dosimeter to the mean signal of the N TLD crystals in the same batch, which were exposed at the same dose (usually in the linear region of the TLD response: 40 to 100 cGy) in a calibration beam. For TLD calibration purposes a ^{60}Co γ-ray or linear accelerator MV x-ray beam calibrated according to international dosimetry protocols[116,117] is used.[23,26,28,118] TLDs are taken from their numbered storage tray and placed in numbered holes, machined as a square or rectangular matrix in a polystyrene or polymethyl methacrylate slab (PMMA, commercially known as Lucite or Plexiglas, see Figure 10.12). Vacuum tweezers should be used so that dosimeter surfaces are not scratched, and

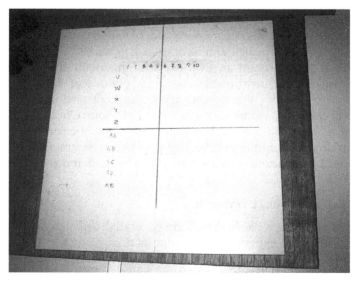

FIGURE 10.12
A rectangular matrix of numbered TLD holes in a PMMA slab.

care should be taken during their transfer from storage trays to the phantom to avoid mixing up with respect to their identifying numbers in the storage trays. The slab bearing the TLDs is sandwiched between slabs of the same material to provide a buildup layer and full backscattering conditions. TLDs are irradiated with a broad beam (usually 20×20 cm^2) and beam inhomogeneity should be verified by simultaneous irradiation of a film used for the quality assurance of the calibration beam. Scanning the film with a laser scanner and assessing the relative dose at the spatial coordinates of each TLD receptor of the calibration phantom can derive appropriate corrections.[30]

The relative detector sensitivity of the *i*th dosimeter at the *k*th experimental irradiation is given by Equation 10.1:

$$S_{i,k} = \frac{TL_{i,k}}{\frac{1}{N} \sum_{i=1}^{N} TL_{i,k}} \tag{10.12}$$

where $TL_{i,k}$ is the TL reading of each dosimeter after background subtraction. Background reading TL_{bkgd} is defined as the average reading of a group of dosimeters from the same batch that were deliberately set aside and were not irradiated.[12,109] Of course, these background TLDs have to undergo the same prereadout, readout, and annealing procedure as the irradiated dosimeters. The relative dosimeter sensitivities S_i are equal to the average value of $S_{i,k}$, resulting from k irradiations and readouts of the TLDs in the batch at the same calibration dose. Usually, this procedure is repeated at least $k = 3$ times to evaluate the reproducibility and stability of each TLD. Dosimeters that do not stabilize, exhibiting a variation of more than a set percentage value (usually 2 to 4%)[60,107,119] are discarded from the batch.

The next procedure is the evaluation of the batch dose response. For this purpose a small set of TLDs is exposed to a dose D_0 lying in the linear response region of 40 to 100 cGy. For absolute absorbed dose measurements an ionization chamber with a calibration traceable to a secondary standard is introduced into the phantom in the central photon beam axis. The dose response at D_0, ε_λ, is given by dividing the mean net (background subtracted) TL reading by the absorbed dose in the calibration phantom medium delivered by the megavoltage or ^{60}Co beam used for calibration and according to

$$\varepsilon_\lambda = \frac{\langle TL_{net} \rangle}{D_0} \tag{10.13}$$

Subsequently, the supralinear response of the TLD detectors to the absorbed dose should be evaluated. It is known that the response of TLD-100 is linear up to 100 cGy for ^{60}Co gamma-rays.[102] At higher doses a supralinearity is observed that is related to the electron trap and recombination center distributions in the crystal lattice, as described by TLD theory.[89,112] It should be noted that the dose beyond which supralinearity of response sets in, is highly dependent on the ionization density of the incident radiation and has been found to appear at lower dose values for high energy photons than for low energy photons[120] or high LET particles.[114] Practically, the TL response may also not be linear for dose values lower than a few cGy due to nonradiation-induced TL and the fact that the TL signal is comparable to the dark current of the reader. A linearity correction may be derived by separating the TLDs of a batch in m groups of $n = 6$–8 dosimeters, which are subsequently irradiated at graded doses ranging from 0.5 to 1000 cGy using a calibration beam, as described above. This procedure is repeated at least one to three times, in order to achieve better precision; an average value of TL signal for the mth TLD group is calculated by

$$\langle TL \rangle_m = \frac{\displaystyle\sum_{i=1}^{l} \frac{\displaystyle\sum_{j=1}^{n} \frac{(TL_j - TL_{bkgd})}{S_j}}{n}}{l} \tag{10.14}$$

For each one of the m TLD groups the dose response, ε_m, is calculated as the ratio of the average TL signal to the absorbed dose:

$$\varepsilon_m = \frac{\langle TL \rangle_m}{D_m} \tag{10.15}$$

At this point it should be noted that doses as low as 0.5 cGy may be delivered using a linear accelerator to irradiate the TLDs at the depth of 50% relative dose by delivering 1 monitor unit: 1 cGy to the dose maximum at the isocenter plane.[109] In that case, however, the monitor-unit end effect, which is significant when delivering doses under 10 cGy, has to be quantified and taken into account.

The TLD responses to the graded doses acquired as above, are finally normalized to the response at D_0 to provide a linearity correction factor according to

$$F_{lin} = \frac{\varepsilon_m(TL_{net})}{\varepsilon_m(TL_0)} \tag{10.16}$$

where $\varepsilon_m(TL_{net})$ is the dose response of the TLDs irradiated to a dose D and $\varepsilon_m(TL_0)$ equals ε_λ, the dose response of the TLDs that were exposed to a dose D_0 chosen for normalization. F_{lin} is plotted against the mean net TLD reading (TL_{net}) of the dosimeter groups, allowing for its evaluation at any subsequent irradiation of the TLD batch to an unknown dose.

10.4.7 Practical Considerations and Experimental Correction Factors

The calibration of TLDs presented in the previous subsection allows for absolute dose determination in the phantom material, m_1, and radiation quality, Q_1, used for the calibration procedure according to

$$D_{m_1} = \frac{TL_{m_1,Q_1}}{S_i F_{lin} \varepsilon_{m_1,Q_1}} = \frac{TL_{m_1,Q_1}}{S_i F_{lin}(TL_{m_1,Q_1}/D_{m_1,Q_1})} \qquad (10.17)$$

where $TL_{m_1,Q1}$ is the net (background subtracted) measured signal of a TLD. In the following the consideration of practical issues and correction factors involved in absolute brachytherapy TLD dosimetry is discussed.

Owing to the steep dose gradient around brachytherapy sources, systematic uncertainty in source dosimeter distance during the experimental procedure can result in significant dosimetric inaccuracy. Achieving dose rate measurements at 1 cm from a source with an accuracy of 2% calls for a source-dosimeter relative positioning accuracy of 100 μm.[11,32] Although experimental dosimetry results should be reported in water, solid phantoms are used so that source and TLD receptors are accurately machined to permit accurate measurements at distances equal to or greater than 1 cm for the derivation of the dosimetric quantities according to the widely used formalism proposed by AAPM TG-43,[1] namely the dose rate constant, radial dose, and anisotropy functions. Owing to the cylindrical symmetry of the source designs, TLD and source receptors are machined in a slab of material which is subsequently sandwiched between slabs of the same material to provide full scatter conditions. The configuration of the TLD receptors, or alternatively, the configuration of loaded receptors during each irradiation should be designed so that possible interference in the dose absorbed by a TLD by any neighboring dosimeters is minimized. In any case, the TLD and source centers must lie in the same plane, while the longitudinal axes of the dosimeters should be parallel to that of the source for dose rate constant and radial dose function measurements and vertical to that of the source for anisotropy function measurements (see Figure 10.13). The irradiation time depends on the phantom configuration and the irradiation scheme and should be adjusted so that the delivered dose is a compromise between linearity of response for dosimeters close to the source and reduced statistical uncertainty for dosimeters at greater distances. A correction factor must be introduced to account for the radioactive decay of the source when long irradiation times T are used. This is given by

$$g(T) = \frac{1}{\int_0^T \exp(-\lambda t)dt} = \frac{\lambda}{1 - \exp(-\lambda T)} \qquad (10.18)$$

where λ is the decay constant of the source.

For the dosimetric characterization of ^{192}Ir brachytherapy sources, experimental phantoms made of polystyrene, PMMA, or solid water may be used since they all closely resemble the dosimetric properties of water, as reported in the literature.[61] This is due to the fact that for the ^{192}Ir energies Compton scattering predominates and no significant difference in absorbed dose exists between materials of comparable electron density. For the

FIGURE 10.13
The solid water dose anisotropy function phantom (GfM).

dosimetry of low energy ^{125}I or [103]Pd sources, where the photoelectric effect predominates, the choice of phantom material should be made with caution. Since the photoelectric effect cross-section is strongly dependent on effective atomic number, slight differences in the exact phantom material composition may result in significant differences in absorbed dose. Polystyrene and PMMA phantoms are not suitable for TLD dosimetry of low energy brachytherapy sources since their dose rate distributions deviate significantly from that of water.[60] This necessitates significant conversion factors that can increase overall experimental uncertainty. Although it has been reported that dose rate distributions around low energy sources in solid water closely resemble that in water,[60] Monte Carlo simulations later revealed that this is not the case.[16,121] Solid water underestimates the dose rate relative to water and an appropriate correction is required.[30] This correction may be calculated using Monte Carlo simulation as the ratio of dose in water to dose in the experimental phantom material at the same position relative to the real brachytherapy source design.

The experimental phantom material, denoted by m_2, may differ from that used for calibration purposes (m_1), and thus a TLD dose response correction should be applied by multiplying ε_{m_1,Q_1} by

$$\frac{(\mathrm{TL}_{m_2,Q_1}/D_{m_2,Q_1})}{(\mathrm{TL}_{m_1,Q_1}/D_{m_1,Q_1})} \qquad (10.19)$$

where TL_{m_i,Q_i} denotes the net reading of a TLD placed in material m_i and exposed to radiation quality Q_i and D_{m_i,Q_i} denotes the corresponding dose of material m_i substituted by the TLD.

For the calibration energies commonly used (^{60}Co and 4 MV or 6 MV linear accelerator beams) absorbed dose does not significantly depend on the exact phantom material composition. Given the predominance of Compton scattering and their comparable electron densities, solid water, polystyrene, and PMMA phantoms are of dosimetric equivalence and this correction factor may be assumed to be unity.

The dose response calibration of the TLDs is valid for the calibration quality Q_1. Since the effective atomic number of TLD-100 (this is 8.4) differs from that of the experimental phantom material m_2, the response of the dosimeter can be expected to vary with experimental irradiation quality, Q_2, relative to the calibration beam quality, Q_1. This is especially true as energy decreases and the photoelectric effect gradually predominates. A calibration of the dosimeters to the quality of the brachytherapy nuclide of the investigated source, Q_2, is feasible.[122] The common practice is to use a relative energy response correction, $E(r)$, according to

$$E(r) = \frac{(TL_{m_2,Q_2}/D_{m_2,Q_2})}{(TL_{m_2,Q_1}/D_{m_2,Q_1})} \tag{10.20}$$

The symbols in Equation 10.20 have the same meaning as those in Equation 10.19. r denotes the distance between source and dosimeter in the experimental phantom, implying that this correction may be position dependent.

Under the fundamental hypothesis that the TL signal of a detector is proportional to the absorbed dose in its sensitive volume, i.e., $TL_{det} = \alpha D_{det}$ where α is the absolute dose response of the TLD which is energy independent,[123] Equation 10.20 may be rewritten as

$$E(r) = \frac{(D_{TLD,m_2,Q_2}/D_{m_2,Q_2})}{(D_{TLD,m_2,Q_1}/D_{m_2,Q_1})} \tag{10.21}$$

According to Burlin's cavity theory,[124] TLDs may be considered as large cavities where dose is deposited by electrons generated by photon interactions within the cavity. This is not only for the low energies of ^{125}I and ^{103}Pd, where the range of secondary electrons in the medium surrounding the TLD is small, but also for the ^{192}Ir energies.[17,24] Moreover, Valicenti et al.[17] who used a modified Burlin cavity theory and Monte Carlo simulation results, showed that TLDs may also be considered as large cavities for the ^{60}Co gamma-ray (or equivalent x-ray photons up to 6 MV) quality to within 1% for TLD-100. Consequently, an approximation of the energy response factor of Equation 10.21 may be performed using the ratios of mass energy absorption of TLD to surrounding medium at radiation qualities Q_1 and Q_2:[22,24,125]

$$E(r) = \frac{(\mu_{en}/\rho)_{m_2,Q_2}^{LiF}}{(\mu_{en}/\rho)_{m_2,Q_1}^{LiF}} \tag{10.22}$$

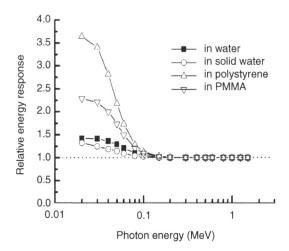

FIGURE 10.14
Relative energy response corrections in water, solid water, polystyrene, and PMMA calculated using the corresponding mass energy absorption coefficients as Equation 10.22. Results are normalized at the ^{60}Co gamma-ray energy.

In Figure 10.14, corresponding results in water, polystyrene, PMMA, and solid water are presented normalized at the ^{60}Co gamma-ray energy. In this figure it can be seen that the relative energy response of TLD in all materials is flat down to approximately 200 keV. Moreover, this figure also indicates that polystyrene or PMMA phantoms are not suitable for TLD dosimetry of low energy emitting brachytherapy sources, as discussed above.

The above approximation assumes homogeneous TLD and material phantoms, thus disregarding the fact that the photon spectrum reaching a TLD in a particular experimental position within a phantom is also characteristic of the surrounding material. The relative energy response of TLDs in a given medium may be determined either experimentally or using Monte Carlo simulation. Experimental determination may be performed using actual brachytherapy sources[24,126] or equivalent x-ray beams chosen by matching the half-value layer-derived effective energy of the beam with the corresponding effective energy of the actual brachytherapy source.[23,40,60] Monte Carlo-aided relative energy response in a medium may be performed either by using calculated photon spectra reaching the center of a TLD at each particular position in the experimental phantom for weighting of the mass energy absorption coefficient ratios in Equation 10.22[24] or by estimating the kerma at each detector volume,[17] always on the basis of the equivalence of TLDs to large cavities, as discussed above.

For ^{125}I sources, various authors[13,40,60,126] have reported values using experimental techniques of 1.41, 1.44, 1.39, and 1.41 relative to ^{60}Co gamma-rays, in water. These results are independent of source-detector distance and in agreement with the approximation of Equation 10.22. This equation value of 1.41 at 30 keV can be seen in Figure 10.14, since the energy spectrum of ^{125}I does not change significantly with distance from the source due to the predominance of photoelectric absorption. Although for the ^{192}Ir energies no relative energy response correction is expected[122] as can be seen in Figure 10.14, a TLD overresponse with increasing distance from the source due to shifting of the energy spectrum to lower energies has been reported.[23] This was of the order of 8.5% at 10 cm distance compared with 1 cm distance. This claim, however, was rejected by other investigators,[127] resulting in an ongoing debate and ^{192}Ir experimental dosimetry results being published with[21] or without[22,24] overresponse corrections. Recently, however,[22] in a study using Monte Carlo-derived photon energy spectra for weighting mass energy absorption coefficients ratios in Equation 10.22, it has been reported that this overresponse is within 3% at radial distances <15 cm from an ^{192}Ir source. Also, experimental work by Pradhan et al.[24] has concluded that this overresponse is, indeed, within experimental uncertainties and remains within 2.5%.

The final correction necessary is associated with the finite dimension of TLD dosimeters. Any TLD positioned with its center at a specific radial distance within the experimental phantom, actually integrates dose over a radial distance range relative to the source center. Therefore, a volume averaging correction factor should be applied to the measured TLD signal representing the ratio of dose at the geometric center to dose over the whole TLD volume. This correction, also known as displacement factor,[3] C_{displ}, is equal to the ratio of the dose to the TLD center to the dose over its volume. It may be computed by subdividing the source into a number of infinitesimal masses, each resembling a point source, and integrating their dose contribution to each infinitesimal mass of the dosimeter.[23]

The segmentation of both the source and a TLD results in infinitesimal volumes $dV_{Source} = \rho_{Source}^{-1} dm_{Source}$ and $dV_{TLD} = \rho_{TLD}^{-1} dm_{TLD}$ (see Figure 10.15). For a point like, infinitesimal source, whose vector to the center of the TLD is $\vec{r}'$ and the connection vector between dV_{Source} and dV_{TLD} is $\vec{r}''$, the ratio of the doses to the arbitrary infinitesimal TLD volume element to that deposited in the TLD center is equal to

$$\frac{D(\vec{r}'')}{D(\vec{r}')} = \frac{\frac{1}{(\vec{r}'')^2}}{\frac{1}{(\vec{r}')^2}} \text{ or } D(\vec{r}'')(\vec{r}'')^2 = D(\vec{r}')(\vec{r}')^2 \qquad (10.23)$$

Given that $D(\vec{r}')$ is the dose deposited in the center of the TLD (D_{center}) and $D = dE/dm$ integrating over the TLD mass, Equation 10.23 results in

$$E_{TLD} = \int dE = D_{center} \rho_{TLD} (\vec{r}')^2 \int_{V_{TLD}} \frac{1}{(\vec{r}'')^2} dV_{TLD} \qquad (10.24)$$

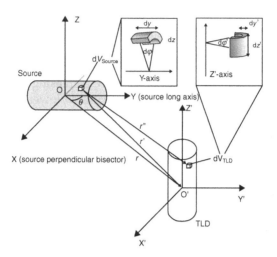

FIGURE 10.15
Geometry set-up of the $F(r,\theta)$ phantom for the finite size dose correction factor (C_{disp}) of both TLD and source core, whose volume centers lie in the same plane.

Assuming that the energy deposited in the TLD mass is equal to the dose in the TLD volume (D_{TLD}) multiplied by the entire mass of the TLD and substituting in Equation 10.24, D_{TLD} is given by Equation 10.25:

$$D_{TLD} = \frac{(\vec{r}')^2 D_{center}}{V_{TLD}} \int_{V_{TLD}} \frac{1}{(\vec{r}'')^2} dV_{TLD}$$

or

$$D_{center} = D_{TLD} V_{TLD} \frac{1}{(\vec{r}')^2} \left[\int_{V_{TLD}} \frac{1}{(\vec{r}'')^2} dV_{TLD} \right]^{-1} \qquad (10.25)$$

and

$$C_{disp}^{Point\ Source} = \frac{D_{center}}{D_{TLD}} = V_{TLD} \frac{1}{(\vec{r}')^2} \left[\int_{V_{TLD}} \frac{1}{(\vec{r}'')^2} dV_{TLD} \right]^{-1} \qquad (10.26)$$

The next step is to average the $C_{disp}^{Point\ Source}$ factor over the source volume according to

$$C_{disp} = \frac{\int_{V_{Source}} C_{disp}^{Point\ Source} dV_{Source}}{V_{Source}}$$

$$= \frac{V_{TLD}}{V_{Source}} \int_{V_{Source}} \left\{ \frac{1}{(\vec{r}')^2} \left[\int_{V_{TLD}} \frac{1}{(\vec{r}'')^2} dV_{TLD} \right]^{-1} \right\} dV_{Source} \qquad (10.27)$$

The value of the integrals depends on the exact active source core and TLD dimensions, and can be calculated using Monte Carlo integration.[128,129]

The Monte Carlo integration of the right-hand side of Equation 10.27 gives

$$C_{disp} = \frac{V_{TLD}}{V_{Source}} V_{Source}(V_{TLD})^1 \frac{1}{N} \sum_{i=1}^{N} \left\{ \frac{1}{(\vec{r}_i')^2} \left[\frac{1}{N} \sum_{j=1}^{N} \frac{1}{(\vec{r}_j'')^2} \right]^{-1} \right\}$$

$$= \sum_{i=1}^{N} \left\{ \frac{1}{(\vec{r}_i')^2} \left[\sum_{j=1}^{N} \frac{1}{(\vec{r}_j'')^2} \right]^{-1} \right\} \tag{10.28}$$

for the case of sampling an equal number of points N, isotropically distributed in both the active source core and TLD volumes.[129] Care should be taken for the exact representation of the geometrical configuration of the TLD and the source in the phantom, since in phantoms utilized for the calculation of radial dose function, $g(r)$, the displacement correction factors are polar angle invariant. This is in contrast to the experimental set-up in phantoms utilized for the calculation of anisotropy function $F(r,\theta)$, where they are polar angle dependent.

An alternative to the above method is the incorporation of this factor in the Monte Carlo simulations for the energy response corrections by calculating the kerma at each detector volume (instead of the kerma at the center of the detector) using an appropriate track length estimator.[17] In any case, the displacement correction factor is significant only for TLDs of large dimensions positioned at relatively small distances from a source. For example, C_{displ} is negligible for TLD-100 minicubes of $1 \times 1 \times 1$ mm^3 dimensions. For TLD-100 rods, however, (6 mm long by 1 mm in diameter), C_{displ} is of the order of 2.7 and 1.2% at distances of 1 and 1.5 cm for the $g(r)$ phantom and 3 and 1.2% for the same radial distances from the source center and all the polar angles in the $F(r,\theta)$ phantom. This becomes insignificant as the distance increases (see Table 10.5). An increased variation in C_{disp} with polar angle θ in the $F(r,\theta)$ phantom is evident at radial distances of 0.5 and 0.6 cm (Table 10.5).

It should also be noted that a difference between the nominal angle, where a TLD is positioned and the effective angle where measurement truly corresponds, exists when using large TLDs positioned at radial distances close to the source center and polar angles close to the longitudinal source axis. For example, nominal-effective polar angle differences of 10° have been reported for TLD rods positioned 1 cm from the center of a source with a 3.4 mm long active core, along its longitudinal axis.[30]

The formula for D_{wat} in absolute TL dosimetry in brachytherapy is given in Equation 10.29:

$$D_{water} = \frac{TL_{m_2,Q_2}}{S_i F_{lin} \varepsilon_{m_1,Q_1} E'(r) g(T)} \tag{10.29}$$

where TL_{m_2,Q_2} denotes the net reading of a TLD in the experimental phantom material and quality and $E'(r)$ is a correction factor in which all the above

TABLE 10.5

Displacement Factor C_{disp} Values Calculated Using the Monte Carlo Integration Method

r (cm)	g(r)	F(r, 0°)	F(r, 15°)	F(r, 30°)	F(r, 60°)	F(r, 90°)
0.5	1.090	1.136	1.135	1.127	1.109	1.102
0.6	1.067	1.091	1.089	1.086	1.076	1.074
0.7	1.051	1.065	1.064	1.062	1.057	1.055
0.8	1.040	1.049	1.048	1.047	1.044	1.043
0.9	1.033	1.037	1.037	1.036	1.035	1.034
1.0	1.027	1.031	1.031	1.029	1.029	1.028
1.2	1.011	1.020	1.020	1.020	1.020	1.020
1.5	1.012	1.013	1.013	1.012	1.012	1.012
1.8	1.009	1.009	1.009	1.009	1.009	1.009
2.0	1.007	1.007	1.007	1.007	1.007	1.007

For both the $g(r)$ and $F(r,\theta)$ phantoms, given for radial distances from the source center of 0.5, 0.6, 0.7, 0.8, 0.9, 1.0, 1.2, 1.5, 1.8, and 2.0 cm.

corrections (except for source decay during irradiation $g(T)$) are bundled according to the notation that is commonly used in reporting brachytherapy TL dosimetry results.[26,28,29] This factor is given by

$$E'(r) = \frac{(TL_{m_2,Q_2}/D_{water,Q_2})}{(TL_{m_1,Q_1}/D_{m_1,Q_1})} C_{displ} \tag{10.30}$$

10.4.8 Uncertainty Estimation

TLD dosimetry results exhibit a degree of uncertainty arising from various sources. Statistical uncertainty can be reduced by performing repeated measurements. Systematic errors, arising mainly from source-detector distance uncertainty in the experimental phantom and the uncertainty in the calibration of the ionization chamber used for absolute calibration beam dosimetry, cannot be controlled and should be taken into account.[27,29,130] Other sources of systematic errors include the statistical uncertainty in the Monte Carlo simulations used to derive the energy dependence correction factor as well as the dose conversion factor from the phantom material to water. In practice, there are two methods used to combine the different systematic, uncorrelated errors.[102] These are simple arithmetic addition of the partial uncertainties or summing them up in quadrature. While the first leads to overestimation, the second underestimates overall uncertainty and the method used should be explicitly reported by investigators.

Statistical errors arise in repeated TLD measurements for calibration as well as experimental purposes[27,29,130] and can be reported as uncertainty in the mean value of the measured quantity in terms of one, two, or three standard deviations. The statistical uncertainty is a reproducibility index not only of the dosimeters but also of relevant instrumentation (TLD reader, annealing oven, and the ionization chamber used for the calibration) and measurement set-up. Therefore, instruments should be checked for reproducibility and consistency over longer periods of time in order to ensure that their impact on the statistical uncertainty of TLD measurements is of minor importance.

A comprehensive analysis of uncertainty estimation as well as methodological recommendations for experimental dosimetry of low energy interstitial brachytherapy sources can be found in the update of the American Association of Physicists in Medicine (AAPM) Task Group No. 43 Report.[2]

10.5 Polymer Gel Dosimetry in Brachytherapy

10.5.1 Introduction

As discussed in more detail in the previous sections of this chapter, experimental dosimetry in brachytherapy must deal with the steep dose gradient around a source which calls for small detector volumes to minimize volume averaging effects. Additionally, the range of photon energies emitted by radionuclides used in brachytherapy (from 20 keV to 1 MeV) necessitates careful selection of phantom and detector materials so that water equivalence of energy attenuation and absorption properties, as well as a well-known (ideally constant) relative energy response are ensured. This is more significant for lower photon energy emitters, such as ^{125}I and ^{103}Pd, and a characteristic example is that of the commonly used experimental set-up of LiF TLD in solid water phantoms. While acceptable for the dosimetry of higher energy photon emitters such as ^{169}Yb, ^{137}Cs, ^{192}Ir, and ^{60}Co, this phantom-detector set-up requires the application of relative energy response and water equivalence corrections of the order of 40% for lower energy sources (see Section 10.4.7 and Figure 10.14).

Potentially, polymer gel dosimetry not only overcomes these problems but also outclasses conventional one-dimensional and two-dimensional dosimeters. The method combines an inherent three-dimensional dose integrating approach with advantages such as fine spatial resolution and arbitrary detector geometry and size. Moreover, gel dosimeters are almost water equivalent and simultaneously comprise both the phantom and the detector material, thus obviating the need for radiation fluence perturbation and significant relative energy response or water equivalence corrections even for low photon energies.[53]

It is due to these advantages that polymer gel dosimetry has gained considerable research interest in the past decade as showcased by numerous publications and the organization of three international conferences dedicated to the subject.[131-133] Although polymer gel dosimetry has not, as yet, been adapted for routine use in a clinical environment, it has been successfully applied to all contemporary radiotherapy applications including brachytherapy, which is of interest to this text.

10.5.2 The Basics of Polymer Gel Dosimetry and Gel Formulations

Polymer gel dosimetry is the latest trend in chemical dosimetry. In 1927, Fricke and Morse[134] proposed dosimetry based on the dose-dependent transformation of ferrous (Fe^{+2}) ions into ferric (Fe^{+3}) ions in irradiated aqueous ferrous sulfate solutions (commonly referred to as Fricke dosimeters). In 1984, Gore et al.[135] proposed the use of nuclear magnetic resonance (NMR) relaxation measurements in Fricke dosimetry. With the increasing availability of magnetic resonance imaging (MRI) facilities and the addition of a gel matrix to the Fricke dosimeters to retain the spatial information of absorbed dose distributions, the term gel dosimetry was soon introduced in radiotherapy.

Besides many successful applications presented in the literature and the ease of Fricke gel manufacture, a number of drawbacks and mainly temporal instability due to ferrous and ferric ion diffusion led to the shift of interest towards three-dimensional dosimetry using polymer gels, introduced in the early 1990s by Maryanski et al.[136,137]

Polymer gels consist of an aqueous solution of vinyl monomers and a gelling agent. Their principle of operation is the radiation-induced radical polymerization of monomer molecules to polymer in a region of interaction of the gel with ionizing radiation. The successive processes distinguished by irradiation of a gel dosimeter are: radical formation from solvent and monomer molecules, initiation propagation and termination of monomer radical polymerization, and cross-linking.

Since polymer gel dosimeters consist of more than 85% water, the main mechanism of radical formation is water radiolysis, a term describing the disassociation of water to ions and highly reactive radicals upon irradiation. The main intermediate radiolysis species to be found in a gel dosimeter shortly after irradiation (10^{-8} sec) are hydroxyls ($^\bullet OH$), hydrated electrons (e_{aq}^-) and hydroxonium ions (H_3O^+). The initiation process involves the reaction of intermediates with vinyl monomer molecules, forming monomer radicals. During the propagation process, the monomer radicals can react with other monomer molecules to form a growing polymer chain. This is followed by termination where "dead," nonradical polymer chains are formed. Growing or final polymer chains readily react with monomers, especially dysfunctional ones (cross-linking), leading to the final formation of a three-dimensional polymer network.

In contrast to Fricke gels, the three-dimensional polymer network formed does not present significant diffusion. It is, therefore, spatially representative of the absorbed dose distribution and its mechanical integrity is ensured by

FIGURE 10.16
(a) A VIPAR gel-filled vial with a 6F plastic catheter fixed through its cap to facilitate irradiation. The catheter is connected to an ^{192}Ir HDR remote afterloader. (b) The same vial as in Figure 10.16a, photographed immediately after irradiation with a single source dwell position that delivered 10 Gy at 10 mm along the source transverse bisector.

the gelling agent (gelatin or agarose). Thus, not only the chemical but also the physical properties of the gel dosimeter are altered so that employing different imaging methods allows for three-dimensional dose measurement (see Section 10.5.3). The most evident alteration of gel properties is shown in Figure 10.16. Figure 10.16a presents an unirradiated, gel-filled vial with a plastic brachytherapy catheter fixed through its cap and connected to a remote afterloader of a HDR ^{192}Ir source. The afterloader was programmed to deliver 10 Gy at 10 mm distance along the transverse source bisector in a single source dwell position. Figure 10.16b shows the irradiated gel where the radiation-induced polymer network has brought about opacity due to light scattering. The transparency of the unirradiated gel volume allows for a first qualitative inspection of the dose distribution pattern.

The degree of polymerization upon radiation exposure (i.e., the rate at which the aforementioned chemical reactions proceed) depends, among other things, on radical concentration, and this presignifies a dependence of polymer gel dose response on radiation characteristics (energy and dose rate). An important factor to be stressed with regard to radical concentration is the presence of oxygen in a gel. Upon radiation exposure oxygen leads to the formation of peroxide radicals which consume other radicals in fast reactions and thus inhibit polymerization initiation.

The degree of polymerization also depends on the choice of monomers in a dosimeter (through their intrinsic reactivity) and their concentration. Thus, the kind and percentage fraction of monomers in a particular gel formulation are characteristic of its dosimetric properties, discussed in Section 10.4: threshold (the lowest detectable dose), sensitivity (reading per unit absorbed dose), and dynamic dose range of response (the dose range from threshold to the dose where saturation sets in).

Table 10.6 summarizes some representative polymer gel dosimeters and their chemical compositions. BANG®-1 (BANG® is a trademark of MGS Research Inc.), whose acronym stands for bisacrylamide, acrylamide, nitrogen, and gelatin,[137] is a variation of the originally proposed PAG (PolyAcrylamide Gelatin).[136] It employs water as solvent and the monomers are dissolved in proportions that depend on the desired sensitivity and dose range.[138] However, a drawback of all PAG-type gels is the use of the highly toxic acrylamide monomer and researchers have long sought to replace it with other less toxic monomers without compromising the performance of

TABLE 10.6

Chemical Compositions of the BANG®-1, -2, VIPAR, PABIG, and MAGIC Polymer Gel Dosimeter Formulations

	Polymer Gel Dosimeter				
Components (% w/w)	BANG®-1	BANG®-2	VIPAR	PABIG	MAGIC
Water	89	88	87.23	87.26	82.8
Gelatin (300 Bloom)	5	5	4.91	4.90	8
N,N'-methylenebisacrylamide (Bis)	3	3	3.93	3.92	—
Acrylamide (Aam)	3	—	—	—	—
Acrylic acid (AA)	—	3	—	—	—
N-vinylpyrrolidone (NVP)	—	—	3.93	—	—
N,N'-methylenebisacrylamide (MAA)	—	—	—	—	9
Poly(ethylene glycol) diacrylate (PEGDA)	—	—	—	3.92	—
Ascorbic acid (AsA)	—	—	—	—	0.0352
Hydroquinone (HQ)	—	—	—	—	0.2
CuSO$_4$·5H$_2$O	—	—	—	—	0.002
NaOH	—	1	—	—	—

the dosimeter. An improvement of the BANG®-1 gel in this respect is BANG®-2,[139] where acrylamide is replaced by acrylic acid and sodium hydroxide resulting in higher sensitivity compared with BANG®-1 and lower toxicity. Likewise, a modification of BANG® was proposed by Murphy et al. (2000).[140] In this formulation acrylamide is replaced with the less toxic monomer sodium methacrylate resulting in a polymer gel of comparable sensitivity with BANG®. Finally, a formulation has been developed in which both acrylamide and bisacrylamide are replaced with methacrylic acid.[141] Although this formulation does not comply with the acronym, it is known as BANG®-3 and only its elemental composition is available in the literature. BANG®-3 has been reported to present the highest sensitivity relative to BANG®-1 and ®-2.[142]

Another modification of the BANG®-1 polymer gel is VIPAR,[143] in which the acrylamide monomer is replaced with N-vinylpyrrolidone (N-VInylPyr-rolidone ARgon). This formulation presents low levels of toxicity but also lower sensitivity compared with BANG-type gels. However, it features a wide dynamic dose range rising to 250 Gy,[144] which is a valuable characteristic in pursuing brachytherapy dosimetry. The PABIG gel (Poly(ethylene glycol)diAcrylate, N,N′-methyleneBIsacrylamide, Gelatin) is another formulation of relatively efficient dynamic dose range where acrylamide is replaced by poly(ethylene glycol) diacrylate (PEGDA) since this macromonomer was found to exhibit an improved sensitivity to radiation[145] which is comparable to that of Bis.[146]

The preparation procedure of all polymer gel dosimeter formulations is quite common and is well described in the literature. Since oxygen presence in the gel inhibits polymerization, a distinctive feature of preparation is the thorough deoxygenating procedure including bubbling of the gel solution with inert gas (nitrogen or argon) and filling polymer gel containers in a glove box under an inert gas atmosphere. The need for anoxic conditions limits the ease of preparation considerably and, along with monomer toxicity, constitutes an obstacle in adapting the method for routine use in a clinical setting.

To address this problem a polymer gel formulation called MAGIC (Methacrylic and Ascorbic acid in Gelatin Initiated by Copper) has been proposed.[147] This formulation incorporates metalloorganic complexes as antioxidant agents capable of binding oxygen dissolved in the gel during preparation, which can therefore be performed in the presence of normal levels of oxygen (hence the name normoxic for this class of gels). Investigations have already been carried out on the use of other antioxidant agents.[148]

Regardless of the polymer gel formulation, however, it is imperative that oxygen does not diffuse into the gel substance after preparation, and prior to, or during, irradiation. To this effect, gel containers must be impermeable to oxygen and air-tight sealed. For brachytherapy polymer gel dosimetry normal glass or Pyrex containers of dimensions appropriate for the application studied can be used. Caution is needed, however, in that plastic catheters introduced in the gel substance to facilitate the introduction of brachytherapy sources for dosimetry purposes, are not impermeable to oxygen. Hence, in the case of prolonged irradiations, necessitated for example in dosimetry of low dose rate sources, oxygen diffusion could occur through the catheter and affect experimental results.[47,49]

10.5.3   Magnetic Resonance Imaging in Polymer Gel Dosimetry

The three-dimensional polymer network, formed as roughly described in the previous section, alters the mobility of surrounding water molecules, which in turn results in a change in spin–lattice ($T1$) and spin–spin ($T2$) relaxation times. The effect is more pronounced for spin–spin relaxation time with the corresponding relaxation rate ($R2 = 1/T2$) being proportional to absorbed dose. Thus, and based on the legacy of Fricke, gel dosimetry using NMR, MRI was the first[136] and most commonly used imaging modality for polymer gel dosimetry.

Since radiation-induced polymerization also alters other physical properties of a gel dosimeter and in view of the need for an existing installation and the time required on an expensive modality such as MRI, other imaging techniques have also been considered. On the basis of the most evident radiation-induced change in a polymerized gel dosimeter, opacity of the gel, optical-computed tomography has been successfully employed.[149–154] Polymerization also induces a density increase which is exploited in studies for the utilization of x-ray-computed tomography[155–159] and ultrasound imaging.[160,161] However, MRI unit availability is increasing and the method provides excellent spatial resolution and signal-to-noise ratio, while recent evidence supports that fast MR scanning sequences employing time-efficient regimes for image formation can significantly reduce the required scanning time.[162,163] These factors make MRI the optimal imaging modality, at least for demanding dosimetry applications such as brachytherapy.

In MRI, the readout of dose distributions within an irradiated gel volume is achieved through the acquisition of proton $T2$ relaxation time maps. The employed imaging sequences involve the acquisition of multiple spin echoes, usually obtained from a Carr-Purcell-Meiboom-Gill (CPMG) RF pulse train in order to minimize the effect of RF refocusing pulse imperfections. Signals measured in successive echo images for a particular voxel are usually fitted to a monoexponential decay curve, the time constant of which represents the proton $T2$ relaxation time of the gel volume contained in that particular voxel. The fitting routines used can either be log-linear or based on nonlinear algorithms, which yield more accurate results (e.g., the Loevenberg-Marquardt method). The result is a voxel-by-voxel representation of $T2$ values found in the imaged plane or volume. Figure 10.17 presents an example of a $T2$ map obtained for the gel irradiated with a single HDR ^{192}Ir source, presented in Figure 10.16b. In this coronal map, the inserted catheter can be clearly discerned and a specific gray-scale

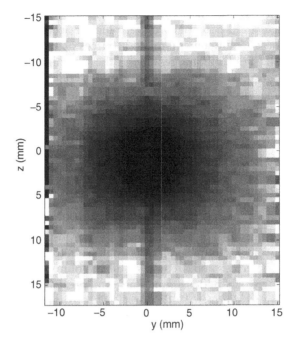

FIGURE 10.17
A coronal $T2$ map of the gel-filled vial in Figure 10.16b as obtained with a multiecho CPMG MRI sequence.

range corresponds to a specific dose range so that qualitative characteristics of the source dose distribution, such as cylindrical symmetry and anisotropy, can be readily observed.

The MR imaging parameters employed in the imaging sequence are of key importance for the accurate determination of $T2$ values of the examined gel. The number and temporal spacing of successive echoes influences the accuracy with which the $T2$ decay of the gel substance within each imaging voxel is sampled, and therefore, the uncertainty with which the $T2$ value corresponding to a particular voxel, is determined. A poor signal-to-noise ratio obtained in the images of individual echoes can also influence the accuracy of the resulting $T2$ map. Furthermore, the repetition time TR of the sequence needs to be long enough (four to five times greater than the $T1$ relaxation time of the unirradiated gel) so that the measured signal is not influenced by $T1$ relaxation. Within this context it is important that MR imaging parameters are optimized,[164,165] since minimization of the uncertainties in the obtained values of the $T2$ map leads to lower uncertainties in the measured dose distributions and therefore to increased dose resolution. This optimization is strongly associated with the type of gel used, its dose response, and the $T1$ and $T2$ relaxation times found in the unirradiated gel.

Besides stochastic uncertainty, MRI involves a number of systematic uncertainties that could potentially undermine the accuracy of the resulting $T2$ values.[166] These include image-related sources of systematic uncertainty, such as magnetic field inhomogeneity,[167] the potential misadjustment of gradient eddy currents (a problem rarely found even in older type, long-bore MR systems),[168] susceptibility artifacts (i.e., close to plastic catheters introduced in the gel to facilitate brachytherapy irradiations,[49] nonuniform temperature as well as possible changes in the gel temperature during MR scanning due to RF heating.[169] Pertinent to brachytherapy, however, another two sources of systematic uncertainty are of increased importance: volume averaging and improper positioning of the imaging slice (or slices) with respect to the irradiated gel volume.

An important factor of the imaging sequence, requiring optimization, is the spatial resolution achieved. The finite voxel size wherein the dose-dependent $T2$ value is sampled (defined by the in-plane resolution and slice thickness of the imaging session) is analogous to the sensitive volume of conventional detectors and it could result in volume averaging effects in applications where steep dose gradients are present (see, for example, Section 10.4.7 and Figure 10.15 for TLD dosimetry). In dosimetry applications where only two-dimensional steep dose gradients exist, such as in the dosimetry of narrow stereotactic beams, an in-plane pixel size of 0.5 mm is adequate for examining the steep falloff region of the penumbra, with less stringent requirements on slice thickness.[170] In contrast, in applications where steep dose gradients exist in all three dimensions, such as dosimetry around a brachytherapy source, it is the voxel rather than the pixel size that is of importance. Ideally, a very small voxel size would minimize dose averaging errors. However, since an increase in spatial resolution is directly associated with a decrease in image signal-to-noise ratio or a significant burden in the total imaging time, a trade-off is required between these three parameters. Monte Carlo integration results have shown[50] that an isotropic voxel size of $0.7 \times 0.7 \times 0.7$ mm introduces volume averaging errors of less than 2% in dosimetry along the transverse bisector of a brachytherapy source even for distances down to 1 mm, constituting an acceptable compromise between measurement accuracy and imaging time. This is shown in Figure 10.18a where an increase of slice thickness to 3 mm, for example, leads to corresponding errors of up to 30% in the millimeter distance range.

Owing to the steep dose gradient around brachytherapy sources, systematic uncertainty in source-dosimeter distance in an experimental procedure can result in significant dosimetric uncertainty. Translated to the polymer gel-MRI dosimetry method this calls for an as accurate as possible registration of individual voxels relative to the source center during irradiation. In polymer gel dosimetry around a single source, one coronal $T2$ map (i.e., a $T2$ map containing the long axis of the source such as that presented in Figure 10.17) is sufficient to provide all the TG-43 dosimetric properties of the source (see Chapter 8). Thus, an important systematic

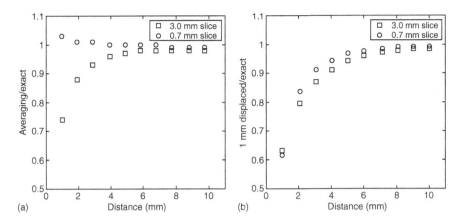

FIGURE 10.18

(a) Monte Carlo integration results quantifying the systematic uncertainty of volume averaging in brachytherapy polymer gel dosimetry. The ratio of values measured along the transverse axis of an ^{192}Ir HDR source in a single coronal $T2$ map to the corresponding exact values is presented for an isotropic voxel of 0.7 mm^3 as well as a voxel of 3 mm slice thickness by 0.43×0.43 mm^2 in-plane resolution. (b) Monte Carlo integration results depicting the systematic uncertainty of mispositioning a single imaging plane. The ratio of measured values along the transverse axis of an ^{192}Ir HDR source in a single coronal $T2$ map to the corresponding exact values is presented for a vertical displacement of 1 mm, for both imaging voxel sizes of Figure 10.18a.

error can arise from the improper positioning of the imaging slice. Monte Carlo simulations have shown that for an isotropic voxel of 0.7 mm^3, a displacement of 1 mm of the imaging plane from the axis of the source may introduce errors of up to 20% in the measurement of radial dose function, for radial distances shorter than 5 mm for the source (Figure 10.18b). Since remedial action for slice mispositioning errors cannot be undertaken after the imaging session, three-dimensional MR imaging sequences, discussed below, can be employed so that the source center during the irradiation can be localized within a single voxel or even subvoxel accuracy. Even in this case, it is advantageous that the scanning volume is aligned with the source. To this end, in the case of irradiations with remotely afterloaded sources, a fast localizing scan may be performed to exploit the cylindrical symmetry of brachytherapy sources (see Section 10.5.5.1 and Figure 10.22). Additionally, in polymer gel dosimetry applications with permanent implant sources, a first localizing scan may be performed employing a multistack and multislice, segmented k-space gradient-echo pulse sequence with very short TE and TR. This sequence is very sensitive to local magnetic field inhomogeneities and susceptibility effects and therefore the source location can be identified by the image artifacts associated with its presence in three orthogonal planes (axial, sagittal, and coronal).

The MR imaging acquisitions can be obtained in single or multiple planes, with an associated increase of imaging time for the latter. Multislice acquisitions offer the advantage of obtaining three-dimensional dose distributions over an extended volume of the irradiated gel, facilitating the reconstruction of isodose contours over any desired plane, and thus allowing the detailed dosimetric investigation of even complex irradiation regimes. However, given the stringent spatial resolution requirements mentioned above, there is an associated significant cost of imaging time required on the MR system, which is normally heavily booked for clinical purposes in a hospital setting. An interesting alternative is the use of a volume selective (three-dimensional) imaging sequence employing Fourier interpolation in the slice selective direction.[171] This method allows the extended coverage of the irradiated gel with the acquisition of thin contiguous slices (about 1 mm slice thickness). Fourier interpolation allows the reconstruction of slice partitions spaced in between the acquired slices, thus providing an increased number of dose measurements along the slice direction. These reconstructed planes provide dose distributions which are the result of interpolation between adjacent slices, without sacrifices in image signal-to-noise ratio or additional expense in imaging time. Furthermore, the reconstruction of isodose contours from the obtained three-dimensional data, set over any desired plane, has been shown to be accurate in the complete three-dimensional dosimetric investigation of irradiation regimes such as interstitial and intravascular brachyther-apy.[171,172] Another promising alternative with regard to significantly reducing the required scanning time without a major compromise in dosimetric accuracy is the use of volume selective turbo spin-echo (TSE) sequences, which involve acquisition of multiple k-space profiles per excitation. The number of profiles measured per excitation is given by the turbo factor (TF), which also equals the reduction factor in the total acquisition time.[162]

10.5.4 Calibration of Polymer Gel Dose Response

Conversion of polymer gel dosimeter $T2$ relaxation time maps obtained with MRI to dose maps requires well-established calibration procedures in which the change in the transverse relaxation rate $R2$ ($R2 = 1/T2$) of the gel is calibrated against delivered radiation dose. An outline of a typical polymer gel dose response curve over its full dose range, with basic features exaggerated to facilitate discussion, is presented in Figure 10.19. This curve of $R2$ vs. absorbed dose is characterized by a linear dose response range, the extent of which is characteristic of a particular gel formulation. A linear fit to data within this dose range according to equation:

$$R2(D) = aD + b \qquad (10.31)$$

yields the values of dose sensitivity, a ($s^{-1}\,Gy^{-1}$) and intercept, b (s^{-1}) for the gel sample. The intercept could differ from the control value, $R2(0)$ in

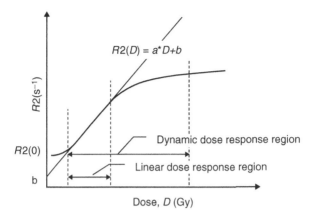

FIGURE 10.19

A graphical representation of polymer gel dose response in the form of MRI measured $R2$ values plotted vs. absorbed dose.

s^{-1}, measured in a separate, nonirradiated gel sample, and therefore the two terms should not be used interchangeably. Besides sensitivity and goodness of fit, the concept of dose resolution has been introduced for characterizing the performance of a polymer gel dosimetry system.[164,173] It is defined as the minimal difference between two absorbed doses that allows them to be distinguished with a specific level of confidence, and it depends on the sensitivity of the gel as well as the statistical uncertainty in $T2$ measurements.[164,173]

Beyond the linear dose response range, a sublinear behavior is shown in Figure 10.19 as saturation gradually sets in due to the consumption of monomers at high absorbed dose values. Sublinearity of response has also been reported in the low dose region, probably due to the presence of chemical inhibitors such as oxygen.[174] This sublinear, inhibition interval[174] extends to greater dose values with increasing oxygen concentration in the gel, without altering the sensitivity, i.e., the calibration curve is shifted to greater dose values and the threshold dose (the lowest detectable dose) increases.[174]

It is quite customary in applications of polymer gel dosimetry to plan the experimental irradiation so that the dose range of interest lies within the linear dose response of the gel. It should be noted, however, that in brachytherapy applications this necessitates gels with an extended linear region given the sharp falloff of dose with distance from a source owing to the geometry factor. If this prerequisite is not met, a nonlinear calibration fit can be employed to exploit the gel dynamic dose range of response (the dose range of dose before saturation). Owing to the regions of sublinear gel response, a biexponential fit has been proposed to describe the form of measured $R2$ vs. dose calibration data presented in Figure 10.19. If focus is on the higher dose range of a polymer gel response, as in the case of dosimetry at short distances from a HDR brachytherapy source, a monoexponential fit has been shown capable of accurately describing calibration data in the corresponding dose range of interest.[144] Polynomial fitting has also been employed in brachytherapy polymer gel dosimetry.[47] However, the dynamic dose range of gel response has to be sufficient for brachytherapy applications, and moreover, a loss of resolution is expected in regions of nonlinear dose response.

The method of choice for obtaining gel calibration data is strongly associated with the employed gel formulation and the range of doses likely to be encountered in the dosimetric application investigated. One calibration method involves the irradiation of a set of gel samples or different regions of a large gel-filled vial at individual known doses using a MV linear accelerator. The measured $R2$ values for each band can then be calibrated against known delivered dose values. This method, although straightforward, has several disadvantages. With regard to HDR brachytherapy dosimetry, the calibration dose rate is different from the range of dose rates encountered around a brachytherapy source. Additionally, the range of calibration doses that can be implemented through this method is limited, thus reducing the ability to explore the full dynamic dose range of a gel composition, which can extend to much greater dose values for certain gel formulations. If different gel samples are used the method is susceptible to intrabatch variability, while implementing this method requires tedious irradiation procedures, the availability of a linear accelerator, and a significant quantity of gel in order to obtain sufficient data points for a reliable calibration curve.

An alternative to this method is the determination of a calibration curve from a gel tube that has been appropriately irradiated in a water bath to record the characteristic depth dose curve of a MV linear accelerator along its long axis.[175] This way, the change of $R2$ values along the long axis direction can be directly calibrated against delivered dose. This method achieves improved accuracy by averaging measured $R2$ values at the same depth along the gel vial long axis and requires a smaller number of calibration gel tubes in order to cover a sufficiently wide dose range. However, it is still hampered by the disadvantages mentioned above.

A third method for obtaining a calibration curve is through the irradiation of a single gel vial with a brachytherapy source of accurately known dosimetric properties, in a single position. The change of $R2$ values with radial distance along the transverse source bisector can then be calibrated vs. delivered dose. This calibration method has been shown to be a time-efficient and accurate source for ample calibration data spreading over a wide dose range.[144] Owing to the cylindrical symmetry of brachytherapy sources, measured R2 values at the same radial distance can be averaged to improve accuracy. Particularly for polymer gel dosimetry in brachytherapy applications, this calibration method is advantageous in that it corresponds to the same energy and dose rate to be investigated. This is important since, although significantly smaller than that for solid water-TLD experimental set-ups, a relative energy response correction is necessary for polymer gel dosimetry of low energy brachytherapy sources relative to calibration data obtained for the ^{60}Co (or equivalently 6 MV linear accelerator x-rays).[53]

Figure 10.20 presents typical calibration curves obtained with the three calibration methods described above, for the VIPAR gel formulation,[143] which exhibits a wide dynamic dose range of response. In the first two methods, a 6 MV linear accelerator was used to deliver discrete doses and depth dose distributions of different maximum doses at d_{max}. For the single source dwell position irradiation, a ^{192}Ir HDR brachytherapy source was employed to deliver 10 Gy at 10 mm radial distance along the transverse source bisector. The uncertainties presented correspond to statistical uncertainty in $T2$ measurement and dose uncertainty due to dose calibration uncertainty of the irradiation modality employed (i.e., 2% absolute dosimetry uncertainty in the linear accelerator and 3% uncertainty in air kerma strength of the brachytherapy source). Brachytherapy data are also

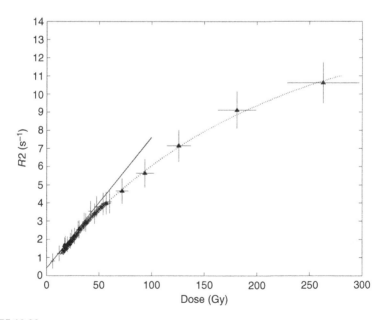

FIGURE 10.20

VIPAR polymer gel calibration data obtained by three different irradiation set-ups. A 6 MV linear accelerator was used to deliver discrete doses in the range 0 to 60 Gy (+) or depth dose distributions up to 60 Gy at d_{max} (×) and an ^{192}Ir HDR brachytherapy source remote afterloader was employed to deliver 10 Gy at 10 mm radial distance along the transverse source bisector (▲). A linear fit to discrete dose data in the linear dose response region as well as a monoexponential fit to the brachytherapy data are also presented.

burdened by a dose uncertainty due to the half-imaging pixel uncertainty of the source center location during irradiation, which was defined with an accuracy of one imaging voxel (see Section 10.5.5.1 and Figure 10.22). In Figure 10.20 it can be seen that the three calibration curves are in relatively good agreement in their common dose range. This implies no significant dependence of ^{192}Ir polymer gel dosimetry results on radiation energy and/or dose rate in the presented range. The curves obtained with the discrete dose and the depth dose methods extend to a limited dose range, mainly due to practical difficulties in obtaining higher dose values with the use of a linear accelerator. In contrast the brachytherapy calibration curve extends over a very wide dose range, rising to 250 Gy.

The characteristics of the calibration curve (sensitivity, intercept, control) for a particular gel formulation and an optimized MRI sequence depend on numerous other factors. These include oxygen presence in the gel,[174] temperature during irradiation and imaging,[166] dose rate[166] and energy of radiation,[53] ongoing polymerization in the first few hours after irradiation,[144,174,176] and ongoing gelation of gelatin for a period of several weeks.[174,177] Much as the study of these factors benefits the optimization of gel formulations, they are not of particular concern in polymer gel dosimetry if a calibration gel is produced, irradiated, stored, and MR scanned along with the experimental gel.[166,178] Minimum precautions include the storage of the irradiated calibration and experimental gels overnight in the MRI room so that the sensitivity stabilizes and the gel temperature equilibrates.

The above constitutes polymer gel formulation, and in particular, the choice and concentration of constituents, the main influence on dose response characteristics. A comprehensive study of the influence of the concentration of different monomers and gelling agents is that of Lepage et al.[179] Figure 10.21 presents calibration data for the VIPAR and PABIG gel formulations obtained with an ^{192}Ir HDR source irradiation in a single dwell position. These two formulations are characterized by a comparable sensitivity which is small compared with other gel formulations. However, this is overcompensated by the extended linear dose region and wide dynamic dose range of response of both gels which are beneficial characteristics for brachytherapy dosimetry. The linear dose range of both gels extends to about 40 Gy while the dynamic dose range rises to 100 Gy for PABIG and at least 250 Gy for VIPAR.

In summary, $T2$ distributions obtained by MRI of an irradiated gel are converted to corresponding $R2$ distributions. These $R2$ distributions can be transformed to absolute dose distributions through an appropriate calibration obtained in the dose range of interest by a calibration gel produced, irradiated, stored, and MR scanned along with the experimental gel. Dose distributions can then be transformed to relative dose distributions. In order to avoid limited dose resolution in sublinear dose response regions one can program the irradiation so that the dose range of interest lies in the linear dose response region. Then, relative dose distributions can be readily

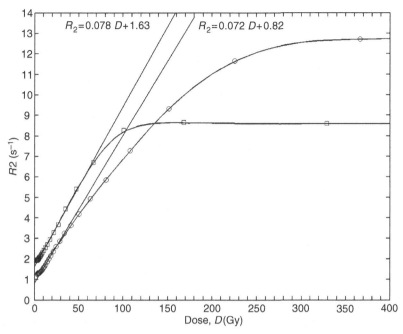

FIGURE 10.21

Calibration curves over the full dynamic dose range of response for the VIPAR and PABIG gels according to data obtained by single source dwell position irradiations with an ^{192}Ir HDR remote afterloader. The curves and corresponding results obtained by a linear fit to data in the linear dose response region are also presented for both gel formulations.

obtained by subtracting the intercept value of the linear fit to calibration data obtained by a calibration gel produced, irradiated, stored, and MR scanned along with the experimental gel (i.e., $R2_{NET}(D) = R2(D) - b$) and normalizing results following equation:

$$D/D_{REF}\% = R2_{NET}(D)/R2_{NET}(D_{REF})100\% \qquad (10.32)$$

It is noted that the control value, $R2(0)$, measured in a separate nonirradiated gel sample could be used instead of the intercept, b, if the two values are in good agreement for a particular gel formulation. This could obviate the need to perform a calibration irradiation for every experiment if dose response variation between different gel preparations could be excluded.

The errors associated with polymer gel dosimetry are calculated using error propagation of the uncertainty of the parameters of the fit to the calibration data and statistical uncertainty in $T2$ measurement,[180] the latter being the predominant source of uncertainty.[173,180]

10.5.5 Polymer Gel Dosimetry Applied in Brachytherapy

10.5.5.1 Single Source

Brachytherapy has been acknowledged as a challenging field of application for polymer gel dosimetry to exhibit its advantages since the early days of the method when Maryanski et al.[137] presented qualitative results for LDR ^{137}Cs and ^{192}Ir sources. Several studies of polymer gel dosimetry have been presented in the literature for LDR ^{137}Cs sources,[45,181] HDR ^{192}Ir sources,[47,49,50,144,171,182] and a beta emitting source of ^{90}Sr/^{90}Y used for intravascular applications.[153]

Even in single source dosimetry applications, the main advantage of the polymer gel method is its ability to obtain measurements in full three dimensions in a single irradiation. As an example of single source dosimetry, consider the gel vial irradiated with a single HDR ^{192}Ir dwell position using a remote afterloader which is shown in Figure 10.16b. In the coordinate system of the AAPM TG-43 formalism[1] distance is measured relative to the source center (see Chapter 8). In view of the steep spatial dose gradient around the source, accurate reporting of dosimetry results calls for the accurate localization of the source center during the irradiation. Assuming that the source long axis lay aligned with the catheter during irradiation and that the MRI session was carefully planned so that the scanned volume is aligned with the catheter, this can be achieved by the utilization of multiple $R2$ profiles along and across the axis of the catheter introduced in the gel to facilitate the irradiation, such as that shown in Figure 10.22a. As shown in Figure 10.22b, owing to the cylindrical symmetry of brachytherapy sources dose, or equivalently measured $R2$, profiles along the longitudinal source axis are symmetrical with respect to the transverse source bisector whereon they present their maxima. Moreover, the two $R2$ profiles along the longitudinal source axis, at equal distance from both sides of it coincide within statistical uncertainty (noise). Additionally, Figure 10.22c shows that dose, or equivalently measured $R2$, profiles across the longitudinal source axis are symmetrical with respect to it and present their maxima when the direction of the profile coincides with the transverse source bisector.

The above method can be used to define the point where the source center resided during irradiation (i.e., the irradiation center) with an accuracy of one imaging voxel. A straightforward approach is to assume that the irradiation center coincides with the center of the defined voxel. Then, a distance uncertainty of half the MR imaging pixel should be ascribed to results. Accordingly, if $R2$ values obtained by the single source irradiation are used for calibration purposes, a dose uncertainty corresponding to that of distance should be added to each data point of the known dose distribution. It is also noted that the standard deviation of $T2$, or equivalently $R2$, results averaged according to cylindrical symmetry, besides statistical uncertainty related to MRI, includes the uncertainty due to the distance

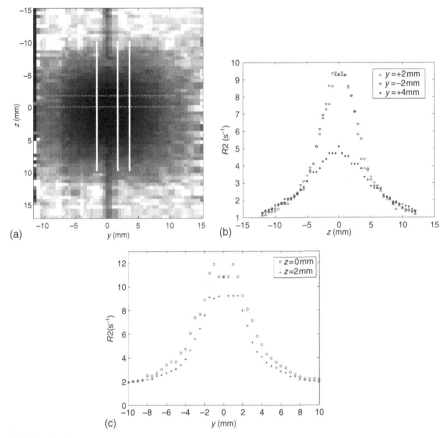

FIGURE 10.22
(a) The coronal *T2* map of Figure 10.17 with five directions marked to facilitate an example on the symmetries of the dose distribution around a single brachytherapy source irradiation. Three directions are shown along the catheter axis at distances $y = -2$ mm, $y = +2$ mm, and $y = +4$ mm from it and two directions are shown across the catheter axis at distances $z = 2$ mm and $z = 0$ mm. (b) The R2 profiles measured at the three directions along the catheter axis shown in Figure 10.22a. (c) The R2 profiles measured at the two directions across the catheter axis shown in Figure 10.22a.

between the actual irradiation center and the center of the imaging voxel where it is supposedly located.

In order to minimize the above uncertainties, the irradiation center could be located in a subpixel scale. To this end, a minimization routine of the standard deviation in R2 values averaged radially on an axial plane (perpendicular to the source long axis) has been proposed.[49] This statistical approach, however, is insensitive to potential misalignment of the source with the scanning volume. In this case, source center localization should rely on the physical properties of the measured distribution and mainly the predominance of the geometry factor.[50]

Besides volume averaging, discussed in Section 10.5.3 and summarized in Figure 10.18, other systematic uncertainties in single source polymer gel dosimetry have also been reported in the literature. These include susceptibility artifacts close to plastic catheters introduced in the gel to facilitate irradiation which affect the first few voxels outside the catheter as well as a dose overshooting effect close to the source.[49] Dose overshooting is attributed to the diffusion of free monomers towards high dose regions where monomers have been depleted and response has reached saturation. This effect, however, depends on the used polymer gel formulation through its characteristic dynamic dose range of response which, if sufficient, helps program irradiations to the distance/dose range of interest to brachytherapy without saturation effects.

A single irradiation of a polymer gel dosimeter is sufficient to provide the full three-dimensional dose distribution around a source such as that shown in Figure 10.23a for the VIPAR gel irradiated with a single dwell position of a HDR ^{192}Ir source (Figure 10.16b). This dose distribution was obtained using a brachytherapy calibration curve measured in a different gel vial, fitted with a monoexponential function. In Figure 10.23a the isodose surface of 45 Gy is presented and the anisotropy of the source dose distribution can be qualitatively observed. Figure 10.23b presents a cutout of the same dose distribution in a color scale to allow for the inspection of different dose levels. It should be noted that points close to the center of the presented dose distribution should not be considered as they correspond to points within the catheter which is depicted in Figure 10.23a.

Given the three-dimensional dose distribution, dose results can be readily obtained on any two-dimensional plane. An example is presented in Figure 10.24 where the two-dimensional dose distribution of an ^{192}Ir HDR source is presented on the central coronal plane containing the source. In this figure the symmetry of the dose distribution with respect to the source longitudinal axis as well as the anisotropy can be observed. Although such two-dimensional dose distributions could be obtained with a comparable spatial resolution by conventional dosimetry methods such as radiochromic film measurements, gel dosimetry presents additional advantages, such as water equivalence and nonsignificant relative energy corrections even for low energy sources such as ^{125}I and [103]Pd.[53]

Polymer gel measured dose distributions can also be readily obtained in the form of one-dimensional profiles in any direction down to short distances from a source and with a spatial resolution superior to that achieved with any conventional detector. Figure 10.25a and Figure 10.25b compare experimental and expected dose values along the transverse bisector and the longitudinal axis of the microSelectron ^{192}Ir HDR source, respectively. Uncertainty of experimental results was calculated by error propagation of the uncertainty in the calibration curve and the standard deviation of $R2$ values averaged according to the source cylindrical symmetry. The uncertainty in reporting distance from the source

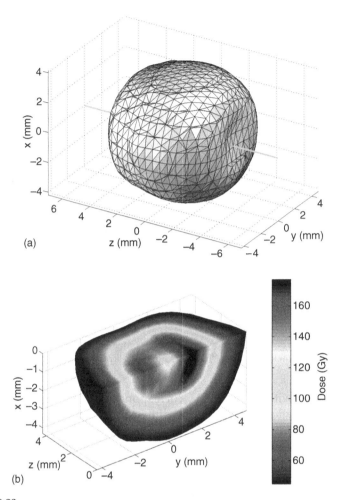

FIGURE 10.23
(a) The 45 Gy, three-dimensional isodose surface for the VIPAR gel irradiated with a single ^{192}Ir HDR source dwell position presented in Figure 10.16. (b) A cutout of the three-dimensional isodose surface in Figure 10.23a presented in a color scale to depict the dose distribution.

corresponds to half the imaging voxel dimension in this plane since the source center was located within an imaging voxel.

Experimental results in Figure 10.25 present satisfying symmetry as well as agreement with the expected dose distribution. The dose underestimation at points closest to the source can be explained by volume averaging effects due to the finite MRI voxel dimensions. A discussion on individual point uncertainties however, although suitable for point detectors, undermines the potential of gel dosimetry which provides complete dose distributions in a single irradiation. Gel dosimetry has received critique with regard to its accuracy in measuring single dose values.[178,183] On the other hand, however, it has been shown to provide brachytherapy source dose rate constant results

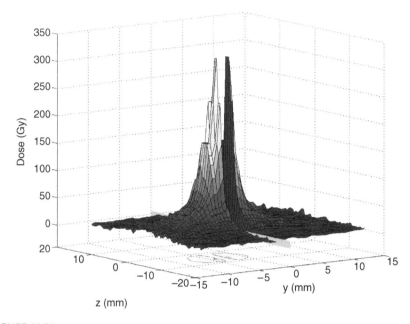

FIGURE 10.24
The two-dimensional dose distribution on the central coronal plane containing the source for the VIPAR gel irradiated with a single ^{192}Ir HDR source dwell position presented in Figure 10.16.

that agree to within 1% with corresponding accurate Monte Carlo calculations[144] when measured one-dimensional dose distributions, such as that presented in Figure 10.25a are employed.

With regard to polymer gel dosimetry of single brachytherapy sources it is worth mentioning that besides its advantages the method was only recently successfully employed for ^{125}I LDR sources. While ^{137}Cs and ^{192}Ir sources are characterized from relatively increased photon energies and from air kerma strengths in the order of tenths to thousands μGy m^2 h^{-1} (1 μGy m^2 h^{-1} = 1 U), interstitial brachytherapy ^{125}I seeds are low energy/LDR sources with air kerma strength in the order of 1 U. Prolonged irradiation times are therefore needed to achieve acceptable measurement accuracy with ^{125}I seeds since delivery of 1 Gy dose to water, at the reference distance of 1 cm along the transverse source bisector, corresponds to an irradiation time in the order of tenths of days for ^{125}I compared with tenths of hours for ^{137}Cs and tenths of seconds for ^{192}Ir.

The first polymer gel dosimetry studies of ^{125}I sources in the literature were those of Ibbot et al.,[46] who used BANG polymer gel calibrated using a 6 MV Linac photon beam, and Heard and Ibbot,[52] who used MAGIC polymer gel calibrated using a well-characterized ^{125}I source to measure the dose distribution of a different type ^{125}I source. However, results presented in both studies were limited, and were either not compared[46] or in disagreement[52] with expected, relative dose distributions. A study using

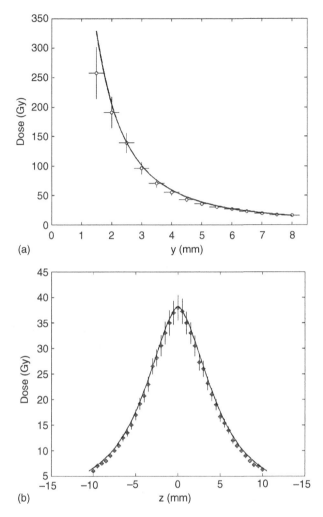

FIGURE 10.25

A comparison of VIPAR gel measured and Monte Carlo calculated dose distributions for the microSelectron ^{192}Ir HDR source (a) along the transverse bisector, and (b) along the longitudinal axis, at a 5 mm distance.

the PABIG gel formulation showed that while linearity is sustained the sensitivity of dose response to the ^{125}I irradiation conditions (LDR and prolonged irradiation) is significantly increased relative to the typical value for [192]Ir (or equivalently ^{60}Co or 6 MV x-rays).[184] This finding is probably attributed to physicochemical processes and suggests that polymer gel dosimetry of ^{125}I or [103]Pd LDR sources requires a calibration curve from a gel that is not only stored and MR scanned along with the experimental gel, but also irradiated in the same energy. Along these lines, the polymer gel-MRI method was employed for the dosimetric characterization of a new ^{125}I low dose rate seed (IsoSeed®model 125.S17).[185] Results in the form of the AAPM TG-43 dosimetric formalism were found to agree with corresponding MC calculations within experimental uncertainties which are smaller for the polymer gel method compared to TLD. It was therefore concluded that the method could be used at least supplementary to the established techniques for the dosimetric characterization of new low energy and low dose rate interstitial brachytherapy seeds.[185]

10.5.5.2 Treatment Plan Dose Verification

Single brachytherapy source irradiation of polymer gel dosimeters is efficient in obtaining calibration data (using a well-characterized source) and, potentially, for measuring the dosimetric properties of new sources. However, the real potential of the method relates to the dose verification of actual brachytherapy treatment plans involving complex irradiation regimes with multiple source positions and/or catheters which are hard, if not impossible, to verify with conventional dosimeters.

An example is brachytherapy of coronary or peripheral vessels following balloon angioplasty. In such intracoronary and intravascular brachytherapy applications acute dose delivery in the vessel, which may be 2 to 5 mm in diameter, is required. Owing to this short distance range of interest relative to conventional brachytherapy the reference distance for dosimetry was shifted from 10 mm[1] to 2 mm.[103] At such short distances, however, measurements with point dosimeters are difficult and may involve large systematic uncertainties while radiochromic film can be used for verification of relative dose distributions in selected planes.

Polymer gel dosimetry has been successfully employed in intravascular HDR brachytherapy for peripheral vessels.[50,171] Figure 10.26a presents the three-dimensional dose distribution measured for an irradiation with a Nucletron microSelectron ^{192}Ir remote afterloader following the PARIS clinical trial protocol. The source was programmed to deliver the protocol dose of 14 Gy at 5 mm away from the long axis of a single catheter fixed in a VIPAR gel phantom, using 15 source dwell positions with a 5 mm source step. This dose distribution allows dosimetry in any selected plane. Figure 10.26b presents the measured dose distribution in a coronal plane containing the long axis of the catheter acquired using a calibration curve obtained from a gel vial irradiated with a single source dwell position. Dose distributions measured with the polymer gel method, such as that shown projected on the yz plane in Figure 10.27b, have been found to compare favorably with corresponding treatment planning system and analytical model calculations thus verifying ^{192}Ir HDR intravascular applications.[50,51]

Another example of a complex brachytherapy irradiation regime where polymer gel dosimetry has been successfully employed for dose verification is ^{192}Ir HDR brachytherapy for dose escalation or as monotherapy of prostate cancer. The technique involves the ultrasound guided, transperineal

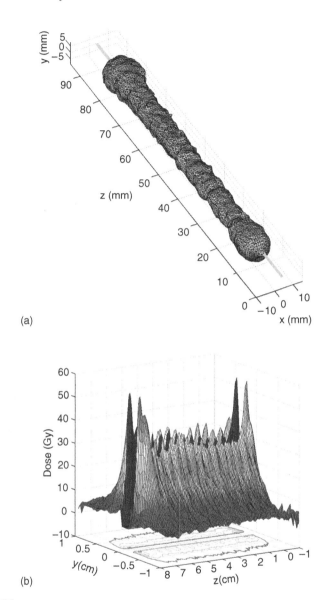

(a)

(b)

FIGURE 10.26
(a) The three-dimensional dose distribution measured for an irradiation of VIPAR gel with a Nucletron microSelectron ^{192}Ir remote afterloader following the PARIS intravascular clinical trial protocol. (b) The two-dimensional dose distribution on the central coronal plane of the three-dimensional dose distribution in Figure 10.26a.

implantation of multiple catheters through a fixed template. Postimplant, computed tomography-based treatment planning is performed with source dwell time and position optimization employed to conform the dose distribution to the prostate while sparing at risk organs in proximity

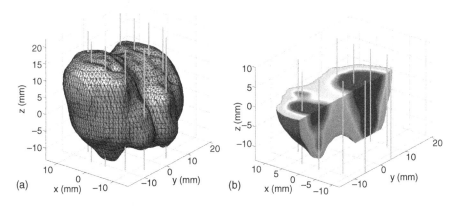

FIGURE 10.27
(a) The 20 Gy, three-dimensional isodose surface measured by a VIPAR gel irradiated with a clinical treatment plan of prostate monotherapy using 39 source dwell positions in 10 catheters which are also depicted. (b) A cutout of the three-dimensional isodose surface in Figure 10.27a presented in a color scale to depict the dose distribution.

(rectum, bladder base) or within the planning target volume (urethra). A fractionated treatment scheme is followed (three or four fractions) to deliver a dose of 9.5 Gy prescribed to the periphery of the gland for monotherapy, or 7 Gy for boost treatment.

Thermoluminescence dosimetry has been evaluated for *in vivo* dose verification of ^{192}Ir HDR prostate brachytherapy using an extra catheter loaded with TLD-100 rods.[31] The *in vivo* character of the method allows for remedial action in subsequent treatment fractions if a significant disagreement is found between the measured one-dimensional dose profile and corresponding treatment planning calculations. The method, however, is invasive and time inefficient to be employed in everyday practice. Most important, measurements of one-dimensional profiles through the planning target volume are only capable of revealing significant delivered dose discrepancies from the treatment plan, such as catheter misregistration.

In contrast, the polymer gel method inherently provides dose results with a fine spatial resolution at any point within a three-dimensional implant. A real clinical treatment plan has been applied to a VIPAR gel phantom.[172] The treatment plan involved a total of 10 catheters and 39 programmed source dwell positions and the prescription dose was set to 20 Gy (instead of 9.5 Gy used in clinical practice) which lies safely within the extended linear dose response range of the gel, since polymer gel-MRI associated statistical uncertainties are reduced in the 10 to 30 Gy region.[172] Figure 10.27a presents the three-dimensional measured dose distribution for the reference dose, also indicating the position of the catheters. Figure 10.27b presents a cutout of the same dose distribution in a color scale to allow for the inspection of different dose levels. Experimental results were qualitatively compared with corresponding treatment planning distributions in the form of three- as well as two-dimensional dose distributions.[172] Furthermore, polymer gel measured and treatment planning calculated dose distribution differences were quantified using the the γ index concept. This method accounts for both the dose difference and the distance to agreement between measured and calculated results on a pixel by pixel basis. The study showed that the majority of points above 50% of the reference dose passed the set 5% dose difference and 3 mm distance to agreement criteria. Systematic offsets were observed at specific segments of the planning target volume that implied uncertainty due to catheter reconstruction, which was however, sustained within the experimental uncertainties. This study concluded that the polymer gel-MRI method is capable of verifying complex brachytherapy irradiation schemes.

References

1. Nath, R., Anderson, L., Luxton, G., Weaver, K., Williamson, J.F., and Meigooni, A.S. Dosimetry of interstitial brachytherapy sources: Recommendations of the AAPM Radiation Therapy Committee Task Group 43, *Med. Phys.*, 22, 209–234, 1995.
2. Rivard, M., Coursey, B., Hanson, W., Hug, S., Ibbot, G., Mitch, M., Nath, R., and Williamson, J. Update of AAPM Task Group No. 43 Report: A revised AAPM protocol for brachytherapy dose calculations, *Med. Phys.*, 31:3, 633–674, 2004.
3. Williamson, J.F. and Meigooni, A.S. Quantitative dosimetry methods in brachytherapy, *Brachytherapy Physics, AAPM Summer School 1994*, J.F. Williamson, B.R. Thomadsen and R. Nath, eds., Medical Physics Publishing Corporation, Madison, Chapter 5, pp. 87–133, 1995.
4. Saylor, W.L. and Dillard, M. Dosimetry of ^{137}Cs sources with the Fletcher-Suit gynecological applicator, *Med. Phys.*, 3:2, 117–119, 1976.
5. Serago, C.C.F., Houdek, P.V., Pisciotta, V., Schwade, J.G., Abitbol, A.A., Lewin, A.A., Poole, D.O., and Marcial-Vega, V. Scattering effects on the dosimetry of iridium-192, *Med. Phys.*, 18:6, 1266–1270, 1991.
6. Siwek, R.A., O'Brien, P.F., and Leung, P.M.K. Shielding effects of Selectron applicator and pellets on isodose distributions, *Radiother. Oncol.*, 20, 132–138, 1991.
7. Baltas, D., Kramer, R., and Loeffler, E. Measurements of the anisotropy of the new Ir-192 source for the microSelectron-HDR, Activity Special Report, No. 3, Nucletron B.V., The Netherlands, 1993.
8. Waterman, F.M. and Holcomb, D.E. Dose distributions produced by a shielded vaginal cylinder using a high-activity iridium-192 source, *Med. Phys.*, 21:1, 101–106, 1994.
9. Verellen, D., De Neve, W., van den Heuvel, F., Storme, G., Coen, V., and Coghe, M. *Med. Phys.*, 21:11, 1677–1684, 1994.
10. Fontenla, A.D.P., Chiu-Tsao, S.T., Chui, C.S., Reiff, J.E., Anderson, L.L., Huang, D.Y.C., and Schell, M.C. Diode dosimetry of models 6711 and 6712 ^{125}I seeds in a water phantom, *Med. Phys.*, 19, 391–399, 1992.

11. Williamson, J.F., Perera, H., Li, Z., and Lutz, W.R. Comparison of calculated and measured heterogeneity correction factors for ^{125}I, ^{137}Cs, and ^{192}Ir brachytherapy sources near localized heterogeneities, *Med. Phys.*, 20, 209–222, 1993.

12. Chiu Tsao, S.T., Anderson, L.L., O'Brien, K., and Sanna, R. Dose rate determination for ^{125}I seeds, *Med. Phys.*, 17, 815–825, 1990.

13. Luxton, G., Astrahan, M.A., Findley, D.O., and Petrovich, Z. Measurement of dose rate from exposure-calibrated ^{125}I seeds, *Int. J. Radiat. Oncol. Biol. Phys.*, 18, 1199–1207, 1990.

14. Nath, R., Meigooni, A.S., and Meli, J.A. Dosimetry on the transverse axes of ^{125}I and ^{192}Ir interstitial brachytherapy sources, *Med. Phys.*, 17, 1032–1040, 1990.

15. Weaver, K.A., Smith, V., Huang, D., Barnett, C., Schell, M.C., and Ling, C. Dose parameters of ^{125}I and ^{192}Ir seed sources, *Med. Phys.*, 16, 636–643, 1990.

16. Williamson, J.F. Comparison of measured and calculated dose rates in water near I-125 and Ir-192 seeds, *Med. Phys.*, 18, 776–786, 1991.

17. Valicenti, R.K., Kirov, A.S., Meigooni, A.S., Mishra, V., Das, R.K., and Williamson, J.F. Experimental validation of Monte Carlo dose calculations about a high intensity ^{192}Ir source of pulsed dose rate brachytherapy, *Med. Phys.*, 122, 821–829, 1994.

18. Kirov, A.S., Williamson, J.F., Meigooni, A.S., and Zhu, Y. TLD diode and Monte Carlo dosimetry of an ^{192}Ir source for high dose rate brachytherapy, *Phys. Med. Biol.*, 40, 2015–2036, 1995.

19. Meigooni, A.S., Mishra, V., Panth, H., and Williamson, J.F. Instrumentation and dosimeter-size artifacts in quantitative thermoluminescence dosimetry of low-dose fields, *Med. Phys.*, 22, 555–561, 1995.

20. Piermattei, A., Azario, L., Rossi, G., Soriani, A., Arcovito, G., Ragona, R., Galelli, M., and Taccini, G. Dosimetry of ^{169}Yb seed model X1267, *Phys. Med. Biol.*, 40, 1317–1330, 1995.

21. Anctil, J.-C., Clark, B.G., and Arsenault, C.J. Experimental determination of dosimetry functions of Ir-192 sources, *Med. Phys.*, 25, 2279–2287, 1988.

22. Karaiskos, P., Angelopoulos, A., Sakelliou, L., Sandilos, P., Antypas, C., Vlachos, L., and Koutsouveli, E. Monte Carlo and TLD dosimetry of an ^{192}Ir high dose rate brachytherapy source, *Med. Phys.*, 25, 1975–1984, 1998.

23. Meigooni, A.S., Meli, J.A., and Nath, R. Influence of the variation of energy spectra with the depth in the dosimetry of ^{192}Ir using LiF TLD, *Phys. Med. Biol.*, 33, 1159–1170, 1988.

24. Pradhan, A.S. and Quast, U. In phantom response of LiF TLD-100 for dosimetry of ^{192}Ir HDR source, *Med. Phys.*, 5, 1025–1029, 2000.

25. Sloboda, R.S. and Menon, G.V. Experimental determination of the anisotropy function and anisotropy factor for model 6711 I-125 seeds, *Med. Phys.*, 27, 1789–1799, 2000.

26. Meigooni, A.S., Sowards, K., and Soldano, M. Dosimetric characteristics of the Intersource ^{103}Palladium brachytherapy source, *Med. Phys.*, 27, 1093–1100, 2000.

27. Meigooni, A.S., Gearheart, D.M., and Sowards, K. Experimental determination of dosimetric characteristics of Best® ^{125}I brachytherapy source, *Med. Phys.*, 27, 2168–2173, 2000.

28. Nath, R. and Yue, N. Dosimetric characterization of a newly designed encapsulated interstitial brachytherapy source of iodine-125- model LS-1 Brachyseed TM, *Appl. Radiat. Isot.*, 55, 813–821, 2001.

29. Patel, N.S., Chiu Tsao, S.T., Williamson, J.F., Fan, P., Duckworth, T., Shasha, D., and Harrison, L.B. Thermoluminescent dosimetry of the Symmetra™ [125]I model I25.S06 interstitial brachytherapy seed, *Med. Phys.*, 28, 1761–1769, 2001.

30. Anagnostopoulos, G., Baltas, D., Karaiskos, P., Sandilos, P., Papagiannis, P., and Sakelliou, L. Thermoluminescent dosimetry of the selectSeed ^{125}I interstitial brachytherapy seed, *Med. Phys.*, 29, 709–716, 2002.

31. Anagnostopoulos, G., Baltas, D., Geretschlaeger, A., Martin, T., Papagiannis, P., Tselis, N., and Zamboglou, N. In vivo TLD dose verification of transperineal ^{192}Ir HDR brachytherapy using CT-based planning for the treatment of prostate cancer, *Int. J. Radiat. Oncol. Biol. Phys.*, 57, 1183–1191, 2003.

32. Perera, H., Williamson, J.F., Li, Z., Mishra, V., and Meigooni, A.S. Dosimetric characteristics, air-kerma strength calibration and verification of Monte Carlo simulation fore a new [169]Yb brachytherapy source, *Int. J. Radiat. Oncol. Biol. Phys.*, 28, 953–970, 1994.

33. Piermattei, A., Azario, L., Monaco, G., Soriani, A., and Arcovito, G. p-type silicon detector for brachytherapy dosimetry, *Med. Phys.*, 22, 835–839, 1995.

34. Flühs, D., Heintz, M., Indenkämpen, F., Wieczorek, C., Kolanoski, H., and Quast, U. Direct reading measurement of absorbed dose with plastic scintillators—The general concept and applications to ophthalmic plaque dosimetry, *Med. Phys.*, 23, 427–434, 1996.

35. Williamson, J.F., Dempsey, J.F., Kirov, S.V., Monroe, J.I., Binns, W.R., and Hedtjärn, H. Plastic scintillator response to low-energy photons, *Phys. Med. Biol.*, 44, 857–872, 1999.

36. Kirov, A.S., Hurlbut, C., Dempsey, J.F., Shrinivas, S.B., Epstein, J.W., Binns, W.R., Dowkontt, P.F., and Williamson, J.F. Towards two dimensional brachytherapy dosimetry using plastic scintillator: New highly efficient water equivalent plastic scintillator materials, *Med. Phys.*, 26, 1515–1523, 1999.

37. Rustgi, S.N. Evaluation of the dosimetric characteristics of a diamond detector for photon beam measurements, *Med. Phys.*, 22, 567–570, 1995.

38. Rustgi, S.N. Application of a diamond detector to brachytherapy dosimetry, *Phys. Med. Biol.*, 43, 2085–2094, 1998.

39. Nakano, T., Suchowerska, N., Bilek, M.M., McKenzie, D.R., Nag, N., and Kron, T. High dose-rate brachytherapy source localisation: Positional resolution using a diamond detector, *Phys. Med. Biol.*, 48, 2133–2146, 2003.

40. Muench, P.J., Meigooni, A.S., Nath, R., and McLaughlin, W.L. Photon energy dependence of the sensitivity of radiochromic film and comparison with silver halide film and LiF TLDs used for brachytherapy dosimetry, *Med. Phys.*, 18, 769–775, 1991.

41. Chiu-Tsao, S.T., de la Zerda, A., Lin, J., and Kim, J.H. High-sensitivity GafChromic film dosimetry for ^{125}I seed, *Med. Phys.*, 21, 651–657, 1994.

42. Hasson, B.F. Chemical dosimetry in the near-zone of brachytherapy sources, *Med. Phys.*, 25, 2076, 1998.

43. Dempsey, J.F., Low, D.A., Kirov, A.S., and Williamson, J.F. Quantitative optical densitometry with scanning-laser film digitizers, *Med. Phys.*, 26, 1721–1731, 1999.

44. Bohm, D., Pearson, D.W., and Das, R.K. Measurements and Monte Carlo calculations to determine the absolute detector response of radiochromic film for brachytherapy dosimetry, *Med. Phys.*, 28, 142–146, 2001.

45. Farajollahi, A.R., Bonnett, D.E., Ratcliffe, A.J., Aukett, R.J., and Mills, J. An investigation into the use of polymer gel dosimetry in low dose rate brachytherapy, *Br. J. Radiol.*, 72, 1085–1092, 1999.

46. Ibbott, G., Maryanski, M., Drogin, A., Gearheart, D., Ranade, M., Painter, T., and Meigooni, A. *Characterization of a New Brachytherapy Source by BANG® Gel Dosimetry*, in Proceedings of the 1st International Workshop on Radiation Therapy Gel Dosimetry (DosGel '99), pp. 196–198, 1999.

47. McJury, M., Tappers, P.D., Cosgrove, V.P., Murphy, P.S., Griffin, S., Leach, M.O., Webb, S., and Oldham, M. Experimental 3D dosimetry around a high-dose-rate clinical ^{192}Ir source using polyacrylamide gel (PAG) dosimeter, *Phys. Med. Biol.*, 44, 2431–2444, 1999.

48. Abramson, D. The measurement of three dimensional dose distribution of a ruthenium- 106 ophthalmological applicator using magnetic resonance imaging of BANG polymer gels, *J. Appl. Clin. Med. Phys.*, 2, 85–89, 2001.

49. De Deene, Y., Reynaert, N., and De Wagter, C. On the accuracy of monomer/polymer gel dosimetry in the proximity of a high-dose-rate ^{192}Ir source, *Phys. Med. Biol.*, 46, 2801–2825, 2001.

50. Papagiannis, P., Pappas, E., Kipouros, P., Angelopoulos, A., Sakelliou, L., Baras, P., Karaiskos, P., Seimenis, I., Sandilos, P., and Baltas, D. Dosimetry close to an ^{192}Ir HDR source using N-vinylpyrrolidone based polymer gels and magnetic resonance imaging, *Med. Phys.*, 28, 1416–1426, 2001.

51. Kipouros, P., Anagnostopoulos, G., Angelopoulos, A., Baltas, D., Baras, P., Drolapas, A., Karaiskos, P., Pantelis, E., Papagiannis, P., Sakelliou, L., and Seimenis, I. Dosimetric calculations and VIPAR polymer gel dosimetry close to the microSelectron HDR, *Zeit. Med. Physik.*, 12, 252–259, 2002.

52. Heard, M.P. and Ibbot, G.S. Measurement of brachytherapy sources using MAGIC gel, *J. Phys.: Conf. Ser.*, 3, 221–223, 2004.

53. Pantelis, E., Karlis, A.K., Kozicki, M., Papagiannis, P., Sakelliou, L., and Rosiak, J.M. Polymer gel water equivalence and relative energy response with emphasis on low photon energy dosimetry in brachytherapy, *Phys. Med. Biol.*, 49, 3495–3514, 2004.

54. Perera, H., Williamson, J.F., Monthofer, S.P., Binns, W.R., Klammen, J.C., Fuller, G.A., and Wong, J.W. Rapid two-dimensional dose measurement in brachytherapy using plastic scintillator sheet: Linearity, signal-to-noise ratio and energy response characteristics, *Int. J. Radiat. Oncol. Biol. Phys.*, 23, 1059–1069, 1992.

55. Williamson, J.F. Recent developments in basic brachytherapy physics in radiation therapy physics, *Medical Radiology Diagnostic Imaging and Radiation Oncology*, A.R. Smith, ed., Series eds.: A.L. Baert, L.W. Brady, H.P. Heilmann, M. Molls, and K. Sartor, Springer, Berlin, pp. 247–302, 1995.

56. Hubbell, J.H. and Seltzer, S.M., *Tables of x-Ray Mass Attenuation Coefficients and Mass Energy-Absorption Coefficients*, Version 1.4 on http://physics.nist.gov/xaamdi. Originally published as NISTIR 5632, National Institute of Standards and Technology, Gaithersburg, 1995.

57. Lymperopoulou, G., Papagiannis, P., Sakelliou, L., Milickovic, N., Giannouli, S., and Baltas, D. A dosimetric comparison of ^{169}Yb versus ^{192}Ir for HDR prostate brachytherapy, *Med. Phys.*, 32, 3832–3842, 2005.

58. Krieger, H. and Baltas, D. eds. Deutsche Gesellschaf für Medizinische Physik, Praktische Dosimetrie in der HDR-Brachytherapie DGMP-Bericht Nr. 13, 1999.

59. Deutsches Institut für Normung e.V. [DIN] Clinical Dosimetry. Part 2. Brachytherapy with Sealed Gamma Sources (Klinische Dosimetrie. Teil 2. Brachytherapie mit umschlossenen gammastrahlenden radioaktiven Stoffen) DIN 6809 (Berlin: DIN), 1993.

60. Meigooni, A.S., Meli, J.A., and Nath, R. A comparison of solid phantoms with water for dosimetry of ^{125}I brachytherapy sources, *Med. Phys.*, 15, 695–701, 1988.

61. Meli, J.A., Meigooni, A.S., and Nath, R. On the choice of phantom material for the dosimetry of ^{192}Ir sources, *Int. J. Rad. Oncol. Biol. Phys.*, 14, 587–594, 1988.

62. Rashid, H., Bjarngard, B.E., Chin, L.M., and Rice, R.K. Dosimetry of ^{125}I sources in a low-density material using scaling, *Med. Phys.*, 20, 765–768, 1993.

63. Meigooni, A.S., Li, Z., and Williamson, J.F. A comparative study of dosimetric properties of plastic water and solid water in brachytherapy applications, *Med. Phys.*, 21, 1983–1987, 1994.

64. Luxton, G. Comparison of radiation dosimetry in water and in solid phantom materials for I-125 and Pd-103 brachytherapy sources: EGS4 Monte Carlo study, *Med. Phys.*, 21, 631–641, 1994.

65. Wallace, R.E. Evaluated phantom material for ^{125}I and ^{103}Pd dosimetry, Poster SUDD-EXH-12, AAPM Annual Meeting, Montreal, Canada, 2002.

66. Hill, R., Holloway, L., and Baldock, C. A dosimetric evaluation of water equivalent phantoms for kilovoltage x-ray beams, *Phys. Med. Biol.*, 50, N331–N344, 2005.

67. Briesmeister, J.F. ed. *MCNPTM—A general Monte Carlo N-particle transport code: version 4C Report LA-13709-M*, Los Alamos National Laboratory, Los Alamos, N.M., 2000.

68. Jayachandran, C.A. Calculated effective atomic number and kerma values for tissue-equivalent and dosimetry material, *Phys. Med. Biol.*, 16, 613–623, 1971.

69. Hawkes, D.J. and Jackson, D.F. An accurate parametrisation of the x-ray attenuation coefficient, *Phys. Med. Biol.*, 25, 1167–1171, 1971.

70. Jackson, D.F. and Hawkes, D.J. x-Ray attenuation coefficients of elements and mixtures, *Phys. Rep.*, 70, 169–233, 1981.

71. Shivaramu, Amutha, R., and Ramprasath, V. Effective atomic numbers and mass attenuation coefficients of some thermoluminescent dosimetric compounds for total photon interaction, *Nucl. Sci. Eng.*, 132, 148–153, 1999.

72. Sidhu, G.S., Singh, P.S., and Mudahar, G.S. A study of energy and effective atomic number dependence of the exposure build-up factors in biological samples, *J. Radiol. Prot.*, 20, 53–68, 2000.

73. Shivaramu, Effective atomic numbers for photon energy absorption and photon attenuation of tissues from human organs, *Med. Dosim.*, 27, 1–9, 2002.

74. Jones, A.K., Hintenlang, D.E., and Bolch, W.E. Tissue-equivalent materials for construction of tomographic dosimetry phantoms in pediatric radiology, *Med. Phys.*, 30, 2072–2081, 2003.

75. Pérez-Calatayud, J., Granero, D., and Ballester, F. Phantom size in brachytherapy source dosimetric studies, *Med. Phys.*, 31, 2075–2081, 2004.

76. Siwek, R.A., O'Brien, P.F., and Leung, P.M.K. Shielding effects of Selectron applicator and pellets on isodose distributions, *Radiother. Oncol.*, 20, 132–138, 1991.

77. Waterman, F.M. and Holcomb, D.E. Dose distributions produced by a shielded vaginal cylinder using a high-activity iridium-192 source, *Med. Phys.*, 21, 101–106, 1994.

78. Verellen, D., De Neve, W., van den Heuvel, F., Storme, G., Cohen, V., and Coghe, M. On the determination of the effective transmission factor for stainless steel avoid shielding and estimation of their shielding efficacy for the clinical use, *Med. Phys.*, 21, 1677–1684, 1994.

79. Selbach, H.J. and Andrássy, M. Experimentally determined TG-43 data for the BEBIG MultiSource ^{192}Ir HDR source, unpublished data, personal communication, 2005.

80. Selbach, H.J. and Andrássy, M. Experimentally determined TG-43 data for the BEBIG MultiSource ^{60}Co HDR source, unpublished data, personal communication, 2005.

81. Pychlau, P. Messung kleiner Aktivitäten mit Teletherapie-Ionisationskammern. Ermittlung des Kalibrierfaktors N_{er}, *Zeit. Med. Physik.*, 1, 194–198, 1991.

82. Deutsches Institut für Normung e.V. [DIN], Procedures of dosimetry with probetype detectors for photon and electron radiation—Part 2 Ionisation dosimetry (Dosismessverafhren nach der Sondenmethode fuer Photonen- und Elektronenstrahlung. Teil 2: ionisationsdosimetrie) DIN 6800 (Berlin: DIN), 1997.

83. Deutsches Institut für Normung e.V. [DIN], Clinical dosimetry—Part 5: Application of x-rays with peak voltages between 100 and 400 kV in radiotherapy (Klinische Dosimetrie. Teil 5: Anwendung von Roentgenstrahlen mit Roehrenspannungen von 100 bis 400 kV in der Starhlentherapie) DIN 6809 (Berlin: DIN), 1996.

84. International Commission on Radiation Units and Measurements, Dosimetry of High-Energy Photon Beams based on Standards of Absorbed Dose to Water ICRU report 64 (Bethesda: ICRU), 2001.

85. IPEMB, The IPEMB code of practice for the determination of absorbed dose for x-rays bellow 300 kV generating potential (0.035 mm Al-4 mm Cu HVL; 10–300 kV generating potential), *Phys. Med. Biol.*, 41, 2605–2625, 1996.

86. International Atomic Energy Agency, Absorbed dose determination in external beam radiotherapy, Technical Reports Series, No. 398, (Vienna: IAEA), 2000.

87. Ma, C.M., Cofefy, C.W., DeWerd, L.A., Liu, C., Nath, R., Seltzer, S.M., and Seuntjens, J.P. AAPM protocol for 40–300 kV x-ray beam dosimetry in radiotherapy and radiobiology, *Med. Phys.*, 28, 868–893, 2001.

88. IPEMB, Addendum to the IPEMB code of practice for the determination of absorbed dose for x-rays below 300 kV generating potential (0.035 mm Al-4 mm Cu HVL), *Phys. Med. Biol.*, 50, 2739–2748, 2005.

89. McKeever, S.W.S. *Thermoluminescence of Solids*, Cambridge University Press, Cambridge, 1985.

90. Wiedemann, E. and Scmidt, G.C. Ueber Lumineszenz, *Ann. Phys. Chem. Neue Folge*, 54, 604, 1895.

91. Curie, M. Radioactive substances, 1904, (English translation of doctoral thesis presented to the Faculty of Science, Paris), Greenwood Press, WestPoint, 1961.

92. Boyle, R. *Register of the Royal Society*, p. 213, 1663.

93. Cameron, J.R., Suntharalingam, N., and Kenney, G.N. *Thermoluminescent Dosimetry*, University of Winsconsin, Madison, WI, 1968.

94. Horowitz, Y.S. *Thermoluminescence and Thermoluminescent Dosimetry*, CRC Press, Boca Raton, 1984.

95. Chen, R. and McKeever, S.W.S. *Theory of thermoluminescence and Related Phenomena*, World Scientific Publishing Co Pte Ltd, Singapore, 1997.

96. Oberhofer, M. and Sharmann, A. *Applied Thermoluminescence Dosimetry ISPRA Courses*, The Commision of the European Communiites, Adam Hilger Ltd, Bristol, 1981.

97. Anderson, L.L., Nath, R., Weaver, K.A., Nori, D., Phillips, T.L., Son, Y.H., Chiu Tsao, S.T., Meigooni, A.S., Meli, J.A., and Smith, V. *Interstitial Brachytherapy, Physical, Biological and Clinical Considerations*, (Interstitial Collaborative Working Group) Raven Press, New York, 1990.

98. Das, R.K., Li, Z., Perrera, H., and Williamson, J.F. Accuracy of the Monte Carlo method in characterizing brachytherapy dosimeter energy response artifacts, *Phys. Med. Biol.*, 41, 995, 1996.

99. Harris, C.K., Elson, H.R., Lamba, M.A.S., and Foster, A.S.A. Comparison of the effectiveness of thermoluminescent crystals LiF: Mg, Ti and LiF:Mg, Cu, P for clinical dosimetry, *Med. Phys.*, 24, 1527, 1997.

100. Horowitz, Y.S. LiF:Mg, Ti versus LiF:Mg, Cu, P the competition heats up, *Rad. Prot. Dosim.*, 47, 135, 1993.

101. Wang, S., Chen, G., Wu, F., Li, Y., Zha, Z., and Zhu, J. Newly developed highly sensitive LiF: (Mg, Cu, P) TL chips with high S/N ratio, *Radiat. Prot. Dosim.*, 14, 223, 1986.

102. Busuoli, G. *Precision and Accuracy of TLD Measurements in Applied Thermoluminescence Dosimetry*, Adam Hilger Ltd, Bristol, UK, Chapter 8, 1981.

103. Nath, R., Amols, H., Coffey, Ch., Whiting, J., Cole, P.E., Crocker, I., and Schwartz, R. Intravascular brachytherapy physics: report of the AAPM radiation therapy committee task group No. 60, *Med. Phys.*, 26, 119, 1999.

104. Petelenz, B., Bilski, P., Walichiewicz, P., Caca, P., and Wilczek, K. Thermoluminescence dosimetry of liquid ^{32}P sources of variable size and composition, *Radiat. Meas.*, 35, 245, 2002.

105. Maruyama, Y., Wierzbicki, J.G., Vigurn, B.M., and Kaneta, K. *Principles and Practices of Brachytherapy*, S. Nag, ed., Futura, Armonk, NY, p. 469, 1997.

106. Rivard, M.J., Wierzbicki, G., Van den Henvel, F., Martin, R.C., and McMahon, R.R. Clinical brachytherapy with neutron emitting 252Cf sources and adherence to AAPM TG-43 dosimetry protocol, *Med. Phys.*, 26, 87, 1999.

107. Cameron, J.R., Zimmermann, D.W., Kenney, G., Buch, R., Bland, R., and Grant, R. Thermoluminescent radiation dosimetry utilizing LiF, *Health Phys.*, 10:20, 1964.

108. Zimmermann, D.W., Rhyner, C., and Cameron, J.R. Annealing effects on the thermoluminescence of LiF, *Health Phys.*, 12, 525, 1966.

109. Yu, C. and Luxton, G. TLD dose measurement: a simplifed accurate technique for the dose range of from 0.5 cGy to 1000 cGy, *Med. Phys.*, 26, 1010, 1999.

110. Driscoll, C.M.H., Barthe, J.R., Oberhofer, M., Busuoli, G., and Hickman, C. Annealimg procedures for commonly used radiothermoluminescent materials, *Radiat. Prot. Dosim.*, 14: 17, 1986.

111. Booth, L.F., Johnson, T.L., and Attix, F.H. Lithium fluoride glow-peak growth due to annealing, *Health Phys.*, 23, 137, 1972.

112. McKnlay, A.F. Thermoluminescence dosimetry, *Medical Physics Handbooks*. Adam Hilger Ltd, Bristol, 1981.

113. Feist, H. Entwicklung der Thermolumineszenzdosimetrie mit LiF zu einer Praezissionsmethode für absolute Energiedosisbestimmungen in der Strahlentherapie mit Photonen- und Elektronenstrahlungen hoher Energie Habilitationsschrift, Ludwig-Maximillian Universität, München, 1992.

114. Mische, E.F. and McKeever, S.W.S. Mechanisms of supralinearity in lithium fuoride thermoluminescence dosimeters, *Radiat. Prot. Dosim.*, 29, 159, 1989.

115. Regulla, D.F. *Dosimetry ISPRA Courses*, The Commission of the European Communities. Adam Hilger Ltd, Bristol, Chapter 7, 1981.

116. Schulz, R.J., Almond, P.R., Cunningham, J.R., Holt, J.G., Loevinger, R., Suntharliningam, N., Wright, K.A., Nath, R., and Lempert, G.D. A protocol for the determination of absorbed dose from high-energy photon and electron beams, *Med. Phys.*, 10, 741, 1983.

117. Andreo, P., Cunningham, J.R., Hohlfeld, K., and Svensson, H. Absorbed Dose determination in photon and electron beams, Technical Reports Series No. 277 IAEA, 1985.

118. Wallace, R.E. Empirical dosimetric characterization of model 1125- SL ^{125}Iodine brachytherapy source in phantom, *Med. Phys.*, 27, 2796, 2000.

119. Weller, M.K., Slessinger, E.D., Wong, J.W., Van Dyke, J., and Leung, P.M.K. A practical method for precise thermoluminescent dosimetry, *Treat. Plan.*, 8, 22, 1983.

120. Horowitz, Y.S. Theory of thermoluminescence gamma dose response: the unified interaction model, *NIM Phys. Res. B*, 184, 68, 2001.

121. Sakelliou, L., Sakellariou, K., Angelopoulos, A., Perris, A., and Zarris, G. Dose rate distributions around ^{60}Co, ^{137}Cs, ^{198}Au, ^{192}Ir, ^{241}Am, ^{125}I (models 6702 and 6711) brachytherapy sources and the nuclide ^{99m}Tc, *Phys. Med. Biol.*, 37, 1859, 1992.

122. Brezovich, I.A., Duan, J., Pareek, P.M., Fiveash, J., and Ezekiel, M. In vivo urethral dose measurements: A method to verify high dose rate prostate treatments, *Med. Phys.*, 27, 2297, 2000.

123. Das, R.K., Li, Z., Perrera, H., and Williamson, J.F. Accuracy of the Monte Carlo method in characterizing brachytherapy dosimeter energy response artifacts, *Phys. Med. Biol.*, 41, 995, 1996.

124. Attix, F.H. *Introduction to Radiological Physics and Radiation Dosimetry*, Wiley, New York, 1986.

125. Mobit, P.N., Nahum, A.E., and Mayles, P. An EGS4 Monte Carlo examination of general cavity theory, *Phys. Med. Biol.*, 42, 1319, 1997.

126. Weaver, K.A. Response of LiF powder to ^{125}I photons, *Med. Phys.*, 11, 850, 1984.

127. Thomason, C. and Higgins, P. Radial dose distributions of Ir-192 and Cs-137 sources, *Med. Phys.*, 16, 254, 1989.

128. Karaiskos, P., Sakelliou, L., Sandilos, P., and Vlachos, L. Limitations of the point and line source approximations for the determination of geometry factors around brachytherapy sources, *Med. Phys.*, 27, 124, 2000.

129. Baltas, D., Giannouli, S., Garbi, A., Diakonos, F., Geramani, K., Ioannidis, G.T., Tsalpatouros, A., Uzunoglou, N., Kolotas, C., and Zamboglou, N. Application of the Monte Carlo integration (MCI) method for calculation of the anisotropy of ^{192}Ir brachytherapy sources, *Phys. Med. Biol.*, 43, 1783, 1998.

130. Gearheart, D.M., Drogin, A., Sowards, K., Meigooni, A.S., and Ibbott, G.S. Dosimetric characteristics of a new ^{125}I brachytherapy source, *Med. Phys.*, 27, 2278, 2000.

131. DosGel'99, *Proceedings of the 1st International Workshop on Radiation Therapy Gel Dosimetry*, Lexington, Kentucky, 21–23 July, 1999.

132. DosGel, *Proceedings of the 2nd International Conference on Radiotherapy Gel Dosimetry, Queensland University of Technology, Brisbane*, 18–21 November, 2001.

133. DosGel, *Proceedings of the Third International Conference on Radiotherapy Gel Dosimetry, Ghent University, Gent*, 13–16 September, 2004.

134. Fricke, H. and Morse, S. The chemical action of roentgen rays on dilute ferrous sulfate solutions as a measure of dose, *Am. J. Roent. Radiat. Ther.*, 18, 430, 1927.

135. Gore, J.C., Kang, Y.S., and Schultz, R.J. Measurement of radiation dose distributions by nuclear magnetic resonance imaging, *Phys. Med. Biol.*, 29, 1189, 1984.

136. Maryanski, M.J., Gore, J.C., Kennan, R.P., and Schulz, R.J. NMR relaxation enhancement in gels polymerised and crosslinked by ionizing radiations: a new approach to 3-D dosimetry by MRI, *Magn. Reson. Imaging*, 11, 253, 1993.

137. Maryanski, M.J., Schulz, R.J., Ibbott, G.S., Gatenby, J.C., Xie, J., Horton, D., and Gore, J.C. Magnetic resonance imaging of radiation dose distributions using a polymer-gel dosimeter, *Phys. Med. Biol.*, 39, 1437, 1994.

138. Maryanski, M.J., Audet, C., and Gore, J.C. Effects of cross-linking and temperature on the dose response of a BANG polymer gel dosimeter, *Phys. Med. Biol.*, 42, 303, 1997.

139. Maryanski, M.J., Ibbot, G.S., Eastman, P., Scultz, R.J., and Gore, J.C. Radiation therapy dosimetry using magnetic resonance imaging of polymer gels, *Med. Phys.*, 23, 699, 1996.

140. Murphy, P.S., Cosgrove, V.P., Leach, M.O., and Webb, S.A. Modifed polymer gel for radiotherapy dosimetry: assessment by MRI and MRS, *Phys. Med. Biol.*, 45, 3213, 2000.

141. Maryanski, M.J. and Barry, M.J. New supersensitive polymer gel dosimeter, *Med. Phys.*, 25, A107, 1998.

142. Maryanski, M.J. *Radiation-Sensitive Polymer gels: Properties and Manufacturing*, Dos Gel'99, p. 63, 1999.

143. Pappas, E., Maris, T., Angelopoulos, A., Paparigopoulou, M., Sakelliou, L., Sandilos, P., Voyiatzi, S., and Vlachos, L. A new polymer gel for magnetic resonance imaging (MRI) radiation dosimetry, *Phys. Med. Biol.*, 44, 2677, 1999.

144. Kipouros, P., Pappas, E., Baras, P., Hatzipanayoti, D., Karaiskos, P., Sakelliou, L., Sandilos, P., and Seimenis, I. Wide dynamic dose range of VIPAR polymer gel dosimetry, *Phys. Med. Biol.*, 46, 2143, 2001.

145. Kozicki, M., Kujawa, P., and Rosiak, J.M. Pulse radiolysis study of diacrylate macromonomer in aqueous solution, *Radiat. Phys. Chem.*, 65, 133, 2002.

146. Kozicki, M., Filipczak, K., and Rosiak, J.M. Reaction of hydroxyl radicals, H atoms and hydrated electrons with N,N'-methylenebisacrylamide in aqueous solution, a pulse radiolysis study, *Radiat. Phys. Chem.*, 68, 827, 2003.

147. Fong, P.M., Keil, D.C., Does, M.D., and Gore, J.C. Polymer gels for magnetic resonance imaging of radiation dose distributions at normal room atmosphere, *Phys. Med. Biol.*, 46, 3105, 2001.

148. De Deene, Y., Hurley, C., Venning, A., Vergote, K., Mather, M., Healy, B.J., and Baldock, C. A basic study of some normoxic polymer gel dosimeters, *Phys. Med. Biol.*, 47, 3441, 2002.

149. Gore, J.C., Ranade, M., Maryanski, M.J., and Schulz, R.J. Radiation dose distributions in three dimensions from tomographic optical density scanning of polymer gels: I. Development of an optical scanner, *Phys. Med. Biol.*, 41, 2695, 1996.

150. Maryanski, M.J., Zastavker, Y.Z., and Gore, J.C. Radiation dose distributions in three dimensions from tomographic optical density scanning of polymer gels: II. Optical properties of the BANG polymer gel, *Phys. Med. Biol.*, 41, 2705, 1996.

151. Oldham, M., Siewerdsen, J.H., Shetty, A., and Jaffray, D.A. High resolution gel-dosimetry by optical-CT and MR scanning, *Med. Phys.*, 28, 1436, 2001.

152. Oldham, M., Siewerdsen, J.H., Kumar, S., Wong, J., and Jaffray, D.A. Optical- CT gel dosimetry I: basic investigations, *Med. Phys.*, 30, 623, 2003.

153. Wuu, C.-S., Schiff, P., Maryanski, M.J., Liu, T., Borzillary, S., and Weinberger, J. Dosimetry study of Re-188 liquid balloon for intravascular brachytherapy using polymer gel dosimeters and laser beam optical CT scanner, *Med. Phys.*, 30, 132, 2003.

154. Xu, Y., Wuu, C., and Maryanski, M.J. Performance of a commercial optical CT scanner and polymer gel dosimeters for 3-D dose verification, *Med. Phys.*, 31, 3024, 2004.

155. Hilts, M., Audet, C., Duzenli, C., and Jirasek, A. Polymer gel dosimetry using x-ray computed tomography: a feasibility study, *Phys. Med. Biol.*, 45, 2559, 2000.

156. Trapp, J.V., Back, S.A.J., Lepage, M., Michael, G., and Baldock, C. An experimental study of the dose response of polymer gel dosimeters imaged with x-ray computed tomography, *Phys. Med. Biol.*, 46, 2939, 2001.

157. Trapp, J.V., Michael, G., De Deene, Y., and Baldock, C. Attenuation of diagnostic energy photons by polymer gel dosimeters, *Phys. Med. Biol.*, 47, 4247, 2002.

158. Audet, C., Hilts, M., Jirasek, A., and Duzenli, C. CT gel dosimetry technique: comparison of a planned and measured 3D stereotactic dose volume, *J. Appl. Clin. Med. Phys.*, 3, 110, 2002.

159. Venning, A.J., Nitschke, K.N., Keall, P.J., and Baldock, C. Radiological properties of normoxic polymer gel dosimeters, *Med. Phys.*, 32, 1047, 2005.

160. Mather, M.L., De Deene, Y., Whitakker, A.K., Simon, G.P., Rutgers, R., and Baldock, C. Investigation of ultrasonic properties of PAG and MAGIC polymer gel dosimeters, *Phys. Med. Biol.*, 47, 4397, 2002.

161. Mather, M.L. and Baldock, C. Ultrasound tomography imaging of radiation dose distributions in polymer gel dosimeters, *Med. Phys.*, 30, 2140, 2003.

162. Bankamp, A. and Schad, L.R. Comparison of TSE, TGSE, and CPMG measurement techniques for MR polymer gel dosimetry, *Magn. Reson. Imaging*, 21, 929, 2003.

163. Baustert, I.C., Oldham, M., Smith, T.A.D., Hayes, C., Webb, S., and Leach, M.O. Optimized MR imaging for polyacrylamide gel dosimetry, *Phys. Med. Biol.*, 45, 847, 2000.

164. Baldock, C., Lepage, M., Back, S.A.J., Murry, P.J., Jayasekera, P.M., Porter, D., and Kron, T. Dose resolution in radiotherapy polymer gel dosimetry: effect of echo spacing in MRI pulse sequence, *Phys. Med. Biol.*, 46, 449, 2001.

165. De Deene, Y. and Baldock, C. Optimization of multiple spin-echo sequences for 3D polymer gel dosimetry, *Phys. Med. Biol.*, 47, 3117, 2002.

166. De Deene, Y. Essential characteristics of polymer gel dosimeters, *J. Phys.: Conf. Ser. 3*: 34, 2004.

167. De Deene, Y., De Wagter, C., De Neve, W., and Achten, E. Artefacts in multi-echo T2 imaging for high-precision gel dosimetry: II. Analysis of B1-field inhomogeneity, *Phys. Med. Biol.*, 45, 1825, 2000.

168. De Deene, Y., De Wagter, C., De Neve, W., and Achtenl, E. Artefacts in multi-echo T2 imaging for high-precision gel dosimetry: I. Analysis and compensation of eddy currents, *Phys. Med. Biol.*, 45, 1807, 2000.

169. De Deene, Y. and De Wagter, C. Artefacts in multi-echo T2 imaging for high-precision gel dosimetry: III. Effects of temperature drift during scanning, *Phys. Med. Biol.*, 46, 2697, 2001.

170. Pappas, E., Seimenis, I., Angelopoulos, A., Georgopoulou, P., Kamariotaki-Paparigopoulou, M., Maris, T., Sakelliou, L., Sandilos, P., and Vlachos, L. Narrow stereotactic beam profile measurements using N-vinylpyrrolidone based polymer gels and magnetic resonance imaging, *Phys. Med. Biol.*, 46, 783, 2001.

171. Baras, P., Seimenis, I., Kipouros, P., Papagiannis, P., Angelopoulos, A., Sakelliou, L., Pappas, E., Baltas, D., Karaiskos, P., Sandilos, P., and Vlachos, L. Polymer gel dosimetry using a three-dimensional MRI acquisition technique, *Med. Phys.*, 29, 2506, 2002.

172. Kipouros, P., Papagiannis, P., Sakelliou, L., Karaiskos, P., Sandilos, P., Baras, P., Seimenis, I., Kozicki, M., Anagnostopoulos, G., and Baltas, D. 3D verification in ^{192}I HDR prostate monotherapy using polymer gels and MRI, *Med. Phys.*, 30, 2031, 2003.

173. Trapp, V., Michael, G., Evans, P.M., Baldock, C., Leach, M.O., and Webb, S. Dose resolution in gel dosimetry: effect of uncertainty in the calibration function, *Phys. Med. Biol.*, 49, N139, 2004.

174. De Deene, Y., Hanselaer, P., De Wagter, C., Achten, E., and De Neve, W. An investigation of the chemical stability of a monomer/polymer gel dosimeter, *Phys. Med. Biol.*, 45, 859, 2000.

175. Oldham, M., McJury, M., Baustert, I.B., Webb, S., and Leach, M.O. Improving calibration accuracy in gel dosimetry, *Phys. Med. Biol.*, 43, 2709, 1998.

176. Lepage, M., Whittaker, A.K., Rintoul, L., Bäck, S.Å.J., and Baldock, C. The relationship between radiation-induced chemical processes and transverse relaxation times in polymer gel dosimeters, *Phys. Med. Biol.*, 46, 1061, 2001.

177. Lepage, M., Whittaker, A.K., Rintoul, L., Bäck, S.Å.J., and Baldock, C. Modeling of post-irradiation events in polymer gel dosimeters, *Phys. Med. Biol.*, 46, 2827, 2001.

178. MacDougall, N.D., Miquel, M.E., Wilson, D.J., Keevil, S.F., and Smith, M.A. Evaluation of the dosimetric performance of BANG3® polymer gel, *Phys. Med. Biol.*, 50, 1717, 2005.

179. Lepage, M., Jayasakera, P.M., Bäck, S.ÅJ., and Baldock, C. Dose resolution optimization of polymer gel dosimeters using different monomers, *Phys. Med. Biol.*, 46, 2665, 2001.

180. Baldock, C., Murry, P., and Kron, T. Uncertainty analysis in polymer gel dosimetry, *Phys. Med. Biol.*, 43, N243, 1999.

181. Fragoso, M., Love, P.A., Verhaegen, F., Nalder, C., Bidmead, A.M., Leach, M., and Webb, S. The dose distribution of low dose rate Cs-137 in intracavitary brachytherapy: comparison of MC simulation, treatment planning calculation and polymer gel measurement, *Phys. Med. Biol.*, 49, 5459, 2004.

182. Ibbot, G.S., Maryanski, M.J., Eastman, P.E., Holcomb, S.D., Zhang, Y., Avison, R.G., Sanders, M., and Gore, J. Three dimensional visualization and measurement of conformal dose distributions using magnetic resonance imaging of PAG polymer gel dosimeters, *Int. J. Radiat. Oncol. Biol. Phys.*, 38, 1097, 1997.

183. MacDougall, N.D., Pitchford, W.G., and Smith, M.A. A systematic review of the precision and accuracy of dose measurements in photon radiotherapy using polymer and Fricke gel dosimetry, *Phys. Med. Biol.*, 47, R107, 2002.

184. Pantelis, E., Lymperopoulou, G., Papagiannis, P., Sakelliou, L., Stiliaris, E., Sandilos, P., Seimenis, I., Kozicki, M., and Rosiak, J.M. Polymer gel dosimetry close to an ^{125}I interstitial brachytherapy seed. *Phys. Med. Biol.*, 50, 4371, 2005.

185. Papagiannis, P., Pantelis, E., Georgiou, E., Karaiskos, P., Angelopoulos, A., Sakelliou, L., Stiliaris, E., Baltas, D., and Seimenis, I. Polymer gel dosimetry for the TG-43 dosimetric characterization of a new ^{125}I interstitial brachytherapy seed. *Phys. Med. Biol.*, 51, 2101, 2006.
186. Low, D.A., Harms, B., Mutic, S., and Purdy, J.A. A technique for the quantitative evaluation of dose distributions, *Med. Phys.*, 25, 656, 1998.

Appendix 1

Data Table of the Selected Nuclides

TABLE A.1.1

Ground State Properties of Natural Nuclides and Selected Radionuclides Including Those of Interest in Brachytherapy

Symbol	Element	Z	A	Standard Atomic Weight	Atomic Mass (u)	Mass Excess (MeV/c²)	Binding Energy (MeV)	Natural Abundance (%)/Decay Mode, Half-Life
H	Hydrogen	1	1	1.00794	1.00782503	7.28897	0.0000	99.9885
			2		2.014101779	13.13572	2.22457	0.0115
			3		3.01604927	14.94979	8.48182	β⁻, 12.33 a
He	Helium	2	3	4.002602	3.01602931	14.93120	7.71806	0.000137
			4		4.00260324	2.42491	28.29567	99.999863
Li	Lithium	3	6	6.941	6.0151223	14.08631	31.99456	7.59
			7		7.0160040	14.90767	39.24453	92.41
Be	Beryllium	4	9	9.012182	9.0121821	11.34758	58.16491	100
B	Boron	5	10	10.811	10.0129370	12.05076	64.75070	19.9
			11		11.0093055	8.66798	76.20480	80.10
C	Carbon	6	12	12.0107	12.0000000	0.000000	92.16175	98.93
			13		13.003354838	3.12501	97.10807	1.07
N	Nitrogen	7	14	14.0067	14.0030740052	2.86342	104.65863	99.632
			15		15.0001088984	0.10144	115.49193	0.368
O	Oxygen	8	16	15.9994	15.9949146221	−4.73699	127.61934	99.757
			17		16.99913150	−0.80900	131.76266	0.038
			18		17.9991604	−0.782064	139.80705	0.205
F	Fluorine	9	19	18.9984032	18.99840320	−1.48741	147.80136	100
Ne	Neon	10	20	20.1797	19.9924401759	−7.04193	160.64485	90.48
			21		20.99384674	−5.73172	167.40596	0.27
			22		21.99138551	−8.02434	177.76991	9.25

(continued)

TABLE A.1.1—*Continued*

Symbol	Element	Z	A	Standard Atomic Weight	Atomic Mass (u)	Mass Excess (MeV/c²)	Binding Energy (MeV)	Natural Abundance (%)/Decay Mode, Half-Life
Na	Sodium	11	22	22.989770	21.9944368	−5.18210	174.14532	EC, 2.6088 a
			23		22.98976967	−9.52948	186.56402	100
Mg	Magnesium	12	24	24.3050	23.9850419	−13.93338	198.25689	78.99
			25		24.98583702	−13.19273	205.58755	10
			26		25.98259304	−16.21448	216.68063	11.01
Al	Aluminium	13	27	26.981538	26.98153844	−17.19683	224.95195	100
Si	Silicon	14	28	28.0855	27.9769265327	−21.49279	236.53688	92.2297
			29		28.97649472	−21.89502	245.01044	4.6832
			30		29.97377022	−24.43288	255.61962	3.0872
P	Phosphorus	15	31	30.973761	30.97376151	−24.44099	262.91668	100
			32		31.97390716	−24.30532	270.85235	β⁻, 14.262 d
S	Sulfur	16	32	32.065	31.97207069	−26.01598	271.78066	94.93
			33		32.97145850	−26.58623	280.42223	0.76
			34		33.96786683	−29.93185	291.83917	4.29
			36		35.96708088	−30.66396	308.71392	0.02
Cl	Chlorine	17	35	35.453	34.96885271	−29.01351	298.20980	75.78
			37		36.96590260	−31.76152	317.10045	24.22
Ar	Argon	18	36	39.948	35.96754628	−30.23044	306.71570	0.3365
			38		37.9627322	−34.71476	327.342670	0.0632
			40		39.962383123	−35.03989	343.81044	99.6003
K	Potassium	19	39	39.0983	38.9637069	−33.80684	333.72371	93.2581
			40		39.96399867	−33.53502	341.52314	0.0117
			41		40.96182597	−35.55887	351.61839	6.7302
Ca	Calcium	20	40	40.078	39.9625912	−34.84611	342.05195	96.941

				42	41.9586183	−38.54676	361.89525	0.647
				43	42.9587668	−38.40844	369.82825	0.135
				44	43.9554811	−41.46909	380.96022	2.086
				46	45.9536928	−43.13491	398.76869	0.004
				48	47.952534	−44.21474	415.99170	0.187
Sc	21	Scandium	44.955910	45	44.9559102	−41.06934	387.84944	100
Ti	22	Titanium	47.867	46	45.9526295	−44.12534	398.19441	8.25
				47	46.9517638	−44.93173	407.07213	7.44
				48	47.9479471	−48.48700	418.69872	73.72
				49	48.9478708	−48.55804	426.84108	5.41
				50	49.9447921	−51.42585	437.78021	5.18
V	23	Vanadium	50.9415	50	49.9471628	−49.21754	434.78955	0.25
				51	50.9439637	−52.19749	445.84083	99.75
Cr	24	Chromium	51.9961	50	49.9460496	−50.25446	435.04412	4.345
				52	51.9405119	−55.41280	456.34510	83.789
				53	52.9406538	−55.28064	464.28427	9.501
				54	53.9388849	−56.92832	474.00327	2.365
Mn	25	Manganese	54.938049	55	54.9380496	−57.70638	482.07030	100
Fe	26	Iron	55.845	54	53.9396148	−56.24841	471.75863	5.845
				56	55.9349421	−60.60100	492.25389	91.754
				57	56.935987	−60.17571	499.89992	2.119
				58	57.9332805	−62.14884	509.94438	0.282
				59	58.9348805	−60.65842	516.52528	β^-, 44.503 d
Co	27	Cobalt	58.933200	59	58.9332002	−62.22361	517.30811	100
				60	59.9338222	−61.64422	524.80005	β^-, 5.2714
Ni	28	Nickel	58.6934	58	57.9353479	−60.22301	506.45384	68.0769
				60	59.9307906	−64.46810	526.84157	26.2231
				61	60.9310604	−64.21678	534.66157	1.1399
				62	61.9283488	−66.74269	545.25881	3.6345

(continued)

TABLE A.1.1—*Continued*

Symbol	Element	Z	A	Standard Atomic Weight	Atomic Mass (u)	Mass Excess (MeV/c²)	Binding Energy (MeV)	Natural Abundance (%)/Decay Mode, Half-Life
Cu	Copper	29	64		63.9279696	−67.09590	561.75466	0.9256
			63	63.546	62.9296011	−65.57616	551.38125	69.17
			65		64.9277937	−67.25972	569.20745	30.83
Zn	Zinc	30	64		63.9291466	−65.99953	559.09359	48.63
			66	65.409	65.9260368	−68.89630	578.13300	27.90
			67		66.9271309	−67.87716	585.18518	4.10
			68		67.9248476	−70.00403	595.38338	18.75
			70		69.925325	−69.55943	611.08142	0.62
Ga	Gallium	31	69	69.723	68.925581	−69.32092	601.98924	60.108
			71		70.924705	−70.13682	618.94779	39.892
Ge	Germanium	32	70	72.64	69.9242504	−70.56032	610.51761	20.84
			72		71.9220762	−72.58556	628.68549	27.54
			73		72.9234594	−71.29714	635.46839	7.73
			74		73.9211782	−73.42201	645.66459	36.28
			76		75.9214027	−73.21289	661.59811	7.61
As	Arsenic	33	75	74.92160	74.9215964	−73.03246	652.56401	100
Se	Selenium	34	74	78.96	73.9224766	−72.21261	642.89048	0.89
			75		74.9225236	−72.16882	650.91801	EC, 119.79 d
			76		75.9192141	−75.25156	662.07208	9.37
			77		76.9199146	−74.55905	669.49089	7.63
			78		77.9173095	−77.02567	679.98884	23.77
			80		79.9165218	−77.75940	696.86521	49.61
			82		81.9167	−77.59344	712.84189	8.73
Br	Bromine	35	79	79.904	78.9183376	−76.06770	686.32011	50.69
			81		80.916291	−77.97436	704.36914	49.31

		Z		A				
Kr	Krypton	36	83.798	78	77.9203863	−74.15970	675.55160	0.35
				80	79.916378	−77.89334	695.43444	2.25
				82	81.9134846	−80.58856	714.27231	11.60
				83	82.914136	−79.98183	721.73690	11.50
				84	83.9115066	−82.43103	732.25743	57.00
				86	85.9106103	−83.26595	749.23498	17.30
Rb	Rubidium	37	85.4678	83	82.915112	−79.07270	720.04541	EC, 86.2 d
				85	84.9117893	−82.16769	739.28305	72.17
				87	86.9091835	−84.59505	757.85305	27.83
Sr	Strontium	38	87.62	84	83.913425	−80.64429	728.90598	0.56
				86	85.9092624	−84.52156	748.92590	9.86
				87	86.9088793	−84.87836	757.35401	7.00
				88	87.9056143	−87.91966	768.46664	82.58
				90	89.9077376	−85.94186	782.63149	β⁻, 28.79 a
Y	Yttrium	39	88.90585	89	88.9058479	−87.70210	775.53805	100
				90	89.9071514	−86.48786	782.39513	β⁻, 64.0 h
Zr	Zirconium	40	91.224	90	89.9047037	−88.76794	783.89285	51.45
				91	90.905645	−87.89113	791.08737	11.22
				92	91.9050401	−88.45456	799.72212	17.15
				94	93.9063158	−87.26629	814.67650	17.38
				96	95.9082757	−85.44065	828.99350	2.80
Nb	Niobium	41	92.90638	93	92.9063775	−87.20874	805.76528	100
Mo	Molybdenum	42	95.94	92	91.906810	−86.80547	796.50832	14.84
				94	93.9050876	−88.41034	814.25584	9.25
				95	94.9058415	−87.70808	821.62490	15.92
				96	95.9046789	−88.79102	830.77916	16.68
				97	96.9060210	−87.54083	837.60030	9.55
				98	97.9054078	−88.11201	846.24280	24.13
				100	99.9074771	−86.18447	860.45791	9.63

(continued)

TABLE A.1.1—*Continued*

Symbol	Element	Z	A	Standard Atomic Weight	Atomic Mass (u)	Mass Excess (MeV/c^2)	Binding Energy (MeV)	Natural Abundance (%)/Decay Mode, Half-Life
Ru	Ruthenium	44	96	101.07	95.907598	−86.07219	826.49563	5.54
			98		97.905287	−88.22447	844.79056	1.87
			99		98.9059393	−87.61696	852.25437	12.76
			100		99.9042197	−89.21879	861.92753	12.60
			101		100.9055822	−87.94958	868.72964	17.06
			102		101.9043495	−89.09785	877.94923	31.55
			104		103.905430	−88.09124	893.08526	18.62
			106		105.907327	−86.32441	907.46108	β^-, 373.59 d
Rh	Rhodium	45	103	102.90550	102.905504	−88.02227	884.16262	100
			106		105.907285	−86.36381	906.71813	β^-, 29.8 sec
Pd	Palladium	46	102	106.42	101.905608	−87.92583	875.21250	1.02
			103		102.906087	−87.47919	882.83718	EC, 16.991 d
			104		103.904035	−89.39089	892.82021	11.14
			105		104.905084	−88.41363	899.99143	22.33
			106		105.903483	−89.90491	909.47687	27.33
			108		107.903894	−89.52173	925.23634	26.46
			109		108.905954	−87.60371	931.38963	β^-, 13.7012 h
			110		109.905152	−88.3500	940.20722	11.72
Ag	Silver	47	107	107.8682	106.905093	−88.40527	915.26620	51.839
			109		108.904756	−88.71965	931.72323	48.161
Cd	Cadmium	48	106	112.411	105.906458	−87.13379	905.14105	1.25
			108		107.9041834	−89.25257	923.40247	0.89
			110		109.9030056	−90.34971	940.64225	12.49
			111		110.9041816	−89.25422	947.61809	12.80

Symbol	Z	Name	A	Atomic weight	Isotopic mass	Mass excess		Abundance / decay
			112		111.9027572	−90.58105	957.01624	24.13
			113		112.9044009	−89.04993	963.55645	12.22
			114		113.9033581	−90.02132	972.59915	28.73
			116		115.9047554	−88.71973	987.44021	7.49
In	49	Indium	113	114.818	112.904061	−89.36638	963.09054	4.29
			115		114.903878	−89.53675	979.40355	95.71
Sn	50	Tin	112	118.710	111.904821	−88.65883	953.52932	0.97
			114		113.902782	−90.55814	971.57127	0.66
			115		114.903346	−90.03263	979.11709	0.34
			116		115.901744	−91.52472	988.68050	14.54
			117		116.902954	−90.39797	995.62507	7.68
			118		117.901606	−91.65310	1004.95152	24.22
			119		118.903309	−90.06719	1011.43693	8.59
			120		119.9021966	−91.10329	1020.54436	32.58
			122		121.9034401	−89.94492	1035.52863	4.63
			124		123.9052746	−88.23610	1049.96246	5.79
Sb	51	Antimony	121	121.760	120.9038180	−89.59290	1026.32294	57.21
			123		122.9042157	−89.22249	1042.09517	42.79
Te	52	Tellurium	120	127.60	119.904020	−89.40488	1017.28124	0.09
			122		121.9030471	−90.31106	1034.33007	2.55
			123		122.9042730	−89.16916	1041.25949	0.89
			124		123.9028195	−90.52307	1050.68472	4.74
			125		124.9044247	−89.02780	1057.26076	7.07
			126		125.9033055	−90.07029	1066.37459	18.84
			128		127.9044614	−88.99364	1081.44058	31.74
			130		129.9062228	−87.35293	1095.94252	34.08
I	53	Iodine	125	126.90447	124.9046241	−88.842019	1056.29264	β⁻, 59.49 d
			127		126.904468	−88.98708	1072.58035	100
			131		130.9061242	−87.44476	1103.32332	β⁻, 8.0207 d

(continued)

TABLE A.1.1—*Continued*

Symbol	Element	Z	A	Standard Atomic Weight	Atomic Mass (u)	Mass Excess (MeV/c^2)	Binding Energy (MeV)	Natural Abundance (%)/Decay Mode, Half-Life
Xe	Xenon	54	124	131.293	123.9058958	−87.65751	1046.25445	0.09
			125		124.9063982	−87.18947	1053.85774	EC, 16.9 h
			126		125.904269	−89.17296	1063.91255	0.09
			128		127.9035304	−89.86081	1080.74304	1.92
			129		128.9047795	−88.69735	1087.6501	26.44
			130		129.9035079	−88.88180	1096.90668	4.08
			131		130.9050819	−88.41561	1103.51181	21.18
			132		131.9041545	−89.27954	1112.44706	26.89
			133		132.905906	−87.64830	1118.88715	β⁻, 5.243 d
			134		133.9053945	−88.12444	1127.43461	10.44
			136		135.907220	−86.42444	1141.18772	8.87
Cs	Cesium	55	131	132.90545	130.905460	−88.06321	1102.37706	EC, 9.689 d
			133		132.905447	−88.07566	1118.53216	100
			137		136.907084	−86.55115	1149.29293	β⁻, 30.07 a
Ba	Barium	56	130	137.327	129.906310	−87.27121	1092.73139	0.106
			132		131.905056	−88.43961	1110.042431	0.101
			134		133.904503	−88.95455	1126.70001	2.417
			135		134.905683	−87.85594	1133.67273	6.592
			136		135.904570	−88.89236	1142.78047	7.854
			137		136.905821	−87.72677	1149.68621	11.232
			138		137.905241	−88.26717	1158.29793	71.698
La	Lathanum	57	138	138.9055	137.907107	−86.52942	1155.77782	0.09
			139		138.906348	−87.23611	1164.55584	99.91
Ce	Cerium	58	136	140.116	135.907140	−86.49519	1138.81860	0.185

Element	Sym	Z	A	Atomic weight	Isotopic mass			
			138		137.905986	−87.57386	1156.03991	0.251
			140		139.905434	−88.08762	1172.6963	88.45
			142		141.909240	−84.54263	1185.29397	11.114
			144		143.9136427	−80.44131	1197.33530	β⁻, 284.91 d
Praseodymium	Pr	59	141	140.90765	140.907648	−86.02558	1177.92324	100
Neodymium	Nd	60	142	144.24	141.907719	−85.95951	1185.14615	27.2
			143		142.909810	−84.01178	1191.26974	12.2
			144		143.910083	−83.75748	1199.08676	23.8
			145		144.912569	−81.44158	1204.84219	8.3
			146		145.913112	−80.935509	1212.40744	17.2
			148		147.916889	−77.41784	1225.03241	5.7
			150		149.920887	−73.69368	1237.45090	5.6
Samarium	Sm	62	144	150.36	143.911995	−81.97637	1195.74094	3.07
			145		144.913406	−80.66214	1202.49804	EC, 340 d
			147		146.914893	−79.27639	1217.25493	14.99
			148		147.914818	−79.34659	1225.39645	11.24
			149		148.917180	−77.14677	1231.26796	13.82
			150		149.917271	−77.06113	1239.25364	7.38
			152		151.919728	−74.77265	1253.10780	26.75
			154		153.922205	−72.46528	1266.94308	22.75
Europium	Eu	63	151	151.964	150.919846	−74.66294	1244.14442	47.81
			153		152.921226	−73.37729	1259.00142	52.19
Gadolinium	Gd	64	152	157.25	151.919788	−74.717010	1251.48755	0.20
			154		153.920862	−73.71631	1266.62940	2.18
			155		154.922619	−72.08011	1273.06453	14.80
			156		155.922120	−72.54515	1281.60090	20.47
			157		156.923957	−70.83388	1287.96094	15.65
			158		157.924101	−70.69989	1295.89827	24.84
			160		159.927051	−67.95190	1309.29294	21.86

(continued)

TABLE A.1.1—*Continued*

Symbol	Element	Z	A	Standard Atomic Weight	Atomic Mass (u)	Mass Excess (MeV/c²)	Binding Energy (MeV)	Natural Abundance (%)/Decay Mode, Half-Life
Tb	Terbium	65	159	158.92534	158.925343	−69.54241	1302.02977	100
Dy	Dysprosium	66	156	162.500	155.924278	−70.53432	1278.2536	0.06
			158		157.924405	−70.41662	1294.05030	0.10
			160		159.925194	−69.68159	13094.5792	2.34
			161		160.926930	−68.06463	1315.91228	18.91
			162		161.926795	−68.19026	1324.10923	25.51
			163		162.928728	−66.38987	1330.38016	24.90
			164		163.929171	−65.97663	1338.03824	28.18
Ho	Holmium	67	165	164.93032	164.930319	−64.90727	1344.25785	100
			166		165.932281	−63.07958	1350.50149	β^-, 26.8 h
Er	Erbium	68	162	167.259	161.928775	−66.34572	1320.69999	0.14
			164		163.929197	−65.95256	1336.44947	1.61
			166		165.930290	−64.93446	1351.57402	33.61
			167		166.932045	−63.29925	1358.01013	22.93
			168		167.932368	−62.99900	1365.78120	26.78
			170		169.935460	−60.11830	1379.04315	14.93
Tm	Thulium	69	169	168.93421	168.934211	−61.28194	1371.35311	100
			170		169.935798	−59.80388	1377.94638	EC + β^-, 128.6 d
Yb	Ytterbium	70	168	173.04	167.933894	−61.57690	1362.79439	0.13
			169		168.935187	−60.372800	1369.66162	EC, 32.015 d
			170		169.934759	−60.77192	1378.13206	3.04
			171		170.936322	−59.31539	1384.74685	14.28
			172		171.9363777	−59.26379	1392.76657	21.83
			173		172.9382068	−57.56003	1399.13414	16.13

			174		173.9388581	−56.95330	31.83
			175		174.9412725	−54.70431	β⁻, 4.185 d
			176		175.942568	−53.49717	12.76
Lu	Lutetium	71	175	174.967	174.9407679	−55.17433	97.41
			176		175.9426824	−53.39099	2.59
Hf	Hafnium	72	174	178.49	173.940040	−55.85224	0.16
			176		175.9414018	−54.58384	5.26
			177		176.9432200	−52.89021	18.60
			178		177.9436977	−52.44522	27.28
			179		178.9458151	−50.47293	13.62
			180		179.9465488	−49.78950	35.08
Ta	Tantalum	73	180	180.9479	179.947466	−48.93542	0.012
			181		180.9479963	−48.44109	99.988
			182		181.950152	−46.43272	β⁻, 114.43 d
W	Tungsten	74	180	183.84	179.946706	−49.64328	0.12
			182		181.948206	−48.24624	26.50
			183		182.9502245	−46.36561	14.31
			184		183.9509326	−45.70603	30.64
			186		185.954362	−42.51133	28.43
			188		187.958487	−38.66915	β⁻, 69.78 d
Re	Rhenium	75	185	186.207	184.9529557	−43.82143	37.40
			186		185.9549865	−41.92977	EC + β⁻, 3.7183 d
			187		186.9557508	−41.21787	62.60
			188		187.9581123	−39.01815	β⁻, 17.004 h
Os	Osmium	76	184	190.23	183.952491	−44.25452	0.02
			186		185.953838	−42.99929	1.59
			187		186.9557479	−41.22053	1.96
			188		187.9558360	−41.13850	13.24
			189		188.9581449	−38.98780	16.15
			190		189.958445	−38.708032	26.26
			192		191.961479	−35.88203	40.78

(continued)

TABLE A.1.1—*Continued*

Symbol	Element	Z	A	Standard Atomic Weight	Atomic Mass (u)	Mass Excess (MeV/c^2)	Binding Energy (MeV)	Natural Abundance (%)/Decay Mode, Half-Life
Ir	Iridium	77	191	192.217	190.960591	−36.70906	1518.09051	37.3
			192		191.962602	−34.83582	1524.28859	EC + β⁻, 73.81 d
			193		192.962924	−34.53635	1532.06044	62.7
			194		193.965076	−32.53186	1538.12727	β⁻, 19.28 h
Pt	Platinum	78	190	195.078	189.959930	−37.32489	1509.85266	0.014
			192		191.961035	−36.29551	1524.96593	0.782
			194		193.962664	−34.77865	1539.59171	32.967
			195		194.964774	−32.81239	1545.69677	33.832
			196		195.964935	−32.66294	1553.61865	25.242
			198		197.967876	−29.92330	1567.02165	7.163
Au	Gold	79	197	196.96655	196.966552	−31.15697	1559.40165	100
			198		197.968225	−29.59799	1565.91399	β⁻, 2.695 d
Hg	Mercury	80	196	200.59	195.965815	−31.84326	1551.23426	0.15
			198		197.966752	−30.97047	1566.50411	9.97
			199		198.968262	−29.56330	1573.16826	16.87
			200		199.968309	−29.52023	1581.19652	23.10

Symbol	Element	Z	A	Atomic weight	Atomic mass	Mass excess	Binding energy	Abundance / half-life
Tl	Thallium	81	201	204.3833	200.970285	−27.67908	1587.42670	13.18
			202		201.970626	−27.36207	1595.18101	29.86
			204		203.973476	−24.70728	1608.66886	6.87
			203		202.972329	−25.77528	1600.88319	29.524
			205		204.974412	−23.83481	1615.08536	70.476
Pb	Lead	82	204	207.2	203.973029	−25.12355	1607.52042	1.4
			206		205.974449	−23.80060	1622.34012	24.1
			207		206.975881	−22.467069	1629.07791	22.1
			208		207.976636	−21.76356	1636.44573	52.4
Bi	Bismuth	83	209	208.98038	208.980383	−18.27289	1640.24403	100
Rn	Radon	86	222	(222)	222.0175705	16.36679	1708.18448	α, 3.8235 d
Ra	Radium	88	226	(226)	226.0254026	23.66232	1731.60953	α, 1600 a
Th	Thorium	90	232	232.0381	232.0380504	35.44368	1766.69141	100
U	Uranium	92	234	238.02891	234.0409456	38.14058	1778.57244	0.0055
			235		235.0439231	40.91406	1783.87029	0.72
			238		238.0507826	47.30366	1801.69465	99.2745
Am	Americium	95	241	(243)	241.0568229	52.93022	1817.93500	α, 432.2 a

The atomic mass,[1] the mass excess,[2] and the binding energy[2] are given for each nuclide along with the natural abundance for natural nuclides or the decay mode and the half-life (in units of a:year, d:day, h:hour) for radionuclides in the last column (see Chapter 3).[3] The atomic weight is also given for each element.

Excess mass is defined as the atomic mass of an atom minus the product of mass number and the unified atomic mass unit (see Chapter 3).

Excess mass is here expressed as energy equivalence using Equation 3.21 in Chapter 3.

References

1. National Institute of Standards and Technology Physics Laboratory Physical Reference Data Electronic Version, available online at http://physics.nist.gov/PhysRefData (Gaithersburg: National Institute of Standards and Technology), August 2004.
2. Nuclear Data Evaluation La, Table of Nuclides (http://atom.kaeri.re.kr/). Korea Atomic Energy Research Institute, 2000.
3. National Nuclear Data Center (NNDC) Brookhaven National Laboratory NUDAT 2.0 Electronic Version, available online at NNDC: www.nndc.bnl.gov/nudat2/, July 2005.

Appendix 2

Unit Conversion Factors and Physical Constants

TABLE A.2.1

Unit Conversion Factors

Physical Quantity	Unit Name	Unit Symbol	Conversion Factor
Distance	Angstrom	A°	1.0×10^{-10} m
Energy	erg	erg	1.0×10^{-7} J
	Electron volt	eV	$1.60217733 \times 10^{-19}$ J
	Calorie[a]	cal	4.18580 J
	Kilocalorie[a]	kcal	4.18580×10^3 J
Mass	Unified atomic mass unit	u	$1.6605402 \times 10^{-27}$ kg
Temperature	Degree celsius	°C	$t_K - 273.15$[b]
	Degree fahrenheit	°F	$(9/5)\, t_K - 459.67$
Pressure	Bar	bar	1.0×10^5 Pa
	Atmosphere	Atm	1.01325×10^5 Pa
	Millimeter of mercury (0°C)	mm Hg	1.33322×10^2 Pa
	Pound force per square inch	psi	6.89476×10^3 Pa
Exposure	Roentgen	R	2.58×10^{-4} C kg^{-1}

[a] Here for the 15°C based definition of calorie/kilocalorie.

[b] t_K is the temperature expressed in kelvin.

Source: From Freim, J. and Feldman, A. *Medical Physics Handbook of Units and Measures*, Medical Physics Publishing, Madison, 1992; International Council of Scientific Unions, Committee on Data for Science and Technology. The 1986 Adjustment of the Fundamental Physical Constants, a report of the CODATA Task Group on Fundamental Constants, CODATA Bulletin 63, Oxford, Pergamon Press 1986; International Organization for Standardization, *ISO Standards Handbook: Quantities and Units*, 3rd ed., ISO 31-0:1992(E), Geneva: ISO, 1993.

TABLE A.2.2

Physical Constants

Physical Quantity	Symbol	Value[a]
Speed of light in vacuum	c	$299\ 792\ 458$ m s^{-1}
Planck constant	h	$6.626\ 069\ 3(11) \times 10^{-34}$ J s
Planck constant reduced	$\hbar = h/2\pi$	$1.054\ 571\ 68(18) \times 10^{-34}$ J s
		$= 6.582\ 119\ 15(56) \times 10^{-22}$ MeV s
Conversion constant	$\hbar c$	$197.326\ 968(17)$ MeV fm
Electron charge magnitude	e	$1.602\ 176\ 53(14) \times 10^{-19}$ C
Fine-structure constant	$\alpha = e^2/4\pi\varepsilon_0\hbar c$	$1/137.035\ 999\ 11(46)$
Unified atomic mass unit	u	$1.660\ 538\ 86(28)\ 10^{-27}$ kg
		$= 931.494\ 043(80)$ MeV/c^2
Electron mass	m_e	$9.109\ 3826(16) \times 10^{-31}$kg
		$= 0.510\ 998\ 918(44)$ MeV/c^2
Proton mass	m_p	$1.672\ 621\ 71(29) \times 10^{-27}$ kg
		$= 938.272\ 029(80)$ MeV/c^2
Neutron mass	m_n	$1.674\ 927\ 28(29) \times 10^{-27}$ kg
		$= 939.565\ 360(81)$ MeV/c^2
Bohr radius ($m_{nucleus} = \infty$)	$\alpha_\infty = 4\pi\varepsilon_0\hbar^2/m_e e^2$	$0.529\ 177\ 2108(18) \times 10^{-10}$ m
Rydberg energy	$m_e\, e^4/2(4\pi\varepsilon_0)^2\hbar^2$	$13.605\ 6923(12)$ eV
Classical electron radius	$r_e = e^2/4\varepsilon_0 m_e c^2$	$2.817\ 940\ 325(28) \times 10^{-15}$ m
Thomson cross-section	$\sigma_T = 8\pi\, r_e^2/3$	$0.665\ 245\ 873(13)$ barn
Avogadro constant	N_A	$6.022\ 1415(10) \times 10^{23}$ mol^{-1}
Boltzmann constant	k	$1.380\ 6505(24) \times 10^{-23}$ J K^{-1}
		$= 8.617\ 343(15) \times 10^{-5}$ eV K^{-1}
Density of dry air at 20 °C and 101 kPa		1.205 kg m^{-3}
Ionization energy[b]	W	$33.97(5)$ eV per ion pair
Ionization constant[b]	W/e	$33.97(5)$ J C^{-1}

[a] The digits in parentheses are the one-standard-deviation (1σ) uncertainty.
[b] Values valid for dry air and electron energies above 1 keV.[5,6]
Source: From The NIST reference on constants, units and uncertainties available online at http://physics.nist.gov/cuu/Constants/index.html.

References

1. Freim, J. and Feldman, A. *Medical Physics Handbook of Units and Measures*, Medical Physics Publishing, Madison, 1992.
2. International Council of Scientific Unions, Committee on Data for Science and Technology. The 1986 Adjustment of the Fundamental Physical Constants, a report of the CODATA Task Group on Fundamental Constants, CODATA Bulletin 63, Oxford, Pergamon Press, 1986.
3. International Organization for Standardization, *ISO Standards Handbook: Quantities and Units*, 3rd ed., ISO 31-0:1992(E), Geneva: ISO, 1993.
4. The NIST reference on constants, units and uncertainties available online at http://physics.nist.gov/cuu/Constants/index.html.

5. Boutillon, M. and Perroche-Roux, A.M. Re-evaluation of the *W* value for electrons in dry air, *Phys. Med. Biol.*, 32/2, 213–219, 1987.
6. Deutsches Institut für Normung e.V. [DIN], *Terms in the Field of Radiological Technique. Part 3. Dose Quantities and Units, Begriffe in der radiologischen Technik. Teil 3. Dosisgrössen und Dosiseinheiten.* DIN, Berlin, DIN 6814-3, 2001.

Appendix 3

TG-43 Tables for Brachytherapy Sources

A.3.1   MicroSelectron HDR Classic [192]Ir Source

TABLE A.3.1

Monte Carlo (MC) Calculated and Experimental Radial Dose Function, $g(r)$

MC Data by Williamson and Li[1]		MC Data by Karaiskos et al.[2]		Experimental (TLD in Polystyrene) Data by Anctil et al.[3]	
Radial Distance r (cm)	$g(r)$	Radial Distance r (cm)	$g(r)$	Radial Distance r (cm)	$g(r) \pm 1 \times$ S.D.
0.1	0.979	0.1	0.990	1.0	1.000
0.2	0.990	0.2	0.993	2.0	1.017 ± 0.025
0.3	0.993	0.3	0.994	3.0	1.002 ± 0.031
0.5	0.997	0.5	0.996	4.0	0.994 ± 0.028
1.0	1.000	0.8	0.998	5.0	0.977 ± 0.026
1.5	1.002	1.0	1.000	6.0	0.993 ± 0.027
2.0	1.003	1.5	1.003	7.0	0.946 ± 0.026
2.5	1.002	2.0	1.004	8.0	0.937 ± 0.025
3.0	1.002	3.0	1.005	9.0	0.923 ± 0.025
4.0	0.997	3.5	1.003	10.0	0.900 ± 0.024
5.0	0.987	4.0	1.000		
6.0	0.973	4.5	0.996		
7.0	0.956	5.0	0.991		
8.0	0.933	6.0	0.979		
9.0	0.904	8.0	0.940		
10.0	0.871	10.0	0.880		
11.0	0.836	12.0	0.800		
12.0	0.795	14.0	0.692		
13.0	0.749	15.0	0.608		
14.0	0.682				

Values for the microSelectron HDR old design, classic, ^{192}Ir source.

TABLE A.3.2

Monte Carlo (MC) Calculated Anisotropy Function, $F(r,\theta)$

Polar Angle θ (°)	Radial Distance r (cm)					
	0.25	0.5	1.0	2.0	3.0	5.0
0.0	0.776	0.671	0.637	0.646	0.663	0.704
1.0	0.777	0.672	0.637	0.647	0.669	0.711
2.0	0.785	0.672	0.640	0.657	0.676	0.717
3.0	0.779	0.669	0.649	0.666	0.686	0.725
5.0	0.776	0.682	0.670	0.687	0.705	0.741
7.0	0.770	0.702	0.695	0.710	0.725	0.759
10.0	0.795	0.737	0.731	0.746	0.757	0.787
12.0	0.823	0.761	0.755	0.770	0.779	0.805
15.0	0.839	0.795	0.790	0.803	0.810	0.832
20.0	0.886	0.844	0.838	0.848	0.852	0.868
25.0	0.920	0.881	0.875	0.884	0.885	0.896
30.0	0.939	0.910	0.905	0.909	0.912	0.917
35.0	0.951	0.933	0.925	0.930	0.931	0.934
45.0	0.974	0.965	0.956	0.960	0.958	0.966
50.0	0.982	0.972	0.970	0.972	0.972	0.977
60.0	0.991	0.988	0.987	0.987	0.988	0.989
75.0	0.995	0.996	0.998	0.998	0.994	1.000
90.0	1.000	1.000	1.000	1.000	1.000	1.000
105.0	0.993	0.996	0.997	0.999	0.992	0.997
120.0	0.991	0.989	0.985	0.985	0.981	0.985
130.0	0.979	0.972	0.969	0.968	0.967	0.970
135.0	0.972	0.966	0.956	0.956	0.955	0.963
145.0	0.951	0.933	0.924	0.925	0.924	0.938
150.0	0.940	0.908	0.900	0.903	0.902	0.918
155.0	0.920	0.877	0.870	0.874	0.876	0.893
160.0	0.884	0.837	0.829	0.833	0.839	0.862
165.0	0.831	0.782	0.770	0.780	0.788	0.817
168.0		0.739	0.721	0.737	0.749	0.783
170.0		0.702	0.683	0.703	0.718	0.758
173.0		0.633	0.616	0.647	0.667	0.712
175.0	Cable	0.611	0.580	0.614	0.635	0.684
177.0		0.603	0.563	0.593	0.614	0.664
178.0		0.605	0.554	0.586	0.606	0.657
179.0		0.605	0.552	0.577	0.601	0.652
180.0		0.605	0.551	0.576	0.596	0.646

Values for the microSelectron HDR old design, classic, ^{192}Ir source. Values related to polar angle θ have been adjusted to fit the TG-43 U1 source coordinate system (Figure 8.6).

Source: From Williamson, J.F. and Li, Z., *Med. Phys.*, 22, 809–819, 1995.

TABLE A.3.3

Monte Carlo (MC) Calculated Anisotropy Factor, $\varphi_{an}(r)$

Radial Distance r (cm)	0.25	0.50	1.00	2.00	3.00	5.00
$\varphi_{an}(r)$	1.173	1.007	0.966	0.959	0.956	0.961

Values for the microSelectron HDR old design, classic, ^{192}Ir source.
Source: From Williamson, J.F. and Li, Z., *Med. Phys.*, 22, 809–819, 1995.

A.3.2 MicroSelectron HDR New ^{192}Ir Source

TABLE A.3.4

Monte Carlo (MC) Calculated Radial Dose Function, $g(r)$

Radial Distance r (cm)	$g(r)$	
	MC Data by Daskalov et al.[4]	MC Data by Papagiannis et al.[5]
0.1	1.004	—
0.2	1.000	—
0.3	1.001	—
0.5	1.000	0.995
1.0	1.000	1.000
1.5	1.003	1.002
2.0	1.007	1.003
2.5	1.008	1.004
3.0	1.008	1.001
4.0	1.004	0.996
5.0	0.995	0.990
6.0	0.981	0.975
7.0	0.964	0.959
8.0	0.940	0.934
9.0	0.913	0.908
10.0	0.882	0.876
11.0	0.844	—
12.0	0.799	—
13.0	0.747	—
14.0	0.681	—

Values for the microSelectron HDR new design ^{192}Ir source.

TABLE A.3.5

Monte Carlo (MC) Calculated Anisotropy Function, $F(r,\theta)$

Polar Angle θ (°)	Radial Distance r (cm)					
	0.25	0.5	1.0	2.0	3.0	5.0
0	0.729	0.667	0.631	0.645	0.660	0.696
1	0.730	0.662	0.631	0.645	0.661	0.701
2	0.729	0.662	0.632	0.652	0.670	0.709
3	0.730	0.663	0.640	0.662	0.679	0.718
4	0.731	0.664	0.650	0.673	0.690	0.726
5	0.733	0.671	0.661	0.684	0.700	0.735
6	0.735	0.680	0.674	0.696	0.711	0.743
7	0.734	0.691	0.687	0.708	0.723	0.753
8	0.739	0.702	0.700	0.720	0.734	0.763
10	0.756	0.727	0.727	0.745	0.758	0.782
12	0.777	0.751	0.753	0.769	0.781	0.804
14	0.802	0.775	0.778	0.791	0.802	0.822
16	0.820	0.797	0.800	0.812	0.822	0.840
20	0.856	0.836	0.839	0.846	0.854	0.872
24	0.885	0.868	0.869	0.874	0.877	0.888
30	0.920	0.904	0.902	0.907	0.906	0.911
36	0.938	0.930	0.929	0.931	0.934	0.933
42	0.957	0.949	0.949	0.955	0.956	0.954
48	0.967	0.963	0.965	0.965	0.969	0.965
58	0.982	0.982	0.982	0.982	0.983	0.978
73	0.994	0.997	0.997	0.998	0.996	0.985
88	0.997	1.001	1.000	1.000	1.000	1.001
90	1.000	1.000	1.000	1.000	1.000	1.000
103	0.995	0.995	1.001	0.999	1.000	0.995
118	0.987	0.987	0.987	0.989	0.989	0.983
128	0.974	0.972	0.976	0.976	0.980	0.979
133	0.969	0.961	0.966	0.965	0.973	0.973
138	0.957	0.949	0.952	0.952	0.959	0.960
143	0.942	0.933	0.935	0.935	0.944	0.941
148	0.924	0.912	0.914	0.915	0.924	0.926
153	0.899	0.886	0.887	0.889	0.899	0.905
158	0.873	0.850	0.850	0.856	0.863	0.870
165	0.806	0.779	0.778	0.791	0.801	0.816
169		0.725	0.723	0.741	0.754	0.785
170		0.710	0.707	0.727	0.742	0.774
172		0.678	0.675	0.697	0.714	0.748
173		0.662	0.657	0.682	0.700	0.733
174		0.642	0.640	0.667	0.686	0.720
175	Cable	0.623	0.624	0.652	0.672	0.707
176		0.605	0.608	0.637	0.658	0.695
177		0.606	0.594	0.624	0.645	0.686
178		0.608	0.586	0.612	0.634	0.675
179		0.609	0.585	0.604	0.624	0.665
180		0.609	0.585	0.603	0.622	0.662

Values for the microSelectron HDR new design ^{192}Ir source.

Source: From Daskalov, G.M., Löffler, E., and Williamson, J.F., *Med. Phys.*, 25, 2200–2208, 1998.

TABLE A.3.6

Monte Carlo (MC) Calculated Anisotropy Factor, $\varphi_{an}(r)$

Radial Distance r (cm)	0.25	0.50	1.00	2.00	3.00	5.00
$\varphi_{an}(r)$	1.155	1.007	0.969	0.960	0.961	0.959

Values for the microSelectron HDR new design ^{192}Ir source.
Source: From Daskalov, G.M., Löffler, E., and Williamson, J.F., *Med. Phys.*, 25, 2200–2208, 1998.

A.3.3   VariSource HDR Old [192]Ir Source

TABLE A.3.7

Monte Carlo (MC) Calculated and Experimental Radial Dose Function, $g(r)$

MC Data by Wang and Sloboda[6]		MC Data by Karaiskos et al.[7]		Experimental (TLD in Solid Water) Data by Meigooni et al.[8]	
Radial Distance r (cm)	$g(r)$	Radial Distance r (cm)	$g(r)$	Radial Distance r (cm)	$g(r) \pm 1 \times$ S.D
0.1	0.952	0.1	0.950	0.5	1.017 ± 0.051
0.2	0.967	0.2	0.969	1.0	1.000
0.3	0.976	0.3	0.981	1.5	1.005 ± 0.050
0.5	0.989	0.5	0.992	2.0	1.007 ± 0.050
1.0	1.000	0.7	0.995	3.0	1.003 ± 0.050
1.5	1.005	1.0	1.000	4.0	1.021 ± 0.051
2.0	1.007	1.5	1.004	5.0	0.987 ± 0.049
2.5	1.006	2.0	1.006	6.0	0.966 ± 0.048
3.0	1.006	2.5	1.007	7.0	0.958 ± 0.048
4.0	1.003	3.0	1.004	8.0	0.933 ± 0.047
5.0	0.998	4.0	1.002	9.0	0.897 ± 0.045
6.0	0.984	5.0	0.993	10.0	0.872 ± 0.044
7.0	0.967	6.0	0.980		
8.0	0.947	8.0	0.941		
9.0	0.920	10.0	0.882		
10.0	0.885	12.0	0.802		
11.0	0.850	14.0	0.694		
12.0	0.807	15.0	0.609		
13.0	0.757				
14.0	0.699				

Values for the VariSource HDR old design ^{192}Ir source.

TABLE A.3.8

Monte Carlo (MC) Calculated Anisotropy Function, $F(r,\theta)$

Polar Angle θ (°)	Radial Distance r (cm)									
	0.25	0.50	1.00	2.00	3.00	4.00	5.00	6.00	7.00	10.00
0.0			0.445	0.410	0.431	0.482	0.531	0.566	0.577	0.571
1.0			0.448	0.439	0.474	0.488	0.562	0.586	0.622	0.659
2.0			0.469	0.472	0.503	0.551	0.583	0.603	0.637	0.669
3.0			0.540	0.545	0.558	0.588	0.628	0.651	0.670	0.719
5.0		0.943	0.637	0.624	0.651	0.671	0.688	0.710	0.724	0.758
7.0		0.944	0.716	0.694	0.714	0.733	0.749	0.765	0.774	0.796
10.0	0.974	0.943	0.787	0.771	0.779	0.788	0.798	0.812	0.822	0.837
12.0	0.974	0.948	0.822	0.804	0.813	0.822	0.831	0.836	0.847	0.863
15.0	0.973	0.953	0.863	0.841	0.849	0.851	0.864	0.868	0.875	0.887
20.0	0.974	0.963	0.904	0.887	0.893	0.892	0.898	0.903	0.908	0.917
25.0	0.979	0.973	0.929	0.917	0.920	0.921	0.922	0.927	0.931	0.936
30.0	0.984	0.980	0.947	0.938	0.940	0.938	0.941	0.943	0.947	0.950
35.0	0.988	0.984	0.961	0.954	0.954	0.951	0.956	0.957	0.960	0.954
45.0	0.993	0.985	0.977	0.973	0.974	0.971	0.974	0.975	0.977	0.986
50.0	0.995	0.988	0.983	0.980	0.981	0.977	0.979	0.981	0.983	0.988
60.0	0.998	0.995	0.991	0.988	0.990	0.987	0.988	0.990	0.993	0.994
75.0	1.000	0.999	0.998	0.995	0.999	0.995	0.997	0.997	0.999	0.998
90.0	1.000	1.000	1.000	1.000	1.000	1.000	1.000	1.000	1.000	1.000
105.0	1.000	0.999	0.998	0.994	0.999	0.994	0.994	0.997	1.000	0.998
120.0	0.998	0.995	0.991	0.989	0.991	0.988	0.989	0.990	0.992	0.995
130.0	0.996	0.992	0.983	0.980	0.981	0.978	0.981	0.982	0.983	0.981
135.0	0.994	0.990	0.975	0.973	0.975	0.971	0.974	0.975	0.977	0.974
145.0	0.989	0.981	0.960	0.952	0.955	0.952	0.955	0.957	0.961	0.960
150.0	0.985	0.973	0.946	0.938	0.940	0.939	0.942	0.944	0.949	0.952
155.0	0.981	0.964	0.930	0.917	0.920	0.920	0.923	0.927	0.931	0.936
160.0	0.976	0.959	0.905	0.887	0.892	0.892	0.899	0.902	0.908	0.914
165.0	0.974	0.954	0.861	0.843	0.851	0.852	0.862	0.869	0.874	0.885
168.0	0.975	0.948	0.824	0.803	0.812	0.819	0.828	0.840	0.847	0.860
170.0	0.973	0.943	0.789	0.765	0.780	0.785	0.800	0.812	0.822	0.842
173.0		0.943	0.712	0.690	0.713	0.722	0.742	0.755	0.771	0.793
175.0		0.942	0.635	0.617	0.645	0.657	0.681	0.709	0.720	0.764
177.0			0.510	0.512	0.544	0.581	0.614	0.645	0.669	0.709
178.0			0.444	0.456	0.494	0.516	0.558	0.594	0.620	0.677
179.0			0.407	0.405	0.454	0.481	0.525	0.555	0.575	0.633
180.0			0.405	0.406	0.441	0.469	0.528	0.524	0.582	0.606

Values for the VariSource HDR old design ^{192}Ir source. Values related to polar angle θ have been adjusted to fit the TG-43 U1 source coordinate system (Figure 8.6). Data not shown correspond to points within the encapsulated source.

Source: From Wang, R. and Sloboda, R.S., *Phys. Med. Biol.*, 43, 37–48, 1998.

A.3.4   VariSource HDR New ^{192}Ir Source

TABLE A.3.9

Radial Dose Function, $g(r)$

Radial Distance r (cm)	$g(r)$
0.1	0.975
0.2	0.985
0.3	0.990
0.5	0.995
0.7	0.998
1.0	1.000
1.5	1.002
2.0	1.005
2.5	1.006
3.0	1.006
4.0	1.002
5.0	0.993
6.0	0.981
8.0	0.941
10.0	0.881
12.0	0.803
14.0	0.693
15.0	0.609

Values for the VariSource HDR new ^{192}Ir source design.
Source: From Monte Carlo calculated data by Angelopoulos, A., Baras, P., Sakelliou, L., Karaiskos, P., and Sandilos, P., *Med. Phys.*, 27, 2521–2527, 2000.

TABLE A.3.10

Monte Carlo (MC) Calculated Anisotropy Function, $F(r,\theta)$

Polar Angle θ (°)	Radial Distance r (cm)								
	0.25	0.50	1.00	2.00	3.00	4.00	6.00	7.00	10.00
0.5		0.564	0.530	0.550	0.616	0.663	0.720	0.736	0.728
1.5		0.574	0.538	0.581	0.642	0.685	0.727	0.748	0.756
2.5		0.588	0.557	0.601	0.657	0.697	0.746	0.760	0.773
3.5		0.620	0.591	0.634	0.687	0.722	0.762	0.777	0.787
4.5		0.646	0.624	0.663	0.706	0.736	0.778	0.790	0.796
5.5		0.675	0.653	0.690	0.730	0.762	0.794	0.805	0.813
7.5	0.849	0.736	0.721	0.745	0.773	0.802	0.824	0.831	0.836
9.5	0.880	0.787	0.766	0.779	0.808	0.827	0.847	0.853	0.860
12.5	0.910	0.837	0.816	0.821	0.841	0.856	0.876	0.882	0.883
14.5	0.925	0.859	0.843	0.845	0.859	0.872	0.884	0.889	0.894
19.5	0.948	0.905	0.890	0.889	0.901	0.907	0.915	0.916	0.918
29.5	0.968	0.949	0.940	0.938	0.943	0.945	0.948	0.952	0.951
39.5	0.983	0.970	0.966	0.965	0.968	0.968	0.971	0.970	0.972
49.5	0.989	0.985	0.982	0.983	0.982	0.983	0.985	0.984	0.984
69.5	0.999	0.998	0.996	0.997	0.995	0.996	0.996	0.995	0.997
89.5	0.999	1.000	1.001	0.999	1.002	1.002	1.000	1.000	0.999
109.5	0.998	0.995	0.995	0.995	0.998	0.997	0.995	0.996	0.998
129.5	0.990	0.986	0.982	0.984	0.985	0.983	0.982	0.984	0.984
139.5	0.982	0.972	0.968	0.968	0.969	0.969	0.971	0.971	0.971
149.5	0.970	0.952	0.945	0.941	0.947	0.949	0.952	0.952	0.953
159.5	0.950	0.910	0.894	0.896	0.902	0.910	0.918	0.922	0.922
164.5	0.929	0.871	0.854	0.856	0.868	0.880	0.895	0.898	0.897
167.5	0.911	0.836	0.816	0.827	0.843	0.857	0.874	0.875	0.879
169.5	0.891	0.807	0.784	0.793	0.814	0.832	0.848	0.860	0.859
172.5	0.844	0.741	0.711	0.725	0.762	0.789	0.808	0.825	0.829
173.5		0.702	0.677	0.706	0.741	0.769	0.802	0.810	0.819
174.5		0.670	0.641	0.671	0.716	0.747	0.778	0.789	0.798
175.5		0.627	0.594	0.632	0.685	0.718	0.756	0.772	0.782
176.5			0.549	0.593	0.646	0.687	0.731	0.753	0.771
177.5				0.544	0.611	0.652	0.710	0.730	0.736
178.5				0.460	0.544	0.596	0.661	0.684	0.711

Values for the VariSource HDR new design ^{192}Ir source. Values related to polar angle θ have been adjusted to fit the TG-43 U1 source coordinate system (Figure 8.6). Data not shown correspond to points within the encapsulated source.

Source: From Angelopoulos, A., Baras, P., Sakelliou, L., Karaiskos, P., and Sandilos, P., *Med. Phys.*, 27, 2521–2527, 2000.

A.3.5 Buchler HDR ^{192}Ir Source

TABLE A.3.11

Radial Dose Function, $g(r)$

Radial Distance r (cm)	$g(r)$
0.2	1.023
0.4	1.004
0.6	1.000
0.8	0.999
1.0	1.000
1.25	1.002
1.5	1.003
1.75	1.004
2.0	1.004
2.5	1.006
3.0	1.008
3.5	1.008
4.0	1.007
5.0	1.002
6.0	0.995
8.0	0.971
10.0	0.941
12.0	0.902
15.0	0.825

Values for the Buchler HDR ^{192}Ir source design. *Source*: From Monte Carlo calculated data by Ballester, F., Pérez-Calatayud, J., Puchades, V., Lluch, J.L., Serrano-Andrés, M.A., Limami, Y., Lliso, F., and Casal, E., *Phys. Med. Biol.*, 46, N79–N90, 2001.

TABLE A.3.12

Monte Carlo (MC) Calculated Anisotropy Function, $F(r,\theta)$

Polar angle θ (°)	Radial distance r (cm)																		
	0.2	0.4	0.6	0.8	1.0	1.25	1.5	1.75	2.0	2.5	3.0	3.5	4.0	5.0	6.0	8.0	10.0	12.0	15.0
0		0.873	0.882	0.881	0.880	0.884	0.887	0.888	0.891	0.896	0.893	0.896	0.896	0.903	0.915	0.920	0.940	0.936	0.952
1		0.871	0.879	0.878	0.878	0.881	0.885	0.885	0.889	0.894	0.892	0.895	0.895	0.903	0.915	0.920	0.938	0.935	0.949
2		0.878	0.885	0.884	0.883	0.885	0.888	0.889	0.891	0.896	0.894	0.897	0.897	0.904	0.916	0.922	0.937	0.937	0.950
3		0.880	0.887	0.886	0.885	0.888	0.890	0.891	0.893	0.899	0.897	0.900	0.901	0.908	0.918	0.925	0.937	0.937	0.947
4		0.882	0.888	0.888	0.887	0.889	0.892	0.893	0.895	0.900	0.899	0.902	0.903	0.910	0.920	0.927	0.937	0.939	0.948
5		0.882	0.888	0.889	0.889	0.891	0.894	0.895	0.897	0.903	0.902	0.905	0.906	0.912	0.921	0.927	0.935	0.939	0.947
6		0.883	0.889	0.890	0.890	0.893	0.895	0.896	0.898	0.904	0.904	0.907	0.909	0.914	0.923	0.929	0.935	0.940	0.947
7		0.884	0.890	0.891	0.891	0.893	0.895	0.897	0.899	0.905	0.905	0.908	0.910	0.915	0.924	0.930	0.935	0.941	0.947
8		0.886	0.892	0.894	0.893	0.896	0.898	0.899	0.901	0.907	0.908	0.910	0.912	0.917	0.925	0.931	0.935	0.942	0.947
9		0.888	0.894	0.896	0.897	0.899	0.901	0.903	0.905	0.911	0.911	0.914	0.916	0.920	0.927	0.932	0.935	0.943	0.949
10		0.889	0.895	0.898	0.899	0.902	0.904	0.905	0.907	0.913	0.914	0.916	0.918	0.921	0.928	0.933	0.935	0.944	0.949
15		0.901	0.906	0.909	0.910	0.912	0.913	0.915	0.916	0.921	0.923	0.924	0.927	0.930	0.936	0.940	0.940	0.949	0.952
20		0.911	0.917	0.919	0.921	0.923	0.924	0.926	0.927	0.931	0.933	0.934	0.937	0.940	0.945	0.948	0.948	0.956	0.957
30	0.910	0.934	0.940	0.942	0.942	0.944	0.946	0.947	0.947	0.949	0.951	0.951	0.954	0.956	0.959	0.962	0.964	0.967	0.969
40	0.939	0.955	0.962	0.964	0.963	0.964	0.967	0.967	0.967	0.967	0.968	0.968	0.970	0.971	0.973	0.975	0.977	0.979	0.981
50	0.964	0.977	0.980	0.981	0.981	0.980	0.983	0.983	0.982	0.983	0.982	0.983	0.983	0.984	0.986	0.987	0.987	0.988	0.990
60	0.976	0.991	0.993	0.994	0.995	0.992	0.995	0.994	0.993	0.994	0.992	0.994	0.993	0.993	0.996	0.995	0.993	0.995	0.995
70	0.989	0.997	1.000	1.001	1.001	0.998	1.000	1.000	0.999	1.000	0.999	1.001	0.999	0.998	1.000	0.998	0.997	0.999	0.998
80	0.997	0.997	1.000	1.001	1.001	1.000	1.001	1.002	1.002	1.002	1.002	1.003	1.002	1.001	1.002	1.000	0.999	1.001	1.000
90	1.000	1.000	1.000	1.000	1.000	1.000	1.000	1.000	1.000	1.000	1.000	1.000	1.000	1.000	1.000	1.000	1.000	1.000	1.000

100	0.998	0.998	1.000	1.000	1.001	1.001	1.002	1.001	1.001	1.002	1.001	1.001	1.000	1.001	1.000	1.000	1.000	1.001
110	0.980	0.996	1.000	1.001	1.002	0.999	1.002	0.999	0.999	1.002	0.998	0.999	0.997	1.000	0.998	0.998	0.998	1.000
120	0.980	0.989	0.993	0.995	0.995	0.993	0.996	0.994	0.994	0.997	0.993	0.995	0.992	0.996	0.994	0.991	0.994	0.995
130	0.957	0.978	0.983	0.983	0.983	0.982	0.984	0.983	0.983	0.985	0.983	0.985	0.983	0.987	0.987	0.982	0.987	0.987
140	0.950	0.957	0.964	0.964	0.965	0.965	0.966	0.967	0.967	0.967	0.968	0.968	0.969	0.971	0.971	0.970	0.974	0.974
150		0.928	0.935	0.937	0.940	0.939	0.939	0.942	0.942	0.941	0.944	0.942	0.948	0.948	0.950	0.954	0.957	0.959
160		0.880	0.893	0.893	0.894	0.894	0.896	0.898	0.897	0.903	0.901	0.905	0.912	0.919	0.922	0.929	0.933	0.937
165		0.855	0.856	0.855	0.858	0.861	0.865	0.864	0.865	0.875	0.870	0.878	0.885	0.895	0.903	0.908	0.916	0.924
170			0.811	0.815	0.820	0.823	0.824	0.823	0.825	0.833	0.827	0.834	0.848	0.859	0.875	0.878	0.892	0.904
171				0.805	0.808	0.812	0.813	0.811	0.815	0.822	0.818	0.823	0.840	0.851	0.868	0.869	0.886	0.898
172				0.788	0.789	0.796	0.799	0.801	0.804	0.809	0.806	0.808	0.830	0.839	0.857	0.860	0.879	0.892
173				0.765	0.770	0.775	0.783	0.785	0.790	0.792	0.791	0.787	0.817	0.825	0.845	0.849	0.871	0.885
174					0.737	0.751	0.758	0.762	0.766	0.772	0.775	0.764	0.804	0.810	0.831	0.837	0.863	0.877
175						0.721	0.727	0.730	0.738	0.747	0.754	0.738	0.787	0.792	0.814	0.824	0.854	0.867
176								0.681	0.698	0.702	0.729	0.713	0.766	0.767	0.791	0.806	0.839	0.856
177									0.610	0.624	0.666	0.678	0.711	0.732	0.763	0.784	0.818	0.840
178											0.558	0.576	0.613	0.659	0.720	0.750	0.778	0.818
179													0.521	0.556	0.605	0.655	0.695	0.752
180															0.541	0.610	0.657	0.725

Values for the Buchler HDR ^{192}Ir source design. Data not shown correspond to points within the encapsulated source.

Source: From Ballester, F., Pérez-Calatayud, J., Puchades, V., Lluch, J.L., Serrano-Andrés, M.A., Limami, Y., Lliso, F., and Casal, E., *Phys. Med. Biol.*, 46, N79–N90, 2001.

TABLE A.3.13

Monte Carlo (MC) Calculated Anisotropy Factor, $\varphi_{an}(r)$

Radial Distance r (cm)	0.25	0.50	1.00	2.00	3.00	4.00	5.00	6.00	8.00	10.00	12.00
$\varphi_{an}(r)$	0.981	0.981	0.981	0.981	0.981	0.981	0.981	0.981	0.981	0.981	0.981

The anisotropy factor values are according to $\varphi_{an}(r) = 0.981 + 0.00034r$.[11]
Values for the Buchler HDR ^{192}Ir source design.
Source: From Ballester, F., Pérez-Calatayud, J., Puchades, V., Lluch, J.L., Serrano-Andrés, M.A., Limami, Y., Lliso, F., and Casal, E., *Phys. Med. Biol.*, 46, N79–N90, 2001.

A.3.6   GammaMed HDR ^{192}Ir Source Models 12*i* and Plus

TABLE A.3.14

Radial Dose Function, $g(r)$

Radial Distance r (cm)	$g(r)$
0.2	0.994
0.4	0.996
0.6	0.997
0.8	0.999
1.0	1.000
1.2	1.001
1.5	1.002
1.8	1.004
2.0	1.004
2.5	1.005
3.0	1.005
3.5	1.004
4.0	1.003
5.0	0.997
6.0	0.989
8.0	0.966
10.0	0.933
12.0	0.893
15.0	0.822

Values for the GammaMed HDR ^{192}Ir source model 12*i* design. $g(r)$ values have been calculated using the third-order polynomial as fitted to the MC data from the original publication by Ballester et al.[12] The same values are valid also for the Plus ^{192}Ir source model design.
Source: From Ballester, F., Puchades, V., Lluch, J.L., Serrano-Andrés, M.A., Limami, Y., Pérez-Calatayud, J., and Casal, E., *Med. Phys.*, 28, 2586–2591, 2001.

TABLE A.3.15

Monte Carlo (MC) Calculated Anisotropy Function, $F(r,\theta)$

Polar Angle θ (°)	Radial Distance r (cm)																		
	0.2	0.4	0.6	0.8	1.0	1.25	1.5	1.75	2.0	2.5	3.0	3.5	4.0	5.0	6.0	8.0	10.0	12.0	15.0
0		0.665	0.654	0.639	0.633	0.639	0.635	0.646	0.637	0.647	0.644	0.676	0.677	0.712	0.711	0.754	0.785	0.806	0.830
1		0.655	0.646	0.635	0.631	0.638	0.635	0.646	0.640	0.650	0.648	0.677	0.677	0.706	0.707	0.747	0.779	0.801	0.828
2		0.694	0.663	0.644	0.640	0.646	0.641	0.652	0.650	0.660	0.661	0.690	0.693	0.719	0.728	0.766	0.790	0.808	0.835
3		0.706	0.669	0.653	0.651	0.656	0.651	0.661	0.661	0.671	0.673	0.700	0.704	0.724	0.734	0.771	0.795	0.813	0.838
4		0.710	0.679	0.664	0.660	0.664	0.659	0.670	0.673	0.682	0.686	0.710	0.715	0.731	0.744	0.778	0.802	0.820	0.843
5		0.718	0.690	0.677	0.672	0.674	0.668	0.678	0.683	0.692	0.699	0.721	0.727	0.742	0.756	0.787	0.807	0.826	0.847
6		0.735	0.700	0.691	0.688	0.688	0.681	0.691	0.696	0.704	0.712	0.731	0.737	0.750	0.765	0.795	0.814	0.830	0.851
7		0.744	0.713	0.701	0.698	0.699	0.694	0.704	0.710	0.717	0.724	0.741	0.749	0.760	0.776	0.803	0.820	0.835	0.856
8		0.755	0.721	0.714	0.713	0.713	0.708	0.717	0.723	0.729	0.738	0.753	0.760	0.770	0.786	0.811	0.826	0.841	0.861
9		0.765	0.734	0.727	0.725	0.725	0.720	0.729	0.734	0.740	0.750	0.763	0.771	0.781	0.796	0.819	0.833	0.848	0.866
10		0.776	0.745	0.739	0.737	0.736	0.733	0.740	0.746	0.751	0.762	0.773	0.781	0.791	0.806	0.826	0.840	0.854	0.871
15		0.829	0.806	0.801	0.797	0.795	0.796	0.800	0.803	0.807	0.815	0.822	0.829	0.839	0.849	0.862	0.871	0.883	0.894
20	0.974	0.876	0.855	0.849	0.845	0.843	0.847	0.849	0.848	0.853	0.856	0.862	0.868	0.875	0.881	0.890	0.897	0.907	0.914
30	0.981	0.931	0.920	0.917	0.914	0.914	0.911	0.914	0.913	0.915	0.915	0.918	0.920	0.922	0.927	0.931	0.936	0.942	0.945
40	0.992	0.965	0.955	0.954	0.952	0.951	0.949	0.950	0.950	0.949	0.950	0.954	0.954	0.955	0.958	0.959	0.962	0.963	0.964
50	0.995	0.980	0.974	0.975	0.974	0.972	0.972	0.973	0.974	0.972	0.973	0.973	0.973	0.975	0.976	0.977	0.978	0.979	0.979
60	0.998	0.990	0.988	0.988	0.987	0.985	0.984	0.988	0.987	0.986	0.986	0.987	0.987	0.988	0.988	0.989	0.989	0.991	0.990
70	0.999	0.995	0.995	0.995	0.995	0.994	0.992	0.995	0.994	0.994	0.994	0.996	0.995	0.995	0.996	0.995	0.995	0.998	0.997
80	1.002	1.000	1.000	0.999	0.999	0.999	0.998	0.998	0.998	0.998	0.998	1.000	1.000	0.999	1.000	0.999	1.000	1.001	1.000

(continued)

TABLE A.3.15—Continued

Polar Angle θ (°)	Radial Distance r (cm)																		
	0.2	0.4	0.6	0.8	1.0	1.25	1.5	1.75	2.0	2.5	3.0	3.5	4.0	5.0	6.0	8.0	10.0	12.0	15.0
90	1.000	1.000	1.000	1.000	1.000	1.000	1.000	1.000	1.000	1.000	1.000	1.000	1.000	1.000	1.000	1.000	1.000	1.000	1.000
100	1.000	0.999	0.998	0.998	0.999	0.998	0.997	0.999	0.998	0.998	0.998	0.997	0.998	0.998	0.998	0.999	0.998	0.998	0.998
110	0.998	0.995	0.994	0.993	0.995	0.993	0.991	0.993	0.993	0.992	0.993	0.992	0.992	0.992	0.993	0.994	0.994	0.995	0.994
120	0.997	0.989	0.987	0.986	0.986	0.985	0.982	0.984	0.984	0.983	0.984	0.985	0.984	0.984	0.986	0.985	0.986	0.988	0.988
130	0.997	0.978	0.971	0.971	0.969	0.969	0.967	0.970	0.968	0.967	0.968	0.970	0.968	0.970	0.974	0.972	0.974	0.977	0.976
140	0.993	0.960	0.949	0.949	0.947	0.946	0.945	0.947	0.946	0.945	0.947	0.948	0.946	0.949	0.954	0.954	0.956	0.959	0.961
150	0.978	0.924	0.912	0.909	0.909	0.906	0.905	0.908	0.909	0.910	0.912	0.913	0.912	0.915	0.921	0.924	0.929	0.933	0.937
160	0.945	0.891	0.841	0.838	0.839	0.836	0.837	0.839	0.842	0.845	0.847	0.854	0.855	0.861	0.868	0.876	0.888	0.895	0.904
165		0.817	0.781	0.782	0.783	0.782	0.783	0.786	0.788	0.795	0.798	0.807	0.811	0.821	0.831	0.845	0.859	0.870	0.879
170		0.765	0.702	0.713	0.703	0.707	0.704	0.713	0.715	0.724	0.730	0.740	0.748	0.762	0.779	0.801	0.817	0.834	0.848
171		0.738	0.701	0.689	0.690	0.687	0.684	0.694	0.696	0.705	0.712	0.723	0.731	0.748	0.766	0.790	0.807	0.825	0.841
172		0.724	0.676	0.676	0.678	0.669	0.664	0.674	0.677	0.684	0.691	0.705	0.713	0.732	0.751	0.777	0.795	0.815	0.833
173			0.651	0.648	0.657	0.627	0.642	0.652	0.655	0.661	0.669	0.685	0.694	0.715	0.734	0.762	0.783	0.803	0.823
174			0.625	0.620	0.618	0.601	0.620	0.628	0.633	0.636	0.645	0.663	0.673	0.695	0.715	0.745	0.769	0.791	0.813
175				0.599	0.575	0.581	0.601	0.600	0.607	0.609	0.618	0.638	0.647	0.672	0.690	0.723	0.752	0.776	0.802
176				0.574	0.560	0.544	0.569	0.588	0.556	0.576	0.588	0.610	0.619	0.648	0.665	0.700	0.733	0.759	0.789
177						0.444	0.492	0.553	0.500	0.527	0.540	0.566	0.577	0.614	0.634	0.671	0.711	0.740	0.774
178								0.506	0.433	0.448	0.479	0.550	0.540	0.583	0.588	0.635	0.682	0.712	0.754
179												0.508	0.510	0.522	0.572	0.584	0.642	0.676	0.725
180																0.551	0.612	0.649	0.706

Values for the GammaMed HDR ^{192}Ir source model 12i design. Data not shown correspond to points within the encapsulated source.

Source: From Ballester, F., Puchades, V., Lluch, J.L., Serrano-Andrés, M.A., Limami, Y., Pérez-Calatayud, J., and Casal, E., *Med. Phys.*, 28, 2586–2591, 2001.

TABLE A.3.16

Monte Carlo (MC) Calculated Anisotropy Function, $F(r, \theta)$

Polar Angle θ (°)	Radial Distance r (cm)																		
	0.2	0.4	0.6	0.8	1.0	1.25	1.5	1.75	2.0	2.5	3.0	3.5	4.0	5.0	6.0	8.0	10.0	12.0	15.0
0		0.666	0.636	0.630	0.608	0.615	0.634	0.625	0.629	0.648	0.654	0.660	0.683	0.702	0.716	0.758	0.789	0.803	0.815
1		0.677	0.643	0.632	0.609	0.614	0.632	0.626	0.633	0.656	0.667	0.676	0.698	0.716	0.727	0.762	0.786	0.794	0.805
2		0.684	0.653	0.640	0.620	0.624	0.638	0.634	0.641	0.661	0.671	0.679	0.697	0.717	0.730	0.764	0.794	0.807	0.821
3		0.684	0.660	0.650	0.634	0.637	0.647	0.646	0.653	0.671	0.682	0.691	0.705	0.726	0.739	0.771	0.798	0.810	0.830
4		0.701	0.666	0.654	0.645	0.652	0.662	0.663	0.668	0.680	0.690	0.697	0.711	0.736	0.750	0.779	0.805	0.814	0.834
5		0.706	0.684	0.673	0.664	0.668	0.674	0.676	0.682	0.693	0.704	0.712	0.724	0.748	0.762	0.788	0.811	0.820	0.842
6		0.709	0.696	0.688	0.681	0.683	0.686	0.691	0.697	0.707	0.718	0.725	0.734	0.757	0.770	0.795	0.817	0.825	0.847
7		0.720	0.705	0.697	0.692	0.697	0.700	0.707	0.712	0.719	0.729	0.736	0.745	0.769	0.780	0.802	0.823	0.831	0.853
8		0.726	0.720	0.715	0.712	0.713	0.713	0.719	0.724	0.731	0.743	0.751	0.758	0.779	0.789	0.810	0.830	0.838	0.860
9		0.738	0.733	0.729	0.726	0.728	0.727	0.734	0.738	0.743	0.754	0.762	0.769	0.790	0.799	0.818	0.837	0.844	0.866
10		0.753	0.743	0.738	0.738	0.741	0.740	0.748	0.752	0.755	0.765	0.772	0.778	0.799	0.808	0.826	0.844	0.851	0.870
15		0.820	0.803	0.801	0.802	0.804	0.802	0.809	0.811	0.813	0.821	0.828	0.829	0.844	0.848	0.863	0.873	0.880	0.895
20	0.962	0.866	0.855	0.854	0.852	0.853	0.852	0.858	0.858	0.862	0.865	0.870	0.869	0.878	0.880	0.893	0.900	0.905	0.914
30	0.968	0.923	0.917	0.916	0.912	0.913	0.912	0.918	0.918	0.918	0.921	0.923	0.923	0.927	0.930	0.933	0.939	0.941	0.945
40	0.979	0.956	0.951	0.952	0.948	0.949	0.946	0.950	0.951	0.950	0.953	0.955	0.954	0.958	0.958	0.959	0.963	0.963	0.963
50	0.987	0.977	0.973	0.973	0.971	0.972	0.971	0.972	0.973	0.973	0.974	0.975	0.974	0.977	0.976	0.978	0.978	0.978	0.979
60	0.993	0.987	0.984	0.985	0.985	0.987	0.985	0.987	0.988	0.989	0.989	0.988	0.988	0.989	0.988	0.988	0.988	0.988	0.988
70	0.994	0.995	0.994	0.994	0.995	0.996	0.993	0.996	0.996	0.996	0.996	0.996	0.996	0.996	0.995	0.995	0.996	0.996	0.996
80	0.998	0.999	0.999	0.998	0.999	1.000	0.998	1.000	1.000	1.000	1.000	1.000	1.000	0.999	0.999	0.999	0.999	0.999	0.999

(continued)

TABLE A.3.16—*Continued*

Polar Angle θ (°)	Radial Distance r (cm)																		
	0.2	0.4	0.6	0.8	1.0	1.25	1.5	1.75	2.0	2.5	3.0	3.5	4.0	5.0	6.0	8.0	10.0	12.0	15.0
90	1.000	1.000	1.000	1.000	1.000	1.000	1.000	1.000	1.000	1.000	1.000	1.000	1.000	1.000	1.000	1.000	1.000	1.000	1.000
100	1.000	0.999	0.999	0.999	0.999	0.998	0.998	0.998	0.998	0.999	0.999	0.999	0.999	1.000	0.999	1.000	0.999	0.999	0.999
110	0.998	0.996	0.994	0.995	0.995	0.995	0.993	0.995	0.994	0.995	0.995	0.995	0.995	0.995	0.994	0.995	0.995	0.995	0.995
120	0.998	0.990	0.985	0.984	0.985	0.986	0.984	0.987	0.987	0.987	0.987	0.988	0.989	0.988	0.987	0.989	0.989	0.990	0.989
130	0.994	0.976	0.973	0.971	0.972	0.972	0.970	0.972	0.973	0.974	0.974	0.974	0.975	0.976	0.977	0.977	0.979	0.979	0.979
140	0.991	0.958	0.950	0.947	0.948	0.947	0.947	0.949	0.950	0.952	0.953	0.954	0.953	0.956	0.958	0.959	0.962	0.963	0.964
150	0.973	0.923	0.914	0.914	0.914	0.914	0.912	0.915	0.916	0.917	0.919	0.921	0.920	0.925	0.927	0.931	0.935	0.939	0.943
160	0.966	0.873	0.851	0.850	0.847	0.850	0.848	0.853	0.856	0.857	0.863	0.867	0.870	0.876	0.878	0.889	0.895	0.904	0.911
165		0.814	0.801	0.798	0.796	0.798	0.801	0.802	0.806	0.809	0.818	0.822	0.829	0.838	0.843	0.860	0.870	0.879	0.889
170		0.756	0.723	0.720	0.725	0.725	0.729	0.730	0.734	0.743	0.754	0.761	0.771	0.784	0.795	0.819	0.831	0.843	0.859
171		0.715	0.700	0.699	0.706	0.706	0.710	0.712	0.716	0.725	0.735	0.744	0.755	0.768	0.782	0.809	0.822	0.834	0.852
172		0.699	0.676	0.679	0.686	0.686	0.689	0.692	0.696	0.705	0.716	0.725	0.738	0.752	0.768	0.798	0.811	0.825	0.845
173			0.642	0.649	0.664	0.663	0.666	0.671	0.675	0.684	0.695	0.705	0.719	0.735	0.752	0.785	0.799	0.814	0.838
174			0.618	0.640	0.643	0.641	0.643	0.650	0.653	0.662	0.672	0.684	0.698	0.715	0.733	0.769	0.784	0.802	0.829
175				0.615	0.611	0.615	0.616	0.625	0.627	0.636	0.646	0.659	0.674	0.692	0.713	0.752	0.769	0.789	0.820
176				0.602	0.566	0.568	0.585	0.597	0.598	0.608	0.618	0.632	0.645	0.666	0.688	0.731	0.750	0.773	0.809
177						0.536	0.500	0.565	0.564	0.575	0.584	0.600	0.610	0.634	0.659	0.706	0.729	0.755	0.796
178								0.518	0.529	0.521	0.533	0.553	0.560	0.591	0.620	0.672	0.702	0.732	0.780
179												0.501	0.527	0.548	0.572	0.621	0.659	0.696	0.755
180																0.589	0.633	0.675	0.741

Values for the GammaMed HDR ^{192}Ir source model plus design. Data not shown correspond to points within the encapsulated source.

Source: From Ballester, F., Puchades, V., Lluch, J.L., Serrano-Andrés, M.A., Limami, Y., Pérez-Calatayud, J., and Casal, E., *Med. Phys.*, 28, 2586–2591, 2001.

A.3.7 BEBIG MultiSource HDR ^{192}Ir Source Model GI192M11

TABLE A.3.17

Monte Carlo (MC) Calculated and Experimental Radial Dose Function, $g(r)$

MC Data by Granero et al.[25]		Experimental (Ionization Chamber in Water Phantom) Data by Selbach and Andrássy[13]		
Radial Distance r (cm)	$g_L(r)$	Radial Distance r (cm)	$g_P(r)$	$g_L(r)$
0.25	0.990	0.6	1.004	1.022
0.5	0.996	0.8	0.997	1.003
0.75	0.998	1.0	1.000	1.000
1.0	1.000	1.2	1.003	1.000
1.5	1.003	1.4	1.004	0.999
2.0	1.003	1.6	1.006	1.000
3.0	1.005	1.8	1.008	1.001
4.0	1.003	2.0	1.012	1.004
5.0	0.999	2.2	1.013	1.006
6.0	0.991	2.4	1.015	1.007
7.0	0.981	2.6	1.016	1.008
8.0	0.968	2.8	1.018	1.010
10.0	0.935	3.0	1.020	1.012
12.0	0.894	3.2	1.020	1.011
15.0	0.821	3.4	1.019	1.010
20.0	0.687	3.6	1.017	1.008
		3.8	1.023	1.014
		4.0	1.017	1.007
		4.2	1.022	1.012
		4.4	1.018	1.009
		4.6	1.017	1.008
		4.8	1.019	1.010
		5.0	1.014	1.004
		5.2	1.015	1.006
		5.4	1.011	1.002
		5.6	1.009	1.000
		5.8	1.011	1.002
		6.0	1.008	0.998
		6.2	1.008	0.999
		6.4	1.004	0.995
		6.6	1.002	0.992
		6.8	1.001	0.992
		7.0	0.997	0.988
		7.2	0.996	0.986
		7.4	0.992	0.982
		7.6	0.989	0.979
		7.8	0.987	0.978
		8.0	0.982	0.972

Values for the BEBIG multiSource HDR ^{192}Ir source model GI192M11 design for the line, $g_L(r)$, and point source, $g_P(r)$, approximation, see also Equation 8.45.

TABLE A.3.18

Monte Carlo (MC) Calculated Anisotropy Function, $F(r,\theta)$

Polar Angle θ (°)	Radial Distance r (cm)															
	0.25	0.5	0.75	1.0	1.5	2.0	3.0	4.0	5.0	6.0	7.0	8.0	10.0	12.0	15.0	20.0
0		0.643	0.610	0.598	0.599	0.605	0.633	0.655	0.679	0.698	0.714	0.728	0.763	0.775	0.801	0.838
1		0.645	0.617	0.608	0.611	0.618	0.644	0.667	0.689	0.707	0.724	0.738	0.769	0.783	0.811	0.837
2		0.651	0.626	0.621	0.626	0.637	0.660	0.683	0.704	0.719	0.739	0.753	0.777	0.797	0.823	0.843
3		0.654	0.633	0.631	0.638	0.649	0.672	0.695	0.714	0.729	0.748	0.761	0.784	0.805	0.828	0.852
4		0.658	0.643	0.643	0.652	0.663	0.685	0.707	0.724	0.741	0.757	0.771	0.793	0.813	0.834	0.856
5		0.667	0.656	0.654	0.662	0.674	0.695	0.717	0.734	0.750	0.765	0.777	0.799	0.818	0.840	0.860
6		0.678	0.667	0.668	0.675	0.686	0.707	0.727	0.744	0.760	0.773	0.784	0.805	0.822	0.843	0.865
8		0.701	0.693	0.695	0.704	0.714	0.733	0.750	0.765	0.779	0.791	0.801	0.820	0.835	0.852	0.872
10		0.728	0.722	0.725	0.732	0.741	0.758	0.772	0.787	0.798	0.808	0.819	0.834	0.849	0.865	0.881
15		0.796	0.790	0.792	0.797	0.803	0.814	0.824	0.834	0.842	0.849	0.856	0.869	0.878	0.890	0.900
20		0.848	0.842	0.843	0.845	0.850	0.857	0.865	0.871	0.877	0.883	0.887	0.896	0.903	0.910	0.918
25		0.887	0.881	0.881	0.883	0.886	0.891	0.896	0.901	0.904	0.909	0.912	0.918	0.923	0.929	0.933
30		0.915	0.911	0.910	0.910	0.912	0.916	0.919	0.923	0.926	0.928	0.931	0.935	0.938	0.943	0.945
40		0.950	0.948	0.948	0.948	0.949	0.951	0.953	0.955	0.956	0.957	0.958	0.960	0.962	0.964	0.965
50		0.972	0.971	0.971	0.971	0.972	0.973	0.973	0.974	0.975	0.976	0.976	0.977	0.978	0.980	0.980
60	0.990	0.985	0.985	0.986	0.985	0.986	0.986	0.986	0.988	0.987	0.988	0.988	0.988	0.989	0.989	0.989
70	0.996	0.994	0.994	0.995	0.995	0.996	0.995	0.995	0.996	0.995	0.995	0.995	0.994	0.995	0.996	0.995

80	0.999	0.998	0.997	0.998	0.998	0.998	0.999	0.999	0.999	0.999	0.999	0.999	0.999	0.999	0.999	0.999
90	1.000	1.000	1.000	1.000	1.000	1.000	1.000	1.000	1.000	1.000	1.000	1.000	1.000	1.000	1.000	1.000
100	0.999	0.998	0.998	0.999	0.999	0.999	0.999	0.998	0.999	0.999	0.999	0.998	0.999	0.999	0.999	0.999
110	0.996	0.993	0.993	0.994	0.994	0.994	0.994	0.995	0.995	0.995	0.995	0.995	0.995	0.995	0.995	0.999
120	0.991	0.985	0.986	0.986	0.986	0.986	0.986	0.986	0.987	0.987	0.988	0.987	0.987	0.988	0.989	0.988
130		0.972	0.971	0.972	0.971	0.972	0.973	0.973	0.974	0.975	0.976	0.977	0.977	0.978	0.979	0.979
140		0.951	0.949	0.949	0.949	0.950	0.952	0.953	0.955	0.956	0.958	0.959	0.960	0.962	0.964	0.965
150		0.916	0.911	0.911	0.912	0.914	0.916	0.919	0.923	0.925	0.927	0.930	0.934	0.937	0.942	0.945
155		0.889	0.883	0.883	0.884	0.887	0.892	0.896	0.901	0.905	0.908	0.911	0.916	0.921	0.927	0.931
160			0.843	0.842	0.843	0.848	0.855	0.862	0.868	0.874	0.880	0.885	0.892	0.900	0.908	0.915
165				0.789	0.792	0.797	0.807	0.817	0.827	0.835	0.843	0.850	0.862	0.872	0.882	0.895
170					0.710	0.719	0.736	0.752	0.766	0.779	0.791	0.801	0.819	0.834	0.851	0.869
172					0.666	0.676	0.698	0.717	0.734	0.750	0.762	0.775	0.796	0.814	0.834	0.855
174						0.621	0.647	0.671	0.692	0.712	0.729	0.744	0.769	0.791	0.814	0.840
175							0.615	0.641	0.664	0.687	0.707	0.723	0.753	0.776	0.802	0.831
176							0.574	0.605	0.632	0.657	0.679	0.698	0.729	0.757	0.788	0.821
177								0.555	0.588	0.618	0.643	0.665	0.701	0.732	0.767	0.808
178										0.560	0.590	0.617	0.662	0.696	0.740	0.787
179														0.654	0.705	0.762

Values for the BEBIG MultiSource HDR ^{192}Ir source model GI192M11 design. Data not shown correspond to points within the encapsulated source.

Source: From Granero, D., Pérez-Calatayud, J., and Ballester, F., *Radiother. Oncol.*, 76, 79–85, 2005.

A.3.8 Ralstron HDR ^{60}Co Source Models Type 1, 2, and 3

TABLE A.3.19

Monte Carlo (MC) Calculated Radial Dose Function, $g(r)$

Radial Distance	$g(r)$		
r (cm)	Type 1 Source	Type 2 Source	Type 3 Source
0.5	0.620	1.001	0.555
1.0	1.000	1.000	1.000
1.5	1.101	0.992	1.168
2.0	1.156	0.984	1.225
2.5	1.173	0.975	1.262
3.0	1.178	0.967	1.281
4.0	1.178	0.949	1.284
5.0	1.166	0.930	1.272
6.0	1.145	0.911	1.253
7.0	1.118	0.890	1.228
8.0	1.092	0.869	1.195
9.0	1.066	0.846	1.163
10.0	1.038	0.823	1.133
11.0	1.007	0.799	1.100
12.0	0.975	0.774	1.067
13.0	0.939	0.746	1.029
14.0	0.900	0.715	0.987
15.0	0.849	0.675	0.932

Values for the Ralstron remote HDR afterloader, type 1, 2, and 3 ^{60}Co source designs. Authors used the point source approximation for the geometry function for deriving the $g(r)$ values and thus $g(r)$ can also be considered as $g_P(r)$.

Source: From Monte Carlo calculated data by Papagiannis, P., Angelopoulos, A., Pantelis, E., Sakelliou, L., Karaiskos, P., and Shimizu, Y., *Med. Phys.*, 29, 712–721, 2003.

TABLE A.3.20

Monte Carlo (MC) Calculated Anisotropy Function, $F(r,\theta)$

Polar Angle					Radial Distance r (cm)								
θ (°)	0.5	1.0	1.5	2.0	3.0	4.0	5.0	6.0	8.0	10.0	12.0	15.0
1		2.601	1.421	1.139	0.966	0.909	0.889	0.878	0.875	0.883	0.878	0.884
3		2.601	1.432	1.145	0.971	0.912	0.885	0.875	0.873	0.868	0.868	0.868
5		2.579	1.430	1.137	0.969	0.910	0.887	0.878	0.873	0.872	0.868	0.870
7		2.558	1.425	1.138	0.973	0.915	0.891	0.882	0.882	0.877	0.876	0.876
9		2.525	1.418	1.147	0.985	0.928	0.905	0.894	0.893	0.892	0.894	0.893
12		2.460	1.432	1.162	1.006	0.952	0.929	0.920	0.917	0.915	0.917	0.918
15		2.386	1.435	1.173	1.022	0.970	0.950	0.939	0.936	0.932	0.932	0.932
18		2.299	1.433	1.180	1.033	0.982	0.961	0.952	0.947	0.943	0.942	0.944
21	5.753	2.207	1.424	1.187	1.041	0.989	0.967	0.958	0.954	0.951	0.950	0.952
25	4.738	2.083	1.398	1.184	1.047	0.998	0.979	0.968	0.964	0.961	0.961	0.961
30	3.818	1.922	1.360	1.178	1.055	1.006	0.986	0.976	0.973	0.969	0.968	0.969
35	3.124	1.759	1.317	1.162	1.059	1.014	0.993	0.983	0.980	0.978	0.978	0.978
40	2.565	1.606	1.267	1.143	1.059	1.019	0.997	0.989	0.987	0.981	0.982	0.984
45	2.131	1.469	1.223	1.121	1.053	1.021	1.002	0.994	0.992	0.985	0.989	0.988
50	1.802	1.355	1.177	1.098	1.047	1.018	1.004	0.997	0.994	0.992	0.990	0.992
55	1.558	1.261	1.139	1.076	1.038	1.016	1.005	0.996	0.998	0.995	0.992	0.993
60	1.350	1.182	1.104	1.058	1.033	1.014	1.005	0.998	0.999	0.996	0.994	0.994
65	1.231	1.121	1.070	1.041	1.023	1.010	1.005	0.999	1.001	0.998	0.996	0.996
70	1.142	1.076	1.047	1.025	1.016	1.007	1.003	1.000	1.000	0.997	0.997	0.997

(continued)

TABLE A.3.20—*Continued*

Polar Angle θ (°)	Radial Distance r (cm)											
	0.5	1.0	1.5	2.0	3.0	4.0	5.0	6.0	8.0	10.0	12.0	15.0
80	1.036	1.020	1.012	1.007	1.004	1.002	1.001	1.000	1.000	0.998	0.998	0.998
90	1.000	1.000	1.000	1.000	1.000	1.000	1.000	1.000	1.000	1.000	1.000	1.000
100	1.026	1.017	1.011	1.007	1.005	1.002	0.999	0.999	1.001	0.999	0.999	0.999
110	1.128	1.067	1.044	1.024	1.016	1.008	1.003	1.000	0.999	0.999	0.998	0.998
120	1.321	1.170	1.098	1.053	1.031	1.015	1.004	1.000	0.998	0.996	0.996	0.997
125	1.518	1.244	1.131	1.074	1.039	1.017	1.006	1.000	0.999	0.996	0.994	0.996
130	1.750	1.336	1.168	1.094	1.046	1.020	1.006	0.998	0.997	0.993	0.991	0.993
135	2.055	1.447	1.213	1.117	1.054	1.021	1.003	0.996	0.994	0.989	0.987	0.988
140	2.465	1.577	1.260	1.141	1.059	1.022	1.001	0.994	0.990	0.986	0.984	0.986
145	2.974	1.728	1.309	1.163	1.064	1.020	0.997	0.988	0.985	0.980	0.979	0.980
150	3.602	1.894	1.356	1.180	1.061	1.010	0.987	0.979	0.976	0.972	0.971	0.973
155	4.401	2.053	1.400	1.189	1.053	1.000	0.978	0.972	0.965	0.961	0.959	0.960
159	5.248	2.183	1.420	1.187	1.041	0.988	0.967	0.956	0.951	0.947	0.946	0.945
162	6.159	2.277	1.429	1.179	1.024	0.973	0.949	0.940	0.936	0.932	0.930	0.930
165		2.361	1.426	1.160	1.006	0.953	0.929	0.919	0.915	0.911	0.911	0.910
168		2.420	1.404	1.129	0.975	0.924	0.898	0.888	0.886	0.882	0.883	0.881
171		2.433	1.348	1.074	0.924	0.868	0.844	0.836	0.835	0.835	0.837	0.837
173			1.287	1.016	0.863	0.811	0.791	0.787	0.787	0.789	0.794	0.794
175				0.915	0.769	0.729	0.712	0.707	0.714	0.716	0.727	0.730
177					0.610	0.574	0.566	0.569	0.583	0.601	0.616	0.622

Values for the Ralstron remote HDR afterloader type 1 ^{60}Co source design. Values have been extracted from the MC calculated dose rate tables using the point source approximated geometry function values. Data not shown correspond to points within the encapsulated source.

Source: From Papagiannis, P., Angelopoulos, A., Pantelis, E., Sakelliou, L., Karaiskos, P., and Shimizu, Y., *Med. Phys.*, 29, 712–721, 2003.

TABLE A.3.21

Monte Carlo (MC) Calculated Anisotropy Function, $F(r,\theta)$

Polar Angle θ (°)	Radial Distance r (cm)											
	0.5	1.0	1.5	2.0	3.0	4.0	5.0	6.0	8.0	10.0	12.0	15.0
1	0.986	0.930	0.920	0.926	0.929	0.929	0.926	0.926	0.951	0.948	0.951	0.953
3	0.979	0.938	0.933	0.928	0.927	0.929	0.930	0.931	0.929	0.937	0.939	0.939
5	0.981	0.942	0.938	0.937	0.938	0.940	0.943	0.942	0.945	0.951	0.947	0.945
7	0.984	0.942	0.935	0.931	0.931	0.933	0.935	0.936	0.943	0.944	0.942	0.944
9	0.982	0.944	0.935	0.933	0.932	0.934	0.938	0.937	0.943	0.946	0.948	0.946
12	0.978	0.944	0.937	0.933	0.935	0.937	0.937	0.938	0.943	0.946	0.944	0.946
15	0.978	0.940	0.935	0.933	0.933	0.936	0.939	0.938	0.945	0.945	0.947	0.948
18	0.980	0.945	0.938	0.935	0.935	0.936	0.939	0.942	0.945	0.950	0.951	0.951
21	0.983	0.946	0.941	0.939	0.939	0.942	0.944	0.947	0.950	0.951	0.952	0.954
25	0.985	0.954	0.950	0.947	0.948	0.950	0.954	0.953	0.958	0.961	0.959	0.961
30	0.993	0.961	0.958	0.956	0.958	0.960	0.963	0.963	0.967	0.969	0.968	0.968
35	0.996	0.969	0.965	0.963	0.965	0.968	0.970	0.970	0.974	0.973	0.975	0.974
40	1.001	0.977	0.974	0.972	0.971	0.972	0.973	0.975	0.978	0.981	0.980	0.981
45	1.005	0.985	0.980	0.979	0.980	0.981	0.981	0.981	0.984	0.985	0.985	0.986
50	1.007	0.989	0.986	0.986	0.985	0.987	0.987	0.987	0.989	0.991	0.992	0.990
55	1.005	0.992	0.991	0.990	0.990	0.991	0.992	0.991	0.993	0.995	0.994	0.995
60	1.004	0.996	0.992	0.993	0.992	0.991	0.992	0.993	0.994	0.997	0.995	0.996
65	1.003	0.997	0.995	0.995	0.995	0.995	0.995	0.995	0.996	0.996	0.996	0.996
70	1.001	0.998	0.998	0.998	0.996	0.996	0.997	0.997	0.997	0.999	0.998	0.998

(continued)

TABLE A.3.21—*Continued*

Polar Angle θ (°)	Radial Distance r (cm)											
	0.5	1.0	1.5	2.0	3.0	4.0	5.0	6.0	8.0	10.0	12.0	15.0
80	1.000	1.001	1.001	1.000	1.000	1.000	1.001	1.000	1.001	1.002	1.000	1.000
90	1.000	1.000	1.000	1.000	1.000	1.000	1.000	1.000	1.000	1.000	1.000	1.000
100	1.000	1.000	0.999	0.999	1.000	1.001	1.001	1.001	1.000	1.000	1.000	1.000
110	1.000	1.000	0.998	0.999	0.998	0.998	0.999	0.999	0.999	1.001	0.998	0.998
120	1.004	0.997	0.993	0.993	0.994	0.995	0.995	0.996	0.995	0.997	0.997	0.998
125	1.008	0.995	0.992	0.992	0.990	0.991	0.991	0.990	0.993	0.994	0.994	0.994
130	1.007	0.991	0.989	0.988	0.989	0.988	0.989	0.989	0.991	0.993	0.992	0.992
135	1.004	0.987	0.983	0.982	0.980	0.981	0.983	0.982	0.986	0.987	0.987	0.987
140	1.004	0.983	0.978	0.978	0.976	0.978	0.979	0.981	0.983	0.985	0.985	0.985
145	1.001	0.976	0.972	0.971	0.972	0.973	0.974	0.973	0.976	0.978	0.977	0.976
150	0.997	0.967	0.963	0.962	0.964	0.966	0.968	0.969	0.972	0.974	0.971	0.971
155	0.986	0.956	0.950	0.950	0.952	0.952	0.955	0.956	0.957	0.958	0.959	0.957
159	0.974	0.942	0.937	0.936	0.935	0.935	0.937	0.937	0.941	0.943	0.942	0.941
162	0.959	0.925	0.922	0.920	0.921	0.921	0.922	0.923	0.927	0.928	0.930	0.929
165		0.905	0.897	0.897	0.895	0.897	0.898	0.898	0.902	0.905	0.905	0.907
168		0.875	0.868	0.865	0.864	0.865	0.867	0.869	0.873	0.877	0.876	0.878
171			0.814	0.810	0.809	0.812	0.814	0.816	0.823	0.829	0.831	0.833
173			0.753	0.751	0.749	0.754	0.758	0.764	0.773	0.782	0.785	0.786
175				0.661	0.663	0.668	0.676	0.681	0.698	0.707	0.716	0.722
177						0.522	0.538	0.550	0.574	0.595	0.609	0.624

Values for the Ralstron remote HDR afterloader type 2 ^{60}Co source design. Values have been extracted from the MC calculated dose rate tables using the point source approximated geometry function values. Data not shown correspond to points within the encapsulated source.

Source: From Papagiannis, P., Angelopoulos, A., Pantelis, E., Sakelliou, L., Karaiskos, P., and Shimizu, Y., *Med. Phys.*, 29, 712–721, 2003.

TABLE A.3.22

Monte Carlo (MC) Calculated Anisotropy Function, $F(r,\theta)$

Polar Angle θ (°)	Radial Distance r (cm)											
	0.5	1.0	1.5	2.0	3.0	4.0	5.0	6.0	8.0	10.0	12.0	15.0
1		4.391	1.721	1.280	1.002	0.920	0.890	0.874	0.875	0.878	0.869	0.857
3		4.333	1.745	1.290	1.012	0.928	0.893	0.876	0.869	0.863	0.860	0.855
5		4.263	1.738	1.287	1.016	0.931	0.896	0.881	0.871	0.869	0.865	0.861
7		4.181	1.740	1.285	1.016	0.937	0.904	0.889	0.882	0.882	0.876	0.874
9		4.051	1.723	1.294	1.036	0.958	0.925	0.910	0.905	0.903	0.902	0.895
12		3.856	1.721	1.311	1.060	0.983	0.949	0.933	0.928	0.924	0.920	0.921
15		3.632	1.714	1.317	1.072	0.999	0.967	0.953	0.948	0.942	0.939	0.934
18		3.399	1.692	1.318	1.078	1.013	0.981	0.965	0.958	0.953	0.949	0.945
21	5.590	3.153	1.663	1.319	1.086	1.016	0.988	0.972	0.965	0.962	0.956	0.954
25	4.633	2.814	1.608	1.306	1.091	1.023	0.997	0.983	0.974	0.971	0.967	0.962
30	3.778	2.439	1.538	1.286	1.095	1.030	1.004	0.989	0.983	0.980	0.975	0.970
35	3.068	2.128	1.464	1.252	1.091	1.035	1.009	0.995	0.990	0.986	0.983	0.981
40	2.499	1.874	1.386	1.218	1.085	1.036	1.011	0.996	0.996	0.992	0.990	0.986
45	2.062	1.661	1.315	1.179	1.076	1.034	1.013	0.999	0.998	0.994	0.991	0.990
50	1.749	1.496	1.250	1.147	1.061	1.030	1.014	1.001	0.999	0.999	0.995	0.994
55	1.495	1.363	1.190	1.114	1.050	1.025	1.012	1.003	0.997	1.000	0.999	0.995
60	1.336	1.257	1.137	1.086	1.036	1.021	1.010	1.003	0.998	1.001	0.998	0.995
65	1.227	1.173	1.096	1.060	1.028	1.015	1.007	1.006	1.000	1.001	1.001	0.996
70	1.139	1.110	1.065	1.040	1.019	1.010	1.005	1.003	0.999	1.000	0.999	0.997

(continued)

TABLE A.3.22—*Continued*

Polar Angle θ (°)	Radial Distance r (cm)											
	0.5	1.0	1.5	2.0	3.0	4.0	5.0	6.0	8.0	10.0	12.0	15.0
80	1.036	1.031	1.017	1.011	1.004	1.003	1.002	1.000	0.999	0.998	1.000	0.999
90	1.000	1.000	1.000	1.000	1.000	1.000	1.000	1.000	1.000	1.000	1.000	1.000
100	1.028	1.025	1.014	1.009	1.004	1.001	1.001	0.999	1.000	1.000	0.999	0.999
110	1.129	1.101	1.058	1.035	1.016	1.010	1.005	1.002	0.999	1.000	1.001	0.999
120	1.311	1.237	1.131	1.081	1.038	1.019	1.012	1.004	1.000	1.001	0.999	0.998
125	1.460	1.337	1.182	1.110	1.048	1.025	1.013	1.003	1.001	1.001	0.998	0.997
130	1.699	1.467	1.237	1.142	1.062	1.027	1.016	1.002	1.001	1.000	0.996	0.995
135	1.987	1.622	1.303	1.177	1.072	1.034	1.013	0.999	0.998	0.998	0.994	0.991
140	2.385	1.829	1.376	1.214	1.083	1.038	1.014	0.999	0.997	0.995	0.991	0.988
145	2.926	2.073	1.453	1.246	1.094	1.040	1.011	0.998	0.992	0.988	0.985	0.982
150	3.552	2.380	1.526	1.285	1.096	1.035	1.005	0.991	0.986	0.981	0.977	0.973
155	4.348	2.742	1.604	1.306	1.094	1.027	0.996	0.982	0.975	0.971	0.966	0.963
159	5.196	3.067	1.656	1.316	1.087	1.016	0.985	0.971	0.964	0.957	0.952	0.947
162		3.327	1.687	1.316	1.074	1.002	0.970	0.952	0.947	0.942	0.937	0.931
165		3.558	1.703	1.300	1.058	0.984	0.949	0.931	0.924	0.922	0.917	0.914
168		3.781	1.694	1.282	1.026	0.950	0.920	0.903	0.895	0.894	0.889	0.884
171		3.937	1.660	1.230	0.974	0.898	0.867	0.851	0.844	0.847	0.844	0.840
173			1.598	1.166	0.912	0.838	0.811	0.797	0.797	0.799	0.797	0.798
175				1.056	0.817	0.754	0.727	0.718	0.721	0.727	0.730	0.732
177						0.597	0.583	0.582	0.592	0.609	0.619	0.624

Values for the Ralstron remote HDR afterloader type 3 ^{60}Co source design. Values have been extracted from the MC calculated dose rate tables using the point source approximated geometry function values. Data not shown correspond to points within the encapsulated source.
Source: From Papagiannis, P., Angelopoulos, A., Pantelis, E., Sakelliou, L., Karaiskos, P., and Shimizu, Y., *Med. Phys.*, 29, 712–721, 2003.

TABLE A.3.23

Monte Carlo (MC) Calculated Anisotropy Factor, $\varphi_{an}(r)$

Radial Distance r (cm)		1.0	1.5	2.0	3.0	4.0	5.0	6.0	8.0	10.0	12.0	15.0
$\varphi_{an}(r)$	Type 1	1.341	1.144	1.072	1.026	1.005	0.994	0.989	0.988	0.985	0.985	0.985
	Type 2	0.980	0.980	0.981	0.981	0.982	0.983	0.983	0.985	0.986	0.986	0.986
	Type 3	1.543	1.212	1.114	1.039	1.014	1.001	0.994	0.991	0.991	0.988	0.986

Values for the Ralstron remote HDR afterloader type 1, 2, and 3 ^{60}Co source designs. The primary values not included in the publication have been provided by the authors.

Source: From Papagiannis, P., Angelopoulos, A., Pantelis, E., Sakelliou, L., Karaiskos, P., and Shimizu, Y., *Med. Phys.*, 29, 712–721, 2003.

A.3.9 BEBIG MultiSource HDR ^{60}Co Source Model GK60M21

TABLE A.3.24

Monte Carlo (MC) Calculated and Experimental Radial Dose Function, $g(r)$

MC Data by Ballester et al.[26]		Experimental (Ionization Chamber in Water Phantom) Data by Selbach and Andrássy[15]		
Radial Distance r (cm)	$g_L(r)$	Radial Distance r (cm)	$g_P(r)$	$g_L(r)$
0.25	1.008	0.6	1.013	1.031
0.5	1.035	0.8	1.004	1.009
0.75	1.014	1.0	1.000	1.000
1.0	1.000	1.2	1.000	0.997
1.5	0.992	1.4	0.998	0.993
2.0	0.984	1.6	0.996	0.990
3.0	0.968	1.8	0.994	0.987
4.0	0.952	2.0	0.991	0.984
5.0	0.935	2.2	0.987	0.979
6.0	0.919	2.4	0.984	0.976
7.0	0.902	2.6	0.981	0.973
8.0	0.884	2.8	0.977	0.969
10.0	0.849	3.0	0.975	0.966
12.0	0.812	3.2	0.971	0.962
15.0	0.756	3.4	0.964	0.955
20.0	0.666	3.6	0.961	0.952
		3.8	0.960	0.951
		4.0	0.956	0.947
		4.2	0.952	0.943
		4.4	0.947	0.938
		4.6	0.942	0.933
		4.8	0.939	0.930
		5.0	0.935	0.926
		5.2	0.931	0.922
		5.4	0.927	0.918
		5.6	0.924	0.915
		5.8	0.922	0.913
		6.0	0.919	0.910
		6.2	0.916	0.907
		6.4	0.914	0.905
		6.6	0.912	0.903
		6.8	0.906	0.897
		7.0	0.899	0.890
		7.2	0.894	0.885
		7.4	0.889	0.880
		7.6	0.885	0.877
		7.8	0.886	0.878
		8.0	0.886	0.877

Values for the BEBIG MultiSource HDR ^{60}Co source model GK60M21 design for the line, $g_L(r)$, and point source, $g_P(r)$, approximation, see also Equation 8.45.

TABLE A.3.25

Monte Carlo (MC) Calculated Anisotropy Function, $F(r, \theta)$

Polar Angle	Radial Distance r (cm)															
θ (°)	0.25	0.5	0.75	1.0	1.5	2.0	3.0	4.0	5.0	6.0	7.0	8.0	10.0	12.0	15.0	20.0
0		0.923	0.951	0.931	0.931	0.934	0.932	0.928	0.926	0.927	0.939	0.936	0.941	0.941	0.944	0.941
1		0.925	0.953	0.934	0.933	0.934	0.935	0.934	0.933	0.934	0.940	0.940	0.943	0.944	0.948	0.949
2		0.924	0.952	0.937	0.937	0.937	0.939	0.940	0.942	0.943	0.945	0.945	0.947	0.949	0.952	0.958
3		0.925	0.953	0.938	0.939	0.939	0.939	0.940	0.942	0.945	0.946	0.947	0.949	0.951	0.952	0.958
4		0.926	0.954	0.938	0.938	0.938	0.938	0.940	0.942	0.944	0.946	0.947	0.949	0.950	0.952	0.957
5		0.927	0.956	0.941	0.940	0.939	0.941	0.942	0.944	0.945	0.946	0.947	0.949	0.951	0.954	0.957
6		0.929	0.958	0.943	0.943	0.942	0.944	0.944	0.946	0.947	0.948	0.949	0.952	0.954	0.956	0.961
8		0.934	0.959	0.946	0.947	0.948	0.949	0.951	0.951	0.952	0.953	0.954	0.958	0.960	0.960	0.963
10		0.938	0.962	0.951	0.952	0.953	0.954	0.955	0.957	0.957	0.959	0.959	0.961	0.961	0.965	0.968
15		0.950	0.969	0.967	0.968	0.967	0.969	0.969	0.969	0.969	0.970	0.971	0.971	0.973	0.974	0.974
20		0.966	0.983	0.977	0.978	0.978	0.979	0.979	0.980	0.979	0.980	0.981	0.980	0.981	0.982	0.983
25		0.983	0.984	0.984	0.984	0.984	0.985	0.985	0.984	0.985	0.985	0.985	0.986	0.986	0.985	0.986
30		0.989	0.992	0.988	0.989	0.989	0.989	0.989	0.989	0.989	0.989	0.989	0.989	0.990	0.989	0.990
40		0.997	0.995	0.993	0.995	0.994	0.994	0.994	0.994	0.994	0.994	0.994	0.994	0.994	0.993	0.994
50	0.918	1.000	1.000	0.997	0.997	0.997	0.998	0.998	0.997	0.997	0.997	0.997	0.997	0.997	0.996	0.997
60	0.957	0.999	1.001	0.998	0.998	0.997	0.998	0.998	0.998	0.998	0.998	0.998	0.998	0.998	0.998	0.999
70	0.974	0.999	1.001	0.999	0.999	0.999	0.999	0.999	0.999	0.999	0.999	0.999	0.999	0.999	0.998	0.999
80	0.990	0.995	0.999	1.000	1.000	1.000	1.000	1.000	1.000	1.000	1.000	1.000	1.000	1.000	1.000	1.000

(continued)

TABLE A.3.25—*Continued*

Polar Angle θ (°)	Radial Distance r (cm)															
	0.25	0.5	0.75	1.0	1.5	2.0	3.0	4.0	5.0	6.0	7.0	8.0	10.0	12.0	15.0	20.0
90	1.000	1.000	1.000	1.000	1.000	1.000	1.000	1.000	1.000	1.000	1.000	1.000	1.000	1.000	1.000	1.000
100	0.990	0.994	0.999	1.000	0.999	1.000	0.999	0.999	1.000	0.999	1.000	0.999	0.999	0.999	0.999	0.999
110	0.975	0.997	1.001	0.999	0.999	1.000	1.000	1.000	0.999	0.999	0.999	0.999	0.999	0.999	0.999	1.000
120	0.956	0.999	1.000	0.999	0.998	0.999	0.999	0.999	0.999	0.999	0.999	0.999	0.999	0.998	0.998	0.998
130		1.001	1.000	0.996	0.997	0.997	0.997	0.997	0.997	0.997	0.996	0.996	0.997	0.996	0.996	0.997
140		0.994	0.995	0.995	0.995	0.995	0.996	0.995	0.996	0.995	0.995	0.995	0.995	0.994	0.994	0.995
150		0.983	0.996	0.991	0.991	0.991	0.990	0.991	0.990	0.990	0.990	0.990	0.990	0.990	0.990	0.990
155		0.971	0.984	0.985	0.985	0.985	0.985	0.986	0.986	0.986	0.986	0.986	0.986	0.986	0.985	0.986
160			0.976	0.980	0.979	0.978	0.978	0.978	0.979	0.978	0.978	0.979	0.978	0.979	0.980	0.981
165				0.967	0.967	0.966	0.966	0.967	0.967	0.967	0.968	0.968	0.968	0.969	0.969	0.971
170					0.938	0.938	0.937	0.938	0.939	0.940	0.940	0.942	0.944	0.945	0.948	0.953
172					0.917	0.917	0.919	0.920	0.921	0.923	0.924	0.926	0.929	0.932	0.935	0.941
174						0.884	0.885	0.887	0.890	0.892	0.896	0.898	0.901	0.906	0.911	0.920
175							0.861	0.864	0.868	0.871	0.875	0.878	0.883	0.890	0.897	0.907
176							0.821	0.827	0.833	0.838	0.842	0.846	0.855	0.864	0.873	0.888
177								0.771	0.779	0.787	0.794	0.799	0.812	0.824	0.836	0.857
178										0.698	0.709	0.719	0.741	0.757	0.775	0.804
179														0.683	0.711	0.743

Values for the BEBIG MultiSource HDR ^{60}Co source model GK60M21 design. Data not shown correspond to points within the encapsulated source.

Source: From Ballester, F., Granero, D., Pérez-Calatayud, J., Casal, E., Agramunt, A., and Cases, R., *Phys. Med. Biol.*, 50, N309–N316, 2005.

A.3.10   Implant Sciences HDR ^{169}Yb Source Model HDR 4140

TABLE A.3.26

Monte Carlo Calculated Radial Dose Function, $g(r)$

Radial Distance r (cm)	$g(r) \pm 1 \times$ S.D
0.5	0.97 ± 0.03
1.0	1.00
1.5	1.04 ± 0.03
2.0	1.07 ± 0.03
2.5	1.10 ± 0.03
3.0	1.12 ± 0.03
3.5	1.14 ± 0.03
4.0	1.15 ± 0.03
4.5	1.16 ± 0.03
5.0	1.17 ± 0.03
5.5	1.17 ± 0.03
6.0	1.16 ± 0.03
6.5	1.15 ± 0.03
7.0	1.15 ± 0.03
7.5	1.14 ± 0.03
8.0	1.12 ± 0.03
8.5	1.12 ± 0.03
9.0	1.09 ± 0.03
9.5	1.07 ± 0.03
10.0	1.05 ± 0.03

Values for the HDR ^{169}Yb source model HDR 4140 design by Implant Sciences Corporation (Wakefield, MA, USA) according to Medich et al. These are $g_L(r)$ values calculated using the line source approximation with an effective source length $L_{eff} = 0.36$ cm according to TG-43 U1, see also Equation 8.45. Values are listed with the calculated uncertainty at $\pm 1 \times$ standard deviation.

Sources: From Medich, C.D., Tries, A.M., and Munro, J.J., *Med. Phys.*, 33, 163–172, 2006; Rivard, M., Coursey, B., Hanson, W., Hug, S., Ibbot, G., Mitch, M., Nath, R., and Williamson, J., *Med. Phys.*, 31/3, 633–674, 2004.

TABLE A.3.27

Monte Carlo (MC) Calculated Anisotropy Function, $F(r,\theta)$

Polar	Radial Distance r (cm)										
Angle θ (°)	0.5	1.0	2.0	3.0	4.0	5.0	6.0	7.0	8.0	9.0	10.0
0	0.500	0.510	0.590	0.620	0.660	0.700	0.720	0.750	0.760	0.780	0.790
10	0.550	0.580	0.640	0.680	0.720	0.750	0.760	0.790	0.800	0.810	0.820
20	0.680	0.710	0.750	0.780	0.800	0.820	0.830	0.840	0.850	0.860	0.860
30	0.790	0.800	0.830	0.840	0.860	0.870	0.880	0.890	0.890	0.900	0.900
40	0.870	0.880	0.880	0.900	0.910	0.910	0.920	0.930	0.930	0.930	0.930
50	0.920	0.930	0.930	0.940	0.940	0.950	0.950	0.950	0.960	0.960	0.960
60	0.960	0.960	0.960	0.970	0.970	0.970	0.970	0.980	0.970	0.980	0.980
70	0.980	0.990	0.980	0.980	0.980	0.990	0.990	0.990	0.990	0.990	0.990
80	0.990	1.000	1.000	0.990	1.000	1.000	1.000	1.000	1.000	1.000	1.000
90	1.000	1.000	1.000	1.000	1.000	1.000	1.000	1.000	1.000	1.000	1.000
100	1.000	1.000	1.000	0.990	1.000	1.000	1.000	1.000	1.000	1.000	1.000
110	0.980	0.980	0.980	0.980	0.980	0.990	0.990	0.990	0.990	0.990	0.990
120	0.960	0.960	0.960	0.960	0.970	0.970	0.970	0.970	0.970	0.970	0.980
130	0.920	0.930	0.930	0.940	0.940	0.950	0.950	0.950	0.950	0.960	0.960
140	0.870	0.870	0.880	0.890	0.900	0.910	0.920	0.920	0.930	0.930	0.930
150	0.790	0.800	0.820	0.840	0.860	0.870	0.870	0.890	0.890	0.890	0.900
160	0.690	0.700	0.740	0.770	0.790	0.810	0.820	0.830	0.850	0.850	0.860
170	0.500	0.540	0.610	0.660	0.690	0.720	0.750	0.770	0.780	0.790	0.800
180	0.410	0.440	0.510	0.570	0.630	0.660	0.690	0.720	0.740	0.750	0.770
$\varphi_{an}(r)$	0.970	0.940	0.940	0.940	0.950	0.950	0.950	0.950	0.960	0.960	0.960

Values and anisotropy factor $\varphi_{an}(r)$ values for the HDR [169]Yb source model HDR 4140 design by Implant Sciences Corporation (Wakefield, MA, USA). The $F(r,\theta)$ value for $r = 8.0$ cm and $\theta = 100°$ from the original publication has been corrected based on the originally published dose rate table.

Source: From Medich, C.D., Tries, A.M., and Munro, J.J., *Med. Phys.*, 33, 163–172, 2006.

A.3.11 MicroSelectron PDR Old and New ^{192}Ir Source Designs

TABLE A.3.28

Monte Carlo (MC) Calculated Radial Dose Function, $g(r)$

	$g(r)$	
Radial Distance r (cm)	PDR Old Source Design, MC Data by Williamson and Li[1]	PDR New Source Design, MC Data by Karaiskos et al.[16]
0.1	1.008	0.948
0.2	0.998	0.983
0.3	0.995	0.990
0.5	0.997	0.995
1.0	1.000	1.000
1.5	1.002	1.003
2.0	1.004	1.005
2.5	1.005	1.006
3.0	1.006	1.006
4.0	1.003	1.002
5.0	0.995	0.994
6.0	0.981	0.982
7.0	0.964	0.964
8.0	0.941	0.943
9.0	0.912	0.915
10.0	0.879	0.884
11.0	0.844	0.848
12.0	0.804	0.803
13.0	0.756	0.754
14.0	0.695	0.694

Values for the microSelectron PDR afterloader, old and new ^{192}Ir source designs. Both author groups considered the point source approximation for extracting the $g(r)$ values and thus $g(r)$ can here be considered as $g_P(r)$.

TABLE A.3.29

Monte Carlo (MC) Calculated Anisotropy Function, $F(r,\theta)$

Polar Angle	Radial Distance r (cm)					
θ (°)	0.25	0.5	1.0	2.0	3.0	5.0
0.0	0.974	0.969	0.969	0.971	0.974	0.977
1.0	0.974	0.969	0.969	0.971	0.975	0.977
2.0	0.975	0.969	0.969	0.973	0.976	0.978
3.0	0.977	0.969	0.971	0.973	0.977	0.979
5.0	0.976	0.971	0.973	0.974	0.978	0.979
7.0	0.973	0.972	0.973	0.975	0.979	0.979
10.0	0.974	0.972	0.973	0.976	0.980	0.979
12.0	0.976	0.972	0.973	0.977	0.981	0.980
15.0	0.976	0.972	0.975	0.979	0.982	0.981
20.0	0.978	0.977	0.981	0.980	0.985	0.985
25.0	0.979	0.982	0.984	0.983	0.988	0.990
30.0	0.983	0.986	0.986	0.985	0.989	0.994
35.0	0.986	0.988	0.988	0.988	0.993	0.994
45.0	0.992	0.991	0.991	0.993	0.996	0.994
50.0	0.990	0.992	0.993	0.996	0.997	0.996
60.0	0.995	0.996	0.997	1.000	0.996	1.001
75.0	0.999	0.999	1.002	1.002	0.998	1.004
90.0	1.000	1.000	1.000	1.000	1.000	1.000
105.0	0.997	0.996	0.997	0.996	1.001	1.002
120.0	0.983	0.983	0.981	0.984	0.984	0.988
130.0	0.963	0.965	0.965	0.968	0.969	0.969
135.0	0.951	0.951	0.954	0.956	0.958	0.959
145.0	0.914	0.912	0.912	0.920	0.922	0.930
150.0	0.886	0.885	0.886	0.895	0.902	0.910
155.0	0.851	0.852	0.856	0.867	0.877	0.887
160.0	0.803	0.813	0.821	0.834	0.848	0.860
165.0	0.761	0.765	0.777	0.794	0.812	0.829
168.0		0.728	0.743	0.764	0.784	0.807
170.0		0.699	0.716	0.741	0.763	0.790
173.0		0.654	0.677	0.707	0.732	0.767
175.0		0.648	0.664	0.692	0.716	0.755
177.0		0.644	0.659	0.686	0.709	0.750
178.0		0.644	0.655	0.685	0.708	0.748
179.0		0.644	0.653	0.680	0.704	0.746
180.0		0.645	0.652	0.679	0.701	0.742

Values for the microSelectron PDR old design ^{192}Ir source. Values related to polar angle θ have been adjusted to fit the TG-43 U1 source coordinate system (Figure 8.6).

Source: From Williamson, J.F. and Li, Z., *Med. Phys.*, 22, 809–819, 1995.

TABLE A.3.30

Monte Carlo (MC) Calculated Anisotropy Function, $F(r,\theta)$

Polar Angle	Radial Distance r (cm)					
θ (°)	0.25	0.5	1.0	2.0	3.0	5.0
0	0.923	0.887	0.882	0.885	0.889	0.905
1	0.924	0.888	0.882	0.887	0.891	0.910
2	0.925	0.888	0.883	0.890	0.894	0.913
3	0.926	0.893	0.890	0.896	0.896	0.914
5	0.928	0.891	0.890	0.890	0.897	0.914
7	0.930	0.902	0.898	0.895	0.899	0.912
9	0.935	0.906	0.897	0.905	0.903	0.916
12	0.939	0.904	0.910	0.904	0.908	0.919
15	0.941	0.916	0.914	0.910	0.914	0.923
20	0.949	0.919	0.927	0.916	0.920	0.927
25	0.957	0.931	0.932	0.928	0.935	0.942
30	0.964	0.940	0.950	0.934	0.939	0.944
35	0.976	0.952	0.958	0.949	0.950	0.953
40	0.983	0.962	0.969	0.962	0.963	0.964
45	0.988	0.970	0.977	0.970	0.973	0.972
50	0.994	0.981	0.982	0.975	0.979	0.981
60	0.995	0.984	0.993	0.984	0.987	0.987
75	0.997	0.991	0.997	0.993	0.993	0.996
90	0.999	1.000	1.000	0.998	0.999	0.999
105	1.000	1.002	1.003	1.002	1.002	1.002
120	1.000	0.999	0.994	1.001	1.002	1.002
130	0.999	0.996	0.985	0.994	0.994	0.995
135	0.997	0.985	0.975	0.984	0.983	0.984
140	0.993	0.980	0.966	0.976	0.977	0.976
145	0.984	0.970	0.951	0.965	0.967	0.966
150	0.978	0.957	0.937	0.953	0.956	0.959
155	0.966	0.946	0.924	0.940	0.943	0.946
160	0.954	0.927	0.897	0.924	0.926	0.928
165	0.926	0.902	0.860	0.894	0.900	0.907
168		0.865	0.828	0.865	0.870	0.881
171		0.834	0.780	0.838	0.843	0.856
173			0.729	0.790	0.804	0.826
175			0.641	0.741	0.758	0.784
177						
178						
179						
180						

Values for the microSelectron PDR new design ^{192}Ir source.

Source: From Karaiskos, P., Angelopoulos, A., Pantelis, E., Papagiannis, P., Sakelliou, L., Kouwenhoven, E., and Baltas, D., *Med. Phys.*, 30, 9–16, 2003.

TABLE A.3.31

Monte Carlo (MC) Calculated Anisotropy Factor, $\varphi_{an}(r)$

Radial Distance r (cm)		0.25	0.5	1.0	2.0	3.0	5.0
$\varphi_{an}(r)$	Old Design MC Data by Williamson and Li[1]	0.969	0.969	0.970	0.973	0.974	0.978
	New Design MC Data by Karaiskos et al.[16]	0.976	0.976	0.977	0.977	0.978	0.979

Values for the microSelectron PDR afterloader, old and new ^{192}Ir source designs.

A.3.12   GammaMed PDR [192]Ir Source Models 12*i* and Plus

TABLE A.3.32

Monte Carlo (MC) Calculated Radial Dose Function, $g(r)$

Radial Distance r (cm)	$g(r)$	
	PDR 12*i* Source Design	PDR Plus Source Design
0.2	1.007	1.010
0.4	1.000	1.004
0.6	0.997	1.001
0.8	0.998	1.000
1.0	1.000	1.000
1.3	1.002	1.002
1.5	1.003	1.004
1.8	1.005	1.006
2.0	1.006	1.008
2.5	1.008	1.011
3.0	1.009	1.012
3.5	1.009	1.013
4.0	1.009	1.013
5.0	1.006	1.011
6.0	1.000	1.005
8.0	0.979	0.983
10.0	0.948	0.951
12.0	0.908	0.910
15.0	0.835	0.838

Values for the GammaMed PDR afterloader, 12*i* and plus ^{192}Ir source designs.

Source: From Pérez-Calatayud, J., Ballester, F., Serrano-Andrés, M.A., Puchades, V., Lluch, J.L., Limami, Y., and Casal, E., *Med. Phys.*, 28, 2576–2585, 2001.

TABLE A.3.33

Monte Carlo (MC) Calculated Anisotropy Function, $F(r, \theta)$

Polar Angle θ (°)	Radial Distance r (cm)																		
	0.2	0.4	0.6	0.8	1.0	1.25	1.5	1.75	2.0	2.5	3.0	3.5	4.0	5.0	6.0	8.0	10.0	12.0	15.0
0		0.944	0.945	0.945	0.945	0.946	0.947	0.948	0.952	0.952	0.953	0.956	0.960	0.960	0.960	0.963	0.969	0.970	0.971
1		0.944	0.945	0.945	0.945	0.946	0.947	0.948	0.952	0.952	0.953	0.956	0.960	0.960	0.961	0.963	0.969	0.970	0.971
2		0.944	0.945	0.945	0.945	0.946	0.947	0.948	0.952	0.952	0.953	0.956	0.960	0.960	0.961	0.965	0.970	0.974	0.975
3		0.944	0.945	0.945	0.945	0.946	0.947	0.948	0.952	0.952	0.953	0.956	0.960	0.960	0.962	0.966	0.970	0.974	0.975
4		0.944	0.945	0.945	0.945	0.946	0.947	0.948	0.953	0.953	0.953	0.956	0.960	0.960	0.962	0.966	0.971	0.974	0.976
5		0.945	0.945	0.945	0.945	0.947	0.947	0.949	0.953	0.953	0.953	0.956	0.960	0.960	0.963	0.966	0.971	0.974	0.977
6		0.946	0.947	0.947	0.947	0.950	0.948	0.950	0.954	0.953	0.954	0.956	0.961	0.960	0.963	0.966	0.971	0.974	0.977
7		0.948	0.949	0.949	0.949	0.951	0.951	0.950	0.955	0.955	0.955	0.956	0.961	0.961	0.964	0.967	0.972	0.974	0.978
8		0.948	0.948	0.949	0.949	0.951	0.951	0.950	0.954	0.955	0.955	0.956	0.962	0.962	0.965	0.968	0.973	0.975	0.978
9		0.947	0.948	0.949	0.950	0.951	0.952	0.951	0.956	0.956	0.956	0.958	0.963	0.963	0.966	0.968	0.973	0.975	0.979
10		0.949	0.950	0.951	0.952	0.953	0.953	0.953	0.957	0.957	0.957	0.959	0.963	0.965	0.967	0.969	0.974	0.976	0.980
15		0.963	0.964	0.965	0.967	0.966	0.966	0.966	0.968	0.968	0.968	0.970	0.971	0.974	0.975	0.976	0.980	0.981	0.984
20	0.970	0.976	0.978	0.979	0.979	0.978	0.979	0.979	0.979	0.979	0.979	0.981	0.980	0.983	0.984	0.984	0.987	0.987	0.989
30	0.975	0.992	0.995	0.995	0.995	0.993	0.994	0.994	0.994	0.993	0.994	0.994	0.994	0.995	0.994	0.995	0.995	0.995	0.995
40	0.987	0.998	0.999	1.001	1.001	1.000	1.000	1.000	1.000	1.000	1.000	0.999	0.999	1.001	0.999	1.000	0.999	0.999	0.999
50	0.993	1.001	1.003	1.004	1.004	1.003	1.004	1.003	1.003	1.002	1.002	1.002	1.002	1.003	1.001	1.002	1.001	1.001	1.001
60	0.996	1.003	1.005	1.006	1.005	1.004	1.005	1.004	1.004	1.003	1.003	1.003	1.003	1.004	1.002	1.002	1.003	1.002	1.002
70	0.996	1.003	1.005	1.005	1.005	1.004	1.005	1.004	1.004	1.003	1.003	1.003	1.003	1.003	1.003	1.002	1.002	1.002	1.002
80	0.995	1.001	1.003	1.002	1.003	1.002	1.002	1.002	1.002	1.002	1.002	1.002	1.002	1.002	1.002	1.002	1.002	1.001	1.001
90	1.000	1.000	1.000	1.000	1.000	1.000	1.000	1.000	1.000	1.000	1.000	1.000	1.000	1.000	1.000	1.000	1.000	1.000	1.000

(continued)

TABLE A.3.33—Continued

| Polar Angle | Radial Distance r (cm) | | | | | | | | | | | | | | | | | | |
θ (°)	0.2	0.4	0.6	0.8	1.0	1.25	1.5	1.75	2.0	2.5	3.0	3.5	4.0	5.0	6.0	8.0	10.0	12.0	15.0
100	0.995	1.001	1.002	1.003	1.003	1.002	1.002	1.002	1.002	1.002	1.001	1.001	1.001	1.001	1.001	1.001	1.001	1.000	1.000
110	0.996	1.002	1.004	1.005	1.004	1.003	1.004	1.003	1.002	1.002	1.002	1.001	1.001	1.002	1.001	1.001	1.000	1.000	1.000
120	0.994	1.001	1.003	1.004	1.003	1.001	1.003	1.001	1.002	1.002	1.001	1.001	1.001	1.001	1.000	1.000	1.000	0.999	0.998
130	0.988	0.997	0.999	1.000	1.000	0.998	1.000	0.998	0.999	0.999	0.998	0.998	0.997	0.998	0.997	0.997	0.997	0.997	0.997
140	0.978	0.990	0.993	0.994	0.994	0.994	0.994	0.994	0.994	0.994	0.994	0.994	0.993	0.993	0.993	0.993	0.993	0.994	0.994
150	0.973	0.981	0.984	0.986	0.986	0.986	0.986	0.985	0.985	0.985	0.985	0.985	0.986	0.986	0.986	0.986	0.986	0.989	0.990
160		0.968	0.968	0.968	0.968	0.968	0.970	0.970	0.970	0.970	0.970	0.971	0.971	0.971	0.971	0.973	0.973	0.977	0.979
165		0.950	0.950	0.950	0.950	0.950	0.951	0.951	0.952	0.953	0.954	0.956	0.957	0.959	0.959	0.961	0.964	0.967	0.970
170		0.946	0.915	0.915	0.915	0.915	0.917	0.917	0.917	0.920	0.921	0.923	0.926	0.932	0.935	0.942	0.946	0.950	0.956
171				0.905	0.905	0.906	0.908	0.908	0.908	0.909	0.910	0.913	0.917	0.923	0.927	0.936	0.941	0.945	0.952
172					0.893	0.893	0.895	0.895	0.896	0.897	0.898	0.901	0.906	0.912	0.916	0.928	0.934	0.939	0.947
173					0.880	0.880	0.881	0.881	0.876	0.882	0.883	0.886	0.892	0.898	0.902	0.918	0.925	0.931	0.940
174						0.856	0.865	0.866	0.857	0.865	0.864	0.867	0.875	0.880	0.884	0.904	0.913	0.922	0.932
175							0.830	0.833	0.833	0.836	0.841	0.844	0.854	0.858	0.862	0.885	0.898	0.909	0.921
176								0.780	0.801	0.806	0.812	0.816	0.828	0.830	0.835	0.861	0.878	0.892	0.906
177										0.738	0.750	0.774	0.789	0.790	0.798	0.827	0.850	0.869	0.886
178													0.700	0.731	0.746	0.778	0.809	0.835	0.859
179																0.704	0.747	0.783	0.820
180																0.660	0.709	0.750	0.795

Values for the GammaMed PDR ^{192}Ir source model 12i design. Data not shown correspond to points within the encapsulated source.

Source: From Pérez-Calatayud, J., Ballester, F., Serrano-Andrés, M.A., Puchades, V., Lluch, J.L., Limami, Y., and Casal, E., *Med. Phys.*, 28, 2576–2585, 2001.

TABLE A.3.34

Monte Carlo (MC) Calculated Anisotropy Function, $F(r,\theta)$

Polar Angle θ (°)	Radial Distance r (cm)																		
	0.2	0.4	0.6	0.8	1.0	1.25	1.5	1.75	2.0	2.5	3.0	3.5	4.0	5.0	6.0	8.0	10.0	12.0	15.0
0		0.940	0.940	0.948	0.951	0.952	0.952	0.952	0.952	0.952	0.952	0.952	0.954	0.954	0.960	0.960	0.965	0.976	0.977
1		0.941	0.941	0.949	0.953	0.953	0.953	0.953	0.953	0.953	0.953	0.953	0.954	0.956	0.961	0.961	0.969	0.976	0.977
2		0.944	0.941	0.949	0.953	0.953	0.953	0.953	0.954	0.955	0.957	0.958	0.958	0.959	0.968	0.970	0.976	0.977	0.978
3		0.945	0.945	0.951	0.953	0.953	0.953	0.954	0.955	0.960	0.960	0.960	0.960	0.962	0.969	0.970	0.976	0.979	0.980
4		0.949	0.949	0.955	0.957	0.957	0.957	0.957	0.958	0.962	0.962	0.961	0.961	0.964	0.969	0.971	0.976	0.979	0.981
5		0.949	0.951	0.956	0.957	0.957	0.957	0.957	0.959	0.962	0.963	0.961	0.961	0.965	0.969	0.971	0.977	0.979	0.981
6		0.949	0.951	0.956	0.957	0.958	0.958	0.958	0.960	0.962	0.963	0.963	0.963	0.968	0.971	0.973	0.977	0.980	0.983
7		0.951	0.955	0.959	0.960	0.962	0.962	0.962	0.962	0.963	0.963	0.963	0.966	0.970	0.972	0.974	0.978	0.981	0.984
8		0.955	0.959	0.962	0.962	0.964	0.964	0.964	0.964	0.964	0.965	0.965	0.968	0.972	0.974	0.976	0.979	0.981	0.984
9		0.955	0.960	0.963	0.962	0.965	0.965	0.965	0.965	0.967	0.967	0.967	0.969	0.974	0.975	0.977	0.980	0.982	0.985
10		0.961	0.965	0.967	0.966	0.969	0.969	0.969	0.969	0.970	0.971	0.970	0.971	0.976	0.977	0.978	0.981	0.982	0.985
15		0.973	0.977	0.978	0.977	0.980	0.979	0.980	0.980	0.980	0.980	0.981	0.981	0.984	0.985	0.986	0.987	0.989	0.991
20	0.973	0.985	0.988	0.990	0.989	0.991	0.989	0.992	0.990	0.989	0.990	0.990	0.990	0.991	0.991	0.992	0.992	0.994	0.994
30	0.979	0.995	0.998	0.999	0.999	0.999	1.000	1.000	1.000	0.999	1.000	1.000	1.000	0.999	0.999	0.999	0.999	0.999	0.999
40	0.985	1.000	1.003	1.004	1.003	1.003	1.004	1.003	1.004	1.003	1.004	1.002	1.002	1.002	1.002	1.002	1.002	1.002	1.001
50	0.990	1.003	1.005	1.006	1.005	1.004	1.004	1.005	1.004	1.004	1.004	1.003	1.004	1.004	1.004	1.003	1.003	1.002	1.001
60	0.992	1.005	1.006	1.007	1.006	1.005	1.006	1.006	1.005	1.005	1.004	1.004	1.005	1.004	1.003	1.003	1.003	1.003	1.002
70	0.993	1.004	1.005	1.006	1.005	1.004	1.006	1.005	1.004	1.004	1.004	1.003	1.004	1.003	1.003	1.002	1.003	1.003	1.002
80	0.999	1.002	1.002	1.003	1.003	1.003	1.004	1.003	1.002	1.002	1.002	1.002	1.002	1.002	1.002	1.002	1.002	1.002	1.001

(continued)

TABLE A.3.34—*Continued*

Polar Angle θ (°)	Radial Distance r (cm)																		
	0.2	0.4	0.6	0.8	1.0	1.25	1.5	1.75	2.0	2.5	3.0	3.5	4.0	5.0	6.0	8.0	10.0	12.0	15.0
90	1.000	1.000	1.000	1.000	1.000	1.000	1.000	1.000	1.000	1.000	1.000	1.000	1.000	1.000	1.000	1.000	1.000	1.000	1.000
100	0.998	1.002	1.003	1.002	1.002	1.002	1.002	1.002	1.002	1.001	1.001	1.001	1.002	1.001	1.001	1.001	1.001	1.000	1.000
110	0.992	1.003	1.005	1.004	1.004	1.003	1.004	1.003	1.003	1.003	1.002	1.001	1.003	1.002	1.001	1.001	1.001	1.001	1.001
120	0.989	1.001	1.003	1.003	1.004	1.003	1.004	1.003	1.002	1.002	1.002	1.001	1.003	1.002	1.001	1.000	1.001	1.000	1.000
130	0.985	0.998	1.001	1.001	1.001	1.001	1.002	1.001	0.999	1.001	1.000	1.000	1.000	0.999	1.000	0.999	0.999	0.999	0.998
140	0.977	0.994	0.997	0.996	0.996	0.996	0.997	0.997	0.997	0.997	0.998	0.996	0.998	0.996	0.997	0.996	0.995	0.996	0.996
150	0.970	0.987	0.990	0.990	0.990	0.989	0.990	0.990	0.991	0.990	0.991	0.991	0.991	0.992	0.991	0.991	0.990	0.991	0.991
160	0.956	0.972	0.975	0.975	0.975	0.977	0.977	0.977	0.977	0.977	0.978	0.978	0.978	0.978	0.978	0.979	0.981	0.983	0.983
165		0.958	0.960	0.960	0.960	0.961	0.962	0.962	0.963	0.965	0.966	0.967	0.969	0.967	0.970	0.970	0.972	0.975	0.976
170		0.937	0.937	0.933	0.933	0.934	0.935	0.936	0.937	0.939	0.939	0.939	0.944	0.944	0.949	0.954	0.956	0.963	0.966
171		0.925	0.925	0.925	0.923	0.921	0.925	0.924	0.925	0.928	0.929	0.929	0.935	0.936	0.941	0.949	0.951	0.959	0.963
172		0.910	0.912	0.912	0.912	0.912	0.912	0.912	0.912	0.916	0.918	0.918	0.925	0.928	0.933	0.942	0.945	0.953	0.958
173			0.898	0.899	0.899	0.899	0.899	0.899	0.899	0.903	0.903	0.903	0.913	0.916	0.921	0.933	0.937	0.946	0.952
174			0.880	0.880	0.881	0.883	0.883	0.883	0.884	0.885	0.887	0.887	0.897	0.903	0.908	0.922	0.929	0.937	0.945
175				0.850	0.859	0.859	0.859	0.859	0.859	0.863	0.865	0.866	0.878	0.887	0.892	0.906	0.917	0.926	0.935
176					0.820	0.822	0.823	0.824	0.831	0.836	0.836	0.840	0.851	0.866	0.872	0.887	0.904	0.912	0.923
177						0.773	0.780	0.780	0.790	0.797	0.798	0.804	0.814	0.836	0.844	0.860	0.884	0.892	0.905
178									0.690	0.724	0.729	0.739	0.756	0.786	0.799	0.816	0.852	0.859	0.878
179															0.779	0.753	0.802	0.810	0.838
180																	0.694	0.750	0.780

Values for the GammaMed PDR ^{192}Ir source model plus design. Data not shown correspond to points within the encapsulated source.

Source: From Pérez-Calatayud, J., Ballester, F., Serrano-Andrés, M.A., Puchades, V., Lluch, J.L., Limami, Y., and Casal, E., *Med. Phys.*, 28, 2576–2585, 2001.

TABLE A.3.35

Monte Carlo (MC) Calculated Anisotropy Factor, $\varphi_{an}(r)$

Radial Distance r (cm)		0.25	0.50	1.00	2.00	3.00	4.00	5.00	6.00	8.00	10.00	12.00
$\varphi_{an}(r)$	12*i*	0.997	0.997	0.997	0.997	0.997	0.997	0.997	0.997	0.997	0.997	0.997
	Plus	0.999	0.999	0.999	0.999	0.999	0.999	0.999	0.999	0.999	0.999	0.999

Values for the GammaMed PDR afterloader, 12*i* and plus ^{192}Ir source designs. The anisotropy factor values are according to $\varphi_{an}(r) = 0.9967 + 0.0000537r$ for the 12*i* source model and $\varphi_{an}(r) = 0.9991 + 0.000033r$ for the plus source model.

Source: From Pérez-Calatayud, J., Ballester, F., Serrano-Andrés, M.A., Puchades, V., Lluch, J.L., Limami, Y., and Casal, E., *Med. Phys.*, 28, 2576–2585, 2001.

A.3.13   LDR ^{137}Cs Source Models

TABLE A.3.36

Monte Carlo (MC) Calculated Radial Dose Function, $g(r)$

			$g(r)$		
Radial Distance r (cm)	Selectron LDR Single Pellet[18]	Amersham CDC Type Source CDC-1[a] (1 × ^{137}Cs Bead)[19]	Amersham CDC Type Source CDC-3[a] (3 × ^{137}Cs Beads)[19]	CDC.K1–K3 Type Sources[a] (1 × ^{137}Cs Bead)[22]	CDC.K4 Type Source[a] (2 × ^{137}Cs Beads)[22]
0.25	1.022	1.042	1.032	1.105	1.057
0.50	1.007	1.010	1.006	1.024	1.020
0.75	1.003	1.003	1.002	1.009	1.006
1.00	1.000	1.000	1.000	1.000	1.000
1.50	0.995	0.995	0.995	0.995	0.995
2.00	0.989	0.989	0.990	0.990	0.990
2.50	0.984	0.984	0.984	0.984	0.984
3.00	0.978	0.978	0.978	0.979	0.978
4.00	0.967	0.965	0.966	0.966	0.966
5.00	0.953	0.951	0.952	0.953	0.952
6.00	0.939	0.936	0.937	0.938	0.937
7.00	0.923	0.921	0.922	0.922	0.922
8.00	0.907	0.904	0.905	0.906	0.905
9.00	0.889	0.887	0.888	0.888	0.887
10.00	0.870	0.869	0.869	0.869	0.869
12.00	0.828	0.830	0.830	0.830	0.829
14.00	0.785	0.788	0.788	0.787	0.787
16.00	0.737	0.744	0.743	0.742	0.742
18.00	0.688	0.697	0.696	0.696	0.697
20.00	0.637	0.649	0.647	0.650	0.650

Values for LDR ^{137}Cs sources. All data are based on Monte Carlo simulations. For the single pellet for the Selectron LDR the point source approximation for the geometry function has been considered.

TABLE A.3.37

Monte Carlo (MC) Calculated Radial Dose Function, $g(r)$

Radial Distance r (cm)	$g(r)$	
	CIS CSM11 Type Source[20]	Amersham CDCS-M Type Source[21]
0.20	1.011	1.005
0.25	1.003	1.005
0.30	1.000	1.005
0.40	0.999	1.004
0.60	1.000	1.003
0.80	1.001	1.001
1.00	1.000	1.000
1.25	0.996	0.998
1.50	0.992	0.996
1.75	0.989	0.994
2.00	0.989	0.992
2.50	0.986	0.987
3.00	0.975	0.982
3.50	0.972	0.976
4.00	0.962	0.970
5.00	0.950	0.957
6.00	0.942	0.943
8.00	0.911	0.911
10.00	0.874	0.874
12.00	0.834	0.834
15.00	0.769	0.768

Values for LDR ^{137}Cs sources. All data are based on Monte Carlo simulations.

TABLE A.3.38

Monte Carlo (MC) Calculated Anisotropy Function, $F(r,\theta)$

Polar Angle	Radial Distance r (cm)																		
θ (°)	0.2	0.4	0.6	0.8	1.0	1.25	1.5	1.75	2.0	2.5	3.0	3.5	4.0	5.0	6.0	8.0	10.0	12.0	15.0
0		0.920	0.929	0.933	0.933	0.936	0.937	0.939	0.940	0.941	0.942	0.944	0.945	0.947	0.949	0.951	0.952	0.954	0.961
1		0.920	0.929	0.933	0.933	0.936	0.937	0.939	0.940	0.941	0.942	0.944	0.945	0.947	0.950	0.951	0.953	0.955	0.961
2		0.920	0.929	0.933	0.933	0.936	0.938	0.939	0.941	0.942	0.943	0.944	0.945	0.948	0.950	0.952	0.953	0.955	0.961
3		0.920	0.929	0.933	0.934	0.937	0.938	0.940	0.941	0.942	0.943	0.945	0.946	0.948	0.950	0.952	0.953	0.955	0.961
4		0.921	0.930	0.933	0.934	0.937	0.939	0.940	0.941	0.943	0.944	0.945	0.946	0.948	0.951	0.953	0.954	0.956	0.961
5		0.922	0.930	0.934	0.934	0.938	0.939	0.940	0.941	0.943	0.944	0.946	0.947	0.949	0.951	0.953	0.954	0.956	0.962
6		0.923	0.931	0.934	0.935	0.938	0.940	0.940	0.942	0.943	0.944	0.946	0.947	0.949	0.952	0.953	0.954	0.957	0.962
7		0.923	0.931	0.935	0.936	0.939	0.940	0.941	0.942	0.944	0.945	0.947	0.948	0.949	0.952	0.954	0.955	0.957	0.962
8		0.924	0.932	0.935	0.936	0.940	0.941	0.942	0.943	0.944	0.945	0.947	0.949	0.950	0.953	0.954	0.955	0.958	0.962
9		0.925	0.932	0.936	0.937	0.941	0.942	0.942	0.944	0.945	0.946	0.948	0.949	0.951	0.953	0.955	0.956	0.959	0.963
10		0.925	0.933	0.937	0.938	0.942	0.943	0.944	0.945	0.946	0.947	0.949	0.950	0.952	0.954	0.956	0.956	0.960	0.963
15		0.929	0.940	0.944	0.945	0.948	0.949	0.950	0.951	0.952	0.953	0.955	0.956	0.957	0.959	0.960	0.961	0.964	0.966
20		0.939	0.947	0.951	0.953	0.955	0.955	0.957	0.957	0.957	0.959	0.960	0.961	0.962	0.964	0.966	0.967	0.969	0.970
30		0.958	0.965	0.968	0.970	0.970	0.972	0.973	0.973	0.973	0.974	0.974	0.975	0.975	0.976	0.976	0.978	0.979	0.980
40	0.921	0.973	0.980	0.982	0.983	0.984	0.984	0.985	0.985	0.985	0.985	0.985	0.985	0.986	0.987	0.986	0.988	0.988	0.988
50	0.949	0.984	0.990	0.992	0.992	0.993	0.993	0.993	0.993	0.992	0.993	0.992	0.992	0.993	0.993	0.993	0.994	0.994	0.994
60	0.966	0.993	0.996	0.997	0.997	0.998	0.997	0.997	0.997	0.997	0.997	0.996	0.997	0.997	0.998	0.997	0.997	0.997	0.997
70	0.985	0.997	0.999	0.999	0.999	0.999	0.999	1.000	1.000	0.999	0.999	0.999	0.999	0.999	0.999	0.999	0.999	0.999	0.999
80	0.997	0.998	1.000	1.000	1.000	1.000	1.000	1.000	1.000	1.000	1.000	1.000	1.000	1.000	1.000	1.000	1.000	1.000	1.000
90	1.000	1.000	1.000	1.000	1.000	1.000	1.000	1.000	1.000	1.000	1.000	1.000	1.000	1.000	1.000	1.000	1.000	1.000	1.000

Values for the Amersham CDC type source CDC-1 ($1 \times {}^{137}$Cs bead). Data not shown correspond to points within the encapsulated source.

Source: From Pérez-Calatayud, J., Ballester, F., Serrano-Andrés, M.A., Llich, J.L., Puchades, V., Limami, Y., and Casal, E., *Med. Phys.*, 29, 538–543, 2002.

TABLE A.3.39

Monte Carlo (MC) Calculated Anisotropy Function, $F(r,\theta)$

Polar Angle θ (°)	Radial Distance r (cm)																		
	0.2	0.4	0.6	0.8	1.0	1.25	1.5	1.75	2.0	2.5	3.0	3.5	4.0	5.0	6.0	8.0	10.0	12.0	15.0
0		0.961	0.965	0.968	0.969	0.970	0.970	0.971	0.972	0.973	0.974	0.974	0.975	0.975	0.975	0.976	0.978	0.980	0.982
1		0.961	0.965	0.968	0.969	0.970	0.970	0.971	0.972	0.973	0.974	0.974	0.975	0.975	0.975	0.976	0.978	0.981	0.982
2		0.961	0.965	0.969	0.969	0.970	0.970	0.971	0.972	0.973	0.974	0.974	0.975	0.975	0.975	0.976	0.978	0.981	0.982
3		0.961	0.965	0.969	0.969	0.970	0.970	0.971	0.972	0.973	0.974	0.974	0.975	0.975	0.975	0.976	0.978	0.981	0.982
4		0.961	0.965	0.969	0.969	0.970	0.970	0.971	0.972	0.973	0.974	0.975	0.975	0.975	0.975	0.976	0.978	0.981	0.983
5		0.961	0.965	0.969	0.969	0.970	0.970	0.971	0.972	0.973	0.974	0.975	0.975	0.975	0.975	0.976	0.978	0.980	0.982
6		0.961	0.965	0.969	0.969	0.970	0.970	0.971	0.972	0.973	0.974	0.975	0.975	0.975	0.976	0.976	0.978	0.981	0.982
7		0.961	0.965	0.969	0.969	0.970	0.970	0.971	0.972	0.973	0.974	0.975	0.975	0.975	0.976	0.976	0.978	0.981	0.982
8		0.962	0.966	0.969	0.970	0.970	0.970	0.971	0.972	0.973	0.974	0.974	0.975	0.976	0.976	0.976	0.977	0.981	0.983
9		0.962	0.966	0.969	0.970	0.970	0.970	0.971	0.972	0.973	0.974	0.975	0.975	0.976	0.976	0.976	0.977	0.981	0.982
10		0.962	0.967	0.969	0.970	0.970	0.971	0.971	0.972	0.973	0.975	0.975	0.976	0.976	0.976	0.977	0.979	0.981	0.983
15		0.963	0.968	0.971	0.972	0.972	0.973	0.974	0.974	0.975	0.976	0.977	0.977	0.977	0.978	0.978	0.979	0.981	0.982	0.984
20		0.966	0.971	0.974	0.975	0.976	0.977	0.978	0.979	0.979	0.980	0.980	0.980	0.981	0.982	0.982	0.984	0.985	0.986
30		0.975	0.980	0.982	0.983	0.983	0.984	0.985	0.985	0.986	0.986	0.987	0.987	0.987	0.988	0.988	0.989	0.989	0.990
40	1.002	0.986	0.987	0.988	0.988	0.989	0.990	0.990	0.991	0.992	0.992	0.992	0.993	0.993	0.993	0.993	0.993	0.993	0.994
50	1.003	0.994	0.994	0.994	0.994	0.994	0.995	0.995	0.995	0.995	0.996	0.996	0.996	0.996	0.996	0.996	0.996	0.996	0.996
60	1.008	0.997	0.997	0.997	0.997	0.997	0.997	0.997	0.998	0.998	0.998	0.998	0.998	0.998	0.998	0.998	0.998	0.998	0.998
70	1.005	0.999	0.999	0.999	0.999	0.999	0.999	0.999	0.999	0.999	0.999	0.999	0.999	0.999	1.000	1.000	1.000	1.000	1.000
80	1.004	1.000	1.000	1.000	1.000	1.000	1.000	1.000	1.000	1.000	1.000	1.000	1.000	1.000	1.000	1.000	1.000	1.000	1.000
90	1.000	1.000	1.000	1.000	1.000	1.000	1.000	1.000	1.000	1.000	1.000	1.000	1.000	1.000	1.000	1.000	1.000	1.000	1.000

Values for the Amersham CDC type source CDC-3 ($3 \times {}^{137}$Cs bead). Data not shown correspond to points within the encapsulated source.

Source: From Pérez-Calatayud, J., Ballester, F., Serrano-Andrés, M.A., Llich, J.L., Puchades, V., Limami, Y., and Casal, E., *Med. Phys.*, 29, 538–543, 2002.

TABLE A.3.40

Monte Carlo (MC) Calculated Anisotropy Function, $F(r,\theta)$

Polar Angle θ (°)	Radial Distance r (cm)																		
	0.2	0.4	0.6	0.8	1.0	1.25	1.5	1.75	2.0	2.5	3.0	3.5	4.0	5.0	6.0	8.0	10.0	12.0	15.0
0			0.887	0.893	0.900	0.907	0.908	0.909	0.913	0.916	0.922	0.924	0.924	0.929	0.929	0.933	0.937	0.939	0.945
1			0.887	0.893	0.900	0.908	0.908	0.910	0.914	0.916	0.922	0.924	0.924	0.929	0.929	0.933	0.937	0.939	0.945
2			0.888	0.893	0.900	0.908	0.908	0.912	0.914	0.916	0.922	0.924	0.924	0.930	0.930	0.933	0.937	0.939	0.945
3			0.888	0.894	0.901	0.910	0.910	0.913	0.915	0.916	0.922	0.924	0.924	0.930	0.930	0.933	0.938	0.939	0.945
4			0.889	0.895	0.901	0.911	0.912	0.915	0.917	0.918	0.923	0.925	0.925	0.930	0.930	0.934	0.938	0.939	0.945
5			0.889	0.899	0.904	0.911	0.912	0.915	0.917	0.919	0.923	0.926	0.926	0.930	0.930	0.934	0.938	0.939	0.945
6			0.891	0.900	0.905	0.913	0.914	0.917	0.919	0.921	0.924	0.927	0.927	0.930	0.930	0.934	0.938	0.940	0.946
7			0.892	0.900	0.906	0.913	0.914	0.917	0.920	0.921	0.924	0.927	0.927	0.930	0.930	0.935	0.939	0.941	0.947
8			0.892	0.901	0.907	0.914	0.916	0.918	0.921	0.922	0.924	0.927	0.927	0.930	0.930	0.935	0.939	0.942	0.947
9			0.893	0.903	0.910	0.915	0.916	0.918	0.921	0.922	0.924	0.927	0.927	0.930	0.932	0.936	0.939	0.942	0.948
10			0.894	0.903	0.909	0.914	0.916	0.918	0.921	0.922	0.925	0.928	0.928	0.931	0.933	0.937	0.940	0.942	0.948
15			0.901	0.912	0.918	0.922	0.925	0.927	0.928	0.930	0.931	0.933	0.935	0.935	0.938	0.942	0.945	0.947	0.952
20			0.907	0.919	0.925	0.928	0.932	0.934	0.935	0.938	0.939	0.941	0.942	0.942	0.946	0.949	0.952	0.953	0.958
30		0.893	0.933	0.942	0.947	0.949	0.953	0.955	0.955	0.957	0.958	0.958	0.960	0.960	0.962	0.963	0.967	0.967	0.970
40		0.931	0.958	0.965	0.968	0.970	0.973	0.974	0.974	0.975	0.975	0.975	0.975	0.975	0.977	0.977	0.979	0.979	0.981
50		0.957	0.978	0.983	0.983	0.985	0.987	0.987	0.987	0.987	0.987	0.987	0.987	0.987	0.988	0.988	0.989	0.989	0.990
60	0.954	0.977	0.991	0.992	0.993	0.994	0.994	0.994	0.994	0.994	0.994	0.994	0.994	0.994	0.994	0.994	0.994	0.995	0.995
70	0.984	0.988	0.998	0.998	0.998	0.998	0.998	0.997	0.997	0.997	0.997	0.997	0.997	0.997	0.997	0.997	0.997	0.997	0.997
80	0.998	0.995	1.000	1.000	1.000	1.000	1.000	1.000	0.999	0.999	0.999	0.999	0.999	0.999	0.999	0.999	0.999	0.999	0.999
90	1.000	1.000	1.000	1.000	1.000	1.000	1.000	1.000	1.000	1.000	1.000	1.000	1.000	1.000	1.000	1.000	1.000	1.000	1.000

Values for the Amersham CDC.K1–K3 type sources ($1 \times {}^{137}$Cs bead). Data not shown correspond to points within the encapsulated source.

Source: From Pérez-Calatayud, J., Ballester, F., Lluch, J.L., Serrano-Andrés, M.A., Casal, E., Puchades, V., and Limami, Y.F., *Phys. Med. Biol.*, 46, 2029–2040, 2001.

TABLE A.3.41

Monte Carlo (MC) Calculated Anisotropy Function, $F(r,\theta)$

Polar Angle θ (°)	Radial Distance r (cm)																		
	0.2	0.4	0.6	0.8	1.0	1.25	1.5	1.75	2.0	2.5	3.0	3.5	4.0	5.0	6.0	8.0	10.0	12.0	15.0
0			0.914	0.935	0.940	0.942	0.944	0.945	0.948	0.948	0.948	0.948	0.952	0.952	0.954	0.957	0.960	0.963	0.965
1			0.914	0.935	0.940	0.942	0.944	0.945	0.948	0.948	0.948	0.948	0.952	0.953	0.954	0.957	0.960	0.963	0.965
2			0.914	0.935	0.940	0.942	0.945	0.945	0.949	0.949	0.949	0.950	0.954	0.954	0.955	0.959	0.962	0.964	0.965
3			0.914	0.935	0.940	0.942	0.945	0.945	0.949	0.949	0.949	0.951	0.954	0.954	0.955	0.959	0.962	0.964	0.965
4			0.914	0.935	0.940	0.942	0.946	0.946	0.949	0.949	0.949	0.951	0.954	0.954	0.955	0.959	0.962	0.964	0.965
5			0.916	0.935	0.940	0.942	0.946	0.946	0.949	0.949	0.949	0.951	0.954	0.955	0.956	0.959	0.962	0.964	0.965
6			0.918	0.936	0.941	0.943	0.946	0.946	0.950	0.950	0.950	0.952	0.954	0.955	0.956	0.959	0.963	0.964	0.965
7			0.919	0.937	0.941	0.944	0.946	0.947	0.950	0.950	0.950	0.952	0.954	0.956	0.957	0.960	0.964	0.964	0.965
8			0.920	0.938	0.941	0.943	0.946	0.947	0.950	0.950	0.951	0.952	0.955	0.956	0.958	0.960	0.964	0.965	0.965
9			0.921	0.938	0.941	0.943	0.946	0.947	0.950	0.950	0.951	0.953	0.955	0.957	0.959	0.961	0.964	0.965	0.966
10			0.921	0.939	0.941	0.943	0.946	0.948	0.950	0.950	0.952	0.953	0.955	0.957	0.959	0.961	0.964	0.966	0.966
15			0.929	0.945	0.946	0.949	0.950	0.952	0.953	0.954	0.956	0.957	0.957	0.960	0.960	0.963	0.967	0.968	0.969
20			0.934	0.948	0.951	0.954	0.955	0.958	0.959	0.960	0.961	0.962	0.962	0.965	0.965	0.967	0.971	0.971	0.972
30			0.953	0.962	0.964	0.967	0.968	0.970	0.970	0.971	0.972	0.972	0.972	0.975	0.976	0.976	0.979	0.979	0.979
40		0.969	0.971	0.978	0.978	0.979	0.981	0.981	0.982	0.982	0.982	0.983	0.983	0.984	0.985	0.985	0.986	0.986	0.986
50		0.985	0.985	0.988	0.988	0.989	0.990	0.990	0.990	0.990	0.990	0.990	0.990	0.990	0.991	0.991	0.992	0.992	0.992
60	1.009	0.995	0.991	0.994	0.994	0.995	0.995	0.995	0.995	0.996	0.996	0.996	0.996	0.996	0.996	0.996	0.996	0.997	0.997
70	1.010	1.000	0.999	0.999	0.997	0.997	0.997	0.997	0.998	0.998	0.998	0.999	0.999	0.999	0.999	0.999	0.999	0.999	0.999
80	1.016	1.001	1.001	1.001	1.001	1.000	1.000	1.000	1.000	0.999	0.999	0.999	0.999	0.999	0.999	1.000	1.000	1.000	1.000
90	1.000	1.000	1.000	1.000	1.000	1.000	1.000	1.000	1.000	1.000	1.000	1.000	1.000	1.000	1.000	1.000	1.000	1.000	1.000

Values for the Amersham CDC.K4 type sources (2 × ^{137}Cs bead). Data not shown correspond to points within the encapsulated source.

Source: From Pérez-Calatayud, J., Ballester, F., Lluch, J.L., Serrano-Andrés, M.A., Casal, E., Puchades, V., and Limami, Y.F., *Phys. Med. Biol.*, 46, 2029–2040, 2001.

TABLE A.3.42

Monte Carlo (MC) Calculated Anisotropy Function, $F(r,\theta)$

Polar Angle θ (°)	Radial Distance r (cm)																
	0.15	0.20	0.25	0.30	0.50	0.75	1.0	1.5	2.0	2.5	3.0	4.0	6.0	8.0	10.0	12.0	15.0
0				0.686	0.785	0.879	0.977	0.885	0.930	0.848	0.948	0.921	0.912	0.926	0.928	0.956	0.945
1				0.693	0.797	0.884	0.923	0.883	0.934	0.908	0.967	0.934	0.928	0.931	0.951	0.958	0.956
2				0.701	0.802	0.874	0.887	0.892	0.943	0.971	0.967	0.941	0.956	0.958	0.966	0.962	0.964
3				0.707	0.790	0.859	0.895	0.910	0.948	0.959	0.941	0.958	0.950	0.981	0.972	0.970	0.982
4				0.715	0.789	0.851	0.882	0.903	0.936	0.936	0.943	0.956	0.939	0.981	0.976	0.966	0.967
5				0.719	0.792	0.831	0.884	0.894	0.919	0.939	0.946	0.943	0.953	0.971	0.961	0.971	0.977
6				0.720	0.791	0.829	0.883	0.909	0.909	0.943	0.945	0.948	0.962	0.951	0.953	0.978	0.979
7				0.730	0.799	0.846	0.872	0.913	0.906	0.929	0.942	0.947	0.959	0.949	0.964	0.958	0.950
8				0.742	0.795	0.853	0.869	0.898	0.908	0.921	0.930	0.931	0.950	0.959	0.966	0.951	0.955
9				0.746	0.785	0.853	0.868	0.898	0.909	0.927	0.930	0.925	0.947	0.973	0.948	0.965	0.963
10				0.736	0.792	0.854	0.869	0.910	0.909	0.932	0.936	0.934	0.945	0.969	0.948	0.973	0.970
20			0.586	0.651	0.806	0.858	0.879	0.912	0.931	0.931	0.951	0.948	0.955	0.968	0.956	0.970	0.982
30		0.690	0.691	0.744	0.826	0.872	0.904	0.921	0.937	0.946	0.958	0.965	0.966	0.970	0.974	0.982	0.983
40	0.709	0.724	0.767	0.793	0.855	0.891	0.911	0.939	0.947	0.962	0.973	0.979	0.974	0.978	0.985	0.986	0.989
50	0.929	0.807	0.825	0.838	0.886	0.916	0.931	0.961	0.960	0.975	0.981	0.980	0.988	0.985	0.991	1.001	0.994
60	0.854	0.866	0.874	0.883	0.914	0.937	0.949	0.973	0.971	0.978	0.987	0.991	0.992	0.985	0.990	0.994	0.993
70	0.918	0.915	0.922	0.927	0.943	0.956	0.963	0.979	0.981	0.982	0.993	0.996	0.989	0.992	0.998	1.000	1.001
80	0.960	0.955	0.961	0.965	0.972	0.978	0.975	0.989	0.992	0.992	0.999	1.004	0.991	0.999	0.990	1.005	1.001
90	1.000	1.000	1.000	1.000	1.000	1.000	1.000	1.000	1.000	1.000	1.000	1.000	1.000	1.000	1.000	1.000	1.000

(continued)

TABLE A.3.42—Continued

Polar Angle θ (°)	Radial Distance r (cm)																
	0.15	0.20	0.25	0.30	0.50	0.75	1.0	1.5	2.0	2.5	3.0	4.0	6.0	8.0	10.0	12.0	15.0
100	1.028	1.036	1.040	1.040	1.032	1.023	1.009	1.016	1.006	1.009	1.018	1.014	1.003	1.001	0.999	1.006	1.004
110	1.059	1.076	1.085	1.086	1.064	1.039	1.030	1.026	1.014	1.015	1.015	1.014	1.005	1.006	1.004	1.005	1.009
120	1.072	1.127	1.142	1.137	1.097	1.063	1.049	1.037	1.024	1.010	1.020	1.009	1.007	1.002	0.995	1.003	1.009
130	1.267	1.181	1.208	1.195	1.130	1.083	1.056	1.042	1.023	1.010	1.024	1.021	1.003	1.006	0.999	1.006	1.000
140	1.152	1.245	1.301	1.280	1.162	1.099	1.066	1.037	1.022	1.020	1.024	1.018	1.004	1.003	0.997	1.002	1.003
150		1.581	1.391	1.374	1.199	1.113	1.073	1.052	1.028	1.018	1.022	1.007	1.002	0.997	0.992	1.004	0.996
160			1.500	1.457	1.215	1.114	1.077	1.034	1.022	1.010	1.010	1.004	1.002	0.992	0.990	0.988	0.997
170				1.695	1.253	1.122	1.076	1.046	1.023	1.011	1.015	1.004	0.992	0.979	0.984	1.000	0.996
171				1.732	1.251	1.127	1.080	1.052	1.026	1.013	1.013	1.001	0.983	0.986	0.996	1.005	0.993
172				1.765	1.258	1.132	1.097	1.065	1.036	1.021	1.012	1.003	0.984	0.988	0.996	1.004	1.001
173				1.802	1.263	1.145	1.098	1.057	1.038	1.027	1.009	1.014	0.997	0.992	0.984	0.999	1.009
174				1.837	1.259	1.145	1.085	1.055	1.025	1.014	1.017	1.022	0.994	0.985	0.978	1.001	1.002
175				1.855	1.266	1.146	1.078	1.071	1.024	1.017	1.025	1.015	0.992	0.978	0.983	1.005	0.995
176				1.880	1.267	1.152	1.080	1.066	1.030	1.019	1.026	1.018	0.987	0.982	0.984	1.020	0.983
177				1.882	1.267	1.155	1.091	1.038	1.029	1.032	1.003	1.027	0.985	0.981	0.975	1.008	0.980
178				1.892	1.274	1.132	1.106	1.022	1.036	1.028	0.987	1.017	1.003	0.977	0.972	1.011	0.968
179				1.907	1.313	1.147	1.118	1.023	1.054	0.986	0.994	1.016	0.998	1.016	0.987	1.022	0.999
180				1.900	1.344	1.185	1.120	1.021	1.067	0.969	1.006	1.020	0.980	1.062	1.004	1.000	1.000

Values for the CIS CSM11 type source design. Data not shown correspond to points within the encapsulated source.

Source: From Ballester, F., Lluch, J.L., Limami, Y., Serrano, M.A., Casal, E., Pérez-Calatayud, J., and Lliso, F., *Med. Phys.*, 27, 2182–2189, 2000.

TABLE A.3.43

Monte Carlo (MC) Calculated Anisotropy Function, $F(r,\theta)$

Polar Angle θ (°)	Radial Distance r (cm)															
	0.15	0.25	0.5	0.75	1.0	1.25	1.5	2.0	3.0	4.0	5.0	6.0	7.0	8.0	10.0	15.0
0						0.892	0.895	0.888	0.875	0.845	0.855	0.860	0.900	0.861	0.867	0.946
1						0.873	0.875	0.866	0.871	0.851	0.861	0.862	0.893	0.857	0.865	0.916
2						0.861	0.861	0.851	0.836	0.853	0.853	0.865	0.871	0.871	0.873	0.897
3						0.853	0.853	0.843	0.833	0.852	0.850	0.860	0.860	0.875	0.876	0.887
4						0.850	0.850	0.839	0.833	0.851	0.852	0.859	0.858	0.875	0.878	0.884
5						0.851	0.850	0.839	0.836	0.852	0.857	0.862	0.863	0.875	0.881	0.886
6						0.854	0.854	0.843	0.842	0.855	0.866	0.868	0.873	0.877	0.886	0.893
7					0.745	0.860	0.860	0.849	0.850	0.861	0.876	0.876	0.885	0.882	0.894	0.903
8					0.890	0.868	0.867	0.858	0.859	0.870	0.888	0.886	0.893	0.892	0.903	0.914
9				0.875	0.890	0.877	0.876	0.867	0.870	0.881	0.897	0.895	0.900	0.906	0.913	0.925
10				0.970	0.898	0.887	0.886	0.878	0.881	0.894	0.904	0.902	0.908	0.912	0.919	0.934
20			1.023	0.980	0.980	0.954	0.952	0.951	0.950	0.954	0.956	0.955	0.956	0.958	0.961	0.965
30		1.022	1.020	0.989	0.980	0.973	0.974	0.974	0.975	0.977	0.978	0.979	0.978	0.979	0.981	0.980
40		0.969	1.044	0.992	0.986	0.983	0.984	0.985	0.987	0.987	0.987	0.989	0.988	0.988	0.990	0.989
50		1.003	1.002	0.996	0.991	0.988	0.989	0.989	0.991	0.992	0.992	0.993	0.992	0.992	0.994	0.994
60	0.756	1.008	1.000	0.998	0.994	0.992	0.992	0.992	0.994	0.995	0.995	0.995	0.995	0.994	0.996	0.997
70	0.901	1.005	1.013	0.999	0.997	0.995	0.995	0.995	0.996	0.997	0.997	0.997	0.997	0.997	0.998	0.998
80	0.977	1.001	1.005	0.999	0.999	0.998	0.998	0.998	0.998	0.999	0.999	0.998	0.999	0.999	0.999	0.999

(continued)

TABLE A.3.43—*Continued*

Polar Angle θ (°)	Radial Distance r (cm)															
	0.15	0.25	0.5	0.75	1.0	1.25	1.5	2.0	3.0	4.0	5.0	6.0	7.0	8.0	10.0	15.0
90	1.000	1.000	1.000	1.000	1.000	1.000	1.000	1.000	1.000	1.000	1.000	1.000	1.000	1.000	1.000	1.000
100	0.974	1.002	1.002	1.002	1.001	1.001	1.002	1.001	1.001	1.000	1.000	1.001	1.000	1.000	1.000	1.000
110	0.896	1.006	1.008	1.004	1.002	1.002	1.003	1.002	1.001	0.999	1.000	1.000	0.999	0.999	1.000	1.000
120	0.754	1.011	1.011	1.006	1.003	1.003	1.004	1.002	1.000	0.999	1.000	0.999	0.998	0.997	0.999	0.999
130		1.008	1.010	1.007	1.005	1.004	1.005	1.001	0.999	0.998	0.998	0.997	0.996	0.995	0.998	0.997
140		1.020	1.014	1.009	1.007	1.005	1.004	0.999	0.996	0.995	0.994	0.993	0.992	0.991	0.994	0.993
150		1.020	1.030	1.016	1.009	1.001	0.999	0.992	0.989	0.986	0.984	0.984	0.984	0.984	0.985	0.984
160			1.040	1.040	0.990	0.988	0.980	0.960	0.968	0.964	0.962	0.964	0.965	0.966	0.968	0.969
170				1.050	1.020	0.948	0.925	0.923	0.910	0.913	0.920	0.921	0.918	0.923	0.932	0.942
171				0.965	1.010	0.938	0.917	0.916	0.904	0.903	0.910	0.910	0.907	0.917	0.927	0.940
172					1.010	0.929	0.909	0.910	0.898	0.893	0.899	0.898	0.896	0.911	0.924	0.937
173					0.761	0.923	0.904	0.905	0.893	0.884	0.888	0.889	0.887	0.906	0.922	0.932
174						0.921	0.902	0.903	0.888	0.877	0.879	0.881	0.882	0.902	0.918	0.926
175						0.924	0.903	0.902	0.885	0.872	0.875	0.878	0.880	0.899	0.911	0.922
176						0.931	0.909	0.903	0.883	0.876	0.875	0.878	0.882	0.898	0.903	0.921
177						0.942	0.920	0.908	0.882	0.884	0.880	0.884	0.884	0.899	0.897	0.924
178						0.954	0.936	0.916	0.882	0.884	0.891	0.896	0.885	0.901	0.900	0.930
179						0.956	0.960	0.927	0.899	0.898	0.892	0.896	0.898	0.906	0.920	0.930
180						0.959	0.991	0.943	0.908	0.900	0.892	0.896	0.895	0.894	0.893	0.894

Values for the Amersham CDCS-M type source design. Data not shown correspond to points within the encapsulated source.

Source: From Casal, E., Ballester, F., Lluch, J.L., Pérez-Calatayud, J., and Lliso, F., *Med. Phys.*, 27, 132–140, 2000.

TABLE A.3.44

Monte Carlo (MC) Calculated Anisotropy Factor, $\varphi_{an}(r)$

	Radial Distance r (cm)																	
$\varphi_{an}(r)$	0.4	0.6	0.8	1.0	1.25	1.5	1.75	2.0	2.5	3.0	3.5	4.0	5.0	6.0	8.0	10.0	12.0	15.0
CDC-1 (1 × 137Cs Bead)[19]	0.990	0.990	0.990	0.990	0.990	0.990	0.990	0.990	0.990	0.990	0.990	0.990	0.991	0.991	0.991	0.992	0.992	0.993
CDC-3 (3 × 137Cs Beads)[19]	1.060	1.027	1.010	1.003	0.995	0.995	0.995	0.995	0.995	0.995	0.995	0.995	0.995	0.996	0.996	0.996	0.996	0.997
CDC.K1-K3 Type Sources (1 × 137Cs Bead)[22]a		0.983	0.983	0.983	0.983	0.983	0.983	0.983	0.983	0.983	0.983	0.983	0.983	0.983	0.983	0.983	0.983	0.983
CDC.K4 Type Source (2 × 137Cs Beads)[22]b				1.074	1.074	1.074	0.991	0.991	0.991	0.991	0.991	0.991	0.991	0.991	0.991	0.991	0.991	0.991

Values for LDR ^{137}Cs sources.

a Anisotropy factor values are according to $\varphi_{an}(r) = 0.98288 + 0.00037r$ as published by the authors.[22]

b Anisotropy factor values are according to $\varphi_{an}(r) = 1.0737 - 0.00017r$ for 0.4 cm $< r <$ 1.26 cm and $\varphi_{an}(r) = 0.9908 + 0.00016r$ for $r >$ 1.26 cm as published by the authors.[22]

A.3.14 LDR ^{125}I Source Models

TABLE A.3.45

Consensus Radial Dose Function, $g(r)$

Radial Distance r (cm)	$g_L(r)$ Amersham 6702[18]	$g_L(r)$ Amersham 6711[19]	$g_L(r)$ BEBIG Symmetra I25.S06	$g_P(r)$ Amersham 6702[18]	$g_P(r)$ Amersham 6711[19]	$g_P(r)$ BEBIG Symmetra I25.S06
0.10	1.020	1.055	1.010	0.673	0.696	0.613
0.15	1.022	1.078	1.018	0.809	0.853	0.760
0.25	1.024	1.082	1.030	0.929	0.982	0.908
0.50	1.030	1.071	1.030	1.008	1.048	1.001
0.75	1.020	1.042	1.020	1.014	1.036	1.012
1.00	1.000	1.000	1.000	1.000	1.000	1.000
1.50	0.935	0.908	0.937	0.939	0.912	0.942
2.00	0.861	0.814	0.857	0.866	0.819	0.863
3.00	0.697	0.632	0.689	0.702	0.636	0.695
4.00	0.553	0.496	0.538	0.557	0.499	0.543
5.00	0.425	0.364	0.409	0.428	0.367	0.413
6.00	0.322	0.270	0.313	0.324	0.272	0.316
7.00	0.241	0.199	0.232	0.243	0.200	0.234
8.00	0.179	0.148	0.176	0.180	0.149	0.178
9.00	0.134	0.109	0.134	0.135	0.110	0.135
10.00	0.0979	0.0803	0.0957	0.0986	0.0809	0.0967

Values for LDR ^{125}I seeds for permanent prostate implants published in TG-43 U1[23] for the line, $g_L(r)$, and point source, $g_P(r)$, approximation, see also Equation 8.45.

TABLE A.3.46

Radial Dose Function, $g(r)$

Radial Distance r (cm)	$g_L(r)$	$g_P(r)$
0.1	1.042	0.643
0.2	1.082	0.905
0.3	1.087	0.998
0.4	1.085	1.036
0.5	1.078	1.049
0.6	1.066	1.049
0.7	1.052	1.042
0.8	1.035	1.030
0.9	1.019	1.017
1.0	1.000	1.000
1.5	0.907	0.912
2.0	0.808	0.814
2.5	0.713	0.719
3.0	0.627	0.632
3.5	0.548	0.553
4.0	0.477	0.481
4.5	0.414	0.418
5.0	0.357	0.360
6.0	0.265	0.267
7.0	0.196	0.198
8.0	0.144	0.145
9.0	0.106	0.107
10.0	0.078	0.079

Values for the nucletron selectSeed model 130.002 ^{125}I seed for permanent prostate implants based on Monte Carlo dosimetry study.[24,25] The data for point source approximation $g_P(r)$ have been extracted from the original publication[24] using the effective source length of 0.34 cm and utilizing Equation 8.45.

TABLE A.3.47

Consensus Values for the Anisotropy Function $F(r,\theta)$ and the Anisotropy Factor $\varphi_{an}(r)$ for the Amersham 6702 ^{125}I -Seed

Polar Angle	Radial Distance r (cm)					
θ (°)	0.5	1	2	3	4	5
0	0.385	0.420	0.493	0.533	0.569	0.589
5	0.413	0.472	0.546	0.586	0.613	0.631
10	0.531	0.584	0.630	0.660	0.681	0.697
15	0.700	0.700	0.719	0.738	0.749	0.758
20	0.788	0.789	0.793	0.805	0.810	0.814
30	0.892	0.888	0.888	0.891	0.892	0.892
40	0.949	0.948	0.944	0.944	0.944	0.944
50	0.977	0.973	0.967	0.967	0.967	0.967
60	0.989	0.985	0.983	0.983	0.983	0.983
70	0.996	0.992	0.990	0.990	0.990	0.990
80	1.000	0.998	0.998	0.998	0.998	0.998
$\varphi_{an}(r)$	**0.986**	**0.960**	**0.952**	**0.951**	**0.954**	**0.954**

Source: From Rivard, M., Coursey, B., Hanson, W., Hug, S., Ibbot, G., Mitch, M., Nath, R., and Williamson, J., *Med. Phys.*, 31/3, 633–674, 2004.

TABLE A.3.48

Consensus Values for the Anisotropy Function $F(r,\theta)$ and the Anisotropy Factor $\varphi_{an}(r)$ for the Amersham 6711 ^{125}I -Seed

Polar Angle	Radial Distance r (cm)					
θ (°)	0.5	1	2	3	4	5
0	0.333	0.370	0.442	0.488	0.520	0.550
5	0.400	0.429	0.497	0.535	0.561	0.587
10	0.519	0.537	0.580	0.609	0.630	0.645
20	0.716	0.705	0.727	0.743	0.752	0.760
30	0.846	0.834	0.842	0.846	0.848	0.852
40	0.926	0.925	0.926	0.926	0.928	0.928
50	0.972	0.972	0.970	0.969	0.969	0.969
60	0.991	0.991	0.987	0.987	0.987	0.987
70	0.996	0.996	0.996	0.995	0.995	0.995
80	1.000	1.000	1.000	0.999	0.999	0.999
$\varphi_{an}(r)$	**0.973**	**0.944**	**0.941**	**0.942**	**0.943**	**0.944**

Source: From Rivard, M., Coursey, B., Hanson, W., Hug, S., Ibbot, G., Mitch, M., Nath, R., and Williamson, J., *Med. Phys.*, 31/3, 633–674, 2004.

TABLE A.3.49

Consensus Values for the Anisotropy Function $F(r,\theta)$ and the Anisotropy Factor $\varphi_{an}(r)$ for the BEBIG Symmetra I25.S06 Seed

Polar Angle θ (°)	Radial Distance r (cm)							
	0.25	0.5	1	2	3	4	5	7
0	0.302	0.429	0.512	0.579	0.610	0.631	0.649	0.684
5	0.352	0.436	0.509	0.576	0.610	0.635	0.651	0.689
10	0.440	0.476	0.557	0.622	0.651	0.672	0.689	0.721
20	0.746	0.686	0.721	0.757	0.771	0.785	0.790	0.807
30	0.886	0.820	0.828	0.846	0.857	0.862	0.867	0.874
40	0.943	0.897	0.898	0.907	0.908	0.913	0.918	0.912
50	0.969	0.946	0.942	0.947	0.944	0.947	0.949	0.946
60	0.984	0.974	0.970	0.974	0.967	0.966	0.967	0.976
70	0.994	0.989	0.988	0.990	0.984	0.985	0.987	0.994
80	0.998	0.998	0.998	1.000	0.994	1.000	0.993	0.999
$\varphi_{an}(r)$	1.122	0.968	0.939	0.939	0.938	0.940	0.941	0.949

Source: From Rivard, M., Coursey, B., Hanson, W., Hug, S., Ibbot, G., Mitch, M., Nath, R., and Williamson, J., *Med. Phys.*, 31/3, 633–674, 2004.

TABLE A.3.50

Values for the Anisotropy Function $F(r,\theta)$ and the Anisotropy Factor $\varphi_{an}(r)$ for the Nucletron SelectSeed Model 130.002 Seed Based on Monte Carlo Dosimetry Study

Polar Angle θ (°)	Radial Distance r (cm)									
	0.3	0.5	0.7	1.0	1.5	2.0	3.0	4.0	6.0	8.0
0.5	0.197	0.205	0.248	0.291	0.344	0.388	0.464	0.515	0.569	0.615
1.5	0.200	0.211	0.251	0.297	0.382	0.441	0.519	0.553	0.620	0.648
2.5	0.201	0.218	0.268	0.354	0.441	0.483	0.552	0.580	0.635	0.655
3.5	0.202	0.246	0.333	0.406	0.463	0.499	0.557	0.591	0.639	0.658
4.5	0.213	0.309	0.373	0.420	0.467	0.508	0.559	0.594	0.641	0.667
5.5	0.240	0.351	0.384	0.425	0.471	0.511	0.562	0.599	0.644	0.671
7.5	0.353	0.371	0.401	0.442	0.495	0.535	0.581	0.616	0.668	0.685
10.5	0.410	0.425	0.460	0.499	0.546	0.580	0.631	0.654	0.699	0.719
12.5	0.469	0.473	0.502	0.540	0.582	0.617	0.657	0.680	0.716	0.731
15.5	0.565	0.543	0.567	0.595	0.633	0.661	0.695	0.715	0.745	0.762
17.5	0.623	0.588	0.602	0.631	0.663	0.689	0.723	0.739	0.763	0.788
20.5	0.700	0.646	0.659	0.679	0.706	0.728	0.754	0.769	0.792	0.810
25.5	0.803	0.728	0.729	0.747	0.766	0.783	0.799	0.814	0.836	0.840
30.5	0.880	0.797	0.790	0.799	0.815	0.828	0.844	0.852	0.867	0.871
35.5	0.937	0.854	0.841	0.844	0.853	0.866	0.879	0.882	0.892	0.903
40.5	0.970	0.902	0.886	0.886	0.889	0.898	0.907	0.908	0.918	0.926
50.5	0.988	0.973	0.955	0.952	0.951	0.956	0.955	0.955	0.958	0.959
60.5	0.966	1.003	1.002	0.998	0.991	0.991	0.993	0.991	0.986	0.987
70.5	0.987	0.981	1.013	1.017	1.016	1.019	1.014	1.008	1.011	1.006
80.5	0.996	0.995	0.995	0.994	1.013	1.019	1.018	1.019	1.018	1.014
$\varphi_{an}(r)$	1.051	0.959	0.938	0.933	0.932	0.936	0.941	0.943	0.946	0.950

Source: From Karaiskos, P., Papagiannis, P., Sakelliou, L., Anagnostopoulos, G., and Baltas, D., *Med. Phys.*, 28, 1753–1760, 2001.

A.3.15 LDR ^{103}Pd Source Models

TABLE A.3.51

Consensus Radial Dose Function, $g(r)$

Radial Distance r (cm)	$g_L(r)$ Theragenics Model 200	$g_L(r)$ NASI MED3633	$g_P(r)$ Theragenics Model 200	$g_P(r)$ NASI MED3633
0.10	0.911		0.494	
0.15	1.210		0.831	
0.25	1.370	1.331	1.154	1.123
0.30	1.380	1.322	1.220	1.170
0.40	1.360	1.286	1.269	1.201
0.50	1.300	1.243	1.248	1.194
0.75	1.150	1.125	1.137	1.113
1.00	1.000	1.000	1.000	1.000
1.50	0.749	0.770	0.755	0.776
2.00	0.555	0.583	0.561	0.589
2.50	0.410	0.438	0.415	0.443
3.00	0.302	0.325	0.306	0.329
3.50	0.223	0.241	0.226	0.244
4.00	0.163	0.177	0.165	0.179
5.00	0.0887	0.098	0.0900	0.099
6.00	0.0482	0.053	0.0489	0.054
7.00	0.0262	0.028	0.0266	0.028
10.00	0.00615		0.00624	

Values for the Theragenics TheraSeed Model 200 and NASI MED3633 LDR ^{103}Pd seeds for permanent prostate implants published in TG-43 U1[23] for the line, $g_L(r)$, and point source, $g_P(r)$, approximation, see also Equation 8.45.

TABLE A.3.52

Consensus Values for the Anisotropy Function $F(r,\theta)$ and the Anisotropy Factor $\varphi_{an}(r)$ for the Theragenics TheraSeed Model 200 [103]Pd-Seed

Polar Angle	Radial Distance r (cm)								
θ (°)	0.25	0.5	0.75	1.0	2.0	3.0	4.0	5.0	7.5
0	0.619	0.694	0.601	0.541	0.526	0.504	0.497	0.513	0.547
1	0.617	0.689	0.597	0.549	0.492	0.505	0.513	0.533	0.580
2	0.618	0.674	0.574	0.534	0.514	0.517	0.524	0.538	0.568
3	0.620	0.642	0.577	0.538	0.506	0.509	0.519	0.532	0.570
5	0.617	0.600	0.540	0.510	0.499	0.508	0.514	0.531	0.571
7	0.579	0.533	0.519	0.498	0.498	0.509	0.521	0.532	0.568
10	0.284	0.496	0.495	0.487	0.504	0.519	0.530	0.544	0.590
12	0.191	0.466	0.486	0.487	0.512	0.529	0.544	0.555	0.614
15	0.289	0.446	0.482	0.490	0.523	0.540	0.556	0.567	0.614
20	0.496	0.442	0.486	0.501	0.547	0.568	0.585	0.605	0.642
25	0.655	0.497	0.524	0.537	0.582	0.603	0.621	0.640	0.684
30	0.775	0.586	0.585	0.593	0.633	0.654	0.667	0.683	0.719
40	0.917	0.734	0.726	0.727	0.750	0.766	0.778	0.784	0.820
50	0.945	0.837	0.831	0.834	0.853	0.869	0.881	0.886	0.912
60	0.976	0.906	0.907	0.912	0.931	0.942	0.960	0.964	0.974
70	0.981	0.929	0.954	0.964	0.989	1.001	1.008	1.004	1.011
75	0.947	0.938	0.961	0.978	1.006	1.021	1.029	1.024	1.033
80	0.992	0.955	0.959	0.972	1.017	1.035	1.046	1.037	1.043
85	1.007	0.973	0.960	0.982	0.998	1.030	1.041	1.036	1.043
$\varphi_{an}(r)$	**1.130**	**0.880**	**0.859**	**0.855**	**0.870**	**0.884**	**0.895**	**0.897**	**0.918**

Source: From Rivard, M., Coursey, B., Hanson, W., Hug, S., Ibbot, G., Mitch, M., Nath, R., and Williamson, J., *Med. Phys.*, 31/3, 633–674, 2004.

TABLE A.3.53

Consensus Values for the Anisotropy Function $F(r,\theta)$ and the Anisotropy Factor $\varphi_{an}(r)$ for the NASI Model MED3633 [103]Pd-Seed

Polar Angle	Radial Distance r (cm)					
θ (°)	0.25	0.5	1.0	2.0	5.0	10.0
0	1.024	0.667	0.566	0.589	0.609	0.733
10	0.888	0.581	0.536	0.536	0.569	0.641
20	0.850	0.627	0.603	0.614	0.652	0.716
30	0.892	0.748	0.729	0.734	0.756	0.786
40	0.931	0.838	0.821	0.824	0.837	0.853
50	0.952	0.897	0.890	0.891	0.901	0.905
60	0.971	0.942	0.942	0.940	0.948	0.939
70	0.995	0.976	0.974	0.973	0.980	0.974
80	1.003	0.994	0.997	0.994	1.000	0.986
$\varphi_{an}(r)$	**1.257**	**0.962**	**0.903**	**0.895**	**0.898**	**0.917**

Source: From Rivard, M., Coursey, B., Hanson, W., Hug, S., Ibbot, G., Mitch, M., Nath, R., and Williamson, J., *Med. Phys.*, 31/3, 633–674, 2004.

References

1. Williamson, J.F. and Li, Z. Monte Carlo aided dosimetry of the microSelectron pulsed and high dose rate ^{192}Ir sources, *Med. Phys.*, 22, 809–819, 1995.
2. Karaiskos, P., Angelopoulos, A., Sakelliou, L., Sandilos, P., Antypas, C., Vlachos, L., and Koutsouveli, E. Monte Carlo and TLD dosimetry of an ^{192}Ir high dose-rate brachytherapy source, *Med. Phys.*, 25, 1975–1984, 1998.
3. Anctil, J.C., Clark, B.G., and Arsenault, C.J. Experimental determination of dosimetry functions of Ir-192 sources, *Med. Phys.*, 25, 2279–2287, 1998.
4. Daskalov, G.M., Löffler, E., and Williamson, J.F. Monte Carlo-aided dosimetry of a new high dose-rate brachytherapy source, *Med. Phys.*, 25, 2200–2208, 1998.
5. Papagiannis, P., Angelopoulos, A., Pantelis, E., Sakelliou, L., Baltas, D., Karaiskos, P., Sandilos, P., and Vlachos, L. Dosimetry comparison of ^{192}Ir sources, *Med. Phys.*, 29, 2239–2246, 2002.
6. Wang, R. and Sloboda, R.S. Influence of source geometry and materials on the transverse axis dosimetry of ^{192}Ir brachytherapy sources, *Phys. Med. Biol.*, 43, 37–48, 1998.
7. Karaiskos, P., Angelopoulos, A., Baras, P., Sakelliou, L., Sandilos, P., Dardoufas, K., and Vlachos, L. A Monte Carlo investigation of the dosimetric characteristics of the VariSource ^{192}Ir high dose rate brachytherapy source, *Med. Phys.*, 26, 1498–1502, 1999.
8. Meigooni, A.S., Kleiman, M.T., Johnson, L.J., Mazloomdoost, D., and Ibbott, G.S. Dosimetric characteristics of a new high-intensity ^{192}Ir source for remote afterloading, *Med. Phys.*, 24, 2008–2013, 1997.
9. Angelopoulos, A., Baras, P., Sakelliou, L., Karaiskos, P., and Sandilos, P. Monte Carlo dosimetry of a new ^{192}Ir high dose rate brachytherapy source, *Med. Phys.*, 27, 2521–2527, 2000.
10. Ballester, F., Pérez-Calatayud, J., Puchades, V., Lluch, J.L., Serrano-Andrés, M.A., Limami, Y., Lliso, F., and Casal, E. Monte Carlo dosimetry of the Buchler high dose rate ^{192}Ir source, *Phys. Med. Biol.*, 46, N79–N90, 2001.
11. J. Venselaar and J. Pérez-Calatayud, eds. *European Guidelines for Quality Assurance in Radiotherapy A Practical Guide to Quality Control of Brachytherapy Equipment Booklet No 8*, ESTRO, Belgium, 2004.
12. Ballester, F., Puchades, V., Lluch, J.L., Serrano-Andrés, M.A., Limami, Y., Pérez-Calatayud, J., and Casal, E. Technical note: Monte Carlo dosimetry of the HDR 12*i* and Plus ^{192}Ir sources, *Med. Phys.*, 28, 2586–2591, 2001.
13. Selbach, H.J. and Andréssy, M. Literature regarding the TG-43 data for the BEBIG MultiSource ^{192}Ir HDR afterloader, unpublished data, private communication, 2005.
14. Papagiannis, P., Angelopoulos, A., Pantelis, E., Sakelliou, L., Karaiskos, P., and Shimizu, Y. Monte Carlo dosimetry of ^{60}Co HDR brachytherapy sources, *Med. Phys.*, 29, 712–721, 2003.
15. Selbach, H.J. and Andréssy, M. Literature regarding the TG-43 data for the BEBIG MultiSource ^{60}Co HDR afterloader, unpublished data, private communication, 2005.
16. Karaiskos, P., Angelopoulos, A., Pantelis, E., Papagiannis, P., Sakelliou, L., Kouwenhoven, E., and Baltas, D. Monte Carlo dosimetry of a new ^{192}Ir pulsed dose rate brachytherapy source, *Med. Phys.*, 30, 9–16, 2003.

17. Pérez-Calatayud, J., Ballester, F., Serrano-Andrés, M.A., Puchades, V., Lluch, J.L., Limami, Y., and Casal, E. Dosimetry characteristics of the Plus and 12*i* GammaMed PDR ^{192}Ir sources, *Med. Phys.*, 28, 2576–2585, 2001.
18. Pérez-Calatayud, J., Granero, D., Ballester, F., Puchades, V., and Casal, E. Monte Carlo dosimetric characterization of the Cs-137 selectron/LDR source: evaluation of applicator attenuation and superposition approximation effects, *Med. Phys.*, 31, 493–499, 2004.
19. Pérez-Calatayud, J., Ballester, F., Serrano-Andrés, M.A., Llich, J.L., Puchades, V., Limami, Y., and Casal, E. Dosimetric characteristics of the CDC-type miniature cylindrical ^{137}Cs brachytherapy sources, *Med. Phys.*, 29, 538–543, 2002.
20. Ballester, F., Lluch, J.L., Limami, Y., Serrano, M.A., Casal, E., Pérez-Calatayud, J., and Lliso, F. A Monte Carlo investigation of the dosimetric characteristics of the CSM11 ^{137}Cs source from CIS, *Med. Phys.*, 27, 2182–2189, 2000.
21. Casal, E., Ballester, F., Lluch, J.L., Pérez-Calatayud, J., and Lliso, F. Monte Carlo calculations of dose rate distributions around the Amersham CDCS-M-type ^{137}Cs source, *Med. Phys.*, 27, 132–140, 2000.
22. Pérez-Calatayud, J., Ballester, F., Lluch, J.L., Serrano-Andrés, M.A., Casal, E., Puchades, V., and Limami, Y.F. Monte Carlo calculation of dose rate distributions around the Walstam CDC.K-Type ^{137}Cs sources, *Phys. Med. Biol.*, 46, 2029–2040, 2001.
23. Rivard, M., Coursey, B., Hanson, W., Hug, S., Ibbot, G., Mitch, M., Nath, R., and Williamson, J. Update of AAPM task group No. 43 report: a revised AAPM protocol for brachytherapy dose calculations, *Med. Phys.*, 31/3, 633–674, 2004.
24. Karaiskos, P., Papagiannis, P., Sakelliou, L., Anagnostopoulos, G., and Baltas, D. Monte Carlo dosimetry of the selectSeed ^{125}I interstitial brachytherapy seed, *Med. Phys.*, 28, 1753–1760, 2001.
25. Granero, D., Pérez-Calatayud, J., and Ballester, F. Monte Carlo calculation of the TG-43 dosimetric parameters of a new BEBIG Ir-192 HDR source, *Radiother. Oncol.*, 76, 79–85, 2005.
26. Ballester, F., Granero, D., Pérez-Calatayud, J., Casal, E., Agramunt, A., and Cases, R. Monte Carlo dosimetric study of the BEBIG Co-60 HDR source, *Phys. Med. Biol.*, 50, N309–N316, 2005.
27. Medich, C.D., Tries, A.M., and Munro, J.J. Monte Carlo characterization of an ytterbium-169 high dose rate brachytherapy source with analysis of statistical uncertainty, *Med. Phys.*, 33, 163–172, 2006.

Appendix 4

Dose Rate Tables for Brachytherapy Sources

A.4.1 MicroSelectron HDR Classic ^{192}Ir Source

TABLE A.4.1

Monte Carlo (MC) Calculated Dose Rate Values Per Unit Source Strength as an Along–Away Dose Rate Table (DRT) for the MicroSelectron HDR Old Design, Classic, ^{192}Ir Source

Distance Along z (cm)	Dose Rate Per Unit Source Strength (cGy h^{-1} U^{-1}) Distance Away y (cm)													
	0.00	0.10	0.25	0.50	0.75	1.00	1.50	2.00	2.50	3.00	4.00	5.00	6.00	7.00
7.00	0.0155	0.0156	0.0157	0.0160	0.0163	0.0165	0.0169	0.0169	0.0167	0.0162	0.0148	0.0130	0.0113	0.0097
6.00	0.0215	0.0217	0.0219	0.0224	0.0228	0.0232	0.0235	0.0234	0.0227	0.0217	0.0192	0.0164	0.0139	0.0116
5.00	0.0313	0.0317	0.0321	0.0329	0.0336	0.0341	0.0344	0.0335	0.0319	0.0298	0.0251	0.0208	0.0169	0.0137
4.00	0.0481	0.0487	0.0498	0.0515	0.0529	0.0537	0.0532	0.0504	0.0463	0.0418	0.0333	0.0260	0.0203	0.0160
3.00	0.0834	0.0849	0.0878	0.0920	0.0948	0.0950	0.0898	0.0804	0.0698	0.0600	0.0436	0.0319	0.0238	0.0182
2.50	0.1188	0.1213	0.1265	0.1338	0.1367	0.1348	0.1218	0.1037	0.0865	0.0718	0.0494	0.0350	0.0256	0.0192
2.00	0.1839	0.1888	0.1995	0.2122	0.2123	0.2025	0.1695	0.1354	0.1073	0.0851	0.0554	0.0379	0.0272	0.0202
1.50	0.3260	0.3370	0.3619	0.3783	0.3598	0.3227	0.2415	0.1766	0.1310	0.0991	0.0610	0.0406	0.0286	0.0209
1.00	0.7401	0.7805	0.8485	0.8124	0.6792	0.5427	0.3412	0.2234	0.1543	0.1119	0.0658	0.0427	0.0296	0.0215
0.75	1.3825	1.4618	1.5506	1.3046	0.9661	0.7058	0.3968	0.2458	0.1646	0.1172	0.0676	0.0434	0.0299	0.0217
0.50	3.4333	3.6550	3.4355	2.2035	1.3624	0.8904	0.4482	0.2648	0.1729	0.1214	0.0690	0.0440	0.0302	0.0218
0.25	27.182	19.893	9.2815	3.5287	1.7728	1.0504	0.4853	0.2772	0.1783	0.1242	0.0698	0.0443	0.0304	0.0219
0.10	—	57.514	14.260	4.1681	1.9294	1.1037	0.4967	0.2810	0.1799	0.1250	0.0701	0.0444	0.0304	0.0220
0.00	—	66.289	15.586	4.3189	1.9638	1.1150	0.4991	0.2818	0.1803	0.1253	0.0702	0.0445	0.0304	0.0220
−0.10	—	57.396	14.246	4.1680	1.9289	1.1033	0.4967	0.2811	0.1799	0.1250	0.0701	0.0444	0.0304	0.0220

(continued)

TABLE A.4.1—*Continued*

Distance Along z (cm)	Dose Rate Per Unit Source Strength (cGy h^{-1} U^{-1}) Distance Away y (cm)													
	0.00	0.10	0.25	0.50	0.75	1.00	1.50	2.00	2.50	3.00	4.00	5.00	6.00	7.00
−0.25	29.109	19.857	9.2742	3.5304	1.7717	1.0495	0.4853	0.2774	0.1782	0.1241	0.0698	0.0443	0.0304	0.0219
−0.50	3.0956	3.5251	3.4217	2.2048	1.3608	0.8888	0.4480	0.2650	0.1728	0.1213	0.0689	0.0439	0.0302	0.0218
−0.75	1.2219	1.3215	1.5300	1.3021	0.9658	0.7044	0.3962	0.2456	0.1644	0.1170	0.0675	0.0434	0.0299	0.0217
−1.00	0.6402	0.6816	0.8218	0.8074	0.6778	0.5418	0.3403	0.2229	0.1538	0.1116	0.0656	0.0426	0.0295	0.0215
−1.50	0.2864	0.2964	0.3372	0.3713	0.3568	0.3211	0.2405	0.1758	0.1301	0.0986	0.0608	0.0405	0.0285	0.0209
−2.00	0.1640	0.1681	0.1821	0.2052	0.2088	0.2004	0.1685	0.1349	0.1067	0.0846	0.0551	0.0378	0.0271	0.0201
−2.50	0.1064	0.1084	0.1143	0.1278	0.1335	0.1329	0.1206	0.1032	0.0863	0.0714	0.0492	0.0348	0.0255	0.0191
−3.00	0.0750	0.0761	0.0791	0.0869	0.0919	0.0933	0.0890	0.0800	0.0696	0.0598	0.0434	0.0318	0.0238	0.0181
−4.00	0.0437	0.0442	0.0452	0.0479	0.0508	0.0524	0.0527	0.0502	0.0463	0.0419	0.0332	0.0258	0.0202	0.0159
−5.00	0.0288	0.0291	0.0294	0.0306	0.0320	0.0331	0.0339	0.0333	0.0318	0.0298	0.0251	0.0207	0.0168	0.0136
−6.00	0.0197	0.0199	0.0201	0.0206	0.0214	0.0222	0.0230	0.0231	0.0226	0.0217	0.0192	0.0164	0.0138	0.0115
−7.00	0.0142	0.0143	0.0144	0.0148	0.0152	0.0157	0.0164	0.0167	0.0166	0.0161	0.0148	0.0131	0.0113	0.0097

The distance along axis z has been adjusted to fit the TG-43 U1 source coordinate system (Figure 8.6). Data not shown correspond to points within the encapsulated source. The dose rate constant, Λ, marked in bold is the dose rate per unit source strength at the reference point, $r = 1.0$ cm and $\theta = 90°$ or in Cartesian coordinates $y = 1.0$ cm and $z = 0.0$ cm.

Source: From Williamson, J.F. and Li, Z., *Med. Phys.*, 22, 809–819, 1995.

A.4.2 MicroSelectron HDR New ^{192}Ir Source

TABLE A.4.2

Monte Carlo (MC) Calculated Dose Rate Values Per Unit Source Strength as an Along–Away Dose Rate Table (DRT) for the MicroSelectron HDR New ^{192}Ir Source Design

Distance Along z (cm)	Dose Rate Per Unit Source Strength (cGy h^{-1} U^{-1}) Distance Away y (cm)													
	0.00	0.10	0.15	0.25	0.35	0.50	0.75	1.00	1.50	2.00	2.50	3.00	5.00	7.00
7.00	0.0164	0.0163	0.0163	0.0164	0.0165	0.0167	0.0170	0.0169	0.0173	0.0172	0.0169	0.0164	0.0132	0.0097
6.00	0.0223	0.0222	0.0223	0.0225	0.0226	0.0230	0.0234	0.0233	0.0238	0.0236	0.0228	0.0219	0.0165	0.0116
5.00	0.0318	0.0319	0.0320	0.0324	0.0326	0.0333	0.0340	0.0341	0.0345	0.0336	0.0319	0.0299	0.0208	0.0137
4.00	0.0483	0.0486	0.0488	0.0496	0.0502	0.0524	0.0530	0.0538	0.0533	0.0504	0.0463	0.0419	0.0259	0.0159
3.00	0.0840	0.0852	0.0859	0.0879	0.0897	0.0926	0.0952	0.0950	0.0899	0.0803	0.0698	0.0598	0.0319	0.0182
2.50	0.1190	0.1220	0.1220	0.1270	0.1300	0.1340	0.1370	0.1340	0.1220	0.1040	0.0864	0.0713	0.0349	0.0192
2.00	0.1830	0.1900	0.1900	0.1980	0.2060	0.2120	0.2120	0.2010	0.1690	0.1350	0.1070	0.0846	0.0379	0.0201
1.50	0.3240	0.3340	0.3430	0.3600	0.3720	0.3770	0.3580	0.3200	0.2390	0.1760	0.1300	0.0985	0.0406	0.0209
1.00	0.7450	0.7810	0.8090	0.8490	0.8540	0.8100	0.6770	0.5400	0.3390	0.2230	0.1540	0.1120	0.0427	0.0215
0.75	1.3570	1.4400	1.5000	1.5390	1.4790	1.3010	0.9630	0.7010	0.3940	0.2460	0.1650	0.1170	0.0435	0.0217
0.50	3.4050	3.6310	3.6910	3.4080	2.9070	2.1850	1.3510	0.8840	0.4460	0.2650	0.1730	0.1210	0.0441	0.0219
0.25	—	19.7100	15.1200	9.1770	5.9680	3.5070	1.7600	1.0420	0.4830	0.2780	0.1780	0.1240	0.0445	0.0219
0.10	—	58.7900	32.1800	14.1900	7.9440	4.1540	1.9170	1.0960	0.4950	0.2820	0.1800	0.1250	0.0446	0.0220
0.00	—	66.3600	36.3600	15.5200	8.4340	4.2990	1.9500	**1.1080**	0.4970	0.2820	0.1800	0.1250	0.0446	0.0220
−0.10	—	58.7900	32.2300	14.2000	7.9520	4.1510	1.9180	1.0970	0.4950	0.2820	0.1800	0.1250	0.0446	0.0220

(continued)

TABLE A.4.2—*Continued*

Distance Along z (cm)	Dose Rate Per Unit Source Strength (cGy h⁻¹ U⁻¹) Distance Away y (cm)													
	0.00	0.10	0.15	0.25	0.35	0.50	0.75	1.00	1.50	2.00	2.50	3.00	5.00	7.00
−0.25	—	19.6800	15.1400	9.1820	5.9760	3.5010	1.7610	1.0420	0.4830	0.2780	0.1780	0.1240	0.0445	0.0219
−0.50	3.1270	3.5590	3.6550	3.4020	2.9090	2.1790	1.3500	0.8830	0.4460	0.2650	0.1730	0.1210	0.0441	0.0219
−0.75	1.2420	1.3790	1.4670	1.5270	1.4760	1.2980	0.9630	0.7010	0.3940	0.2460	0.1650	0.1170	0.0435	0.0217
−1.00	0.6680	0.7390	0.7830	0.8370	0.8480	0.8060	0.6770	0.5390	0.3390	0.2230	0.1540	0.1120	0.0427	0.0215
−1.50	0.3010	0.3140	0.3270	0.3500	0.3660	0.3730	0.3570	0.3210	0.2400	0.1760	0.1300	0.0987	0.0406	0.0209
−2.00	0.1700	0.1790	0.1800	0.1900	0.2020	0.2100	0.2110	0.2000	0.1690	0.1350	0.1070	0.0847	0.0379	0.0201
−2.50	0.1120	0.1150	0.1150	0.1220	0.1270	0.1320	0.1360	0.1340	0.1210	0.1040	0.0861	0.0714	0.0349	0.0192
−3.00	0.0790	0.0803	0.0814	0.0840	0.0869	0.0904	0.0940	0.0943	0.0896	0.0802	0.0695	0.0597	0.0319	0.0182
−4.00	0.0455	0.0470	0.0466	0.0473	0.0494	0.0509	0.0528	0.0532	0.0529	0.0502	0.0461	0.0418	0.0259	0.0159
−5.00	0.0303	0.0305	0.0307	0.0311	0.0316	0.0322	0.0333	0.0334	0.0342	0.0344	0.0318	0.0298	0.0207	0.0137
−6.00	0.0212	0.0213	0.0214	0.0216	0.0219	0.0222	0.0228	0.0229	0.0236	0.0234	0.0228	0.0217	0.0165	0.0116
−7.00	0.0156	0.0156	0.0157	0.0158	0.0160	0.0162	0.0165	0.0166	0.0171	0.0171	0.0168	0.0163	0.0132	0.0097

Data not shown correspond to points within the encapsulated source. The dose rate constant, Λ, marked in bold is the dose rate per unit source strength at the reference point, $r = 1.0$ cm and $\theta = 90°$ or in Cartesian coordinates $y = 1.0$ cm and $z = 0.0$ cm.

Source: From Daskalov, G.M., Löffler, E., and Williamson, J.F., *Med. Phys.*, 25, 2200–2208, 1998.

A.4.3 VariSource HDR Old ^{192}Ir Source

TABLE A.4.3

Monte Carlo (MC) Calculated Dose Rate Values Per Unit Source Strength as an Along–Away Dose Rate Table (DRT) for the VariSource HDR Old ^{192}Ir Source Design

Distance Along z (cm)	Dose Rate Per Unit Source Strength (cGy h^{-1} U^{-1}) Distance Away y (cm)														
	0.00	0.10	0.25	0.50	0.75	1.00	1.50	2.00	2.50	3.00	4.00	5.00	6.00	7.00	10.00
10.00	0.0057	0.0065	0.0066	0.0071	0.0074	0.0077	0.0080	0.0082	0.0082	0.0081	0.0077	0.0071	0.0065	0.0058	0.0038
7.00	0.0129	0.0136	0.0142	0.0155	0.0166	0.0173	0.0180	0.0181	0.0177	0.0172	0.0155	0.0136	0.0116	0.0101	0.0059
6.00	0.0175	0.0180	0.0193	0.0217	0.0234	0.0243	0.0250	0.0248	0.0240	0.0228	0.0199	0.0170	0.0142	0.0118	0.0067
5.00	0.0241	0.0253	0.0282	0.0320	0.0343	0.0358	0.0364	0.0353	0.0334	0.0310	0.0260	0.0212	0.0172	0.0139	0.0075
4.00	0.0347	0.0372	0.0440	0.0522	0.0557	0.0570	0.0563	0.0528	0.0482	0.0433	0.0340	0.0264	0.0206	0.0162	0.0083
3.00	0.0558	0.0647	0.0825	0.0967	0.1018	0.1021	0.0949	0.0837	0.0720	0.0614	0.0441	0.0323	0.0241	0.0184	0.0089
2.50	0.0780	0.0969	0.1237	0.1444	0.1489	0.1452	0.1280	0.1076	0.0888	0.0730	0.0499	0.0354	0.0258	0.0194	0.0092
2.00	0.1238	0.1614	0.2073	0.2340	0.2321	0.2175	0.1770	0.1393	0.1091	0.0860	0.0559	0.0383	0.0274	0.0203	0.0095
1.50	0.2292	0.3173	0.4144	0.4337	0.3976	0.3463	0.2495	0.1794	0.1319	0.0998	0.0614	0.0409	0.0287	0.0211	0.0097
1.00	0.6676	0.9841	1.1420	0.9742	0.7502	0.5706	0.3449	0.2239	0.1547	0.1123	0.0661	0.0429	0.0298	0.0217	0.0098
0.75	1.8060	2.5951	2.3897	1.5627	1.0378	0.7225	0.3962	0.2449	0.1646	0.1175	0.0679	0.0437	0.0302	0.0219	0.0099
0.50	—	15.435	5.7097	2.4301	1.3746	0.8779	0.4418	0.2622	0.1724	0.1216	0.0692	0.0443	0.0304	0.0220	0.0099
0.25	—	28.207	8.9179	3.2147	1.6503	0.9981	0.4733	0.2735	0.1774	0.1242	0.0701	0.0446	0.0306	0.0222	0.0099
0.10	—	29.225	9.5971	3.4506	1.7406	1.0347	0.4827	0.2768	0.1787	0.1251	0.0700	0.0448	0.0307	0.0221	0.0100
0.00	—	29.418	9.7037	3.4958	1.7563	1.0432	0.4842	0.2777	0.1793	0.1247	0.0705	0.0447	0.0307	0.0221	0.0100
−0.10	—	29.268	9.5966	3.4502	1.7394	1.0342	0.4833	0.2768	0.1788	0.1245	0.0701	0.0448	0.0307	0.0222	0.0100
−0.25	—	28.252	8.9090	3.2076	1.6511	0.9978	0.4725	0.2734	0.1770	0.1239	0.0701	0.0445	0.0306	0.0220	0.0099

(continued)

TABLE A.4.3—Continued

Dose Rate Per Unit Source Strength (cGy h^{-1} U^{-1})

Distance Along z (cm)	Distance Away y (cm)														
	0.00	0.10	0.25	0.50	0.75	1.00	1.50	2.00	2.50	3.00	4.00	5.00	6.00	7.00	10.00
-0.50	—	15.424	5.7044	2.4362	1.3730	0.8778	0.4411	0.2619	0.1721	0.1214	0.0692	0.0443	0.0304	0.0220	0.0099
-0.75	—	2.6069	2.3937	1.5630	1.0377	0.7228	0.3960	0.2446	0.1644	0.1175	0.0677	0.0437	0.0302	0.0219	0.0099
-1.00	0.6072	0.9784	1.1398	0.9753	0.7501	0.5711	0.3452	0.2240	0.1549	0.1124	0.0660	0.0429	0.0300	0.0217	0.0098
-1.50	0.2218	0.3094	0.4123	0.4345	0.3979	0.3458	0.2494	0.1794	0.1320	0.0998	0.0614	0.0407	0.0288	0.0211	0.0097
-2.00	0.1227	0.1524	0.2059	0.2346	0.2321	0.2176	0.1774	0.1394	0.1091	0.0861	0.0559	0.0382	0.0274	0.0203	0.0095
-2.50	0.0789	0.0926	0.1232	0.1442	0.1487	0.1450	0.1281	0.1077	0.0889	0.0730	0.0500	0.0353	0.0258	0.0194	0.0092
-3.00	0.0570	0.0633	0.0815	0.0969	0.1021	0.1021	0.0950	0.0838	0.0721	0.0614	0.0443	0.0323	0.0241	0.0184	0.0089
-4.00	0.0338	0.0349	0.0434	0.0514	0.0554	0.0569	0.0563	0.0529	0.0482	0.0433	0.0340	0.0264	0.0206	0.0162	0.0083
-5.00	0.0239	0.0245	0.0274	0.0317	0.0343	0.0357	0.0364	0.0353	0.0334	0.0311	0.0259	0.0212	0.0172	0.0139	0.0075
-6.00	0.0162	0.0172	0.0192	0.0215	0.0231	0.0242	0.0250	0.0247	0.0239	0.0228	0.0200	0.0169	0.0142	0.0118	0.0067
-7.00	0.0130	0.0129	0.0138	0.0154	0.0165	0.0173	0.0180	0.0180	0.0178	0.0172	0.0155	0.0136	0.0117	0.0100	0.0059
-10.00	0.0061	0.0063	0.0066	0.0071	0.0074	0.0076	0.0080	0.0081	0.0081	0.0081	0.0077	0.0071	0.0065	0.0058	0.0038

The distance along axis z has been adjusted to fit the TG-43 U1 source coordinate system (Figure 8.6). Data not shown correspond to points within the encapsulated source. The dose rate constant, Λ, marked in bold is the dose rate per unit source strength at the reference point, $r = 1.0$ cm and $\theta = 90°$ or in Cartesian coordinates $y = 1.0$ cm and $z = 0.0$ cm.

Source: From Wang, R. and Sloboda, R.S., Phys. Med. Biol., 43, 37–48, 1998.

A.4.4 VariSource HDR New ^{192}Ir Source

TABLE A.4.4

Monte Carlo (MC) Calculated Dose Rate Values Per Unit Source Strength as an Along–Away Dose Rate Table (DRT) for the VariSource HDR New ^{192}Ir Source Design

Distance Along z (cm)	Dose Rate Per Unit Source Strength (cGy h^{-1} U^{-1}) Distance Away y (cm)													
	0.00	0.10	0.25	0.50	0.75	1.00	1.50	2.00	2.50	3.00	4.00	5.00	6.00	7.00
7.00	0.014	0.016	0.014	0.017	0.017	0.018	0.018	0.018	0.018	0.017	0.016	0.014	0.012	0.010
6.00	0.020	0.022	0.020	0.023	0.023	0.024	0.025	0.024	0.024	0.023	0.020	0.017	0.014	0.012
5.00	0.028	0.028	0.030	0.035	0.035	0.036	0.036	0.035	0.034	0.031	0.026	0.021	0.017	0.014
4.00	0.042	0.044	0.048	0.053	0.053	0.057	0.056	0.053	0.048	0.043	0.034	0.026	0.020	0.016
3.00	0.071	0.075	0.087	0.098	0.100	0.100	0.095	0.084	0.072	0.061	0.044	0.033	0.024	0.018
2.50	0.101	0.110	0.126	0.144	0.144	0.142	0.126	0.107	0.089	0.073	0.050	0.035	0.026	0.019
2.00	0.154	0.172	0.204	0.229	0.221	0.213	0.175	0.139	0.109	0.086	0.055	0.038	0.027	0.020
1.50	0.275	0.326	0.383	0.412	0.373	0.336	0.245	0.179	0.132	0.099	0.062	0.041	0.029	0.021
1.00	0.646	0.801	0.926	0.877	0.699	0.556	0.342	0.225	0.156	0.113	0.066	0.043	0.030	0.021
0.75	1.209	1.589	1.716	1.394	0.984	0.714	0.398	0.247	0.166	0.119	0.068	0.044	0.030	0.022
0.50	3.375	4.669	3.850	2.288	1.350	0.893	0.448	0.265	0.174	0.122	0.069	0.044	0.031	0.022
0.25	—	29.623	9.600	3.505	1.736	1.044	0.481	0.276	0.179	0.125	0.070	0.045	0.031	0.022
0.10	—	49.775	13.012	4.033	1.873	1.091	0.492	0.281	0.180	0.125	0.070	0.045	0.031	0.022
0.00	—	52.157	13.770	4.147	1.900	1.101	0.496	0.281	0.180	0.125	0.070	0.045	0.031	0.022
-0.10	—	49.859	13.028	4.029	1.872	1.091	0.493	0.281	0.180	0.126	0.070	0.045	0.031	0.022
-0.25	—	29.677	9.596	3.494	1.736	1.037	0.479	0.276	0.178	0.125	0.070	0.044	0.031	0.022

(continued)

TABLE A.4.4—*Continued*

Distance Along z (cm)	Dose Rate Per Unit Source Strength (cGy h^{-1} U^{-1}) Distance Away y (cm)													
	0.00	0.10	0.25	0.50	0.75	1.00	1.50	2.00	2.50	3.00	4.00	5.00	6.00	7.00
−0.50	—	4.710	3.889	2.297	1.358	0.887	0.447	0.264	0.173	0.122	0.069	0.044	0.031	0.022
−0.75	—	1.600	1.727	1.391	0.978	0.718	0.396	0.246	0.166	0.118	0.068	0.043	0.030	0.022
−1.00	—	0.816	0.938	0.878	0.699	0.554	0.344	0.224	0.156	0.112	0.066	0.043	0.030	0.021
−1.50	—	0.306	0.381	0.415	0.373	0.333	0.246	0.180	0.132	0.100	0.061	0.041	0.029	0.021
−2.00	—	0.161	0.200	0.231	0.223	0.209	0.176	0.138	0.109	0.086	0.056	0.038	0.027	0.020
−2.50	—	0.098	0.123	0.146	0.146	0.143	0.126	0.106	0.088	0.073	0.050	0.035	0.026	0.019
−3.00	—	0.068	0.083	0.099	0.101	0.102	0.092	0.083	0.071	0.061	0.044	0.032	0.024	0.018
−4.00	—	0.039	0.045	0.053	0.055	0.057	0.056	0.052	0.048	0.043	0.034	0.026	0.021	0.016
−5.00	—	0.024	0.028	0.033	0.035	0.036	0.036	0.035	0.033	0.031	0.026	0.021	0.017	0.014
−6.00	—	0.017	0.021	0.022	0.022	0.024	0.025	0.025	0.024	0.023	0.020	0.017	0.014	0.012
−7.00	—	0.012	0.015	0.017	0.016	0.018	0.018	0.018	0.018	0.017	0.015	0.014	0.012	0.010

The primary values not included in the publication have been provided by the authors. The distance along axis z has been adjusted to fit the TG-43 U1 source coordinate system (Figure 8.6). Data not shown correspond to points within the encapsulated source. The dose rate constant, Λ, marked in bold is the dose rate per unit source strength at the reference point, $r = 1.0$ cm and $\theta = 90°$ or in Cartesian coordinates $y = 1.0$ cm and $z = 0.0$ cm.

Source: From Angelopoulos, A., Baras, P., Sakelliou, L., Karaiskos, P., and Sandilos, P., *Med. Phys.*, 27, 2521–2527, 2000.

A.4.5 Buchler HDR ^{192}Ir Source

TABLE A.4.5

Monte Carlo (MC) Calculated Dose Rate Values Per Unit Source Strength as an Along–Away Dose Rate Table (DRT) for the Buchler HDR ^{192}Ir Source Design

Distance Along z (cm)	Dose Rate Per Unit Source Strength (cGy h^{-1} U^{-1}) Distance Away y (cm)																	
	0.03	0.20	0.40	0.60	0.80	1.00	1.25	1.50	1.75	2.00	2.50	3.00	3.50	4.00	5.00	6.00	8.00	10.00
10.00	0.0085	0.00861	0.00980	0.00979	0.00977	0.00972	0.00966	0.00963	0.00954	0.00947	0.00928	0.00907	0.00881	0.00851	0.00786	0.00720	0.00586	0.00467
8.00	0.0131	0.0136	0.0156	0.0156	0.0156	0.0155	0.0154	0.0153	0.0151	0.0149	0.0145	0.0140	0.0134	0.0128	0.0115	0.0102	0.00784	0.00595
6.00	0.0279	0.0281	0.0281	0.0282	0.0280	0.0278	0.0274	0.0271	0.0266	0.0260	0.0248	0.0234	0.0219	0.0204	0.0174	0.0147	0.0104	0.00742
5.00	0.0403	0.0405	0.0404	0.0404	0.0401	0.0396	0.0390	0.0383	0.0373	0.0362	0.0339	0.0313	0.0287	0.0262	0.0216	0.0176	0.0119	0.00817
4.00	0.0629	0.0632	0.0630	0.0627	0.0620	0.0610	0.0594	0.0577	0.0555	0.0532	0.0482	0.0433	0.0385	0.0341	0.0267	0.0209	0.0133	0.00890
3.50	0.0818	0.0821	0.0819	0.0813	0.0801	0.0785	0.0760	0.0730	0.0695	0.0660	0.0585	0.0513	0.0447	0.0389	0.0295	0.0227	0.0141	0.00925
3.00	0.113	0.112	0.112	0.110	0.107	0.104	0.0998	0.0944	0.0887	0.0829	0.0714	0.0610	0.0519	0.0443	0.0326	0.0245	0.0147	0.00955
2.50	0.160	0.160	0.159	0.156	0.151	0.145	0.136	0.126	0.116	0.106	0.0878	0.0724	0.0600	0.0500	0.0356	0.0262	0.0154	0.0098
2.00	0.250	0.249	0.246	0.237	0.226	0.211	0.193	0.173	0.154	0.137	0.108	0.0856	0.0687	0.0559	0.0384	0.0277	0.0159	0.0101
1.75	0.325	0.324	0.317	0.302	0.284	0.261	0.233	0.204	0.179	0.156	0.119	0.0926	0.0731	0.0587	0.0398	0.0284	0.0162	0.0102
1.50	0.442	0.440	0.426	0.399	0.367	0.329	0.284	0.243	0.208	0.177	0.131	0.0995	0.0773	0.0614	0.0410	0.0290	0.0164	0.0103
1.25	0.638	0.630	0.599	0.547	0.486	0.421	0.349	0.289	0.240	0.200	0.143	0.106	0.0813	0.0639	0.0421	0.0296	0.0165	0.0103
1.00	0.994	0.971	0.894	0.781	0.658	0.545	0.429	0.341	0.274	0.223	0.155	0.112	0.0849	0.0661	0.0430	0.0300	0.0167	0.0104
0.80	1.558	1.491	1.307	1.072	0.8491	0.669	0.502	0.386	0.302	0.242	0.163	0.117	0.0873	0.0675	0.0436	0.0303	0.0168	0.0104
0.60	2.776	2.561	2.044	1.513	1.0998	0.815	0.579	0.428	0.327	0.257	0.170	0.120	0.0893	0.0687	0.0441	0.0306	0.0169	0.0105
0.40	6.318	5.283	3.400	2.122	1.3846	0.959	0.648	0.465	0.348	0.270	0.175	0.123	0.0906	0.0695	0.0444	0.0307	0.0169	0.0105
0.20	24.78	13.81	5.555	2.780	1.635	1.073	0.698	0.489	0.361	0.278	0.179	0.124	0.0915	0.0700	0.0446	0.0308	0.0169	0.0105
0.00	—	27.62	6.942	3.089	1.738	1.115	0.715	0.497	0.366	0.280	0.180	0.125	0.0918	0.0702	0.0448	0.0308	0.0169	0.0105
−0.20	—	13.74	5.551	2.784	1.640	1.073	0.698	0.489	0.362	0.278	0.179	0.124	0.0915	0.0700	0.0447	0.0308	0.0169	0.0105

(continued)

TABLE A.4.5—*Continued*

Dose Rate Per Unit Source Strength (cGy h^{-1} U^{-1})

Distance Along z (cm)	Distance Away y (cm)																	
	0.03	0.20	0.40	0.60	0.80	1.00	1.25	1.50	1.75	2.00	2.50	3.00	3.50	4.00	5.00	6.00	8.00	10.00
−0.40	—	5.239	3.412	2.124	1.390	0.961	0.649	0.465	0.348	0.270	0.176	0.123	0.0907	0.0695	0.0444	0.0307	0.0169	0.0105
−0.60	—	2.500	2.047	1.514	1.105	0.816	0.580	0.429	0.328	0.258	0.170	0.120	0.0892	0.0687	0.0441	0.0305	0.0169	0.0105
−0.80	—	1.434	1.296	1.075	0.856	0.673	0.504	0.386	0.303	0.242	0.163	0.117	0.0873	0.0676	0.0437	0.0303	0.0168	0.0104
−1.00	—	0.892	0.872	0.777	0.660	0.547	0.431	0.342	0.275	0.224	0.155	0.112	0.0849	0.0660	0.0430	0.0300	0.0167	0.0104
−1.25	—	0.568	0.573	0.539	0.484	0.422	0.350	0.290	0.241	0.201	0.143	0.106	0.0813	0.0639	0.0421	0.0296	0.0166	0.0103
−1.50	—	0.387	0.403	0.390	0.363	0.329	0.285	0.244	0.208	0.178	0.131	0.0996	0.0773	0.0614	0.0410	0.0290	0.0164	0.0103
−1.75	—	0.280	0.297	0.293	0.279	0.260	0.233	0.205	0.180	0.157	0.120	0.0926	0.0731	0.0587	0.0398	0.0284	0.0162	0.0102
−2.00	—	0.209	0.227	0.227	0.220	0.209	0.192	0.173	0.155	0.138	0.108	0.0857	0.0687	0.0558	0.0384	0.0277	0.0159	0.0101
−2.50	—	0.129	0.145	0.146	0.144	0.141	0.134	0.125	0.115	0.106	0.0879	0.0725	0.0600	0.0500	0.0355	0.0261	0.0154	0.00983
−3.00	—	0.0926	0.0982	0.101	0.102	0.101	0.0978	0.0932	0.0881	0.0827	0.0714	0.0610	0.0519	0.0442	0.0325	0.0244	0.0148	0.00955
−3.50	—	0.0639	0.0701	0.0745	0.0752	0.0749	0.0736	0.0713	0.0685	0.0653	0.0583	0.0513	0.0447	0.0389	0.0295	0.0227	0.0141	0.00924
−4.00	—	0.0486	0.0545	0.0570	0.0579	0.0579	0.0573	0.0561	0.0545	0.0525	0.0480	0.0432	0.0384	0.0341	0.0266	0.0209	0.0133	0.00889
−5.00	—	0.0334	0.0345	0.0363	0.0371	0.0372	0.0371	0.0368	0.0363	0.0355	0.0335	0.0311	0.0286	0.0261	0.0215	0.0176	0.0119	0.00815
−6.00	—	0.0203	0.0235	0.0248	0.0255	0.0256	0.0257	0.0257	0.0256	0.0253	0.0243	0.0231	0.0217	0.0203	0.0174	0.0147	0.0104	0.00740
−8.00	—	0.0116	0.0128	0.0135	0.0139	0.0141	0.0143	0.0144	0.0145	0.0144	0.0141	0.0137	0.0132	0.0126	0.0114	0.0102	0.00782	0.00592
−10.00	—	0.00745	0.00796	0.00828	0.00860	0.00859	0.00890	0.00898	0.00902	0.00900	0.00892	0.00878	0.00859	0.00834	0.00777	0.00713	0.00583	0.00465

Data not shown correspond to points within the encapsulated source. The dose rate constant, Λ, marked in bold is the dose rate per unit source strength at the reference point, $r = 1.0$ cm and $\theta = 90°$ or in Cartesian coordinates $y = 1.0$ cm and $z = 0.0$ cm.

Source: From Ballester, F., Pérez-Calatayud, J., Puchades, V., Lluch, J.L., Serrano-Andrés, M.A., Limami, Y., Lliso, F., and Casal, E., *Phys. Med. Biol.,* 46, N79–N90, 2001.

A.4.6 GammaMed HDR ^{192}Ir Source Models 12*i* and Plus

TABLE A.4.6

Monte Carlo (MC) Calculated Dose Rate Values Per Unit Source Strength as an Along–Away Dose Rate Table (DRT) for the GammaMed HDR ^{192}Ir Source Model 12*i* Design

Distance Along z (cm)	Dose Rate Per Unit Source Strength (cGy h^{-1} U^{-1}) Distance Away y (cm)																	
	0.03	0.20	0.40	0.60	0.80	1.00	1.25	1.50	1.75	2.00	2.50	3.00	3.50	4.00	5.00	6.00	8.00	10.00
10.00	0.00827	0.00822	0.00830	0.00835	0.00844	0.00846	0.00853	0.00857	0.00859	0.00859	0.00857	0.00849	0.00833	0.00814	0.00763	0.00705	0.00579	0.00463
8.00	0.0129	0.0128	0.0131	0.0132	0.0134	0.0134	0.0136	0.0136	0.0137	0.0137	0.0136	0.0133	0.0129	0.0123	0.0113	0.0101	0.00780	0.00592
6.00	0.0221	0.0226	0.0230	0.0235	0.0239	0.0242	0.0244	0.0246	0.0245	0.0243	0.0236	0.0226	0.0214	0.0200	0.0172	0.0146	0.0104	0.00740
5.00	0.0321	0.0325	0.0330	0.0339	0.0344	0.0349	0.0352	0.0352	0.0348	0.0343	0.0327	0.0306	0.0283	0.0260	0.0214	0.0176	0.0118	0.00815
4.00	0.0482	0.0497	0.0518	0.0531	0.0543	0.0547	0.0546	0.0541	0.0529	0.0512	0.0473	0.0428	0.0383	0.0339	0.0267	0.0209	0.0133	0.00891
3.50	0.0628	0.0648	0.0675	0.0695	0.0709	0.0713	0.0706	0.0692	0.0669	0.0641	0.0576	0.0509	0.0446	0.0388	0.0295	0.0227	0.0141	0.00925
3.00	0.0819	0.0866	0.0915	0.0949	0.0964	0.0964	0.0943	0.0910	0.0865	0.0815	0.0709	0.0609	0.0520	0.0443	0.0326	0.0245	0.0148	0.00958
2.50	0.118	0.125	0.132	0.137	0.138	0.136	0.131	0.123	0.114	0.105	0.0876	0.0726	0.0601	0.0501	0.0356	0.0262	0.0154	0.00986
2.00	0.182	0.197	0.209	0.215	0.212	0.204	0.189	0.171	0.154	0.137	0.108	0.0859	0.0689	0.0560	0.0386	0.0279	0.0160	0.0101
1.75	0.242	0.257	0.276	0.280	0.271	0.255	0.230	0.204	0.179	0.156	0.120	0.0930	0.0734	0.0590	0.0400	0.0286	0.0163	0.0102
1.50	0.323	0.353	0.380	0.378	0.356	0.325	0.283	0.243	0.208	0.178	0.132	0.100	0.0777	0.0617	0.0412	0.0292	0.0165	0.0103
1.25	0.470	0.522	0.549	0.528	0.479	0.420	0.350	0.290	0.241	0.201	0.144	0.107	0.0818	0.0643	0.0424	0.0298	0.0166	0.0104
1.00	0.740	0.842	0.856	0.771	0.659	0.547	0.432	0.343	0.276	0.225	0.156	0.113	0.0854	0.0665	0.0433	0.0303	0.0168	0.0105
0.80	1.19	1.36	1.30	1.08	0.856	0.673	0.505	0.387	0.304	0.243	0.164	0.118	0.0880	0.0681	0.0440	0.0306	0.0169	0.0105
0.60	2.27	2.52	2.09	1.53	1.11	0.818	0.581	0.430	0.329	0.259	0.172	0.121	0.0900	0.0693	0.0445	0.0308	0.0170	0.0105
0.40	6.15	5.70	3.50	2.14	1.39	0.963	0.651	0.468	0.350	0.272	0.177	0.124	0.0915	0.0702	0.0449	0.0310	0.0171	0.0106
0.20	—	15.1	5.48	2.75	1.63	1.07	0.701	0.493	0.364	0.280	0.181	0.126	0.0924	0.0707	0.0451	0.0311	0.0171	0.0106
0.00	—	23.0	6.60	3.04	1.73	1.118	0.719	0.502	0.369	0.283	0.182	0.126	0.0927	0.0709	0.0451	0.0311	0.0171	0.0106

(continued)

TABLE A.4.6—*Continued*

| | Dose Rate Per Unit Source Strength (cGy h⁻¹ U⁻¹) | | | | | | | | | | | | | | | | | |
| Distance Along z (cm) | Distance Away y (cm) | | | | | | | | | | | | | | | | | |
	0.03	0.20	0.40	0.60	0.80	1.00	1.25	1.50	1.75	2.00	2.50	3.00	3.50	4.00	5.00	6.00	8.00	10.00
−0.20	—	15.2	5.48	2.76	1.63	1.07	0.700	0.492	0.364	0.280	0.180	0.126	0.0923	0.0707	0.0451	0.0311	0.0171	0.0106
−0.40	—	5.71	3.49	2.14	1.39	0.962	0.650	0.467	0.350	0.272	0.177	0.124	0.0914	0.0701	0.0448	0.0310	0.0171	0.0105
−0.60	—	2.48	2.08	1.53	1.11	0.818	0.581	0.431	0.329	0.259	0.171	0.121	0.0898	0.0692	0.0444	0.0308	0.0170	0.0105
−0.80	—	1.33	1.29	1.08	0.855	0.673	0.504	0.387	0.303	0.243	0.164	0.117	0.0877	0.0679	0.0439	0.0305	0.0169	0.0105
−1.00	—	0.810	0.849	0.771	0.657	0.546	0.430	0.342	0.275	0.224	0.156	0.113	0.0853	0.0664	0.0433	0.0302	0.0168	0.0104
−1.25	—	0.493	0.542	0.523	0.472	0.419	0.349	0.289	0.240	0.201	0.144	0.107	0.0816	0.0642	0.0423	0.0297	0.0166	0.0104
−1.50	—	0.329	0.373	0.373	0.352	0.324	0.282	0.242	0.207	0.177	0.132	0.0998	0.0777	0.0617	0.0412	0.0291	0.0164	0.0103
−1.75	—	0.237	0.268	0.277	0.268	0.254	0.229	0.203	0.178	0.156	0.120	0.0926	0.0731	0.0588	0.0398	0.0285	0.0162	0.0102
−2.00	—	0.176	0.202	0.212	0.211	0.203	0.188	0.171	0.153	0.137	0.108	0.0856	0.0687	0.0559	0.0385	0.0278	0.0160	0.0101
−2.50	—	0.108	0.127	0.134	0.137	0.135	0.130	0.123	0.114	0.105	0.0875	0.0724	0.0599	0.0500	0.0356	0.0262	0.0154	0.00984
−3.00	—	0.0728	0.0853	0.0915	0.0943	0.0950	0.0937	0.0899	0.0863	0.0811	0.0706	0.0603	0.0517	0.0441	0.0325	0.0245	0.0148	0.00956
−3.50	—	0.0536	0.0620	0.0666	0.0690	0.0702	0.0702	0.0687	0.0663	0.0635	0.0571	0.0504	0.0442	0.0387	0.0294	0.0227	0.0141	0.00924
−4.00	—	0.0406	0.0467	0.0503	0.0524	0.0535	0.0540	0.0536	0.0523	0.0507	0.0469	0.0424	0.0380	0.0338	0.0265	0.0209	0.0133	0.00890
−5.00	—	0.0264	0.0297	0.0318	0.0330	0.0338	0.0345	0.0347	0.0345	0.0339	0.0323	0.0304	0.0282	0.0258	0.0213	0.0175	0.0118	0.00815
−6.00	—	0.0183	0.0205	0.0218	0.0227	0.0233	0.0238	0.0241	0.0241	0.0239	0.0233	0.0223	0.0212	0.0199	0.0172	0.0145	0.0104	0.00738
−8.00	—	0.0103	0.0114	0.0120	0.0125	0.0128	0.0131	0.0133	0.0134	0.0134	0.0134	0.0131	0.0127	0.0123	0.0112	0.0100	0.00776	0.00589
−10.00	—	0.00684	0.00728	0.00760	0.00780	0.00799	0.00816	0.00826	0.00833	0.00839	0.00842	0.00836	0.00825	0.00806	0.00758	0.00700	0.00578	0.00462

Data not shown correspond to points within the encapsulated source. The dose rate constant, Λ, marked in bold is the dose rate per unit source strength at the reference point, $r = 1.0$ cm and $\theta = 90°$ or in Cartesian coordinates $y = 1.0$ cm and $z = 0.0$ cm.

Source: From Ballester, F., Puchades, V., Lluch, J.L., Serrano-Andrés, M.A., Limami, Y., Pérez-Calatayud, J., and Casal, E., *Med. Phys.*, 28, 2586–2591, 2001.

TABLE A.4.7

Monte Carlo (MC) Calculated Dose Rate Values Per Unit Source Strength as an Along–Away Dose Rate Table (DRT) for the GammaMed HDR ^{192}Ir Source Model Plus Design

Distance Along z (cm)	Dose Rate Per Unit Source Strength (cGy h^{-1} U^{-1}) Distance Away y (cm)																	
	0.03	0.20	0.40	0.60	0.80	1.00	1.25	1.50	1.75	2.00	2.50	3.00	3.50	4.00	5.00	6.00	8.00	10.00
10.00	0.00836	0.00834	0.00841	0.00848	0.00848	0.00856	0.00854	0.00857	0.00860	0.00860	0.00857	0.00849	0.00834	0.00814	0.00765	0.00707	0.00580	0.00464
8.00	0.0129	0.0131	0.0131	0.0132	0.0134	0.0135	0.0136	0.0137	0.0137	0.0137	0.0136	0.0133	0.0129	0.0124	0.0113	0.0101	0.00779	0.00591
6.00	0.0225	0.0227	0.0233	0.0236	0.0239	0.0243	0.0245	0.0247	0.0246	0.0245	0.0237	0.0227	0.0214	0.0201	0.0173	0.0146	0.0104	0.00741
5.00	0.0321	0.0325	0.0335	0.0341	0.0347	0.0352	0.0354	0.0354	0.0351	0.0346	0.0329	0.0308	0.0296	0.0267	0.0215	0.0176	0.0118	0.00818
4.00	0.0488	0.0497	0.0517	0.0533	0.0540	0.0550	0.0550	0.0543	0.0531	0.0515	0.0473	0.0429	0.0389	0.0340	0.0260	0.0209	0.0134	0.00892
3.50	0.0623	0.0643	0.0673	0.0699	0.0712	0.0717	0.0712	0.0695	0.0671	0.0643	0.0577	0.0510	0.0446	0.0383	0.0296	0.0227	0.0141	0.00927
3.00	0.0837	0.0867	0.0919	0.0952	0.0969	0.0970	0.0950	0.0914	0.0875	0.0818	0.0711	0.0610	0.0519	0.0443	0.0326	0.0245	0.0148	0.00956
2.50	0.119	0.125	0.133	0.138	0.139	0.138	0.131	0.124	0.115	0.105	0.0878	0.0727	0.0601	0.0501	0.0357	0.0262	0.0154	0.00986
2.00	0.181	0.197	0.211	0.216	0.214	0.204	0.189	0.172	0.156	0.137	0.108	0.0859	0.0689	0.0560	0.0386	0.0278	0.0160	0.0101
1.75	0.235	0.260	0.279	0.283	0.273	0.258	0.231	0.204	0.179	0.156	0.120	0.0931	0.0735	0.0590	0.0400	0.0285	0.0162	0.0102
1.50	0.322	0.356	0.382	0.379	0.357	0.326	0.284	0.244	0.208	0.178	0.132	0.100	0.0777	0.0617	0.0412	0.0292	0.0165	0.0103
1.25	0.454	0.526	0.553	0.528	0.478	0.420	0.350	0.290	0.241	0.201	0.144	0.107	0.0817	0.0642	0.0423	0.0297	0.0167	0.0104
1.00	0.716	0.845	0.861	0.773	0.656	0.546	0.432	0.343	0.276	0.225	0.156	0.113	0.0854	0.0665	0.0433	0.0302	0.0168	0.0104
0.80	1.19	1.37	1.30	1.08	0.854	0.674	0.506	0.387	0.304	0.243	0.164	0.118	0.0879	0.0681	0.0440	0.0306	0.0169	0.0105
0.60	2.23	2.52	2.10	1.53	1.11	0.818	0.583	0.431	0.330	0.259	0.172	0.121	0.0900	0.0693	0.0445	0.0308	0.0170	0.0105
0.40	6.05	5.68	3.51	2.14	1.39	0.963	0.652	0.467	0.351	0.272	0.177	0.124	0.0915	0.0702	0.0449	0.0310	0.0170	0.0106
0.20	—	15.0	5.50	2.76	1.64	1.08	0.702	0.493	0.365	0.280	0.180	0.126	0.0923	0.0707	0.0451	0.0311	0.0171	0.0106
0.00	—	23.2	6.63	3.05	1.73	1.118	0.720	0.501	0.370	0.283	0.182	0.126	0.0926	0.0707	0.0451	0.0311	0.0171	0.0106
−0.20	—	15.1	5.50	2.77	1.64	1.08	0.701	0.493	0.365	0.280	0.180	0.124	0.0913	0.0701	0.0448	0.0310	0.0171	0.0105
−0.40	—	5.70	3.51	2.15	1.40	0.965	0.652	0.468	0.351	0.272	0.177	0.121	0.0899	0.0693	0.0445	0.0308	0.0170	0.0105
−0.60	—	2.52	2.09	1.53	1.11	0.820	0.582	0.431	0.330	0.259	0.171	0.118	0.0878	0.0693	0.0445	0.0306	0.0170	0.0105
−0.80	—	1.36	1.30	1.08	0.857	0.676	0.506	0.388	0.304	0.243	0.164	0.113	0.0853	0.0665	0.0440	0.0302	0.0169	0.0105
−1.00	—	0.833	0.858	0.776	0.659	0.549	0.431	0.343	0.276	0.225	0.156	0.113	0.0853	0.0665	0.0433	0.0298	0.0168	0.0104
−1.25	—	0.512	0.552	0.531	0.481	0.421	0.350	0.290	0.241	0.201	0.144	0.107	0.0817	0.0643	0.0424	0.0292	0.0166	0.0104
−1.50	—	0.343	0.381	0.380	0.359	0.328	0.284	0.244	0.208	0.178	0.132	0.1001	0.0776	0.0618	0.0413	0.0287	0.0164	0.0103

(continued)

TABLE A.4.7—*Continued*

| | Dose Rate Per Unit Source Strength (cGy h^{-1} U^{-1}) | | | | | | | | | | | | | | | | | |
| | Distance Away y (cm) | | | | | | | | | | | | | | | | | |
Distance Along z (cm)	0.03	0.20	0.40	0.60	0.80	1.00	1.25	1.50	1.75	2.00	2.50	3.00	3.50	4.00	5.00	6.00	8.00	10.00
-1.75	—	0.246	0.276	0.283	0.274	0.257	0.231	0.204	0.179	0.157	0.120	0.0929	0.0733	0.0589	0.0400	0.0285	0.0162	0.0102
-2.00	—	0.184	0.208	0.216	0.214	0.205	0.189	0.171	0.154	0.137	0.108	0.0859	0.0689	0.0561	0.0386	0.0278	0.0160	0.0101
-2.50	—	0.114	0.130	0.137	0.139	0.137	0.132	0.123	0.115	0.106	0.0879	0.0726	0.0601	0.0501	0.0356	0.0262	0.0154	0.00985
-3.00	—	0.0772	0.0885	0.0941	0.0965	0.0970	0.0953	0.0915	0.0869	0.0820	0.0710	0.0608	0.0518	0.0442	0.0326	0.0245	0.0148	0.00958
-3.50	—	0.0563	0.0641	0.0685	0.0706	0.0713	0.0711	0.0695	0.0672	0.0645	0.0577	0.0510	0.0446	0.0389	0.0295	0.0227	0.0141	0.00925
-4.00	—	0.0428	0.0488	0.0519	0.0537	0.0546	0.0549	0.0544	0.0531	0.0513	0.0474	0.0428	0.0382	0.0340	0.0266	0.0209	0.0133	0.00891
-5.00	—	0.0269	0.0308	0.0328	0.0340	0.0348	0.0353	0.0354	0.0351	0.0345	0.0329	0.0308	0.0284	0.0260	0.0215	0.0176	0.0118	0.00816
-6.00	—	0.0192	0.0211	0.0225	0.0234	0.0239	0.0243	0.0245	0.0245	0.0243	0.0237	0.0226	0.0215	0.0201	0.0173	0.0146	0.0104	0.00740
-8.00	—	0.0110	0.0120	0.0125	0.0129	0.0133	0.0135	0.0135	0.0136	0.0136	0.0135	0.0133	0.0129	0.0124	0.0113	0.0101	0.00779	0.00591
-10.00	—	0.00706	0.00751	0.00777	0.00802	0.00816	0.00834	0.00842	0.00848	0.00852	0.00855	0.00846	0.00833	0.00814	0.00765	0.00706	0.00581	0.00464

Data not shown correspond to points within the encapsulated source. The dose rate constant, Λ, marked in bold is the dose rate per unit source strength at the reference point, $r = 1.0$ cm and $\theta = 90°$ or in Cartesian coordinates $y = 1.0$ cm and $z = 0.0$ cm.

Source: From Ballester, F., Puchades, V., Lluch, J.L., Serrano-Andrés, M.A., Limami, Y., Pérez-Calatayud, J., and Casal, E., *Med. Phys.*, 28, 2586–2591, 2001.

A.4.7 BEBIG MultiSource HDR ^{192}Ir Source Model GI192M11

TABLE A.4.8

Monte Carlo (MC) Calculated Dose Rate Values Per Unit Source Strength as an Along–Away Dose Rate Table (DRT) for the BEBIG MultiSource HDR ^{192}Ir Source Model GI192M11 Design

Distance Along z (cm)	Dose Rate Per Unit Source Strength (cGy h^{-1} U^{-1}) Distance Away y (cm)														
	0.0	0.25	0.5	0.75	1.0	1.5	2.0	2.5	3.0	4.0	5.0	6.0	8.0	10.0	14.0
14.0	0.00384	0.00393	0.00398	0.00397	0.00398	0.00400	0.00400	0.00400	0.00399	0.00394	0.00380	0.00363	0.00322	0.00276	0.00192
10.0	0.00800	0.00813	0.00825	0.00830	0.00832	0.00844	0.00847	0.00848	0.00839	0.00806	0.00757	0.00699	0.00574	0.00459	0.00284
8.0	0.01234	0.01280	0.01296	0.01308	0.01323	0.01350	0.01353	0.01344	0.01316	0.01231	0.01118	0.01001	0.00772	0.00586	0.00337
6.0	0.0215	0.0224	0.0230	0.0234	0.0238	0.0243	0.0241	0.0234	0.0224	0.01984	0.01708	0.01449	0.01028	0.00733	0.00392
5.0	0.0304	0.0320	0.0328	0.0337	0.0345	0.0349	0.0340	0.0325	0.0304	0.0257	0.0213	0.01741	0.01174	0.00810	0.00416
4.0	0.0462	0.0493	0.0513	0.0529	0.0541	0.0536	0.0509	0.0469	0.0424	0.0337	0.0264	0.0207	0.01322	0.00883	0.00439
3.0	0.0795	0.0866	0.0918	0.0951	0.0955	0.0903	0.0808	0.0703	0.0602	0.0438	0.0323	0.0243	0.01465	0.00950	0.00458
2.5	0.1116	0.1243	0.1333	0.1372	0.1352	0.1222	0.1042	0.0868	0.0718	0.0496	0.0353	0.0260	0.01530	0.00978	0.00466
2.0	0.1715	0.1957	0.211	0.212	0.203	0.1698	0.1357	0.1072	0.0850	0.0555	0.0382	0.0276	0.01585	0.01002	0.00473
1.5	0.304	0.357	0.378	0.360	0.322	0.241	0.1761	0.1306	0.0991	0.0612	0.0409	0.0289	0.01632	0.01022	0.00477
1.0	0.692	0.845	0.813	0.680	0.542	0.340	0.223	0.1542	0.1120	0.0659	0.0430	0.0299	0.01665	0.01036	0.00481
0.75	1.289	1.546	1.304	0.966	0.703	0.395	0.245	0.1645	0.1173	0.0677	0.0437	0.0303	0.01678	0.01042	0.00482
0.5	3.310	3.430	2.190	1.355	0.885	0.446	0.263	0.1727	0.1215	0.0690	0.0443	0.0306	0.01687	0.01045	0.00483
0.25	—	9.210	3.510	1.763	1.043	0.483	0.276	0.1779	0.1240	0.0699	0.0446	0.0308	0.01691	0.01047	0.00484
0.0	—	15.500	4.300	1.954	1.108	0.497	0.280	0.1797	0.1249	0.0702	0.0447	0.0308	0.01694	0.01048	0.00484
−0.25	—	9.210	3.510	1.764	1.044	0.483	0.276	0.1779	0.1240	0.0699	0.0445	0.0308	0.01693	0.01047	0.00484

(continued)

TABLE A.4.8—*Continued*

Dose Rate Per Unit Source Strength (cGy h^{-1} U^{-1})

Distance Along z (cm)	Distance Away y (cm)														
	0.0	0.25	0.5	0.75	1.0	1.5	2.0	2.5	3.0	4.0	5.0	6.0	8.0	10.0	14.0
−0.5	—	3.440	2.190	1.356	0.885	0.446	0.263	0.1727	0.1215	0.0691	0.0443	0.0306	0.01686	0.01044	0.00484
−0.75	—	1.547	1.307	0.966	0.703	0.395	0.245	0.1646	0.1173	0.0677	0.0437	0.0303	0.01679	0.01040	0.00482
−1.0	—	0.838	0.815	0.681	0.542	0.340	0.223	0.1542	0.1121	0.0659	0.0429	0.0300	0.01668	0.01035	0.00481
−1.5	—	0.344	0.377	0.360	0.323	0.241	0.1762	0.1305	0.0990	0.0612	0.0409	0.0289	0.01632	0.01020	0.00477
−2.0	—	0.1825	0.209	0.212	0.203	0.1700	0.1357	0.1072	0.0851	0.0556	0.0382	0.0276	0.01585	0.01003	0.00472
−2.5	—	0.1122	0.1306	0.1363	0.1352	0.1223	0.1043	0.0868	0.0719	0.0496	0.0353	0.0260	0.01529	0.00977	0.00466
−3.0	—	0.0758	0.0887	0.0940	0.0953	0.0905	0.0809	0.0703	0.0602	0.0438	0.0322	0.0243	0.01464	0.00951	0.00458
−4.0	—	0.0412	0.0484	0.0519	0.0534	0.0536	0.0509	0.0468	0.0424	0.0336	0.0264	0.0208	0.01323	0.00884	0.00439
−5.0	—	0.0261	0.0304	0.0326	0.0338	0.0346	0.0340	0.0325	0.0304	0.0257	0.0212	0.01741	0.01173	0.00810	0.00416
−6.0	—	0.01813	0.0209	0.0223	0.0232	0.0240	0.0240	0.0234	0.0224	0.01985	0.01708	0.01448	0.01028	0.00733	0.00391
−8.0	—	0.01029	0.01159	0.01228	0.01271	0.01321	0.01341	0.01336	0.01313	0.01229	0.01118	0.00999	0.00772	0.00586	0.00338
−10.0	—	0.00666	0.00729	0.00769	0.00793	0.00823	0.00837	0.00840	0.00834	0.00803	0.00757	0.00698	0.00575	0.00459	0.00284
−14.0	—	0.00335	0.00353	0.00367	0.00376	0.00385	0.00392	0.00396	0.00397	0.00390	0.00379	0.00362	0.00321	0.00276	0.00192

Data not shown correspond to points within the encapsulated source. The dose rate constant, Λ, marked in bold is the dose rate per unit source strength at the reference point, $r = 1.0$ cm and $\theta = 90°$ or in Cartesian coordinates $y = 1.0$ cm and $z = 0.0$ cm.

Source: From Granero, D., Pérez-Calatayud, J., and Ballester, F., *Radiother. Oncol.*, 76, 79–85, 2005.

A.4.8 Ralstron HDR ^{60}Co Source Models Type 1, 2, and 3

TABLE A.4.9

Monte Carlo (MC) Calculated Dose Rate Values Per Unit Source Strength as an Along–Away Dose Rate Table (DRT) for the Ralstron Remote HDR Afterloader Type 1 ^{60}Co Source Design

Distance Along z (cm)	Dose Rate Per Unit Source Strength (cGy h^{-1} U^{-1}) Distance Away y (cm)											
	0.50	0.75	1.00	1.50	2.00	2.50	3.00	4.00	5.00	6.00	7.00	10.00
7.00	0.017	0.017	0.018	0.018	0.018	0.017	0.016	0.015	0.013	0.011	0.009	0.006
6.00	0.027	0.024	0.025	0.024	0.024	0.023	0.022	0.019	0.016	0.013	0.011	0.006
5.00	0.036	0.037	0.037	0.036	0.034	0.032	0.029	0.024	0.020	0.016	0.013	0.007
4.00	0.059	0.060	0.059	0.055	0.051	0.046	0.041	0.032	0.024	0.019	0.015	0.008
3.00	0.109	0.108	0.107	0.093	0.081	0.069	0.058	0.041	0.030	0.022	0.017	0.008
2.50	0.168	0.162	0.153	0.127	0.105	0.085	0.069	0.046	0.033	0.024	0.018	0.008
2.00	0.274	0.257	0.232	0.177	0.135	0.104	0.081	0.052	0.035	0.025	0.019	0.009
1.50	0.514	0.440	0.372	0.247	0.172	0.125	0.094	0.057	0.037	0.026	0.019	0.009
1.00	1.281	0.846	0.600	0.331	0.211	0.145	0.104	0.061	0.039	0.027	0.020	0.009
0.75	2.033	1.122	0.726	0.371	0.228	0.153	0.109	0.063	0.040	0.028	0.020	0.009
0.50	2.616	1.316	0.820	0.403	0.241	0.159	0.112	0.064	0.040	0.028	0.020	0.009
0.25	2.423	1.354	0.867	0.422	0.250	0.163	0.114	0.065	0.041	0.028	0.020	0.009
0.10	2.217	1.348	0.874	0.428	0.253	0.165	0.114	0.065	0.041	0.028	0.020	0.009
0.00	2.179	1.334	0.878	0.430	0.254	0.165	0.115	0.065	0.041	0.028	0.020	0.009
−0.10	2.229	1.347	0.873	0.427	0.252	0.165	0.115	0.065	0.041	0.028	0.020	0.009
−0.25	2.410	1.360	0.865	0.424	0.250	0.164	0.114	0.065	0.041	0.028	0.020	0.009

(continued)

TABLE A.4.9—*Continued*

Distance Along z (cm)	Dose Rate Per Unit Source Strength (cGy h^{-1} U^{-1}) Distance Away y (cm)											
	0.50	0.75	1.00	1.50	2.00	2.50	3.00	4.00	5.00	6.00	7.00	10.00
−0.50	2.619	1.316	0.824	0.402	0.242	0.159	0.112	0.064	0.041	0.028	0.020	0.009
−0.75	2.031	1.120	0.727	0.373	0.229	0.154	0.109	0.063	0.040	0.028	0.020	0.009
−1.00	1.298	0.852	0.601	0.332	0.212	0.145	0.105	0.061	0.039	0.027	0.020	0.009
−1.50	0.516	0.440	0.374	0.246	0.173	0.125	0.093	0.057	0.037	0.026	0.019	0.009
−2.00	0.275	0.257	0.234	0.177	0.136	0.105	0.081	0.052	0.035	0.025	0.019	0.009
−2.50	0.162	0.162	0.154	0.128	0.105	0.085	0.069	0.047	0.032	0.024	0.018	0.008
−3.00	0.105	0.108	0.107	0.094	0.082	0.069	0.058	0.041	0.030	0.022	0.017	0.008
−4.00	0.052	0.057	0.058	0.056	0.051	0.046	0.041	0.032	0.024	0.019	0.015	0.008
−5.00	0.030	0.034	0.035	0.035	0.034	0.032	0.029	0.024	0.020	0.016	0.013	0.007
−6.00	0.020	0.022	0.023	0.024	0.024	0.023	0.022	0.019	0.016	0.013	0.011	0.006
−7.00	0.014	0.016	0.016	0.017	0.017	0.017	0.016	0.015	0.013	0.011	0.009	0.006

Data not shown correspond to points within the encapsulated source. The dose rate constant, Λ, marked in bold is the dose rate per unit source strength at the reference point, $r = 1.0$ cm and $\theta = 90°$ or in Cartesian coordinates $y = 1.0$ cm and $z = 0.0$ cm.

Source: From Papagiannis, P., Angelopoulos, A., Pantelis, E., Sakelliou, L., Karaiskos, P., and Shimizu, Y., *Med. Phys.*, 29, 712–721, 2003.

TABLE A.4.10

Monte Carlo (MC) Calculated Dose Rate Values Per Unit Source Strength as an Along–Away Dose Rate Table (DRT) for the Ralstron Remote HDR Afterloader Type 2 ^{60}Co Source Design

Distance Along z (cm)	Dose Rate Per Unit Source Strength (cGy h^{-1} U^{-1}) Distance Away y (cm)											
	0.50	0.75	1.00	1.50	2.00	2.50	3.00	4.00	5.00	6.00	7.00	10.00
7.00	0.019	0.019	0.019	0.018	0.017	0.016	0.016	0.014	0.013	0.011	0.009	0.006
6.00	0.026	0.026	0.025	0.025	0.023	0.022	0.021	0.018	0.015	0.013	0.011	0.006
5.00	0.038	0.038	0.037	0.035	0.033	0.031	0.029	0.024	0.019	0.015	0.013	0.007
4.00	0.060	0.059	0.057	0.054	0.049	0.045	0.040	0.031	0.024	0.019	0.014	0.008
3.00	0.106	0.105	0.100	0.090	0.078	0.067	0.057	0.040	0.029	0.022	0.016	0.008
2.50	0.153	0.151	0.139	0.121	0.101	0.083	0.067	0.046	0.032	0.023	0.017	0.008
2.00	0.239	0.231	0.206	0.168	0.131	0.101	0.080	0.051	0.035	0.025	0.018	0.008
1.50	0.411	0.389	0.324	0.237	0.169	0.124	0.092	0.056	0.037	0.026	0.019	0.009
1.00	0.843	0.754	0.539	0.330	0.213	0.146	0.105	0.061	0.039	0.027	0.019	0.009
0.75	1.326	1.104	0.692	0.384	0.235	0.156	0.110	0.062	0.040	0.027	0.020	0.009
0.50	2.196	1.631	0.871	0.434	0.252	0.163	0.113	0.064	0.040	0.028	0.020	0.009
0.25	3.529	2.273	1.026	0.469	0.265	0.169	0.116	0.064	0.041	0.028	0.020	0.009
0.10	4.223	2.547	1.079	0.480	0.268	0.170	0.117	0.065	0.041	0.028	0.020	0.009
0.00	4.382	2.609	**1.101**	0.483	0.269	0.171	0.118	0.065	0.041	0.028	0.020	0.009
−0.10	4.218	2.543	1.080	0.481	0.268	0.170	0.117	0.064	0.041	0.028	0.020	0.009
−0.25	3.526	2.273	1.025	0.469	0.264	0.168	0.116	0.064	0.041	0.028	0.020	0.009
−0.50	2.190	1.637	0.872	0.433	0.252	0.163	0.114	0.064	0.040	0.028	0.020	0.009
−0.75	1.329	1.104	0.695	0.384	0.234	0.156	0.110	0.062	0.040	0.027	0.020	0.009
−1.00	0.841	0.755	0.538	0.330	0.213	0.146	0.105	0.061	0.039	0.027	0.019	0.009
−1.50	0.401	0.387	0.325	0.236	0.170	0.123	0.093	0.056	0.037	0.026	0.019	0.009
−2.00	0.223	0.224	0.205	0.168	0.131	0.102	0.080	0.051	0.035	0.025	0.018	0.009

(continued)

TABLE A.4.10—*Continued*

Dose Rate Per Unit Source Strength (cGy h^{-1} U^{-1})

Distance Along z (cm)	Distance Away y (cm)											
	0.50	0.75	1.00	1.50	2.00	2.50	3.00	4.00	5.00	6.00	7.00	10.00
−2.50	0.139	0.143	0.139	0.121	0.101	0.083	0.068	0.046	0.032	0.023	0.017	0.008
−3.00	0.093	0.097	0.097	0.089	0.078	0.067	0.056	0.040	0.029	0.022	0.017	0.008
−4.00	0.052	0.052	0.054	0.053	0.049	0.045	0.040	0.031	0.024	0.019	0.015	0.008
−5.00	0.032	0.033	0.033	0.034	0.033	0.031	0.028	0.024	0.019	0.016	0.013	0.007
−6.00	0.021	0.022	0.022	0.023	0.023	0.022	0.021	0.018	0.016	0.013	0.011	0.006
−7.00	0.015	0.016	0.016	0.017	0.016	0.016	0.016	0.014	0.012	0.011	0.009	0.006

Data not shown correspond to points within the encapsulated source. The dose rate constant, Λ, marked in bold is the dose rate per unit source strength at the reference point, $r = 1.0$ cm and $\theta = 90°$ or in Cartesian coordinates $y = 1.0$ cm and $z = 0.0$ cm.

Source: From Papagiannis, P., Angelopoulos, A., Pantelis, E., Sakelliou, L., Karaiskos, P., and Shimizu, Y., *Med. Phys.*, 29, 712–721, 2003.

TABLE A.4.11

Monte Carlo (MC) Calculated Dose Rate Values Per Unit Source Strength as an Along–Away Dose Rate Table (DRT) for the Ralstron Remote HDR Afterloader Type 3 ^{60}Co Source Design

Distance Along z (cm)	Dose Rate Per Unit Source Strength (cGy h^{-1} U^{-1}) Distance Away y (cm)											
	0.50	0.75	1.00	1.50	2.00	2.50	3.00	4.00	5.00	6.00	7.00	10.00
7.00	0.018	0.018	0.018	0.018	0.018	0.017	0.016	0.014	0.013	0.011	0.009	0.006
6.00	0.025	0.025	0.025	0.025	0.024	0.023	0.022	0.019	0.016	0.013	0.011	0.006
5.00	0.037	0.037	0.038	0.037	0.035	0.032	0.029	0.024	0.020	0.016	0.013	0.007
4.00	0.060	0.061	0.060	0.057	0.052	0.046	0.041	0.032	0.024	0.019	0.015	0.008
3.00	0.115	0.114	0.110	0.097	0.083	0.070	0.058	0.041	0.029	0.022	0.017	0.008
2.50	0.178	0.169	0.159	0.132	0.107	0.085	0.070	0.046	0.032	0.023	0.018	0.008
2.00	0.299	0.273	0.245	0.186	0.138	0.105	0.081	0.052	0.035	0.025	0.018	0.009
1.50	0.601	0.490	0.396	0.256	0.175	0.125	0.093	0.056	0.037	0.026	0.019	0.009
1.00	1.508	0.918	0.624	0.337	0.212	0.144	0.104	0.060	0.039	0.027	0.020	0.009
0.75	2.269	1.157	0.728	0.372	0.227	0.152	0.108	0.062	0.039	0.027	0.020	0.009
0.50	2.479	1.256	0.789	0.397	0.238	0.157	0.111	0.063	0.040	0.027	0.020	0.009
0.25	2.016	1.218	0.805	0.414	0.245	0.161	0.113	0.063	0.040	0.028	0.020	0.009
0.10	1.819	1.181	0.807	0.417	0.248	0.162	0.114	0.064	0.040	0.027	0.020	0.009
0.00	1.778	1.177	**0.800**	0.417	0.248	0.162	0.114	0.064	0.040	0.028	0.020	0.009
−0.10	1.816	1.184	0.804	0.419	0.247	0.162	0.114	0.064	0.040	0.028	0.020	0.009
−0.25	2.020	1.218	0.805	0.412	0.245	0.160	0.113	0.064	0.040	0.027	0.020	0.009
−0.50	2.480	1.258	0.790	0.399	0.238	0.158	0.112	0.063	0.040	0.027	0.020	0.009
−0.75	2.266	1.160	0.732	0.373	0.227	0.152	0.109	0.062	0.040	0.027	0.020	0.009
−1.00	1.505	0.919	0.620	0.338	0.211	0.144	0.104	0.060	0.039	0.027	0.020	0.009
−1.50	0.610	0.490	0.394	0.256	0.174	0.124	0.093	0.056	0.037	0.026	0.019	0.009
−2.00	0.296	0.277	0.245	0.184	0.138	0.105	0.081	0.051	0.035	0.025	0.018	0.009

(continued)

TABLE A.4.11—*Continued*

Distance Along z (cm)	Dose Rate Per Unit Source Strength (cGy h⁻¹ U⁻¹) Distance Away y (cm)											
	0.50	0.75	1.00	1.50	2.00	2.50	3.00	4.00	5.00	6.00	7.00	10.00
−2.50	0.168	0.170	0.162	0.133	0.107	0.086	0.069	0.046	0.032	0.023	0.018	0.008
−3.00	0.107	0.111	0.111	0.098	0.083	0.070	0.058	0.041	0.029	0.022	0.017	0.008
−4.00	0.053	0.057	0.059	0.057	0.052	0.047	0.041	0.032	0.024	0.019	0.015	0.008
−5.00	0.030	0.034	0.035	0.036	0.035	0.032	0.029	0.024	0.020	0.016	0.013	0.007
−6.00	0.019	0.022	0.023	0.024	0.024	0.023	0.022	0.019	0.016	0.013	0.011	0.006
−7.00	0.013	0.015	0.016	0.017	0.017	0.017	0.016	0.015	0.013	0.011	0.009	0.006

Data not shown correspond to points within the encapsulated source. The dose rate constant, Λ, marked in bold is the dose rate per unit source strength at the reference point, $r = 1.0$ cm and $\theta = 90°$ or in Cartesian coordinates $y = 1.0$ cm and $z = 0.0$ cm.

Source: From Papagiannis, P., Angelopoulos, A., Pantelis, E., Sakelliou, L., Karaiskos, P., and Shimizu, Y., *Med. Phys.*, 29, 712–721, 2003.

A.4.9 BEBIG MultiSource HDR ^{60}Co Source Model GK60M21

TABLE A.4.12

Monte Carlo (MC) Calculated Dose Rate Values Per Unit Source Strength as an Along–Away Dose Rate Table (DRT) for the BEBIG MultiSource HDR ^{60}Co Source Model GK60M21 Design

| Distance Along z (cm) | Dose Rate Per Unit Source Strength (cGy h^{-1} U^{-1}) Distance Away y (cm) | | | | | | | | | | | | | | |
	0.0	0.25	0.5	0.75	1.0	1.5	2.0	2.5	3.0	4.0	5.0	6.0	8.0	10.0	14.0
14.0	0.00406	0.00410	0.00411	0.00412	0.00410	0.00409	0.00406	0.00401	0.00398	0.00385	0.00368	0.00349	0.00307	0.00262	0.00185
10.0	0.00874	0.00882	0.00881	0.00878	0.00872	0.00867	0.00860	0.00841	0.00824	0.00774	0.00717	0.00653	0.00529	0.00421	0.00263
8.0	0.01417	0.01430	0.01428	0.01422	0.01420	0.01398	0.01369	0.01335	0.01287	0.01175	0.01047	0.00923	0.00702	0.00531	0.00309
6.0	0.0259	0.0264	0.0262	0.0261	0.0260	0.0253	0.0244	0.0232	0.0218	0.01874	0.01586	0.01326	0.00926	0.00658	0.00354
5.0	0.0380	0.0387	0.0384	0.0382	0.0378	0.0365	0.0344	0.0321	0.0294	0.0242	0.01964	0.01589	0.01055	0.00724	0.00375
4.0	0.0606	0.0612	0.0608	0.0602	0.0591	0.0557	0.0511	0.0458	0.0407	0.0315	0.0243	0.01888	0.01186	0.00787	0.00394
3.0	0.1105	0.1105	0.1094	0.1070	0.1034	0.0925	0.0801	0.0681	0.0574	0.0409	0.0296	0.0220	0.01311	0.00844	0.00410
2.5	0.1613	0.1604	0.1579	0.1528	0.1445	0.1238	0.1026	0.0837	0.0683	0.0462	0.0324	0.0236	0.01366	0.00868	0.00417
2.0	0.254	0.252	0.246	0.232	0.213	0.1701	0.1324	0.1027	0.0806	0.0516	0.0350	0.0250	0.01414	0.00889	0.00422
1.5	0.459	0.452	0.427	0.383	0.331	0.238	0.1707	0.1246	0.0935	0.0568	0.0374	0.0262	0.01456	0.00906	0.00427
1.0	1.061	1.020	0.878	0.700	0.544	0.332	0.214	0.1466	0.1055	0.0611	0.0393	0.0271	0.01486	0.00918	0.00430
0.75	1.979	1.808	1.381	0.982	0.698	0.384	0.235	0.1562	0.1105	0.0628	0.0400	0.0274	0.01498	0.00924	0.00431
0.5	4.740	3.830	2.250	1.362	0.872	0.433	0.252	0.1638	0.1143	0.0640	0.0405	0.0277	0.01506	0.00926	0.00431
0.25	—	9.450	3.560	1.753	1.021	0.467	0.264	0.1688	0.1167	0.0648	0.0409	0.0279	0.01512	0.00929	0.00432
0.0	—	15.100	4.310	1.923	1.084	0.480	0.268	0.1706	0.1176	0.0651	0.0409	0.0279	0.01512	0.00930	0.00433
− 0.25	—	9.440	3.560	1.756	1.021	0.467	0.264	0.1689	0.1167	0.0648	0.0408	0.0278	0.01510	0.00928	0.00433
− 0.5	—	3.800	2.260	1.360	0.871	0.432	0.252	0.1639	0.1143	0.0640	0.0405	0.0277	0.01506	0.00925	0.00432

(continued)

TABLE A.4.12—*Continued*

Distance Along z (cm)	Dose Rate Per Unit Source Strength (cGy h^{-1} U^{-1}) Distance Away y (cm)														
	0.0	0.25	0.5	0.75	1.0	1.5	2.0	2.5	3.0	4.0	5.0	6.0	8.0	10.0	14.0
−0.75	—	1.798	1.378	0.982	0.698	0.384	0.235	0.1563	0.1105	0.0628	0.0400	0.0275	0.01497	0.00924	0.00431
−1.0	—	1.018	0.878	0.701	0.544	0.332	0.214	0.1467	0.1055	0.0611	0.0393	0.0271	0.01486	0.00918	0.00430
−1.5	—	0.444	0.428	0.383	0.332	0.239	0.1707	0.1246	0.0935	0.0568	0.0374	0.0262	0.01456	0.00906	0.00427
−2.0	—	0.242	0.245	0.232	0.213	0.1703	0.1326	0.1027	0.0805	0.0516	0.0350	0.0250	0.01415	0.00890	0.00423
−2.5	—	0.1491	0.1563	0.1528	0.1446	0.1239	0.1026	0.0838	0.0682	0.0462	0.0324	0.0236	0.01366	0.00868	0.00417
−3.0	—	0.0999	0.1074	0.1067	0.1033	0.0926	0.0801	0.0682	0.0575	0.0409	0.0296	0.0220	0.01310	0.00844	0.00410
−4.0	—	0.0524	0.0583	0.0594	0.0589	0.0557	0.0510	0.0459	0.0407	0.0315	0.0243	0.01887	0.01186	0.00788	0.00394
−5.0	—	0.0316	0.0358	0.0372	0.0374	0.0364	0.0344	0.0320	0.0294	0.0242	0.01967	0.01588	0.01055	0.00724	0.00376
−6.0	—	0.0207	0.0239	0.0251	0.0255	0.0253	0.0244	0.0232	0.0218	0.01877	0.01587	0.01328	0.00928	0.00658	0.00354
−8.0	—	0.01055	0.01252	0.01328	0.01366	0.01383	0.01366	0.01333	0.01285	0.01172	0.01047	0.00924	0.00704	0.00530	0.00309
−10.0	—	0.00634	0.00747	0.00802	0.00825	0.00847	0.00849	0.00837	0.00823	0.00774	0.00715	0.00654	0.00530	0.00422	0.00263
−14.0	—	0.00296	0.00336	0.00362	0.00376	0.00390	0.00394	0.00396	0.00393	0.00383	0.00368	0.00348	0.00306	0.00262	0.00185

Data not shown correspond to points within the encapsulated source. The dose rate constant, Λ, marked in bold is the dose rate per unit source strength at the reference point, $r = 1.0$ cm and $\theta = 90°$ or in Cartesian coordinates $y = 1.0$ cm and $z = 0.0$ cm.

Source: From Ballester, F., Granero, D., Pérez-Calatayud, J., Casal, E., Agramunt, A., and Cases, R., *Phys. Med. Biol.*, 50, N309–N316, 2005.

A.4.10 MicroSelectron PDR Old and New ^{192}Ir Source Designs

TABLE A.4.13

Monte Carlo (MC) Calculated Dose Rate Values Per Unit Source Strength as an Along–Away Dose Rate Table (DRT) for the MicroSelectron PDR Old Design ^{192}Ir Source

Distance Along z (cm)	Dose Rate Per Unit Source Strength (cGy h^{-1} U^{-1}) Distance Away y (cm)													
	0.00	0.10	0.25	0.50	0.75	1.00	1.50	2.00	2.50	3.00	4.00	5.00	6.00	7.00
7.00	0.0217	0.0217	0.0217	0.0216	0.0215	0.0213	0.0207	0.0200	0.0192	0.0183	0.0162	0.0140	0.0119	0.0101
6.00	0.0300	0.0300	0.0300	0.0299	0.0296	0.0292	0.0283	0.0271	0.0256	0.0241	0.0207	0.0174	0.0144	0.0120
5.00	0.0438	0.0438	0.0438	0.0435	0.0429	0.0422	0.0403	0.0380	0.0353	0.0324	0.0266	0.0216	0.0174	0.0141
4.00	0.0690	0.0690	0.0689	0.0681	0.0669	0.0652	0.0610	0.0558	0.0502	0.0446	0.0345	0.0267	0.0208	0.0163
3.00	0.1229	0.1230	0.1225	0.1202	0.1165	0.1117	0.0997	0.0865	0.0738	0.0624	0.0448	0.0326	0.0243	0.0185
2.50	0.1764	0.1765	0.1754	0.1707	0.1634	0.1540	0.1321	0.1100	0.0902	0.0740	0.0505	0.0357	0.0260	0.0195
2.00	0.2750	0.2749	0.2719	0.2609	0.2436	0.2232	0.1798	0.1412	0.1103	0.0869	0.0564	0.0386	0.0276	0.0205
1.50	0.4873	0.4865	0.4764	0.4430	0.3958	0.3439	0.2501	0.1808	0.1331	0.1005	0.0619	0.0412	0.0289	0.0212
1.00	1.0930	1.0867	1.0347	0.8889	0.7145	0.5604	0.3476	0.2265	0.1562	0.1132	0.0666	0.0432	0.0299	0.0218
0.75	1.9403	1.9134	1.7621	1.3701	0.9940	0.7191	0.4021	0.2485	0.1663	0.1185	0.0683	0.0440	0.0303	0.0220
0.50	4.3590	4.2047	3.5403	2.2318	1.3810	0.9014	0.4529	0.2670	0.1744	0.1226	0.0696	0.0445	0.0305	0.0221
0.25	17.513	15.158	8.9081	3.5887	1.8022	1.0638	0.4895	0.2791	0.1796	0.1252	0.0704	0.0448	0.0307	0.0222
0.10	110.78	56.163	15.450	4.3226	1.9680	1.1177	0.5004	0.2826	0.1811	0.1260	0.0707	0.0449	0.0307	0.0222
0.00	—	113.74	17.980	4.4985	2.0023	1.1280	0.5024	0.2832	0.1814	0.1261	0.0707	0.0449	0.0307	0.0222
−0.10	—	53.842	15.347	4.3127	1.9640	1.1156	0.4997	0.2823	0.1811	0.1260	0.0707	0.0449	0.0307	0.0222
−0.25	13.683	12.717	8.5435	3.5490	1.7903	1.0588	0.4878	0.2784	0.1795	0.1253	0.0705	0.0448	0.0307	0.0222
−0.50	2.9015	3.1072	3.1057	2.1445	1.3533	0.8892	0.4493	0.2657	0.1743	0.1228	0.0697	0.0445	0.0305	0.0221
−0.75	1.2985	1.3263	1.4481	1.2559	0.9569	0.7015	0.3966	0.2464	0.1660	0.1188	0.0683	0.0439	0.0303	0.0219

(continued)

TABLE A.4.13—*Continued*

Distance Along z (cm)	Dose Rate Per Unit Source Strength (cGy h^{-1} U^{-1}) Distance Away y (cm)													
	0.00	0.10	0.25	0.50	0.75	1.00	1.50	2.00	2.50	3.00	4.00	5.00	6.00	7.00
−1.00	0.7355	0.7469	0.8140	0.7824	0.6660	0.5395	0.3405	0.2237	0.1553	0.1132	0.0666	0.0432	0.0299	0.0217
−1.50	0.3343	0.3375	0.3532	0.3689	0.3509	0.3176	0.2408	0.1765	0.1312	0.0996	0.0616	0.0411	0.0289	0.0212
−2.00	0.1923	0.1938	0.1977	0.2094	0.2085	0.1990	0.1683	0.1358	0.1075	0.0853	0.0558	0.0383	0.0275	0.0204
−2.50	0.1252	0.1262	0.1275	0.1339	0.1361	0.1337	0.1209	0.1036	0.0869	0.0719	0.0497	0.0353	0.0258	0.0194
−3.00	0.0884	0.0892	0.0896	0.0930	0.0954	0.0951	0.0894	0.0801	0.0699	0.0601	0.0438	0.0322	0.0241	0.0184
−4.00	0.0510	0.0513	0.0515	0.0523	0.0535	0.0541	0.0531	0.0502	0.0463	0.0420	0.0333	0.0260	0.0204	0.0161
−5.00	0.0333	0.0335	0.0336	0.0337	0.0342	0.0345	0.0345	0.0335	0.0318	0.0298	0.0252	0.0208	0.0169	0.0138
−6.00	0.0228	0.0229	0.0230	0.0230	0.0232	0.0235	0.0237	0.0234	0.0227	0.0217	0.0192	0.0165	0.0139	0.0116
−7.00	0.0165	0.0165	0.0166	0.0166	0.0167	0.0168	0.0171	0.0170	0.0167	0.0162	0.0148	0.0131	0.0114	0.0097

The distance along axis z has been adjusted to fit the TG-43 U1 source coordinate system (Figure 8.6). Data not shown correspond to points within the encapsulated source. The dose rate constant, Λ, marked in bold is the dose rate per unit source strength at the reference point, $r = 1.0$ cm and $\theta = 90°$ or in Cartesian coordinates $y = 1.0$ cm and $z = 0.0$ cm.

Source: From Williamson, J.F. and Li, Z., *Med. Phys.*, 22, 809–819, 1995.

TABLE A.4.14

Monte Carlo (MC) Calculated Dose Rate Values Per Unit Source Strength as an Along–Away Dose Rate Table (DRT) for the MicroSelectron PDR New Design ^{192}Ir Source

Distance Along z (cm)	Dose Rate Per Unit Source Strength (cGy h^{-1} U^{-1}) Distance Away y (cm)													
	0.00	0.10	0.25	0.50	0.75	1.00	1.50	2.00	2.50	3.00	4.00	5.00	6.00	7.00
7.00	0.021	0.020	0.020	0.021	0.021	0.021	0.020	0.019	0.018	0.018	0.016	0.014	0.012	0.010
6.00	0.027	0.028	0.028	0.027	0.027	0.027	0.027	0.026	0.024	0.023	0.020	0.017	0.014	0.012
5.00	0.039	0.041	0.040	0.041	0.040	0.040	0.038	0.036	0.034	0.031	0.026	0.021	0.017	0.014
4.00	0.061	0.065	0.063	0.063	0.063	0.062	0.058	0.053	0.048	0.043	0.034	0.027	0.021	0.016
3.00	0.108	0.114	0.114	0.112	0.109	0.105	0.095	0.084	0.072	0.061	0.044	0.032	0.024	0.018
2.50	0.156	0.162	0.164	0.159	0.154	0.146	0.128	0.107	0.088	0.072	0.050	0.035	0.026	0.019
2.00	0.242	0.254	0.251	0.245	0.234	0.214	0.174	0.138	0.108	0.086	0.056	0.038	0.027	0.020
1.50	0.431	0.458	0.445	0.418	0.381	0.334	0.246	0.178	0.131	0.099	0.061	0.041	0.029	0.021
1.00	0.960	1.006	0.977	0.855	0.706	0.552	0.343	0.224	0.155	0.113	0.066	0.043	0.030	0.021
0.75	1.723	1.798	1.662	1.321	0.977	0.708	0.398	0.247	0.165	0.118	0.067	0.043	0.030	0.022
0.50	3.917	4.017	3.400	2.189	1.372	0.894	0.450	0.264	0.173	0.121	0.069	0.044	0.030	0.022
0.25	16.310	15.307	8.778	3.552	1.783	1.055	0.485	0.277	0.178	0.124	0.070	0.044	0.031	0.022
0.10	—	59.317	15.270	4.276	1.954	1.109	0.497	0.280	0.179	0.125	0.070	0.044	0.031	0.022
0.00	—	106.25	17.740	4.460	1.995	1.121	0.500	0.282	0.180	0.125	0.070	0.045	0.031	0.022
-0.10	—	59.405	15.324	4.272	1.958	1.107	0.496	0.280	0.179	0.124	0.070	0.044	0.031	0.022
-0.25	—	14.973	8.790	3.570	1.796	1.055	0.485	0.277	0.178	0.124	0.070	0.044	0.030	0.022
-0.50	—	3.687	3.348	2.188	1.362	0.896	0.450	0.265	0.174	0.121	0.069	0.044	0.030	0.022
-0.75	—	1.535	1.608	1.318	0.979	0.711	0.399	0.245	0.165	0.118	0.068	0.043	0.030	0.022
-1.00	—	0.794	0.918	0.844	0.707	0.550	0.343	0.224	0.155	0.113	0.066	0.043	0.029	0.022
-1.50	—	0.311	0.395	0.402	0.374	0.332	0.245	0.178	0.132	0.099	0.061	0.041	0.029	0.021
-2.00	—	0.160	0.213	0.230	0.227	0.210	0.174	0.138	0.108	0.086	0.055	0.038	0.027	0.020
-2.50	—	0.094	0.132	0.146	0.149	0.144	0.126	0.106	0.088	0.073	0.050	0.035	0.026	0.019

(continued)

TABLE A.4.14—*Continued*

Dose Rate Per Unit Source Strength (cGy h^{-1} U^{-1})

Distance Along z (cm)	Distance Away y (cm)													
	0.00	0.10	0.25	0.50	0.75	1.00	1.50	2.00	2.50	3.00	4.00	5.00	6.00	7.00
−3.00	—	0.062	0.088	0.100	0.102	0.101	0.093	0.083	0.072	0.061	0.044	0.032	0.024	0.018
−4.00	—	0.033	0.047	0.055	0.057	0.059	0.057	0.053	0.048	0.043	0.034	0.026	0.020	0.016
−5.00	—	0.021	0.029	0.034	0.036	0.037	0.037	0.035	0.033	0.031	0.026	0.021	0.017	0.014
−6.00	—	0.014	0.019	0.023	0.024	0.026	0.025	0.025	0.024	0.023	0.020	0.017	0.014	0.012
−7.00	—	0.011	0.014	0.017	0.017	0.018	0.019	0.018	0.018	0.017	0.016	0.014	0.012	0.010

Data not shown correspond to points within the encapsulated source. The dose rate constant, Λ, marked in bold is the dose rate per unit source strength at the reference point, $r = 1.0$ cm and $\theta = 90°$ or in Cartesian coordinates $y = 1.0$ cm and $z = 0.0$ cm.

Source: From Karaiskos, P., Angelopoulos, A., Pantelis, E., Papagiannis, P., Sakelliou, L., Kouwenhoven, E., and Baltas, D., *Med. Phys.*, 30, 9–16, 2003.

A.4.11   GammaMed PDR ^{192}Ir Source Models 12*i* and Plus

TABLE A.4.15

Monte Carlo (MC) Calculated Dose Rate Values Per Unit Source Strength as an Along–Away Dose Rate Table (DRT) for the GammaMed PDR ^{192}Ir Source Model 12*i* Design

Distance Along z (cm)	Dose Rate Per Unit Source Strength (cGy h^{-1} U^{-1}) Distance Away y (cm)																	
	0.00	0.20	0.40	0.60	0.80	1.00	1.25	1.50	1.75	2.00	2.50	3.00	3.50	4.00	5.00	6.00	8.00	10.00
10.00	0.0103	0.0103	0.0103	0.0103	0.0102	0.0102	0.0102	0.0101	0.0100	0.00993	0.00974	0.00951	0.00924	0.00893	0.00824	0.00751	0.00608	0.00481
8.00	0.0165	0.0165	0.0165	0.0165	0.0164	0.0163	0.0162	0.0161	0.0159	0.0157	0.0153	0.0147	0.0141	0.0134	0.0120	0.0106	0.00809	0.00609
6.00	0.0299	0.0299	0.0299	0.0298	0.0295	0.0293	0.0290	0.0286	0.0281	0.0275	0.0261	0.0246	0.0229	0.0213	0.0181	0.0152	0.0107	0.00756
5.00	0.0433	0.0434	0.0431	0.0428	0.0425	0.0421	0.0414	0.0405	0.0395	0.0383	0.0357	0.0329	0.0300	0.0272	0.0223	0.0181	0.0121	0.00832
4.00	0.0680	0.0678	0.0674	0.0667	0.0658	0.0648	0.0631	0.0611	0.0587	0.0561	0.0506	0.0451	0.0399	0.0352	0.0274	0.0214	0.0136	0.00905
3.50	0.0880	0.0881	0.0874	0.0863	0.0851	0.0832	0.0804	0.0770	0.0733	0.0692	0.0611	0.0532	0.0462	0.0400	0.0302	0.0231	0.0143	0.00938
3.00	0.120	0.120	0.118	0.117	0.114	0.111	0.106	0.0997	0.0933	0.0868	0.0743	0.0630	0.0534	0.0453	0.0332	0.0249	0.0150	0.00970
2.50	0.172	0.171	0.169	0.165	0.160	0.153	0.143	0.133	0.121	0.110	0.0909	0.0745	0.0614	0.0510	0.0361	0.0265	0.0156	0.0100
2.00	0.268	0.266	0.261	0.252	0.239	0.224	0.202	0.181	0.160	0.142	0.111	0.0875	0.0700	0.0568	0.0390	0.0281	0.0161	0.0102
1.75	0.348	0.346	0.337	0.322	0.301	0.276	0.244	0.213	0.185	0.161	0.122	0.0943	0.0743	0.0596	0.0403	0.0288	0.0164	0.0103
1.50	0.474	0.468	0.452	0.425	0.387	0.346	0.296	0.252	0.213	0.181	0.134	0.101	0.0785	0.0623	0.0416	0.0294	0.0166	0.0104
1.25	0.682	0.668	0.636	0.578	0.509	0.440	0.361	0.297	0.245	0.204	0.146	0.108	0.0824	0.0648	0.0426	0.0299	0.0168	0.0105
1.00	1.06	1.030	0.951	0.821	0.685	0.564	0.441	0.349	0.279	0.227	0.157	0.114	0.0859	0.0668	0.0435	0.0304	0.0169	0.0105
0.80	1.66	1.590	1.390	1.120	0.879	0.688	0.513	0.392	0.306	0.244	0.165	0.118	0.0882	0.0683	0.0441	0.0307	0.0170	0.0106
0.60	2.95	2.730	2.150	1.560	1.120	0.830	0.588	0.434	0.331	0.260	0.172	0.121	0.0901	0.0694	0.0446	0.0309	0.0171	0.0106
0.40	6.66	5.530	3.500	2.160	1.410	0.973	0.656	0.469	0.351	0.272	0.177	0.124	0.0914	0.0701	0.0449	0.0311	0.0171	0.0106
0.20	—	14.000	5.610	2.810	1.660	1.080	0.703	0.492	0.364	0.280	0.180	0.125	0.0922	0.0706	0.0451	0.0312	0.0172	0.0106
0.00	—	28.200	7.000	3.110	1.750	**1.120**	0.719	0.500	0.368	0.283	0.181	0.126	0.0924	0.0708	0.0452	0.0312	0.0172	0.0106
−0.20	—	13.900	5.600	2.810	1.650	1.080	0.702	0.492	0.364	0.280	0.180	0.125	0.0922	0.0706	0.0451	0.0312	0.0172	0.0106
−0.40	—	5.480	3.490	2.160	1.410	0.971	0.655	0.469	0.351	0.272	0.177	0.124	0.0914	0.0701	0.0449	0.0311	0.0171	0.0106

(continued)

TABLE A.4.15—*Continued*

Dose Rate Per Unit Source Strength (cGy h^{-1} U^{-1})

Distance Along z (cm)	Distance Away y (cm)																	
	0.00	0.20	0.40	0.60	0.80	1.00	1.25	1.50	1.75	2.00	2.50	3.00	3.50	4.00	5.00	6.00	8.00	10.00
-0.60	—	2.700	2.130	1.550	1.120	0.828	0.587	0.433	0.331	0.260	0.172	0.121	0.0900	0.0693	0.0446	0.0309	0.0171	0.0106
-0.80	—	1.560	1.380	1.110	0.875	0.685	0.512	0.391	0.306	0.244	0.165	0.118	0.0882	0.0682	0.0441	0.0307	0.0170	0.0106
-1.00	—	0.998	0.942	0.816	0.681	0.561	0.439	0.347	0.278	0.226	0.157	0.114	0.0858	0.0668	0.0435	0.0304	0.0169	0.0105
-1.25	—	0.638	0.628	0.574	0.506	0.437	0.360	0.296	0.244	0.203	0.145	0.108	0.0823	0.0647	0.0426	0.0299	0.0168	0.0105
-1.50	—	0.437	0.445	0.420	0.384	0.343	0.294	0.250	0.212	0.181	0.133	0.101	0.0784	0.0622	0.0415	0.0294	0.0166	0.0104
-1.75	—	0.317	0.329	0.319	0.298	0.274	0.242	0.212	0.184	0.160	0.122	0.0941	0.0742	0.0596	0.0403	0.0287	0.0164	0.0103
-2.00	—	0.237	0.252	0.249	0.237	0.222	0.201	0.179	0.159	0.141	0.110	0.0873	0.0699	0.0567	0.0390	0.0281	0.0161	0.0102
-2.50	—	0.148	0.161	0.162	0.158	0.152	0.142	0.132	0.121	0.110	0.0903	0.0742	0.0612	0.0509	0.0361	0.0265	0.0156	0.00996
-3.00	—	0.101	0.111	0.113	0.112	0.109	0.105	0.0989	0.0926	0.0862	0.0738	0.0627	0.0532	0.0452	0.0331	0.0248	0.0150	0.00969
-3.50	—	0.0725	0.0803	0.0828	0.0830	0.0818	0.0794	0.0763	0.0726	0.0687	0.0606	0.0529	0.0460	0.0399	0.0301	0.0231	0.0143	0.00938
-4.00	—	0.0548	0.0608	0.0632	0.0637	0.0634	0.0622	0.0604	0.0581	0.0556	0.0502	0.0449	0.0397	0.0351	0.0273	0.0213	0.0136	0.00904
-5.00	—	0.0338	0.0381	0.0399	0.0407	0.0408	0.0405	0.0399	0.0390	0.0378	0.0354	0.0327	0.0299	0.0271	0.0221	0.0180	0.0121	0.00831
-6.00	—	0.0230	0.0258	0.0272	0.0279	0.0282	0.0283	0.0280	0.0276	0.0271	0.0258	0.0243	0.0228	0.0211	0.0179	0.0151	0.0106	0.00755
-8.00	—	0.0126	0.0141	0.0148	0.0153	0.0155	0.0157	0.0156	0.0156	0.0155	0.0151	0.0146	0.0140	0.0133	0.0119	0.0105	0.00803	0.00607
-10.00	—	0.00806	0.00871	0.00916	0.00941	0.00959	0.00968	0.00973	0.00971	0.00971	0.00958	0.00938	0.00913	0.00884	0.00818	0.00746	0.00603	0.00478

Data not shown correspond to points within the encapsulated source. The dose rate constant, Λ, marked in bold is the dose rate per unit source strength at the reference point, $r = 1.0$ cm and $\theta = 90°$ or in Cartesian coordinates $y = 1.0$ cm and $z = 0.0$ cm.

Source: From Pérez-Calatayud, J., Ballester, F., Serrano-Andrés, M.A., Puchades, V., Lluch, J.L., Limami, Y., and Casal, E., *Med. Phys.*, 28, 2576–2585, 2001.

TABLE A.4.16

Monte Carlo (MC) Calculated Dose Rate Values Per Unit Source Strength as an Along–Away Dose Rate Table (DRT) for the GammaMed PDR ^{192}Ir Source Model *Plus* Design

| Distance Along z (cm) | Dose Rate Per Unit Source Strength (cGy h^{-1} U^{-1}) | | | | | | | | | | | | | | | | | |
| | Distance Away y (cm) | | | | | | | | | | | | | | | | | |
	0.00	0.20	0.40	0.60	0.80	1.00	1.25	1.50	1.75	2.00	2.50	3.00	3.50	4.00	5.00	6.00	8.00	10.00
10.00	0.0103	0.0104	0.0104	0.0103	0.0103	0.0103	0.0102	0.0102	0.0101	0.0100	0.00983	0.00957	0.00930	0.00898	0.00828	0.00754	0.00609	0.00482
8.00	0.0166	0.0166	0.0167	0.0166	0.0165	0.0165	0.0163	0.0162	0.0161	0.0159	0.0154	0.0148	0.0142	0.0135	0.0121	0.0107	0.00810	0.00610
6.00	0.0298	0.0302	0.0301	0.0300	0.0298	0.0296	0.0292	0.0288	0.0283	0.0277	0.0263	0.0247	0.0230	0.0213	0.0181	0.0152	0.0107	0.00757
5.00	0.0432	0.0435	0.0434	0.0433	0.0429	0.0425	0.0417	0.0408	0.0398	0.0386	0.0359	0.0330	0.0301	0.0273	0.0223	0.0181	0.0121	0.00832
4.00	0.0675	0.0680	0.0676	0.0670	0.0664	0.0654	0.0636	0.0614	0.0591	0.0563	0.0508	0.0453	0.0401	0.0353	0.0274	0.0214	0.0136	0.00905
3.50	0.0879	0.0889	0.0882	0.0873	0.0860	0.0841	0.0811	0.0775	0.0737	0.0696	0.0612	0.0534	0.0463	0.0401	0.0302	0.0231	0.0143	0.00939
3.00	0.120	0.121	0.120	0.118	0.115	0.112	0.106	0.100	0.0938	0.0871	0.0744	0.0631	0.0535	0.0453	0.0332	0.0249	0.0150	0.00970
2.50	0.172	0.173	0.171	0.168	0.162	0.155	0.144	0.133	0.122	0.111	0.0910	0.0746	0.0615	0.0510	0.0361	0.0265	0.0156	0.0100
2.00	0.268	0.268	0.264	0.255	0.242	0.225	0.203	0.181	0.161	0.142	0.111	0.0876	0.0701	0.0568	0.0390	0.0280	0.0161	0.0102
1.75	0.351	0.349	0.341	0.325	0.303	0.277	0.245	0.213	0.185	0.161	0.122	0.0944	0.0743	0.0596	0.0403	0.0287	0.0164	0.0103
1.50	0.477	0.473	0.457	0.428	0.390	0.348	0.297	0.252	0.214	0.182	0.134	0.101	0.0785	0.0622	0.0415	0.0294	0.0166	0.0104
1.25	0.685	0.677	0.643	0.582	0.511	0.440	0.362	0.297	0.245	0.204	0.146	0.108	0.0824	0.0647	0.0426	0.0299	0.0168	0.0105
1.00	1.07	1.050	0.959	0.825	0.687	0.564	0.441	0.349	0.279	0.227	0.157	0.114	0.0859	0.0669	0.0435	0.0303	0.0169	0.0105
0.80	1.66	1.610	1.400	1.120	0.881	0.688	0.513	0.392	0.306	0.245	0.165	0.118	0.0882	0.0683	0.0441	0.0307	0.0170	0.0106
0.60	2.94	2.760	2.150	1.560	1.130	0.829	0.588	0.433	0.331	0.260	0.172	0.121	0.0900	0.0694	0.0445	0.0309	0.0171	0.0106
0.40	6.64	5.570	3.510	2.160	1.410	0.971	0.655	0.469	0.351	0.272	0.177	0.124	0.0914	0.0702	0.0449	0.0310	0.0171	0.0106
0.20	—	14.000	5.640	2.810	1.650	1.080	0.703	0.493	0.364	0.280	0.180	0.125	0.0922	0.0706	0.0451	0.0311	0.0171	0.0106
0.00	—	28.200	7.000	3.110	1.750	1.120	0.718	0.500	0.368	0.283	0.181	0.126	0.0924	0.0708	0.0451	0.0312	0.0172	0.0106
-0.20	—	14.000	5.610	2.810	1.650	1.080	0.703	0.493	0.364	0.280	0.180	0.125	0.0921	0.0706	0.0451	0.0311	0.0172	0.0106
-0.40	—	5.510	3.500	2.160	1.400	0.970	0.654	0.469	0.351	0.272	0.177	0.124	0.0914	0.0701	0.0445	0.0310	0.0171	0.0106
-0.60	—	2.720	2.140	1.550	1.120	0.827	0.587	0.433	0.330	0.260	0.172	0.121	0.0899	0.0693	0.0444	0.0309	0.0171	0.0106
-0.80	—	1.580	1.380	1.120	0.875	0.686	0.512	0.391	0.306	0.244	0.165	0.118	0.0881	0.0682	0.0441	0.0307	0.0170	0.0106
-1.00	—	1.010	0.947	0.818	0.682	0.562	0.440	0.348	0.279	0.227	0.157	0.114	0.0857	0.0668	0.0435	0.0304	0.0169	0.0105
-1.25	—	0.645	0.633	0.577	0.507	0.438	0.360	0.296	0.245	0.204	0.145	0.108	0.0823	0.0647	0.0426	0.0299	0.0168	0.0105
-1.50	—	0.445	0.449	0.423	0.385	0.345	0.295	0.251	0.213	0.181	0.134	0.101	0.0784	0.0623	0.0415	0.0294	0.0166	0.0104

(continued)

TABLE A.4.16—*Continued*

Dose Rate Per Unit Source Strength (cGy h^{-1} U^{-1})

Distance Along z (cm)	Distance Away y (cm)																	
	0.00	0.20	0.40	0.60	0.80	1.00	1.25	1.50	1.75	2.00	2.50	3.00	3.50	4.00	5.00	6.00	8.00	10.00
−1.75	—	0.323	0.334	0.321	0.300	0.275	0.243	0.212	0.185	0.160	0.122	0.0942	0.0742	0.0596	0.0403	0.0288	0.0164	0.0103
−2.00	—	0.245	0.256	0.251	0.239	0.223	0.202	0.180	0.160	0.141	0.111	0.0873	0.0699	0.0568	0.0389	0.0281	0.0161	0.0102
−2.50	—	0.153	0.164	0.164	0.160	0.153	0.143	0.132	0.121	0.110	0.0905	0.0743	0.0613	0.0509	0.0361	0.0265	0.0156	0.00996
−3.00	—	0.104	0.113	0.115	0.113	0.110	0.105	0.0993	0.0930	0.0865	0.0740	0.0628	0.0533	0.0452	0.0331	0.0248	0.0150	0.00968
−3.50	—	0.0751	0.0820	0.0843	0.0843	0.0830	0.0800	0.0767	0.0730	0.0690	0.0608	0.0530	0.0461	0.0399	0.0301	0.0231	0.0143	0.00938
−4.00	—	0.0570	0.0626	0.0645	0.0648	0.0642	0.0627	0.0607	0.0584	0.0559	0.0504	0.0450	0.0398	0.0351	0.0273	0.0213	0.0135	0.00904
−5.00	—	0.0362	0.0395	0.0409	0.0414	0.0414	0.0410	0.0402	0.0392	0.0381	0.0355	0.0328	0.0299	0.0272	0.0222	0.0180	0.0121	0.00831
−6.00	—	0.0248	0.0270	0.0280	0.0284	0.0287	0.0286	0.0283	0.0278	0.0273	0.0260	0.0245	0.0229	0.0212	0.0180	0.0151	0.0106	0.00755
−8.00	—	0.0134	0.0147	0.0154	0.0156	0.0158	0.0159	0.0158	0.0156	0.0155	0.0152	0.0147	0.0141	0.0134	0.0120	0.0106	0.00805	0.00608
−10.00	—	0.00861	0.00913	0.00949	0.00961	0.00973	0.00982	0.00984	0.00982	0.00978	0.00965	0.00945	0.00919	0.00889	0.00821	0.00747	0.00605	0.00479

Data not shown correspond to points within the encapsulated source. The dose rate constant, Λ, marked in bold is the dose rate per unit source strength at the reference point, $r = 1.0$ cm and $\theta = 90°$ or in Cartesian coordinates $y = 1.0$ cm and $z = 0.0$ cm.

Source: From Pérez-Calatayud, J., Ballester, F., Serrano-Andrés, M.A., Puchades, V., Lluch, J.L., Limami, Y., and Casal, E., *Med. Phys.*, 28, 2576–2585, 2001.

A.4.12   LDR ^{137}Cs Source Models

TABLE A.4.17

Monte Carlo (MC) Calculated Dose Rate Values Per Unit Source
Strength as a Radial Dose Rate Table (DRT) for the Selectron
LDR Single Pellet

Radial Distance r (cm)	Dose Rate Per Unit Source Strength (cGy h^{-1} U^{-1})
0.25	18.10
0.50	4.459
0.75	1.973
1.00	**1.107**
1.50	0.4893
2.00	0.2738
2.50	0.1742
3.00	0.1203
4.00	0.06687
5.00	0.04220
6.00	0.02886
7.00	0.02086
8.00	0.01568
9.00	0.01215
10.00	0.009628
12.00	0.006369
14.00	0.004433
16.00	0.003188
18.00	0.002350
20.00	0.001763

The dose rate constant, Λ, marked in bold is the dose rate per unit source
strength at the reference point, $r = 1.0$ cm and $\theta = 90°$. Due to the spherical
symmetry of this source only the radial distance r is needed.
Source: From Pérez-Calatayud, J., Granero, D., Ballester, F., Puchades, V.,
and Casal, E., *Med. Phys.*, 31, 493–499, 2004.

TABLE A.4.18

Monte Carlo (MC) Calculated Dose Rate Values Per Unit Source Strength as an Along–Away Dose Rate Table (DRT) for the Amersham CDC Type Source CDC-1 (1 × ^{137}Cs bead)

Distance Along z (cm)	Dose Rate Per Unit Source Strength (cGy h^{-1} U^{-1}) Distance Away y (cm)																	
	0.00	0.20	0.40	0.60	0.80	1.00	1.25	1.50	1.75	2.00	2.50	3.00	3.50	4.00	5.00	6.00	8.00	10.00
10.00	0.00923	0.00925	0.00925	0.00922	0.00920	0.00916	0.00911	0.00905	0.00897	0.00889	0.00873	0.00850	0.00824	0.00797	0.0074	0.00673	0.00546	0.00434
8.00	0.01497	0.01503	0.01501	0.01498	0.01492	0.01482	0.01470	0.01455	0.01438	0.01419	0.01374	0.01323	0.01267	0.01207	0.0108	0.00955	0.00729	0.00550
6.00	0.02749	0.02754	0.02744	0.02732	0.02715	0.02691	0.02655	0.02611	0.02559	0.02505	0.02376	0.02234	0.02083	0.01933	0.0164	0.01377	0.00965	0.00684
5.00	0.0401	0.0402	0.0400	0.0397	0.0394	0.0389	0.0382	0.0373	0.0362	0.0352	0.0327	0.0301	0.02748	0.02496	0.0204	0.01651	0.01099	0.00753
4.00	0.0636	0.0635	0.0631	0.0625	0.0616	0.0605	0.0587	0.0566	0.0543	0.0520	0.0468	0.0417	0.0369	0.0325	0.0252	0.01963	0.01237	0.00820
3.50	0.0837	0.0834	0.0829	0.0816	0.0799	0.0779	0.0750	0.0719	0.0682	0.0646	0.0569	0.0496	0.0430	0.0372	0.0279	0.02129	0.01305	0.00852
3.00	0.1146	0.1142	0.1127	0.1106	0.1077	0.1043	0.0991	0.0935	0.0875	0.0815	0.0696	0.0591	0.0499	0.0423	0.0308	0.02291	0.01368	0.00881
2.50	0.1659	0.1648	0.1619	0.1575	0.1517	0.1452	0.1356	0.1253	0.1147	0.1044	0.0857	0.0702	0.0577	0.0478	0.0336	0.02452	0.01426	0.00906
2.00	0.2599	0.2579	0.2512	0.2410	0.2282	0.2130	0.1925	0.1722	0.1528	0.1351	0.1054	0.0829	0.0661	0.0534	0.0364	0.02597	0.01479	0.00929
1.75	0.339	0.336	0.325	0.309	0.2879	0.2640	0.2333	0.2039	0.1772	0.1537	0.1163	0.0896	0.0703	0.0561	0.0377	0.02665	0.01501	0.00938
1.50	0.463	0.456	0.437	0.407	0.371	0.332	0.2851	0.2427	0.2055	0.1745	0.1278	0.0963	0.0744	0.0588	0.0388	0.02725	0.01520	0.00946
1.25	0.666	0.656	0.615	0.557	0.491	0.425	0.350	0.2876	0.2369	0.1967	0.1394	0.1027	0.0782	0.0611	0.0399	0.02779	0.01538	0.00953
1.00	1.045	1.012	0.922	0.796	0.666	0.549	0.429	0.338	0.2704	0.2194	0.1506	0.1087	0.0816	0.0632	0.0408	0.02822	0.01552	0.00959
0.80	1.636	1.555	1.350	1.093	0.860	0.673	0.502	0.382	0.2974	0.2369	0.1588	0.1129	0.0840	0.0647	0.0414	0.02853	0.01562	0.00963
0.60	2.917	2.673	2.105	1.534	1.108	0.815	0.576	0.423	0.322	0.2525	0.1658	0.1163	0.0859	0.0658	0.0418	0.02875	0.01569	0.00966
0.40	6.610	5.450	3.470	2.143	1.392	0.958	0.644	0.459	0.343	0.2648	0.1710	0.1188	0.0872	0.0666	0.0422	0.02893	0.01575	0.00968
0.20	—	13.860	5.610	2.799	1.641	1.070	0.693	0.484	0.356	0.2729	0.1745	0.1205	0.0882	0.0672	0.0424	0.02903	0.01577	0.00970
0.00	—	29.090	7.030	3.110	1.746	**1.113**	0.711	0.493	0.361	0.2755	0.1757	0.1210	0.0884	0.0673	0.0425	0.02908	0.01580	0.00970

Data not shown correspond to points within the encapsulated source. The dose rate constant, Λ, marked in bold is the dose rate per unit source strength at the reference point, $r = 1.0$ cm and $\theta = 90°$ or in Cartesian coordinates $y = 1.0$ cm and $z = 0.0$ cm.

Source: From Pérez-Calatayud, J., Ballester, F., Serrano-Andrés, M.A., Llich, J.L., Puchades, V., Limami, Y., and Casal, E., *Med. Phys.*, 29, 538–543, 2002.

TABLE A.4.19

Monte Carlo (MC) Calculated Dose Rate Values Per Unit Source Strength as an Along–Away Dose Rate Table (DRT) for the Amersham CDC Type Source CDC-3 (3 × ^{137}Cs bead)

Dose Rate Per Unit Source Strength (cGy h^{-1} U^{-1})

Distance Along z (cm)	Distance Away y (cm)																	
	0.00	0.20	0.40	0.60	0.80	1.00	1.25	1.50	1.75	2.00	2.50	3.00	3.50	4.00	5.00	6.00	8.00	10.00
10.00	0.00937	0.00950	0.00948	0.00946	0.00942	0.00939	0.00932	0.00927	0.00918	0.00908	0.00888	0.00864	0.00837	0.00808	0.00744	0.00678	0.00548	0.00434
8.00	0.01537	0.01541	0.01535	0.01531	0.01522	0.01513	0.01500	0.01485	0.01466	0.01444	0.01397	0.01344	0.01285	0.01220	0.01089	0.00960	0.00730	0.00550
6.00	0.02823	0.02827	0.02813	0.02800	0.02749	0.02749	0.02711	0.02660	0.02606	0.02546	0.02409	0.02261	0.02105	0.01947	0.01647	0.01381	0.00965	0.00684
5.00	0.0414	0.0414	0.0411	0.0408	0.0403	0.0398	0.0390	0.0379	0.0369	0.0357	0.0331	0.0304	0.02770	0.02509	0.02039	0.01653	0.01098	0.00753
4.00	0.0656	0.0656	0.0650	0.0642	0.0631	0.0618	0.0598	0.0576	0.0551	0.0526	0.0472	0.0420	0.0371	0.0326	0.02519	0.01963	0.01235	0.00819
3.50	0.0861	0.0860	0.0852	0.0838	0.0820	0.0798	0.0766	0.0730	0.0692	0.0653	0.0573	0.0498	0.0430	0.0372	0.02789	0.02126	0.01302	0.00850
3.00	0.1181	0.1179	0.1163	0.1137	0.1105	0.1065	0.1009	0.0949	0.0885	0.0822	0.0700	0.0591	0.0499	0.0422	0.0307	0.02289	0.01365	0.00878
2.50	0.1706	0.1702	0.1672	0.1619	0.1557	0.1480	0.1376	0.1266	0.1156	0.1051	0.0859	0.0702	0.0576	0.0477	0.0335	0.02447	0.01423	0.00904
2.00	0.2678	0.2663	0.2593	0.2472	0.2331	0.2166	0.1949	0.1735	0.1535	0.1355	0.1054	0.0828	0.0660	0.0533	0.0363	0.02593	0.01474	0.00926
1.75	0.352	0.349	0.337	0.317	0.2936	0.2678	0.2354	0.2048	0.1776	0.1539	0.1163	0.0894	0.0701	0.0560	0.0375	0.02658	0.01497	0.00935
1.50	0.481	0.475	0.453	0.418	0.378	0.336	0.2869	0.2427	0.2055	0.1745	0.1277	0.0961	0.0742	0.0586	0.0387	0.02718	0.01517	0.00944
1.25	0.698	0.684	0.638	0.571	0.499	0.429	0.351	0.2874	0.2367	0.1965	0.1392	0.1025	0.0780	0.0609	0.0398	0.02771	0.01534	0.00950
1.00	1.103	1.066	0.956	0.814	0.674	0.552	0.430	0.338	0.2700	0.2189	0.1502	0.1084	0.0813	0.0630	0.0407	0.02816	0.01549	0.00957
0.80	1.755	1.654	1.399	1.114	0.867	0.675	0.501	0.381	0.2965	0.2361	0.1582	0.1125	0.0837	0.0644	0.0413	0.02844	0.01558	0.00961
0.60	3.23	2.886	2.178	1.553	1.111	0.815	0.575	0.422	0.321	0.2514	0.1650	0.1160	0.0856	0.0656	0.0417	0.02867	0.01566	0.00963
0.40	8.06	6.059	3.556	2.146	1.385	0.953	0.641	0.457	0.341	0.2636	0.1703	0.1186	0.0870	0.0664	0.0421	0.02883	0.01570	0.00965
0.20	—	15.386	5.541	2.755	1.621	1.061	0.688	0.481	0.355	0.2715	0.1736	0.1202	0.0878	0.0669	0.0423	0.02894	0.01574	0.00967
0.00	—	24.392	6.679	3.036	1.720	**1.103**	0.705	0.490	0.359	0.2742	0.1748	0.1207	0.0881	0.0671	0.0423	0.02895	0.01574	0.00968

Data not shown correspond to points within the encapsulated source. The dose rate constant, Λ, marked in bold is the dose rate per unit source strength at the reference point, $r = 1.0$ cm and $\theta = 90°$ or in Cartesian coordinates $y = 1.0$ cm and $z = 0.0$ cm.

Source: From Pérez-Calatayud, J., Ballester, F., Serrano-Andrés, M.A., Llich, J.L., Puchades, V., Limami, Y., and Casal, E., *Med. Phys.*, 29, 538–543, 2002.

TABLE A.4.20

Monte Carlo (MC) Calculated Dose Rate Values Per Unit Source Strength as an Along–Away Dose Rate Table (DRT) for the Amersham CDC.K1–K3 Type Sources (1 × ^{137}Cs Bead)

Distance Along z (cm)	Dose Rate Per Unit Source Strength (cGy h^{-1} U^{-1}) Distance Away y (cm)																	
	0.03	0.20	0.40	0.60	0.80	1.00	1.25	1.50	1.75	2.00	2.50	3.00	3.50	4.00	5.00	6.00	8.00	10.00
10.00	0.00896	0.00896	0.00896	0.00895	0.00893	0.00889	0.00885	0.00881	0.00875	0.00866	0.00848	0.00827	0.00803	0.00777	0.00720	0.00658	0.00536	0.00427
8.00	0.0145	0.0145	0.0145	0.0145	0.0144	0.0144	0.0143	0.0142	0.0140	0.0138	0.0134	0.0129	0.0124	0.0118	0.0106	0.0094	0.00717	0.00542
6.00	0.0266	0.0266	0.0266	0.0265	0.0263	0.0261	0.0258	0.0254	0.0249	0.0244	0.0232	0.0218	0.0204	0.0189	0.0161	0.0136	0.00954	0.00676
5.00	0.0394	0.0392	0.0389	0.0385	0.0381	0.0376	0.0369	0.0361	0.0352	0.0342	0.0319	0.0295	0.0270	0.0245	0.0201	0.0163	0.0109	0.00746
4.00	0.0620	0.0620	0.0615	0.0608	0.0597	0.0585	0.0568	0.0549	0.0528	0.0505	0.0458	0.0409	0.0363	0.0320	0.0249	0.0194	0.0122	0.00812
3.50	0.0813	0.0812	0.0805	0.0792	0.0776	0.0756	0.0728	0.0698	0.0665	0.0629	0.0558	0.0487	0.0423	0.0366	0.0276	0.0210	0.0129	0.00844
3.00	0.111	0.111	0.109	0.107	0.105	0.101	0.0963	0.0911	0.0854	0.0797	0.0684	0.0581	0.0492	0.0417	0.0304	0.0227	0.0135	0.00872
2.50	0.159	0.159	0.157	0.153	0.148	0.141	0.132	0.122	0.112	0.102	0.0843	0.0692	0.0570	0.0472	0.0333	0.0243	0.0141	0.00898
2.00	0.250	0.250	0.244	0.234	0.222	0.208	0.188	0.169	0.150	0.133	0.104	0.0819	0.0653	0.0528	0.0360	0.0257	0.0146	0.00920
1.75	0.325	0.325	0.316	0.300	0.280	0.258	0.228	0.200	0.174	0.151	0.115	0.0886	0.0695	0.0556	0.0373	0.0264	0.0149	0.00929
1.50	0.447	0.443	0.424	0.395	0.362	0.325	0.280	0.239	0.203	0.172	0.126	0.0953	0.0736	0.0582	0.0385	0.0270	0.0151	0.00938
1.25	0.645	0.636	0.597	0.543	0.481	0.418	0.345	0.284	0.234	0.194	0.138	0.102	0.0774	0.0606	0.0395	0.0275	0.0152	0.00945
1.00	1.009	0.986	0.896	0.779	0.655	0.541	0.424	0.335	0.268	0.217	0.149	0.108	0.0808	0.0627	0.0404	0.0280	0.0154	0.00952
0.80	1.583	1.518	1.314	1.073	0.848	0.665	0.496	0.378	0.295	0.235	0.157	0.112	0.0832	0.0641	0.0410	0.0283	0.0155	0.00955
0.60	2.834	2.616	2.064	1.513	1.095	0.806	0.571	0.420	0.320	0.250	0.164	0.115	0.0851	0.0652	0.0415	0.0285	0.0156	0.00958
0.40	—	—	3.439	2.122	1.380	0.950	0.639	0.455	0.340	0.263	0.170	0.118	0.0865	0.0660	0.0419	0.0287	0.0156	0.00960
0.20	—	—	5.595	2.774	1.627	1.061	0.688	0.480	0.353	0.271	0.173	0.120	0.0874	0.0666	0.0421	0.0288	0.0156	0.00961
0.00	—	—	7.037	3.089	1.732	**1.106**	0.706	0.489	0.358	0.274	0.174	0.120	0.0877	0.0667	0.0421	0.0288	0.0157	0.00961

Data not shown correspond to points within the encapsulated source. The dose rate constant, Λ, marked in bold is the dose rate per unit source strength at the reference point, $r = 1.0$ cm and $\theta = 90°$ or in Cartesian coordinates $y = 1.0$ cm and $z = 0.0$ cm.

Source: From Pérez-Calatayud, J., Ballester, F., Lluch, J.L., Serrano-Andrés, M.A., Casal, E., Puchades, V., and Limami, Y.F., *Phys. Med. Biol.*, 46, 2029–2040, 2001.

TABLE A.4.21

Monte Carlo (MC) Calculated Dose Rate Values Per Unit Source Strength as an Along–Away Dose Rate Table (DRT) for the Amersham CDC.K4 Type Sources ($2 \times {}^{137}$Cs Bead)

Distance Along z (cm)	Dose Rate Per Unit Source Strength (cGy h^{-1} U^{-1}) Distance Away y (cm)																	
	0.03	0.20	0.40	0.60	0.80	1.00	1.25	1.50	1.75	2.00	2.50	3.00	3.50	4.00	5.00	6.00	8.00	10.00
10.00	0.00922	0.00922	0.00922	0.00919	0.00916	0.00913	0.00907	0.00901	0.00893	0.00885	0.00865	0.00843	0.00818	0.00791	0.00731	0.00667	0.00541	0.00429
8.00	0.0150	0.0150	0.0150	0.0149	0.0149	0.0148	0.0146	0.0145	0.0143	0.0141	0.0136	0.0131	0.0126	0.0120	0.0107	0.0095	0.00721	0.00544
6.00	0.0273	0.0273	0.0273	0.0271	0.0270	0.0268	0.0264	0.0260	0.0255	0.0249	0.0236	0.0222	0.0207	0.0191	0.0162	0.0136	0.00955	0.00677
5.00	0.0403	0.0403	0.0400	0.0396	0.0392	0.0387	0.0379	0.0370	0.0360	0.0349	0.0324	0.0299	0.0273	0.0247	0.0201	0.0163	0.0109	0.00746
4.00	0.0640	0.0638	0.0632	0.0622	0.0612	0.0600	0.0582	0.0562	0.0540	0.0515	0.0464	0.0414	0.0366	0.0322	0.0249	0.0194	0.0122	0.00813
3.50	0.0832	0.0832	0.0826	0.0813	0.0798	0.0778	0.0748	0.0715	0.0678	0.0640	0.0563	0.0490	0.0425	0.0367	0.0276	0.0210	0.0129	0.00844
3.00	0.114	0.114	0.113	0.110	0.108	0.104	0.0987	0.0929	0.0868	0.0807	0.0690	0.0584	0.0494	0.0418	0.0304	0.0227	0.0135	0.00872
2.50	0.167	0.166	0.163	0.157	0.152	0.145	0.135	0.124	0.114	0.103	0.0849	0.0695	0.0571	0.0473	0.0333	0.0243	0.0141	0.00897
2.00	0.263	0.260	0.252	0.241	0.228	0.212	0.192	0.171	0.151	0.134	0.104	0.0820	0.0653	0.0528	0.0360	0.0257	0.0146	0.00919
1.75	0.340	0.339	0.328	0.310	0.288	0.263	0.232	0.202	0.176	0.152	0.115	0.0887	0.0695	0.0556	0.0373	0.0264	0.0149	0.00928
1.50	0.470	0.464	0.442	0.409	0.371	0.331	0.283	0.240	0.203	0.173	0.127	0.0954	0.0736	0.0582	0.0384	0.0270	0.0150	0.00937
1.25	0.681	0.667	0.624	0.560	0.491	0.423	0.347	0.285	0.235	0.195	0.138	0.102	0.0773	0.0605	0.0395	0.0275	0.0152	0.00944
1.00	1.082	1.046	0.939	0.803	0.666	0.546	0.426	0.335	0.268	0.217	0.149	0.108	0.0807	0.0626	0.0404	0.0280	0.0154	0.00949
0.80	1.736	1.634	1.382	1.103	0.859	0.669	0.497	0.378	0.294	0.234	0.157	0.112	0.0830	0.0640	0.0409	0.0283	0.0155	0.00953
0.60	3.235	2.888	2.171	1.544	1.103	0.808	0.570	0.419	0.319	0.250	0.164	0.115	0.0850	0.0651	0.0414	0.0285	0.0155	0.00956
0.40	—	—	3.575	2.131	1.375	0.946	0.635	0.453	0.339	0.262	0.169	0.118	0.0864	0.0660	0.0418	0.0287	0.0156	0.00958
0.20	—	—	5.516	2.723	1.606	1.052	0.682	0.477	0.352	0.270	0.172	0.119	0.0872	0.0665	0.0419	0.0287	0.0156	0.00960
0.00	—	—	6.577	3.001	1.699	**1.092**	0.699	0.485	0.356	0.273	0.173	0.120	0.0875	0.0667	0.0421	0.0288	0.0156	0.00959

Data not shown correspond to points within the encapsulated source. The dose rate constant, Λ, marked in bold is the dose rate per unit source strength at the reference point, $r = 1.0$ cm and $\theta = 90°$ or in Cartesian coordinates $y = 1.0$ cm and $z = 0.0$ cm.

Source: From Pérez-Calatayud, J., Ballester, F., Lluch, J.L., Serrano-Andrés, M.A., Casal, E., Puchades, V., and Limami, Y.F., *Phys. Med. Biol.*, 46, 2029–2040, 2001.

TABLE A.4.22

Monte Carlo (MC) Calculated Dose Rate Values Per Unit Source Strength as an Along–Away Dose Rate Table (DRT) for the CIS CSM11 Type Source Design

Dose Rate Per Unit Source Strength (cGy h^{-1} U^{-1})

Distance Along z (cm)	Distance Away y (cm)																	
	0.00	0.20	0.40	0.60	0.80	1.00	1.25	1.50	1.75	2.00	2.50	3.00	3.50	4.00	5.00	6.00	8.00	10.00
10.00	0.00932	0.00931	0.00930	0.00927	0.00924	0.00921	0.00915	0.00906	0.00899	0.00889	0.00870	0.00846	0.00822	0.00794	0.00734	0.00669	0.00542	0.00430
8.00	0.0152	0.0152	0.0151	0.0151	0.0150	0.0149	0.0147	0.0145	0.0143	0.0141	0.0137	0.0132	0.0126	0.0120	0.0108	0.00948	0.00724	0.00545
6.00	0.0276	0.0276	0.0275	0.0274	0.0272	0.0269	0.0265	0.0260	0.0254	0.0249	0.0236	0.0222	0.0206	0.0191	0.0162	0.0136	0.00956	0.00678
5.00	0.0399	0.0399	0.0397	0.0394	0.0390	0.0385	0.0377	0.0368	0.0358	0.0347	0.0323	0.0298	0.0272	0.0247	0.0202	0.0163	0.0109	0.00747
4.00	0.0626	0.0625	0.0622	0.0616	0.0607	0.0596	0.0578	0.0558	0.0536	0.0511	0.0462	0.0411	0.0364	0.0321	0.0249	0.0194	0.0122	0.00814
3.50	0.0847	0.0836	0.0821	0.0805	0.0787	0.0767	0.0739	0.0706	0.0672	0.0634	0.0559	0.0488	0.0423	0.0366	0.0276	0.0210	0.0129	0.00845
3.00	0.116	0.114	0.111	0.108	0.105	0.102	0.0968	0.0916	0.0859	0.0799	0.0683	0.0579	0.0490	0.0416	0.0304	0.0227	0.0136	0.00874
2.50	0.167	0.163	0.159	0.154	0.148	0.142	0.132	0.122	0.112	0.102	0.0839	0.0689	0.0568	0.0470	0.0332	0.0242	0.0141	0.00899
2.00	0.258	0.251	0.242	0.232	0.220	0.206	0.187	0.167	0.149	0.132	0.103	0.0813	0.0650	0.0526	0.0360	0.0257	0.0147	0.00922
1.75	0.332	0.326	0.314	0.297	0.277	0.255	0.226	0.198	0.172	0.150	0.114	0.0879	0.0692	0.0554	0.0372	0.0264	0.0149	0.00932
1.50	0.447	0.439	0.419	0.390	0.355	0.319	0.275	0.234	0.199	0.170	0.125	0.0946	0.0733	0.0580	0.0385	0.0270	0.0151	0.00941
1.25	0.638	0.624	0.586	0.531	0.469	0.407	0.336	0.278	0.230	0.192	0.137	0.101	0.0771	0.0604	0.0395	0.0276	0.0153	0.00947
1.00	0.985	0.957	0.868	0.749	0.630	0.523	0.413	0.328	0.263	0.214	0.148	0.107	0.0806	0.0625	0.0405	0.0280	0.0154	0.00953
0.80	1.518	1.461	1.262	1.023	0.810	0.640	0.482	0.370	0.290	0.232	0.156	0.111	0.0830	0.0640	0.0411	0.0283	0.0155	0.00958
0.60	2.638	2.462	1.931	1.421	1.042	0.777	0.556	0.412	0.315	0.248	0.163	0.115	0.0850	0.0652	0.0416	0.0286	0.0156	0.00961
0.40	6.156	4.886	3.127	1.975	1.312	0.918	0.624	0.449	0.337	0.261	0.169	0.118	0.0865	0.0661	0.0419	0.0287	0.0156	0.00963
0.20	—	12.18	5.028	2.614	1.570	1.037	0.677	0.475	0.351	0.269	0.173	0.120	0.0875	0.0666	0.0421	0.0288	0.0157	0.00966

0.00	—	23.30	6.585	3.008	1.707	**1.096**	0.702	0.487	0.358	0.273	0.174	0.120	0.0878	0.0668	0.0422	0.0289	0.0157	0.00964
−0.20	—	18.582	5.845	2.827	1.643	1.071	0.693	0.483	0.356	0.272	0.174	0.120	0.0877	0.0667	0.0421	0.0288	0.0157	0.00965
−0.40	11.518	7.521	4.004	2.302	1.447	0.982	0.653	0.463	0.344	0.265	0.171	0.119	0.0870	0.0664	0.0420	0.0288	0.0157	0.00964
−0.60	3.959	3.380	2.437	1.681	1.177	0.849	0.591	0.430	0.325	0.254	0.166	0.116	0.0858	0.0656	0.0417	0.0286	0.0156	0.00963
−0.80	2.040	1.874	1.544	1.201	0.919	0.706	0.518	0.390	0.302	0.239	0.160	0.113	0.0839	0.0646	0.0413	0.0284	0.0156	0.00959
−1.00	1.243	1.180	1.040	0.873	0.713	0.578	0.445	0.347	0.276	0.223	0.152	0.109	0.0817	0.0632	0.0407	0.0282	0.0155	0.00956
−1.25	0.765	0.742	0.685	0.608	0.526	0.448	0.364	0.296	0.242	0.200	0.141	0.103	0.0784	0.0612	0.0398	0.0277	0.0153	0.00949
−1.50	0.505	0.515	0.479	0.440	0.396	0.351	0.298	0.250	0.211	0.178	0.130	0.097	0.0747	0.0589	0.0388	0.0272	0.0152	0.00943
−1.75	0.374	0.369	0.355	0.333	0.307	0.279	0.244	0.211	0.182	0.158	0.118	0.0905	0.0707	0.0563	0.0377	0.0267	0.0150	0.00936
−2.00	0.284	0.281	0.273	0.260	0.244	0.225	0.202	0.179	0.158	0.139	0.107	0.0838	0.0665	0.0536	0.0364	0.0260	0.0148	0.00926
−2.50	0.181	0.178	0.174	0.168	0.161	0.153	0.141	0.130	0.118	0.107	0.0875	0.0712	0.0582	0.0481	0.0337	0.0245	0.0142	0.00904
−3.00	0.122	0.122	0.120	0.117	0.113	0.109	0.103	0.0970	0.0905	0.0838	0.0712	0.0600	0.0505	0.0426	0.0309	0.0230	0.0137	0.00879
−3.50	0.0884	0.0881	0.0873	0.0860	0.0841	0.0818	0.0784	0.0746	0.0706	0.0664	0.0582	0.0505	0.0435	0.0375	0.0281	0.0213	0.0130	0.00852
−4.00	0.0675	0.0673	0.0667	0.0658	0.0647	0.0632	0.0611	0.0587	0.0562	0.0534	0.0480	0.0425	0.0374	0.0329	0.0254	0.0197	0.0124	0.00820
−5.00	0.0426	0.0424	0.0422	0.0418	0.0412	0.0406	0.0397	0.0386	0.0374	0.0362	0.0335	0.0308	0.0280	0.0253	0.0206	0.0166	0.0110	0.00754
−6.00	0.0287	0.0287	0.0286	0.0284	0.0282	0.0279	0.0275	0.0270	0.0264	0.0257	0.0244	0.0229	0.0212	0.0196	0.0166	0.0139	0.00968	0.00685
−8.00	0.0155	0.0155	0.0155	0.0154	0.0154	0.0153	0.0151	0.0150	0.0148	0.0146	0.0141	0.0135	0.0129	0.0123	0.0110	0.00966	0.00733	0.00551
−10.00	0.00955	0.00954	0.00953	0.00951	0.00948	0.00944	0.00939	0.00934	0.00926	0.00915	0.00893	0.00869	0.00842	0.00812	0.00749	0.00681	0.00550	0.00435

Data not shown correspond to points within the encapsulated source. The dose rate constant, *A*, marked in bold is the dose rate per unit source strength at the reference point, $r = 1.0$ cm and $\theta = 90°$ or in Cartesian coordinates $y = 1.0$ cm and $z = 0.0$ cm.

Source: From Ballester, F., Lluch, J.L., Limami, Y., Serrano, M.A., Casal, E., Pérez-Calatayud, J., and Lliso, F., *Med. Phys.*, 27, 2182–2189, 2000.

TABLE A.4.23

Monte Carlo (MC) Calculated Dose Rate Values Per Unit Source Strength as an Along–Away Dose Rate Table (DRT) for the Amersham CDCS-M Type Source Design

Distance Along z (cm)	Dose Rate Per Unit Source Strength (cGy h^{-1} U^{-1}) Distance Away y (cm)													
	0.00	0.15	0.30	0.50	0.75	1.00	2.00	2.50	3.00	4.00	5.00	6.00	8.00	10.00
10.00	0.00850	0.00851	0.00845	0.00850	0.00844	0.00849	0.00851	0.00843	0.00828	0.00786	0.00730	0.00667	0.00539	0.00428
8.00	0.0137	0.0137	0.0136	0.0137	0.0138	0.0138	0.0137	0.0135	0.0131	0.0120	0.0108	0.00949	0.00721	0.00542
6.00	0.0250	0.0251	0.0248	0.0247	0.0250	0.0255	0.0247	0.0236	0.0223	0.0193	0.0163	0.0136	0.00953	0.00675
5.00	0.0368	0.0368	0.0363	0.0365	0.0373	0.0375	0.0351	0.0328	0.0302	0.0249	0.0202	0.0164	0.0109	0.00744
4.00	0.0593	0.0588	0.0581	0.0585	0.0598	0.0596	0.0526	0.0473	0.0419	0.0324	0.0250	0.0194	0.0122	0.00811
3.00	0.110	0.108	0.107	0.109	0.110	0.107	0.0832	0.0704	0.0591	0.0418	0.0304	0.0227	0.0135	0.00869
2.50	0.165	0.159	0.158	0.163	0.161	0.153	0.106	0.0862	0.0700	0.0473	0.0332	0.0242	0.0141	0.00894
2.00	0.275	0.262	0.266	0.267	0.255	0.230	0.136	0.105	0.0823	0.0528	0.0359	0.0256	0.0146	0.00917
1.50	0.565	0.539	0.542	0.507	0.438	0.363	0.174	0.127	0.0949	0.0578	0.0382	0.0269	0.0150	0.00935
1.00	—	1.990	1.630	1.180	0.810	0.582	0.215	0.147	0.106	0.0621	0.0402	0.0278	0.0153	0.00948
0.75	—	6.990	3.300	1.810	1.070	0.714	0.233	0.156	0.111	0.0637	0.0408	0.0282	0.0155	0.00953
0.50	—	12.270	5.000	2.420	1.320	0.834	0.248	0.163	0.115	0.0649	0.0413	0.0284	0.0156	0.00957
0.25	—	13.590	5.700	2.800	1.490	0.917	0.257	0.167	0.117	0.0656	0.0417	0.0286	0.0156	0.00958

0.00	—	13.590	5.880	2.910	1.540	**0.946**	0.261	0.169	0.117	0.0659	0.0418	0.0287	0.0156	0.00959
−0.25	—	13.610	5.730	2.800	1.490	0.920	0.258	0.167	0.117	0.0657	0.0417	0.0286	0.0156	0.00961
−0.50	—	12.670	5.070	2.460	1.340	0.842	0.249	0.163	0.115	0.0651	0.0415	0.0285	0.0155	0.00956
−0.75	—	7.620	3.450	1.860	1.100	0.725	0.234	0.157	0.111	0.0638	0.0409	0.0282	0.0155	0.00953
−1.00	—	2.260	1.730	1.210	0.831	0.595	0.217	0.148	0.107	0.0622	0.0402	0.0279	0.0154	0.00950
−1.50	0.620	0.579	0.568	0.519	0.448	0.371	0.176	0.128	0.0953	0.0580	0.0384	0.0270	0.0151	0.00937
−2.00	0.294	0.282	0.278	0.274	0.258	0.234	0.138	0.106	0.0828	0.0530	0.0361	0.0258	0.0146	0.00915
−2.50	0.175	0.169	0.168	0.168	0.164	0.155	0.107	0.0871	0.0707	0.0477	0.0334	0.0244	0.0141	0.00898
−3.00	0.1171	0.113	0.111	0.112	0.112	0.109	0.0843	0.0712	0.0597	0.0422	0.0306	0.0228	0.0136	0.00874
−4.00	0.0622	0.0611	0.0597	0.0605	0.0612	0.0607	0.0533	0.0479	0.0424	0.0327	0.0252	0.0196	0.0123	0.00816
−5.00	0.0384	0.0383	0.0374	0.0376	0.0381	0.0381	0.0359	0.0334	0.0307	0.0253	0.0205	0.0165	0.0110	0.00750
−6.00	0.0264	0.0261	0.0256	0.0256	0.0258	0.0259	0.0251	0.0239	0.0224	0.0194	0.0164	0.0137	0.00957	0.00677
−8.00	0.0142	0.0142	0.0141	0.0140	0.0139	0.0140	0.0139	0.0136	0.0132	0.0121	0.0108	0.00951	0.00725	0.00547
−10.00	0.00867	0.00880	0.00875	0.00870	0.00860	0.00862	0.00870	0.00858	0.00839	0.00792	0.00734	0.00670	0.00541	0.00429

Data not shown correspond to points within the encapsulated source. The dose rate constant, Λ, marked in bold is the dose rate per unit source strength at the reference point, $r = 1.0$ cm and $\theta = 90°$ or in Cartesian coordinates $y = 1.0$ cm and $z = 0.0$ cm.

Source: From Casal, E., Ballester, F., Lluch, J.L., Pérez-Calatayud, J., and Lliso, F., *Med. Phys.*, 27, 132–140, 2000.

References

1. Williamson, J.F. and Li, Z. Monte Carlo aided dosimetry of the microSelectron pulsed and high dose rate ^{192}Ir sources, *Med. Phys.*, 22, 809–819, 1995.

2. Daskalov, G.M., Löffler, E., and Williamson, J.F. Monte Carlo-aided dosimetry of a new high dose-rate brachytherapy source, *Med. Phys.*, 25, 2200–2208, 1998.

3. Wang, R. and Sloboda, R.S. Influence of source geometry and materials on the transverse axis dosimetry of ^{192}Ir brachytherapy sources, *Phys. Med. Biol.*, 43, 37–48, 1998.

4. Angelopoulos, A., Baras, P., Sakelliou, L., Karaiskos, P., and Sandilos, P. Monte Carlo dosimetry of a new ^{192}Ir high dose rate brachytherapy source, *Med. Phys.*, 27, 2521–2527, 2000.

5. Ballester, F., Pérez-Calatayud, J., Puchades, V., Lluch, J.L., Serrano-Andrés, M.A., Limami, Y., Lliso, F., and Casal, E. Monte Carlo dosimetry of the Buchler high dose rate ^{192}Ir source, *Phys. Med. Biol.*, 46, N79–N90, 2001.

6. Ballester, F., Puchades, V., Lluch, J.L., Serrano-Andrés, M.A., Limami, Y., Pérez-Calatayud, J., and Casal, E. Technical note: Monte Carlo dosimetry of the HDR 12*i* and Plus ^{192}Ir sources, *Med. Phys.*, 28, 2586–2591, 2001.

7. Granero, D., Pérez-Calatayud, J., and Ballester, F. Monte Carlo calculation of the TG-43 dosimetric parameters of a new BEBIG Ir-192 HDR source, *Radiother. Oncol.*, 76, 79–85, 2005.

8. Papagiannis, P., Angelopoulos, A., Pantelis, E., Sakelliou, L., Karaiskos, P., and Shimizu, Y. Monte Carlo dosimetry of ^{60}Co HDR brachytherapy sources, *Med. Phys.*, 29, 712–721, 2003.

9. Ballester, F., Granero, D., Pérez-Calatayud, J., Casal, E., Agramunt, A., and Cases, R. Monte Carlo dosimetric study of the BEBIG Co-60 HDR source, *Phys. Med. Biol.*, 50, N309–N316, 2005.

10. Karaiskos, P., Angelopoulos, A., Pantelis, E., Papagiannis, P., Sakelliou, L., Kouwenhoven, E., and Baltas, D. Monte Carlo dosimetry of a new ^{192}Ir pulsed dose rate brachytherapy source, *Med. Phys.*, 30, 9–16, 2003.

11. Pérez-Calatayud, J., Ballester, F., Serrano-Andrés, M.A., Puchades, V., Lluch, J.L., Limami, Y., and Casal, E. Dosimetry characteristics of the Plus and 12*i* GammaMed PDR ^{192}Ir sources, *Med. Phys.*, 28, 2576–2585, 2001.

12. Pérez-Calatayud, J., Granero, D., Ballester, F., Puchades, V., and Casal, E. Monte Carlo dosimetric characterization of the Cs-137 selectron/LDR source: evaluation of applicator attenuation and superposition approximation effects, *Med. Phys.*, 31, 493–499, 2004.

13. Pérez-Calatayud, J., Ballester, F., Serrano-Andrés, M.A., Llich, J.L., Puchades, V., Limami, Y., and Casal, E. Dosimetric characteristics of the CDC-type miniature cylindrical ^{137}Cs brachytherapy sources, *Med. Phys.*, 29, 538–543, 2002.

14. Pérez-Calatayud, J., Ballester, F., Lluch, J.L., Serrano-Andrés, M.A., Casal, E., Puchades, V., and Limami, Y.F. Monte Carlo calculation of dose rate distributions around the Walstam CDC.K-Type ^{137}Cs sources, *Phys. Med. Biol.*, 46, 2029–2040, 2001.

15.   Ballester, F., Lluch, J.L., Limami, Y., Serrano, M.A., Casal, E., Pérez-Calatayud, J., and Lliso, F. A Monte Carlo investigation of the dosimetric characteristics of the CSM11 ^{137}Cs source from CIS, *Med. Phys.*, 27, 2182–2189, 2000.

16.   Casal, E., Ballester, F., Lluch, J.L., Pérez-Calatayud, J., and Lliso, F. Monte Carlo calculations of dose rate distributions around the Amersham CDCS-M-type ^{137}Cs source, *Med. Phys.*, 27, 132–140, 2000.

Index

Printed in the United States
by Baker & Taylor Publisher Services